Living Machines

Living Machines
A Handbook of Research in Biomimetic and Biohybrid Systems

Edited by

Tony J. Prescott

Nathan F. Lepora

Paul F. M. J. Verschure

Assistant Editor: Michael Szollosy

OXFORD
UNIVERSITY PRESS

OXFORD
UNIVERSITY PRESS

Great Clarendon Street, Oxford, OX2 6DP,
United Kingdom

Oxford University Press is a department of the University of Oxford.
It furthers the University's objective of excellence in research, scholarship,
and education by publishing worldwide. Oxford is a registered trade mark of
Oxford University Press in the UK and in certain other countries

Published in the United States of America by Oxford University Press
198 Madison Avenue, New York, NY 10016, United States of America

British Library Cataloguing in Publication Data

Data available

Library of Congress Control Number: 2017949321

ISBN 978–0–19–967492–3

Printed and bound by
CPI Group (UK) Ltd, Croydon, CR0 4YY

Foreword

Terrence J. Sejnowski

This handbook brings together a wide range of knowledge from many different areas of science and engineering toward the realization of what is called "living machines"— man-made devices having capabilities shared with creatures that evolved in nature. Living machines are not easy to build, for many reasons. First, nature evolved smart materials that are assembled at the molecular level, compared with current technology that is assembled from macroscopic parts; second, bodies are built from the inside out, turning food into materials that are used to replace all the proteins inside cells, which themselves are periodically replaced; third, some of the most difficult problems in seeing, hearing, and moving around have not been solved, and our intuition on how nature has solved them is misleading. Take, for example, vision.

The MIT AI Lab was founded in the 1960s and one of its first projects was a substantial grant from a military research agency to build a robot that could play ping pong. I once heard a story that the principal investigator forgot to ask for money in the grant proposal to build a vision system for the robot, so he assigned the problem to a graduate student as a summer project. I once asked Marvin Minsky whether or not it was a true story. He snapped back that I had it wrong: "We assigned the problem to undergraduate students." A document from the archives at MIT confirms his version of the story (Papert, 1966).

The intuition that it would be easy to write a vision program is based on what we find easy to do. We are all experts at vision because it is important for our survival and evolution had millions of years to get it right. This misguided the early AI pioneers into thinking that writing a vision program would be easy. Little did anyone know in the 1960s that it would take more than 50 years and a million times increase in computer power before computer vision would begin to reach human levels of performance (Sejnowski, 2018).

A closer look at vision reveals that we may not even understand what computational problems nature solved. It has been almost universally assumed by researchers in computer vision that the goal of vision is to create an internal model of the outside world. However, a complete and accurate internal model of the world may not be necessary for most practical purposes, and might not even be possible given the low sampling rate of current video cameras. Based on evidence from psychophysics, physiology, and anatomy, Patricia Churchland, V. S. Ramachandran, and I came to the conclusion that the brain represents only a limited part of the world, only what is needed at any moment to carry out the task at hand (Churchland et al., 1994). We have the illusion of high resolution everywhere because we can rapidly reposition our eyes. The apparent modularity of vision also is an illusion. The visual system integrates information from other streams, including signals from the reward system indicating the values of objects in the scene, and the motor system actively seeks information by repositioning sensors, such as moving eyes and, in some species, ears to gather information that may lead to rewarding actions. Sensory systems are not ends in themselves; they evolved to support the motor system, to make movements more efficiently and survival more likely.

Writing computer programs to solve problems was the dominant method used by AI researchers until recently. In contrast, we learn to solve problems through experience with the world. We have many kinds of learning systems, including declarative learning of explicit events and unique objects, implicit perceptual learning, and motor learning. Learning is our special power. Recent advances in speech recognition, object recognition in images, and language translation in AI have been made by deep learning networks inspired by general features found in brains, such as massively parallel simple processing units, a high degree of connectivity between them, and the connection strengths learned through experience (Sejnowski, 2018). There is much more we can learn from nature by observing animal behavior and looking into brains to find the algorithms that are responsible.

Adaptability is essential for surviving in a world that is uncertain and nonstationary. Reinforcement learning, which is found in all invertebrates and vertebrates, is based on exploring the various options in the environment, and learning from their outcomes. The reinforcement learning algorithm found in the basal ganglia of vertebrates, called temporal difference learning, is an online way to learn a sequence of actions to accomplish a goal (Sejnowski et al., 2014). Dopamine cells in the midbrain compute the reward prediction error, which can be used to make decisions and update the value function that predicts future rewards. As learning improves, exploration decreases, eventually leading to pure exploitation of the best strategy found during learning.

One of the problems with reinforcement learning is that it is very slow, since the only feedback from the world is whether or not reward is received at the end of a long sequence of actions. This is called the temporal credit assignment problem. The limitation of vision discussed earlier, attending to only one object at a time, makes it easier for reinforcement learning to narrow down the number of possible sensory inputs that contribute to obtaining rewards, making this a feature rather than a bug. Reinforcement learning coupled with deep learning networks recently produced AlphaGo, a neural network that defeated the world's best Go players (Sejnowski, 2018). Go is an ancient board game of great complexity and, until recently, the best Go programs were far beneath the capability of humans. What this remarkable success has revealed is that multiple synergistic learning systems, each of which has limitations, working together can achieve complex behaviors.

Building bodies has proven an even more difficult problem than vision. In the 1980s, a common argument for ignoring nature was that it is foolish to study birds if you want to build a flying machine. But the Wright Brothers, who built the first flying machine, studied gliding birds with great interest and incorporated their observations into the design of wings, as Prescott and Verschure point out in Chapter 1 of *Living Machines*. Soaring birds often rely on ascending thermal plumes in the atmosphere as they search for prey or migrate across large distances (Figure F1). Recently, reinforcement learning was used to teach a glider to soar in turbulent thermals in simulations (Reddy et al., 2016) and in the field (Reddy et al., in press). The control strategy that was learned produced behavior similar to that of soaring birds (Figure F2). This is another

Figure F1 Soaring birds have wingtip feathers that break up the air-flow from below. They also curve these feathers upward to create arrays of small winglets.
Photo courtesy of John Lienhard.

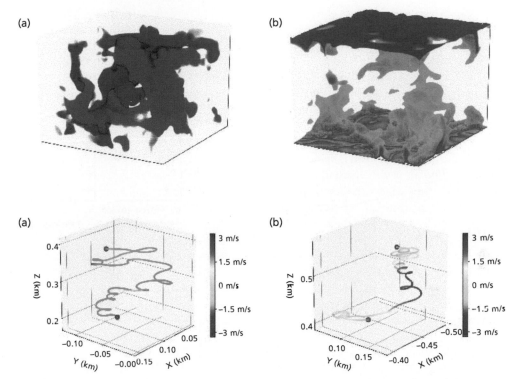

Figure F2 Simulations of a glider learning to soar in a thermal. *Lower panels*: Snapshots of the vertical velocity (a) and the temperature fields (b) in our numerical simulations of 3D Rayleigh–Bénard convection. For the vertical velocity field, the red and blue colors indicate regions of large upward and downward flow, respectively. For the temperature field, the red and blue colors indicate regions of high and low temperature, respectively. *Upper panels*: (a) Typical trajectories of an untrained and (b) a trained glider flying within a Rayleigh–Bénard turbulent flow. The colors indicate the vertical wind velocity experienced by the glider. The green and red dots indicate the start and the end points of the trajectory, respectively. The untrained glider takes random decisions and descends, whereas the trained glider flies in characteristic spiraling patterns in regions of strong ascending currents, as observed in the thermal soaring of birds and gliders.

Reproduced from Gautam Reddy, Antonio Celani, Terrence J. Sejnowski and Massimo Vergassol, Learning to soar in turbulent environments, *Proceedings of the National Academy of Sciences of the United States of America*, 113 (33), pp. E4877–E4884, Figure 1, doi.org/10.1073/pnas.1606075113, a, 2016.

example of how AI is beginning to tackle problems in the real world. Have you noticed that jet planes are sprouting winglets at the tips of wings, similar to the wingtips found on soaring birds (Figure F1)? These save millions of dollars of jet fuel. Evolution has had millions of years to optimize flying and it would be foolish to ignore nature.

Agility is another holy grail for living machines. Hummingbirds are renowned for their ability to maneuver with great speed and precision. Maneuverability requires the rapid integration of sensory information into control systems to avoid obstacles and reach goals, such as the nectar in flowers. A recent study of over 25 species of hummingbird revealed that fast rotations and sharp turns evolved by recruiting change in wing morphology, whereas acceleration maneuvers evolved by recruiting changes in muscle capacity (Figure F3) (Dakin et al., 2018).

Humans invented the wheel for efficient transportation, an engineering achievement that seems to have eluded nature. The reality is that nature was far cleverer and instead invented

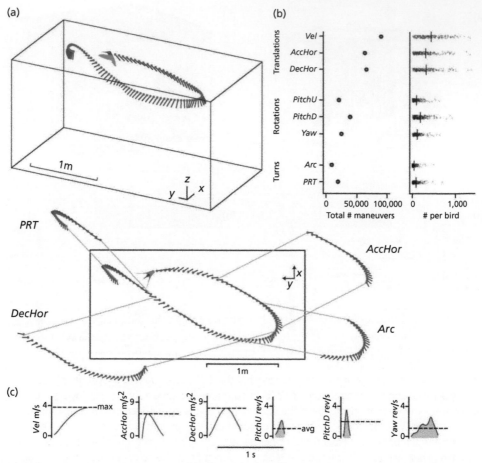

Figure F3 Flight performance of hummingbirds. (a) A tracking system recorded body position (blue dots) and orientation (red lines) at 200 frames per second. These data were used to identify stereotyped translations, rotations, and complex turns. The sequence in (a) shows a bird performing a pitch-roll turn (PRT) followed by a deceleration (DecHor), an arcing turn (Arc), and an acceleration (AccHor).The sequence duration is 2.5 s, and every fifth frame is shown. (b) The number of translations, rotations, and turns recorded in 200 hummingbirds. (c) Example translations and rotations illustrating the performance metrics.

From Roslyn Dakin, Paolo S. Segre, Andrew D. Straw, Douglas L. Altshuler, Morphology, muscle capacity, skill, and maneuvering ability in hummingbirds, *Science*, 359 (6376), pp. 653–657, doi: 10.1126/science.aao7104, © 2018 The Authors. Reprinted with permission from AAAS.

walking. Take a wheel with a 6-foot radius. As it turns, one spoke after another is in contact with the ground, supporting all the other spokes that are off the ground. A far more efficient design only needs two spokes: the one on the ground and the next one to hit the ground, exchanging them along the way. The bipedal wheel even works well over uneven ground. This illustrates Orgel's second rule: Evolution is cleverer than you are. Leslie Orgel was a chemist who studied the origin of life at the Salk Institute.

Sensorimotor coordination in living machines is a dynamical process that requires the integration of sensory information with control systems that have a layered architecture: Reflexes in the spinal cord provide rapid correction to unexpected perturbations, but are inflexible; another

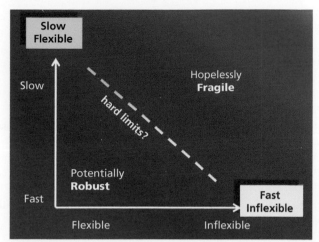

Figure F4 Trade-off between speed and flexibility. Systems can be designed that are slow and flexible, fast and inflexible, or along the line between these extremes (dashed line). Control systems outside the envelope are fragile and those inside are robust.
Courtesy of John Doyle.

layer of control in the midbrain is able to adjust the reflexes in the spinal cord on a longer time scale to handle more slowly changing conditions, such as slippery terrain; longer term planning based on integrating visual and auditory information from the cerebral cortex is more flexible, albeit much slower. John Doyle has made the tradeoff between flexibility and speed a central principle driving layered control systems (Figure F4). Central to this view of control and communication is the mitigation of time delays, which can be accomplished with predictive coding (Rao and Ballard, 1999; Adams et al., 2013).

The challenge for the designer of living machines is autonomy, a *sine qua non* for nature. For example, bacteria have adapted to extreme environments that range from hydrothermal vents in the ocean to sheets of ice in Antarctica and your gut, which harbors thousands of species. Bacteria like *E. coli* have developed algorithms for swimming up gradients toward food sources. Because they are too small (a few micrometers across) to sense the gradient directly, they use chemotaxis, which consists in periodically tumbling and setting off in a random direction (Berg, 2003). This may seem counterproductive, but by adjusting the swim times to be longer at higher concentrations, they can reliably climb up the gradient. Bacteria are smarter than the smartest biologists, who have not yet figured out how they manage to survive in such a wide range of environments.

The prospect for creating living machines has never been better, judging by all the advances that are summarized in this handbook. Nature remains a rich source of inspiration. Just as general principles from birds gave rise to flying machines that can fly much faster than any bird, today's deep learning networks based on general principles of brains are beginning to outperform human brains in limited domains. As materials become more sophisticated, as homeostatic mechanisms become more robust, and as we discover new principles for self-organization, autonomous living machines are rising, slowly at first, eventually taking their place in the world beside their creators.

References

Adams, R.A., Shipp, S. and Friston, K.J. (2013). Predictions not commands: active inference in the motor system. *Brain Structure and Function*, 218(3), 611–43.

Berg, Howard C. (2003). E. coli *in Motion*. New York: Springer.

Churchland, P.S., Ramachandran, V.S. and Sejnowski, T.J. (1994). A Critique of Pure Vision. In: C. Koch & J.D. Davis (eds). *Large-Scale Neuronal Theories of the Brain*. Cambridge, Massachusetts: MIT Press, 23–60.

Dakin, R., Segre, P.S., Straw, A.D. and Altshuler, D.L. (2018). Morphology, muscle capacity, skill, and maneuvering ability in hummingbirds. *Science*, 359, 653–7.

Papert, S.A. (1966). The Summer Vision Project, AI Memo AIM-100. July 1, DSpace@MIT. https://dspace.mit.edu/handle/1721.1/6125.

Rao, R.P.N. and Ballard, D.H. (1999). Predictive coding in the visual cortex: a functional interpretation of some extra-classical receptive-field effects. *Nature Neuroscience*, 2, 79–87.

Reddy, G., Celani, A., Sejnowski, T.J. and Vergassola, M. (2016). Learning to soar in turbulent environments. *Proceedings of the National Academy of Sciences of the United States of America*, 113(33), E4877–E84.

Reddy, G., Ng, J.W., Celani, A., Sejnowski, T.J. and Vergassola, M. (in press). Soaring like a bird via reinforcement learning in the field. *Nature*.

Sejnowski, T.J. (2018). *The Deep Learning Revolution*. Massachusetts: MIT Press.

Sejnowski, T.J., Poizner, H., Lynch, G., Gepshtein, S. and Greenspan, R.J. (2014). Prospective Optimization. *Proceedings of the Institute of Electrical and Electronic Engineering*, 102:799–811.

Acknowledgments

This book has come about through the contributions of a great many people. We are particularly grateful to the chapter authors for their incredible research in living machines, their inspired writing, and their patience and understanding with the rather long and complex process of assembling the manuscript for this handbook (unfortunately for us, it did not self-assemble as some living machines are able to do!). Beyond those who have directly contributed, there is also a very much larger set of researchers and students, at all career stages—particularly, those that have participated in our conferences and summer schools—who have inspired the ideas in this handbook and have kept us motivated by challenging our preconceptions and delighting us with their energy and enthusiasm. We would particularly like to thank the following for their advice, ideas and encouragement: Ehud Ahissar, Joseph Ayers, Mark Cutkosky, Terrence Deacon, Mathew Diamond, Paul Dean, Marc Desmulliez, Peter Dominey, Frank Grasso, José Halloy, Leah Krubitzer, Maarja Kruusmaa, David Lane, Uriel Martinez, Barbara Mazzolai, Bjorn Merker, Giorgio Metta, Ben Mitchinson, Edvard Moser, Martin Pearson, Roger Quinn, Peter Redgrave, Scott Simon, and Stuart Wilson.

The editors have benefited enormously from the help of staff in their own institutions. Michael Szollosy, who we have acknowledged as an assistant editor, both contributed to the final section of the book "Perspectives" and edited and wrote the introduction to that section. Anna Mura has made a colossal contribution through her tireless work in planning and promoting the Living Machines (LM) conference series, now in its seventh edition, and the Barcelona Summer School on Cognition, Brain, and Technology, now in its eleventh year; we also wish to acknowledge all the local organizers of the LM conferences and the LM International Advisory Board. We would like to thank our co-investigators in the two Convergence Science Network (CSN) coordination action projects, who helped conceive of the LM Handbook, particularly Stefano Vassanelli who also co-ordinated two schools on neurotechnology. The editors are grateful to the support teams in their own institutions, particularly, Carme Buisan and Mireia Mora from the SPECS group at the Department of Information and Communication Technologies at Universitat Pompeu Fabra, and Ana MacIntosh and Gill Ryder from the University of Sheffield and Sheffield Robotics.

The possibility for this handbook has also come about through projects funded by the European Union Framework Programme. The CSN coordination action (grant numbers 248986 and 601167) has been the key integrator, but other collaborative projects also stand behind many of the ideas and technologies seen in this book. Specific ones that we have participated in and benefited from include BIOTACT: BIOmimetic Technology for vibrissal ACtive Touch (FP7 ICT-215910), EFAA: Experimental Functional Android Assistant (FP7 ICT-270490), WYSIWYD: WhatYouSayIsWhatYouDid (FP7 ICT-612139), SF: Synthetic Forager (FP7 ICT-217148), cDAC: The role of consciousness in adaptive behavior: A combined empirical, computational and robot based approach (ERC-2013-ADG-341196), CEEDS: The Collective Experience of Empathic Data Systems (FP7-ICT-258749), CA-RoboCom: Robot Companions for Citizens (FP7-FETF-284951) and the European FET Flagship Programme through the Human Brain Project (HBP-SGA1 grant agreement 720270).

We would like to thank the editorial staff at Oxford University Press for their help in putting the handbook together. Finally, we would like to express our gratitude to our families and friends for their patience, understanding, and tireless love and support.

Contents

Contributors

Iain A. Anderson
Auckland Bioengineering Group, The
University of Auckland, New Zealand

Minoru Asada
Graduate School of Engineering, Osaka
University, Japan

Joseph Ayers
Marine Science Center, Northeastern
University, USA

Belén Rubio Ballester
SPECS, Institute for Bioengineering of
Catalonia (IBEC), the Barcelona Institute
of Science and Technology (BIST),
Barcelona, Spain

Sliman J. Bensmaia
Department of Organismal Biology and
Anatomy, University of Chicago, USA

Theodore W. Berger
Department of Biomedical Engineering,
University of Southern California, Los
Angeles, USA

Josh Bongard
Department of Computer Science, University
of Vermont, USA

Frédéric Boyer
Automation, Production and Computer
Sciences Department, IMT Atlantique
(former Ecole des Mines de Nantes), France

Dieter Braun
Systems Biophysics, Center for Nanoscience,
Ludwig-Maximilians-Universität München,
Germany

Joanna J. Bryson
Department of Computer Science, University
of Bath, UK

Gregory S. Chirikjian
Department of Mechanical Engineering,
Johns Hopkins University, USA

Roberto Cingolani
Istituto Italiano di Tecnologia, Genoa, Italy

Emily C. Collins
Sheffield Robotics, University
of Sheffield, UK

Holk Cruse
Faculty of Biology, Universität Bielefeld,
Germany

Mark R. Cutkosky
School of Engineering, Stanford
University, USA

Terrence W. Deacon
Anthropology Department, University of
California, Berkeley, USA

Piotr Dudek
School of Electrical & Electronic Engineering,
The University of Manchester, UK

Uğur Murat Erdem
Department of Mathematics, North Dakota
State University, USA

Martin S. Fischer
Institute of Systematic Zoology and
Evolutionary Biology with Phyletic Museum,
Friedrich-Schiller-Universität Jena,
Germany

Toshio Fukuda
Institute for Advanced Research, Nagoya
University, Japan

Ulrich Gerland
Theory of Complex Biosystems,
Technische Universität München,
Garching, Germany

John Greenman
Bristol Robotics Laboratory, University of the
West of England, UK

David J. Gunkel
Department of Communication, Northern
Illinois University, USA

José Halloy
Paris Interdisciplinary Energy Research
Institute (LIED), Université Paris
Diderot, France

Yasuhisa Hasegawa
Department of Micro-Nano
Systems Engineering, Nagoya
University, Japan

Michael E. Hasselmo
Department of Psychological and Brain
Sciences, Center for Systems Neuroscience,
Boston University, USA

Anders Hedenström
Department of Biology, Lund
University, Sweden

Ivan Herreros
SPECS, Institute for Bioengineering of
Catalonia (IBEC), the Barcelona Institute
of Science and Technology (BIST),
Barcelona, Spain

James Hughes
Institute for Ethics and Emerging
Technologies, Boston, USA; and University of
Massachusetts, Boston, USA

Ioannis A. Ieropoulos
Centre for Research in Biosciences,
University of the West of England, UK

Auke Jan Ijspeert
Biorobotics Laboratory, EPFL,
Switzerland

Akio Ishiguro
Research Institute of Electrical
Communication, Tohoku University, Japan

Hoon Eui Jeong
School of Mechanical and Advanced Materials
Engineering, Ulsan National Institute of
Science and Technology, South Korea

Moritz Kreysing
Max Planck Institute of Molecular Cell
Biology and Genetics, Dresden, Germany

Leah Krubitzer
Centre for Neuroscience, University of
California, Davis, USA

Maarja Kruusmaa
Centre for Biorobotics, Tallinn University of
Technology, Estonia

Vincent Lebastard
Automation, Production and Computer
Sciences Department, IMT Atlantique
(former Ecole des Mines de Nantes), France

Pablo Ledezma
Bristol Robotics Laboratory, UK, and
Advanced Water Management Centre,
University of Queensland, Australia

Chanseok Lee
School of Mechanical and Aerospace
Engineering, Seoul National University,
South Korea

Torsten Lehmann
School of Electrical Engineering and
Telecommunications, University of New
South Wales, Australia

Joel Z. Leibo
Google DeepMind and McGovern Institute
for Brain Research, Massachusetts Institute
of Technology (MIT), USA

Charles Lenay
Philosophy and Cognitive Science, University
of Technology of Compiègne, France

John J. Leonard
Computer Science and Artificial Intelligence
Laboratory, Massachusetts Institute of
Technology, USA

Nathan F. Lepora
Department of Engineering Mathematics and
Bristol Robotics Laboratory, University of
Bristol, UK

Hannah Maslen
Oxford Uehiro Centre for Practical Ethics,
University of Oxford, UK

Christof Mast
Systems Biophysics, Ludwig-Maximilians-
Universität München, Germany

Barbara Mazzolai
Center for Micro-BioRobotics, Istituto
Italiano di Tecnologia, Italy

Chris Melhuish
Bristol Robotics Laboratory, University of the West of England, UK

Giorgio Metta
Istituto Italiano di Tecnologia, Genoa, Italy

Abigail Millings
Department of Psychology, University of Sheffield, UK

Ben Mitchinson
Department of Psychology, University of Sheffield, UK

Friederike Möller
Systems Biophysics, Ludwig-Maximilians-Universität München, Germany

Matthew S. Moses
Applied Physics Laboratory, Johns Hopkins University, USA

Anna Mura
SPECS, Institute for Bioengineering of Catalonia (IBEC), the Barcelona Institute of Science and Technology (BIST), Barcelona, Spain

Masahiro Nakajima
Center for Micro-nano Mechatronics, Nagoya University, Japan

Stefano Nolfi
Laboratory of Autonomous Robots and Artificial Life, Institute of Cognitive Sciences and Technologies (CNR-ISTC), Rome, Italy

Benjamin M. O'Brien
Auckland Bioengineering Institute, The University of Auckland, New Zealand

Benedikt Obermayer
Berlin Institute for Medical Systems Biology, Max Delbrück Center for Molecular Medicine, Berlin, Germany

Changhyun Pang
School of Chemical Engineering, SKKU Advanced Institute of Nanotechnology, Sungkyunkwan University, South Korea

Tim C. Pearce
Department of Engineering, University of Leicester, UK

Tomaso Poggio
Department of Brain and Cognitive Sciences, McGovern Institute for Brain Research, Massachusetts Institute of Technology, USA

Girijesh Prasad
School of Computing, Engineering and Intelligent Systems, Ulster University, Londonderry, UK

Tony J. Prescott
Sheffield Robotics and Department of Computer Science, University of Sheffield, UK

Holger Preuschoft
Institute of Anatomy, Ruhr-Universität Bochum, Germany

Roger D. Quinn
Mechanical and Aerospace Engineering Department, Case Western Reserve University, USA

Roy E. Ritzmann
Biology Department, Case Western Reserve University, USA

Nicholas Roy
Computer Science and Artificial Intelligence Laboratory, Massachusetts Institute of Technology, USA

Julian Savulescu
Oxford Uehiro Centre for Practical Ethics, University of Oxford, UK

Giacomo Scandroglio
Bristol Robotics Laboratory, University of the West of England, UK

Cornelius Schilling
Biomechatronics Group, Technische Universität Ilmenau, Germany

Malte Schilling
Center of Excellence for Cognitive Interaction Technology, Universität Bielefeld, Germany

Severin Schink
Theory of Complex Biosystems, Technische Universität München, Garching, Germany

Allen Selverston
Division of Biological Science, University of California, San Diego, USA

Wolfgang Send
ANIPROP GbR, Göttingen, Germany

Anil K. Seth
Sackler Centre for Consciousness Science, University of Sussex, UK

Leslie S. Smith
Department of Computing, Science and Mathematics, University of Stirling, UK

Dong Song
Department of Biomedical Engineering, University of Southern California, USA

Kahp-Yang Suh
School of Mechanical and Aerospace Engineering, Seoul National University, South Korea

Michael Szollosy
Sheffield Robotics, University of Sheffield, UK

Masaru Takeuchi
Department of Micro-Nano Systems Engineering, Nagoya University, Japan

Matthieu Tixier
Institut Charles Delaunay, Université de Technologie de Troyes, France

Barry Trimmer
Biology Department, Tufts University, USA

Takuya Umedachi
Graduate School of Information Science and Technology, The University of Tokyo, Japan

André van Schaik
Bioelectronics and Neuroscience, MARCS Institute for Brain, Behaviour and Development, Western Sydney University, Australia

Stefano Vassanelli
Department of Biomedical Sciences, University of Padova, Italy

Paul F. M. J. Verschure
SPECS, Institute for Bioengineering of Catalonia (IBEC), the Barcelona Institute of Science and Technology (BIST), and Catalan Institute of Advanced Studies (ICREA), Spain

Julian Vincent
School of Engineering, Heriot-Watt University, UK

Danja Voges
Biomechatronics Group, Technische Universität Ilmenau, Germany

Vasiliki Vouloutsi
SPECS, Institute for Bioengineering of Catalonia (IBEC), the Barcelona Institute of Science and Technology (BIST), Barcelona, Spain

Stuart P. Wilson
Department of Psychology, University of Sheffield, UK

Hartmut Witte
Biomechatronics Group, Technische Universität Ilmenau, Germany

Robert H. Wortham
Department of Computer Science, University of Bath, UK

Section I

Roadmaps

Chapter 1

Living Machines: An introduction

Tony J. Prescott[1] and Paul F. M. J. Verschure[2]

[1] Sheffield Robotics and Department of Computer Science,
University of Sheffield, UK
[2] SPECS, Institute for Bioengineering of Catalonia (IBEC), the Barcelona
Institute of Science and Technology (BIST), and Catalan Institute of Advanced
Studies (ICREA), Spain

Homo sapiens has moved beyond its evolutionary inheritance by progressively mastering nature through the use of tools, the harnessing of natural principles in technology, and the development of culture. Quality of life in our modern industrialized societies is far removed from the conditions in which our early ancestors existed and we have reached a stage where we ourselves have become the main driving force of global change, recognized now as a distinct geological era— the *Anthropocene*. However, our better health, longer life expectancy, greater prosperity, ubiquitous user-friendly devices, and expanding control over nature are creating challenges in multiple domains—personal, social, economic, urban, environmental, and technological. Globalization, the demographic shift, the rise of automation and of the digital life, together with resource depletion, climate change, income inequality, and armed conflict with autonomous weaponry are creating unprecedented hazards.

For the ancient Greek philosophers, such as Socrates, Aristotle, and Epicurus, the major question was how to find *eudaemonia* or "the good life." We face the same question today, but in a technologically advanced, rapidly changing, and hyperconnected world. We are also faced with a paradox—the prosperity that we have achieved through science and engineering risks becoming a threat to our quality of life. Where in the nineteenth and early twentieth century technological advances seemed to be automatically coupled to a general improvement in the quality of life, now this relationship is not so obvious anymore. Hence, how can we assure that our species will continue to thrive? The answer must involve greater consideration for each other and respect for the ecosystems that support us, but the future development of science and technology will also play a crucial role.

The gap between the natural and the human-made is narrowing as this book will show. Beyond the pressures that we have already noted, humanity is faced with an even stranger and unprecedented future—not only will our artifacts become increasingly autonomous, we may also be transitioning towards a post-human or biohybrid era in which we incrementally merge with the technology that we have created.

As we move towards this future we believe there is a need to review and reconsider how we develop, deploy, and relate to the artifacts that we are creating and that will help define what we are becoming. Our specific approach is biologically grounded. We seek to learn from the natural world about how to create sustainable systems that support and enhance life, and we see this as a way to make better technologies that are more human-compatible.

The approach we take is also convergent—including science, technology, humanities, and the arts—as we look to deal with both the physics and metaphysics of the new entities, interfaces, and implants we are developing, and to discover what this research tells us about the human condition.

Figure 1.1 Mimicking the appearance or principles of flight. (a) The steam-driven 'pigeon' designed by Archytas of Tarentum (428–347 bc), considered to be the founder of mathematical mechanics. The artificial bird was apparently suspended by wires but nevertheless astonished watching crowds. (b) The Wright brothers designed and built the world's first successful flying machine, the *Kitty Hawk*, which took to the air in 1903. Their breakthrough was achieved through a careful study

We call our field "living machines" because we consider that a single unifying science can help us to understand both the biological things that we are and the kinds of synthetic things that we make. In doing so, we knowingly dismiss a Cartesian dualism that splits minds from bodies, humans from animals, and organisms from artifacts, and we celebrate the richness of our own evolved biological sentience that is enabling us, through our science and technology, to make better sense of ourselves.

The living machines opportunity

Contemporary technology is still far removed from the versatility, robustness, and dependability of living systems. Despite tremendous progress in material science, mechatronics, robotics, artificial intelligence, neuroscience, and related fields, we are still unable to build systems that are comparable with insects. There are a number of reasons for this, but two stand out as fundamental and intrinsic limitations. On one hand, artifacts are still very limited in autonomously dealing with the real world, especially when this world is populated by other agents. On the other, we face fundamental challenges in scaling and integrating the underlying component technologies. The approach to addressing these two bottlenecks, followed in this handbook, is to translate into engineering the fundamental principles underlying the still unsurpassed ability of natural systems to operate and sustain themselves in a myriad of different real world niches. Nature provides the only example of sustainability, adaptability, scalability, and robustness that we know. Harnessing these principles should render a radically new class of technology that is renewable, adaptive, robust, self-repairing, potentially social, moral, perhaps even conscious (Prescott et al. 2014; Verschure, 2013). This is the realm of living machines.

Within the domain of living machines we distinguish two classes of entities—*biomimetic* systems that harness the principles discovered in nature and embody them in new artifacts, and *biohybrid* systems that couple biological entities with synthetic ones in a rich and close interaction so forming a new hybrid bio-artificial entity.

Biomimetics is an approach that is as ancient as civilization itself. In the fourth century BC the Greek philosopher Archytas of Tarentum built a self-propelled flying machine made from wood and powered by steam that was structured to resemble a bird (Figure 1.1, top). As history progresses we can observe an important shift from mimicking the form of natural processes, such as the shape and movement of bird wings, to the extraction of principles of lift and drag that, for instance, allowed the Wright brothers to realize the first flying machines (Figure 1.1, middle). Hence, the domain of biomimetics has learned the important lesson to look beyond the surface of living systems and understand and re-use the principles they embody. Today, we are able to mimic the winged flight of birds with good accuracy (see Hedenström, Chapter 32, this

---

of bird flight and aerodynamics using a home-built air tunnel that uncovered some of the key principles of flight. Or, in the words of Orville Wright, "learning the secret of flight from a bird was a good deal like learning the secret of magic from a magician." (McCullough, 2015). (c) *SmartBird* (described in detail in Send, Chapter 46, this volume) is one of a number of biomimetic robots created by the German components manufacturer Festo, who have also built a robotic elephant trunk, swimming robot manta rays, and a flying robot jellyfish. *SmartBird* demonstrates how an understanding of the principles of flight is now enabling researchers to build models of natural systems that capture their specific capabilities.

(b) Image: Library of Congress. (c) Image: Uli Deck/DPA/PA Images.

volume), the Festo *SmartBird* (Figure 1.1, bottom) being a good example (see Send, Chapter 46, this volume), through an understanding of those aspects of bird design that are critical to generating the right balance between lift and thrust.

Research in biomimetics is flourishing. Promising areas include system design and structure, self-organization and cooperativity, biologically inspired active materials, self-assembly and self-repair, learning, memory, control architectures and self-regulation, movement and locomotion, sensory systems, perception, cognition, control, and communication. The list goes on and is explored throughout the chapters in Sections II–IV of this volume. In all of these areas, biomimetic systems offer the potential for more effective, adaptive, and robust technologies than we have at present, and this is being shown through the construction and demonstration of a wide range of different biomimetic devices and animal-like robots (see Section V).

Biohybrid systems is a relatively new field, with exciting and largely unknown potential, but one that is likely to shape the future of humanity. Examples of current research in biohybrid systems (see Section VI) include brain–machine interfaces—where neurons are connected to microscopic sensors and actuators—and various forms of intelligent prostheses from sensory devices like artificial retinas, to life-like artificial limbs, wearable exoskeletons, and virtual reality-based rehabilitation approaches and neuroprosthetics that incorporate fundamental principles underlying the organization of the brain.

One key defining feature of both biomimetic and biohybrid research is that it establishes a bi-directional positive feedback loop between the basic scientific study or *analysis* of living systems, and the engineering disciplines, where the natural principles underlying the body, behavior, and the brain are *synthesized* into novel technologies. At the same time, the construction of artificial systems also constitutes a direct method to validate our hypotheses about the importance of the discovered biological principles via an existence proof (see Figure 1.2). The artifact instantiates our understanding of the living system, or, to put it another way, the machine *is* the theory (see Verschure and Prescott, Chapter 2, this volume, and Verschure 2012).

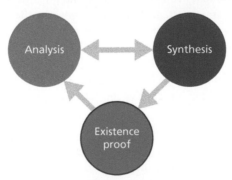

Figure 1.2 Since analysis (natural science) and synthesis (engineering) proceed by different methods, they are complementary approaches that can unlock challenges previously seen as beyond our reach (Braitenberg, 1986; Simon, 1969). Further, as synthetic approaches result in real-world technologies, they can stand as powerful existence proofs for the validity of the insights, drawn from biology, but translated into engineering. This is especially relevant for the study of complex multi-scale non-linear systems that elude effective analytic description, such as living systems and complex artifacts.

The living machines paradigm can, in principle, extend to all fields of biological research, from physiology and molecular biology to ecology, and from zoology to botany to agriculture. Biomimetic research, particularly at the nano-scale, should also lead to important advances in component sustainability, miniaturization, self-configuration, self-repair, and energy efficiency. Another key focus is on complete behaving systems in the form of biologically grounded robotic systems that can operate on the different substrates of sea, land, and air. A further central theme is the physiological basis for advanced adaptive behaviour as explored through the modeling of embodied neural systems. Exciting emerging topics within this field include the development of neuromimetic controllers in hardware, termed neuromorphics, and within the control architectures of robots, sometimes termed neurorobotics. One of the challenges of science is to better understand both human capabilities and limitations; this is fundamental in order to advance technologies that can expand human skill while compensating for any failings. The pressing need to develop assistive technologies for individuals with impairments can be related to that of developing bionic technologies for healthy individuals, where the Living Machine's focus is to advance those technologies that augment in a symbiotic manner, rather than replace, human capability.

A short history of living machines

As noted earlier, the ambition to understand and apply insights from nature has been with us since ancient times. The great Renaissance polymath and inventor, Leonardo da Vinci, developed many designs for machines inspired by his own observations of natural systems and their internal mechanisms. A century later, Francis Bacon, today seen as the founder of scientific method, described the natural world as a pyramid, with observations at the bottom and the laws of nature (what we are calling biological principles) at the top—there to be discovered by carefully crafted empirical research. For his Italian contemporary, Galileo Galilei, the world, and the universe it sits in, comprise a vast mechanical system that could be understood through mathematically describable laws. Thomas Hobbes extended Galileo's vision to the human condition, regarding it as likewise one of matter in motion, with bodies as machines and the different elements of our mental life—deliberation, imagination, emotion—forms of interior motion. The notion that living things were biological machines was also shared by Rene Descartes, inspired by the clockwork automata of his day. Famously, of course, Descartes was unwilling to extend this idea to his own species, forcing him to assert and defend a mind–body dualism that he applied to humans but not animals, and whose legacy in Western culture extends to this day. It was left to the physician and philosopher Julien Offray de La Mettrie to directly challenge the orthodox eighteenth-century view that humans incorporate an otherworldly soul—a brave step at a time when such a heresy could still result in censorship or prosecution. De La Mettrie argued for the continuity between animals and humans, that animals were living machines, meaning that we too should be understood as complex, but entirely mechanistic, entities. De La Mettrie's book *L'Homme Machine* (1748), hurriedly written, and more rhetorical than scientific, also introduced the important idea that machines—natural or otherwise—can be dynamic, autonomous, and purposive entities. Around the same time, the famous automatons of the French inventor Jacques de Vaucanson (see Figure 1.3)—the ancestors of modern theme-park animatronics—were emblematic of this emerging view of nature.

With the rise of cybernetics in the first half of the twentieth century, it became clear that there was the possibility to create inventions that would realize this vision of machines that were both autonomous and purposive. The term "biomimetics" was introduced by Otto

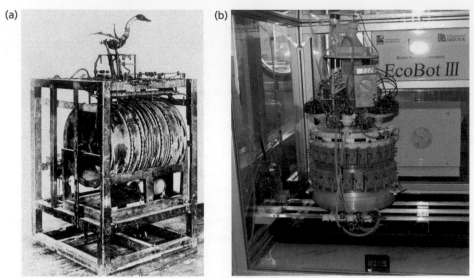

Figure 1.3 (a) Jacques de Vaucanson's *Canard Digérateur*, or Digesting Duck was designed to appear as though it could consume and digest kernels of corn, turning them into pellets of excrement. The duck did not actually digest but was fitted with a concealed container that retained the food whilst a trap-door released the artificial dropping. The duck, which could also flap its wings and drink water, had over four hundred moving parts but was sadly destroyed in a museum fire at the end of the nineteenth century. In 2006, Wim Delvoye created an automaton called the "Cloaca Machine" that could actually digest food, turning it into fecal matter similar to human excrement; packs of the latter were sold to collectors as works of art. (Image: The automaton Digesting Duck by French inventor Jacques de Vaucanson (1709–1782), created in 1739/Bridgeman Images.) (b) Living machines now exist, such as Bristol Robotics Lab's *EcoBot III* pictured here, that can extract energy by breaking down organic matter to do useful work (see Ieropoulos et al., Chapter 6, this volume).

Schmitt during the 1950s, and "bionics" by Jack Steel (popularized in Daniel Halacy's book *Bionics: the science of "living" machines* (Halacy, 1965)). These fields represented a growing movement in engineering that sought to build better ties with the biological sciences and to make progress through the "reverse engineering" of natural systems. The biological sciences had also taken up the notion of cells and organisms as living machines, with an important movement growing around systems theory (see Prescott, Chapter 4, this volume) that saw the need to understand the natural world at multiple levels of scale (Figure 1.4). However, if organisms are mechanical then it is clear that they are also a very different kind of machine to man-made devices such as clockwork ducks, internal combustion engines, or even modern computers. An important focus of recent times has been on autonomy, with the consequence that we are now entering a new era of (relatively) autonomous machines, such as self-driving cars and assistive robots. Despite this progress, some, arguably more fundamental, differences remain to be captured, notably that organisms, unlike most machines, are self-organizing and self-creating systems (Maturana and Varela, 1973/1980). An important theme in the science and engineering described in this book is to recognize and rise to the challenge of understanding these essential properties of living entities that are currently missing in our artifacts and only partially explored in our science.

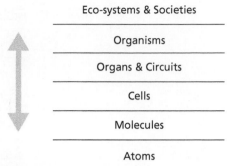

Eco-systems & Societies

Organisms

Organs & Circuits

Cells

Molecules

Atoms

Figure 1.4 Multiple levels of organization in living systems. In line with the systems approach in biology (Bertalanffy 1969), and the emergentist view in complexity theory, our approach advocates an understanding of living machines, natural and human-made, at multiple interacting levels of organization. Reciprocal interactions between elements of a given scale of complexity give rise to emergent structure at the higher scale that in turn can structure lower levels, suggesting a hierarchy of complexity (Rasmussen et al. 2001; Verschure et al. 2003).

Since the start of this century there has been explosive growth in biomimetic research, with the number of published papers doubling every two to three years (Lepora et al. 2013). In 2012 a new international conference series *Living Machines: Biomimetic and Biohybrid Systems* (LM) held its first meeting in Antoni Gaudi's bio-inspired building *La Pedrera* in Barcelona (Prescott et al. 2012). The second edition, in 2013, included an exhibition of biomimetic and biohybrid artifacts and art at the London Science Museum (Figure 1.5), a theme of combining science and engineering with art and design that permeates the field (an earlier example is the biomimetic sentient space Ada that was realized for the Swiss national exhibition Expo.02 (Eng et al. 2003)). The conference series has provided a valuable forum for this dialogue between nature and technology and the arts, a place where people can discuss the biomimetic and biohybrid machines of tomorrow, and where they can consider what this mean for understanding the biological machines—humans and animals—of today. The current book is the result of talks and discussions from the conference series, from the *Barcelona Summer School in Brain Cognition and Technology* (BCBT) (running since 2008), and from the wider activities of *The Convergent Science Network for Biomimetic and Biohybrid Systems* (CSN, www.csnetwork.eu), an international scientific network sponsored by the European Framework 7 Programme between 2010 and 2016. The work of the network continues, as does the Living Machines conference series, which held its 6th annual meeting in Stanford, California in July 2017 (Mangan et al. 2017).

Ethical and societal issues

As with all technology development, it is important to balance enthusiasm for the development of future biomimetic and biohybrid technologies with forethought about their potential societal impacts (Prescott et al. 2014). Nevertheless, we believe that, given the speed at which our civilization is careering towards resource depletion, continuation with the status quo is not an option. Our existing advanced technologies make heavy use of rare metals and metalloids for which supplies are running out. They are also difficult to dismantle and recycle, leading to stockpiles of waste materials and urban mines that are dangerously polluting throughout the world, but particularly in developing countries. We have much to learn from nature about

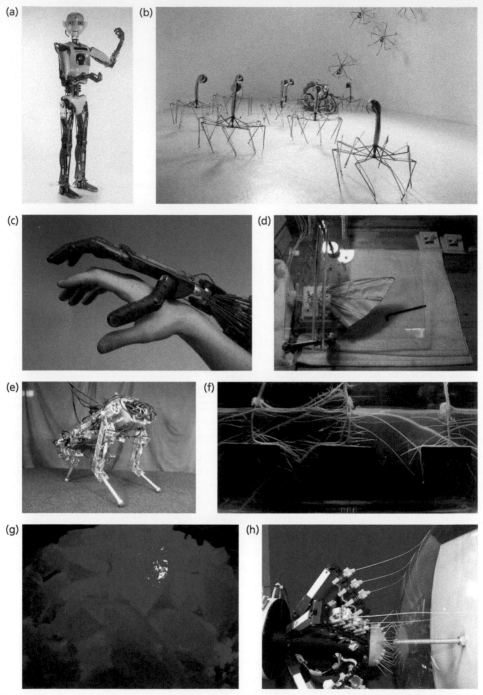

Figure 1.5 Exhibits at the Second International Conference on Living Machines in London, 2013. (a) Robothespian—an interactive multilingual humanoid created by *Engineered Arts* (© Engineered Arts Ltd, 2017). (b) Arachnoids—this artwork by Konstantinos Grigoriadis reflects the imagining of a biomechanical, post-apocalyptic world (© Konstantinos Grigoriadis, 2013). (c) The Shadow dextrous hand—an anthropomorphic robot hand pictured here above a human hand (© Shadow Robot Company, 2013). (d) A robotic fish fin with a sense of touch from Drexel University (© Jeff Kahn and James Tangorra, 2017). (e) HyQ—a quadruped robot mammal developed by the Advanced Robotics Group at the Istituto Italiano di Tecnologia (IIT) (© Claudio Semini, IIT, 2013). (f) Biological roots move and grow—this exhibit showed how principles from the study of plant movement can be translated into robotics (see also Mazzolai, Chapter 9, this volume) (© Center

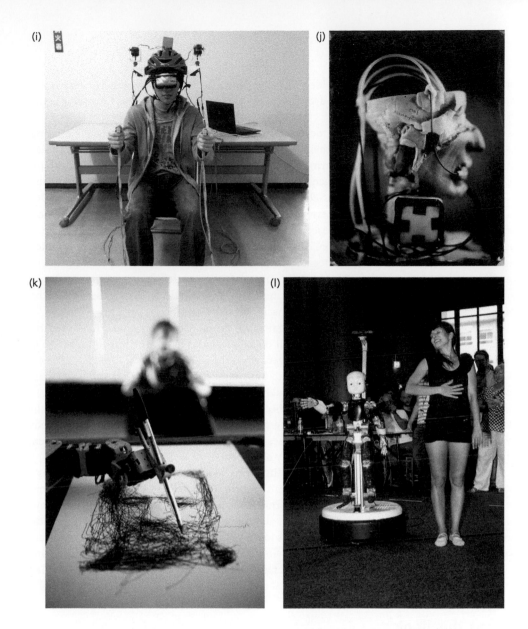

for Micro-BioRobotics, IIT, 2017). (g) Dreamscape concrete—an exploration of materials that react to human presence by the artist Sam Conran (© Sam Conran, 2017). (h) The BIOTACT Sensor—a mammal-like whiskered touch system developed by Bristol Robotics Laboratory together with Sheffield Robotics (see also Prescott, Chapter 45, this volume) (Image courtesy of Charlie Sullivan. © Bristol Robotics Laboratory, 2013). (i) The virtual chameleon—chameleons have independently moveable eyes that can be directed in different directions. This exhibit from Tohoku Institute of Technology allowed the user to control two cameras by hand each of which presents its current field of view to one eye (© Fumio Mizuno, 2017). (j) AnTon—an animatronic talking head designed and built by Robin Hofe; one of a series of wet-plate collodion images of biomimetic robots by the photographer Guy J Brown (© Guy J. Brown, 2017). (k) Paul, an artificial drawing entity—a robot that draws portraits created by the artist Patrick Tresset (photo by Tommo © Patrick Tresset, 2017). (l) The iCub humanoid—in the exhibited work the iCub robot, developed by a European consortium lead by IIT (see also Metta and Cingolani, Chapter 47, this volume), performed a repertoire of dance-like movements in dialogue with the dancer Anuska Fernandez (© Sarah Prescott Photography, 2017).

building complex machines out of widely available, intrinsically-less-hazardous materials that have natural life-cycles of growth, decomposition and re-use (see Halloy, Chapter 65, this volume).

A growing concern is that human society could lose control of the technologies it is creating, as the latter rush past us in terms of their capabilities through a so-called intelligence explosion and towards a "technological singularity." This scenario is often put forward by people with the misapprehension that human intelligence is a unitary phenomenon. In fact, there are undoubt-edly multiple intelligences, as the psychologist Howard Gardner has argued (Gardner 2006), grounded in a number of brain processes (Hampshire et al. 2012), and despite recent advances in machine learning, there is good reason to be sceptical that machines will match human intel-ligence in all of its facets anytime soon. Also we have to question whether the construct of intel-ligence that emerged in the late nineteenth century, and was given prominence by the artificial intelligence movement of the mid-twentieth century, is the key defining property of humans and animals to pursue in Living Machines. Alternatively, we should focus on other integrative properties such as cognition, emotion and consciousness.

Irrespective of how we measure humans against machines, what seems more likely, regard-ing our shared future, is that our species will do as we always have, which is to make use of new and smarter artifacts to boost our intelligence and the collective capabilities of our society (Clark 2003; Prescott 2013); this scenario of an interconnected global network of humans and machines is sometimes described as the "global brain" (Russell 1982).

We might also anticipate that people will become more biohybrid themselves, through the development of personalized systems that are more-and-more closely interfaced to the body (Prescott 2013; Wu et al. 2013). This also raises ethical issues of course, though of a different kind, for instance around privacy and equality of access.

More generally, and in order for advances in Living Machine technologies to be seen as a benefit rather than a threat, we should ensure that progress is made with broad societal con-sent (see e.g. Prescott and Verschure 2016). This will require engagement with a wide range of stakeholders, including the general public, about our scientific aims and the technological advances we wish to pursue, and a willingness to adapt those aims to address concerns that are well-founded. As a step in this direction, the final section of this book is devoted to exploring future applications of living machines, from human augmentation through to robot–human relationships, and some of their possible ethical and societal impacts.

About this book

Whilst these are undoubtedly exciting times for research in biomimetic and biohybrid systems, at the same time, research is currently very fragmented with respect to the disciplines involved, the problems addressed, and the different types of end uses; this despite the commonality in many of the underlying approaches and aims. Arguably this has held back the visibility of the research and weakened the position of the community pursuing this approach in mobilizing support from institutions, funding bodies, and governments. In order to advance the field, we consider that there is a pressing need to (i) increase the amount of shared knowledge, (ii) estab-lish a clearer picture of the current state-of-the-art, (iii) identify shared methodologies, and (iv) collectively consider our visions of the future. The success of any research endeavor also crit-ically depends on capturing the enthusiasm of young researchers and providing them with the tools and skills to pursue their own ideas. These are some of the goals that have motivated us to organize conferences, summer schools, and exhibitions, and to bring a review and synthesis of these ideas together in this book.

Our handbook seeks to provide a survey of the field that, whilst not complete (that would require multiple volumes), we hope is representative of many of the major themes and approaches. We have included a mixture of shorter and longer contributions, written by known leaders in their fields, and targeting a wide audience from the lay reader through to the informed researcher. We have also included some additional introductory material about the Living Machines methodology and some ideas from the community about how the field of biomimetics might develop in the next few years. In most chapters we try to balance a description of fundamental principles gleaned from research on natural systems with a discussion of the biomimetic or biohybrid technologies they have given rise to. In addition, to assist the reader in following up on the work described here, many chapters include pointers for further reading and specific ideas for future research.

Our book is organized in seven main sections—*Roadmaps, Life, Building blocks, Capabilities, Biomimetic systems, Biohybrid systems*, and *Perspectives*.

The remainder of Section I, *Roadmaps*, brings together some of the key ideas and themes of the book in two longer chapters written by the book's editors. The first of these, *The Living Machines approach to the sciences of mind and brain*, explores some of the core methodologies that we believe should underpin the science of Living Machines, focusing on the research domain we know best—that of the mind and brain—and relating these to the past history of the field and current research landscape. A key emphasis is on the need for a renewed commitment to multi-scale approaches, to the discovery of general principles, and to the value of physical models in advancing and integrating scientific understanding. The second chapter in this section, *A roadmap for biomimetics*, brings together some insights into the future of research in biomimetics gathered from field leaders; we also summarize some general findings about the current state-of-the-art and provide pointers to relevant chapters throughout the rest of the handbook.

Section II, *Life*, focuses on understanding and emulating some of the fundamental properties of living systems such as their capacity to self-organize, to metabolize, to grow, and to reproduce, and the materials out of which they construct themselves. This section also includes a proposal for a unified theory of evolution in natural and artificial systems.

Section III, *Building blocks*, is concerned with principles and technologies relating to specific system elements—for example, we include chapters on the different sensory systems, including proprioception, on muscles and skin, and on the neural system underlying oscillations and rhythms.

Section IV, *Capabilities*, is concerned with the integration of components, such as those described in Section III, into systems that deliver some of the key behavioral competencies that we have identified in natural systems and that we wish to see in our artifacts. Contributions here focus on cognitive capacities such as perception, attention, learning, and memory; sensorimotor capabilities such as reach and grasp, locomotion, and flight; and integrative capacities such as emotions and consciousness. Two chapters on brain architecture discusses how components can be combined to support a range of robust life-like behaviours in living machines.

Section V, *Biomimetic systems*, considers examples of how these ideas are being used to build complete systems—a (partial) "phylogeny" of current efforts to devise biomimetic artifacts. We begin with a contribution describing attempts to generate laboratory conditions similar those that gave rise to the origin of life on Earth with the aim of creating new chemical living machines that resemble the first replicating microorganisms. This is followed by chapters on artificial slime molds, and robot models of soft-bodied invertebrates, insects, fish, mammals, birds, and humans.

Section VI, *Biohybrid systems*, explores how we are moving towards different forms of biohybridicity, through human machine–brain interfaces, sensory, motor, and cognitive prostheses. We also include a chapter about building biohybrid robots using tissue from animals and techniques from synthetic biology.

The final section, *Perspectives*, includes invited contributions from philosophers and other commentators on the potential impacts of this research on society and on how we see ourselves.

References

Bertalanffy, L. v. (1969). *General system theory: foundations, development, applications.* New York: Braziller.

Braitenberg, V. (1986). *Vehicles: experiments in synthetic psychology.* Cambridge, MA: MIT Press.

Clark, A. (2003). *Natural-born cyborgs: minds, technologies and the future of human intelligence.* Oxford: Oxford University Press.

Eng, K., Klein, D., Bäbler, A., Bernardet, U., Blanchard, M., Costa, M., ... Verschure, P.F.M.J. (2003). Design for a brain revisited: the neuromorphic design and functionality of the interactive space 'Ada'. *Rev. Neurosci.,* **14**(1–2), 145–80.

Gardner, H. (2006). *Multiple Intelligences: New Horizons.* New York: Basic Books.

Halacy, D.S. (1965). *Bionics, the science of "living" machines.* New York: Holiday House.

Hampshire, A., Highfield, R.R., Parkin, B.L., and Owen, A.M. (2012). Fractionating human intelligence. *Neuron,* **76**(6), 1225–37.

Lepora, N.F., Verschure, P., and Prescott, T.J. (2013). The state of the art in biomimetics. *Bioinspiration and Biomimetics,* **8**(1), 013001. doi:10.1088/1748-3182/8/1/013001

Mangan, M., Cutkosky, M., Mura, A., Verschure, P. F. M. J., Prescott, T. J., and Lepora, N. (eds), *Biomimetic and Biohybrid Systems: 6th International Conference, Living Machines* 2017, Stanford, CA, USA, July 26–28, 2017, Proceedings (pp. 86–94). Cham: Springer International Publishing.

Maturana, H., and Varela, F. (1973/1980). *Autopoiesis: The Organization of the Living.* Dordrecht, Holland: D. Reidel Publishing.

McCullough, D. (2015). *The Wright Brothers.* New York: Simon & Schuster.

Prescott, T. J. (2013). The AI singularity and runaway human intelligence. *Biomimetic and Biohybrid Systems; Second International Conference on Living Machines,* Lecture Notes in Computer Science, vol. **8064**, pp. 438–40.

Prescott, T.J., Lepora, N., and Verschure, P.F.M.J. (2014). A future of living machines?: International trends and prospects in biomimetic and biohybrid systems. *Proc. SPIE 9055, Bioinspiration, Biomimetics, and Bioreplication 2014,* 905502. doi: 10.1117/12.2046305

Prescott, T.J., Lepora, N.F., Mura, A., and Verschure, P.F.M.J. (2012). *Biomimetics and Biohybrid Systems: First Iinternational Cconference on Living Machines.* Lectures Notes in Computer Science, vol. 7375. Berlin: Springer-Verlag.

Prescott, T.J., and Verschure, P.F.M.J. (2016). Action-oriented cognition and its implications: Contextualising the new science of mind. In: A.K. Engel, K. Friston, and D. Kragic (eds.), *Where's the Action? The Pragmatic Turn in Cognitive Science.* Cambridge, MA: MIT Press for the Ernst Strüngmann Foundation, pp. 321–31.

Rasmussen, S., Baas, N.A., Mayer, B., Nilsson, M., and Olesen, M.W. (2001). *Ansatz* for dynamical hierarchies. *Artificial Life,* 7(4), 329–53. doi:10.1162/106454601317296988

Russell, P. (1982). *The Awakening Earth: The Global Brain.* London: Ark.

Simon, H. A. (1969). *The Sciences of the Artificial.* Cambridge, MA: MIT Press.

Verschure, P.F.M.J. (2012). The distributed adaptive control architecture of the mind, brain, body nexus. *Biologically Inspired Cognitive Architectures,* 1(1), 55–72.

Verschure, P.F.M.J. (2013). Formal minds and biological brains II: from the mirage of intelligence to a science and engineering of consciousness. *IEEE Expert,* 28(5), 33–6.

Wu, Z., Reddy, R., Pan, G., Zheng, N., Verschure, P.F.M.J., Zhang, Q., ... Müller-Putz, G.R. (2013). The Convergence of Machine and Biological Intelligence. *IEEE Intelligent Systems,* 28(5), 28–43. doi:10.1109/MIS.2013.137

Chapter 2

A Living Machines approach to the sciences of mind and brain

Paul F. M. J. Verschure[1] and Tony J. Prescott[2]

[1] SPECS, Institute for Bioengineering of Catalonia (IBEC), the Barcelona Institute of Science and Technology (BIST), and Catalan Institute of Advanced Studies (ICREA), Spain
[2] Sheffield Robotics and Department of Computer Science, University of Sheffield, UK

How do the different sciences of mind and brain—from neuroscience and psychology, through to cognitive science and artificial intelligence (AI)—stand in relation to each other at this moment in the twenty-first century?

Our aim, in this chapter, is to persuade you that despite the fact that our knowledge is expanding at ever-accelerating rates, our understanding of the relationship between mind and brain is, in some important sense, becoming less and less. An explanatory gap is building that, for us, can only be bridged by a kind of multi-tiered and integrated theoretical framework that recognizes the value of developing explanations at different levels of description of mind and brain and combining these into cross-level integrated theories.

A second goal of this chapter is to show that, in bridging this explanatory gap, we can directly contribute to advancing new technologies that improve the human condition. Indeed, our view is that the development of technologies that instantiate principles gleaned from the study of the mind and brain, or biomimetic technologies, is a key part of the validation process for the scientific theories that we present.

We call this strategy for the integration of science and engineering a Living Machines approach as it recognizes the symbiotic relationship between these domains. The strategy has many antecedents, but one that should be more widely known is the work of the eighteenth century Neapolitan philosopher Giambattista Vico who famously proposed that we can only understand that which we create: *Verum et factum reciprocantur seu convertuntur*[1]. We will argue that following the creative path proposed by Vico can lead not only to better science (understanding), and useful engineering (new life-like technologies in form and function), but can also guide us towards a richer view of human experience and of the boundaries and relationships between science, engineering and art (Verschure 2012).

Matter over mind

To begin, let us consider some concrete examples of how the science of the mind and brain is currently being pursued. Since mind emerges from the brain, an important trend that we have

---

[1] Famously echoed by the physicist Richard Feynman, whose blackboard at the time of his death held the statement "What I cannot create, I do not understand."

noticed is an increasing focus of resources and efforts towards the brain side of the mind–brain duality, seemingly in the hope that this will unlock the secrets of both. We call this trend "matter over mind" because we feel that it is drawing attention towards physical things that can be directly measured—brain processes—but in a manner that risks losing sight of what those processes collectively achieve—instantiating the mind.

Two concrete and significant examples of this trend are as follows. In 2013, the European Commission initiated the Human Brain Project (HBP)—a decade-long, €1 billion effort to understand and emulate the human brain. In the same year, the US Government announced the BRAIN Initiative—projected to direct funding of $3 billion to brain research over a similar ten-year period. With this level of investment and enthusiasm we can hope that great advances in brain science are surely just around the corner. Knowing so much more about the brain, we should surely also know much more about the mind and thus about ourselves.

However, although this increased international enthusiasm for brain science is exciting and in many ways welcome, there are some niggles. Looking at these flagship projects we are struck by how both initiatives are convinced that an understanding of the brain, and hence the mind, will proceed from a very large-scale, systematic approach to measuring the brain and its physical properties. More precisely both of these projects are leveraging powerful twenty-first century technologies—such as the latest human brain imaging, nanotechnology, and optogenetic methods—that can make the connectivity and activity of the brain more apparent. They will then apply the tools of "big data," such as automated classification, reconstruction, and machine learning approaches, powered by the accelerating power and capacity of computers, to help make sense of what will amount to a tsunami of new measurements.

While all of this is well and good, we see a significant gap. Will we know ourselves and our brains once we have all of these "facts" in our database? Where are the theories of brain function that are going to explain all of this new anatomical and physiological detail? Do we need a theory at all? How are we going to make the connection between the understanding of the brain at a tissue level and the understanding of mind at a psychological level? In this fascination with the brain as the physiologically most complex organ in the human body, are we losing sight of what is needed to understand and explain the role of the brain in guiding and generating behavior and shaping experience?

While the proponents of these big data projects have argued that we need better data to drive theory building (Alivisatos et al. 2012; Markram 2012), we contend that there is already a mountain of unexplained high-quality data about mind and brain, and what is needed are better theories for trying to make sense of it all.

Part of the solution to the challenge of connecting the brain to behavior is computational modeling, either using computer simulation, or, to understand the link between brain and behavior more directly, by embedding brain models in robots. Naturally these large-scale projects that are exploring the human brain will apply and extend current computational neuroscience models and methods so, on the surface, all seems well. In particular, they will develop computer simulations that seek to capture rich new data sets at an unprecedented level of accuracy using hugely powerful massively parallel machines, in an attempt to show how interactions among the microscopic elements that constitute a brain can give rise to the global properties that we associate with the mind or its maladies. One of the goals of these projects is to better understand/treat mental illness, thus showing societal relevance. Indeed, this program of brain simulation has ambitions to match those of the corresponding endeavor of brain measurement, so why are we still worried?

Well, our concern lies in the observation that the technical possibility of amassing new data itself risks becoming the main driving force. The analogy is often made with the human genome whose decoding has unlocked new avenues for research in biology and medicine. However, whilst the genome is large (3 billion base pairs) it is finite, linear, and discrete (each pair can only be one of a fixed number of known patterns), and the case could be (and was) made that deciphering it would concretely and permanently address a key bottleneck for research. The brain, on the other hand, has no equivalent target to the genome—there is no template for brain design that once we have described it we can say we are finished. There will always be another level of description and accuracy that we can strive for and which, for some, will be the key to unlocking the brain's secrets, from microtubules and CAM kinase to gamma range oscillations and blood flow dynamics. Further, whilst the new tools of twenty-first century brain science are attractive in terms of their greater accuracy and power, what we see with many of their results is confirmation of observations that had already been made in previous decades albeit in a more piecemeal fashion. To unlock the value of these new data sets, we believe, will require the development of multi-tiered explanations of brain function (as discussed in this chapter). Data analysis tools will help, but the multi-scale theory-building activity itself will largely be a human endeavor of which interpretation and abstraction will remain an important part.

A key point for us is that description and measurement, whilst vital to doing good research, are not the ultimate goal of science. Rather, we describe in order to prepare the grounds for explanation. As the physicist David Deutsch (1998) noted, there are an infinite number of facts that we could collect about the natural world (and this will include a countless number of brain facts), but this kind of knowledge is not, by itself, what we would call understanding. The latter comes when we are able to explain the amassed data by uncovering powerful general principles. In astronomy, for instance, Ptolemy, followed by Copernicus, Galileo, Newton, and then Einstein all developed theories that sought to explain observations of the motion of stars and planets. Each new theory succeeded in explaining more of the assembled data and did so more accurately and more succinctly. For instance, Copernicus explained data that had been problematic for Ptolemy's geocentric cosmology by replacing the Earth with the Sun as the center point around which the planets turn. Einstein showed that Newton's law of gravitation breaks down when gravity becomes very strong, and was thus able to better (or more succinctly) explain some data on planetary orbits. In physics the search for a theory with more explanatory power than general relativity continues, with the hope to one day explain the origin of everything, beginning with and including the Big Bang, according to a single set of over-arching principles.

In comparison to astronomy, how far have we come in developing theories for understanding brain data? The answer is not very far yet. With the current focus on large-scale data sets one would assume that there is an interest in discovering principles, but there also seems to be an expectation that these will bubble up autonomously through the accumulation of observations; a process of induction if you will, powered by the tools of data mining and computational modeling. Moreover, in place of striving for the kind of compact theoretical description seen in physical science, there is an increasing focus on descriptions that can capture more and more of the potentially relevant detail. The boundary becomes blurred between capturing principles and what can become, in the end, an exercise in function fitting. In our admiration of the elegance and beauty of brain data, the technology to capture it, and with the power of modern ICT systems to simulate it, we can come to believe that the best model of the brain is the most exact model. Following this path, however, can only lead to the conclusion that brain is its own best explanation, an idea satirized by Rosenbleuth and Wiener in their comment that "the best material model for a cat is another, or preferably the same cat" (Rosenblueth and Wiener 1945), and reminiscent of Borge's famous story of the cartographical institute whose best map was identical to the landscape it described and thus lost its usefulness. This story was in turn an elaboration of Lewis Carroll's *Sylvie and Bruno Concluded*, where a map was produced with a scale of 1:1, but because of the practical difficulties of such a map one of the characters exclaims: "we now use the country itself, as its own map, and I assure you it does nearly as well."

A second way of summarizing the concern about data-oriented initiatives is that the zeitgeist seems to favor more reductionist descriptions rather than holistic theoretical explanations. The logic appears to go that we still don't know enough of the key facts about the brain—cells, circuits, synapses, neurotransmitters, and so forth—therefore let's go and find out these details. Once we know these things we will necessarily better understand both brain and mind. Thus in an era where data itself has become a commodity the trend is to mimic the complexities found in nature starting at the lowest level of description. The challenges such an approach faces are manifold. First, it makes the unsubstantiated assumption, already questioned by the seventeenth century philosopher Thomas Hobbes, that the observations that provide scientific data are theoretically unbiased. Rather the instrumentation we construct to obtain data already includes some biases as to how we should inspect reality. The idea that "truth" will emerge from data once critical mass has been reached is therefore already questionable. Second, it assumes a uni-directional bottom-up causality where lower levels of organization "cause" phenomena at higher levels thus ignoring the tight coupling between the many levels of organization found in biological systems. Third, it fails to distinguish structural and functional levels of description. Fourth, it seems ignorant of the simple fact that mimicry does not imply understanding, as the so-called cargo cults of the Pacific Ocean had to discover at their own peril—in trying to physically copy what they thought were important parts of a plane, they demonstrated that you cannot build flying machines without first understanding the principles of flight.

However, while neuroscience tilts towards more data gathering, it is interesting to note that other areas of biology have become more holistic in their approach, adopting what is often described as a "systems" view (see Prescott, Chapter 4, this volume). Indeed, in systems biology, explanations are sought that go across levels from the molecular through the cellular, the organismic, and the ecological. No one level of explanation (or description) is privileged, and understanding at each level informs and constrains understanding at the levels above and below it. In much the same way, and within the sciences of the mind, parallel complementary explanations can be sought at the psychological level (mind) and at the biological level (brain), and we can allow that there may be other useful explanatory levels between these two. Indeed, we contend, as do many others, that useful theories of mind and brain can be motivated that abstract away from the biological details of the brain but at the same time capture regularities at a level below that of our direct intuitions (what some have called "folk psychology"), and that in this area some of the most powerful explanatory ideas might lie.

Cognitive science turf wars

The notion of a multi-tiered understanding of the mind and brain is of course nothing new. Indeed, in many ways it is captured in a research program that since the mid twentieth century has gone by the name of cognitive science (Gardner 1987) and that has largely replaced psychology as the dominant program for the study of mind. Acting in many ways as a kind of scientific umbrella, cognitive science has fostered interdisciplinary dialogues across the sciences of the mind and brain for the last seventy years, promoting the complementarity of explanations emanating from neuroscience, psychology, linguistics, philosophy, and computer science. At the same time, however, cognitive science has never really succeeded in building a consensus around a core set of scientific principles. Although a valiant attempt was mounted to elevate the computer metaphor to the status of a theory of mind (Newell 1976; Putnam 1960), instead it has seen struggles between different communities as to what should be the preferred level of description of mind and brain and it has hosted heated debates over the meaning and relevance of central concepts such as representation and computation. Perhaps this is the nature of a healthy science; however, unlike neuroscience, for instance, which holds a successful annual conference for around 30,000 delegates, the focus of cognitive scientists is dispersed across dozens of events each favoring a particular perspective or approach. Moreover, despite its potential relevance to society—for instance in the scientific understanding of mental disorders and the development of new smart technologies, it has surrendered much of its ground on the former to neuroscience and on the latter to AI. Finally, whilst neuroscience as a community has been able to mobilize support at the highest levels for endeavors such as HBP and the Brain Initiative, funding for cognitive science appears to be flagging, at least momentarily. The European Union, for example, scrapped its *Robotics and Cognitive Systems* program in 2014 in favor of one solely focused on robotics, partly due to the failure, as they saw it, of cognitive systems to address society-relevant challenges. This action was disappointing but not surprising—publicly funded science, particularly on this scale, must seek to make the world a better place; impactful science is also better science (see Prescott and Verschure 2016).

Standing back for a moment, we wonder if the current resurgence of a more reductionist brain science program is, at least in part, due to the failure of cognitive science to really capitalize on the great start that it made more than half a century ago. A commitment to interdisciplinarity has till now failed to lead to powerful interdisciplinary theories that command broad assent, leaving a vacuum to be filled by explanations couched at only one level of description. But then again, if we define the neuroscience challenge as gathering data and the cognitive science one as developing theories bridging the explanatory gap of mind and brain, we must acknowledge that the former is easier than the latter.

A further way to look at the current status of the field is to recognize that, in terms of the sociology of science as described by Thomas Kuhn (1962/1970), cognitive science sometimes appears to be "pre-paradigmatic." For Kuhn, work within any given domain of science begins with multiple competing general theories, or "paradigms," but then progresses to a point where one of these is clearly more successful than the rest, comes to dominate the field, and attracts more and more supporters to work within it—this then is the normal state for a mature scientific field. According to this narrative, it is possible for an alternate paradigm to arise by building on any weaknesses in the current dominant general theory, such as a failure to adequately explain key data, and by questioning some of its core precepts. If sufficiently persuasive, such an alternative can provoke a "scientific revolution" in which the current dominant paradigm is overthrown and replaced by a new orthodoxy. Whilst a scientific revolution can come about because the new paradigm is more explanatory, a key element of the Kuhnian analysis is that trends in scientific research are partly determined by social and political forces rather than purely scientific ones (see also Feyerabend 1975). The dominant paradigm might crumble, for instance, not simply because it is weaker but because it has become unfashionable; conversely, an alternate paradigm might fail to thrive not because it does not offer better explanations but simply because it fails to attract enough supporters, or resources, to mount a serious challenge; as in politics, the incumbent can have power and influence that allows them to suppress contenders at least for a while. Hence, the new paradigm might not be the best from the perspective of scientific understanding but rather in terms of its ability to convince those that control the resources that fund science (for instance, in the rise of nuclear physics in the 1940s and of artificial intelligence in the 1950s).

The Kuhnian narrative appears to work well in physics, a domain that Kuhn was trained in, and where the Newtonian view succeeded the Galilean view, then to be replaced by Einstein's (special and general) theories of relativity. Applied to the sciences of the brain and mind, however, the picture looks more complicated.

From one view, cognitive science stands as a distinct scientific domain still looking to find its feet (i.e. pre-paradigmatic), with symbolic AI, connectionism, dynamic systems, enactivism, machine learning, and perhaps cognitive neuroscience all vying as competing paradigms within it. Within the field, the navel-gazing continues. For optimists, there is still the expectation that eventually a consensus will emerge and cognitive science will have come of age. More pessimistically, and in view of the strongly fractured nature of the field, we wonder if many have already given up the effort to find a general theoretical understanding of the mind and brain, in favor of a focus on sub-domains to which preferred levels of explanation and methods can be more easily applied.

From an alternative perspective, cognitive science is itself a paradigm competing within the broader domain of the natural sciences to be *the* approach to understanding the mind and brain. According to this view, and to which we subscribe, cognitive science replaced behaviorism as the dominant paradigm in the mid-twentieth century and has succeeded in holding its ground until recently, despite a lack of consensus and internal division. If so, we might ask if cognitive science is now at risk, in Kuhn's sense, of being overthrown and, if so, who would be the contender? Surveying the landscape, does the new breed of assertive reductionist neuroscience have, as its ambition, the desire to replace the cognitive science consensus (or lack thereof) and to transform the notion of multidisciplinary explanations into unidisciplinary reduction? Could neuroscience potentially succeed in eliminating cognitivist theories, and all their conceptual intermediaries, in favor of descriptions couched directly in terms of brain states and dynamics? Will a future retelling of the history of science conclude that cognitive science was a useful approximation—like the descriptions of Tycho Brahe of a pre-Newtonian astronomy—attempting to plug the explanatory gap left by predecessors such as behaviorism, but ultimately not as powerful as a fully formed (Newtonian or relativist) neuroscientific theory of the relationship between mental phenomena and brain activity?

We describe this scenario not because we think it is inevitable, or because we think an eliminativist neuroscience really "has the legs" to become the paradigm in the science of mind and brain. However, we recognize, with Kuhn, that science is a societal activity, and that the field of cognitive science could wane, and perhaps is already waning. We would like to reinvigorate cognitive science—move its focus away from turf wars about privileged levels of explanation, and get back to its core agenda of building powerful multi-tiered theories of the mind and brain. The alternate reductionist agenda, that sees the brain as the best theory of itself, is actually a retreat from properly advancing the sciences of the mind (or any science for that matter). Like behaviorism seventy years ago, which declared the organism to be a black box instructed by its environment, the brain again becomes a box whose contents are ultimately unanalyzable; this time we can describe what is inside, but we surrender the hope of a theoretical explanation of the emergence of the mind in favor of the aspiration that if we copy it accurately enough we will somehow also replicate or understand interesting aspects of mental function.

Toward a multi-tiered theoretical framework

So what should a twenty-first-century approach to understanding the mind and brain look like? In the study of mind and brain there have been few general theories and the last concerted attempt to define one came to a halt in the early 1950s with Clark Hull's theory of the behaving system which followed the logical-positivist school. Since then, it has gone relatively quiet in terms of attempts to postulate theories that show how a physical system like the brain can give rise to all of mind and behavior; at best we have seen micro-theories that are highly specialized, as we exemplified earlier, or theories that specialize (or are most applicable) in specific areas of cognition. This is the explanatory gap that needs to be filled—a general theory, or framework, connecting brain and mind. It is important to emphasize that this challenge of linking the physical and the mental is a unique scientific endeavor that is unparalleled by any other natural science. In this case we have to go beyond solely describing and explaining the physical world to address what, following David Chalmers, we might call the "hard problem" of finding a scientific third person perspective of subjective first person states: a science of human experience.

In broader terms what should we look for in such a theory of mind and brain? First, as we already noted, a theory must *explain*, in this case, the scientific observations that constitute the relevant "facts" of empirical science concerning measurement of the brain and behavior, and this data must be interpreted in such a way that it provides an explanation of human experience. Second, a theory must make testable *predictions* that can be validated with available methods and technologies (making predictions that require measurements to be made with science fiction technologies cannot be taken too seriously). Third, a scientific theory must be able to *control* natural phenomena. This means, for instance, to be able to define a set of manipulations that constitute an experiment or the principles on the basis of which a useful artifact can be constructed. In addition to these primary requirements, we can include that a scientific theory must be supported by a broad base of observations, generate multiple predictions in a range of areas, and display a form of continuity with pre-existing knowledge and theories, even if they conflict. Furthermore, it must follow Adelard of Bath's dictum, formulated in the twelfth century, that nature is a closed system and all natural phenomena must be explained as caused by other natural agents, combined with Occam's razor, from the fourteenth century, which asks for parsimony in scientific theories.

Our commitment to the value of models, particularly those that make useful impact in the world, derive from Vico's dictum "verum et factum"—we understand by making. This epistemological model has been echoed by a number of scientists such as Lord Kelvin (in his rejection of Maxwell's theory of electromagnetism) or Richard Feynman; it stood at the heart of the short-lived cybernetics revolution, and was directly followed—probably unknowingly—by Watson and Crick in their discovery of the structure of DNA. Another way of expressing this idea is that "the machine *is* the theory," concretely embodied and observable, hence, a Living Machines approach to the science of mind and brain.

This approach naturally enough, has a rich history. Indeed, Harvey relied on a hydraulic mechanism metaphor to interpret and explain his anatomical observations of the heart and blood circulation, exploiting available artifacts, whilst the discovery by Watson and Crick of the structure of deoxyribonucleic acid was based on the active construction of a mechanical model, a purposefully built artifact. Despite the insight this artifact-driven model rendered, many aspects of the nature of circulation and of DNA were fully unclear at the time and are still being discovered today. Hence, the model facilitated the scientific discovery process without actually mimicking biological reality; rather it provided a suitably defined abstraction.

Theories and models face the problem of being under-constrained. From this view, there are many possible ways to interpret observations. We can think of the model as a fit of a curve through a cloud of data points. There is a practically infinite number of lines we can draw; which ones to retain and which to ignore? In the study of mind and brain we consider that we can reduce this search space by imposing the requirements that theories of mind and brain must be able to relate to multiple levels of description—minimally the structure and function of the brain, or its anatomy and physiology, and the behavior it generates. This method is called convergent validation (Verschure 1996). Convergent validation is a strategy which renders theories as models that are empirically adequate rather than "true" (van Fraassen 1980). In more practical terms this strategy is to build models to emulate the brain's anatomy and physiology and to embody these models using interfaces to the physical world (for instance, via a robot) capturing pertinent constraints from behavior and embodiment. In this form our theory as a model can *explain* anatomy, physiology, and behavior, make *predictions* at multiple levels of description, and *control* physical devices, i.e. the three criteria for scientific theories. In addition, it instantiates what we might call "Vico's loop"—we have made an artifact with life-like capabilities that can potentially be deployed in useful tasks and have societal impact (Prescott and Verschure 2016; Verschure 2013).

An additional form of constraint and a critical aspect of understanding the mind, and not just its parts, is that we need to develop theoretical frameworks that have the potential to inclusively explain all of the interesting capabilities of mind and brain including, but not limited to, perception, sensorimotor control, affect, personality, memory, learning, language, imagination, creativity, planning, consciousness, etc. In short, work within particular sub-domains—and most research in cognitive science is necessarily of this character—must have the potential to be incorporated within the bigger picture. This means that theories of components must be integrated in architectures that show how components are integrated into a complete embodied and situated system. This evolving idea of identifying the full architecture of the mind and brain should reveal how underlying principles operate across these sub-domains at the same time as identifying that specific sub-domains may also have their own specialisms.

Naturally, such a framework should not adopt an a priori view about privileged levels of explanation; rather, it should stress the importance of finding links between them. Indeed, it is a requirement of our commitment to convergent validation that we should generate models at multiple levels of abstraction—some very close to mechanisms revealed through the microscope of neuroscience, others very high-level and pertaining to psychology or connecting to principles relevant to engineering or computer science. Finally, we recognize the value of different methodologies for acquiring explanatory concepts. For instance, we can work inductively and bottom-up, using the powerful data analysis tools now being developed, to identify the "good tricks" (Dennett 1995), that have been discovered in the evolution of nervous systems. Likewise, we can work top-down and deductively, going from the three-tiered computational analyses of the functions of mind proposed by David Marr (1982) to ideas about the underlying computations and the mechanisms that can implement them. Here advances made in engineering and AI can furnish us with candidate principles that could be instantiated by the brain and mind. We do not prefer bottom-up or top-down approaches, but rather strive for the completeness of our theory, and to have elements of both in order to have strong constraints in these two directions. However, convergent validation holds for all of these approaches. Theories with no explanatory and predictive power are not theories but rather noise. Only in this inclusive yet methodologically disciplined perspective can we hope to unravel the multi-scale relationship of mind and brain.

The approach we advocate stands in contrast to the early twentieth century idea of a unity of science going back to Cartesian epistemology, where all higher levels of description of the *scala naturae* will be reduced to one fundamental level of description: *res extensa* of physics. Only the conscious mind was reserved a special place as *res cogitas*. Life is more interesting than that! However, how do we capture a non-reductionist view in an understanding of physical and mental life? More specifically what should be the ingredients of a framework that would facilitate the development of a theory of mind and brain?

We do not propose to define a "new" science with the negating of preceding paradigms that this implies. Rather, it is vital that we build on previous scientific attempts at a general theory of the mind and brain.

For example, many key ideas originated with the invention of control and information theory in engineering, digital computers in ICT, and systems theory in biology. Amongst these we would highlight the insights of Norbert Wiener, Warren McCulloch, Rosh Ashby, and William Grey Walters, cyberneticians who combined the theory of feedback-based control from engineering, with the notion of homeostasis from biology, to produce a theory of the brain as a mechanism for maintaining balance. This can be applied to diverse areas of brain function from autonomic function (the regulation of bodily processes such as breathing, circulation, and metabolism) to motor control, to cognitive processes such as learning and memory (see Cordeschi (2002) and Herreros, Chapter 26, this volume).

During the emergence of cognitive science in the 1950s, leading figures were also concerned with general theories of the mind. For instance, building on attempts to understand decision making, Alan Newell and John Anderson both elaborated general theories of cognitive architecture using if-then rules (productions) as their primary building block (see Verschure, Chapter 35, this volume). These models explored and demonstrated the power of a simple principle, recursively applied, in generating mind-like properties, but they also revealed some of the limitations of prematurely settling on a specific level of analysis, or computational primitive. Indeed, this model of the brain as a symbol system, which emerged from computer science, rather unfortunately declared the independence of theories of the computational mind from theories of the analog brain. This approach, which then established a trend for cognitive scientists to declare that some levels of explanation are preferable to others, also illustrates a normativity often expressed by a high-level, or "engineering," view on mind and brain where it is conceptualized as an evolutionary suboptimal kludge that actually *should* perform in a specific way and conform to current engineering standards. Instead, we should perhaps admit that we actually do not know how the brain operates, or how it realizes the mind, and should thus be careful in declaring our current engineering state-of-the-art as the norm against which to measure the mind.

In a reaction to the perceived failure of the AI program, connectionist models were proposed by David Rumelhart, Jay McClelland, Geoffrey Hinton, Terry Sejnowski and others (e.g. Rumelhart et al. 1986), who looked much more at the specific structural properties of the brain, particularly its massively distributed nature, as inspiration for their models of the mind and brain. The excitement around these models, which showed the capacity to explain facets of learning and memory that were problematic for symbol-based approaches, encouraged claims to be made that this now constituted the privileged level of explanation at which theories of mind should be couched (e.g. Smolensky 1988). This is therefore a further example, in cognitive science, where success in applying one particular kind of approach in a number of sub-domains led to premature conclusions about its explanatory power. Looking back, thirty years on from the connectionist "revolution," this approach has also succumbed to the same criticism but this time from computational neuroscience—that the more abstract version of brain architecture favored by the connectionists overlooked critical details while also failing to scale-up to high-level cognitive functions. But the answer is not to go deeper and deeper to find the one perfect model but to recognize that different scientific questions can be addressed at many different levels (Churchland and Sejnowski 1992), the challenge becoming how to link these different levels of description together in comprehensive system level theories.

A dynamic systems view, emerging in the 1990s, and championed by Scott Kelso, Esther Thelen, Tim van Gelder, and others, attempted a synthesis between connectionist theory and systems biology (e.g. Port and van Gelder 1995). However, once again, the effort to distinguish itself from what had gone before, in this instance by declaring itself to be non-computational, limited the impact of the approach. More recently, new bandwagons have emerged based on the notion of the brain as a machine for Bayesian inference (Doya 2007), and predictive coding (minimizing its own ability to be surprised by the world) (reviewed in Clark 2013), which largely stem from popular engineering tools and approaches that are being deployed to stem the data deluge in computer science and industry. The most ambitious versions of these theories hope to be full accounts of how mind emerges from brain. The attraction that we have as scientists to the possibility of uncovering core principles that succinctly explain many of the things we want to understand, as the physicists managed to do in explaining the motion of the stars and planets over many generations from Ptolemy to Einstein, must be tempered by the recognition that such notions have so far only captured a small fraction of the competencies of the mind and the intricacies of the brain. These recent approaches therefore have a very long way to go before they can make any claim to theoretical completeness (following the proposed criteria of explain, predict, control). As an evolved system that must solve many different types of challenges in order to survive and thrive we must also be open to the possibility that there is no single principle, or even a small cluster of principles, that will explain the mind/brain. We certainly hope and expect a theory that is much simpler than the brain itself, but we expect, nevertheless, that it will be a long journey and an answer that will be significantly more complex than "42."

Conclusion

As we have explained in this chapter, we consider that an important element of theory development and testing is its instantiation as an embodied model or machine. This step allows us to achieve a level of completeness and explicitness not possible at the purely theoretical level or even in simulation, and allows us to test theories whose complexity we cannot easily entertain in our own minds as thought experiments or using the formal tools the mind has at its disposal. More emphatically, we consider, following Vico, that a mark of a good theory of the human mind and brain is that it can be physically instantiated, and an advantage of this approach is that it can also lead to the development of new, biologically grounded technologies that have direct value to society because they solve problems that are beyond what contemporary engineered systems can do. This strategy, in other words, takes seriously the notion of "living machine" as applied to the understanding of the biology of mind and brain, and its emulation in real-world artifacts. Further, as many of the chapters in this book illustrate, this approach is beginning to lead to a range of technologies, including rehabilitation, prosthetics, education, and assistive technologies, that show how a multi-tiered science of the mind and brain, validated through Vico's loop, can lead to useful innovation and ultimately to broad societal benefit, the ultimate benchmark for any science.

References

Alivisatos, A. P., Chun, M., Church, G. M., Greenspan, R. J., Roukes, M. L., and Yuste, R. (2012). The brain activity map project and the challenge of functional connectomics. *Neuron*, 74(6), 970–4.

Churchland, P. S., and Sejnowski, T. J. (1992). *The computational brain*. Cambridge, MA: Bradford Books.

Clark, A. (2013). Whatever next? predictive brains, situated agents, and the future of cognitive science. *Behavioural and Brain Sciences*, 36(3), 181–204.

Cordeschi, R. (2002). *The discovery of the artificial: behavior, mind and machines before and beyond cybernetics*. Dordrecht : Kluwer.

Dennett, D. C. (1995). *Darwin's Dangerous Idea*. London: Penguin Books.

Deutsch, D. (1998). *The fabric of reality*. London: Penguin Books.

Doya, K. (2007). *Bayesian brain: Probabilistic approaches to neural coding.* Cambridge, MA: MIT Press.

Feyerabend, P. K. (1975). Against method: outline of an anarchistic theory of knowledge. *Philosophia*, 6(1), 156–77.

Gardner, H. (1987). *The Mind's New Science: A history of the cognitive revolution.* New York: Basic Books.

Kuhn, T. S. (1962/1970). *The Structure of Scientific Revolutions* (2nd ed.). Chicago: University of Chicago Press.

Markram, H. (2012). The human brain project. *Scientific American*, **306**(6), 50–5.

Marr, D. (1982). *Vision: A Computational Investigation into the Human Representation and Processing of Visual Information.* New York: Freeman.

Newell, A. S. H. A. (1976). Computer science as empirical inquiry: Symbols and search. *Communications of the ACM*, **19**, 113–26.

Port, R. F., and van Gelder, T. (1995). *Mind as Motion: Explorations in the Dynamics of Cognition.* Cambridge, MA: Bradford Books.

Prescott, T. J., and Verschure, P. F. M. J. (2016). Action-oriented cognition and its implications: Contextualising the new science of mind. In: A. K. Engel, K. Friston, and D. Kragic (eds.), *Where's the Action? The Pragmatic Turn in Cognitive Science.* Cambridge, MA: MIT Press for the Ernst Strüngmann Foundation, pp. 321–31.

Putnam, H. (1960). Minds and Machines. In S. Hook (ed.), *Dimensions of Mind.* New York: New York University Press, pp. 20–33.

Rosenblueth, A., and Wiener, N. (1945). The role of models in science. *Philosophy of Science*, **12**(4), 316–21.

Rumelhart, D. E., McClelland, J. L., and the PDP Research Group (1986). *Parallel Distributed Processing: Explorations in the Microstructure of Cognition. Volume 1: Foundations.* Cambridge, MA: Bradford Books.

Smolensky, P. (1988). On the proper treatment of connectionism. *Behavioural and Brain Sciences*, **11**, 1–74.

van Fraassen, B. (1980). *The Scientific Image.* Oxford: Oxford University Press.

Verschure, P. F. M. J. (1996). Connectionist explanation: taking positions in the mind–brain dilemma. In: G. Dorffner (ed.), *Neural Networks and a New Artificial Intelligence.* London: Thompson, pp. 133–88.

Verschure, P. F. M. J. (2012). The distributed adaptive control architecture of the mind, brain, body nexus. *Biologically Inspired Cognitive Architectures*, **1**(1), 55–72.

Verschure, P. F. M. J. (2013). Formal minds and biological brains ii: from the mirage of intelligence to a science and engineering of consciousness. *IEEE Expert*, **28**(5), 33–36.

Chapter 3

A roadmap for Living Machines research

Nathan F. Lepora[1], Paul F. M. J. Verschure[2], and Tony J. Prescott[3]

[1] Department of Engineering Mathematics and Bristol Robotics Laboratory, University of Bristol, UK
[2] SPECS, Institute for Bioengineering of Catalonia (IBEC), the Barcelona Institute of Science and Technology (BIST), and Catalan Institute of Advanced Studies (ICREA), Spain
[3] Sheffield Robotics and Department of Computer Science, University of Sheffield, UK

In the first two decades of this millennium there has been an explosive growth in research on biomimetics and biohybrid systems that comprise the subject areas underlying the Living Machines approach described in this handbook. A bibliographic analysis by the authors of this chapter (Lepora et al. 2013) found that the number of papers in biomimetics published per annum had doubled every two to three years (Figure 3.1). From a relatively small field in the mid-1990s of just ten or so papers per year, biomimetics has expanded exponentially thereafter to reach a critical mass of several hundred papers per year by 2003–2005. By 2010 over 1500 papers were being published every year in biomimetic science, engineering, and technology. Based on this analysis we can safely claim that there is a boom in the biologically grounded and inspired research of living machines. Since then this growth as continued. One new contributor is the *Living Machines* conference series that began in 2012 (Prescott et al. 2012) and has just passed its sixth edition (Mangan et al. 2017) organized by the *Convergent Science Network*. Meanwhile, there has also been steady growth in established journals, such as the IOP journal *Bioinspiration & Biomimetics*, whose impact factor has increased by 60% since 2010 and also recent journals that address the field such as *Advanced Biosystems*, *Biomimetics* and *Soft Robotics*.

Our 2013 bibliographic study examined the popular themes in this publication growth and showed that research in biomimetics now spans a diverse range of applications (Figure 3.2). Leading concepts include *robot* followed by *control*, which indicates a strong focus of research on the development and application of biomimetic and neurorobotics. We observe that research interests are broad and include terms such as *tissue, cell*, and *bone*, reflecting the large impact of the biomimetic approach in the field of biomedical science. Also evident are the abilities and characteristics of biological organisms that principles are being taken from, such as *vision, walking, learning*, and *development*. In addition, concepts from control engineering and artificial intelligence are also represented, including *networks, adaptive, algorithm*, and *optimization*.

These findings indicate that the study of living machines is entering a renaissance, becoming a significant paradigm for engineering and the life sciences alike. The impacts of this increased emphasis on biomimetics and biohybrids can be expected to include major new discoveries in many areas of the physical and biological sciences, and revolutionary new technologies with significant societal and economic impact.

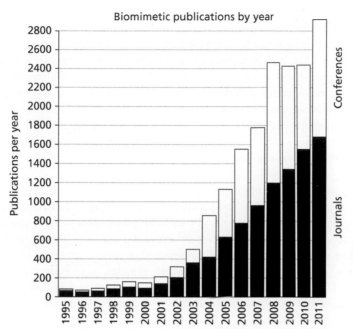

Figure 3.1 Growth of biomimetic research. The bar chart plots the number of journal and conference papers published each year in biomimetics starting from 1995.

Adapted from Nathan F. Lepora, Paul F. M. J. Verschure, and Tony J. Prescott, The state of the art in biomimetics, *Bioinspiration & Biomimetics*, 8, (013001), p. 5, Figure 4, doi:10.1088/1748-3182/8/1/013001, © 2013, IOP Publishing.

The goal of this roadmap is to identify current trends in biomimetic systems together with their implications for future research. Important questions include the scale at which these systems are defined, the types of biological systems addressed, the kinds of principles sought, the role of biomimetics in the understanding of living systems, relevant application domains, common benchmarks, the relationship to other fields, and further developments on the horizon. In this chapter we also provide pointers to where these issues are further elaborated, in terms of biological principles, biomimetic methods, and exemplar technologies, throughout this handbook.

Figure 3.2 Popular terms in biomimetics displayed using a word cloud. The size of the word reflects the relative frequency of occurrence.

Reproduced from Nathan F. Lepora, Paul F. M. J. Verschure, and Tony J. Prescott, The state of the art in biomimetics, *Bioinspiration and Biomimetics*, 8, (013001), p. 6, Figure 5, doi:10.1088/1748-3182/8/1/013001, © 2013, IOP Publishing.

Methodology

The initial draft for this roadmap was developed around a road-mapping exercise centred at a biomimetics week for the 2011 Barcelona Cognition, Brain, and Technology Summer School organized by the Convergent Science Network. A group of experts on biomimetics who were lecturing at the school attended a workshop where general questions about biomimetics were debated, including ideas about how best to conduct information gathering for a roadmap. Based on this workshop, we constructed a standardized questionnaire to be completed during one-on-one interviews lasting about an hour each. These questionnaires were then edited by the interviewer (Lepora) and fed back to the interviewees to check the accuracy of the note-taking.

This roadmap is based on collating the answers from these experts, and covers a broad range of areas in Living Machines. The authors of this roadmap constructed the breakdown into areas of research, based partly on the results of their earlier state-of-the-art survey of biomimetics (Lepora et al. 2013), but also by assigning the responses to research themes evident from the questionnaire results. This resulted in eleven domains: self-assembly and micro-machines; micro air vehicles; underwater vehicles; insect-inspired robotics; soft robotics; bipedal and quadrupedal robotics; social robotics and human–machine interaction; brain-based systems; olfaction and chemosensing; audition and echolocation; and touch and hands.

Since our initial roadmap discussions, bio-inspired approaches have been making headlines in the fields of artificial intelligence (AI) and machine learning, owing partly to the success of "deep learning" algorithms in solving significant challenges in areas such as automatic speech recognition and computer vision. In this roadmap we have focused on physically embodied systems, also known as cyber-physical systems (e.g. Lee, 2008); however, the impacts and prospects for biomimetics in AI and machine learning are also considered, in some detail, in this handbook. For example, Leibo and Poggio (Chapter 25) explore the links between neuroscience and the development of convolutional neural networks to solve pattern recognition problems in perception; Herreros (Chapter 26) relates animal learning theories to machine learning; and Lepora (Chapter 28) shows how theories of optimal decision making in animals and machines have evolved together through the interplay between biology and AI. Verschure (Chapters 35 and 36) places the development of architectures for living machines in the context of historic trends and research in Artificial Intelligence and Cognitive Science.

In the following sections we briefly review the goals and risks for each of the eleven research domains related to embodied living machines.

Domain 1: Self-assembly and micro-machines

Self-assembly is a biomimetic approach that allows multi-component systems to collectively configure themselves to address a challenge (see Moses and Chirikjian, Chapter 7). When the collective entity is composed of multiple distinct homogeneous agents, this is typically referred to as a "swarm" (Figure 3.3 and Nolfi, Chapter 43). Agents can also join together to form a new entity: examples from nature include cellular multiplication and growth (e.g. Mazzolai, Chapter 9), and the ability of some animals, such as the single-celled slime molds, to form themselves into temporary multicellular aggregates that move and behave as one (see Ishiguro and Umedachi, Chapter 40).

As robotic applications move away from structured environments, such as factory assembly lines, and into unstructured scenarios such as the world at large, a critical requirement is to interact robustly with their surroundings. Interactions with the environment inherently involve both

Figure 3.3 Swarm robots form a symbiotic multi-robot organism and collectively pass a barrier. Demonstration from the European Union REPLICATOR project. See Nolfi (Chapter 43) for more on cooperative behavior in collective systems.

Reproduced from Serge Kernbach, Eugen Meister, Florian Schlachter, Kristof Jebens, Marc Szymanski, Jens Liedke, Davide Laneri, Lutz Winkler, Thomas Schmickl, Ronald Thenius, Paolo Corradi, and Leonardo Ricotti, Symbiotic Robot Organisms: REPLICATOR and SYMBRION Projects, *Proceedings of the 8th Workshop on Performance Metrics for Intelligent Systems*, p. 3, Figure 1c, doi: 10.1145/1774674.1774685 Copyright © 2008, Association for Computing Machinery, Inc.

microscopic forces (friction, molecular attraction, and capillary forces) and macroscopic forces (from inertia, gravity, and aerodynamics). In nature, these requirements are addressed with hierarchical systems and structures exhibiting complex geometry and functionality over multiple length scales. Specification of such systems involves a complex mix of genetic, phenotypic, and environmental constraints, and construction relies on self-organising developmental processes (see Wilson, Chapter 5, and Prescott and Krubitzer, Chapter 8). Conversely, artificial manufacturing utilizes top-down processes that operate primarily on bulk materials using a precise specification of the final product.

When human engineers design processes, they generally consider just one length scale. Biomimetic robotic systems offer the opportunity to be at the forefront of current challenges and opportunities by operating over many scales while fulfilling multiple objectives. Biology also presents prime examples of the integration of nanotechnology with macroscopic fabrication methods, which may ultimately overturn the current top-down manufacturing approach with a fully integrative methodology, such as "growing" an entire product (see Fukuda et al., Chapter 52).

Experts in this research area identified several goals for self-assembly:

Rapid prototyping of inhomogenous structures. This includes the ability to build structures with spatially varying structural properties. Most man-made things are homogenous, whereas nature uses continuously varying material structures, such as from hard to elastic and back into a hard material.

Fabrication at the micro- and nano-scale. Organisms are incredibly complex at the micro-scale, for example neural systems, muscle, skin, etc. Conventional fabrication at the micro- and nano-scale is specialized and very different from nature.

Self-assembly at the micro- and nano-scale. There is significant interest in this area, for example the pop-up mechanism developed by Wood (2014) for the fabrication of a synthetic "robot bee". Devices are getting too small or too complex to assemble manually, so there is a need for manufacturing techniques involving self-assembly.

Exploitation of micro- and nano-scale properties. Material properties can simplify control, confer robustness, and provide extra stability and energy efficiency. For example, the micro-structure of animal skin can have adhesive properties that support gripping, climbing, locomotion and so on (see Pang et al., Chapter 22). Another example is to have sensing and processing at the micro-scale, as in sensory receptors and neurons. Working at the micro-scale becomes essential when developing interfaces for biohybrid systems, for instance to the nervous system (e.g. Vassanelli, Chapter 50, Lehmann and van Schaik, Chapter 54, and Song and Berger, Chapter 55).

Identified risks include:

Following nature too closely. This could also be called "blind mimicry." Nature has other priorities that may not be relevant to the engineered solution. Moreover, to mimic does not imply to understand. It can also lead to a copying of surface features that are not relevant to the underlying targeted function or property (see also Pfeifer and Verschure, 1995).

Identifying the appropriate principles from nature. It can be difficult in biology, because of its complexity, to identify the underlying principles to be utilized biomimetically; in particular, it can be hard to set up a controlled experiment in biology to isolate the principle you are interested in. Engineering a biomimetic solution can help this process, because these principles can be much easier to validate in artifacts such as robots.

Domain 2: Micro air vehicles

A micro air vehicle (MAV) is a small unmanned aerial vehicle, typically of the order of tens of centimeters in scale. Modern developments are driving further miniaturization, with insect-sized aircraft expected in the near future. Existing MAVs have multiple applications in commercial, research, government, and military purposes. A recent focus in aerial robotics is to take inspiration from flying insects or birds to achieve a new state-of-the-art in flight capabilities (see Figure 3.4), with biomimetic MAVs forming an important class of current approaches (see Ward 2017). The principles underlying flapping-wing flight are increasingly well understood (see Hedenström, Chapter 32) and there are emerging exemplar technologies at a number of scales from bird (Send, Chapter 46) to insect size (e.g. Wood 2014). Coordinated agile MAVs would have a wide range of new applications, from pollinating crops as artificial bees, to exploring hazardous and constrained environments, to surveillance and security.

Experts in this research area identified several goals for biomimetic micro air vehicles:

Development of an autonomous flying insect-like robot. Achieving autonomous flight will require compact high-energy power sources and associated electronics, integrated into the body of the micro air vehicle.

Progress in low-powered devices. Low power is a combination of engine efficiency and energy source density, and is needed for the device to achieve autonomy for a significant period of time.

Colony behavior and swarm robotics. To mimic the sophisticated behavior of biological swarms and flocks will require development of novel coordination algorithms and communication methods (i.e. the ability for individual machines to "talk" to one another and the hive) (see Wortham and Bryson, Chapter 33).

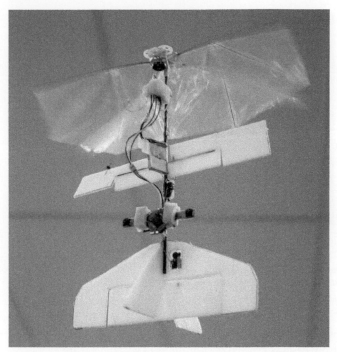

Figure 3.4 The "Delfly", a flapping wing robot designed at Delft University in The Netherlands.

Reproduced from G.C.H.E. de Croon, M. Perçin, B.D.W. Remes, R. Ruijsink, and C. De Wagter, *The DelFly: Design, Aerodynamics, and Artificial Intelligence of a Flapping Wing Robot*, doi: 10.1007/978-94-017-9208-0, © Springer Science+Business Media Dordrecht, 2016.

Identified risks include:

Threats to privacy. One application of micro-flying bugs is for spying, if equipped with a mini-ature camera or microphone and transmitter. This would create threats to privacy if the tech-nology were widely available.

Micro flying robots in warfare. Applications of micro flying robots in warfare include intelli-gence gathering and as anti-personnel devices (if fitted with a deadly payload, such as a bio-logical agent). This latter use is of considerable concern, particularly if the technology could be mass-produced, and hence it is likely that the availability of this technology would need to be controlled. The US defense agency recently acknowledged that it was developing weapons to counter the threat of MAVs.

Domain 3: Underwater vehicles

Development of autonomous underwater vehicle (AUVs) is being driven by applications in several industries. For example, the offshore oil and gas sector use them to map the seafloor before sub-sea construction for drilling. Marine scientists rely on AUVs to research the ocean and its floor. Other civilian applications include clearing up marine pollution and mining the deep ocean. Military uses of AUVs include identifying and neutralizing underwater mines and monitoring protected areas (e.g. harbors) for potential threats. A recent trend in under-water robotics has been to create AUVs based on the form and function of marine life (e.g. Figure 3.5 and Kruusmaa, Chapter 44). Although most of these biomimetic AUVs are still

Figure 3.5 Biomimetic fish-like robot; see also Kruusmaa (Chapter 44).

Reproduced from Jin-Dong Liu and Huosheng Hu, Biologically inspired behaviour design for autonomous robotic fish, *International Journal of Automation Computing*, 3(4), pp. 336–47. https://doi.org/10.1007/s11633-006-0336-x, Copyright © Institute of Automation, Chinese Academy of Sciences, 2006.

under development, they offer the potential to make a step-change in maneuverability by taking inspiration from designs evolved over millions of years in nature.

Experts in the research area identified several goals for biomimetic AUVs:

Modeling and controlling compliance. Existing simulation methods for complex, non-linear materials are not suitable for real-time control of underwater robotics. The ideal solution would be a compromise between empirical and physical approaches.

Novel materials. For example, the development of novel low-power actuation systems would greatly benefit the capabilities of small underwater robots.

New sensing technologies. There is a need for new sensing technologies for underwater devices, both to survey the nearby environment and to control the AUV. Model biological sensor systems include the electroreceptive capability of some marine organisms (see Boyer and Lebastard, Chapter 19), the lateral line organ in fish and the whiskers of pinnipeds (both of which can detect hydrodynamic flow patterns), and the echolocation capabilities of dolphins and whales.

Identified risks include:

Dependence on interests of biological researchers. Engineers using a biomimetic approach are reliant on biologists to be interested in research that would benefit the development of the new technology; however, often biologists have their own priorities and may not answer the right questions for bio-inspiration.

Underwater security. Widespread availability of cheap autonomous underwater vehicles capable of long-range travel could generate a number of security issues from drug smuggling to autonomous bomb delivery. Therefore it is likely this technology will need to be controlled.

Domain 4: Insect-inspired robotics

Insects are an excellent source of inspiration for robotics because of the compact nature of their morphologies, sensing, and nervous systems, compared with larger animals, that can nevertheless support a range of adaptive and robust behaviors. In insects, integrated behavioral patterns can often be identified as under the control of a network of identified neurons containing the entire sensory–motor pathways from sensation to action (see Selverston, Chapter 21, and Ayers, Chapter 51). For example, the principles underlying insect walking behavior and neural architecture form a good basis for designing legged robots and their control systems that can then be used to test hypotheses about the model animal (see Quinn and Ritzmann, Chapter 42). The Living Machines approach advanced in this handbook also argues that in many cases where morphology, control and environment come together, this synthetic route might be the only option for a controlled decomposition and validation of the different factors underlying function and their interactions (Verschure and Prescott, Chapter 2). The ethology and physiology of the cockroach, cricket, stick insect, ant, and honeybee have been studied in great detail and are thus the most commonly used insect species for inspiring robot and control technologies (see Figure 3.6).

Experts in this research area identified several goals for insect-based robotics:

Biomimetic approaches to navigation. The machine learning algorithm SLAM is the leading approach to solving autonomous navigation based on map building. However, at present, the computational demands can prevent SLAM from being an effective solution in complex, natural environments. Biomimetic studies of how insects such as ants and bees navigate offer alternative approaches for solving this problem; for instance, by showing that insect navigation can be well explained without relying on maps per se (Mathews et al. 2009).

Miniaturization, energy-efficiency, and costs. We would like future robots to be smaller, less energy hungry, more robust, and cheaper. Biomimetic fabrication methods (as discussed in domain 1) are relevant to miniaturization, as are neuromorphic and biohybrid approaches that could result in improved energy efficiency by exploiting biological principles such as spike-based coding and event-based processing (see Dudek, Chapter 14 for examples related to insect-like vision, and Ayers, Chapter 51 for steps towards biohybrid robots).

Figure 3.6 Hector is a six-legged walking robot, with a design based on the scaled-up body geometry of a stick insect, that incorporates several biomimetic concepts. These include a decentralized walking controller, muscle-like actuation, and vision processing based on neural mechanisms found in flying insects (Schneider at al. 2012; Cruse and Schilling, Chapter 24). Image: Sina Gantenbrink/CITEC.

New applications for swarm robotics. Applications that could benefit from miniature swarms include distributed sensing inside machinery. Swarm robotics can also be suited to tasks that demand cheap designs and disposable robots, for example search and rescue in disaster sites. The simplicity of individual team members motivates an approach to achieve meaningful behavior at swarm level instead of at the individual level (see Nolfi, Chapter 42).

New experimental techniques for biological research on insects. Biomimetics would benefit from advances in techniques for *in vivo* electrophysiological neural recordings in freely moving insects in natural environments. Also developments in insect tracking technology are needed, such as low-weight, cheap GPS sensors combined with physiological data acquisition.

Domain 5: Soft robotics

Most robotic systems are hard, being composed of rigid metallic or plastic structures with joints around bearings. The development of new types of soft robotic structures such as artificial muscles (see Anderson and O'Brien, Chapter 20) and fully soft robotic bodies (see Trimmer, Chapter 41), along with materials and methods for their fabrication, offers a range of new opportunities based on a close collaboration between organic chemists, materials scientists, and roboticists. Although the areas of biomimetics involved in soft robotics are still in their infancy, scientists are making good progress (see Figure 3.7) and this research is having substantial impact. Nevertheless, there is a long way to go to approach the benchmarks set by biological models such as the elephant trunk or the octopus arm.

Soft robotics has a range of applications, from dextrous manipulation in hazardous or unstructured environments, to applications in healthcare and assistive robotcs (see also Rubio Ballester, Chapter 59). They are also a relatively safe technology to use around humans as their naturally compliant nature makes them intrinsically safer for human physical interaction compared to conventional systems.

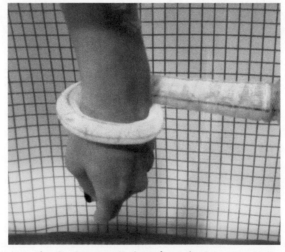

Figure 3.7 Soft robot arm, the OctArm manipulator from the European Union OCTOPUS project.

Reprinted from *Advanced Robotoics*, 26(7), Cecilia Laschi, Matteo Cianchetti, Barbara Mazzolai, Laura Margheri, Maurizio Follador, and Paolo Dario, Soft Robot Arm Inspired by the Octopus, pp. 709-27, doi: 10.1163/156855312X626343, Copyright © 2012 Taylor & Francis and Robotics Society of Japan. Reprinted with permission of Taylor & Francis Ltd, www.tandfonline.com on behalf of Taylor & Francis and Robotics Society of Japan.

Experts in the research area identified several goals for soft robotics:

Improved actuators. Present-day artificial muscles are often built around electro-active polymer (EAP) materials that are limited in their strength and safety. Development of EAPs, and of new kinds of bio-inspired actuator, in different configurations (e.g. fibers and fiber bundles) would be beneficial in order to increase the range and efficiency of soft robotic actuation (see Anderson and O'Brien, Chapter 20).

Development of a true muscular hydrostat robot. The muscular hydrostat— sealed, liquid-filled units that are acted on by muscles—are one of nature's solutions to the problem of soft, mor-phable, low-power, high-strength manipulation and are seen in many invertebrate animals, and in structures such as the octopus arm and the vertebrate tongue. To achieve a synthetic hydrostat (see Trimmer, Chapter 41, for some work in this direction), there is a need to develop better soft actuator technology as noted above. Modular hydrostats could also be plugged together to create assemblies of soft actuators in useful configurations. For example, a struc-ture such as a tongue contains many such units, held together by a flexible scaffold or matrix, that can together provide near infinite degrees of freedom for movement (hyper-redundancy).

Understanding "squishy" cognition. This research goal includes both how to control soft robot arms in an ecological context and the development of theoretical models for this research. At present there is a lack of theory behind such devices, which are often phenomenologically rather than the-oretically driven. The study of identified circuits in invertebrate animals, such as those underly-ing pattern-generation (e.g. Selverston, Chapter 21), provides an exciting prospect for developing low-power real-time systems that are computationally efficient and capable of robust behavior.

Identified risks include:

Potential for human injury. Whilst soft robotic systems are intrinsically safer than hard ones, they can still pose risks to humans, for instance through compression. Since many potential applications for soft robotics involve close physical interactions with people (for instance, in a care setting) attention to human safety will be important. Some current EAP technologies also face the limitation that they require high voltages to operate, which could restrict their usefulness around people.

Animal care. The United Kingdom and now the European Union have instituted animal care protocols for cephalopods that place limits on experiments. In principle, biomimetics could offer an alternative to animal research in the development of appropriate hydrostatic plat-forms for testing biological hypotheses. Simpler marine animals than the octopus can also provide useful biological models (see Trimmer, Chapter 41).

Domain 6: Bipedal and quadrupedal robots

Robot locomotion has traditionally involved wheels or treads. Legged robots, unlike wheeled robots, have the potential to access nearly all of the Earth's surface, and can access vertical surfaces by climb-ing, enabling robotic applications in areas where more conventional robots would not be suitable. Improved understanding of the principles underlying human and animal locomotion, which often exploits passive mechanisms to minimize power consumption (see Figure 3.8), has led to recent suc-cess in creating quadrupedal and bipedal robots capable of walking, running, and climbing. Reptile-inspired quadrupeds can support stable and fast locomotion. Mammal-inspired systems are more difficult to control but could prove to be more energy efficient (Witte et al., Chapter 31). Bipedal robots may be useful in spaces designed for humans (Metta and Cingolani, Chapter 47). Robust bio-mimetic controls systems for locomotion can be built around models of central pattern generator systems (see Cruse and Schilling, Chapter 24) and adaptive control (Herreros, Chapter 26).

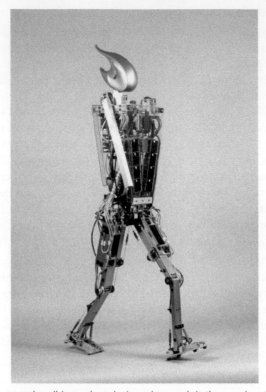

Figure 3.8 Flame, a powered walking robot designed to exploit the passive dynamics of human-like legs leading to a more energy-efficient gait.

© 2008 IEEE. Reprinted, with permission, from Daan Hobbelen, Tomas de Boer, and Martijn Wisse, System overview of bipedal robots Flame and TUlip: Tailor-made for Limit Cycle Walking, 2008 IEEE/RSJ *International Conference on Intelligent Robots and Systems*, doi: 10.1109/IROS.2008.4650728.

Experts in this research area identified several goals for bipedal and quadrupedal robots:

Robot legged locomotion. Highlighted topics include locomotion across unstructured terrain and improved mechanics for locomotion—for example, adding more compliance and improving energy efficiency. Another goal is to combine locomotion and reaching/manipulation so that both tasks are carried out simultaneously on the same platform as animals typically do (for example, running to catch a ball).

Common platforms for research. Research would benefit from the availability of more types of standard platform allowing research groups to co-develop for a single platform rather than developing their own hardware each time. Ideally, platforms should be open source. A leading example of this is the *iCub* platform (see Metta and Cingolani, Chapter 47).

Identified risks/limitations include:

Energy efficiency. Wheels are simpler and more efficient than legs for moving on flat or shallow-slopped surfaces. Applications that require efficiency alongside the ability to manage different terrain types could benefit from hybrid wheel/leg solutions such as "whegs" (see Quinn and Ritzmann, Chapter 42).

Research openess. As elsewhere in biomimetics, research that is funded by the military or by commercial organizations is often not openly published, limiting its value to the wider research community and slowing progress in the field.

Domain 7: Social robotics and human–machine interaction

As the introduction of robots into our daily life becomes a reality, the social compatibility of robots is gaining importance (e.g. Figure 3.9; see also Millings and Collins, Chapter 60). In order to meaningfully interact with humans, robots could benefit from advances in real-world social intelligence based on the perceptual, behavioral, emotional, motivational, cognitive, and communicative capabilities discussed in section IV of this handbook. A social robot is one that interacts and communicates with significant others in accordance with the rules for social behavior attached to its role. Wortham and Bryson (Chapter 33) discuss some of the requirements for successful communication, comparing and contrasting biomimetic approaches based on animal and human communication. The ultimate benchmark for human-like robot communication skills is often considered to be the Turing Test, although it may be of limited practical use in devising useful systems. For this reason, Verschure (Chapter 35) proposes a more stringent benchmark for cognitive architectures building on an earlier proposal by Newell (1990).

To match human sociality, robot behavior will need to be regulated by an understanding of values, norms, and standards; Vouloutsy and Verschure (Chapter 34) describe how we can think of this in terms of emotional appraisal and self-regulation. Isaac Asimov's Three Laws of Robotics have sometimes been proposed for governing robot decision-making but these rules are too weak to sufficiently constrain behavior, while, at the same time, require a capacity for consequential reasoning that may be difficult to attain. Lepora provides an overview of available decision-making methods for living machines (Chapter 28) while Verschure looks at this issue in the context of embodied cognitive architectures (Chapter 36). Appropriate behavior also implies a requirement for cultural sensitivity, since norms and values vary across human societies, which will require advanced communication and language abilities for living machines (Wortham & Bryson, Chapter 33). Another domain in which living machines will become relevant is the application of symbiotic human–machine health systems for rehabilitation (Rubio Ballester, Chapter 59).

Figure 3.9 A lifelike android (of the physicist Albert Einstein) created by Hanson robotics. The development of increasingly life-like humanoid robots implies the possibility of closer human–robot relationships but also raises societal and ethical issues (see Millings and Collins, Chapter 60).

(Image: Flickr/Erik (HASH) Hersman. Terms of Use: This work is licensed under a Creative Commons Attribution 2.0 Generic License (CC BY 2.0). It is attributed to Erik (HASH) Hersman and the original version can be found at https://www.flickr.com/photos/whiteafrican/albums/72157613341134707 .)

Experts in the research area identified several goals for social robotics:

A robot representation of self. To interact effectively with people a robot should have good awareness of the external world, and of itself, and significant others, as embodied agents and social actors—essentially a synthetic sense of "self" (Verschure, Chapter 36; Prescott, 2015). Internal models can provide awareness of physical embodiment (Bongard, Chapter 11; Metta and Cingolani, Chapter 47), body pose and position (Asada, Chapter 18), localization within the environment (Erdem et al., Chapter 29), inner states (Vouloutsi and Verschure, Chapter 34; Seth, Chapter 37), motor control (Herreros, Chapter 26), and the ability to reflect on the past and plan for the future (Verschure, Chapters 35 and 36).

Verbal and non-verbal communication. Spoken language will play a crucial role in guiding social interactions with robots, which will involve acquiring a shared vocabulary of meaning. Whilst there is progress towards building speech recognition systems suitable for robots, understanding the meaning of utterances in context remains a critical research challenge. As language learning in humans builds on substrates of joint attention and non-verbal communication, it seems likely that similar strategies will be required for social robots that could be advanced by approaches in developmental robotics (Wortham and Bryson, Chapter 33).

State perception of others and social attention. This requires the development of machine learning and biomimetic algorithms for perceiving the state and intentions of others, and a social attention system, that builds on control systems that direct orienting, for appropriate social interaction (see Mitchinson, Chapter 27; Verschure, Chapter 36).

Identified risks include:

Deception. A number of researchers have considered ethical issues around the development of social robots, and some have argued that such robots are potentially, or even intrinsically, deceptive, since they can never be truly social according to accepted norms of human sociality (see e.g. Boden et al. 2017). However, this view is not universal; indeed, the Living Machines approach sees a continuity between humans and machines (Prescott and Verschure, Chapter 1) that allows for the possibility that future robots could have all the attributes required to be a truly social agent (see also Prescott 2017).

Robot rights. A related issue is the possibility that we could create new kinds of agents towards whom humans could owe some form of moral obligation. Gunkel (Chapter 63) provides a philosophical perspective on this question, arguing that the historical extension of rights to different classes of living systems suggests that we may eventually decide that we have some obligations towards synthetic living machines.

Domain 8: Brain-based systems

Biomimetic approaches to neuroscience, including brain-based robots and neuromorphic engineering, seek to advance the scientific understanding of the neural mechanisms underlying movement, perception, cognition, and learning in animals, and, at the same time, take inspiration from animals to design new control and sensing methods for robotics. In neuroscience and neuroethology, robotics gives an opportunity to test biological hypotheses that would otherwise not be accessible using traditional techniques from biology, by embodying these principles in robots that provide an accessible and measurable physical model (e.g. Figure 3.10; Verschure and Prescott, Chapter 3; Verschure, Chapter 36; and Prescott et al. 2016b).

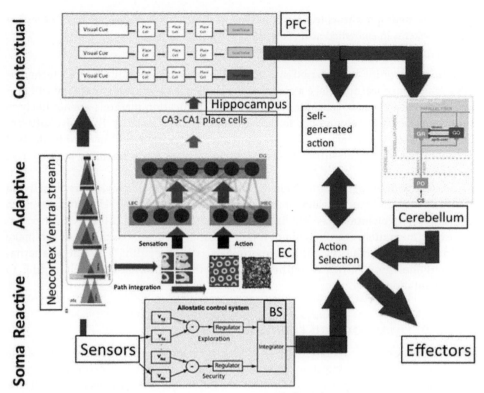

Figure 3.10 NeuroDAC: core elements of the Distributed Adaptive Control (DAC) architecture for robot control mapped to the rodent brain (see Verschure, Chapter 36).

Reprinted from *Biologically Inspired Cognitive Architectures*, 1, Paul F. M. J. Verschure, Distributed Adaptive Control: A theory of the mind, brain, body nexus, pp. 55–72, doi: 10.1016/j.bica.2012.04.005, Copyright © 2012 Elsevier B.V. All rights reserved. with permission from Elsevier.

Experts in this research area identified several goals for brain-based systems:

Robots as scientific tools for biology. At present, computational neural simulation is a standard tool for biology, as in the field of computational neuroscience. Many people concentrate on specific brain subsystems or circuits. However, this approach can be too focused to answer wider research questions in biology: there is a need to consider these components as part of a larger system. Models gain validity by satisfying different forms of constraint, such as from anatomy, physiology, and behavior (see Verschure and Prescott, Chapter 2). Animal-like robotics could be the next standard tool for investigating the interaction between brains, bodies, and the environment.

Biomimetic companion and assistive robots. Brain-based robotics provides a path toward companion and assistive robots whose cognitive and social capabilities are better matched to humans than those developed by conventional engineering (which are more likely to under- or over-shoot the human benchmark). Metta and Cingolani (Chapter 47) discuss some recent progress towards incorporating brain-based control and neuromorphic hardware in humanoid robots.

Prototyping implant technologies. In the domain of biohybrids, brain-based systems have the long-term potential to replace or augment human cognitive function. Work towards implantable neuroprostheses (e.g. Song and Berger, Chapter 55) can benefit by prior testing of these systems in embodied neurobotic model systems.

Identified risks include:

Safety and verifiability. Brain-based systems, along with other brain-inspired technologies such as deep learning, face a considerable challenge in verifying that they will perform appropriately and safely. Standard software verification methods are not likely to work with these technologies due to the lack of analytic tractability. Extensive software simulation could be a partial solution, though testing regimes will need to be devised that identify and provide sufficient tests of the system for low probability but high risk scenarios.

Dual-use issue with biological data. Neuroscientific data is collected invasively from living tissue. Ethical and moral issues in animal research are therefore relevant to research in brain-based biomimetics even if synthetic methods do not directly involve animal experimentation. One contribution that research on brain-based robots can make to reducing the number of animal experiments is by providing a synthetic alternative methodology in which to refine hypotheses through physical modeling prior to experimentation.

Domain 9: Olfaction and chemosensing

Artificial olfaction (or chemosensation) is an emerging requirement in engineering where there are an increasing number of applications that require measuring the presence or concentration of a particular chemical in a solid, liquid, or gaseous medium. These include: quality control in food processing; detection and diagnosis in medicine; detection of drugs, explosives, and dangerous or illegal substances; military and law enforcement such as against chemical warfare agents; disaster response such as to toxic industrial chemicals; and environmental monitoring of pollutants. The successful design of pattern analysis systems for artificial olfaction requires careful consideration of the various issues involved in processing multivariate data—signal pre-processing, feature extraction, feature selection, classification, regression, clustering, and validation—with biomimetic inspiration featuring strongly due to its potential for robustness and accuracy (see Figure 3.11 and Pearce, Chapter 17).

Experts in this research area identified several goals for artificial olfaction:

Comparison between neuromorphic and natural performance. Machine learning is good at some aspects of odor classification but struggles on segmentation of individual classes, whereas the mammalian olfactory system, composed of the olfactory bulb and olfactory cortex, is good at both. Active sensing (sniffing) could be useful too. The chemosensory systems of other animals such as bees, moths, and lobsters have also been targets for neuromorphic modeling and could provide useful clues.

Multi-modal perception with olfaction. The goal here is to combine artificial olfaction with other senses, for example using complementary information from flow sensors.

Improved artificial sensors. Neuromorphic chemosensors have been constrained by the quality of sensors: at present, there is not sufficient sensitivity or diversity of receptor types. Biohybridicity might be part of the answer, for example, Ayers (Chapter 51) describes a chemosensory system for controlling taxis behavior in a prototype biohybrid robot.

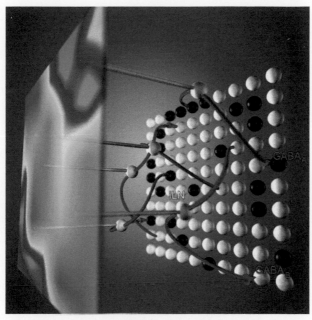

Figure 3.11 Biomimetic approaches, such as this insect-inspired neural network model, could lead to more accurate and robust solutions to machine olfaction (see also Pearce, Chapter 17).

Reproduced from Marco, A. Gutiérrez-Gálvez, A. Lansner, D. Martinez, J. P. Rospars, R. Beccherelli, A. Perera, T. C. Pearce, P. F. M. J. Verschure, and K. Persaud, A biomimetic approach to machine olfaction, featuring a very large-scale chemical sensor array and embedded neuro-bio-inspired computation, *Microsystem Technologies*, 20(4), pp. 729–42, S. © Springer-Verlag Berlin Heidelberg, 2013. With permission of Springer.

Domain 10: Biomimetic audition and echolocation

Audition is the ability to make sense of the world by interpreting pressure waves at different frequencies emanating from different directions (see Smith, Chapter 15). In nature, auditory sensing is often critical to survival and some animals have evolved far superior capabilities, in comparison with current technology, with regard to resolution, object identification, and material characterization (Assous et al. 2012). For example, bats can resolve acoustic pulses more efficiently than current technology, whilst dolphins are capable of discriminating different materials based on acoustic energy; these animals also demonstrate excellent acoustic focusing characteristics. These animal capabilities have inspired biomimetic robots, such as the "bat-bot" (Schillebeeckx et al. 2010), illustrated in Figure 3.12, that can wriggle its ears, an active sensing technique often used by bats to modulate the characteristics of the echo. Biomimetic echolocation in both air and water has many applications, not least in the development of improved sonar navigation and object detection/classification for aerial and marine robots.

Experts in this research area identified several goals for artificial audition:

Improved sensing technologies. Acoustic technologies with greater resolution and imaging would allow enhanced physical characterization of materials, processes, and structures.

Novel applications of sensor technologies. For example, cochlear implants could utilize sound to localize direction which is a challenge for current technology—see Lehmann and van Schaik (Chapter 54) for a discussion of the state of the art.

Figure 3.12 The CIRCE robot bat head with biomimetic sonar and moveable artificial pinnae can perform binaural 3D localization.
© Herbert Peremans.

Low-power active sensors. There is a need to reduce the peak power needed to achieve a given resolution in active sensors that emit energy (such as sound waves) to resolve physical systems.

Improved signal-to-noise capabilities. In audition, as in other sensory modalities, we can learn from the strategies used by animals to reduce the signal-to-noise problems in sensing, such as cancellation of self-produced noise.

Domain 11: Touch and hands

Tactile sensors can be used to sense a diverse range of stimuli ranging from detecting the presence or absence of a grasped object to a complete tactile image (see Lepora, Chapter 16). A tactile sensor can consist of an array of touch-sensitive sites, called taxels, analogous to sensory mechanoreceptors. The measurements obtained by a tactile sensor should convey a large amount of information about the state of a grip, for example within manipulation, and the properties of the contacted object. Evolution has devised examples of many different types of touch sensor to inspire artificial devices, such as the human fingertip, the tactile whiskers of rodents, and the tactile star of the star-nosed-mole. An important application of tactile sensing is to help reproduce the grasping and manipulation capabilities of humans (see Cutkosky, Chapter 30). Even though many robotic devices have been developed, from very simple grippers to very complex anthropomorphic robotic hands (Figure 3.13), their usability and reliability in unstructured environments still lags far behind human capabilities. A further important application domain

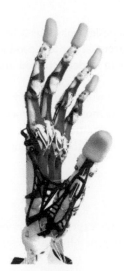

Figure 3.13 A strongly anthropomorphic robot hand from the University of Washington, Seattle. See Cutkosky (Chapter 30) for a discussion of grasp and manipulation strategies for biomimetic hands.

© 2016 IEEE. Reprinted, with permission, from Zhu Xu and Emanuel Todorov (personal copyright). See Zhe Xu and Emanuel Todorov, Design of a highly biomimetic anthropomorphic robotic hand towards artificial limb regeneration, *2016 IEEE International Conference on Robotics and Automation*, doi: 10.1109/ICRA.2016.7487528.

is to create touch-sensitive surfaces for prosthetic limbs (see Bensmaia, Chapter 53). The *Scholarpedia of Touch* (Prescott et al. 2016a) is a recent volume that collects many articles on biological touch systems and their emulation in artifacts via biomimetics.

Experts in this research area identified several goals for biomimetic artificial touch:

Robot cognition for touch. The utility and safety of robots can be improved by developing and integrating biomimetic systems that implement reflexive reactions to tactile events, can flexibly represent spatially and temporally distributed patterns of physical contact, and utilize tactile data to construct a body schema (Asada, Chapter 18) and support physical interaction.

Human–robot interactions via touch. Biomimetic systems using tactile feedback can improve human–robot interaction capabilities; for instance, by stimulating the C-tactile fiber system in human hairy skin thought to underlie aspects of affective and social touch.

Virtual skin for delivering prosthetics and telehaptics. The use of tactile sensing in prosthetic limbs has not extended far beyond prototype systems (see Bensmaia, Chapter 53), but will have clear societal benefit in terms of improving the useability of prosthetics for grasping and manipulation of objects. A further application is to deliver haptic feedback to human operators in contexts such as teleoperation and robotic surgery. Both applications require a tactile sensing capacity and a display capacity so that the signals generated can be relayed to the user in an easily intelligible way.

Dextrous manipulation. Human-level object manipulation is an unsolved problem in robotics. Tactile sensing is likely to be key, as is machine learning and artificial hand-eye coordination. In all of these aspects, brain-based modeling could provide useful insights. There are many high-impact application domains, including fully automated assembly lines, agricultural robots (particularly harvesting), and healthcare (particularly physical robot–human interaction).

Identified risks include:

Limitations of conventional robot hands. The adoption of design solutions inherited from conventional mechanics is, in part, responsible for the current failure to reproduce human-level manipulation capabilities in robots. Recent progress in under-actuated and soft grippers may provide a partial answer.

Threat to human workforce. A considerable fraction of the world's labour force is involved in hand-assembling products for human consumption. As more of this process is automated there will be significant societal and economic impacts. Political actions may be needed to ensure that the advances in productivity brought about by these technologies are for the benefit of everyone.

The Living Machines approach

In addition to comments about particular research areas, the experts we interviewed were also asked to comment about Living Machines research more generally. Some common concerns and lessons can be summarized as follows:

Good practice in Living Machines research. Biomimetic methods must remain anchored in the natural and biomedical sciences, and linked to best practice and knowledge in engineering, or else the field risks isolation and the susceptibility to fads, fashions, and the irrelevance that this brings. Often in biomimetics the links between engineers and scientists can be tenuous, although there can be good links between modelers and scientists and between modelers and engineers. To better straddle the divide we need more researchers who have some research training in both engineering *and* science. We also need to identify biomimetic methodologies that can translate across domains and to have better paths to disseminate good practice. The international landscape for interdisciplinary training relevant to Living Machines research, and some strategies to improve it, are considered by Mura and Prescott (Chapter 64).

Beyond bio-inspiration. Related to the above, a beginner's mistake is to try to do something that is a bit scientific and a bit engineering with no clear goal. You should either have a defined engineering challenge, or, if you wish to contribute to biology, understand the relevant sub-field sufficiently well to be sure where you are adding new knowledge. It may be possible to do both at once, but it is easy to fall down the middle.

The limits of biomimetics. The Living Machines approach is not a panacea for solving all engineering challenges. We can learn from nature's ingenuity, but we should be critical rather than just following blindly. The Panglossian assumption that everything in nature is optimized is patently false (Gould and Lewontin 1979). The focus should be on using biomimetics to identify and prototype candidate engineering solutions. Engineering can improve on nature, but it often needs to catch up first.

Public perceptions. Support for Living Machines research, including better funding, requires better communication about our goals, methods, and achievements. Biomimetic and biohybrid technologies entail some societal risk, as does any new technology, but this field is particularly susceptible to inciting fear of monsters (Szollosy, Chapter 61), or criticism that it is promoting individual human enhancement over broader public good (Hughes, Chapter 57). Perceptions of research endeavors can sour rapidly if the public see scientists as guilty of hubris or as overly committed to commercial interests. Secrecy, which can be brought about by an over-reliance on commercial or military funding, can also create a justifiable atmosphere of distrust.

Science and engineering are not value-neutral, and it is important that the research field of Living Machines openly and honestly appraises its goals and exposes them to public criticism (Prescott and Verschure, 2016).

Attention to real-world challenges. There is a mistaken view that the best science is pure science. If we follow the example of Pasteur[1] we can see that research is often most effective, important, and impactful when applied questions intersect with basic science, for example where challenges in stroke rehabilitation are addressed through advances in brain theory (Ballester, Chapter 59). Fortunately, the Living Machines ethos of "understanding through making" (Verschure and Prescott, Chapter 2) is well-positioned to have positive impacts on the world, but it must be recognized that the step of turning proof-of-concept into societal benefit requires additional effort. In our view, there is a special need for Living Machines technologies at this time due to the growing crisis in the sustainability of our modern technological society (Halloy, Chapter 65). Unless we can learn from nature how to maintain our way of life, without destroying the environment that sustains it, then we risk going the way of past human civilizations (see Diamond, 2005), except this time on a global scale.

Conclusion

Our roadmapping survey confirms that Living Machines research has entered a phase of rapid growth with broad impact across a wide range of science and technology. Taken together, these findings indicate that biomimetic and biohybrid approaches are becoming increasingly important in robotics, materials science, neuroscience, and other disciplines, with the potential for significant societal and economic impact over the coming decades.

It is often said that future scientific discoveries are hard to predict. This is not necessarily the case in biomimetics. There are plenty of examples surrounding us in the natural world. The future will produce artificial devices with these abilities, from mass-produced flying micro devices based on insects, to robotic manipulators based on the human hand, to swimming robots based on fish. Less certain is what they will do to our society, economy, and way of life, or to the human condition as we become more biohybrid. In exploring this exciting world we must tread carefully with due attention to the impacts our footsteps might bring.

Acknowledgments

The surveys and events mentioned in this chapter were sponsored by the European Union 7th Framework through the Future Emerging Technologies (FET) programe and Coordination Actions "Convergent Science Network for Biomimetics and Biohybrid Systems (CSNI)" (ICT-248986) and "Convergent Science Network of Biomimetics and Neurotechnology (CSNII)" (ICT-601167). We are grateful to the many speakers and participants in CSN events and conferences who have helped to craft and inspire the Living Machines approach or, in one way or another, have contributed ideas to this roadmap.

---

[1] Pasteur spent much of his scientific career working on practical problems such the diseases of domestic animals and the fermentation of sugar; nevertheless, his research led to important discoveries such as the germ theory of disease, and to the birth of microbiology. See Stokes (1992) for the concept of "Pasteur's Quadrant" as the sweet spot for important, societally relevant research that combines basic science with applied concerns.

Contributors

Robert Allen, Professor in Biodynamics and Control, University of Southhampton, UK

Joseph Ayers, Professor of Marine and Environmental Sciences, Northeastern University, MA, USA

Dieter Braun, Professor in Systems Biophysics, Ludwig Maximilians University, München, Germany

Yoseph Bar-Cohen, Senior Research Scientist, Jet Propulsion Laboratory, California Institute of Technology, CA, USA

Mark Cutkovsky, Fletcher Jones Professor of Mechanical Engineering, Stanford University, CA, USA

Yiannis Demeris, Professor in Human-Centred Robotics, Imperial College, London, UK

Frank Grasso, Associate Professor in Psychology, Brooklyn College, NY, USA

Mitra Hartmann, Professor of Biomedical Engineering and Mechanical Engineering, Northwestern University, IL, USA

Auke Ijspeert, Professor, Head of Biorobotics Laboratory, École polytechnique fédérale de Lausanne, Switzerland

William Kier, Professor in Marine Biology, Biomechanics of musculoskeletal systems, University of North Carolina, NC,USA

Danica Kragic, Professor of Computer Science, KTH Royal Institute of Technology, Sweden

Maarja Kruusma, Professor of Biorobotics, Tallinn University of Technology, Estonia

David Lane, Professor of Autonomous Systems Engineering, Heriot-Watt University, Edinburgh, UK

Nathan Lepora, Associate Professor in Engineering Mathematics, University of Bristol and Bristol Robotics Laboratory, UK

David Lentink, Assistant Professor in Mechanical Engineering, Stanford University, CA, USA

Tim Pearce, Reader in Bioengineering, University of Leicester, UK

Giovanni Pezzulo, Researcher at the National Research Council of Italy, Institute of Cognitive Sciences and Technologies (ISTC-CNR), Rome, Italy

Andrew Phillipides, Reader in Informatics, University of Sussex, UK

Barry Trimmer, Henry Bromfield Pearson Professor of Natural Sciences, Tufts University, MA, USA

Tony Prescott, Professor of Cognitive Robotics, Sheffield University, UK

Paul Verschure, Professor of Cognitive science and Neurorobotics at the Technology Department, Universitat Pompeu Fabra, Spain

Ian Walker, Professor of Electrical and Computer Engineering, Clemson University, SC, USA

David Zipser, Professor Emeritus of Cognitive Science, University of California, CA, USA

References

Assous, S., Lovell, M., Linnett, L., Gunn, D., Jackson, P. and Rees, J., 2012. A novel bio-inspired acoustic ranging approach for a better resolution achievement. In: S. Bourennane (ed.), *Underwater Acoustics*. Rijeka, Croatia: InTech. doi: 10.5772/32924

Boden, M., Bryson, J., Caldwell, D., Dautenhahn, K., Edwards, L., Kember, S., Newman, P., Parry, V., Pegman, G., Rodden, T., Sorrell, T., Wallis, M., Whitby, B., and Winfield, A. (2017). Principles of robotics: regulating robots in the real world. *Connection Science*, 29(2), 124–9. doi:10.1080/09540091.2016.1271400

Cianchetti, M., Calisti, M., Margheri, L., Kuba, M., and Laschi, C. (2015). Bioinspired locomotion and grasping in water: the soft eight-arm OCTOPUS robot. *Bioinspiration & Biomimetics (Special Issue on Octopus-inspired robotics)*, 10, 035003.

de Croon, G.C.H.E., Ruijsink, R., Perçin, M., De Wagter, C., and Remes, B.D.W. (2016). *The DelFly: Design, Aerodynamics, and Artificial Intelligence of a Flapping Wing Robot.* Dordrecht: Springer Science and Business Media. doi: 10.1007/978-94-017-9208-0

Diamond, J. (2005). *Collapse: How Societies Choose to Fail or Succeed.* New York: Viking Press.

Gould, S. J., and Lewontin, R. C. (1979). The spandrels of San Marco and the Panglossian paradigm: a critique of the adaptationist programme. *Proc. R Soc. Ldn B: Biol. Sci.*, **205**(1161), 581–98.

Hobbelen, D., de Boer, T., and Wisse, M. (2008). System overview of bipedal robots Flame and TUlip: Tailor-made for limit cycle walking. In: *2008 IEEE/RSJ International Conference on Intelligent Robots and Systems.* IEEE., pp. 2486–91.

Kernbach, S., Meister, E., Schlachter, F., Jebens, K., Szymanski, M., Liedke, J., Laneri, D., Winkler, L., Schmickl, T., Thenius, R., and Corradi, P. (2008). Symbiotic robot organisms: REPLICATOR and SYMBRION projects. In: *Proceedings of the 8th workshop on performance metrics for intelligent systems.* ACM, pp. 62–9.

Lee, E. A. (2008). Cyber Physical Systems: Design Challenges. *11th IEEE International Symposium on Object and Component-Oriented Real-Time Distributed Computing (ISORC)*, Orlando, FL, 2008, pp. 363–9. doi: 10.1109/ISORC.2008.25.

Lepora, N. F., Verschure, P., and Prescott, T. J. (2013). The state of the art in biomimetics. *Bioinspiration & Biomimetics*, **8**(1), 013001.

Liu, J., and Hu, H. (2006). Biologically inspired behaviour design of autonomous robotic fish. *International Journal of Automation and Computing*, **3**(4), 336–47.

Mangan, M., Cutkosky, M., Mura, A., Verschure, P. F. M. J., Prescott, T. J., and Lepora, N. (eds), *Biomimetic and Biohybrid Systems: 6th International Conference, Living Machines* 2017, Stanford, CA, USA, July 26–28, 2017, Proceedings (pp. 86–94). Cham: Springer International Publishing.

Martinez, D., and Montejo, N. (2008). A model of stimulus-specific neural assemblies in the insect antennal lobe. *PLoS Comput. Biol.*, **4**(8): e1000139

Mathews, Z., Lechon, M., Calvo, J. M., Dhir, A., Duff, A., Bermudez i Badia, S., and Verschure, P. F. M. J. (2009). Insect-like mapless navigation based on head direction cells and contextual learning using chemo-visual sensors. In: *IEEE/RSJ International Conference on Intelligent Robots and Systems.* IEEE, pp. 2243–2250. doi: 10.1109/IROS.2009.5354264.

Newell, A. (1990). *Unified Theories of Cognition.* Cambridge, MA: Harvard University Press.

Pfeifer, R., and Verschure, P. (1995). The challenge of autonomous agents: Pitfalls and how to avoid them. In: L. Steels and R. Brooks (eds.), *The artificial life route to artificial intelligence: building embodied, situated agents.* Hillsdale, New Jersey: Lawrence Erlbaum Associates, pp. 237–263.

Prescott, T. J. (2015). Me in the Machine. *New Scientist*, 21 March 2015, 36–9.

Prescott, T. J. (2017). Robots are not just tools. *Connection Science*, **29**(2), 142–9. doi:10.1080/09540091.2017.1279125

Prescott, T. J., Ahissar, E, and Izhikevich, E. (2016a). *Scholarpedia of Touch.* Amsterdam: Atlantic Press.

Prescott, T. J., Ayers, J., Grasso, F. W., and Verschure, P. F. M. J. (2016b). Embodied Models and Neurorobotics. In: M. A. Arbib and J. J. Bonaiuto (eds), *From neuron to cognition via computational neuroscience.* Cambridge, MA: MIT Press, pp. 483–512.

Prescott, T. J., Lepora, N. F., Mura, A., and Verschure, P. F. M. J. (eds) (2012). *Biomimetic and Biohybrid Systems: First International Conference, Living Machines 2012*, Barcelona, Spain, July 9–12, 2012, Proceedings (Vol. 7375). Basel: Springer.

Schneider, A., Paskarbeit, J., Schaeffersmann, M., and Schmitz, J. (2012). Hector, a new hexapod robot platform with increased mobility-control approach, design and communication. In: U. Rückert, J. Sitte, and F. Werner (eds), *Advances in Autonomous Mini Robots.* Berlin Heidelberg: Springer, pp. 249–64. .

Schillebeeckx, F., De Mey, F., Vanderelst, D., and Peremans, H. (2010). Biomimetic sonar: binaural 3D localization using artificial bat pinnae. *The International Journal of Robotics Research*, **30**(8), 975–987.

Stokes, D. E. (1992). *Basic Science and Technological Innovation*. Washington: Brookings Institution Press.

Verschure, P.F. (2012). Distributed adaptive control: a theory of the mind, brain, body nexus. *Biologically Inspired Cognitive Architectures*, **1**, 55–72.

Ward, T. A., Fearday, C. J., Salami, E., and Soin, N. B. (2017). A bibliometric review of progress in micro air vehicle research. *International Journal of Micro Air Vehicles*, **9**(2), 146–65. doi:10.1177/1756829316670671

Wood, R. J. (2014). The challenge of manufacturing between macro and micro. *American Scientist*, **102**(2), 124.

Xu, Z., and Todorov, E. (2016). Design of a highly biomimetic anthropomorphic robotic hand towards artificial limb regeneration. In: *2016 IEEE International Conference on Robotics and Automation (ICRA)*, Stockholm, 2016. IEEE, pp. 3485–92.

Section II

Life

Chapter 4

Life

Tony J. Prescott
Sheffield Robotics and Department of Computer Science,
University of Sheffield, UK

A core tenet of the emerging field of research in living machines is that biological entities that live and act—organisms—have much in common with certain kinds of man-made entities—machines—that can display autonomous behaviour. But underlying this parallel, which is made even more persuasive by observing the life-like behaviour of many of the artifacts described in this book, are a host of critical, and still only partially answered questions. Perhaps the most fundamental of these is "what is life?"

The chapters in this section of our handbook delve into this core question, exploring some of the most fundamental properties of living systems, such as their capacity to self-organize, to evolve, to metabolize, to grow, to self-repair, and to reproduce. But perhaps the best way to begin is with the answer given to the question of life by the biologist Ludwig von Bertalanffy, one of the sharpest and most forward-thinking minds of the twentieth century.

Von Bertalanffy (1932, 1969) argued that although an organism can be viewed as an aggregate of a very large number of processes that can be described and explained in terms of their physics and chemistry, this description alone does not get to heart of the question of what it means to be alive or not. What makes an organism alive rather than dead requires a different kind of explanation, and it is the presence of *order*[1]. He went on to explain that order comes in different forms and he illustrated this by reference to the very idea that is at the heart of our handbook—that is of the "living machine" (Bertalanffy, 1969). From the clockwork machines of the seventeenth century, through the thermodynamic machines of the days of steam, to the self-regulating machines of cybernetics, he saw patterns of order emerging at different scales, and from very different chemical and physical substrates, in carefully designed artificial systems. The machine metaphor, he suggested, allows us to look again at organisms and to see them as made up of lots of little molecular machines all intricately coupled and working to support and maintain one another.

But von Bertalanffy also saw problems in the analogy between the organism and the machine, or at least unanswered questions. One of these is the problem of origin. Since organisms are evolved and machines are made, how can theories about machines help unravel the mystery of the origin of life? In particular, can machines make themselves? We will see in the chapters of this book that they are beginning to and that the answer to the origins question may be that it was living machines all the way back to the dawn of life.

---

[1] Von Bertalanffy also emphasized the wholeness of living entities, influenced, in part by the notion of *Gestalt* (from Köhler (1924)). The emphasis on the whole being more than sum of the parts permeates the multi-scale approach in biology that von Bertalanffy helped to establish.

A second problem for von Bertalanffy concerned regulation. He was aware, following Turing, that many complex processes can be resolved into a series of steps (an algorithm) that can be reproduced by an automaton. Nevertheless, he worried that the number of steps that would underlie such processes as are required for life might be too immense to be computable. With the benefit of more than sixty years of advances in computing it is now easier to be reassured on this point, particularly whilst the trend towards faster computation at smaller and smaller scales continues. Perhaps as importantly though, the science of Living Machines is also showing that the principles underlying the regulation of life can be abstracted, and that it should be possible to recreate the dynamic patterns that von Bertalanffy saw to be so critical to natural living systems without the need for all of life's intricately rich biochemistry.

Finally, von Bertalanffy wondered about the fundamental characteristic of the organism as a system engaged in the continuous exchange of components, the process we call metabolism. If metabolism is what creates the organism then higher-level descriptions of the organism as a machine, such as its cybernetic capability, do not explain life. Rather, order is already needed to bring the organism into existence, and to allow it to maintain its existence[2], before we can explore all these further interesting analogies with our man-made automata.

This brings us to the core idea that von Bertalanffy considered to underpin life—the notion of the organism as an *open system*. An open system, as compared with a closed one such as a chemical reaction inside a test-tube, is one that continuously exchanges material with its environment, and so succeeds, at least temporarily, in thwarting the second law of thermodynamics—the tendency towards entropy or maximum disorder. Living things, in staving off entropy, achieve and maintain a steady state that is equilibrium-like but that is far from a true equilibrium. In order to maintain this state they do work, and in working, all of the marvellous properties that we see in living systems such as behaviour and cognition can come about.

The idea of living things as open self-sustaining systems has further evolved over the past century. The physicist Erwin Schrödinger (1944) popularized the notion that living organisms extract order from the environment by feeding on "negative entropy", while the chemist Ilya Prigogine emphasized the irreversibility of the dynamical processes that sustain life, describing living systems as structures that thrive on a throughput of energy that then dissipates into the environment (Prigogine and Stengers 1984). A formal definition of a living machine, intended to capture the self-generating nature of life and the ability of living organisms to acquire and distribute the materials they need to stabilize themselves within a boundary, was proposed by Humberto Maturana and Francisco Varela (1973/1980) in their notion of an *autopoietic machine*. Whilst there have been various attempts to create such machines as computational models (e.g. Agmon et al. 2016; McMullin 2004), it is much harder to build a physical device that instantiates these ideas; with the exception of some of the machines described in this volume, most of today's robots enforce a separation between fixed physical hardware and mutable software such that robotic embodiment is typically very different from autopoietic embodiment (Ziemke 2004).

The chapters in this section of the book further explore these fundamental questions of the essence of life and some of the steps that are being taken to understand them through building living machines.

We begin with the notion of emergent order itself, which in recent decades has increasingly been described using the term *self-organization*. Wilson (Chapter 5) explores the fundamental

---

[2] Von Bertalanffy wrote: "A machine-like structure of the organism cannot be the ultimate reason for the order of life processes because the machine itself is maintained in an ordered flow of processes. The primary order therefore must lie in the process itself." (von Bertalanffy, 1969, p. 141)

nature of self-organization, its relationship to chaos, and its role in the evolution of biological organisms.

Next we look at the question of metabolism and steps towards the development of living machines that generate their own energy by harvesting and metabolizing food. Specifically, Ieropoulos and colleagues (Chapter 6) explore the symbiotic relationship between a microbial fuel cell and the robotic body in which it is embedded, and the lessons that can be learned from biological homeostasis about designing and building self-sustaining living machines.

Moses and Chirikjian (Chapter 7) investigate the key challenge of machine reproduction beginning with von Neumann's design for a cellular automaton, called the Universal Constructor, which was devised both to allow machine replication and to support open-ended growth. Moses and Chirikjian track the history of attempts to build physical instantiations of such a system up to and including recent modular robotic systems inspired by cell molecular biology.

Biological organisms are products of the twin processes of evolution and development, and contemporary systems biology increasingly recognizes the interdependence of these two in giving rise to the complexity and variety of forms that we find in nature. Artificial evolution is widely explored in synthetic systems; however, attempts to combine evolution and development— artificial *evo-devo*—are fewer and far between. Prescott and Krubitzer (Chapter 8) trace some of the key ideas in the modern biological synthesis, particularly in relation to the evolution and development of the nervous system, and show how these are being investigated to allow evo-devo for living machines.

Living systems that move and grow are the focus of Mazzolai's contribution to our handbook (Chapter 9), focusing particularly on what we can learn from the study of plants as living machines that operate in diverse and often hostile environmental conditions. Mazzolai explains how the biology of plants, and particularly of their root systems, is inspiring the design of a new class of plant-inspired actuator systems.

Vincent (Chapter 10) addresses the broad topic of biological materials noting their composite nature, their durability, and multifunctional capabilities. Vincent considers how we can learn from biology to manufacture low-energy, high-performance materials that can self-assemble.

The adaptability of living systems arises in part from their capability to self-monitor and adapt or self-repair in response to change or damage. Bongard (Chapter 11) discusses how living machines can use self-models to understand their own morphology, adapt to change, learn from others, or metamorphose from one body plan to another.

We conclude this section by returning to the core question of the nature of life, and to the differences between living and man-made systems that intrigued von Bertalanffy and many others. Building on Darwin's theory of evolution, von Neumann's Universal Constructor, and theories of self-organization, Deacon (Chapter 12) attempts a general theory of evolution, intended to straddle both the organic and inorganic domains and to resolve the challenges left aside by Darwin's theory of natural selection—the origin of organisms that are capable of self-replication, a question that is explored empirically, via the physics and chemistry of self-replicating molecular open systems, by Mast et al. (Chapter 39) later in this book. In setting out this theory, Deacon rejects the notion that organisms are "mere mechanisms" and seeks to reinject a notion of agency into our definition of living systems. Specifically, he argues that living entities are both self-specifying and self-determining, and that they create their own inherent teleology or purpose, towards which their dynamics evolves.

If Deacon is right then the developers of living machines may need to embed more of the core "formative powers" of biological systems into their artifacts in order to bring them truly to life.

Beyond this, however, I would follow Maturana and Varela in being sceptical of further efforts to move the goal-posts:

> To the extent that the nature of the living organization is unknown, it is not possible to recognize when one has at hand, either as a concrete synthetic system or as a description, a system that exhibits it. Unless one knows which is the living organization, one cannot know which organization is living. In practice, it is accepted that plants and animals are living but their characterization as living is done through the enumeration of their properties. Among these, reproduction and evolution appear as determinant, and for many observers the condition of living appears subordinated to the possession of these properties. However, when these properties are incorporated in a concrete or conceptual man-made system, those who do not accept emotionally that the nature of life can be understood, immediately conceive of other properties as relevant, and do not accept any synthetic system as living by continuously specifying new requirements. (Maturana and Varela, 1973/1980, p. 83)

References

Agmon, E., Gates, A.J., Churavy, V., and Beer, R.D. (2016). Exploring the space of viable configurations in a model of metabolism–boundary co-construction. *Artificial Life*, **22**(2), 153–71. doi:10.1162/ARTL_a_00196

Bertalanffy, L. v. (1932). *Theoretische Biologie: Band 1: Allgemeine Theorie, Physikochemie, Aufbau und Entwicklung des Organismus*. Berlin: Gebrüder Borntraeger.

Bertalanffy, L. v. (1969). *General system theory: foundations, development, applications*. New York: Braziller.

Köhler, W. (1924). *Die physischen Gestalten in Ruhe und im stationären Zustand*. Braunschweig: Vieweg.

Maturana, H., and Varela, F. (1973/1980). *Autopoiesis: the organization of the living*. Dordrecht, Holland: D. Reidel Publishing.

McMullin, B. (2004). Thirty years of computational autopoiesis: A review. *Artificial Life*, **10**(3), 277–95.

Prigogine, I., and Stengers, I. (1984). *Order out of chaos: Man's new dialogue with nature*. New York: Bantam Books.

Schrödinger, E. (1944). *What is Life?* Cambridge, UK: Cambridge University Press.

Ziemke, T. (2004). Are robots embodied? In: C. Balkenius, J. Zlatev, H. Kozima, K. Dautenhahn, and C. Breazeal (eds), *Proceedings of the First International Workshop on Epigenetic Robotics: Modeling Cognitive Development in Robotic Systems*. Lund University Cognitive Studies, **85**. Lund: LUCS.

Chapter 5

Self-organization

Stuart P. Wilson

Department of Psychology, University of Sheffield, UK

Self-organization describes a dynamic in a system whereby local interactions between individuals collectively yield global order, i.e. patterns unobservable in their entirety to the individuals. By this working definition, self-organization is intimately related to chaos, i.e. global order in the dynamics of deterministic systems that are locally unpredictable. A useful distinction is that a small perturbation to a chaotic system causes a large deviation in its trajectory, i.e. the butterfly effect, whereas self-organizing patterns are robust to noise and perturbation. For many, self-organization is as important to the understanding of biological processes as natural selection. For some, self-organization explains where the complex forms that compete for survival in the natural world originate from. This chapter outlines some fundamental ideas from the study of simulated self-organizing systems, before suggesting how self-organizing principles could be applied through biohybrid societies to establish new theories of living systems.

Biological principles

Perhaps the most influential description of self-organization can be found in the system of equations considered by Turing (1952). Turing described *morphogenesis* (pattern formation) in chemical systems as emergent properties of local interactions between *activators*, i.e. chemicals that increase the rate of a reaction, and *inhibitors*, i.e. chemicals that decrease the rate of a reaction. Turing's equations describe how the balance and spatial distribution of activators and inhibitors affect the two-dimensional distribution of reactions across a continuous medium, e.g. a substance in a petri dish. Turing considered the influence of activators and inhibitors to be local and homogeneous, so his equations essentially define the iterative application of a 2D kernel across the medium. He showed that different constraints on the form of this kernel can lead to dynamics resulting in the emergence of different patterns of reaction. Crucially, constraints yielding regular patterns involved activators with more local influence than inhibitors. Under such constraints, any slight disturbances in the initial distribution of chemicals cause reactions to build up around points of elevated activator concentration, and for adjacent concentrations to separate out across the medium into regular patterns determined by the form of the kernel and the shape of the boundary (e.g. the petri dish). Similar to how a small number of modes become selectively amplified when a string held at two ends is vibrated, different kernels selectively amplify particular modes during *Turing instabilities*, leading to the emergence of regularly spaced spots (like those of a leopard), stripes (like those of a zebra), fingerprint patterns (like those of primates), Dirichlet domains (like cortical columns in rodent brains), etc. An important insight is that while different random initial chemical distributions will lead to the emergence of different patterns, their overall qualitative form will be determined by the kernel, e.g. spots may emerge in different locations, but their size, shape, and spatial frequency across the medium will be equivalent.

As demonstrated through Turing morphogenesis, self-organization is often described as the spontaneous emergence of global order from a system of individuals interacting without a plan (e.g. Camazine et al. 2001). In some systems, a useful extension is to describe self-organization as the emergence of a plan (or schema), i.e. an organization with implications for function, and thus for fitness. This extension is most clearly explained through self-organizing map algorithms, which demonstrate how statistical structure in the world external to a network can become consolidated in the functional organization of the network over time. A concrete example is the neural network model of von der Malsburg (1973).

Von der Malsburg (1973) constructed a model of how the mammalian primary visual cortex (V1) develops spatially organized patterns of neuronal preferences for the orientation of edges that appear on the retina. The network comprised a two-dimensional sheet of artificial photoreceptors corresponding to the retina and sheets of excitatory and inhibitory neurons corresponding to V1. When photoreceptors were activated in patterns resembling images of oriented lines, V1 neurons responded by exciting each other if they were nearby and inhibiting if they were further apart, creating a recurrent dynamic like that studied by Turing, in which initially random patterns of activity across the V1 sheet collected into localized "bubbles" of activity. These dynamics were consolidated by Hebbian learning at connections from the retina to V1, which increases future influences of the active photoreceptors on neurons in proportion to their activities after the dynamics have settled. Neurons contained by emergent activity bubbles would thus be more strongly active for a second image if it involved photoreceptors active in the first, but otherwise activity would more likely originate (and therefore ultimately settle) around neurons in other locations across V1. After iterating for many oriented patterns, neurons each develop selective responses to specific orientations, and von der Malsburg found that similar orientations became represented by adjacent neurons, to create patterns of orientation preference across V1 resembling those in animals. The patterns reflect across the tissue the (periodic) topology of the space of possible orientations, and thus may be considered an emergent plan of the orientation domain.

As a description of self-organization, the key principles common to the Turing and von der Malsburg models are as follows; i) individuals interact locally, directly influencing only neighbors within some finite distance, ii) individuals interact by a balance of cooperation (short-range chemical activators or synaptic excitation) and competition (long-range chemical inhibitors, or synaptic inhibition), and iii) individuals had no direct access to the plan, i.e. to generate camouflaging patterns, or to represent periodic topologies, as we might describe these apparent plans *post hoc*.

Our working definition of self-organization as a dynamic in a system whereby local interactions between individuals collectively yield global order can therefore be fleshed out a little: self-organization can describe the dynamic emergence of a plan from a balance of local cooperative and competitive interactions between individuals.

The emergence of a plan does not necessarily constitute the emergence of function. For example, it has been argued that emergent patterns in the anatomical or physiological properties of self-organized neural maps may not in themselves be useful (Purves et al. 1992; Wilson and Bednar 2015). Maps may well be *spandrels*, that is, essentially useless by-products of developmental mechanisms selected for their efficiency in some non-computational capacity. But self-organized plans may provide a starting point for evolution by natural selection. And the ability of models to self-organize into naturally occurring patterns can be used to validate theories of how forms of local interaction have been selected in natural systems. This is an essential part of the *synthetic* scientific approach (Braitenberg 1984). For example, a recent Turing-style model of orientation map development showed that only long-range inhibitory interactions between the neural correlates of similar orientations guarantee the emergence of mammalian orientation maps, which suggests that mammalian brain evolution has favoured such interactions (Kaschube et al. 2010).

Self-organization provides a starting point for evolution. When we understand how ordered patterns can emerge from disorder, the door is opened for evolution by natural selection to then explain how the fittedness of those patterns to their environments is evaluated by competition between forms. This view of self-organization, as a necessary prerequisite for evolution, can be understood through the ideas of Kauffman (1993).

Kauffman considered networks of interacting nodes, which he called *NK* networks, where N is the number of nodes, and K is the average number of other nodes affecting each. Kauffman considered networks of interacting chemicals but we might think of nodes as genes or neurons or animals etc. Imagine 20 nodes, inside a hat, initially unconnected. Pick out a pair from the hat at random and connect them, for example by adding a synaptic connection between two neurons, and then replace them in the hat. Now pick another pair at random and connect them. It is unlikely that by chance you will have picked one of the first two. Repeat the process, adding a new connection each time and replacing the nodes in the hat, and consider that each connection increases the chance that by randomly picking a node you will also lift several connected nodes from the hat. The network you are constructing is called a random graph.

As the ratio of connections to nodes in a random graph increases beyond 0.5 (e.g. 10 connections between 20 nodes), the chances that picking out one node will mean lifting all others too increases suddenly. Plotting the size of the largest cluster lifted against the number of connections made reveals a sigmoidal (S-shaped) curve, indicating a *phase transition*; gradual change in the control parameter (the ratio of connections) causes a rapid change in behaviour. When there are as many connections as nodes, pathways of connections of all lengths become possible, hence activation of one node can affect all others. Kauffmann considered also a second control parameter. If nodes can be either on or off and each is connected, for example, to three others, then there are $2^3 = 8$ combinations of input that could affect each node. If every combination turns the node on (or off) the activity in the network will be uninteresting; however, another phase transition occurs when around half of the input combinations turn a given node on. The two control parameters, the connections/nodes ratio and the proportion of activating interactions, define a two-dimensional space wherein network dynamics are either *subcritical*, e.g. inputs fix all nodes as either on or off, or *supercritical*, e.g. the dynamics explode into chaos. In this space, Kauffman described the *edge of chaos*, a boundary defined by the two phase transitions between sub- and super-criticality, along which patterned dynamics are self-organizing and self-sustaining. Kauffman's combinatorial approach led him to propose i) that life (i.e. self-sustaining dynamics in metabolic networks) could emerge spontaneously through a phase transition when enough interacting chemicals interact, and ii) that evolution by natural selection describes a tendency for natural systems to remain poised at the edge of chaos, in a regime where self-organizing dynamics prevail.

Note that the phase transition is a hallmark of self-organization, owing to a direct analogy that can be drawn between the control parameter of a system and the thermodynamic concept of temperature, and thus note also that the concepts of self-organization, thermodynamics, and life, are thought to be closely intertwined (see Schrödinger, 1967).

Biomimetic or biohybrid systems

Considering how self-organization and natural selection may interact, an idea with important implications for the design of synthetic systems is *the Baldwin effect*. Not to be confused with Lamarckian evolution (the inheritance of acquired characteristics), which assumes that information gathered during a lifetime can be passed directly from phenotype to genotype, the central idea of Baldwin was that evolution could be sped up through interactions with the environment during the lifetime that change the fitness landscape for future generations. The most concrete example is through the model of Hinton and Nolan (1987). They imagined a population of organisms each with a genotype comprising 20 binary states, and specified a fitness landscape in which just one target configuration of the states was optimal; searching 2^{20} possible states for this isolated fitness peak is almost impossible, like trying to find a needle in a haystack. However, when a proportion of the binary states could be randomly flipped many times to simulate individual lifetimes, and hence increase the chance of finding the target configuration, a genetic algorithm could solve the problem quickly. A phase transition occurred such that when one in the population stumbled on the target, passing on its configuration of correct and flippable states to the next population rapidly increased the fitness for subsequent generations. This smoothing of the otherwise rugged fitness landscape occurred because Hinton and Nolan defined a cost to adaptation; initial genetic conditions from which fewer flips were required to reach the target configuration were survived in the next generation with greater probability. Crucially, whether the flippable states should be one or zero was not inherited directly (so the acceleration was not Lamarckian), rather the capacity to modify state during the lifetime reduced the problem space searched by the genetic algorithm. Hinton and Nolan (1987) described the capacity to flip states as the capacity to learn, hence their paper "How Learning Can Guide Evolution" has been highly influential in formalizing Baldwin's idea and related concepts such as *adaptability* and *evolvability* (albeit not without criticism).

The following statement from their discussion encapsulates what may now be called *evo-devo* theory (from evolutionary developmental biology; see Carroll, 2005; and Prescott and Krubitzer, Chapter 8, this volume); the idea that evolution operates not by specifying phenotypic forms directly, but by specifying the self-organizing algorithms by which phenotypic forms develop through interactions with the environment:

> […] the same combinatorial argument can be applied to the interaction between evolution and development. Instead of directly specifying the phenotype, the genes could specify the ingredients of an adaptive process and leave it to this process to achieve the required end result. An interesting model of this kind of adaptive process is described by Von der Malsburg and Willshaw [1977]. Waddington suggested this type of mechanism to account for the inheritance of acquired characteristics within a Darwinian framework. There is selective pressure for genes which facilitate the development of certain useful characteristics in response to the environment. In the limit, the developmental process becomes *canalized*: The same characteristic will tend to develop regardless of the environmental factors that originally controlled it. Environmental control of the process is supplanted by internal genetic control. Thus, we have a mechanism which as evolution progresses allows some aspects of the phenotype that were initially specified indirectly via an adaptive process to become more directly specified.

Strong support for this idea comes from a Turing-style model capturing the self-organizing principles of the von der Malsburg network (Kaschube et al. 2010), from confirmation of its strong prediction that singularities in V1 orientation maps should be spaced by π cortical columns, which holds in mammalian species that diverged more than 65 million years ago. This is a remarkable demonstration of canalization in natural self-organizing systems, and of the application of the synthetic approach.

As the designers of Living Machines, we might therefore consider adopting evo-devo design principles, not by engineering solutions directly, but through a synthetic approach, by specifying the local self-organizing rules through which robust solutions to the challenges of varying and variable environments self-organize.

Future directions

An algorithm that is closely related to that of von der Malsburg, and is therefore useful for exploring evo-devo design principles in living machines, is the self-organizing map (SOM) by Kohonen (see Ritter et al. 1992; see also Figure 5.1). The key difference is that the SOM algorithm substitutes the recurrent interactions in the earlier model by applying learning only to nodes surrounding that which responds maximally to a given input. The SOM algorithm therefore violates a strict definition of self-organization as an emergent property of only local interactions, because identification of the maximally active node invokes a *global supervisor* with access to the state of all individuals in the system. Nonetheless the substitution of truly local interactions for a global supervisor in SOM allows maps to be generated efficiently. SOM has therefore been widely applied in data

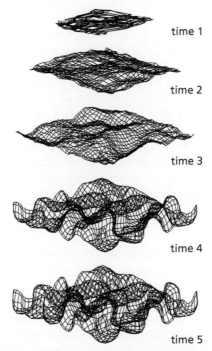

time 1

time 2

time 3

time 4

time 5

Figure 5.1 Map self-organization. The SOM algorithm (see text) was used to generate a 3D functional organization in a 2D network. Each network node was driven by three input units, whose activities varied uniform randomly in the ranges $x \in \pm 1$, $y \in \pm 1$, and $z \in \pm\frac{1}{3}$. Over time (i.e. after training on 1 to 5 hundred random input patterns) the network self-organizes to smoothly cover the 3D space in which the inputs varied. Each point in these "fishnet plots" represents a node, and the 3D location of each shows the strength of its association to the three input units. Lines connect nodes that are adjacent in the 2D network. An initially random organization "unfolds" first to smoothly cover the x–y plane, which explains most input variance, and subsequently folds further to represent variability also in the z dimension. The initial strengths of association from input units to network nodes were random, hence this smooth projection of a 3D space onto a 2D network topology captures the essential dynamics of map self-organization.

mining and pattern recognition problems, and as a mechanism for generating maps of the statistical structures of multisensory inputs to robots that are attributable to their own morphologies and movements, i.e. a rudimentary robot body schema (Ritter et al. 1992; Hoffmann et al. 2010).

Nature works without supervision (Bak 1996). Thus, a promising approach for future work on the application of self-organizing principles to the design of artificial systems, is to strip away from the systems we create all global supervisory mechanisms. An example is in the development of self-organizing models of aggregation patterns in groups of interacting animals. Previous models had explained emergent aggregation patterns in animals that huddle together to keep warm, e.g. huddling penguins, using the idea of self-organization to describe the dynamics, but implicitly assuming global supervision by including in the algorithms selection of the warmest and coolest animals or locations in the group. A recent model could instead explain thermoregulatory aggregation patterns (in rodents) as a self-organizing system based on entirely local interactions, and this model led to a mathematical description of the group as a *super-organism* (Glancy et al. 2015, 2016; Wilson 2017a, 2017b, 2017c).

An intriguing possibility is that by instantiating the local rules by which animals interact in a robot, a natural system might adopt the imposter into its own self-organizing dynamics, or even be steered by the plan of the imposter (see Halloy et al. 2007). In the other direction, the principle of the Baldwin effect could perhaps be used to genetically imprint onto a robot control system the local rules of interaction obeyed by animals comprising a self-organizing aggregation.

With models describing the self-organization of *macro-scale* (group-level) dynamics from *micro-scale* interactions between organisms, comes the possibility to generate a new type of scientific model, the *biohybrid society*, i.e. a system of locally interacting artificial and biological individuals that collectively sustain their collective dynamics.

Let us end with a thought experiment. If living systems are those sustaining edge-of-chaos self-organizing dynamics, which emerge like all others from simple interactions between individuals, and organisms are iteratively replaced by machines, then at what ratio of machines to organisms does the biohybrid society cease to be a living system? One to many; many to many; many to one; all to none? Addressing this question empirically may enable us to sidestep philosophical questions about whether or not the systems we create can ever be considered living.

Learning more

For a comprehensive text on self-organization in natural systems see Camazine et al. (2001). For Turing's final and perhaps greatest gift to science see Turing (1952), which establishes the basic principles of reaction-diffusion models. For an essential text on simulating and analyzing self-organizing maps see Ritter et al. (1992), and for critical reviews of the idea that self-organized patterns in the brain are functionally useful see Wilson and Bednar (2015) and Bednar and Wilson (2015). For a highly influential body of work on the relationship between self-organization and natural selection that is still ahead of its time see Kauffman (1993), or for a more accessible overview see Kauffman (1995). The notion of evo-devo in natural and synthetic systems is explored further in this handbook by Prescott and Krubitzer (Chapter 8, this volume). For a collection of thoughtful chapters on the Baldwin effect(s) see Weber and Depew (2003), in particular for the chapters by Dennett and Deacon. The books of Carroll (2005) on evo-devo, Braitenberg (1984) on synthetic psychology, Gleick (1988) on chaos, Bak (1996) on self-organized criticality, and Schrodinger (1967) on life are also inspiring.

Acknowledgments

Thanks to students in the Department of Psychology at the University of Sheffield taking the *cooperative models of mind* course from 2013 to 2017, interactions with whom have helped shape the ideas presented in this chapter. And thanks to Jim Bednar at the University of Edinburgh.

References

Bak, P. (1996). *How nature works: The science of self-organized criticality*. Kraków: Copernicus Press.

Bednar, J. A., and **Wilson, S. P.** (2015). Cortical maps. *The Neuroscientist*, **22**(6), 604–17.

Braitenberg, V. (1984). *Vehicles, experiments in synthetic psychology*. Cambridge, MA: MIT Press.

Camazine, S., Deneubourg, J. L., Franks, N. R., Sneyd, J., Theraulaz, G., and **Bonabeau, E.** (2001). *Self-organisation in biological systems*. Princeton: Princeton University Press.

Carroll, S. (2005). *Endless Forms Most Beautiful: The New Science of Evo Devo and the Making of the Animal Kingdom*. New York: W. W. Norton & Company.

Glancy, J., Groß, R., Stone J. V., and **Wilson, S. P.** (2015). A self-organizing model of thermoregulatory huddling. *PLoS Comput Biol*, **11**(9), e1004283.

Glancy, J., Stone J. V., and **Wilson, S. P.** (2016). How self-organisation can guide evolution. *Royal Society Open Science*, **3**, 160553.

Gleick, J. (1988). *Chaos: Making a new science*. London: Penguin Books.

Halloy, J., Sempo, G., Caprari, G., Rivault, C., Asadpour, M., Tche, F., Sad, I., Durier, V., Canonge, S., Am, J. M., Detrain, C., Correll, N., Martinoli, A., Mondada, F., Siegwart, R., and **Deneubourg, J. L.** (2007). Social integration of robots into groups of cockroaches to control self-organized choices. *Science*, **318**(5853), 1155–8.

Hinton, G., and **Nolan, S.** (1987). How learning can guide evolution. *Complex systems*, **1**, 495–502.

Hoffmann, M., Marques, H., Hernandez Arieta, A., Sumioka, H., Lungarella, M., and **Pfeifer, R.** (2010). Body schema in robotics: A review. *Autonomous Mental Development, IEEE Transactions on*, **2**(4), 304–24.

Kaschube, M., Schnabel, M., Lwel, S., Coppola, D. M., White, L. E., and **Wolf, F.** (2010). Universality in the evolution of orientation columns in the visual cortex. *Science*, **330**(6007), 1113–16.

Kauffman, S. (1995). *At home in the universe: The search for laws of complexity*. London: Penguin Books.

Kauffman, S. A. (1993). *Origins of order: Self-organization and selection in evolution*. New York: Oxford University Press.

Purves, D., Riddle, D. R., and **LaMantia, A. S.** (1992). Iterated patterns of brain circuitry (or how the cortex gets its spots). *Trends Neurosci.*, **15**(10), 362–368.

Ritter, H., Martinez, T., and **Schulten, K.** (1992). *Neural computation and self- organizing maps: An introduction*. New York: Addison-Wesley.

Schrödinger, E. (1967) *What is life?* Cambridge, UK: Cambridge University Press.

Turing, A. M. (1952). The chemical basis of morphogenesis. *Philosophical Transactions of the Royal Society of London. Series B, Biological Sciences*, **237**(641), 37–72.

von der Malsburg, C. (1973). Self-organization of orientation sensitive cells in the striate cortex. *Kybernetik*, **14**(2), 85–100.

von der Malsburg, C., and **Willshaw, D. J.** (1977). How to label nerve cells so that they can interconnect in an ordered fashion I. *Proc. Natl Acad. Sci. U.S.A.*, **74**, 5176–8.

Weber, B. H. and **Depew, D. J.** (2003). *Evolution and learning: The Baldwin effect reconsidered*. Cambridge, MA: MIT Press.

Wilson, S. P., and **Bednar, J. A.** (2015). What, if anything, are topological maps for? *Developmental Neurobiology*, **75**(6), 667–81.

Wilson, S. P. (2017a). A self-organizing model of thermoregulatory huddling. *PLoS Comput Biol*, **13**(1), e1005378.

Wilson, S.P. (2017b). Self-organising thermoregulatory huddling in a model of soft deformable littermates. *Proceedings of Living Machines 6, Biomimetic and Biohybrid Systems (LNCS)*, **10384**, 487–96.

Wilson, S.P. (2017c). Modelling the emergence of rodent filial huddling from physiological huddling. *Royal Society Open Science*, **4**, 170885.

Chapter 6

Energy and metabolism

Ioannis A. Ieropoulos[1,2], Pablo Ledezma[1],
Giacomo Scandroglio[1], Chris Melhuish[1], and
John Greenman[1,2]

[1] Bristol Robotics Laboratory, UK
[2] Centre for Research in Biosciences, University of the West of England, UK

Living Machines is a relatively new approach in the fields of robotics and artificial intelligence that endorses a biomimetic and biohybrid strategy for developing intelligent machines and systems (Prescott and Verschure, Chapter 1, this volume). In the current chapter the focus is on living machines that are literal hybrids between living entities and mechatronic systems and introduces the novel concept of *symbiotic biohybrid systems*, whereby the living entity is reliant on the mechatronic/artificial hardware for survival and continuous operation of the system as a whole. The unique feature of such a symbiotic association is that it is rooted in energy transduction and exchange, which serves as the currency for survival.

McFarland and Spier (1997, pp.179–190) were the first to introduce the concept of self-sustainability, by discussing utility cycles and "lethal limits" in robots that would potentially be autonomous. The vast majority of robot examples describe systems in simulation—and similarly where they have been implemented in practice—where energy is supplied by batteries or mains. For a truly autonomous living machine, the energy must be sustainable and collectable without human intervention. This could be in the form of renewable energy from the sun, wind, and waves, mechanical transduction of vibrations or movement, and heat conversion. Within the confines of each environment, an appropriately equipped robot can be considered to be autonomous and self-sustainable—this has in fact been the definition of living machines. However, none of the previous examples entails a real living entity, integrated in a hybrid manner for operation, and in this sense it would perhaps be more accurate to define a living machine as one including real life. This artificial agent will be truly bio-based as well as bio-inspired and the term "living" has a literal, rather than a metaphorical sense. This chapter describes work that is both biologically inspired and biohybrid in practice—a robot as a multicellular creature—operating through the interaction with living microbes and their electrochemical transformations. The work encompasses living machines in the context of natural and artificial metabolism for energy management and it touches upon artificial life using hardware, software, and wetware to produce functional embodied robotic systems. Research on self-sustaining machines that metabolize their own energy can be viewed as part of a broader effort to create future robotic and autonomous systems that are more sustainable in terms of their environmental impact, as discussed by Halloy (Chapter 65, this volume).

Symbiotic robots (SymBots) as living machines

Artificial life (A-life) imitates traditional biology by attempting to *recreate* some aspects of biological phenomena (Brooks and Maes1996; Bedau et al. 2013). The term "artificial intelligence" is often used to specifically refer to soft *in silico* A-life, being a system entirely confined to a digital environment. However, hardware based A-life has also been proposed, mainly with robots programmed or designed to fulfil specific functions of interest to humans, with some degree of autonomy. Biochemical-based "wet" A-life has also been proposed and implemented, and it can take two main forms: (a) synthetic biology, which according to the UK Synthetic Biology Roadmap (Clarke 2012) is: " ... *the design and engineering of biologically based parts, novel devices and systems as well as the redesign of existing, natural biological systems* ... ";[1] (b) non-synthetic (natural) biochemical-based living systems. The latter can include self-organized networks of real neurons connected to electronic circuits (Demarse et al. 2001), and also "soft" networks where the circuits are not fixed but easily reconfigurable. These include bio-electronic hybrid architectures such as the dynamic circuits made from the slime mold *Physarum polycephalum* (Tsuda et al. 2006) or "chemical brains" based on collisions between chemical waves in the Belousov–Zhabotinsky reaction (Adamatzky and Lacy Costello 2002).

"SymBot" is a term that refers to the beneficial integration between a live part and a mechatronic part, embodied in a robotic platform. A SymBot is therefore a robot that can potentially demonstrate artificial symbiosis, which is a proto-cooperative association between two entities. Proto-cooperation is the result of the integration between the hardware and the live microbes in the form of biofilms. The microorganisms, as part of their normal activities, would be supplying electrical energy to the robot that in turn would be performing various tasks, one of which would be to collect feedstock for the microbes. This allows the robot to be self-sustainable, provided that the robot can collect fresh nutrients from the environment and get rid of waste by-products, thus maintaining physicochemical stability and homeostasis. Functional autonomy then becomes feasible for a robot that includes energy collection, distribution, expenditure, and overall management in its behavioural repertoire and physical embodiment. The latter is particularly important since microorganisms need to be contained and allowed to feed on their natural substrate, in order to efficiently convert it to electricity directly. The potential of using biohybrid devices to sense the physicochemical environment, or replace microcontrollers entirely, would move the science further and allow conventional electronics to be reserved for higher levels of control. This implies the use of natural or unconventional computing, all effected by living machines. The technology that allows for all of the above to take place onboard a mobile robotic platform is known as the Microbial Fuel Cell (MFC), which is discussed in detail in the sections to follow.

Natural (unconventional) computation

Conventional computing as a science is generally based on the application of binary element programming and the systematic study of algorithmic processes and programmes for implementation into applications. Most modern computing is based on the idea of a theoretically perfect Turing machine (a Universal Turing Machine) that can be efficiently automated (Denning et al. 1989). Modern computers are close to being Universal Turing Machines, and can run programs and sets of instructions so that the main processor may compute local functions and support efficient processing. Turing machines, recursive functions, and formally equivalent models rely on notions of symbols and symbol manipulation, which are fundamentally human mental constructs. For computation, a thought process must be expressible as software.

---

[1] For example, Mast et al. (Chapter 39, this volume) describe a synthetic biology approach to the creation of minimal systems capable of self-reproduction as a means to understand the evolution of the first life forms.

According to Abbott (2006, pp.41–56), a process in nature may be considered computation only when it is used as a way to work with externalized thought. Almost all processes, whether natural or artificial, will proceed depending on the environment. Computing occurs when contingencies are interpreted and controlled, so that the resulting process can be used to work with one's thoughts. Furthermore, processes must be subject to alterations for computation and there must be something contingent about the environment within which they operate, which both determines how it proceeds as well as how it is interpreted, to give meaning to the agent in terms of the results. From a robot perspective, autonomy and survival are contingent environmental pressures. Its functionality requires it to be sensing environmental parameters so that it can seek food or shelter and respond to environmental variations in order to survive and function. It should be noted that real mammalian brains and neuronal systems do not use binary algorithms in order to process.

Homeostasis and metabolic computing

Metabolism includes all the metabolic processes, anabolic as well as catabolic that occur in a cell. In heterotrophic microbes (e.g. *E. coli*) there are over 2000 distinct metabolic reactions (Neidhardt et al. 1990). These reactions include four broad categories of pathway: (1) fuelling reactions where electron-rich substrates are metabolized into more oxidized products with the generation of energy in the form of reducing power (NADH, NADPH), proton-motive force, and ATP which the cell uses for growth and motility; (2) biosynthetic reactions to synthesize fatty acids, cell-wall sugars, amino acids, and nucleotides; (3) polymerization reactions, of fatty acids to lipids, of sugars to peptidoglycan, glycogen, or capsular polysaccharides, and of amino acids to proteins; and (4) assembly reactions where the cell assembles functional structures including ribosomes, envelope, flagella, and pili. All the aforementioned reactions employ 100s of enzymes, many of which are susceptible to some form of regulatory feedback control that allows the cell to conserve energy by down-regulating non-essential pathways and up-regulating those that are rate-limiting to produce vital commodities. Two fundamental distinctions can be made in regulation; whether they change the rate of catalysis of proteins already present in the cell or whether they alter the net rate of synthesis (i.e. cell concentration) of protein (Neidhardt et al. 1990). Methods which control metabolism by modifying the activity of existing proteins work very quickly in the cell (within seconds) whilst methods that control the DNA switching (transcription) may take at least 20 minutes before the change is manifested by the cell. Bacteria have evolved multiple ways to modulate gene expression with control based on a plethora of mechanisms that allow the cell to best adapt to its environment.

The property of an open system such as a living organism to regulate its internal environment so as to maintain a stable condition is described as homeostasis. Multiple dynamic equilibrium adjustments and regulation mechanisms within the cell achieve homeostasis. It is this feature of the cell that can be exploited for computing. A combination of different families of microbe on board a living machine may indeed allow running some forms of rudimentary responses without recourse to silicon circuits. The following section describes the biofilm electrodes (BE), which are the living entities with microbes inside Microbial Fuel Cells that serve as live engines onboard the EcoBot living machines.

Biohybrid Microbial Fuel Cell Biofilm Electrodes (MFC-BE) as live engines

Microbial Fuel Cells are bio-electrochemical transducers, which generate electricity directly from the metabolic reactions of the constituent microorganisms, feeding on a wide variety of organic substrates (Bennetto 1990). Continuous culture of biofilm microbes has been previously described (Helmstetter and Cummings 1963), but the nearest equivalent to biofilm electrodes is the "Sorbarod" matrix perfusion model (Hodgson et al. 1995; Spencer et al. 2007; Taylor and Greenman 2010). The biofilm electrode simply replaces the cellulosic substratum (the Sorbarod) with a conductive carbon-veil electrode and places it into the anodic chamber of a MFC (Figure 6.1).

The idea and feasibility of robots powered by MFCs has been previously demonstrated (Ieropoulos et al. 2010). The microbes around the MFC biofilm electrode (BE) supply electrical energy to the robot in the form of electrons as part of their normal metabolic activity. The robot then performs tasks, one of which is to collect more organic "fuel" from which the microbes can continue growing and extracting energy. The ideal robot would thus be both energetically and computationally autonomous in terms of performing "intelligent" functions and extracting energy in unstructured environments with minimum human intervention. EcoBot-III used conventional silicon circuits and sensors to control the system (Ieropoulos et al. 2010), thus creating interest within the artificial life community (Lowe et al. 2010; Montebelli et al. 2010). The symbiotic relationship between microbes and artifacts can go a lot further than merely producing electrical power and this may be extended to include *live sensors* and biofilm-electrode-based artificial neuronal circuits for other specialized functions.

Living sensors and artificial neurons need to be resilient to the normal range of physicochemical conditions they are likely to operate within and need to exhibit longevity, stability, and reliability. It has previously been shown that dynamic steady states can be achieved with MFC-BEs, for constant physicochemical environmental conditions, in terms of electrical output, with one variable parameter at any given time (Greenman et al. 2008). Such findings have already been demonstrated (Greenman et al. 2008; Ledezma et al. 2012). Figure 6.1 shows the conceptual and schematic diagrams of MFC-BEs.

Under certain conditions of continuous flow, when anodic units are supplied with buffer medium and where the electrode is the *only* electron acceptor in the system, the microbes colonize the electrode giving a stable output. The type of microbes selected for this work (called anodophiles) can only metabolize and grow if they have electrical contact (via conducting pili or transmembrane cytochrome complexes) with the anode electrode surface. This results in the population being *non-accumulative but fully viable* to respond to changes in the environment. Once mature, biofilm electrodes grow and shed daughter cells at a rate equal to their production and a dynamic steady state is produced (Figure 6.1 inset).

If physicochemical conditions change, the output from the electrode goes through transition and reaches a new steady state. Some transitions can be relatively quick whilst others (involving expression of de novo proteins) may take minutes to hours. Electrode units may therefore hold thousands (possibly as high as 10^{12}) potentially different, yet accessible physiological-metabolic states. Made-to-be-identical electrode units should behave in identical ways even through quite wide changes in environmental parameters, and should all recover when settings return to null. Long-term stability and repeatability requires testing but in theory the full responsive dynamic state of the electrode should remain stable for as long as comfortable conditions remain (years or decades). In theory, an array, matrix, or group of biofilm electrode units can be configured to produce logic gates or embody associative learning (Greenman et al. 2006; Greenman et al. 2008).

Figure 6.2 (top) shows the effects of switching different values of MFC-BE load resistor. In time sequence, the data show the responses starting at steady-state output on a 5kΩ resistor switched to a suboptimal load (1kΩ). This produced a rapid initial response towards higher power, but then the output slowly decayed. When a 2.5kΩ load was connected, which was matching the

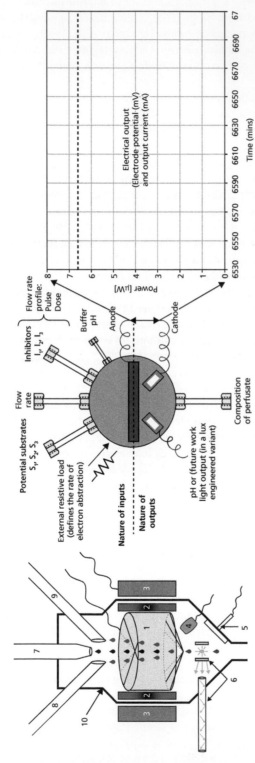

Figure 6.1 Left: Schema of an integrated prototype: (1) Anode biofilm electrode; (2) Ion exchange interface; (3) Cathode electrode; (4) pH probe; (5) Bioluminescence light path—transparent window plus fibre optic cable; (6) Main feedstock input; (7) Additive stimulator; (8) Additive inhibitor; (9) MFC casing. Right: Schematic representation of the biofilm-electrode system showing the diversity of physicochemical parameters that can be used to induce a different response state for the microorganisms resulting in a new output state. Inset: Example of real metabolic steady state for *Geobacter* under defined conditions, measured as electrical output.

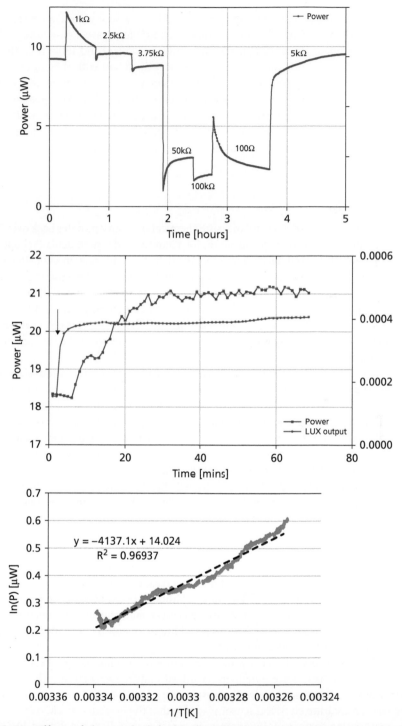

Figure 6.2 Top: Effects of changes (switches) in the MFC-BE external resistance values (starting from 5kΩ). Middle: Electrical power and lux output response from a continuous flow MFC, when the nutrient flow rate was increased (arrow) from 7.2 to 50.8 mL.h^{-1}. Bottom: Relationship between temperature and MFC-BE output with data expressed in terms of Arrhenius plot.

system's internal resistance, the initial response was within a few seconds and a new steady state was quickly established; similarly the connection of a 3.75kΩ load gave a new steady state. The same pattern was observed when using higher value resistors (50kΩ and 100kΩ). In the case of a heavier load of 100Ω the initial burst of power was very rapid, but with slow recovery (deceleration lasting over an hour). Upon returning to 5kΩ there was concomitant recovery of power output to the initial level. Some transitions (e.g. between very high or very low resistance values) appear to be rapid upon initial switching, but hit a slow phase before reaching a new steady state. The transitional responses following stimuli reflect the metabolic and physiological state of the microorganisms, particularly with regard to the various homeostatic mechanisms that are operating within the biofilm.

Live bio-sensing

The use of microorganisms to detect and measure chemicals in solutions dates back over 75 years with the development of microbiological assays for vitamins and amino acids (Schopfer 1935; Sandford 1943; Snell 1945; Henderson and Snell 1948). Such assays rely on choosing a species that can only grow in the presence of the target analyte. In the assay, the natural growth of the existing viable cells relates in a quantitative manner to the concentration of the target vitamin. This is a fine example of the early use (all prior to 1960) of coupling growth and metabolism to the quantification of specific chemicals in solution and thus gene expression (Thellier et al. 2006; Norris et al. 2011); the assay was not originally interpreted within the modern paradigm of computing (bactocomputing).

Results using MFC-BE colonized by *Shewanella* monocultures (Figure 6.2 top) show consistent transitions and establishment of new steady states, for as long as the new conditions are held. The data support the theory that the biofilm is actively growing, and that new daughter cells are released continuously, leaving a non-accumulating yet highly active living layer of fixed population numbers. Perturbation results in a transitional state that gives rise to a new steady state for this new setting of the physicochemical parameters. It therefore demonstrates that it is both a controlled and controllable dynamic multi-state system.

Using a genetically engineered strain of *Shewanella* to express lux luminescence, Figure 6.2 (middle), illustrates the electrical output response (red colour) after an increase in the fuel supply rate, which was nicely reflected in the measured light output of the biofilm daughter cells, as they were being shed out of the system. The microbial cell progeny released by the MFC-BE were shown to increase with the increase in the flow rate of feedstock. Both electrode power output and lux luminescence increased when the nutrient flow rate was increased. The power output was quick to respond to the new conditions, but the increase in bioluminescence to a new steady state was a much slower response (20–30 minutes). Although lux luminescence was used to verify that the growth rate and production of new cells was constant, it further demonstrates the feasibility of using light channels on a robot as an additional method of monitoring, communication and connectivity of the MFC-BE units.

Figure 6.2 (bottom) shows the effects of temperature on a stack of four MFCs connected fluidically in cascade; these were employing a mixed culture consortium. The exposure to temperatures above ambient were only of relatively short duration, and were performed by co-measuring with an accurate electrical sensor placed within the insulated stack. As the temperature increases there is a corresponding increase in power output. Between 20 and 30°C there is ca. a two-fold increase in power, which is in line with the Arrhenius prediction (Laidler 1987). These findings reveal the metabolic profile of the biofilm electrode, because it is assumed that the brief exposure times did not significantly affect the ecological stability. Higher temperatures and/or longer exposure periods would normally kill heat-sensitive species and change the ecology irreversibly with time (Michie et al. 2011). Such responses can serve thermo-tactic robots and, like biological agents, change their behaviour and direction, based on temperature.

Previous work has already demonstrated the ability of photo-MFCs to respond to light when the anode or cathode are colonized by microalgae or phototrophic bacteria (Cao et al. 2009; Powell et al. 2011). The rate of response is rather slow but is one that could potentially be of some functional use to a robot; for example to detect light and give the robot the ability to switch on or off "daytime" and "night-time" behavioural patterns or schema.

Future directions

MFC-BEs are stable, resilient, and robust, provided they are operating within the comfort zone (biotic limits) of the constituent species. They can be used as living sensors, as exemplified by the temperature effects on MFC and the photo-MFC being able to detect visible light radiation. It is therefore possible to use (photonic) channels on a robot as an additional method of monitoring, communication, and connectivity of the MFC-BE units. Mixed culture communities may be more compliant with real robots due to the higher risk of monoculture contamination and the energy costs of fuel sterilization. If mixed culture microcosms are used, there may be some physicochemical conditions that should be avoided since they may give rise to irreversible variations and lack fidelity with time. The precise types of connection between nodes of artificial neurons, sensors, and actuator controllers have yet to be determined by experimentation.

In simple terms, the block diagram of Figure 6.3 illustrates how the MFC technology can come together as building blocks for a non-silicon-based living machine. A light sensitive MFC can be linked to an interneuron-MFC, which in turn can activate a collective stack to energize an actuator. The connections allow the light sensor to modulate the actuator in a quantitative manner, in proportion to the intensity of light being sensed at any time. The interneuron can be modulated by further circuits (not shown). The general idea is to have numerous autonomous, i.e. homeostatic, channels replicated onboard the robot, with cross-channel interneurons coordinating the whole. For example, six or more complete circuits (six sensors and six actuators), arranged around the perimeter of a robot, may be configured to show positive or negative phototaxis.

We envisage MFC nodes as the propagating force for information channels (akin to neuronal circuits in mammals) to transport or receive information and differentially activate the living machine into functional purposes. There are three main levels of complexity onboard biohybrid

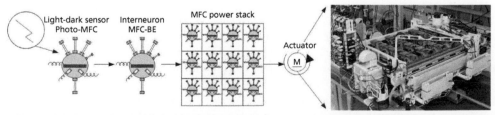

Figure 6.3 Left: Implementation of MFC sensor connected to an interneuron whose output can modulate an MFC stack to power an actuator. The interneuron allows other channels of information to influence the information going from the sensor to the stack and actuator. Right: Photo of the latest EcoBot-IV, which implements the MFC stack technology, showing the actuators and the peripheral maintenance system. See Figure 1.3b for a picture of the previous generation Ecobot-III.

robots; single cells, cell collectives (the biofilm community as a whole), and collectives of MFC-BE nodes. This in turn is dictated by the nature of the feedstock, soluble chemicals, and electron abstraction rate determined by the internal and external resistances.

A future biohybrid living machine robot may therefore possess the SymBot equivalent of sense-organs by adopting living sensors and artificial neurons, and the whole regarded as a new level of complexity, showing robotic properties that could never be guessed at from the component elemental blocks, i.e. the living cells. EcoBot-IV (Figure 6.3) is the latest generation of SymBots, and although it does not employ MFCs for information processing, it is operating in a more interactive manner with its environment. This is the platform that will allow further development of MFCs as sensing and information processing units, as well as energy supply systems, and will therefore move closer to a true biohybrid living machine.

Learning more

Autonomy in robotics can have different meanings, depending on the context of the experiments/ mission and the environment. There is a plethora of robots that can operate with minimum human intervention, with the majority being powered by conventional batteries and photovoltaics. Many of these can harvest natural energy from the surrounding environment, and for as long as they continue operating within the limits of that environment, their operation can be classed as autonomous. For example, underwater autonomous vehicles have been reported to be operating on solar or wave power (Davis et al.1991; Yuh 2000; Jalbert et al. 2003; Bergbreiter and Pister 2007; Hine et al. 2009; Slesarenko and Knyazhev 2012). Likewise, unmanned solar-powered air vehicles have also been implemented, at different scales (Noth et al. 2007; Richter and Lipson 2011), and the Mars Rovers have demonstrated longevity of solar-powered operation in an alien atmosphere (Squyres et al. 2003). Perhaps the first example of a mobile robot involving MFCs and organic fuel was Gastrobot, which employed the MFCs to recharge the onboard batteries that were used to drive the actuators (Wilkinson 2001).

Acknowledgments

This work has been partly funded by the EPSRC, grant no. EP/H046305/, EP/L002132/1 and the Bill & Melinda Gates Foundation, grant numbers OPP1094890 and OPP1149065. Ioannis Ieropoulos has been funded by EPSRC, Career Acceleration Fellowship grant no. EP/I004653/1.

References

Abbott, R. (2006). If a tree casts a shadow is it telling the time? In: C.S. Calude, M.J. Dinneen, G. Păun, G. Rozenberg, S. Stepney (eds). *Unconventional Computation*. Heidelberg: Springer, pp.41–56.

Adamatzky, A., and Lacy Costello, B. (2002). Collision-free path planning in the Belousov–Zhabotinsky medium assisted by a cellular automaton. *Naturwissenschaften*, **89**, 474–8.

Bedau, M.A., McCaskill, J.S., Packard, N.H., Parke, E.C., and Rasmussen, S.R. (2013). Introduction to Recent Developments in Living Technology, *Artificial Life*, **19**, 291–8.

Bennetto, H.P. (1990). Electricity generation by microorganisms. *Biotechnology Education*, **1**, 163–8.

Bergbreiter, S. and Pister, K.S.J. (2007). Design of an Autonomous Jumping Microrobot. In: *Proceedings of the International Conference on Robotics and Automation*, 447–53, Rome, Italy.

Brooks, R.A., and Maes, P. (eds.) (1996). *Artificial Life: Proceedings of the Fourth International Workshop on the Synthesis and Simulation of Living Systems*. Cambridge, Massachusetts: MIT Press.

Cao, X.X., Huang, X., Liang, P., Boon, N., Fan, M.Z., Zhang, L., and Zhang, X.Y. (2009). A completely anoxic microbial fuel cell using a photo-biocathode for cathodic carbon dioxide reduction. *Energy & Environmental Science*, **2**, 498–501.

Clarke, L. (2012). Synthetic Biology Roadmap, [Online] 21 Nov. 2013, Available at: http://www.rcuk. ac.uk/documents/publications/SyntheticBiologyRoadmap.pdf

Davis, R.E., Webb, D.C., Regier, L.A., and Dufour, J. (1991). The Autonomous Lagrangian Circulation Explorer (ALACE). *Journal Atmospheric and Oceanic Technology*, **9**, 264–85.

Demarse, T.B., Wagenaar, D.A., Blau, A.W., and Potter, S.M. (2001). The neurally controlled Animat: biological brains acting with simulated bodies. *Autonomous Robots*, **11**, 305–10.

Denning, P.J., Comer, D.E., Gries, D., Mulder, M.C., and Tucker, A. (1989). Computing as a discipline. *Communications ACM*, **32**, 9–23.

Greenman, J., Ieropoulos, I. and Melhuish, C. (2008). Biological computing using perfusion anodophile biofilm electrodes (PABE). *International Journal of Unconventional Computing*, **4**, 23–32.

Greenman, J., Ieropoulos, I., McKenzie, C., and Melhuish, C. (2006) Microbial Computing using Geobacter Biofilm Electrodes: Output Stability and Consistency. *International Journal of Unconventional Computing*, **2**, 249–65.

Helmstetter, C.E., and Cummings, D.J. (1963). Bacterial synchronization by selection of cells at division. *Proc. Natl Acad. Sci. USA*, **56**, 707–74.

Henderson, L.M. and Snell, E.E. (1948). A uniform medium for determination of amino acids with various micro-organisms. *Journal of Biological Chemistry*, **172**, 15–29.

Hine, R., Willcox, S., Hine, G., and Richardson, T. (2009). The Wave Glider: A wave-powered autonomous marine vehicle. In: Proceedings of the OCEANS 2009, MTS/IEEE Biloxi—Marine Technology for Our Future: Global and Local Challenges, 1–6, Biloxi, MS, US.

Hodgson, A.E., Nelson, S.M., Brown, M.R.W, and Gilbert, P. (1995). A simple *in vitro* model for growth control of bacterial biofilms. *Journal of Applied Bacteriology*, **79**, 87–93.

Ieropoulos, I., Greenman, J., Melhuish, C. and Horsfield, I. (2010). EcoBot-III: a Robot with Guts. In: H. Fellermann, M. Dörr, M. M Hancz, LL. Laursen, S. Maurer, D. Merkle, P-A. Monnard, K. Støy & S. Rasmussen (eds). *Artificial Life XII*. Cambridge, Massachusetts: MIT Press, pp.733–40.

Jalbert, J., Baker, J., Duchesney, J., et al. (2003). A solar-powered autonomous underwater vehicle. In: Proceedings of the OCEANS 2003, **2**, 1132–40, San Diego, CA.

Laidler, K.J. (1987). *Chemical Kinetics*, 3rd Edition, New York, Harper & Row.

Ledezma, P., Greenman, J., and Ieropoulos, I. (2012). Maximising electricity production by controlling the biofilm specific growth rate in microbial fuel cells. *Bioresource Technology*, **32**, 1228–40.

Lowe, R., Montebelli, A., Ieropoulos, I., Greenman, J., Melhuish, C., and Ziemke, T. (2010). Grounding motivation in energy autonomy: a study of artificial metabolism constrained robot dynamics. In: H. Fellermann, M. Dörr, M. M Hancz, LL. Laursen, S. Maurer, D. Merkle, P-A. Monnard, K. Støy & S. Rasmussen (eds). *Artificial Life XII*. Cambridge, Massachusetts : MIT Press, pp.725–32.

McFarland, D. and Spier, E. (1997). Basic cycles utility and opportunism in self-sufficient robots. *Robotics & Autonomous Systems*, **20**, 179–90.

Michie, I.S., Kim, J.R., Dinsdale, R.M., Guwy, A.J., and Premier, G.C. (2011). Operational temperature regulates anodic biofilm growth and the development of electrogenic activity. *Applied Microbiology and Biotechnology*, **92**, 419–30.

Montebelli, A., Lowe, R., Ieropoulos, I., Greenman, J., Melhuish, C., and Ziemke, T. (2010). Microbial fuel cell driven behavioural dynamics in robot simulations. In: H. Fellermann, M. Dörr, M. M Hancz, LL. Laursen, S. Maurer, D. Merkle, P-A. Monnard, K. Støy & S. Rasmussen (eds). *Artificial Life XII*. Cambridge, Massachusetts : MIT Press, pp.749–56.

Neidhardt, F.C., Ingraham, J.L., and Schaechter, M. (1990). *Physiology of the bacterial cell: A molecular Approach*. Sunderland, Massachusetts : Sinauer Associates Inc.

Norris, V., Zemirline, A., Amar, P., et al. (2011). Computing with bacterial constituents, cells and populations: from bioputing to bactoputing. *Theory in Biosciences*, **130**, 211–28.

Noth, A., Siegwart, R., and Engel, W. (2007). Autonomous Solar UAV for Sustainable Flights. In: K.P. Valavanis (ed.). *Advances in Unmanned Aerial Vehicles, State of the art and the road to autonomy.* Dordrecht, The Netherlands: Springer-Verlag, pp. 377–406.

Powell, E.E., Evitts, R.W., Hill, G.A., and Bolster, J.C. (2011). A Microbial Fuel Cell with a photosynthetic microalgae cathodic half cell coupled to a yeast anodic half cell. *Energy Sources Part a -Recovery Utilization and Environmental Effects*, **33**, 440–8.

Richter, C. and Lipson, H. (2011). Untethered Hovering Flapping Flight of a 3D-Printed Mechanical Insect. *Artificial Life*, **17**, 73–86.

Sandford, M. (1943). Historical development of microbiological methods in vitamin research. *Nature* **152**, 374–6.

Schopfer, W.H. (1935). Standardization and possible uses of a plant growth test for vitamin B. *Bulletin of the Society of Chimique Biol*, **17**, 1097–109.

Slesarenko, V.V. and Knyazhev, V.V. 2012. Energy sources for autonomous unmanned underwater vehicles. In: *Proceedings of the 22nd International Offshore and Polar Engineering Conference*, 538–42, Rhodes, Greece.

Snell, E.E. (1945). The microbiological assay of amino acids. In: M.L. Anson and J.T. Edsall (eds.). *Advances in Protein Chemistry.* New York: Academic Press, pp. 85–116.

Spencer, P., Greenman, J., McKenzie, C., Gafan, G. Spratt, D. and Flanagan, A. (2007). *In vitro* biofilm model for studying tongue flora and malodour. *Journal of Applied Microbiology* **103**, 985–92.

Squyres, S.W., Arvidson R.E., Baumgartner, E.T., et al. (2003). Athena Mars rover science investigation, *Journal of Geophysical Research*, **108**, 8062.

Taylor, B. and Greenman, J. (2010). Modelling the effects of pH on tongue biofilm using a sorbarod biofilm perfusion system *Journal of Breath Research*, **4**, 017107.

Thellier, M., Legent, G., Amar, P., Norris, V., and Ripoll, C. (2006). Steady-state kinetic behaviour of functioning-dependent structures. *Federation of European Biochemical Societies, Journal*, **273**, 4287–99.

Tsuda, S., Zauner, K-P., and Gunji, Y-P. (2006). Robot control: from silicon circuitry to cells. In: A.J. Ijspeert, T. Masuzawa, and S. Kusumoto (eds.). *Lecture Notes in Computer Science.* Heidelberg: Springer, pp. 20–32.

Wilkinson, S. (2001). Hungry for success—future directions in gastrobotics research. *Industrial Robot*, **28**, 213–19.

Yuh, J. (2000). Design and Control of Autonomous Underwater Robots: A Survey. *Autonomous Robots*, **8**, 7–24.

Chapter 7

Reproduction

Matthew S. Moses and Gregory S. Chirikjian

Johns Hopkins University, USA

John von Neumann introduced the concept of a Universal Constructor as part of his effort to develop a mathematical theory describing living organisms. He described a Universal Constructor as a kinematic machine able to manipulate and assemble primitive building blocks. Von Neumann showed how this hypothetical constructor, being itself composed of the same primitive blocks, could self-replicate and evolve. Remarkably, this model system pre-dates the discovery of the genetic code, yet it applies to molecular biology of the cell equally well as to man-made machines. As early as 1957, researchers built physical instantiations of von Neumann's building blocks, assembling them into simple constructors capable of varying levels of construction and self-replication. This work continues today. Modern demonstrations typically involve the automatic assembly of a handful of robotic modules, each module comprising an intricate arrangement of parts. The use of complex prefabricated building blocks limits the space of machines that may be constructed. Development of simpler building blocks is thus key to increasing the evolutionary power of artificial self-replicating machines. This chapter reviews recent progress toward this goal, and discusses how knowledge of cell molecular biology guides and informs the design of these prototype self-replicating machines. Readers may also be interested to consult Mast et al. (Chapter 39, this volume) which explores a complementary approach that seeks to construct new kinds of biochemical replicators using synthetic biology.

Biological principles

It is difficult to agree upon precise criteria of what something must do in order to be "alive" (see Prescott, Chapter 4, this volume), but there is general consensus that a living organism must pursue six key activities: metabolizing, maintaining homeostasis, responding to stimulus, developing through growth, reproducing, and adapting to the environment through natural selection.

Reproduction occurs at many levels: sub-cellular, cellular, and organism. Reproduction of an organism is, of course, essential for adaptation through natural selection. In multicellular organisms, cell reproduction is key for growth and development. Reproduction at these higher organizational levels is dependent on the molecular self-reproduction processes that take place within a cell. In fact, even the reproduction of viruses (which are generally considered to be "somewhat but not entirely" alive) relies on the molecular machinery of self-reproduction at the sub-cellular level. Sub-cellular self-reproduction is of central importance to all living organisms.

A very simple picture of the sub-cellular reproduction processes is shown in Figure 7.1. A prokaryotic cell (a bacterium or archaeon) has no membrane-bound internal organelles, and therefore the processes outlined in Figure 7.1 take place more or less within the entirety of the cell. In a eukaryotic cell (a single- or multi-celled organism with membrane-bound organelles)

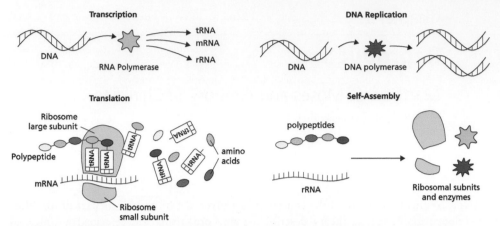

Figure 7.1 The core processes of subcellular self-reproduction in a prokaryote. A network of molecular machines is able to collectively reproduce all of its constituent components. Information stored in DNA can be translated into arbitrary polypeptide sequences, which can then self-assemble into a multitude of functional macromolecules. A key similarity between the cell's functions and von Neumann's model of self-reproduction is the use of a separate means for storing, replicating, and interpreting the information describing the system's construction.

the process is much more complex, with different stages of the process confined to different organelles within the cell. Even in the relatively simple prokaryote cell, the process shown in Figure 7.1 is assisted and regulated by a host of metabolic enzymes, transcription factors, initiation factors, and other functional macromolecules. These additional molecules are created either by large biomolecule assemblies called ribosomes, or created by things created by ribosomes. Hence, for the purposes of our discussion, the simplified picture of the prokaryotic process in Figure 7.1 is of sufficient detail.

Transcription

The primary mode of storing genetic information in a cell is in the sequence of nucleotide bases in DNA. There is a complex interplay between messages coded in segments of DNA and the surrounding molecular machinery of the cell. One of the most important classes of coded messages are called "genes." Genes are segments of DNA containing information that can be converted into sequences of RNA and then translated into polypeptides, which then can form useful molecular machinery. Transcription is the process by which DNA is converted into RNA. It is the first step in synthesizing proteins. Transcription need not take place inside a living cell—it can occur in a vial provided the necessary molecular components are kept under suitable conditions and given a source of chemical energy.

An enzyme called RNA polymerase is responsible for transcription. RNA polymerase (with assistance from smaller enzymes called "sigma factors") will recognize special codes in a stretch of DNA as initiation sites, and then bind to the DNA and proceed to synthesize a complementary strand of RNA, until reaching a special termination code in the DNA, at which point the transcription stops. Both DNA and RNA use a similar complementary four-character alphabet (A,T,G,C for DNA and A,U,G,C for RNA) and so there is a one-to-one relation between a symbol in the DNA strand and a symbol in the newly created RNA strand. It is for this reason the process is called "transcription" as there is no encoding or decoding involved in the conversion.

Once transcribed, a strand of RNA may serve many different purposes. Messenger RNAs (mRNA) serve as coded instructions to ribosomes for specifying a certain sequence of amino acids in a polypeptide. Ribosomal RNAs (rRNA) self-assemble into chemically active structures that form parts of ribosomes themselves. Transfer RNAs (tRNA) self-assemble into special "adapter molecules" that help match amino acids with their corresponding instruction (called a "codon") on an mRNA during polypeptide synthesis.

DNA replication

Double-stranded DNA can be replicated by a family of enzymes known as DNA polymerases. The process within a living cell is somewhat involved, requiring the interaction of many enzymes with the proper initiation sites on the DNA to be replicated but, like RNA transcription, DNA replication can take place outside a living cell. For example, polymerase chain reaction (PCR) is routinely used to replicate DNA strands in a laboratory environment.

Translation

A single strand of messenger RNA (mRNA) contains the information to produce a linear chain of amino acids called a polypeptide. This process is performed by the ribosome, a large biomolecule containing many subunits composed of rRNA in addition to polypeptide chains. Information in mRNA is stored in a non-overlapping, three-character code. Three consecutive nucleotide bases in mRNA make up a "codon" which represents one of a possible 20 standard amino acids. Each of the 20 amino acids bind to a unique tRNA molecule. The tRNA acts as an adapter, in that one end binds to an amino acid, and the other contains a complementary three-nucleotide "anti-codon" that will bind to a mRNA codon. Enzymes called aminoacyl-tRNA synthetases are responsible for matching up and attaching a given amino acid to its corresponding tRNA molecule.

Translation begins with a strand of mRNA threaded through the ribosome (see Figure 7.1). As translation proceeds, the ribosome creates a corresponding sequence of amino acids by matching the correct tRNA to each exposed codon in the mRNA bound to the active sites within the ribosome. With a tRNA and its amino acid payload properly positioned, the ribosome catalyzes the formation of a peptide bond between the new amino acid and the growing polypeptide chain (see Figure 7.2, center). This proceeds until a special "stop codon" is read in the mRNA, at which point the ribosome releases both mRNA and the newly formed polypeptide sequence.

Self-assembly

Many biomolecules are long polymers of repeated subunits. These molecules are highly flexible. Under normal biological conditions (in water, at room temperature) random thermal motion will cause these molecules to undergo vigorous changes in shape, in effect rapidly searching through the space of possible molecular configurations. The free energy of the biomolecule depends crucially on its shape. The many subunits of the biomolecule present opportunities for weak interatomic bonds to form—either between subunits within a polymer, between subunits and surrounding water molecules, or even between separate biomolecules. For example, amino acids can be hydrophilic or hydrophobic; weak bonds can form between amino acids themselves; hydrogen bonds form between complementary nucleotides (A–T, G–C) in DNA and (A–U, G–C) in RNA.

Rapid conformational change due to thermal motion helps the biomolecule to quickly settle, fold, or "assemble" into shapes that minimize the free energy of the molecule itself and the surrounding water. A simple change in temperature or pH may alter the energy balance,

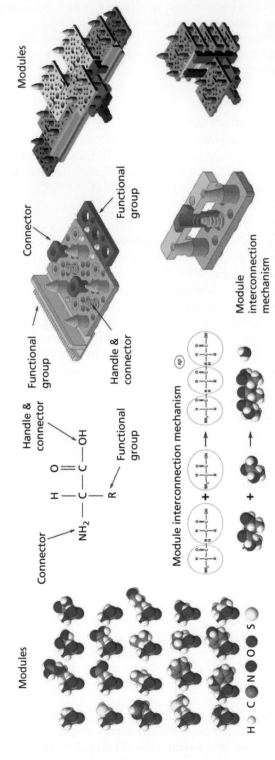

Figure 7.2 Shown on the left side are space-filling models of the standard 20 amino acids which are coded in mRNA. The ribosome adds amino acids to a growing polypeptide by repeating the same reaction over and over. This reaction combines an amino group (NH_2) with a carboxyl (COOH) group to release a water molecule and form a peptide bond. This motif can be mimicked in a system of macroscopic components. Like amino acids, each component has a basic structure with common "handles" and "connectors" to which are added various functional groups. The modules can be recombined into many variations of functional structures.

causing an assembled system to unravel or "denature". Remarkably, in many cases this process is entirely reversible: complex three-dimensional structures may denature into individual unfolded strands at high temperature, and then spontaneously reform into complex structures as temperature is reduced or pH returns to normal.

The linear polypeptide chains that are produced during translation typically become active shortly after their creation by self-assembling into functional forms. In the cell, self-assembly is often aided by enzymes called "molecular chaperones". These enzymes greatly increase the speed of self-assembly, but self-assembly can also occur outside the cell in laboratory apparatus.

All of the essential molecular machinery associated with the processes in Figure 7.1, including ribosomes, has been shown to self-assemble from denatured molecules in laboratory experiments. Further, the four different processes have all been demonstrated in laboratory apparatus outside of a living cell. The four pieces of this complex sub-cellular reproduction process (transcription, translation, DNA replication, and self-assembly) form a complete analog to von Neumann's theoretical Universal Constructor.

Biomimetic systems

Let us consider again life's six principal pursuits. Most machines pursue some subset of these activities as part of their normal operation. They might perform these functions automatically— a century-old steam engine with self-regulating speed control satisfies metabolism, homeostasis, and response to stimulus. Or they might perform these functions while closely supervised by a human operator—networks of machines and manufacturing tools are routinely directed to grow, develop, and reproduce (albeit with intensive intervention of engineers and technicians). These machines "evolve," becoming (in some sense) better adapted to their environment, because successive generations are refined and improved by their human designers. Machine evolution without actual reproduction is possible as well. For example, genetic algorithms can be used to design real machines, by performing simulated evolution on virtual machines reproducing within an artificial environment.

If humans are removed from the picture, the number of examples of machine reproduction is dramatically reduced. Von Neumann's work was the first serious investigation of *automatic* machine self-reproduction. Von Neumann proposed a collection of three hypothetical machines: A, a constructing machine that can assemble components in arbitrary arrangements given a tape T of instructions; B, a tape copying machine; and C, a control unit that coordinates the activities of A and B. The reproduction process is represented by the expression

$$(A + B + C) + T(A + B + C) \rightarrow (A + B + C) + (A + B + C) + T(A + B + C) + T(A + B + C)$$

where T(A+B+C) represents a tape containing instructions for the creation of machines A, B, and C. (In the 1940s and 1950s it was common for automatic machine tools to be controlled by commands encoded in a strip of paper with holes punched in it called a "punched tape".) Von Neumann's model is closely analogous to the sub-cellular reproduction processes of Figure 7.1. The tape T corresponds to DNA, the tape copying device B to DNA polymerase, the constructing machine A to the ribosome, and the controller C to the associated enzymes that help control and regulate the process.

In 1957, directly inspired by von Neumann's work, the geneticist L.S. Penrose built a set of wooden modules containing a variety of interlocking hook mechanisms. These blocks could be confined in a box and shaken by hand in order to replicate patterns in a manner roughly similar to DNA replication. It wasn't until much later that machine systems with a capability more resembling the full sub-cellular reproduction system were realized.

Modular self-reconfigurable robots

In the early 1990s, beginning with T. Fukuda's work on the CEBOT, researchers began to investigate modular self-reconfigurable robotics (MSRR). Typically a MSRR system is a collection of simple robots, each containing sensors, actuators, and small computers (Yim et al. 2007). Often the robots are identical, or of a small number of types, and each runs identical control software. The behavior of modules in a collection is somewhat analogous to cells in a multicellular organism. The modules can connect to one another, transmit signals, and share energy sources. Collectively, many modules can self-organize and cooperate to perform tasks. Reproduction is usually not an important goal for these systems—usually it is envisioned that traditional manufacturing is used to produce vast numbers of identical modules, which can then be used interchangeably. However, the work in modular reconfigurable robots laid the foundation for a resurgence in interest in machine self-reproduction which began in the late 1990s and early 2000s.

Track-following robots

In 2003, a mobile modular robot composed of several subunits was shown to reproduce by following a track (Suthakorn et al. 2003). This was one of the first demonstrations of a general purpose programmable robot that could assemble components into its own duplicate. The key feature of both von Neumann's Universal Constructor and sub-cellular reproduction—namely, a means for manipulating information separate from the means for arranging physical components, was provided by the track. While this machine did not replicate the track, the same track could be utilized by multiple machines. In effect, this system demonstrates a ribosome analog, in which the robots are the ribosomes, the disassembled modules are the amino acids, and the track is analogous to RNA. More recent versions of track-based systems are described in Liu et al. (2007) and Lee et al. (2008).

Self-reproduction in MSRR

In 2005, a tower-like assembly of four modular robot components was shown to self-reproduce when provided with suitably placed individual modules (Zykov et al. 2005). Each module is a cube containing a controller, a single motor, and electromagnetic connectors on each face. This work was an important step toward a general purpose programmable construction robot that is capable of self-reproduction.

One measure of a self-reproducing system's "manufacturing power" is the loosely defined notion of a ratio between the internal complexity of the modules and the complexity of the arrangements of modules the system can produce. A ribosome can construct sequences down to almost atomic detail—individual amino acids are under its control. In contrast, a modular robotic system may be composed of up to a few dozen robotic modules, but the modules themselves may contain hundreds, thousands, or even more internal components (especially if the complexity of the microcontrollers is accounted for). In an effort to reduce the complexity ratio, a special MSRR with very simple components was developed (Moses et al. 2014). Figure 7.2 shows how these components are designed with sub-cellular reproductive processes in mind. In particular each component is kept as simple as possible, preferably with no moving parts. In addition, like an amino acid, the components have standard "handles" where a constructing machine can grasp them, and universal "connectors" that allow them to be assembled using the same process for all components. Figure 7.3 shows an example of a ribosome-like constructor made from these components.

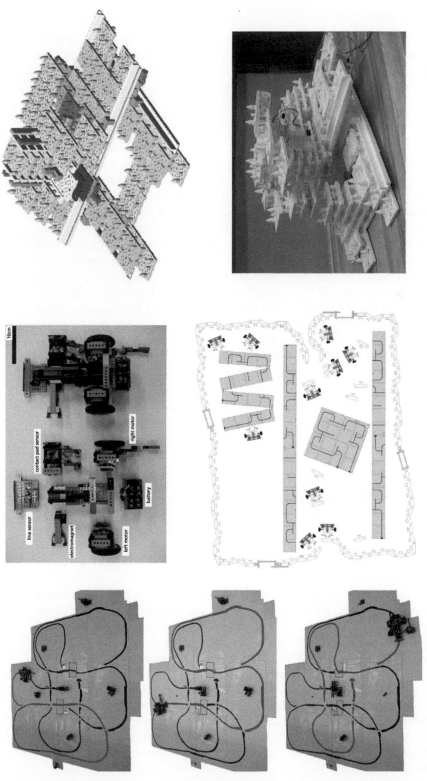

Figure 7.3 Top center: a mobile track-following robot made of seven simple modules. The standard handle for each module is a metal plate which can be grasped by an electromagnet. The standard connector is made of complementary permanent magnets attached on each mating face of the components.

Left: reproduction sequence for the track-following robot.

Right: CAD drawing and photograph of a ribosome-like constructing robot. The machine can handle and assemble the same type of components it is constructed from.

Bottom center: conceptual design for a membrane-bound artificial cell based on a combination of modular, track-following, and self-assembling components.

© 2007 IEEE. Reprinted, with permission, from Andrew Liu, Matt Sterling, Diana Kim, Andrew Pierpont, Aaron Schothauer, Matt Moses, Kiju Lee, and Greg Chirikjian, A Memoryless Robot that Assembles Seven Subsystems to Copy Itself, ISAM'07. *IEEE International Symposium on Assembly and Manufacturing*, pp. 264–269, doi: 10.1109/ISAM.2007.428848.

Robotic self-assembly

Advances in self-reconfigurable robots were accompanied by additional improved demonstrations of self-assembly. In Griffith et al. (2005) simple plastic tiles that resemble nucleotide bases were shown to replicate in a manner similar to DNA when moving randomly on a low-friction surface. Klavins (2007) shows a number of methods for controlling the geometry of self-assembled robotic structures.

Future directions

Beginning in about 2005, the RepRap and Fab@Home projects initiated efforts to build low cost 3D printers capable of producing some of their own components. This started an exciting new direction for self-reproduction research. The creators of RepRap liken their machine to an organism in symbiotic relation with its human users (Jones et al. 2011). The machine produces parts of itself that would be hard to procure otherwise, and the human user performs the assembly. An interesting possibility is to combine the fabrication powers of a low cost 3D printer with the assembly ability of a modular self-reconfigurable robot. The constructing machine shown in Figure 7.3 is one possible avenue for this. The collection of modules form a machine with three axes of motion, similar to a machine tool. In principle, this device can pick up a variety of material deposition tools and build modules using free form fabrication, and then assemble these modules into a growing grid to make new machines. A computer simulation by Stevens (2011) provides an example of how such a machine could go so far as to reproduce its own controller and memory from logic gates embedded into individual components.

Another avenue for future work is to extend the capabilities of track-following robots. Figure 7.3 shows a proposed system, built of heterogeneous components that act analogous to the cell membrane, enzymes, and DNA/RNA. These diverse components each perform simple specialized functions. Collectively, an ensemble of these devices could form a robotic system that can reconfigure, self-replicate, and evolve.

Learning more

There are a multitude of good textbooks available on cell and molecular biology: Karp et al. (2015) is one of them, currently in its eighth edition. A very thorough review of artificial self-replicating machines can be found in Freitas and Merkle (2004). This book begins with von Neumann's lectures, and covers most of the experimental work in this field up to 2004, including Penrose's early 1957 experiments with simple wooden modules, as well as modern work with modular robots and computer simulations.

Experimental demonstrations of machine reproduction more recent than 2004 can be found in Griffith et al. (2005), Zykov et al. (2005), Liu et al. (2007), Lee et. al. (2008), and MacCurdy et al. (2014). A good review of self-reconfigurable modular robotics is Yim et al. (2007), which focuses primarily on non-stochastically assembled devices. A good overview of stochastically driven modular robot self-assembly is Klavins (2007). More information on the simplified modular robot components and the constructor made from them (Figure 7.2) can be found in Moses et al. (2014). Details of several track-based self-reproducing robots can be found in Suthakorn et al. (2003), Liu et al. (2007), and Lee et al. (2008). A formal academic description of the RepRap project can be found in Jones et al. (2011). An extensive computer simulation of a near-physical self-reproducing system with the capacity for general purpose construction is described in Stevens (2011). Programmable self-assembly in a 1000-robot swarm is described in Rubenstein et al. (2014).

Acknowledgments

The authors' work was partially supported by NSF Grant IIS 0915542 "Robotic Inspection, Diagnosis, and Repair".

References

Freitas, R. A., and Merkle, R. C. (2004). *Kinematic self-replicating machines*. Georgetown, TX: Landes Bioscience.

Griffith, S., Goldwater, D., and Jacobson, J. M. (2005). Self-replication from random parts. *Nature*, 437(7059), 636.

Jones, R., Haufe, P., Sells, E., Iravani, P., Olliver, V., Palmer, C., and Bowyer, A. (2011). Reprap—the replicating rapid prototyper. *Robotica*, 29(1), 177–91.

Karp, G., Iwasa, J., and Marshall, W. (2015). *Cell and Molecular Biology*, 8th edition. Chichester, UK: Wiley.

Klavins, E. (2007). Programmable Self-assembly. *Control Systems Magazine*, 24(4), 43–56.

Lee, K., Moses, M., and Chirikjian, G.S. (2008). Robotic Self-Replication in Partially Structured Environments: Physical Demonstrations and Complexity Measures. *International Journal of Robotics Research*, 27(3–4), 387–401.

Liu, A., Sterling, M., Kim, D., Pierpont, A., Schlothauer, A., Moses, M., and Chirikjian, G. (2007). A memoryless robot that assembles seven subsystems to copy itself. In *ISAM'07. IEEE International Symposium on Assembly and Manufacturing, 2007*, pp. 264–9.

MacCurdy, R., McNicoll, A., and Lipson, H. (2014). Bitblox: Printable digital materials for electromechanical machines. *The International Journal of Robotics Research*, 33(10), 1342–60.

Moses, M. S., Ma, H., Wolfe, K. C., and Chirikjian, G. S. (2014). An architecture for universal construction via modular robotic components. *Robotics and Autonomous Systems*, 62(7), 945–65.

Rubenstein, M., Cornejo, A., and Nagpal, R. (2014). Programmable self-assembly in a thousand-robot swarm. *Science*, 345(6198), 795–9.

Stevens, W. M. (2011). A self-replicating programmable constructor in a kinematic simulation environment. *Robotica*, 29(1), 153–76.

Suthakorn, J., Cushing, A. B., and Chirikjian, G. S. (2003). An autonomous self-replicating robotic system. In *AIM 2003. Proceedings. 2003 IEEE/ASME International Conference on Advanced Intelligent Mechatronics, 2003.* (Vol. 1, pp. 137–42).

Yim, M., Shen, W.-M., Salemi, B., Rus, D., Moll, M., Lipson, H., Klavins, E., and Chirikjian, G. S. (2007). Modular self-reconfigurable robots [Grand Challenges of Robotics]. *IEEE Robotics & Automation*, 14(1), 43–52.

Zykov, V., Mytilinaios, E., Adams, B., and Lipson, H. (2005). Self-reproducing machines. *Nature*, 435(7039), 163–4.

Chapter 8

Evo-devo

Tony J. Prescott[1] and Leah Krubitzer[2]

[1] Sheffield Robotics and Department of Computer Science, University of Sheffield, UK
[2] Centre for Neuroscience, University of California, Davis, USA

Technology in the areas of artificial intelligence, cognitive systems, and robotics has already made many significant advances. We have constructed bipedal robots that walk and run, quadrupeds that negotiate uneven ground, cars that drive themselves, and micro-scale hover-bots that fly in formation. We have developed sensing systems that read hand-writing, recognize faces, and parse human speech; and we have devised planning, reasoning, and inference systems that integrate terabytes of information to coordinate traffic, handle the logistics of large organizations, optimize complex financial transactions, and mine scientific data sets. In some areas we are close to mimicking the achievements of biological systems, in others we have already outstripped them. But we are reaching a limit in the design of these systems and this is reflected in their brittleness in the face of unexpected challenges and the increasing difficulty and cost of updating these systems to keep pace with the changing world. "Evolvability" is therefore recognized as an important challenge for the design of complex artifacts (Lehman and Belady 1985; Mannaert et al. 2012) and technologists have been inspired by biological systems to make artificial systems that are more dynamic and evolvable (Fortuna et al. 2011; Le Goues et al., 2010).

But what can we learn from nature about evolvability? In the biological sciences, evolution has traditionally been considered through the lens of selection—preferring organisms that thrive over ones that fail. Selection is inarguably one of the most powerful mechanisms operating in biology that has made possible the evolution of complex organisms with rich repertoires of behavior. However, selection can only work if there is suitable variation within the population from which to select. Put it another way, whilst selection is the mechanism through which evolution operates it is not what makes organisms evolvable. The study of biological evolvability points to the sources of natural variability (e.g. Figure 8.1) as critical to understanding and replicating the power of natural evolution (Carroll 2012; Kirschner and Gerhart 2006).

A general definition of biological evolvability (e.g. Wagner and Altenberg 1996) might consider two sources of variability in the population. The first is concerned with how gene sequences vary within a population to give rise to variations in phenotype. Mechanisms that affect this form of variability include processes that directly alter chromosomal DNA such as mutation, crossover, and recombination, as well as population effects such as migration, translocation, and so forth. A second important source of variability, less widely known, derives from the developmental processes through which this genetic infrastructure produces a specific phenotype, as beautifully illustrated by the differences in patterning of the butterfly wings in Figure 8.1. This kind of phenotypic diversity does *not* require variability in the underlying gene sequences but instead relates to variation in the way that genes are expressed during development.

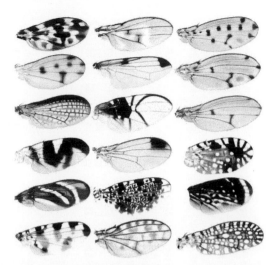

Figure 8.1 Variability in natural systems. Understanding how evolution operates requires that we go beyond the processes that allow *selection* of genetic material and decode the mechanisms that give rise to *variation*. For example, the stunning variety of patterns in these butterfly wings is thought to arise through changes in the way that regulatory gene networks operate without requiring any changes in the underlying gene sequences.

Reproduced from *Proceedings of the National Academy of Sciences of the United States of America*, Emerging principles of regulatory evolution, 104 (Supplement 1), pp. 8605–8612, Figure 1, doi:10.1073/pnas.0700488104, Benjamin Prud'homme, Nicolas Gompel, and Sean B. Carroll, Copyright (2007) National Academy of Sciences, USA.

To understand variability due to development we first have to appreciate that genes do not specify the organism directly—rather, there is a rich set of mechanisms operating inside the cell and across the embryo that, amongst other things, determine how genes are transcribed into proteins and messenger molecules, and the probability with which they are transcribed. The operation of these *epigenetic mechanisms* determines whether any given cell becomes a neuron or a white blood cell, and they resolve the mystery of how bodies can form that have a myriad of different parts despite the fact that all cells share the same DNA. Developmental biologists have now dissected many of the underlying molecular mechanisms that regulate gene expression. Interestingly, we are now realizing that these mechanisms can also be affected by the environment, within and outside the embryo, in a manner that can produce non-genetic (epigenetic), but heritable, trans-generational change.

In recent decades the *evo-devo*—evolutionary and developmental—approach (see, e.g. Carroll 2012; Muller 2007) has become central to our current understanding of biological evolvability. A key starting pointing was the discovery that similar networks of regulatory genes underlie body patterning in all modern multi-celled animals from sea urchins to humans (see Figure 8.2 and Raff 1996; Swalla 2006). This surprising finding demonstrated the flexibility of these networks to generate different body plans and also drove the realization that the manner in which genes determine phenotypic outcomes depends on multiple interactions across different organizational levels—the adult organism is the outcome of a series of genetic cascades modulated in time and space by the wider embryological, bodily, and environmental context. Selection can operate on phenotypic variability that arises at any stage in this process.

What we have also learned since these remarkable discoveries is that the toolbox of the genetic-developmental system is both sensitive to small changes and yet remarkably robust. For example, slight alterations in the spatial and temporal patterning of gene expression can lead to large changes in the developmental outcomes, as we illustrate with respect to nervous system evolution below.

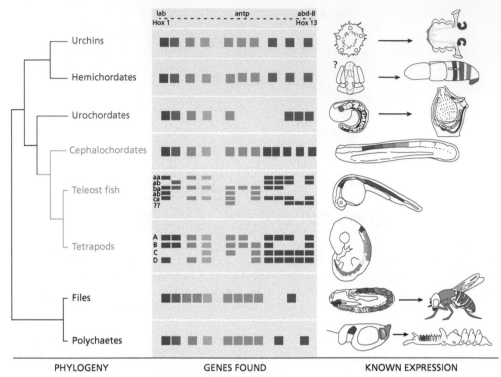

PHYLOGENY   GENES FOUND   KNOWN EXPRESSION

Figure 8.2 The evo-devo of biological systems. In the first half of the twentieth century evolutionary biology was dominated by a reductionist emphasis on the role of genes. A renewed interest in the contribution of development began with the discovery that the homeotic gene clusters, involved in body patterning in *Drosophila*, were strongly conserved across all multi-celled animals including vertebrates, despite obvious differences in the morphological outcomes they produced. This finding forced a re-examination of the role of developmental processes in specifying body and brain designs. This figure shows how the same regulatory gene cluster (Hox) is involved in specifying body patterning in all the different groups of modern multi-celled animals (our own species is a member of the class of tetrapods).

Reprinted by permission from Macmillan Publishers Ltd: *Heredity*, 97(3), Building divergent body plans with similar genetic pathways, B J Swalla, pp. 235–243, Figure 6, doi:10.1038/sj.hdy.6800872 Copyright © 2006, Nature Publishing Group.

However, at the same time, the wider system is able to respond flexibly to these shifting patterns in a manner that still results in a viable organism. This takes us far from traditional notions of evolution as exploring random changes in form. Understanding how developmental dynamics resists damaging perturbations and promotes convergence to viable outcomes could allow us to crack the puzzle of biological evolvability and uncover principles that can be usefully applied to artificial systems.

Our particular interest is in understanding the evolution and development of one specific complex biological system—the mammalian brain and nervous system—and in the possibility of designing brain-like adaptable architectures that could control biomimetic robots. In the next section, we give some additional consideration to particular mechanisms involved in brain evo-devo and to the evolutionary history of brains including our own. In the third part of the chapter we then explore how the principles underlying natural evo-devo have inspired, and are continuing to inspire, the design of artificial living machines.

Evo-devo of the mammalian nervous system

As a control system becomes more complex, interconnected, and integrated, it becomes more likely that any structural variation will result in degraded function. So how did animal nervous systems retain the capacity to evolve as they became larger and more sophisticated? One answer is by avoiding over-specification of the system in the genome and by exploiting developmental mechanisms that, as we have discussed above, promote compensation for disruptive change (Deacon 2010; Katz 2011). More generally, evo-devo research in neurobiology is now beginning to identify the toolbox of developmental processes that can give rise to useful forms of variability in nervous system organization, recognizing that this set of mechanisms has itself been the target of strong selection pressure during evolution (Charvet et al. 2011).

The toolbox for brain evolvability

So what's inside the toolbox for brain evolvability? First, like other developing systems the complex wiring of the brain is the result of a much simpler (though still very complex!) set of growth rules, implemented through genetic cascades, and interacting within a self-assembling structure, first over-specifying and then using intrinsic mechanisms, such activity-based pruning, to sculpt out useful network topographies. One set of tools thus relates to the *generative*, or self-organizing, mechanisms that permit the brain to wire during early development by tuning in to internal neural signals (see also Wilson, Chapter 5, this volume). The system is first directed to incoming sensory signals, and then to the correlations between inputs from different modalities (see e.g. Krubitzer and Kaas 2005).

Second, there is a set of tools that instantiate *adaptive* mechanisms (which in this chapter we use to mean those involved in learning as opposed to general adaptation). Learning operates to promote the selection of circuits that support behavioral capacities that are well matched to the environment in which the individual develops. The development of cortical areas, for instance, is driven in significant part by specific patterns of incoming sensory inputs that can support tasks such as object recognition and detection, decision-making, path-planning, and motor control. These circuits are continually modified and honed by experience in the world, using different forms of internal and externally-mediated feedback (see Herreros, Chapter 26, this volume).

A third set of tools relates to systems components and architectural principles as discussed in Sections Four (building blocks) and Five (capabilities) of this book. For example, oscillators that can create patterned rhythms, decision circuits that can resolve competitions, and sensorimotor circuits that can implement predictive control. Architectural principles include *layered control* whereby lower tier pathways that link sensing with action can operate in the absence of higher ones (Prescott et al. 1999), and *redundancy*, where multiple substrates provide alternative means to support a given function (Deacon 2010). These architectural features create robustness and thus provide protection against catastrophic local change—smoothing out the fitness landscape by providing more options for exploring useful variability (Kauffman 1990). An important feature that we see in brain evolution, and that can underlie layering and redundancy, is the duplication of existing structures. Making copies of working sub-circuits can lead to system components that are under-used and that can then adjust to take on new functions (Deacon 2010; Whitacre and Bender 2010).

From phenotypic variability to hopeful monsters

Biologists have long been puzzled by the difficulty of reconciling rapid changes in animal lineages with the notion of gradual change brought about by genetic operators such as mutation. In the 1940s, the biologist Richard Goldschmidt coined the phrase "*hopeful monster*" (Goldschmidt 1940) to capture the idea of an organism that represented a large leap in design-space (also termed macroevolution). This idea was deemed controversial by many biologists of the day who reasoned that any change in the genome that brought about a radical reorganization of the organism would be likely to be lethal. Nevertheless, evolution does seem to have proceeded by taking a number of larger steps in addition to many smaller ones. How might this be possible? Evolvability again appears to be key and the new understanding engendered by the evo-devo approach is resurrecting Goldschmidt's hopeful monster hypothesis. Specifically, genetic mutations, or consistent non-genetic departures from normality, that bring about radical changes can be compensated for by pattern-forming processes that steer development towards building a viable animal. Moreover, where a single population is affected by a significant disruptive event, such as abrupt environmental change, similar changes might be triggered in multiple individuals. Where this happens successful large strides within design space could be made.

We can illustrate this idea using studies of brain evolution that show a number of dramatic changes in the size and organization of the mammalian cortex through evolutionary history, perhaps most remarkably, in terms of speed of change, in the sequence of primate and hominid forms leading to modern humans. This is illustrated in Figure 8.3 which shows approximate shape and areal structure of the cortex in different groups of modern mammals.

Experimental manipulations in brain development are beginning to suggest how such changes may have come about. For example, Chenn and Walsh (2002) have shown that the impact of a single

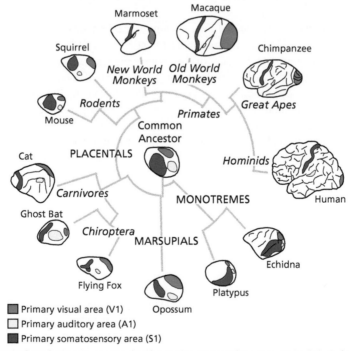

Figure 8.3 Cortical evolution in mammals. The major mammalian groups and their family relationships illustrating how the shape of the cortical mantle, and of some of its primary sensory areas, vary relative to an inferred common ancestor.

Reproduced from *Proceedings of the National Academy of Sciences of the United States of America*, Cortical evolution in mammals: The bane and beauty of phenotypic variability, 109 (Supplement 1), pp. 10647–10654, Figure 1, doi: 10.1073/pnas.1201891109, Leah A. Krubitzer and Adele M. H. Seelke, Copyright (2012) National Academy of Sciences, USA.

change in the sequence of developmental events involved in early neurogenesis in the mouse can lead to a change from a smooth rodent-like neocortex to a convoluted primate-like one, bringing about an expanded cortical surface without an increase in thickness. Although these particular mice have brain abnormalities that make them non-viable, what is impressive is that the over-production of neural progenitor cells triggers changes later in the developmental cascade that at least partially compensate for this perturbation (see Rakic (2009) for further discussion). This study demonstrates that changes that occur early in development can have particularly profound effects (thus larger strides). Other experiments, summarized by Krubitzer and Seelke (2012), have used knock-out mice to explore the role of different transcription factors in regulating arealization—the size, shape, and position of different cortical areas. As illustrated in Figure 8.4, eliminating any one of the four factors Emx2, COUP-TF1, Pax6, or Sp8 radically alters the size and shape of multiple fields—enlarging some, reducing others, whilst preserving their overall topological relationships.

In a final illustrative experiment, by Fukuchi-Shimogori and Grove (2001), an electrically induced overexpression of a growth factor in a specific locus in the mouse forebrain caused the animal to develop a second complete barrel field (a representation of the facial whiskers) that was a mirror-image of the normal one. This result indicates the relative ease with which develop-mental processes can generate useful redundancy in the form of substantial and well-organized structures that can then be co-opted to serve new roles.

Krubitzer and Seelke (2012) summarize the main differences in mammalian cortex that can be observed by comparing across species, as shown in Figure 8.3. These include (i) the size of the cortical sheet, (ii) the relative amount of space allocated to different sensory domains (and more generally, to the different cortical fields), (iii) cortical magnification of behaviorally relevant body parts, (iv) the addition of new modules, (v) the overall number of cortical fields, and (vi) the connections between cortical fields. Interestingly, many of the same changes can be seen when comparing between individuals of a given species but to a much less dramatic extent than when comparing across species. This illustrates that the phenotypic variability required to allow selection to direct cortical evolution in any of these different directions may be present in most species most of the time.

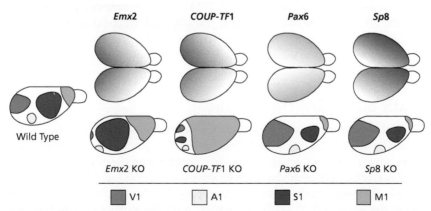

Figure 8.4 The effect of removing specific transcription factors (via genetically-modified 'knock-out' mice) on arealization in mouse cortex. The size and shape of major cortical areas—primary visual (v1), auditory (a1), somatosensory (s1), and motor (m1) cortices—is radically disrupted by each knock-out whilst the overall topology is maintained.

Conservation of basic design principles

Whilst exploring the paths to effective change through biological evolution it is also important to recognize that some things have stayed the same. Indeed, the story of biological evolution can be regarded as one of conservation of basic design principles, that were discovered early on, and that instantiated evolvable control architectures (Kirschner and Gerhart 2006). For example, fossil evidence shows a remarkable explosion of animal forms during the Cambrian period (~541–485 million years ago) in which all of the major bilaterian phyla were represented, despite being absent, for the most part, from the fossil record at the end of the previous Ediacaran period (see also Prescott, Chapter 38, this volume). Current hypotheses suggest that the last common ancestor of all modern bilaterian animals, the *Urbilateria*, evolved some time before the Cambrian boundary, and is thought to have possessed the "essential bilaterian toolbox" (Erwin and Davidson 2002) of regulatory genes, including the homeotic gene clusters that underlie cell differentiation and body patterning in all modern bilaterians (as in Figure 8.2). Vertebrate-like animals also appear earlier than was once thought—finds from Chengjiang in China (the Chinese "Burgess shale") show the presence of fish-like creatures (craniates) in the early Cambrian (Mallatt and Chen 2003). This evidence implies a rapid evolution of complex nervous systems, as part of the general evolution of new body plans (Gabor Miklos et al. 1994).

Analysis of comparative brain architecture indicates some startling conservation of nervous system structure across all vertebrate classes. Notably, all vertebrate brains comprise a layered architecture with spinal, hindbrain, midbrain, and forebrain components; an integrative core, comprised of the medial reticular formation and the basal ganglia; and a number of specialized learning/memory systems—hippocampus, striatum, cortex, and cerebellum (Prescott et al. 1999). Of course there have been considerable changes in the size, shape, number of sub-divisions (parcellation), and microstructural organization of many brain regions, as well as the addition of new cell types, and the migration of cell clusters within the brain. Connectional changes have included axonal invasion of new areas and increased differentiation of local areas through selective connection loss. Nevertheless, comparative neurobiology indicates that the scope for alteration of the basic vertebrate brain plan may be remarkably limited (Charvet et al. 2011). Vertebrates have a control architecture that is evolvable but only within a set of constraints that were already laid down within the nervous system of a 500 million-year-old fish. If we could better understand this constrained but evolvable architecture—that carried generations of owners from sea to land, from four legs to bipedal walking, into the air, even to explore space—this should make for a powerful model for the design of artificial living machines.

Biomimetic and biohybrid systems

To learn from the evo-devo of natural systems it is clearly necessary to identify, simulate, and abstract the evolutionary and developmental mechanisms that gave rise to their evolvable nature. This approach could result in a toolbox of generative, adaptive, and selective mechanisms that can be used to formulate a design methodology to be applied to complex bio-inspired systems. Such a framework will also require an abstracted understanding of body plans and control architectures since it is clear that evolution acts to select evolvable and flexible control systems and then operates within some limits determined by those designs (and by the laws of physics).

In computational neuroscience, artificial intelligence, and robotics, researchers have explored generative methods, such as self-organization, or selective methods, such as genetic algorithms and genetic programming, and adaptive methods, such as reinforcement and supervised learning, for several decades. A smaller, but still significant, community have investigated how generative, selective, and adaptive mechanisms, or some combination of these, can operate together.

Much of this literature has followed a biomimetic approach. There is insufficient space here to provide a comprehensive review, therefore we briefly survey the history of this field and highlight some examples that were particularly groundbreaking or insightful.

Foundations of artificial evo-devo

The attempt to create artificial systems using the methodologies of evolution, development, and learning is driven by the recognition that engineered systems cannot match either the complexity or the robustness of biological life, and moreover, that nature has invented some remarkable techniques for encapsulating the specification for such complex entities in highly compact codes. For example, the human brain is estimated to have 100 trillion neural connections and yet our genome has less than 30,000 active genes. One of nature's most important tools, then, is this indirect but superbly productive mapping of the genome onto the phenotype.

One analogy that occurs repeatedly is that of a language—with a small vocabulary, and a clutch of grammatical rules saying how words can combine, a language can generate millions of varying but valid sentences. In computer science, the productivity of a formal language, defined as set of rules that operate over an identified set of symbols, has been understood since Chomsky's work in the 1950s, and this general idea was first applied to model biological development by Lindenmayer (1968) to make branching shapes similar in structure to threads of algae (e.g. Figure 8.5). This approach, now known as L-systems (a contraction of Lindenmayer-systems), continues to be widely used. For instance Torben-Nielsen et al. (2008) describe a

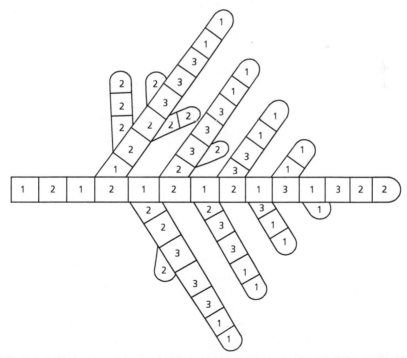

Figure 8.5 A plant-like structure generated by an L-system grammar for modeling biological development.

Reprinted from *Journal of Theoretical Biology*, 18(3), Aristid Lindenmayer, Mathematical models for cellular interactions in development I. Filaments with one-sided inputs, pp. 280–99, doi: 10.1016/0022-5193(68)90079-9, Copyright © 1968, Elsevier Ltd., with permission from Elsevier.

modeling system, EVoL-Neuron, that builds models of single neuron arborization, using genetic algorithms to tune these to match experimentally measured cells.

The L-systems approach was also generalized by Kitano (1990) to the construction of artificial neural networks. Whereas Lindenmayer's goal was to show that a small grammar could give rise to differentiated and life-like structures, Kitano sought to demonstrate a specific advantage, in terms of coding efficiency, of a grammar-based system compared to a more direct encoding approach. Moreover, Kitano used L-systems in combination with selective (genetic algorithm) and adaptive (back-propagation) methods, that is, as part of a toolbox approach, and sought to show that this combination of methods could be usefully applied to difficult computational problems (he chose decoding, a classic problem in computer science). Kitano demonstrated that, compared to a direct encoding approach, a grammar-based generative model converged faster and scaled better as the size of the target network increased—a clear win for evo-devo. He also showed that larger networks could be specified without growing the size of the genetic coding—a path to concise and scalable developmental codes for specifying useful artificial complex systems was beginning to emerge.

Whilst grammars beautifully capture some of the power of a compact encoding to recursively express a developmental program, it is not obvious how such a system can map onto the chemical and mechanical systems inside the cell. Historically, two other approaches have provided important leads.

The first begins with Turing's (1952) paper on the chemical basis of morphogenesis (the development of body form). Now seen as a foundational paper for understanding self-organizing systems in general (see Wilson, Chapter 5, this volume), Turing was particularly concerned with the question of how the genes of a fertilized egg enable it to develop into a multicellular organism with asymmetric differentiated structure. Drawing on the idea that development can be guided by diffusing chemicals, that he termed *morphogens*, Turing provided a mathematical model of interacting chemical gradients, termed *reaction-diffusion* processes, that could create elegant patterns such as dappling, stripes, and whorls. Models based on Turing's idea are able to capture some of the more analog and non-local character of processes that happen inside the developing organism. For instance, Lewis (2008) reviewed half a century of work, building on Turing's idea, that has used mathematical models to understand the role of chemical gradients in gene expression, with examples from patterning in plant meristems to the dorsoventral organization of frog embryos. Fleischer and Barr (1994), in following a toolbox approach, combined reaction-diffusion modeling based on differential equations with a grammar-based model of growth through cell-division and a mechanical model of cell-to-cell interactions to create a 2D model of a developing cell sheet. The resulting system produced rich life-like patterns, such as those illustrated in Figure 8.6, that have found applications in areas such as computer graphics. For a contemporary toolbox for synthetic development capable of defining multicellular "soft robots" see Doursat and Sánchez (2014).

A third starting point was the realization that chromosomal DNA, and the chemical processes that surround it, instantiate a rich dynamical system with compressed degrees of freedom and emergent order, and for which it might be possible to define a simpler analog with similar dynamical properties. Kauffman (1969a, 1969b) attempted just this using randomly wired Boolean networks as models of regulatory gene systems. He showed that, given the right degree of connectivity between network nodes, a Boolean network of a size mapped to a specific animal genome will enter dynamically stable states that can predict the number of cell types in that organism, behavior cycles that can predict cell division times, and responses to noise that look like behavior switching. That Boolean networks could be a useful abstraction of regulatory gene systems was further demonstrated by Dellaert and Beer (1996) who embedded one within a model cell, and used the state of the network, as it varied over time, to regulate cycles of cell division and differentiation. Using a genetic algorithm to configure the initial network, they were able to grow

Figure 8.6 Fleischer and Barr (1994) applied their developmental tool-box, inspired by Turing's reaction-diffusion model and Lindenmayer's L-systems, to the challenge of generating computer graphics such as these spheres coated with animal-like scales.

Reproduced from Fleischer, K. W. "A multiple-mechanism developmental model for defining self-organizing geometric structures," Figure 7, PhD thesis, California Institute of Technology, Pasadena, 1995.

multicellular 2D model organisms, with a mix of sensor and actuator cells, that could be evolved to follow curved lines. Dellaert and Beer also built a second model organism more closely modeled on the chemistry of genetic regulatory networks; however, configuring such a model to be evolvable proved to be less tractable than for the simpler Boolean system. The status of Boolean networks as simplified models of cellular regulatory networks is discussed in Bornholdt (2008), who argues that Boolean network dynamics can provide useful insights into how cells "compute." Giacomantonio and Goodhill (2010) have applied Boolean networks to the problem of understanding the genetic regulatory networks underlying arealization in mammalian cerebral cortex, as shown in Figure 8.3, and involving all of the genetic transcription factors shown in Figure 8.4 plus one more (Fgf8). Exhaustively simulating all possible network configurations of these five genes, they found that only 0.1% of possible networks could reproduce experimentally observed expression patterns. Moreover, the networks that worked tended to have certain kinds of within-network interactions and not others, thus giving clues as how these gene networks may operate during brain development.

Reviewing several decades of work, Stanley and Miikkulainen (2003) proposed a taxonomy of model evo-devo, or to use their term *artificial embryogeny*, systems that identified five major dimensions of design choice: (1) cell fate—the mechanism(s) through which cell type is determined; (2) targeting—how cells connect with each other; (3) heterochrony—how the timing of development events is regulated; (4) canalization—how systems are made robust to variation in the genotype; and (5) complexification—how the genome (and the phenotype) become more complex over time.

Various further proposals have been made as to how these different design mechanisms can be abstracted from biological systems to lead to more compact and scaleable codings (and hence to improvements in evolvability). For example, Stanley (2007) has suggested that the core role of local interactions within developing organisms is to provide cells with information about their location in space. If so, a useful abstraction of the developmental process could be to provide an explicit coordinate frame allowing development at each location to be specified by the composition of a series of parameterized functions. This process can be visualized as a directed graph or "compositional pattern producing network" (CPPN). Removing local interaction from the mix also allows this method to collapse time, that is to compute the final outcome at each location in a single step. Applying artificial evolution to CPPNs shows that they are flexible and evolvable ways of encoding patterns that exhibit phenomena characteristic of natural developing systems such as symmetry and repetition with variation. Other approaches have explored the utility of redundancy, via duplication, of network components (Calabretta et al. 2000); still others have explored the benefits of modularity within evolved controlled systems (e.g. Bongard 2002), and in the emergence of functionally distinct sub-networks in mammalian cortex (Calabretta 2007). Garcia-Bernardo and Eppstein (2015) have recently described an approach to complexification that involves pruning dense model networks to find minimal configurations that retain required functionality, then using these more compact circuits as fixed building blocks within larger systems.

Many model systems have included adaptive mechanisms (learning) as a key step in building a working model organism (e.g. Sendhoff and Kreutz, 1999). Recognized as the "Baldwin effect" (see Wilson, Chapter 5, this volume), a learning capacity can make a system more evolvable by smoothing out the fitness landscape around peaks (if learning can reliably bring the model system to the fitness peak then it should be sufficient for evolution and development to place the system somewhere in the vicinity of the peak). Adding learning capacity is thus another means for reducing code-size for evolving systems whilst making good solutions easier to find (reducing search time). Of course, this comes at the cost of the time needed for lifetime learning and a period of reduced fitness while the system adapts.

Scaling-up

An important goal of research in artificial evo-devo has been to show that model solutions can be scaled to cope with real-world complexity. A useful stepping-stone for this has been the use of simulated 3D worlds that include real-time physics engines adapted from applications in computer graphics and gaming. A well-known example of this approach is Karl Sims (1994) "creatures." Sims used genetic algorithms to select directed graphs (analogous to an L-system grammar) that specify solutions to the problem of being a digital creature built of collections of blocks, linked by powered flexible joints, and controlled by circuits. Sims evolved a neural network control system for his creatures alongside their physical morphology and designed fitness functions for model aquatic and terrestrial environments that selected for proficiency at tasks such as swimming speed, swimming after a light source, moving across a surface, jumping on a surface, and attempting to possess a square block in contest with another creature—a model evolutionary arms race (see Figure 8.7). These experiments produced a fascinating array of creatures, some with familiar morphologies and behavior reminiscent of that of actual animals such as snakes, tadpoles, or crabs. Others, often equally effective at their tasks, accomplished them with less familiar or bizarre patterns of movement and body-form. The resulting panoply of digital organisms has been compared to the Cambrian explosion of early life.

The success of Sims' approach depended on the power of a generative encoding to build working model organisms out of parameterized simple elements (blocks), and the ability of genetic algorithms to exploit physics in a manner that simplified the challenge of control. Lipson and Pollack (2000) took this idea closer to physical reality by combining evolution in 3D simulation with additive manufacturing (3D printing) of physical robots for the most successful designs. To perform evo-devo in an actual physical system is more difficult, although some work in reconfigurable robots is moving in this direction. For example, the Eyebot, shown in Figure 8.8, developed by Lichtensteiger and Eggenberger (1999) and modeled on the insect compound eye, applied

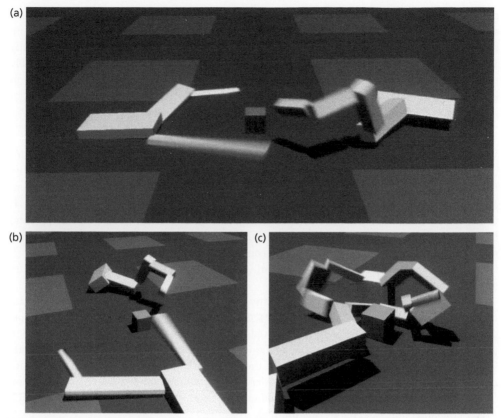

Figure 8.7 A simulated evolutionary arms race. Evolved artificial creatures fight for possession of a block in one of the experiments performed by Karl Sims (1994).

Reproduced from Karl Sims, Evolving virtual creatures, *Proceedings of the 21st annual conference on computer graphics and interactive techniques*, pp. 15–22, doi:10.1145/192161.192167 © 1994, Association for Computing Machinery, Inc. Reprinted by permission.

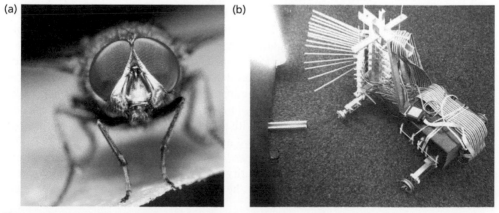

Figure 8.8 *Eyebot* (b), a reconfigurable robot inspired by the insect compound eye (a), designed to apply artificial evolution to robot morphology.

(a): ©Denis Vesely/Shutterstock.com (b) Reprinted by permission from Macmillan Publishers Ltd: *Nature*, 406 (6799), Automatic design and manufacture of robotic lifeforms, Hod Lipson and Jordan B. Pollack, pp. 974–978, Figure 5, doi:10.1038/35023115, Copyright © 2000, Nature Publishing Group.

genetic algorithms to tune the position of model ommatidia; Gomez et al. (2004) extended this approach to a robot hand-eye system to explore parallel development of control systems and morphology (degrees of freedom) in a model of human hand-eye coordination. Most recently Vujovic et al. (in press) have used additive manufacturing, together with automated assembly using a robot arm, to evolve and grow robots with variable morphology, testing their ability to locomote across a flat arena.

Future directions

We are reaching a point in time where the widespread availability of super-fast computers, combined with increased sharing of open source toolboxes, could allow a much wider community to get going with artificial evo-devo. On the other hand, a limiting factor is education—current approaches require mastery of a range of different tools and, if mining of biological principles is to be successful, understanding of one of the most daunting scientific literatures—developmental biology—with its complicated vocabulary and detailed data sets. To overcome these problems an emphasis on multi-scale modeling must continue (see Verschure and Prescott, Chapter 2, this volume) alongside interdisciplinary training opportunities for young scientists (see Mura and Prescott, Chapter 64, this volume).

The field of artificial evo-devo also has many territories still to explore. At the most fundamental level the identification of the principles required for living systems—such as being viable and self-sustaining—can be effectively explored through simulation (e.g. Agmon et al. 2016) in parallel with efforts to create new artificial living systems from organic chemistry (Mast et al., Chapter 39, this volume). At the other end of the complexity spectrum, there is also a community seeking to apply this approach to the understanding of human development and cognition (Parisi 1997) through the medium of embodied robotic modeling (e.g. Cangelosi et al. 2015; Metta and Cingolani, Chapter 47, this volume).

One strategy to take these methods into the real world is to try to circumvent some of the bottlenecks that have set limits to natural evolution, such as the need to compress the information required to created new life-forms into a single cell, to limit inheritance to family lines, and to start each experiment afresh with a new organism built from that single cell. For instance, we can explore the retention of acquired adaptations, mimicking Lamarckian evolution (or memetic cultural evolution) (Le et al. 2009), perhaps directly copying useful adaptations from one experiment to the next (much as biologists now do with genetic tools). As already noted, experiments can run in simulated bodies and worlds, where reasonably realistic model systems can now be simulated faster than real time. Model populations can be tuned to optimize evolvability, as theories of this improve, for instance by choosing additional selection criteria that promote phenotypic diversity (Lehman et al. 2016). For physical systems we can repeatedly adapt and re-use modifiable and modular robot hardware whilst taking advantage of auto-fabrication methods. Further, we can more effectively utilize our understanding of biological evolution to start with pre-structured models that are not created from scratch. Nature has identified many "forced moved and good tricks" (Dennett 1995) for the construction of animal bodies and of complex control architectures and conserves these as species evolve (Prescott 2007). We can identify design features that promote robustness and evolvability and either build these in from the beginning or promote them via selection criteria.

To apply the lessons of evo-devo to technology will require a continued effort to identify the evo-devo principles that gave rise to biological life, to determine appropriate levels of abstraction away from the biological detail, and to define design methodologies that allow these principles to be applied to artificial systems. Current toolbox approaches show great promise, although, in the spirit of evo-devo, these systems are now getting sufficiently complex that they themselves could be candidates for automated redesign.

Learning more

Several books have captured the excitement around the fusion of evolutionary and developmental approaches in modern biology—Raff (1996), Carroll (2012), and Kirschner and Gerhart (2006) are all good places to start. The application of the evo-devo approach to neurobiology is still relatively new with most aspects of brain architecture still unexplored. Charvet et al. (2011) discuss some of the key developmental mechanisms in brain evo-devo and consider the balance between constancy and change in vertebrate brain evolution, Deacon (2010) and Katz (2011) discuss a number of neural generative, selective, and adaptive mechanisms that operate to make behavior evolvable, whilst Krubitzer and Seelke (2012) explore the bases for phenotypic diversity in relation to the evolution of mammalian cortex. Much of the research that applies evo-devo methodologies to artifacts has emerged under the umbrella of "Artificial Life," and a number of classic contributions in this field are provided in Langton (1995); Parisi (1996) also provides a conceptual overview and introduction to much of the foundational work, whilst Stanley and Miikkulainen (2003) provide a useful synthesis and taxonomy. Nolfi and Floreano (2000) review much of the classic work in evolutionary robotics, whilst Pfeifer and Bongard (2006) explore the application of some ideas from evo-devo in robots with particular emphasis on embodiment. Haddow and Tyrrell (2011) provide a review and critical assessment of the application of evo-devo methodologies to electronic systems. Finally, Downing (2015) considers many of the topics introduced in this chapter with an overall emphasis on the emergent nature of biological intelligence, and with similar enthusiasm for re-purposing nature's evo-devo toolbox to create new kinds of Living Machines.

References

Agmon, E., Gates, A. J., Churavy, V., and Beer, R. D. (2016). Exploring the space of viable configurations in a model of metabolism–boundary co-construction. *Artificial Life*, 22(2), 153–71. doi:10.1162/ARTL_a_00196

Bongard, J. (2002). Evolving modular genetic regulatory networks. In: *Proceedings of the 2002 Congress on Evolutionary Computation*, CEC '02, Honolulu, HI, 2002, pp. 1872–7. doi: 10.1109/CEC.2002.1004528

Bornholdt, S. (2008). Boolean network models of cellular regulation: prospects and limitations. *Journal of The Royal Society Interface*, 5(Suppl 1), S85–S94. doi:10.1098/rsif.2008.0132.focus

Calabretta, R. (2007). Genetic interference reduces the evolvability of modular and non-modular visual neural networks. *Philosophical Transactions of the Royal Society B: Biological Sciences*, 362(1479), 403–10. doi:10.1098/rstb.2006.1967

Calabretta, R., Nolfi, S., Parisi, D., and Wagner, G. P. (2000). Duplication of modules facilitates the evolution of functional specialization. *Artificial Life*, 6(1), 69–84.

Cangelosi, A., Schlesinger, M., and Smith, L. B. (2015). *Developmental robotics: from babies to robots*. Cambridge, MA: MIT Press.

Carroll, S. B. (2012). *Endless forms most beautiful: the new science of evo devo and the making of the Animal Kingdom*. London: Quercus.

Charvet, C. J., Striedter, G. F., and Finlay, B. L. (2011). Evo-devo and brain scaling: candidate developmental mechanisms for variation and constancy in vertebrate brain evolution. *Brain Behavior and Evolution*, 78(3), 248–57.

Chenn, A., and Walsh, C. A. (2002). Regulation of cerebral cortical size by control of cell cycle exit in neural precursors. *Science*, **297**(5580), 365–9. doi:10.1126/science.1074192

Deacon, T. W. (2010). Colloquium paper: a role for relaxed selection in the evolution of the language capacity. *Proc. Natl Acad. Sci. USA*, **107** (Suppl 2), 9000–06. doi:0914624107 [pii] 10.1073/pnas.0914624107

Dellaert, F., and Beer, R. D. (1996). A developmental model for the evolution of complete autonomous agents. In: P. Maes, M. J. Mataric, J.-A. Meyer, J. Pollack, and S. W. Wilson (eds), *From Animals to Animats 4: Proceedings of the Fourth International Conference on Simulation of Adaptive Behavior*. Cambridge, MA: MIT Press, pp. 394–401.

Dennett, D. C. (1995). *Darwin's dangerous idea*. London: Penguin Books.

Doursat, R., and Sánchez, C. (2014). Growing fine-grained multicellular robots. *Soft Robotics*, **1**(2), 110–21.

Downing, K. L. (2015). *Intelligence emerging: adaptivity and search in evolving neural systems*. Cambridge, MA: MIT Press.

Erwin, D. H., and Davidson, E. H. (2002). The last common bilaterian ancestor. *Development*, **129**(13), 3021–32.

Fleischer, K. W. (1995). *A multiple-mechanism developmental model for defining self-organizing geometric structures*. PhD thesis, California Institute of Technology, Pasadena.

Fleischer, K. W., and Barr, A. (1994). A simulation testbed for the study of multicellular development: the multiple mechanisms of morphogenesis. In: C. G. Langton (ed.), *Artificial Life III*. Boston, MA: Addison-Wesley.

Fortuna, M. A., Bonachela, J. A., and Levin, S. A. (2011). Evolution of a modular software network. *Proceedings of the National Academy of Sciences of the United States of America*, **108**(50), 19985–89.

Fukuchi-Shimogori, T., and Grove, E. A. (2001). Neocortex patterning by the secreted signaling molecule FGF8. *Science*, **294**(5544), 1071–4. doi:10.1126/science.1064252

Gabor Miklos, G. L., Campbell, K. S. W., and Kankel, D. R. (1994). The rapid emergence of bio-electronic novelty, neuronal architectures, and organismal performance. In: R. J. Greenspan and C. P. Kyriacou (eds), *Flexibility and Constraint in Behavioral systems*. New York: John Wiley and Sons, pp. 269–93.

Garcia-Bernardo, J., and Eppstein, M. J. (2015). Evolving modular genetic regulatory networks with a recursive, top-down approach. *Systems and Synthetic Biology*, **9**(4), 179–89. doi:10.1007/s11693-015-9179-5

Giacomantonio, C. E., and Goodhill, G. J. (2010). A Boolean model of the gene regulatory network underlying mammalian cortical area development. *PLOS Computational Biology*, **6**(9), e1000936. doi:10.1371/journal.pcbi.1000936

Goldschmidt, R. (1940). *The material basis of evolution*. Yale: Yale University Press.

Gomez, G., Lungarella, M., and Eggenberger, H. (2004). Simulating development in a real robot: on the concurrent increase of sensory, motor, and neural complexity. In: L. Berthouze, et al. (eds), *Proceedings of the Fourth International Workshop on Epigenetic Robotics: Modeling Cognitive Development in Robotic Systems*.Lund University Cognitive Studies, **117**. Lund: LUCS, pp. 119–22. doi:citeulike-article-id:549800

Haddow, P. C., and Tyrrell, A. M. (2011). Challenges of evolvable hardware: past, present and the path to a promising future. *Genetic Programming and Evolvable Machines*, **12**(3), 183–215.

Katz, P. S. (2011). Neural mechanisms underlying the evolvability of behaviour. *Philosophical Transactions of the Royal Society B: Biological Sciences*, **366**(1574), 2086–99.

Kauffman, S. A. (1969a). Homeostasis and differentiation in random genetic control networks. *Nature*, **224**(5215), 177–8.

Kauffman, S. A. (1969b). Metabolic stability and epigenesis in randomly constructed genetic nets. *Journal of Theoretical Biology*, **22**(3), 437–67. doi:http://dx.doi.org/10.1016/0022-5193(69)90015-0

Kauffman, S. A. (1990). Requirements for evolvability in complex-systems—orderly dynamics and frozen components. *Physica D*, **42**(1–3), 135–52.

Kirschner, M. W., and Gerhart, J. C. (2006). *The plausibility of life: resolving Darwin's dilemma*. Yale: Yale University Press.

Kitano, H. (1990). Designing neural networks using genetic algorithms with graph generation system. *Complex Systems*, **4**, 461–76.

Krubitzer, L. A., and Kaas, J. (2005). The evolution of the neocortex in mammals: how is phenotypic diversity generated? *Current Opinion in Neurobiology*, **15**(4), 444–53. doi:10.1016/J.Conb.2005.07.003

Krubitzer, L. A., and Seelke, A. M. H. (2012). Cortical evolution in mammals: the bane and beauty of phenotypic variability. *Proc. Natl Acad. Sci. USA*, **109**(Supplement 1), 10647–54. doi:10.1073/pnas.1201891109

Langton, C. G. (1995). *Artificial Life*. Reading, MA: Addison-Wesley.

Le Goues, C., Forrest, S., and Weimer, W. (2010). The case for software evolution. In: G.-C. Roman (ed.), *FoSER '10 Proceedings of the FSE/SDP workshop on Future of software engineering research*. New York, NY, USA: ACM.

Le, M. N., Ong, Y.-S., Jin, Y., and Sendhoff, B. (2009). Lamarckian memetic algorithms: local optimum and connectivity structure analysis. *Memetic Computing*, **1**(3), 175. doi:10.1007/s12293-009-0016-9

Lehman, J., Wilder, B., and Stanley, K. O. (2016). On the critical role of divergent selection in evolvability. *Frontiers in Robotics and AI*, **3**, 45.

Lehman, M. M., and Belady, L. A. (1985). *Program evolution: processes of software change*. San Diego, CA: Academic Press Professional, Inc.

Lewis, J. (2008). From signals to patterns: space, time, and mathematics in developmental biology. *Science*, **322**(5900), 399–403. doi:10.1126/science.1166154

Lichtensteiger, L., and Eggenberger, P. (1999). Evolving the morphology of a compound eye on a robot. In: *The Third European Workshop on Advanced Mobile Robots, (Eurobot '99)*, Zurich, pp. 127–34. doi: 10.1109/EURBOT.1999.827631

Lindenmayer, A. (1968). Mathematical models for cellular interaction in development: Parts I and II. *Journal of Theoretical Biology*, **18**, 280–315.

Lipson, H., and Pollack, J. B. (2000). Automatic design and manufacture of robotic lifeforms. *Nature*, **406**(6799), 974–8. doi:http://www.nature.com/nature/journal/v406/n6799/suppinfo/406974a0_S1.html

Mallatt, J., and Chen, J. Y. (2003). Fossil sister group of craniates: predicted and found. *Journal Of Morphology*, **258**(1), 1–31.

Mannaert, H., Verelst, J., and Ven, K. (2012). Towards evolvable software architectures based on systems theoretic stability. *Software: Practice and Experience*, **42**(1), 89–116.

Muller, G. B. (2007). Evo-devo: extending the evolutionary synthesis. *Nat. Rev. Genet.*, **8**(12), 943–9.

Nolfi, S., and Floreano, D. (2000). *Evolutionary robotics*. Cambridge, MA: MIT Press.

O'Leary, D. D., and Sahara, S. (2008) Genetic regulation of arealization of the neocortex. *Current Opinion in Neurobiology*, **18**(1), 90–100. doi:10.1016/j.conb.2008.05.011

Parisi, D. (1996). Computational models of developmental mechanisms. In: R. Gelman and T. K, Au (eds), *Perceptual and cognitive development*. San Diego, CA: Academic Press, pp. 373–412.

Parisi, D. (1997). Artificial life and higher level cognition. *Brain and Cognition*, **34**(1), 160–84.

Pfeifer, R., and Bongard, J. C. (2006). *How the body shapes the way we think: a new view of intelligence*. Cambridge, MA: MIT Press.

Pfeifer, R., and Gómez, G. (2009). Morphological computation—connecting brain, body, and environment. In: B. Sendhoff, E. Körner, O. Sporns, H. Ritter, and K. Doya (eds), *Creating brain-like intelligence: from basic principles to complex intelligent systems*. Berlin, Heidelberg: Springer, pp. 66–83.

Prescott, T. J. (2007). Forced moves or good tricks in design space? Landmarks in the evolution of neural mechanisms for action selection. *Adaptive Behavior*, **15**(1), 9–31.

Prescott, T. J., Redgrave, P., and Gurney, K. N. (1999). Layered control architectures in robots and vertebrates. *Adaptive Behavior*, **7**(1), 99–127.

Prud'homme, B., Gompel, N., and Carroll, S. B. (2007). Emerging principles of regulatory evolution. *Proc. Natl Acad. Sci.USA*, **104**(Suppl 1), 8605–12. doi:10.1073/pnas.0700488104

Raff, R. A. (1996). *The shape of life: genes, development and the evolution of animal form.* Chicago: Chicago University Press.

Rakic, P. (2009). Evolution of the neocortex: perspective from developmental biology. *Nature Reviews Neuroscience*, **10**(10), 724–35. doi:10.1038/nrn2719

Sendhoff, B., and Kreutz, M. (1999). A model for the dynamic interaction between evolution and learning. *Neural Processing Letters*, **10**(3), 181–93.

Sims, K. (1994). Evolving virtual creatures. In: D.Schweitzer, A. Glassner, and M. Keeler (eds), *Proceedings of the 21st annual conference on Computer graphics and interactive techniques.* Association for Computing Machinery, Inc., pp. 15–22. doi:10.1145/192161.192167

Stanley, K. O. (2007). Compositional pattern producing networks: A novel abstraction of development. *Genetic Programming and Evolvable Machines*, **8**(2), 131–62. doi:10.1007/s10710-007-9028-8

Stanley, K. O., and Miikkulainen, R. (2003). A taxonomy for artificial embryogeny. *Artificial Life*, **9**(2), 93–130.

Swalla, B. J. (2006). Building divergent body plans with similar genetic pathways. *Heredity*, **97**(3), 235–43. doi:10.1038/Sj.Hdy.6800872

Torben-Nielsen, B., Tuyls, K., and Postma, E. (2008). EvOL-Neuron: Neuronal morphology generation. *Neurocomputing*, **71**(4–6), 963–72. doi:http://dx.doi.org/10.1016/j.neucom.2007.02.016

Turing, A. M. (1952). The chemical basis of morphogenesis. *Philosophical Transactions of the Royal Society of London Series B: Biological Sciences*, **237**(641), 37.

Vujovic, V., Rosendo, A., Brodbeck, L., and Iida, F. (in press). Evolutionary developmental robotics: improving morphology and control of physical robots. *Artificial Life*.

Wagner, G. P., and Altenberg, L. (1996). Perspective—complex adaptations and the evolution of evolvability. *Evolution*, **50**(3), 967–76.

Whitacre, J., and Bender, A. (2010). Degeneracy: a design principle for achieving robustness and evolvability. *Journal of Theoretical Biology*, **263**(1), 143–53.

Chapter 9

Growth and tropism

Barbara Mazzolai

Center for Micro-BioRobotics, Istituto Italiano di Tecnologia (IIT), Italy

Plants have conquered almost all surfaces on our planet; they were the first settlers in hostile environments, making way for habitats that could then be settled by nearly all living beings. Plants, as sessile organisms, spend their entire lives at the site of their seed germination. Consequently, they require a suite of strategies to survive extremely diverse environmental conditions and stresses. Some may think that plants do not actually move unless they grow; however, this is incorrect. Plants move a great deal, although they typically move in a different time frame than animals. Plants are able to move their organs, often in response to stimuli, and they have developed a variety of motion systems, primarily as a result of water absorption and transport. Despite their exceptional evolutionary and ecological success, plants have rarely been a source of inspiration for robotics and artificial intelligence, likely due to misconceptions of their capabilities and because of their radically different operational principles compared with other living organisms. This chapter will describe several plant features that can be translated to technological solutions and, thus, open up new horizons in engineering by offering inspiration for developing novel principles to create growing, adaptable robots and smart-actuation systems.

Biological principles

One of the fundamental characteristics of plants is the correlation between growth and movement. The growth process is mostly associated with development in animals, although growth can occur for both development and movement in plants. Movements in plants can be classified on the basis of reversibility or irreversibility; of their active (triggered by action potentials) or passive (primarily based on already dead tissues) nature (Burgert and Fratzl 2009); of their independency from (nastic response) or influence by (tropic response or tropism) the spatial direction of stimulus; or of the different timescales of water transport in the cellulosic wall (Dumais and Forterre 2012).

Tropisms play a crucial role in plant fitness because they allow plants to use limited resources and avoid unfavourable growing conditions. In tropic responses, the direction of stimulation is extremely important and can either be positive (i.e. towards the stimulus) or negative (i.e. away from the stimulus; Esmon et al. 2005). There is a wide range of tropisms in plants, including phototropism (light), gravitropism (gravity), thigmotropism (touch), thermotropism (temperature), chemotropism (chemicals), and hydrotropism (water/humidity gradient).

Behavior in plant roots. To fulfil its primary needs and survive, an organism must exhibit adequate behavior with respect to the environment. This behavior is the result of the interaction between an organism and its environment. To understand the behavior, specific attention should be paid to morphological properties: interaction is primarily guided by the morphological characteristics that define a certain range of movements and the ability to explore the environment.

Morphological adaptation plays a primary role in plant fitness, in particular, the (i) modularity and redundancy of structural and functional units, (ii) shape alteration in the root tip, and (iii) rich and distributed sensing, which are the basis of high-performance exploration.

The root system architecture of plants varies substantially between species and also shows extensive natural variation within species. As a consequence, the extension in space and time of the root system is governed by genetically driven developmental rules, which are modulated by environmental conditions. As a result, genetically identical plants can have extremely different root architectures. A general plant root system consists of a primary root, lateral roots, adventitious roots, and root hairs. Plant roots perform multi-faceted roles, of which two functions are foremost: (i) firmly anchor the plant into the soil, and (ii) supply water and nutrients to the aerial parts of the plant.

Living roots are capable of penetrating a variety of soils, including rocks and dense ground, because of their peculiar growth strategy: roots penetrate the soil by elongating the meristematic area (the region situated a few mm behind the tip) by cell mitosis and cell expansion (Bengough et al. 1997). In this way, plants avoid generating high pressures associated with high-energy consumption, which would be necessary for a penetration system operating from the surface, and prevent damage due to friction processes, which would occur during advancement. The cell expansion in the elongation region produces an axial pressure, which is dissipated when the soil cavity expands and from the frictional resistance of the soil during the root advancement.

Another important characteristic of plant root growth is the reduction in soil–root friction by producing mucus and releasing cells from the tip. In addition, plant roots grow by extending hairs into existing pore spaces into the soil and moving soil particles aside in the radial direction, increasing the anchoring capabilities of the root to the soil. (See Figure 9.1 A, B.)

Actuation in plants. The most familiar *biological actuator* is the muscle. Muscle movement results from an arrangement of micro-metric contractile filaments. However, muscles are not the only actuators in nature. Many living organisms perform a variety of movements despite lacking muscles; this is the case for plants. The lack of muscles has led plants to exploit alternative strategies to perform movements essential for their survival. Plant cells are surrounded by a thin, stiff cell wall made of cellulose microfibrils embedded in a pectin matrix. This stiffness is the basis of the capability of plant cells to sustain a large, internal hydrostatic pressure, known as turgor. This pressure originates from an osmotic gradient between the cell cytoplasm and environment (Dumais and Forterre 2012). Osmosis plays an unquestionable, key role in plant actuation systems. In particular, cell turgor is generally responsible for slow, small-scale movements. In fast, large-scale movements, osmosis acts more as a "trigger" of a cascade sequence of elastic instability movements of smart, natural engineered structures. An analysis of these mechanisms reveals that the engineering of soft, non-muscular, hydraulically actuated systems for rapid movement requires either a small size or enhanced motion on large scales via elastic instabilities (Sinibaldi et al. 2013). Thus, the elastic release of energy acts as a "speed-booster". A few examples of these fast and slow movements include the rapid (∼100 ms) closure of *Dionaea muscipula* (Venus' flytrap), partly actuated by an abrupt decrease in internal tissue pressure (Burgert and Fratzl 2009); leaves that close by touch stimuli in *Mimosa pudica*, which occurs within 20 ms from when they are touched (Martone et al. 2010); the impressive pollination mechanism of *Stylidium*, in which the gynostaemium flips rapidly (∼25 ms) to hit and pollinate insects (Hill and Findlay 1981); and the remarkable 4.5-MPa actuation pressure exhibited by stomatal guard cells during the closing phase, whose objective is to prevent water loss (Hill and Findlay 1981).

An improved understanding of these movements will push developments forward, both in applied sciences and engineering and, in particular, in creating novel biomimetic actuation

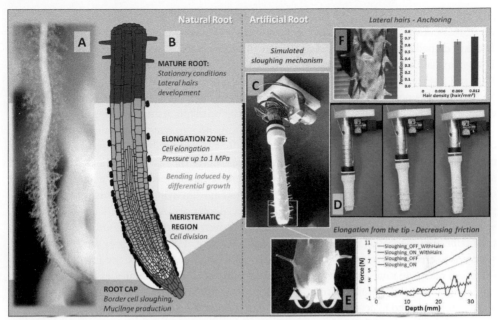

Figure 9.1 An overview of growing mechanisms in plant roots and their artificial implementation counterpart. A) Plant root apex of *Zea mays*. B) Growing mechanism in plant roots: growing from the tip (cells sloughing); bending from differential elongation; root anchorage by lateral hairs. C) Prototype of an artificial root using the sloughing mechanism for soil penetration. D) How the artificial sloughing mechanism works. E) Detail of the flexible skin released from the tip for implementing the sloughing mechanism: this mechanism greatly decreases the force required by the artificial root to penetrate into the soil. F) Anchorage exploited by lateral hairs: greater hair density leads to higher penetration performances.

strategies characterized by high-energy efficiency and low power consumption (Burgert and Fratzl 2009; Martone et al. 2010).

Biomimetic or biohybrid systems

Robotic solutions inspired by plant roots. The development of robotic artifacts capable of moving autonomously and efficiently exploring an environment in a non-destructive way, as plant roots do, poses new and bold challenges in robotics and materials research. The mechanical properties of plant roots and the morphology of their structure have been considered in developing the first artificial root apex, based on the plant-inspired actuation principle (i.e. osmotic actuation; Mazzolai et al. 2011). This artificial apex has a conical shape to favour penetration and includes two sensors for gravity and humidity detection and behavior control. The prototype was made from acrylic material and is actuated hydraulically. Sensory feedback is used to achieve proper steering during hydrotropism and gravitropism, which were tested and validated both in air and in soil.

More recently, a robotic system inspired by low-friction penetration strategies in plant roots has been proposed. Cell growth at the root tip deforms soil, while sloughing cells in the cap create an interface between the root and soil to reduce root–soil friction during penetration. A simple prototype, inspired by these root features, is based on a tubular shaft and a soft continuum skin

(Sadeghi et al. 2013; see Figure 9.1 C–F). The skin slides from the inner to the external part of the shaft. This outward movement of the skin opens the soil in front of the tip and penetrates the soil. The skin embeds soft hairs at its surface that provide the prototype with self-anchorage capabilities. The performance of this robotic system was characterized during penetration in granular soils. The skin–soil interaction was fundamental for 1) displacing the soil in front of the tip and 2) preventing backward movements of the robot by anchoring the posterior body to the soil. An increased hair density (0.012 hairs/mm^2) resulted in a higher penetration depth of the robot (approximately 30%).

Plant-inspired actuators. Actuation represents a bottleneck in many current engineering applications, including bio-robotics. The available actuators are primarily electromagnetic, and their performance is far from that achieved by natural actuators. The main limitations involve inertia and back-drivability, stiffness control, and power consumption. Nevertheless, new and promising technologies are emerging from study of and inspiration from living organisms, which offer new possibilities to fill the gap between natural and artificial solutions. Among these biological models, several plant features, such as hydraulic movements (driven by osmotic or humidity gradients) and the material properties and geometry of the cell wall, have already been proposed as innovative solutions to create new actuators characterized by remarkable energy efficiency and high-actuation force. An adequate understanding of plant osmotic-driven actuation strategies has already led to modeling a new concept of an osmotic actuator (Sinibaldi et al. 2013; Figure 9.2). The proposed approach has helped to determine explicit scaling laws for the actuation figures of merit, namely, characteristic time, maximum force, peak power, power density, cumulative work, and energy density. These expressions, which are based on analogous properties in plants, required several basic design considerations, i.e. preliminary dimensioning of the envisaged osmotic actuator based on design targets/constraints (e.g. assigned actuation characteristic time and/or maximum force). The role of the volume-to-surface aspect ratio was elucidated in terms of actuator performance.

An example of a forward osmosis actuator based on the analysis of plant movements and on the osmotic actuation modeling previously described is reported in Sinibaldi et al. (2014). The

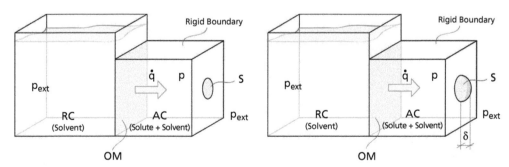

Figure 9.2 Schematic of the osmotic actuator concept. Left: The osmotic actuator is composed of a reservoir chamber (RC), an actuation chamber (AC), and an osmotic membrane (OM). The osmotic pressure generated by the difference in solute concentration between the solvent (RC) and solution (AC) induces a solvent flux (ϕ) through the osmotic membrane. The deformable bulging membrane (S) exploits the energy associated with the osmotic process into an actuation. At the beginning, the bulging membrane is flat. Right: After a while, the solvent flux induces the bulging of the membrane (displacement δ).

Adapted from E Sinibaldi, G L Puleo, F Mattioli, V Mattoli, F Di Michele, L Beccai, F Tramacere, S Mancuso, and B Mazzolai, Osmotic Actuation Modeling for Innovative Biorobotic Solutions Inspired by the Plant Kingdom, *Bioinspiration and Biomimetics*, 8(2), 025002, doi: 10.1088/1748-3182/8/2/025002. © IOP Publishing. Reproduced by permission of IOP Publishing. All rights reserved.

system achieves actuation timescale (2–5 minutes) comparable to that of a typical plant cell, and produces forces above 20 N, while containing the power consumption (in the order of 1 mW).

Another promising approach with the objective of overcoming several of the limitations of current actuation solutions is based on so-called smart materials. Smartness is a feature typically referring to biological systems that are capable of interacting adaptively with the environment, and plants are the best example in nature of this capability. Analogously, the "smart materials" term is applied to those materials that can reversibly modify one or more (functional or structural) properties responding to external stimuli or to changes in surrounding conditions.

The fibrillar organizations of the plant cell wall have inspired the design of a fluidic, flexible matrix composite (F^2MC) cell concept, a novel adaptive structural element for new actuators (Li and Wang 2012). This novel, bio-inspired system consists of two F^2MC cells with different fibre angles connected through internal fluid circuits. A non-dimensional dynamic model was developed to identify crucial constitutive parameters for the F^2MC cellular design.

Shape memory alloys (SMA) have been used to develop a Venus' flytrap-inspired robot: the system mimics the fast snap-through motion observed in *Dionaea muscipula* (Figure 9.3 A) using

Figure 9.3 Venus' flytrap-inspired mechanisms. (a) Image of *Dionaea muscipula* (Venus' flytrap). (b) A bio-mimetic flytrap robot. The robot uses a bistable asymmetrically laminated carbon fibre-reinforced prepreg (CFRP) structure, which has a bistable mechanism similar to the Venus' flytrap's passive elastic mechanism. By embedding shape memory alloy (SMA) springs, a large deformation is induced, and the bistable structure can be triggered to snap-through. This concept of the flytrap robot can be applied to rapid grippers of various sizes.

(a) © Stefano Mancuso, 2017. (b) © 2014 IEEE. Reprinted, with permission, from Seung-Won Kim, Je-Sung Koh, Jong-Gu Lee, Junghyun Ryu, Maenghyo Cho and Kyu-Jin Cho, Flytrap-inspired robot using structurally integrated actuation based on bistability and a developable surface, *Bioinspiration and Biomimetics*, 9(3), p. e036004, doi: 10.1088/1748-3182/9/3/036004.

bistable laminate embedded with an SMA spring actuator to generate the bending moment. This structure is called BIMAC: Bistable Intelligent Morphing Active Composite plate. The robot can open and close its bistable laminate leaves repeatedly with a closure time of approximately 100 ms, which is similar to its natural counterpart (Kim et al. 2010) (Figure 9.3).

The observation and investigation of plant movements have revealed a plethora of physical and mechanical principles that are extremely interesting for creating smart actuators. The exploitation in artificial systems of osmosis or of the hierarchical structure of the cellulose layer in cell walls will allow the implementation of movements that are primarily driven by changes in environmental conditions and, thus, do not require further control and energy supply, creating new perspectives in engineering.

Future directions

The study of plant features to take advantage of their biological and engineering tools and methods can lead to the development of new soft plant-inspired robots that are able to grow by adding new materials while they search for specific targets. These robots will require the development of new actuation solutions for steering and elongation of the robotic root, soft sensing, control and robotic architectures (distributed control, coordination of many degrees-of-freedom), and kinematics models.

Interestingly, concerning plant root behavior, it seems that root apices, which include sensing mechanisms, act as decision-making centres, which is supported by the observation that the growth pattern of the root system (e.g. root swarming or proliferation in favourable patches) is not random and is actually coordinated and efficiently shaped to exploit soil resources and to avoid hazards (Ciszak et al. 2012). Within this perspective, a group of apices in a given plant could be considered a "colony" whose individual elements work for the benefit of the colony, akin to colonies of animals such as social insects (see Nolfi, Chapter 43, this volume). Similar to a flock of birds, it has been demonstrated that collectively, roots can be influenced by their neighbours and can align their growth directions (Ciszak et al. 2012). The occurrence of this swarming behavior may be an advantage for exploratory purposes. Thus, because plant roots address multiple mechanical, sensory, and communication constraints as they grow and react to the environment, future directions in science and engineering will be in the development of novel, bio-inspired methods of sensor fusion and collective decision-making in decentralized structures with local computation, simple communication, and a high-level of control of a new generation of growing, soft robots inspired by plant roots.

Applications for technologies inspired by plants include soil monitoring, penetration, and exploration in on-Earth and extraterrestrial environments, and could also include medical and surgical applications. For example, new flexible endoscopes, able to steer and grow, could be envisioned for safer operations on delicate structures, such as the central nervous system.

Learning more

An interesting view of how plants perform actuation, including systems activated without any metabolism, can be found in Burgert and Fratzl (2009). This work also revealed that actuation in plants can offer several insights into the design of innovative, bio-inspired devices. A review to better understand the principles of the basis of fluid mechanics that allow plants to achieve movements has been given by Dumais and Forterre (2012). Details on the first attempt to develop a mechatronic system that behaves as a plant root apex for gravitropism and hydrotropism are described in Mazzolai et al. (2011). Recent works on soft plant-inspired robots endowed with bending and growing abilities are proposed in Sadeghi et al. (2016), Laschi et al. (2016), and Sadeghi et al. (2017).

Many interesting details can be found in Darwin's pioneering works on plant movements and behavior, which represent important milestones in plant research. More information is available in other works by the authors cited in this chapter and from works of other scientists that are active in plant biomechanics and behavior.

Acknowledgments

The author's work reported here was funded in part by the Future and Emerging Technologies (FET) programme within the 7th Framework Programme for Research of the European Commission, under the PLANTOID project, FET-Open grant number 293431. The author thanks Dr. Lucia Beccai and Dr. Virgilio Mattoli for helpful comments on the manuscript.

References

Bengough, A. G., Croser, C. and Pritchard, J. (1997). A biophysical analysis of root growth under mechanical stress. *Plant and Soil* **189**, 155–64.

Burgert, I. and Fratzl, P. (2009). Actuation systems in plants as prototypes for bioinspired devices. *Philos. Trans. R. Soc. Lond. A* **367**, 1541–57.

Ciszak, M., Comparini, D., Mazzolai, B., et al. (2012). Swarming behavior in plant roots. *PLoS One*, 7(1), e29759.

Dumais, J. and Forterre, Y. (2012). "Vegetable Dynamicks:" The Role of Water in Plant Movements *Annu. Rev. Fluid. Mech.* **44**, 453–78.

Esmon, C. A., Pedmale, U. V., and Liscum, E. (2005). Plant tropisms: providing the power of movement to a sessile organism. *International Journal of Developmental Biology* **49**, 665.

Hill, B. S. and Findlay, G. P. (1981). The power of movement in plants: the role of osmotic machines. *Q. Rev. Biophys.* **14**, 173–222.

Kim, S.-W., Koh, J.-S., Cho, M. and Cho K.-J. (2010). Towards a bio-mimetic flytrap robot based on a snap-through mechanism. *Proceedings of the 2010 3rd IEEE RAS & EMBS International Conference on Biomedical Robotics and Biomechatronics*, The University of Tokyo, Tokyo, Japan, September 26–29.

Laschi, C., Mazzolai, B., and Cianchetti, M. (2016). Soft robotics: Technologies and systems pushing the boundaries of robot abilities. *Science Robotics* 1(1), eaah3690.

Li, S. and Wang, K. W. (2012). On the dynamic characteristics of biological inspired multicellular fluidic flexible matrix composite structures. *Journal of Intelligent Material Systems and Structures* 23(3), 291–300.

Martone, P. T., Boller, M. and Burgert, I. et al. (2010). Mechanics without muscle: biomechanical inspiration from the plant world. *Integr. Comp. Biol.* **50**, 888–907.

Mazzolai, B., Mondini, A. and Corradi, P. et al. (2011). A miniaturized mechatronic system inspired by plant roots for soil exploration. *IEEE/ASME Trans. Mechatronics* **16**, 201–12.

Sadeghi, A., Tonazzini, A., Popova, L., and Mazzolai, B. (2013). Robotic mechanism for soil penetration inspired by plant roots. *Proceedings of the 2013 IEEE International Conference on Robotics and Automation (ICRA)*, Karlsruhe, May 6–10.

Sadeghi, A., Mondini, A., Del Dottore, E., et al. (2016). A plant-inspired robot with soft differential bending capabilities. *Bioinspiration and Biomimetics* 12(1), 015001.

Sadeghi, A., Mondini, A., and Mazzolai, B. (2017). Towards self-growing soft robots inspired by plant roots and based on additive manufacturing technologies. *Soft Robotics*, in press.

Sinibaldi, E., Puleo, G.L., Mattioli, F., et al. (2013). Osmotic actuation modelling for innovative biorobotic solutions inspired by the plant kingdom. *Bioinspiration&Biomimetics* 8(2), 025002.

Sinibaldi, E., Argiolas, A., Puleo, G.L., and Mazzolai, B. (2014). Another lesson from plants: the forward osmosis-based actuator. *PLoS One* 9(7): e102461.

Chapter 10

Biomimetic materials

Julian Vincent

School of Engineering, Heriot-Watt University, UK

Biological materials, if we take their density into account, are as stiff and strong as man-made materials (Wegst et al. 2015). Although they cannot take the very high stresses and temperatures of some high-performance ceramics and metal composites, they are more durable, more tolerant of damage, made at ambient temperatures, and can easily be recycled. More important, they are invariably adapted to their particular function, partly through genetic control and partly through responses to their working conditions. They therefore tend to be inhomogeneous, changing from point to point within the structure of which they are a part.

Biological materials work at a number of levels. Most obviously they provide mechanical support and integrity. Very often they are multifunctional, so that new wood transports water and becomes a support when it dies; the centre of mammalian bones doesn't contribute to bending stiffness but it's hollow, houses bone marrow, and produces red blood cells. The materials which cover organisms have to be resistant to mechanical and chemical damage, are often skeletal in some way, and have specialized secondary functions such as drag reduction (shark skin, feathers), water repellence (most plant surfaces), camouflage, or thermal control.

In performing all these functions, biological materials and structures are constrained by their evolutionary origins. These constraints emerge as specialization, and commonly reduce the organism's reliance on other systems such as nervous or muscular control.

Biological materials and principles

Biological materials are nearly all composites of fibers embedded in a more-or-less organized matrix that transfers force from fiber to fiber. The literature on man-made composites is huge simply because composites represent such a successful way of assembling a material. Biological composites conform to the same rules as man-made composites, but are often more subtle and versatile. In vertebrate animals the main fiber is collagen, itself a hierarchical composite. The basic molecule is tropocollagen, which has no particular conformation. When combined with two other molecules of tropocollagen it forms a stable triple helix that is sufficiently stiff in bending to show liquid crystallinity and self-assembly. A collagen fiber has a Young's modulus of between 2 and 10 GPa, but the stiffness of collagenous materials depends on the proportion of collagen present (volume fraction) and its orientation. The collagen is embedded in a matrix of hydrated polysaccharides. Collagenous materials are characterized by their high toughness. Examples are mammalian skin, tendon, and the egg capsule of elasmobranchs (sharks, rays, and skates). Collagen is often found in conjunction with a rubbery protein—elastin—that is found in skin, arteries, and ligaments. The outer layer of skin is composed of alpha keratin, an intracellular protein, so the material is formed of fiber-filled cells closely stuck together. Alpha keratin forms other dermal structures such as hoof, horn, baleen, and hair. Its Young's modulus when dry is about 10 GPa.

Invertebrates in general use a different fiber in their composites—the polysaccharide fiber, chitin. The strongly linear molecules of chitin are tightly H-bonded to form crystallites with 19 or more chains, about 3 nm in diameter and 100 nm or more long. Opinion differs as to the stiffness of these crystallites, which might be as high as 150 GPa (comparing it with cellulose, a very similar molecule which has a stiffness of about 135 GPa) or only 40 GPa (a measured value). Chitin is typically found in the external coverings of arthropods, embedded in a protein matrix that may also be calcified (in crustaceans). The bonding between the protein and chitin is very specific; there are areas of silk-like beta-sheet on the protein that closely match the spacing of the hydroxyl groups on the chitin crystallite. The protein may or may not be hydrated, which governs the stiffness of the composite. Hydration is very probably controlled by hydrophobic phenolics such as Dopa, which drive the proteins to form cooperative structures, extruding the water. A very similar chemistry occurs throughout invertebrates, yielding extracellular materials that are coloured brown or black. Mussel byssus thread (Waite et al. 2005) and dogfish egg-case are typical. Some very hard and durable materials can result from this chemistry; the sharp tip of the beak of squid and octopus is an example that is as hard as metal (Miserez et al. 2010).

More durable and harder materials result from the addition of calcium salts. Bone contains hydroxyapatite (HAp), a hydrated phosphate of calcium. It is present as small platelets that, individually, are almost impossible to break. Up to a volume fraction of about 0.3 the platelets are associated with regular gaps in the collagenous structure; at higher volume fractions they are formed outside this structure. This increase in volume fraction tends to reduce the space available for water, so although antler, with a low volume fraction of HAp, can be somewhat flexible and very tough, the increased internal connectivity of stiffer bone also makes it brittle, since cracks can move more easily across the internal structures. This trade-off between stiffness and toughness is a recurring theme in materials science, but biological materials have developed numerous ways of circumventing it. Bone mostly does it by allowing microcracks to exist within the structure that (crudely put) intercept and divert larger cracks (Zioupos et al. 2008). A crack needs strain energy to propagate, and biological materials have evolved methods by which that energy can be used up remote from a primary site of fracture or blocked from reaching the fracture site. Parallel fibers in a softer matrix stop the development of a stress concentration. In horn and hoof the intermediate filaments are highly oriented; this allows fracture between the fibers with relative ease, but across the fibers it is very difficult to propagate a crack since the stress cannot be transmitted through the softer matrix. Thus the bony horn core of bovids is well protected in a fight; a crack in the horn sheath rarely becomes large enough to cross into the bone. In mollusc shells the uneven surface texture of the platelets of mother of pearl (nacre), combined with the lubrication provided by the thin layers of protein between them, allows the platelets in the stretched nacre to move (absorbing energy) but to lock up before fracture strains are reached (Barthelat and Rabiei 2011). Another part of the strained nacre then deforms and locks up in turn, so that eventually the entire piece of shell has absorbed energy and nowhere has the energy been sufficiently large or concentrated to be able to start a crack. The strongest material seems to be in the radular teeth of snails (specifically the limpet) that are made of chitin crystallites in a tanned protein matrix with mineral filler. A micromachined test piece had a tensile strength of about 5 GPa and stiffness of up to 180 GPa. This is comparable with high performance carbon fiber and is better than silk. Clearly biological materials can display impressive mechanical properties, and typically have a density significantly less than half that of many metals.

Plant materials are equally impressive. Their primary construction is of water-containing cells surrounded by a wall of cellulose fibrils. The generation of morphology is largely controlled by the local orientation of the cellulose, allowing the cell to deform in various ways under the influence of the internal turgor pressure. This pressure is typically 5 to 10 bar. Thus the cell wall is under continual strain; the stiffness of an herbaceous (unlignified) plant is generated by the resistance of this pre-stressed structure to external forces. Tensile loads are well within the strength limit of the cellulose cell walls; compressive loads are paid off against the prestrain. If the plant is long-lived (i.e. biannual or perennial) the turgor system is usually replaced by lignification (similar to phenolic tanning). This stabilizes the cell walls by providing a stiff matrix for the cellulose fibers, and also tends to drive out the water, encouraging cooperative interactions. The ultimate result is wood, which has long been lauded as the material of choice for the highest flagpole (Greenhill 1881). It does this in part because it is relatively lightweight, having lost all its water, and is cellular (50–80% empty space). The long cells of the xylem in wood give extremely good fracture control (Gordon and Jeronimidis 1980), transmitting forces away from the primary fracture site and increasing the nominal surface area of a crack (hence the energy for fracture) by a factor of about 200. Many plants have fibers arranged parallel in a weak matrix of parenchyma cells, which defeats the formation of a stress concentration. Examples are liana when in its tension phase of life, and the g-fibers of angiosperm trees that form tension wood.

Differentiation between material and structure is impossible in biology. The ability of a biological material to perform properly within a macrostructure such as a skeleton is largely due to its microstructure. At static equilibrium, total tension and compression must be equal and oppositely signed vectors. In bending both tension and compression are experienced, as is shear. Since shear is more difficult to resist, either in bending or in torsion, it is mostly avoided. Longitudinal compression of isolated members can induce instability (Euler buckling) that can lead to failure. These considerations suggest that the simplest and most reliable structures have a tensile membrane (a stable, well-defined, component) surrounding a compressed fluid (less well defined, both in shape and lability), nearly always aqueous (and thus easily obtained). This is the arrangement seen in the majority of animals, especially the "lower" animals and larval forms, and in the non-lignified parts of plants. If the structure is to move (or be moved) the forces to be accommodated are less predictable, and it is advantageous for the structure to be able to redistribute the forces as they impinge on it. This leads to a concept rather like Buckminster Fuller's tensegrity (although this term is currently being misused by many people) where struts (which take compression) are isolated in a web of elastic tensile elements. If the tensile elements can also contract (muscles) then the deformability of the entire structure can be intimately controlled and do work. Since the struts are not attached to each other they experience no bending moments and can move relatively freely. Tensegrity structures are ideal candidates for deployability.

Deployability enables a structure to be kept safe until needed, when it can be extended or enlarged. Examples are wings of birds and insects (though note that many primitive insects such as some dragonflies can't fold their wings and so can't hide in the undergrowth; they are "confined" to open spaces), leaves of many plants, and the male intromittent organ.

Finally, biological materials frequently have sensory function built in. The sensory receptor has to be in contact with the nervous system, but the signal it transduces can be conditioned or amplified by structures that are otherwise insensate. This is found best in arthropods, which have to transmit information across the sensory shield of a cuticular exoskeleton. There are many types of bristle that can respond to direct displacement, or to movements of air or water, or resonate in response to external vibrations (Humphrey et al. 2007). They can be tuned to respond to a narrow range of stimulus so that the receptor cell simply has to respond to an off/on signal. An assembly of receptive hairs each responding to a slightly different range of a stimulus can thus generate a digital response, much simplifying the requirements of the nervous system. Another form of sensory conditioner is controlled compliance of areas of the cuticular exoskeleton by the introduction of more or less oval holes (e.g. campaniform sensilla in insects) or slits (lyriform organ or slit-sensilla of arachnids; Young et al. 2014). The sensory signal is a change in the aspect ratio of the hole as it is deformed. The change is rotated through 90 degrees by an arched membrane sealing off the hole. As the hole is deformed the membrane moves up and down and this displacement is sensed. A displacement of the order of nm can be detected. In conjunction with the orientation of the fibrous component of the cuticle (Skordos et al. 2002), the compliance of the cuticle can be made very directional, so that a field of orientated holes or slits of different sizes can pre-process a signal in many ways.

Biomimetic materials

What is the advantage, if any, of biological, or biomimetic, materials over more conventional ones? We have great skills and industries based on metals and plastics and can produce and manipulate them with ease and accuracy. What can biology offer? It helps to consider the constraints inherent in biological and artificial materials. Is there any complementarity that could indicate advantage?

Biological materials have complex and variable properties. In part this is due to their responsive mode of synthesis—at least where the material is still within the metabolic pool and has cells which can sense and respond to demands of use by means of modification and repair. This adaptiveness does not occur throughout nature; heart-wood, antler bone (after the velvet has sloughed away), nail, hair, wings of many insects (e.g. flies and bees) are all materials which are not in contact with living cells and therefore are not repaired. One supposes that evolution has built in a larger safety factor where necessary, though this cannot nullify eventual obsolescence. In animals with a well-developed nervous system, accuracy of manufacture of skeletal structures is needed only at the articulations, which must be able to take proper loads, move properly, etc. For the rest of it we learn how to use our bodies as they are provided to us, adapting movement and control to produce the desired output. This reduces the need for initial quality control of the shape of the skeletal elements. By contrast animals such as insects rely very much on the intrinsic properties of their materials and structures; I have not found evidence of locusts needing to learn how to walk or fly. All the same, this highlights the importance of feedback from sensors that can be an intrinsic part of the material. Arthropods are probably the best paradigm since their skeleton is exterior and made of a (relatively!) simple fibrous composite. The campaniform sensillum has been modelled both theoretically and practically. Its main advantages are:

a) simplicity (it's a carefully designed hole);

b) adaptability (the hole can be any shape you like, but is usually oval, and so can be orientated to amplify strains in one main direction);

c) versatility (a group of sensilla orientated in two or three primary directions can detect strains in any direction by comparison of their responses);

d) robustness (in context: although the sensilla tend to be placed where strains—and therefore stresses—are the greatest they must not act as stress concentrators although their function is to concentrate strains. This can be achieved by controlling the size of the hole and the orientation and continuity of fibers in the surrounding material); and

e) sensitivity (the deformation of the hole can be measured in a number of ways, including remotely by laser).

Attempts to use arthropodan bristles of various sorts seem to have been limited by the insistence of using stiff materials for the articulation, which rather negates the concept of using the movement of the bristle as the sensing action. The ideal would seem to be a compliant material (insects use resilin, a rubbery form of cuticle). The performance of these sensors can be tuned not just by their geometry, but by anisotropy and non-uniform distribution of mass within the material.

Robots mostly move. Their internal skeleton can be made of almost anything, depending on weight, strength, and ease of fabrication. Tools such as the Ashby Materials Selector can provide the essential information for the designer. If a skin is needed, for protection, containment, etc., things get more difficult since the combination of toughness and flexibility presents problems. The most obvious solution is overlapping armour plates suspended in a flexible material. Examples in nature are fish that can have dermal bones (Osteolepids) and sea-horses (Porter et al. 2013), scaly fish, armadillos (Chintapalli et al. 2014) and woodlice. However, on the whole flexible armour is not so effective as a flexible skin properly toughened with fibers. Probably the crucial design feature is that the fibers should be connected so that they can distribute, and thus dissipate, local stresses, and that the fibers should be relatively unconstrained by the matrix, which must therefore be rather compliant. A body that is more or less cylindrical is best dressed with a net made of two counter-wound helices rather like a fish-net stocking. If the fibers are connected only by a soft matrix they can move and accommodate to a variety of shapes. A prime example is the skin of sharks that doubles in duty as an exotendon (Wainwright et al. 1978). The basic design is therefore a scaffold made of fibers impregnated with a low modulus matrix. This general approach has been successful in the design of skin analog for the arm of a robotic octopus (Hou et al. 2012). Commercial tights are made of knitted nylon; a 10-denier nylon was chosen as the scaffold. It was impregnated with a mix of silicone rubbers yielding a tube made of a thin and waterproof membrane. The Young's moduli of the membrane at low extensions were 0.08 MPa (longitudinal) and 0.13 MPa (transverse), which are much lower than those of the octopus skin and allowed the robotic arm to extend easily. Strength and fracture toughness are higher than those of real octopus skin. Modern knitting designs are capable of creating remarkably complex three-dimensional structures, such that the entire external shape of a robot can be realized in a single component. This can then be impregnated with matrices of differing stiffnesses, which can give a range of properties over the structure. In general, our understanding of handling fibers and assembling them into knitted or woven structures is hardly utilized in robotics.

With a fibrous scaffold controlling the ultimate deformability, and holding all the components together, the possibility emerges of a multifunctional exoskeletal structure with lots of space inside. Such an exoskeleton can be powered by compressed air, which is much lighter than a solid component. If the arrangement of fibers in the scaffold is asymmetric, then the actuated scaffold can be made to bend and turn and perform complex manoeuvres without the need for hinges and bearing surfaces. Such soft actuation more nearly conforms to current concepts of how end effectors should work. There are many possibilities. A few have been realized by Festo, but they have confined themselves to the crossed-helices of the McKibben actuator. Asymmetrically arranged fibers also occur in plants, where they can limit the extensibility or deformability of a part of a structure. If the matrix around the fibers expands on hydration, then the structure can change shape or position as surrounding moisture changes. Thus the fibrous asymmetry allows the structure to become a chemical engine, transducing hydration into work. If the matrix is suitably chosen, heat or light can have the same effect. This sort of effect is used in the self-erecting "origami" robots recently developed in MIT. Shape change can be used for direct actuation, locomotion, or as part of a switch mechanism. Associated with strain energy storage in a bistable structure, it can also be used to achieve power amplification without the need for an extra energy source (Forterre 2013; Hayashi et al. 2009).

In biology, once the basic molecules have been synthesized, the generation of materials is nearly all assembly at ambient temperatures. Such self-assembled structures tend to be stable due to extensive and cooperative H-bonding (melting temperature around 60 °C) rather than the relatively few high-energy bonds that typify plastics (melting temperature 120 °C or higher). Properly assembled, such materials can have very impressive properties considering the apparently weak molecules from which they are constructed. Their durability comes from the cooperative effects that well-directed assembly can engender. The bonds employed in biological materials are relatively low energy, and therefore require less energy to make and to break individually, although there are more of them. Thus breakdown and recycling can be at ambient temperatures. In a world where energy and material resources are ever more expensive, this low-energy approach is important. Note that when materials are broken down and recycled in nature, there are restrictions that may not apply to engineering. For instance, the breakdown products become the food of the next generation, but unless those products are broken down to small molecules (an amino acid, for instance, with MW ~ 120 Da) they might retain enough structure to be immunogenic or poisonous. Such restrictions need not apply to technical uses, so molecules could be designed to retain more of their functionality for recycling.

Self-assembly requires a template or mold of some sort that for our purposes can be manufactured using traditional methods such as knitting, weaving, felting, and paper-making. Introducing a second phase into this scaffold depends on diffusion and the usual problems associated with avoiding internal voids. A very durable analog of nacre, with very similar mechanical properties, has been made using particles of clay (montmorillonite) assembled using paper-making technology or coating (Liu and Berglund 2012; Walther et al. 2010). This ticks nearly all the boxes, being durable, made of easily-sourced components, recyclable (if the adhesive polymer is xyloglucan, from plant cell walls, or similar), and assembled at ambient temperature. An exciting signpost for low-energy high-performance materials.

An increasingly significant method for the production of structures which can use recyclable materials such as poly(lactic acid) is 3D printing—additive manufacturing or rapid prototyping—which allows the generation of complex shapes, some impossible to make any other way, at half the weight. With multiple printing heads it is possible to deposit a range of materials with different properties so that, for instance, a rubbery hinge can be interposed between two stiff components in the way that a skeleton might be articulated. Additive manufacturing can also be cheaper and quicker, requiring less machinery. However, it has the disadvantage that the materials are amorphous and relatively weak. At the very least the technique needs to be able to incorporate fibers into the plastic matrix. An elegant way to do this would be to use block copolymers that can separate into phases either on deposition or by post-processing such as annealing. The morphology of the phases can be controlled by volume fractions of the starting components, and their geometry by post-processing under applied strain. This would be closely analogous to much of the self-assembly of biological materials, where the phase separation is based on hydrophobic interactions within a hydrating environment. A nice paradigm is the deposition of insect cuticle, where the components (protein and chitin) are secreted into a hydrated zone (the Schmidt layer) where they have the freedom to self-associate and assemble at a concentration of about 10% solids. They are then incorporated into the bulk of the developing exoskeleton where they can be further processed and dehydrated.

Future directions

Biological materials are to a large extent soft and adaptive. They retain their coherence and durability by means of a network of fibers that are stiff in tension but flexible—like rope—that is intimately bonded with a compliant matrix. Even stiff materials such as bone and nacre have a matrix that has, or can have, more of the qualities of a lubricant than of a solid or glue. Such soft materials are unusual in technology, possibly because animals and plants usually yield to high loads rather than fighting them. This way they need to invest less material in their structure and can be more durable. They can do this in part because they can repair themselves, but mainly because they can sense the loads. This sensitivity allows them to reduce energy required to sequester and metabolize materials, and also reduces the payload that has to be transported.

There is another aspect to softness and adaptiveness. Although collagen (and other fibrous materials) is stiff in tension, it is flexible. Thus it can be bent and twisted with no harm done. With an appreciative nod to Jim Gordon, a good model of a skeleton is the rigging of a sailing ship. The compressive members are few and massive; the tensile members are many, distributed, and much smaller. The aim is to reduce bending loads on the main members by adjusting the tensile members. The tensile members thus need to have some way of changing their prestrain within the system. The crucial factor is actuation. In an animal, a postural muscle can change its stiffness against a load and hold itself immobile, giving the onlooker the impression that the system of which the muscle is part is infinitely stiff. Additionally, by adjusting its prestrain, a muscle can ensure that a compressive load is confined to the stiff part of the skeleton, preferably down the centre. In fact you don't always need a muscle to do this. In echinoderms (sea urchins, starfish, etc.) collagen takes on the role of postural muscle, being able to soften when a part has to move, and stiffen again when that part has to be held steady against a force. The Crown-of-Thorns Starfish, for example, can soften the collagen holding the ossicles that constitute its hard skeleton, adjust its shape, and stiffen the collagen again, having in the process redesigned its morphology. The nearest we have which could deliver such changes is a fibrous composite, probably made of knitted or woven fibers, in a matrix made of an electrorheological liquid. This could then be softened, actuated, and stiffened again.

Learning more

For a general introduction to the mechanical analysis of organisms, a rather old book by Steve Wainwright and others (Wainwright et al. 1976) is still the most useful. It combines biology, materials science, structural engineering, and composite theory in a manner that has not since been repeated. It has been reprinted by Princeton University Press. This Press has a number of other books that should be on your list. Vincent's book on biological materials (Vincent 2012) brings parts of Wainwright's book up to date, and Jim Gordon's books (also reprinted by Princeton UP) complete the essential background (Gordon 1976, 1978). Steve Vogel (Vogel 2003) gives an exhaustive account of the physical context of living organisms.

Mike Ashby and his associates have produced numerous papers and books giving methods for selection and use of materials in engineering design, with comparisons of technical and biological materials (Ashby 2005, 2011). In the current handbook, Fukuda et al. (Chapter 52, this volume) discuss micro- and nanofabrication methods that could prove useful for assembling biomimetic materials. Halloy (Chapter 65, this volume) discusses the broader challenge of creating sustainable technologies for the twenty-first century highlighting the need for new kinds of materials that are recyclable and have low environmental impact.

References

Ashby, M. F. (2005). *Materials selection in mechanical design*. Oxford: Butterworth–Heinemann.

Ashby, M. F. (2011). Hybrid materials to expand the boundaries of material-property space. *Journal of the American Ceramic Society*, **94**, S3–S14.

Barthelat, F., and Rabiei, R. (2011). Toughness amplification in natural composites. *Journal of the Mechanics and Physics of Solids*, **59**, 829–40.

Chintapalli, R. K., Mirkhalaf, M., Dastjerdi, A. K., and Barthelat, F. (2014). Fabrication, testing and modeling of a new flexible armor inspired from natural fish scales and osteoderms. *Bioinspiration and Biomimetics*, **9**, 036005.

Forterre, Y. (2013). Slow, fast and furious: understanding the physics of plant movements. *Journal of Experimental Botany*, **64**, 4745–60.

Gordon, J. E. (1976). *The new science of strong materials, or why you don't fall through the floor.* Harmondsworth: Penguin.

Gordon, J. E. (1978). *Structures, or why things don't fall down.* Harmondsworth: Penguin.

Gordon, J. E., and Jeronimidis, G. (1980). Composites with high work of fracture. *Philosophical Transactions of the Royal Society A*, **294**, 545–50.

Greenhill, A. C. (1881). Determination of the greatest height consistent with stability that a vertical pole or mast can be made, and of the greatest height to which a tree of given proportions may grow. *Proceeding of the Cambridge Philosophical Society*, **4**, 65–73.

Hayashi, M., Feilich, K. L., and Ellerby, D. J. (2009). The mechanics of explosive seed dispersal in orange jewelweed (*Impatiens capensis*). *Journal of Experimental Botany*, **60**, 2045–53.

Hou, J., Bonser, R. H. C., and Jeronimidis, G. (2012). Developing skin analogues for a robotic octopus. *Journal of Bionic Engineering*, **9**, 385–90.

Humphrey, J. A. C., Barth, F. G., Casas, J., and Simpson, S. J. (2007). Medium flow-sensing hairs: biomechanics and models. In: J. Casas and S. Simpson (eds), *Advances in Insect Physiology, Volume 34: Insect Mechanics and Control*. New York: Academic Press, pp. 1–80.

Liu, A., and Berglund, L. A. (2012). Clay nanopaper composites of nacre-like structure based on montmorrilonite and cellulose nanofibers: Improvements due to chitosan addition. *Carbohydrate Polymers*, **87**, 53–60.

Miserez, A., Rubin, D., and Waite, J. H. (2010). Cross-linking chemistry of squid beak. *Journal of Biological Chemistry*, **285**, 38115–24.

Porter, M. M., Novitskaya, E., Castro-Cesea, A. B., Meyers, M. A., and McKittrick, J. (2013). Highly deformable bones: Unusual deformation mechanisms of seahorse armor. *Acta Biomaterialia*, **9**, 6763–70.

Skordos, A., Chan, C., Jeronimidis, G., and Vincent, J. F. V. (2002). A novel strain sensor based on the campaniform sensillum of insects. *Philosophical Transactions of the Royal Society A*, **360**, 239–54.

Vincent, J. F. V. (2012). *Structural Biomaterials*. Princeton: Princeton University Press.

Vogel, S. (2003). *Comparative Biomechanics*. Princeton and Oxford: Princeton University Press.

Wainwright, S. A., Biggs, W. D., Currey, J. D., and Gosline, J. M. (1976). *The mechanical design of organisms*. London: Arnold.

Wainwright, S. A., Vosburgh, F., and Hebrank, J. H. (1978). Shark skin: function in locomotion. *Science*, **202**, 747–9.

Waite, J. H., Andersen, N. H., Jewhurst, S., and Sun, C. (2005). Mussel adhesion: finding the tricks worth mimicking. *The Journal of Adhesion*, **81**, 297–317.

Walther, A., Bjurhager, I., Malho, J.-M., Ruokolainen, J., Berglund, L. A., and Ikkala, O. (2010). Supramolecular control of stiffness and strength in lightweight high-performance nacre-mimetic paper with fire-shielding properties. *Angew. Chem. Int. Ed.*, **49**,1–7.

Wegst, U. G. K., Bai, H., Saiz, E., Tomsia, A. P., and Ritchie, R. R. (2015). Bioinspired structural materials. *Nature Materials*, **14**, 23–36.

Young, S. L., Chyasnavichyus, M., Erko, M., Barth, F. G., Fratzl, P., Zlotnikov, I., Politi, Y., and Tsukruk, V. V. (2014). A spider's biological vibration filter: Micromechanical characteristics of a biomaterial surface. *Acta Biomaterialia*, **10**, 4832–42.

Zioupos, P., Hansen, U., and Currey, J. D. (2008). Microcracking damage and the fracture process in relation to strain rate in human cortical bone tensile failure. *Journal of Biomechanics*, **41**, 2932–9.

Chapter 11

Modeling self and others

Josh Bongard

Department of Computer Science, University of Vermont, USA

The body plan of an animal or robot constrains and provides opportunities for how it generates behavior. For example a legged robot will require a more complex control policy than a wheeled robot, but more energy-efficient movement may be possible by exploiting the momentum of the leg during the swing phase (Collins et al. 2005). This chapter explores one particular relationship between a robot's body plan and adaptive behavior: the ability to learn and mentally simulate the topology of its own (or another's) morphology. It will be shown that such ability enabled a robot to recover from unanticipated mechanical damage (Bongard et al. 2006). In another experiment it was shown that this ability to create self-models can be used to find appropriate teachers: the robot creates models of other robots in its environment; ignores those that are morphologically different from itself; and learns from morphologically similar teachers (Kaipa et al. 2010). These experiments draw on three theories from neuroscience: forward and inverse models, the neuronal replicator hypothesis, and the brain as a hierarchical prediction machine. This chapter concludes with a discussion of how the ability to mentally simulate self and others may enable future biomimetic machines to scale up to adaptive behaviors beyond the current state of the art.

Biological principles

Forward and inverse models in the brain

Internal models have long been a central concept in motor control: internal models allow an animal or robot to "mentally rehearse" an action before attempting it in reality. Two forms of internal models have been proposed. Forward models take as input a motor command and produce as output the predicted sensory feedback. Inverse models, as the name implies, invert this relationship: for some desired sensory state, an inverse model generates the motor command that is likely to produce that state. Forward models enable an agent to predict the consequences of a given action; inverse models enable it to find a motoric means to a desired sensory end. However, animals are able to generate a seeming infinitude of action variants that compensate for changes in the environmental context in which the action takes place: a human can maintain balance while holding a bag regardless of whether the bag is empty or full, or whether the human is walking or standing still. This suggests that the brain maintains multiple forward and inverse models that are responsible for one or a few contexts in which a desired action can be executed. Evidence for this modular organization of internal models has been reported in numerous papers including (Wolpert and Kawato 1998). However, it is not yet clear what form these models take, how many there are, how they are learned, the span of contexts for which they are responsible, and how much overlap there is between them.

The neuronal replicator hypothesis

The prospect of multiple internal models cooperating to help find appropriate actions and predict the result of them promotes a population-based approach to understanding brain function. The idea of multiple models (or memories, or ideas) co-existing in the brain can be dated back to Gerald Edelman's concept of neuronal group selection (popularly known as neural Darwinism after the publication of his book of the same name; Edelman 1987). The possibility that selection in the Darwinian sense is occurring in the brain has great appeal, as it concords with other selective systems that seem to operate in the body including the immune system, carcinogenesis, and competition between cells. It has been pointed out that Edelman's theory lacks a necessary condition for Darwinian selection to act: replication. A more recent theory, coined the neuronal replicator theory, proposes neurophysical processes that could allow for replication and mutation of neural activity patterns (Fernando et al. 2010). If such patterns encode forward and inverse models, then the neuronal replicator theory may provide an explanation for how an animal may learn (1) different forward models that successfully predict the result of specific actions in specific contexts, and (2) different inverse models that successfully provide actions which result in desired sensor states in specific contexts. In the next section a robot is described that continuously evolves forward and inverse models which enable it to autonomously recover from mechanical damage.

Hierarchical predictive coding

As mentioned earlier, it is not yet clear what form forward and inverse models take in the brain. Recently, the hierarchical predictive coding (HCP) account of brain function has gained much popularity as it seems to unite perception and action (for a review see Clark 2013). HCP promotes the view that much of the brain is organized into a particular hierarchy with signals propagating "upward" and "downward." In brain regions responsible for perception, signals propagating downward predict sensory signals; signals propagating upward from the sensory systems report errors between actual and predicted sensor signals. Motor regions determine what actions to perform to meet the predictions flowing down from upper levels of this hierarchy or, in other words, to reduce any errors occurring between predicted and actual sensory signals.

Thus, this hierarchy may implicitly encode forward and inverse models at different levels of abstraction. Upper levels of the hierarchy make vaguely defined predictions about the sensory repercussions of whole-body but ill-defined actions (a kind of forward model) and may also generate global, vaguely defined actions that will produce desired sensory signals (a kind of inverse model). Lower levels of the hierarchy would thus encode forward and inverse models related to local and specific sensations and actions.

One is left with the impression from the HPC hypothesis that the sole function of the brain is to reduce error. However as is demonstrated in the next section, it is often useful to generate actions that *increase* the prediction errors of forward models: this provides increased fodder for learning and allows an animal or robot to learn about hitherto unknown possible interactions with the world.

Biomimetic or biohybrid systems

This section presents two biomimetic systems. The first is a robot that generates self-models using an internal Darwinian process and uses this ability to diagnose and recover from unanticipated damage. The second is a social robot that generates models of itself and other robots in its environment, and uses this ability to find a teacher that is sufficiently similar to itself than it can successfully learn from it.

The Resilient Machines project

Bongard et al. (2006) reported an autonomous robot that demonstrated resiliency: the ability to generate a qualitatively different behavior to compensate for an unexpected situation (see Figure 11.1).

The project that led to this robot drew on the concepts of forward and inverse models, Darwinian processes in the brain, and hierarchical prediction coding. The resilient machine in Figure 11.1 maintains three Darwinian processes. After performing a random action (Figure 11.1A) and recording the sensor result, the first Darwinian process evolves a population of self-models, which are encoded as hypothetical body plans (two such models are shown in Figure 11.1B, C). It does so by actuating each self-model in the population with each action; scoring each model by how well it reproduces the sensory repercussions of each action; deleting the robots with low scores; and replacing them with randomly modified copies of the remaining higher scoring models. By repeating this process an accurate self-model is obtained. Periodically the robot performs a new action (Figure 11.1D). This action is generated with a second Darwinian process: hypothetical actions animate each of the self-models in the current population, and the resulting disagreement among the models' resulting sense data is used to score each action: an action receives a higher score if it induces more prediction disagreement among the models. The best of these evolved actions is performed by the robot and the self-models must now evolve to account for the original actions in addition to the new one (Figure 11.1E, F).

Once an accurate self-model is achieved (Figure 11.1G), a third Darwinian process is activated: this process evolves control policies on the best self-model such that it exhibits some desired behavior, which in this case is legged locomotion (Figure 11.1H, I). The best policy, when executed by the robot, results in locomotion (Figure 11.1). If the robot is damaged (Figure 11.1M) it re-evolves its self-models to reflect the damage (Figure 11.1N, O), re-evolves a compensating control policy (Figure 11.1P–R), and executes it (Figure 11.1S–U).

The self-models are evolved within a three-dimensional physics engine. This model/physics engine pair can be viewed as a forward model: when the virtual robot is actuated by an action or control policy, it generates virtual sensor data, which is the model's prediction. In addition, the third Darwinian process that generates control policies can be seen as an inverse model: it takes as input some desired sensation (e.g. the "feel" of moving forward) and generates actions (encoded as a control policy) that achieve that sensation.

Although these three Darwinian processes do not take the form of a monolithic hierarchical network capable of predictions and error correction at different spatial and temporal scales, there are striking similarities between the two approaches to adaptive behavior. If the robot performs a familiar action and obtains an expected sensory result, there is no change in the self-models or control policies. If the robot performs an action and receives a slightly unexpected result, it may evolve slight changes to a few of the self models and control policies. If the robot however experiences a truly unexpected event (such as after the loss of a leg), this triggers major changes in all of the self models and outright replacement of the current control policies with new ones.

Modeling of self and others

In Kaipa et al. (2010), the above method was adapted to enable a robot to observe other robots in its environment and create models of them. This robot maintained five Darwinian processes: the

Figure 11.1 A resilient machine. The robot first evolves self models that describe its body (a–f); then uses such a self model to evolve the ability to locomote (g–l); if damaged (m) it re-evolves a model of its damaged self (n,o) and uses the new model to evolve a compensating control policy (p–u).

first and second evolved self-models and exploratory actions as described above (Figure 11.2a–d). The third evolved models of a potential teacher robot in its visual field (Figure 11.2e–h). In short, the student robot uses proprioceptive information to evolve a model of self and visual information to evolve a model of the other.

Despite the differences in sensor modality, these "other models" take the same form as the self models: they are three-dimensional dynamical models of the potential teacher's morphology. The fourth Darwinian process is specific to the student robot's perception of the potential teacher. The student evolves a single-pixel position within its visual field at which the current models of the teacher disagree, and then the color of that pixel is queried from the camera. This new datum generally increases the prediction error of the current teacher models, causing them to evolve into more accurate models. The fifth and final Darwinian process enables the robot to imitate the potential teacher: populations of control policies (Figure 11.2i, l) are evaluated on the student model (Figure 11.2j), and their error is determined as the difference between the motions of the student model and the observed motions of the teacher (Figure 11.2k, m). If this error can be significantly reduced then the student robot can imitate the teacher robot. If this Darwinian process fails to reduce the error significantly, the student robot concludes that the potential teacher is inappropriate as it can generate movements that the student robot cannot because it has a different body plan (Figure 11.2n).

Future directions

The above two experiments demonstrate that a robot which mentally maintains multiple Darwinian processes can be capable of autonomously diagnosing and recovering from damage, or discovering an appropriate teacher from whom to learn. This work was inspired by the prospect that the mammalian brain generates multiple neural firing patterns that evolve to explain the organism's experience and propose actions. As compelling as it is, this theory of neural Darwinism (Edelman 1987)—or its more recent instantiation, the neuronal replicator hypothesis (Fernando et al. 2010)—has little neurophysiological data to support it: such evidence would greatly alter our thinking about brain function, and suggest novel approaches to autonomous robotics.

The hierarchical predictive coding theory of brain function (Clark 2013(is also gaining much attention in the neuro- and cognitive science communities. Deep belief networks (Hinton et al. 2006) are a class of computational models based on this approach, and have begun to demonstrate their power in the robotics domain (Hadsell et al. 2008). DBNs however currently lack an account for how and when agents should generate novel actions that increase rather than decrease error among the network's predictive models. It is anticipated that future work which integrates neural Darwinism with DBNs may lead to biomimetic systems that act to experience novelty and learn from it, as well to reduce uncertainty.

Learning more

Clark's recent (2013) review of the hierarchical predictive coding paradigm provides a good overview, and is accompanied by several responses which argue its limitations or forward alternative theories of brain function. Hawkins and Blakeslee (2005) provide a convincing argument in support of HPC and capture well the often non-intuitive accounts of human behavior generated by this theory. Asada (Chapter 18, this volume) discusses the human body schema and how it arises from multi-modal sensations including proprioception. Seth (Chapter 37, this volume) builds on the predictive coding hypothesis to create a framework for understanding the first person self.

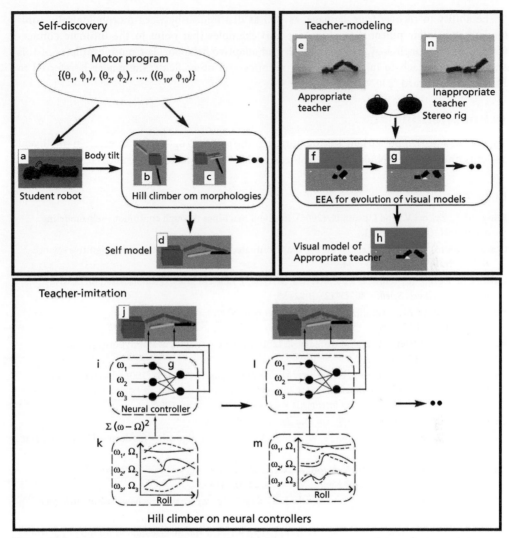

Figure 11.2 A student robot learning to distinguish between an inappropriate and appropriate teacher. First the student robot (a) evolves a set of increasingly accurate self models using proprioceptive sensor data (b–d). Then, it observes a potential teacher robot (e) and evolves increasingly accurate models of it using visual data (f–h). Then, the student robot attempts to imitate the potential teacher: it generates a random control policy (i) and uses it to actuate its self model (j). This leads to large errors between the observed motions of the teacher (solid lines in (k)) and its own motions (dashed lines in (k)). This policy is evolved into a more accurate policy (l) that enables the self model to reproduce the observed motions of the teacher (m). If the robot observes a teacher with a dissimilar body plan (n), it evolves a model of it but fails to evolve a control policy that enables it to reproduce that robot, so it ignores it.

Reprinted from *Neural Networks*, 23 (8-9), Krishnanand N. Kaipa, Josh C. Bongard, Andrew N. Meltzoff, Self-discovery enables robot social cognition: Are you my teacher?, pp. 1113–24, doi.org/10.1016/j.neunet.2010.07.009. Copyright © 2010 Elsevier Ltd.

The ability to recover from damage, as well as distinguish between morphologically similar and dissimilar potential teachers, are two examples that point to the intimate connection between morphology, neural control, and adaptive behavior. For more examples of this embodied approach to intelligence from robotics and other fields, the reader is referred to Pfeifer and Bongard (2007).

Acknowledgments

The author's work reported here was supported by the NASA Program for Research in Intelligent Systems under grant NNA04CL10A, NSF grant DMI 0547376, and NSF grant SGER 0751385.

References

Bongard, J., Zykov, V., and Lipson, H. (2006). Resilient machines through continuous self-modeling. *Science*, 314(5802), 1118–21.

Clark, A. (2013). Whatever next? Predictive brains, situated agents, and the future of cognitive science. *Behav. Brain Sci.* 36(3), 181–204.

Collins, S., Ruina, A., Tedrake, R., and Wisse, M. (2005). Efficient bipedal robots based on passive-dynamic walkers. *Science*, 307(5712), 1082–85.

Edelman, G. M. (1987). *Neural Darwinism: The Theory of Neuronal Group Selection.* New York: Basic Books.

Fernando, C., Goldstein, R., and Szathmáry, E. (2010). The neuronal replicator hypothesis. Neural Computation, 22(11), 2809–57.

Hadsell, R., Erkan, A., Sermanet, P., Scoffier, M., Muller, U., and LeCun, Y. (2008). Deep belief net learning in a long-range vision system for autonomous off-road driving. In *Proceedings of the IEEE/RSJ International Conference on Intelligent Robots and Systems*, pp. 628–33.

Hawkins, J., and Blakeslee, S. (2005). *On intelligence*. Owl Books.

Hinton, G. E., Osindero, S., and Teh, Y. W. (2006). A fast learning algorithm for deep belief nets. *Neural Computation* 18(7): 1527–54.

Kaipa, K. N., Bongard, J. C., and Meltzoff, A. N. (2010). Self discovery enables robot social cognition: Are you my teacher? *Neural Networks*, 23(8), 1113–24.

Pfeifer, R., and Bongard, J. C. (2007). *How the body shapes the way we think: a new view of intelligence.* Cambridge: MIT Press.

Wolpert, D. M., and Kawato, M. (1998). Multiple paired forward and inverse models for motor control. *Neural Networks*, 11(7), 1317–29.

Chapter 12

Towards a general theory of evolution

Terrence W. Deacon

Anthropology Department, University of California, Berkeley, USA

Machines are designed artifacts, and as such they are the physical exemplars of abstract principles. By metaphoric extension researchers often describe cellular structures as molecular machines, and engineers often use living processes as models for the mechanisms they design. But the design process is teleological. Teleological accounts begin with an aim to achieve a type of end without specifying how that end is to be produced, then proceed to explore options to accomplish this. This is also one of the defining features of teleological explanations. But teleological explanations have been systematically shunned by the natural sciences throughout most of the last century. This exclusionary stance is neither surprising nor problematic. For the most part teleological explanations rest on "black box" mechanisms.

There have however been efforts to extend the concept of teleology so that it is consistent with biological function. For example, building on the philosophy of C. S. Peirce, the philosopher T. L. Short (2002) has argued that Darwinism is a teleological theory because natural selection produces certain general outcomes—adaptation for general functions like flight or thermoregulation—but doesn't specify what mechanism must be developed to do this. Nevertheless, individual instances of an adaptation are consequences of modifications to specific mechanisms in such a way that these ends are achieved. For Short, this generality of what constitutes an adaptation of a specific type is all that is required to classify natural selection as teleological. However, the process of natural selection does not exhibit another feature generally assumed to define a teleological process: the organization of behavior with respect to the representation of a yet-to-be-achieved general goal. Although the incorporation of the concept of representation into this end-directed context makes it intrinsically semiotic, many biologists and philosophers of biology would deny that organisms embody a represented or implicit target state, except for organisms with complex brains that embody proximate adaptive ends.

Clearly the extreme anti-teleological and the semiotic conceptions of organisms are incompatible. Defining teleology in Darwinian terms, as Short has attempted, does not really resolve this incompatibility. It simply renames the terms of the argument. Moreover, it effectively reduces teleology to a reflective property that exists by virtue of observer judgment that classifies diverse specific physical outcomes as similar (e.g. by ignoring innumerable specific physical differences) and makes them tokens of a type of adaptation. The question remains, whether teleology in a stronger semiotic sense must also be invoked and explained. In other words, the question is whether there are anything like organisms' semiotic representations of adaptive outcomes (whether instantiated in molecular processes or brain processes) and, if so, whether they are doing any of the work of determining these ends. If so, then semiosis must be considered central to evolutionary theory, even if the process of evolution itself is non-teleological, because organism teleology is what generates the forms and behaviors upon which "blind" selection operates.

In the following short essay I argue that for these reasons natural selection theory is fundamentally incomplete and that a non-trivial non-reducible form of teleological and semiotic causality must be introduced in order to complete the synthesis. However, this does not necessarily undermine the standard conception of evolution by natural selection as an *undirected* process, even though it leaves open the possibility that semiotic processes can enter indirectly into its workings (e.g. via niche construction effects).

Completing Darwin's "one long argument"

> "There is grandeur in this view of life, with its several powers, having been originally breathed into a few forms or into one; and that, whilst this planet has gone cycling on according to the fixed law of gravity, from so simple a beginning endless forms most beautiful and most wonderful have been, and are being, evolved." — Charles Darwin (1859)

Charles Darwin ended his paradigm-changing book with the above poetic reflection on the origin of life and the initiation of the evolutionary process. In this lone published hint about the origin of life, Darwin implicitly recognizes that the process of natural selection cannot account for the origin of life, and indeed that it depends on life's "several powers" coming into existence first, in order to begin. These "powers," such as the generation of new component structures, the maintenance of non-equilibrium conditions, and the repair and reproduction of the whole integrated organism, are prerequisites for natural selection. Of course Darwin was not interested in explaining the nature of life itself, only in explaining how species have adapted to their distinct niches and have transmuted into new forms over evolutionary timescales. Thus his intention was to describe a mechanistic logic that could account for functional design in living nature by spontaneous natural processes—processes which were blind to consequent function and eventual outcome and thus did not depend upon prior purpose or representation. This mechanism—natural selection—has been understood by many to represent the ultimate triumph of materialist science over forms of supernatural design and teleological processes.

The genius of the theory of natural selection was that it inverted the logic of classical mechanistic causality in one important respect and yet was consistent with the strictures of this non-teleological paradigm. It demonstrated that causal consequences—ends, not just prior conditions—could play a constitutive role in determining how living things are organized, collectively and individually. Most importantly, it demonstrated a way that this could occur irrespective of any anticipatory, representational, or even functional feedback mechanism. But as Darwin's closing line demonstrates, he nevertheless recognized that the origin of the core distinguishing powers of life itself are not explained by this mechanism. It is the availability of these "several powers" that makes natural selection possible. This passage, along with Darwin's reticence to even discuss the origins of life question, suggests that he was under no illusions about this distinction. There is no natural selection without the prior availability of self-repair and reconstitution, reproduction, and inheritance. So natural selection is necessarily an incomplete explanation for the evolution of life.

In this respect, neoDarwinian theories, including such abstract versions as described as artificial life, replicator selection (e.g. Dawkins 1976), and universal Darwinism (e.g. Dennett 1996), must make non-trivial and quite complex assumptions concerning the prerequisites for this process to occur. For this reason, I describe neoDarwinism as a "special theory" of evolution. It is "special" because it makes non-trivial assumptions about the context in which it occurs and the boundary conditions that it depends on. Unless these conditions are satisfied there is no evolution. Specifically, natural selection theory leaves nearly all dynamical detail out of the account in order to focus on what is special about a Darwinian selection process compared to other natural and living processes. To provide a more complete and thus "general theory" of evolution, capable of being applied beyond the narrow confines of biology, we must fully account for the dynamics that makes natural selection possible and explain how these atypical processes arise in the context of more typical thermodynamic processes found in the inanimate world.

In this respect, a general theory of evolution amounts to a bridge theory. It connects the end-directed dynamics of self-reproducing physical systems (e.g. organisms) that are susceptible to natural selection to the undirected physics of non-living chemistry. In answer to Darwin's implied question about these "several powers" the general theory of evolution must explain the spontaneous emergence of dynamical systems capable of resisting the ravages of thermodynamics and of producing replicas of themselves, each inheriting this same general capability, despite each exhibiting some degree of structural and dynamical variability. In other words, the general theory of evolution must begin with an account of the processes by which Darwin's "several powers" could have arisen spontaneously in a lifeless context.

The generality of the theory is not, however, provided by an account of life's ultimate origin, but by the implication that these processes, which constitute the necessary substrate for evolution, are therefore also its ever-present handmaids. Those "special powers" which emerged with life's origin (or before, depending on how we define life) did not cease to operate or diminish in importance once natural selection became a robust part of terrestrial dynamics. Biological evolution (and likely evolution in general in all its many forms) is necessarily a higher-order emergent consequence of these processes, which are neither merely chemical nor entirely biological.

Blind variation and multiple-realizability

In the classical account of natural selection, organism structures and processes, and the variations of these in different individuals, are, by assumption, generated irrespective of any eventual functional consequence. And yet, despite this agnosticism concerning the mechanisms involved in generating the various forms that become subject to selection, Darwin showed that differential preservation of certain of these variant forms with respect to others could lead to progressively better fittedness of these forms to surrounding conditions. With only this much specified, Darwin was nevertheless justified in concluding that adaptive changes would tend to follow. Although Darwin did not exclude many possible mechanisms for the generation of these living forms—e.g. including those advocated by Lamarck, Spencer, or Haeckel—it was Darwin's genius to recognize that end-directed evolutionary mechanisms were unnecessary to account for these adaptive consequences.

This disjunction between the production of variant organism forms and their eventual preservation or elimination within a lineage allows for the widest diversity of mechanisms to contribute to the evolutionary process. Significantly, the partial independence of functional analysis from mechanistic analysis allows for what philosophers of science have called multiple-realizability. This seemingly imposing term describes a familiar fact of our everyday lives, but one that doesn't easily fit into standard physical theories. It is, for example, the essential property of money. A dollar can be realized in coins, by a paper bill, or even by an electronic funds transfer. In biology multiple-realizability is a defining property of functions and adaptations. There are typically multiple evolutionary routes to the same general type of adaptation, such as flight, oxygen metabolism, or chemotaxis. Natural selection is a process that can only be defined with respect to outcomes, and most outcomes can be achieved via any number of alternative means. This makes them general types of outcomes, not specific mechanical consequences (Short 2002).

For this same reason, natural selection theory cannot be properly described as a mechanism. Natural selection theory provides an abstract description of how an open-ended class of mechanisms can lead to outcomes that converge on some likewise open-ended class of dynamical relationships between organisms and their environments. Natural selection is a logic (or algorithm) that describes what will occur should a certain class of organism processes fit with the local environment in such a way that the disposition to produce similar adaptive processes is also maintained in its life and in future progeny. The question that is left unaddressed by this analysis is whether one needs to understand this general disposition as anything other than a chance constellation of mechanisms. An incautious focus on replication alone—as captured by the phrase "blind variation with selective retention"—often presumes that mere accident is sufficient to explain natural selection. However, this ignores the fact that *what* varies is necessarily a physically embodied structure or dynamical process that requires specific physical work to be produced.

If these generative processes are not already in place, accidental variation contributes no more than thermodynamic decay. Accident is the source of the novel variations that feed natural selection, but this already assumes some functional substrate that is susceptible to variations—an organ or process that already contribute directly or indirectly to survival and reproduction. So first we must ask: what kind of causal dynamic is sufficient to constitute the sort of dispositions (i.e. several powers) that characterize an organism such that it can autonomously compensate for spontaneous and extrinsic tendencies to disrupt its coherence and can under the right conditions reproduce an approximate replica of itself embodying these same powers? Secondarily, we must also ask how such an atypical dynamical process could have been spontaneously initiated, despite the fact that it tends to run counter to the ubiquitous and unavoidable second law of thermodynamics.

Although Darwin took for granted those processes of life which exhibit the prerequisite "powers" on which natural selection depends—because they were everywhere in evidence in the living world and unanimously accepted by his contemporaries—a complete evolutionary theory cannot take them for granted. They are a critical part of the explanation of what it is that physically occurs to generate this process, and yet are anomalous dispositions to change when contrasted with the tendencies to change that are ubiquitous in the non-living world.

Replication isn't free

Because of the anomalous physical nature of these "several powers" of life, being able to evolve by natural selection is not a typical physical property like weight or a capacity like flammability. In simple terms it is a consequence of the ability to maintain systemic integration, to reproduce, and to reproduce these capacities despite slight variations of form in the process. But this simple statement belies the physical complexity of what is necessary to achieve these capabilities.

An early insightful abstract analysis of the physical requirements for a system to be considered capable of self-reproduction came from the work of the mathematician John von Neumann, one of the pioneers of modern computer design. He was interested in the logical and physical problem of machine self-reproduction (Neumann and Burks 1966). He argued that the most basic criterion for self-reproduction is that the system in question must be able to construct a replica of itself which includes this same capability as well. Specifying how to construct a mechanism with this capability turns out to be much more difficult than this simple recipe might suggest. Von Neumann eventually abandoned the effort to specify the physical requirements of such a system much less build one. He settled instead on the effort to precisely describe the logical (i.e. computational) requirements of self-replication. This work ultimately contributed to the development of cellular automata theory. In the decades since von Neumann outlined the problem, many software simulations of natural selection processes have been developed (see

Moses and Chirikjian, Chapter 7, this volume). The field that has grown up around this work is often described as artificial life (or A-life). But the goal of physically implementing machine self-reproduction, as opposed to merely simulating it, has not been achieved.

This distinction is analogous to narrowly defining evolution in terms of the replication of DNA alone, and ignoring the physical demands of producing a body capable of replicating its DNA and reproducing itself. The missing element in von Neumann's explanation is, of course, the special means to produce the physical work required in both biological and machine reproduction. He simply described this mechanism as a "universal constructor" for which the functional details were omitted. In A-life, cellular automata, and evolutionary computing the production of computer code is entirely parasitic on the physical work done by the computer mechanism. The reproduction of such a code is not, then, self-reproduction in the full sense of the word. It is only a set of instructions that causes the machine to make another copy of those same instructions. But physically embodied self-reproduction requires one physical system to construct another like itself. And this requires quite complex forms of physical work to modify and organize materials. In the terms of von Neumann's proposal, in order to complete a general theory of machine evolution we need to specify the organization and functioning of the universal constructor mechanism. So, what must it accomplish? In order to reproduce itself as well as this same disposition a system must:

1. acquire the necessary substrates to generate all its critical system components;
2. transform substrates into critical system components;
3. assemble all components in a way that maintains the critical dynamical relationships of these components that make 1 and 2 possible;
4. be sufficiently tolerant of structural–functional degeneracy to allow some degree of component replacement and dynamical variation without losing capacities 1–3; and
5. perform these operations in sufficient excess of what is needed to maintain system integrity so as to generate a redundant whole self-replica that also satisfies requirements 1–4.

This is only a list of what must be accomplished, it is not yet an account of how it *can* be accomplished. It is like the many descriptions of what needs to be accomplished in order to contain and control nuclear fusion in order to produce energy. Knowing what's needed is no guarantee that we know how to do it, or even that it is possible. Nevertheless, this need not be a specific mechanistic account because of the multiply realized nature of the process. It need only specify the kind of mechanisms required. But without this further degree of explanation we lack a complete general theory.

Von Neumann's partitioning of the machine reproduction problem into a physical and a logical problem, respectively, and then abandoning the effort to deal with the physical problem, thus provides an insightful parallel to most mainstream contemporary accounts of the evolutionary process. However, von Neumann made an explicit choice to temporarily set the mechanistic challenge aside, fully aware that it would unavoidably need to be addressed. Perhaps because he was approaching this as an engineering challenge, in which physical and practical details eventually need to be worked out, this physical problem seemed obvious. That it is often ignored or taken for granted in evolutionary theorizing probably derives from the fact that life deals with this for us. Likewise for theorists developing computer simulations of Darwinian processes. The pattern generating and copying capacities that are built in as a basic operating principles of computing devices by the engineers and system programmers that created them, make it possible for A-life and evolutionary computing theorists to also approach their subjects oblivious to the physical dynamics.

But like von Neumann, in order to completely account for the special powers that make physical reproduction possible we are forced to pay attention to these difficulties. This is unavoidable if our intention is to produce a fully generalizable explanation of the evolutionary process. It's time to face up to von Neumanns's problem.

The role of self-organization

In an effort to provide a philosophically coherent account of the teleological features of living organisms, the philosopher Immanuel Kant (1790) focused on two critical features which seemed to set organisms apart from machines: causal circularity and the generation of form. In an oft-quoted passage he says:

> "... a machine has solely motive power, whereas an organized being possesses inherent formative power, and such, moreover, as it can impart to material devoid of it — material which it organizes. This, therefore, is a self-propagating formative power ..."

and later,

> "... in which, every part is reciprocally both end and means."

In other words, organisms physically create themselves and do so in a circular or reciprocal manner. He argues that these distinctive properties of organisms constitute what he describes as "intrinsic finality" (alluding to Aristotle's notion of final cause, or that for the sake of which something occurs). Although he later argues that this organizational feature of organisms is only understandable as teleological from an outside perspective, so to speak—one that enables a sufficiently global perspective to include more than just each isolated cause and effect interaction. For this he turns to a transcendental agency. But recasting Kant's logic in modern non-equilibrium thermodynamic terms we can begin to make good on a critical promissory note in this account. This pertains to the concept of "formative power" which Kant explicitly contrasts to "motive" (or mechanistic) power.

Despite the ubiquitous and unavoidable increase in entropy and the breakdown of order that is made explicit by the second law of thermodynamics, living organisms persistently generate orderly structures and processes (forms). This physically atypical property of organism dynamics was elevated to a defining property of life by the quantum physicist Erwin Schrödinger in his influential book *What is Life?* (Schrödinger 1944). Though also anticipating the discovery of the critical features of DNA molecules that enables them to serve as the medium of genetic inheritance, it was his description of life as "feeding on negentropy" that became its most enigmatic claim. In the generations that followed, researchers trying to make sense of this thermodynamic riddle eventually found a critical clue in the analysis of thermodynamic systems that persist far from equilibrium. Ilya Prigogine was awarded a Nobel Prize for his expansion of thermodynamic theory to incorporate the special properties exhibited by such far-from-equilibrium systems, and particularly their tendency to locally increase their regularity and orderliness. This property has generally been described as "self-organization" because the orderliness is not imposed from without, but rather emerges due to the recursive interactions among components (see Prigogine and Stengers 1984).

The concept of self-organization has become a key principle in complex systems theories and in systems biology in general (see Wilson, Chapter 5, this volume). It also offers a way to begin to reframe Kant's concept of "formative power" precisely because it provides an opposed dynamical tendency to the simple close-to-equilibrium thermodynamic tendency that characterizes the common conception of mechanism. Specifically, mechanistic processes have a linear causal character whereas self-organizing processes (even though they are also composed of individual mechanistic interactions) have a non-linear (i.e. circular or recursive) causal character. This is a significant advance over the modeling of biological processes as complexes of chemical mechanisms. Self-organization is particularly critical for embryogenesis. The differentiation of tissue types, body segments, limbs, and organs from undifferentiated precursors during embryogenesis depends extensively on such form-generating processes. And it is not too much of a caricature to say that it is self-organization that generates the forms that natural selection operates upon (see Prescott and Krubitzer, Chapter 8, this volume). Of course, genetic information is also critical, and effectively provides the network of constraints that serve as the boundary conditions that enable specific self-organizing dynamics to take place in specific contexts. In many respects genetic information doesn't instruct embryogenesis but rather constrains the spontaneous self-organizing tendencies that emerge spontaneously in molecular and cellular interactions.

This seeming "negentropic" character of self-organizing processes has led many to assume that a living organism can be characterized as a complex self-organizing process. This is both true and misleading. Both at the level of intracellular molecular architecture and at the level of embryogenesis, self-organizing dynamics play an important role in the production of complex structures. And, given the fact that organisms build themselves, it would seem appropriate to describe an organism as a self-organized system. It would be a mistake, however, to confuse the term "self" as used in these two senses. With respect to self-organization the term merely indicates that the orderliness of a physical process is not directly imposed from without but emerge due to the way extrinsic perturbations induce internal interactions among components to spontaneously regularize. In contrast, when we describe an organism as differentiating, repairing, defending, or reproducing itself the term "self" takes on a teleological meaning in the sense that these are target states that are in some sense prefigured within the organism almost entirely irrespective of extrinsic perturbation or influence. In other words organisms exhibit a characteristic locus of autonomy and a reflexive form of causality that is missing in non-living phenomena.

In my recent book *Incomplete Nature* (Deacon 2012a) I show how this form of autonomous agency can, however, arise even in an exceedingly simple molecular system called an autogenic system (also "autocell" in Deacon 2006) which results from a reciprocally reinforcing linkage between complementary self-organizing molecular processes. This occurs because, like the more complex reciprocal constellations of complementary self-organizing processes that constitute simple organisms, the form-generating dynamics of the component self-organizing processes can reciprocally generate each others' supportive boundary conditions; i.e. the constraints that predispose these otherwise intrinsic form-generating dynamics. This reciprocal interdependence of constraint-generation processes is ultimately the source of autonomy. It provides a persisting locus of the specific global constraints required to channel energy in a way that does the work of continually reconstituting and preserving this very capacity. To state this somewhat enigmatically: organism *self* emerges from a special form of higher-order self-organization of chemical self-organizational processes. Mast et al. (Chapter 39, this volume) provide a discussion of recent efforts in synthetic biology in this direction.

This intrinsically maintained self-specification can be considered to be a representation interpreted by the dynamics of organism development and metabolism. It is a higher-order constraint that is intrinsically capable of channeling the work that determines the production of the specific target form of the mature body. It is, in this sense, both self-referential and self-determinative. In semiotic terms this form of higher-order constraint is a sign vehicle that is dynamically interpreted when it brings about the form of a new physical system in which it will again become embedded, complete with this same future capacity. It is a form that informs. What makes this form of constraint more than a mere restriction, structure, or regularity is that its most distinctive property is an existence *in futuro*, so to speak. It specifies and preserves its own form into the future, despite changes in its physical-chemical embodiment. It is in some sense *outside* the specifics of its necessary physical embodiment.

This is the ultimate basis for all biosemiosis. It is why we cannot escape assuming teleological concepts when talking about living processes, even when only considering the individual molecules or chemical reactions of life. These structures and processes exist and assume their distinctive forms because they are serving functions for the sake of the system that they partially constitute. Thus in every respect organisms are semiotic. First, they embody their own self-reference. Second, they embody in their form and dynamics a kind of negative image of their *Umwelt*—i.e. those aspects of their environment that they either depend on or must avoid. And third, they modify their form under the influence of complex constraints—sign vehicles, genetic and extragenetic—that invariably determine development toward a culminative form—a state of "completeness" sufficient to support reproduction. So organisms aren't mere mechanisms, nor is the appearance of teleological organization merely derivative of natural selection. And they are teleological in the full sense of the word, even if not mentalistically based, because the core organism dynamics are preserved in the form of a self-representation.

Elements of the general theory

In conclusion, Darwin's "several powers" are ultimately derived from a thermodynamic basis, and specifically from processes that use the tendency for entropy to increase against itself. In order for organismic processes to locally thwart the breakdown of dynamical and structural constraints due the relentless increase of entropy, work must be constantly performed to continually repair and replace component features. The ability to locally generate and preserve forms (constraints) faster than thermodynamic breakdown can degrade them is the prerequisite to reproduction. In this respect, the competitive aspect of natural selection is also grounded in thermodynamics. It is this requirement of constant work that necessitates nearly uninterrupted access to resources and persistent maintenance of the web of interdependent constraints that appropriately channel this work.

Evolution has not only been initiated by the coalescence of complementary self-organizing processes, the evolution of new adaptive forms has also gravitated toward the recruitment of self-organizing dynamics whenever available to ease demands on precise constraint maintenance and energy use. Thus not only do living forms converge toward ever closer correspondence with the environmental factors that they depend upon, they also tend to converge toward incorporating increasing numbers and forms of self-organizing dynamics in their molecular, physiological, and developmental processes.

To be evolvable, a type of dynamical system must also maintain the higher-order self-referential constraint complex that constitutes self. The generation of new material embodiments of these constraints at a rate sufficient to also enable the emergence of this higher-order self-organization of component self-organizing processes is thus also critical to reproduction. And reproduction is the ultimate hedge against the second law of thermodynamics.

In general, then, the several powers that make natural selection possible are processes that generate and preserve forms, and are susceptible to selection with respect to minimizing the amount of work required to produce and maintain organic forms. Thus, intrinsic spontaneous constraint-generating dynamics tend to predominate in the living world because of the competitive thermodynamic advantages they provide. More importantly, the role played by reciprocal form-generation dynamics provides the basis for explaining the semiotic nature of life. This is because the autonomy that emerges due to the reciprocal closure of constraint-generation is implicitly self-referential. In other words, the higher-order constraint complex that constitutes the reciprocal closure of such a system of self-organizing processes effectively instructs the representation and regulates the reproduction of a self-replica that can either be referred to for self-repair or for the generation of progeny. This is a quintessential semiotic relationship, even if "sender" and "receiver" are not distinct individual interpreters but are rather component phases of the self-creation process. This is also why systems constituted by interdependent complementary self-organizing processes (autogenic systems) are also the most likely precursors to evolvable life and the origins of semiosis.

These considerations are important to keep in mind as we attempt to extend the evolutionary paradigm beyond the confines of biology and its molecular substrates. Whether applying this paradigm logic to computational processes, neurological processes, or artifact design, it is important to be able to identify the "special powers" that are relevant in these domains. To ignore these underlying formative processes risks falling prey to overly simple analogies. Ignoring this makes evolution appear to be ungrounded, with neither teleological nor semiotic character, when in fact it is the very expression of these properties. So a general theory of evolution that can apply both to organic and inorganic systems must specify what Darwin alluded to as life's special powers.

Acknowledgment

This chapter is an amended version of one previously published as: Is semiosis one of Darwin's "several powers"? In Timo Maran, Kati Lindström, Riin Magnus, and Morten Tønnessen (eds.) *Semiotics in the Wild*, pp. 71–77 (Deacon 2012b). It is republished here by permission of University of Tartu Press.

References

Darwin, Charles (1971 [1859]). *On the Origin of Species by Means of Natural Selection or the Preservation of Favored Races in the Struggle for Life.* (London: J. M. Dent & Sons.)

Dawkins, R. (1976). *The Selfish Gene.* New York: Oxford University Press.

Deacon, T. (2006). Reciprocal linkage between self-organizing processes is sufficient for self-reproduction and evolvability. *Biological Theory,* 1(2), 136–49.

Deacon, T. (2012a). *Incomplete Nature: How Mind Emerged from Matter.* New York: W.W. Norton & Co.

Deacon, T. (2012b). Is semiosis one of Darwin's "several powers"? In T. Maran, K. Lindström, R. Magnus, and M. Tønnessen (eds.), *Semiotics in the Wild.* Tartu: University of Tartu Press, pp. 71–77.

Dennett, D. (1996). *Darwin's Dangerous Idea: Evolution and the Meanings of Life.* New York: Simon & Schuster.

Kant, I. (1790, 1951). *Critique of Judgement.* J. H. Bernard, trans. New York: Hafner Press.

Neumann, J. v., and **Burks, A. W.** (1966). *Theory of Self-Reproducing Automata.* Champaign/Urbana: University of Illinois Press.

Prigogine, I., and **Stengers, I.** (1984). *Order Out of Chaos: Man's New Dialogue with Nature.* New York: Bantam Books.

Schrödinger, E. (1944). *What is Life? The Physical Aspect of the Living Cell.* Cambridge, UK: Cambridge University Press.

Short, T. L. (2002). Darwin's concept of final cause: neither new nor trivial, *Biology and Philosophy,* 17, 323–340.

Section III

Building blocks

Chapter 13

Building blocks

Nathan F. Lepora

Department of Engineering Mathematics and Bristol Robotics Laboratory,
University of Bristol, UK

At their most basic level, animals can be thought of as systems that move and act in response to sensory information about their surroundings. Thus living machines based on animals may similarly be considered as sensory–motor systems. In animals, the way in which sensory information is turned into an action can vary from the very simple, such as in-built reflexes, to the very sophisticated, using memories across our entire lifetimes. Therefore there are three basic components that all animals and their corresponding living machines must have: sensors that take in information about the world, actuators that move the body to act upon the world, and a computation/memory system that interprets the sensory information to control the actuators.

The chapters in this "building blocks" section of the *Handbook of Living Machines* explore the individual sensory and motor components that when pieced together can comprise a complete biological or artificial system. Each biological sense organ, such as our eyes, ears, and skin, and aspect of biological actuation, such as muscles and their control, has led to biomimetic counterparts that have been applied technologically, such as in robotics. Moreover, these biomimetic artifacts provide excellent models of their biological counterparts, allowing us to ask and answer questions about the biological systems that cannot be addressed through experiment alone.

The first part of this section concerns sensing. Living machines such as robots need to be able to sense their surroundings otherwise they would be unable to react to the world. Therefore, most robots need sensors that translate information about the surroundings into electrical signals. These signals can then be interpreted by the robot's controllers to tell the actuators how to move the robot. While many sensors have been developed in engineering, the Living Machines approach focuses on biomimetic sensors and often on *neuromorphic* technologies that use very large-scale integration (VLSI) methods to build electronic sensor arrays that mimic some of the capacities of sensory cells to transduce stimuli into patterns of simulated neural activity.

In ancient times it was thought we have only five senses—sight, hearing, touch, taste, and smell. The first four chapters of this section mirror that classical thinking, with taste and smell considered together as chemosensing. However, it is now appreciated that we have many more senses. An important example of another human sense is proprioception, which allows us to feel where the parts of our body are relative to each other. Animals mostly have the same senses as us, although there are senses that some animals have that we do not—for example, the sensing of electric fields in some species of fish.

We start with biomimetic vision. Dudek (Chapter 14) explains the biological principles underlying the function of vertebrate and insect eyes, including the processing of visual information in the neural circuitry of the eye. These principles have led to a range of "neuromorphic" artificial vision systems, from devices based on the compound eyes of insects and vertebrate retinas, to more abstract dynamic vision sensors and vision chips.

Next, we progress to biomimetic audition. Smith (Chapter 15) reviews the mammalian auditory system, connecting its principles with the problems it must solve: locating sound sources and interpreting those important for the animal. The best-known technological translations of the auditory system are hearing aids, in particular cochlear implants, which are discussed elsewhere in this handbook (Lehmann and van Schaik, Chapter 54). Approaches include implementing principles of audition in software (CPU-based) and in neuromorphic hardware such as silicon cochlear.

Lepora (Chapter 16) next describes three principles of biological tactile sensing: that touch attains hyperacuity (superresolution), is actively controlled, and uses a lexicon of exploratory procedures to characterize object properties. These principles have been implemented in artificial devices to result in superior sensing capabilities, including systems that mimic the human fingertip and hand (cutaneous touch) and the rodent whisker system (vibrissal touch).

Pearce (Chapter 17) describes how biological chemosensing, encompassing also the olfactory and gustatory senses, is extremely sensitive with an enormous dynamic range, benefitting from the population code in olfactory sensory neurons and the structure of the olfactory mucosa. His article discusses three technological translations from biology: biohybrid chemosensors, neural-inspired processing architectures, and neuromorphic implementations of the animal olfactory system.

Biomimetic proprioception is another key sense in humans and other animals. Asada (Chapter 18) explains the biological principles underlying human proprioception and body representation, in relation to the spatial–temporal integration of the senses over the body. This chapter follows on from Bongard (Chapter 11) who provides a more general introduction to the challenge of learning a body model. Asada considers the applications of body schema in robotics and concludes by discussing a series of constructive approaches attacking the mystery of body representation.

The last sense considered here is biomimetic electrosensing, as found in some species of fish. Boyer and LeBastard (Chapter 19) describe how the African fish *Gnathonemus petersii* can sense the perturbations in electric fields emitted in its tail as a form of active electrolocation. This biological sense has motivated a biomimetic implementation of electrosensing in underwater robots.

The second part of this section is broadly concerned with actuation. Robotic living machines are only useful if they can perform tasks, such as picking up objects or travelling from one place to another. Likewise, animals can only live if they forage for food and water, and do the many other tasks they need to survive. In a human, the control system is our brain and the actuators are our muscles coupled with our skeleton, which operate together to interact with our surroundings. In a robot, the controller is a computer and the actuators are the various motors, gears, wheels, and structures that function together as the robot's muscles and skeleton.

Our three chapters on the biomimetics of animal movement take three very different perspectives.

Biomimetic muscles are artificial devices that mimic aspects of biological muscle. Anderson and O'Brien (Chapter 20) describe how natural muscle is a soft lightweight material whose speed, strain and pressure for low density is unmatched by any man-made alternative. For artificial muscles to be used appropriately, they should mimic mechanisms for biological muscle actuation and also incorporate self-sensing feedback for their control.

Although there are challenges in reverse engineering biological control systems, invertebrate circuits where individual neurons can be more easily isolated and studied provide a suitable starting point. In animals, many important motor behaviors are rhythmic, for example locomotion, chewing, and breathing. Selverston (Chapter 21) explains the biological mechanisms that underlie rhythmogenesis, in particular collections of nerve cells in invertebrate central nervous systems that produce rhythmic behaviors autonomously. These biological principles can be incorporated into biomimetic and hybrid devices using electronic models of spiking neurons and have been demonstrated in the control of biomimetic and biohybrid robots (see Ayers, Chapter 51).

We finish this section with a key aspect of biological movement: how skin adheres to surfaces. Pang and colleagues (Chapter 22) describe how the unique structural features of gecko foot hairs and attachment systems in beetle feet lead to remarkable adhesive properties. Nanofabrication methods are beginning to mimic these properties, with applications to robots that can climb vertical surfaces, and to a wide range of applications from clean transportation to biomedical patches.

Chapter 14

Vision

Piotr Dudek

School of Electrical & Electronic Engineering, The University of Manchester, UK

For many animal species, vision is the primary sense through which they perceive their environment. It is also the one to which they dedicate the largest proportion of their brain resources. The processing of visual information starts early, inside the eye. From insects to vertebrates, eyes are not only sensory organs, but also include neural circuitry providing pre-processing of the visual signal right next to the photoreceptors. Massively parallel near-sensor computation allows rapid extraction of information from the visual signal, compression of redundant or less relevant data, and the creation of signal representations that are more informative, more useful, and easier to process further than raw light intensity images. This highly efficient scheme has been an inspiration for constructing artificial vision sensors. In this chapter, the principles of operation and key features of the early stages of biological vision systems are reviewed. Several examples are presented, ranging from devices inspired by the compound eyes of insects, through silicon circuits mimicking the structure and operation of vertebrate retinas, to more abstract interpretations of biological principles by dynamic vision sensors and vision chips with pixel-parallel array processors. Key future research directions in this area are identified.

Biological principles

The sense of vision starts with signal transduction and processing of the information inside the eyes. All vertebrates, and many invertebrate species, have complex image-forming eyes, with the basic optical arrangement not dissimilar to that of a camera (Figure 14.1a). The light enters the eye through the cornea, is further focused by the lens, and falls onto the retina, creating an image there. This image is sensed by a film of light-sensitive cells—photoreceptors—which include proteins that deform in response to incident photons, triggering a sequence of biochemical reactions that result in a change of electrical potential across the cell membrane and the rate at which the photoreceptor cells release a neurotransmitter substance. This electrochemical signal is then processed in several layers of neural cells in the retina, before being sent down the bundle of axon fibres in the optic nerve to other structures in the brain.

In vertebrates, the retinal photoreceptor cells are subdivided into *rods*, which have superior light sensitivity (capable of detecting single photons, and thus providing night vision), but saturate in bright light, and *cones*, which provide daylight vision, with different spectral sensitivities (e.g. primates have three types of cones, with peak sensitivities in the red, blue, and green parts of the light spectrum) that allow the perception of colour. The wide range of responsivity of the photoreceptors, together with adaptive circuitry that moderates the retinal outputs based on the past and surrounding image intensities, result in a very wide dynamic range of image intensities that can be detected.

The photoreceptor concentrations are not uniform across the retina. In primates, a small central region called the *fovea* has very tightly packed cones, providing a high resolution image in a very small region of the visual field (a few degrees), while the periphery is populated mostly by rods, with fewer cones, and density of photoreceptors decreasing away from the center region. Consequently, to resolve the fine detail in a scene, gaze has to be constantly shifting towards the locations of interest in a sequence of rapid eye movements called *saccades*, facilitated by the eye muscles, with peripheral vision identifying the regions of interests for the saccades. Gaze shifts can be also accomplished by movements of the head, and the entire body.

Insect eyes have evolved a different optical structure. The compound eyes of these animals consist of a large number of *ommatidia*, direction-sensitive light-sensing units, each consisting of a micro-lens covering a small number of photoreceptor cells; the image is formed across the compound eye as the individual ommatidia are arranged on a curved surface, each one pointing in a slightly different direction. This has the advantage of providing a large field of view in a very

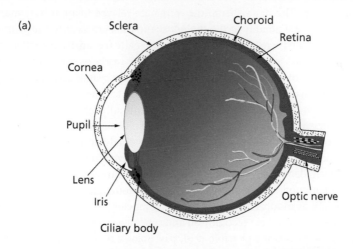

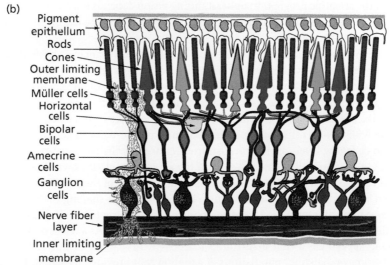

Figure 14.1 Vertebrate eye: (a) cross-section of the eye, (b) structure of the retina.

Figures reproduced from http://webvision.umh.es/webvision/index.html

compact optical system. Other, simpler optical structures, such as concave mirrors and pinhole cameras, or even more basic eyes with clusters of photosensitive cells providing rudimentary sensitivity to the intensity and direction of light can be found in less advanced animal species.

The detection of light, however, is only the first stage in the visual pathway. Eyes are not simply "cameras" that capture and transmit the images to be then processed by some distant circuitry in the brain—it is inside the eyes where the actual processing of the visual information starts. Fruit flies have very small brains (about 10^5 neurons), yet perform sophisticated vision processing that allows them to maneuver at high flight speeds, avoid obstacles, and precisely land on a surface. Much of this ability can be attributed to the image processing carried out right next to the sensors in the eye, using specialized neural circuitry to determine the speed and direction of motion of objects in the fly's visual field.

The human brain is far more complex (about 10^{11} neurons), and of course, capable of a far more sophisticated interpretation of visual information. Visual perception, including the ability to recognize objects and their relations in the environment, is a demanding task, which engages large parts of the entire brain. The signals from the retina are relayed in an orderly manner to the other parts of the brain, most notably to the *thalamus* and then onwards to the *primary visual cortex*. While the processing there is organized retinotopically (i.e. adjacent cortical regions correspond to adjacent locations on the retina), the complex networks of cortical neurons start to integrate the various types of available information over larger spatial and temporal scales. It is also there that the signals from both eyes are put together for the first time. From there, information flows to other brain regions, and back, being continually refined and interpreted in the context of other sensory information and past knowledge and experience. The full details of this process are still unknown, but neuroscience is making a steady progress towards unraveling its mysteries. Basic neural pathways and brain regions involved in various tasks that serve visual perception have been identified. Models of neural systems carrying out lower-level vision tasks such as feature, orientation, and motion selectivity, contour integration, colour perception, stereopsis, etc., as well as models integrating these competencies toward the accomplishment of higher-level tasks such as object recognition and space awareness, are being continually developed and tested through various physiological, psychological, and computational methods. Much of the scientific progress in understanding vision depends on, but also feeds into, our overall ability to understand the information processing principles in the brain (see Leibo and Poggio, Chapter 25, this volume).

In this chapter, however, our main focus is on how biological principles have influenced the development of engineered vision systems at the level of sensors and early sensory information processing. The vertebrate retina develops as part of the brain, and contains neural circuitry that performs sophisticated computations. The basic structure of the retina is schematically depicted in Figure 14.1b. The information captured by the photoreceptors is processed in parallel channels, starting with *bipolar* cells that respond with increased or reduced activity, depending on their type, to changing levels of neurotransmitter released by the photoreceptors. The bipolar cells connect to *ganglion* cells, which then send the action-potentials down their axons, which are bundled to form the optic nerve that transmits the data out of the eye. This direct pathway is modulated by *horizontal* cells, which influence the bipolar cells shaping their receptive fields through the addition of information from neighbouring photoreceptors, and *amacrine* cells providing further spatial and temporal processing of ganglion signals. As a result of this processing, a typical ganglion cell has a centre-surround receptive field that, for instance, emphasizes the "edges" in the images. Distinct populations of ganglion cells provide ON signals (increased firing in response to a light patch on a dark background) and OFF signals (detecting dark patches on a light background). Colour information is similarly represented by differences in signals rather than their absolute values. Furthermore, ganglion cells respond on various timescales, with some being activated by an onset or disappearance of a light stimulus (hence detecting rapid appearance of objects and motion in the visual field), and some responding more to a sustained light stimulation. An attempt can be made at describing the operation of these circuits through combinations of spatial and temporal filters, although prevalent non-linearities and feedback paths, non-uniform physical distribution of rods and various types of cones and their different processing circuits, together with the complex spatio-temporal characteristics of the images impinging on the photoreceptors, caused by saccades and micro-saccades (tiny oscillatory eye movements that are always present between the gaze shifts), make for a very complicated system. And, in addition to several relatively well understood neural circuits in the retina, many more exist, formed by the plethora of amacrine, horizontal, and ganglion cell types, communicating through a combination of electrical and biochemical signalling, from direct electrical couplings, through rapid release of neurotransmitters in response to cellular potentials, to slow diffusion of neuromodulators across larger distances. Cells have been identified that, for example, provide a direction-of-motion response (i.e. only fire if the image on the retina moves in a particular direction), and others with a looming response (signalling an expanding object image and thus directly detecting an approaching object) (Münch et al. 2009). Furthermore, the neural code used to represent the information extracted by the retina also appears complex. Ganglion cells have been found that appear to encode both the intensity and edge information using a combination of firing rate and time-to-first spike (Gollisch and Meister 2010).

While a comprehensive description of the circuits and the function of the retina still eludes scientists, it is clear that retinas perform sophisticated pre-processing on the sensory signals, before sending the information through to the rest of the brain. Undoubtedly, all this helps to relieve the cortex of some of the workload, facilitating the brain's amazing ability to evaluate complex visual situations in a short time. Retinal computation, together with a region-of-interest processing strategy implied by the saccades, help to reduce the volume of sensory data that needs to be processed by the cortex, and hence to make more efficient use of the limited physical resources available to carry out computations in the brain.

Biomimetic systems

The extraction of features of the visual signal right at the sensor level, using massively parallel networks of locally interconnected analog processing units operating in continuous time and placed right next to the sensors, is a key feature of biological vision systems. The computations are localized, with the neural circuitry processing the information originating from the adjacent photoreceptors, and neurons communicating information to other neurons within a small neighbourhood. Only the results of these computations, providing more informative data than the raw light intensity information obtained by the photosensors, are transmitted onwards from the eye for further processing. This highly efficient scheme has provided cues to several engineered approaches that try to optimize performance, power efficiency, and size of the vision system beyond what is possible when using a traditional setup of a conventional video camera feeding a central processing unit. The following section presents several examples of such systems.

Mimicking insect eyes

Insect eyes have long been an inspiration for designing vision sensors. The intricate structure of the compound eye has captured the imagination of engineers who have attempted to reproduce the physical organization of this system in hardware (Figure 14.2). Recent systems developed by

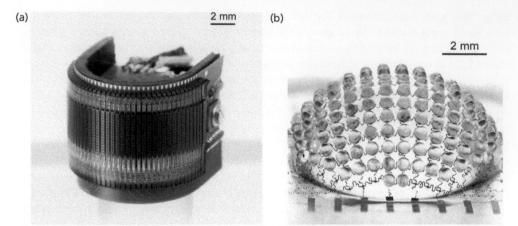

Figure 14.2 Insect-inspired vision sensors: (a) the CURVACE sensor. (b) Sensor developed by Song et al. (2013).

(a) Photo courtesy of Professor Dr Dario Floreano, LIS-EPFL. (b) Photo courtesy of Professor J. Rogers, University of Illinois, USA.

Floreano et al. (2013) and Song et al. (2013) integrate arrays of polymer micro-lenses, and silicon photodetector circuits on flexible substrates, which are then mechanically curved to provide the varying direction of the optical axes and thus covering the wide-angle visual field with over-lapping direction-sensitive receptive fields of individual sensor elements, much like the insect ommatidia. But it is not just the physical arrangement of the insect eye, with its compact yet wide field of view, that is intriguing from the engineering perspective. Its processing capabilities are no less remarkable. Researchers working on miniature autonomous flying robots are faced with the challenge of providing enough computational power to carry out visual information processing required for obstacle avoidance and navigation, given the extremely stringent weight and power consumption budgets of micro air vehicles. The simple yet effective way in which insects are able to extract the direction of motion information directly at the sensor level has provided the inspiration for attempting to solve this problem. The work of Barrows et al. (2003) and others has demonstrated that through the implementation of the algorithms mimicking motion processing in the neural circuitry of insect eyes, either in conventional hardware or through specialized electronic circuitry, sufficient information can be obtained to implement basic flight maneuvering without the need for a powerful central processor or a high-resolution imaging system.

Mimicking vertebrate retinas

The neural circuitry supporting the near-sensory processing of visual information in insects is relatively simple, and engineered systems may attempt to mimic their functionality in elec-tronic hardware with reasonably high fidelity. Vertebrate retinas present a far more complex picture. Nevertheless, retinas were providing models for some of the very first *neuromorphic* VLSI (Very Large Scale of Integration) silicon integrated circuits developed in the late 80s and early 90s, first in Carver Mead's lab at Caltech and later in other labs worldwide. The full reper-toire of retinal responses could never be captured by these devices, and the technological limi-tations of the day meant they never provided more than proof-of-concept studies, rather than any practical solution to building artificial vision systems. Nonetheless, designs by Mahowald, Indiveri, Delbruck, Liu, Boahen, Andreou, and others have pioneered the implementation of microelectronic circuits mimicking neural circuits, usually focusing on providing an electronic circuit equivalent of one particular aspect of a simplified retina model. Constructed in a manner similar to image sensor arrays, which are nowadays ubiquitous in digital cameras, and implemented in industry-standard CMOS chip technologies, these devices used pixelated photosensor (e.g. photodiode or phototransistor) matrices to transform incoming light into electrical signals. However, rather than directly outputting thus obtained image data, they placed signal processing transistor circuitry right next to the photosensors, so that image pixels captured by the sensor could be immediately processed *in situ*, with the complete silicon retina chip outputting not raw light-intensity images, but instead results of processing the images with corresponding filters, according to the specific circuit design and the implemented model. The work by Zaghloul and Boahen (2006) typifies this approach, with arguably the most comprehensive silicon retina to date (Figure 14.3). Thirteen different neuron cell types have been modeled in electronic circuitry, providing spiking responses resembling those generated by four distinct types of retinal ganglions. The authors speculate that such circuits might find applications in retinal prosthetics. In such applications, the biological realism of the produced outputs might indeed be of great importance, although it remains to be seen how much of the retinal complexity needs to be replicated in order to provide a viable artificial replacement for the organ.

From the perspective of constructing artificial vision systems, however, the detailed modeling of the neurophysiology of the retinal tissue in electronic circuits would only make sense if the rest of the system could efficiently use this information to improve the overall performance. Given the complexity, as well as the still incomplete understanding, of not only retinal function but also the rest of the biological visual system, this is not the case today. Nevertheless, the basic principle of performing computations right next to the sensors, and outputting not raw images but only data relevant to further processing by the system, has very clear practical advantages, as compared with a more conventional way of sensing, digitizing the images, and then processing the data on inherently sequential (single-core or even many-core) computers. The massive parallelism—thousands of hardware processing units, each processing only the immediate neighbourhood of a pixel—offers huge computational speedups. The reduction in overall data transfer requirements of the system, and efficiency with which individual pixel-level computations can be performed by dedicated circuitry, lead to dramatic power consumption savings.

Dynamic vision sensors

In can be argued that functional abstractions of retinal function, at a level that well matches the capabilities of the rest of the system, are needed to fully exploit the potential suggested by the basic organizational principles of biological vision. An excellent example of how this can be achieved is provided through the work of Lichtsteiner et al. (2008), and related developments of Dynamic Vision Sensors (DVS). The retinal function has been abstracted here to a very simple operation—these chips produce instantaneous binary ON and OFF responses (asynchronously firing ganglion cells) in pixels that increase or decrease in light intensity (Figure 14.4a, b). Logarithmic sensing ensures that a large dynamic range of light intensities is handled internally, and the output is generated by sensing temporal contrast at individual image locations, with no lateral communication between the pixels. Most notably, the output from the chip, in the form of a stream of addresses (array coordinates) of pixels that change, is entirely input-driven. Conventional vision systems operate with video frames, i.e. discrete images acquired at sampling frequencies dictated by the system's frame rate. For example, when processing a typical video stream of 25 fps (frames per second), a new value is obtained for each pixel every 40 ms. In contrast, the DVS system produces an asynchronous stream of pixel change "events," with latencies in the range of 1 μs, outputting only those pixels that actually change. This reduces the

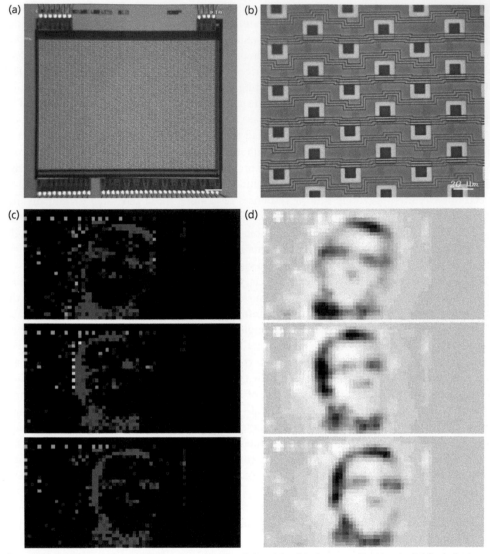

Figure 14.3 Silicon retina: (a) microphotograph of the chip; (b) close-up showing individual pixels; (c) output from the chip: three frames of a sequence are shown with colours (red, blue, green, yellow) corresponding to firing of four distinct retinal ganglion cell types, black cells are not firing; (d) reconstructed grayscale images.

sensor/processor bandwidth, and facilitates, for example, high-speed motion tracking. Several robotic applications of DVS devices have been demonstrated (Figure 14.4c). In each case, the bulk of the vision processing relies on algorithms running on a conventional microprocessor, but whose computational load is greatly reduced by the pre-processing carried out by the DVS chip.

Figure 14.4 Dynamic vision sensor: (a) output of the DAVIS chip (Brandli et al. 2014), white pixels represent ON events, black pixels represent OFF events, generated within a short time window as the camera is panned from right to left; (b) corresponding gray-level light intensity image output from the chip; (c) a DVS-based camera (pointed by the arrow) mounted on a line-following robot. Images courtesy of T.Delbruck, INI Zurich.

Vision chips with pixel-parallel processor arrays

Retina-inspired vision chips implementing simplified functional models can provide practical solutions for building artificial vision systems, but applications of devices that carry out only one particular aspect of early visual processing (e.g. temporal contrast, detection of edges, direction of motion, etc.) are limited. The circuits for computing various functions could be integrated on a single chip, but the circuit area in a silicon pixel is limited if practical image resolutions are to be obtained. To relieve this tension between versatility and space, a multitude of specialized continuous-time circuits (one per function) can be replaced with a programmable processing element, which contains an execution unit capable of a basic set of elementary operations and a memory. Following the Universal Turing Machine concept, the functionality of such a processor is then fully described by the code it executes. This does not mean that the desirable features of the biological blueprint must be foregone. Like in retinas, computations can still be carried out in massively parallel fashion, and processing elements can be tightly integrated in physical proximity to the image sensors. Indeed, it is even possible to continue carrying out the computations in the analog domain—while the software-programmable processor concept has been fundamental to modern digital computing, it can be also applied to a sampled-data analog system. The vision chip described by Carey et al. (2013) demonstrates this approach (Figure 14.5). Each processing element contains a photosensor, arithmetic/logic unit, and memory. A mixture of digital and analog circuits is used to achieve a balance between the robustness of digital computation and energy efficiency of analog computing. The system operates as a SIMD (Single Instruction Multiple Data) machine, with an array of 65,536 parallel processors, one per image pixel, executing a sequence of instructions provided by a controller. Lateral connectivity between pixels allows the implementation of spatial neighbourhood filters, while in-pixel memory provides a means to execute temporal operations. The chip can thus be programmed to compute spatio-temporal filters akin to retinal pre-processing, as well as more abstract maps (e.g. edge orientation, feature detection, and optic flow), in some sense also emulating early vision processing in parallel networks of the visual cortex. While the computations across the array are synchronized, and pixel data is sampled, the high speed afforded by the massively parallel processing permits very high frame rates (thousands of frames per second, if required). The information is not transmitted out of the chip at this rate; like in a retina the information can be extracted by the near-sensor circuitry, and only relevant data transmitted onwards to further processing stages. High frame rates permit data-driven, event-based output, continually generating neural "spikes" at detected locations of interest in the visual field.

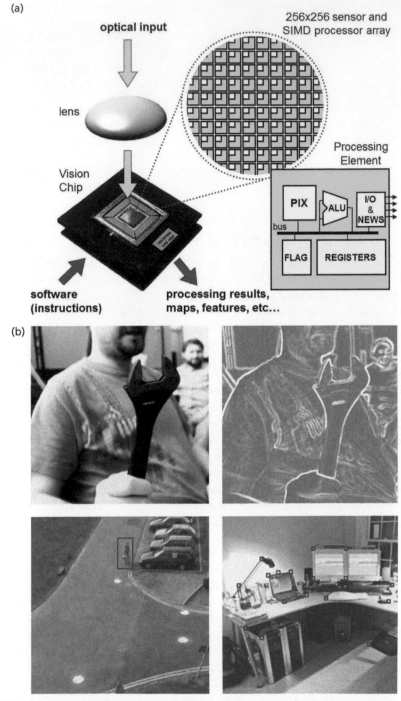

Figure 14.5 Vision sensor with a pixel-parallel SIMD processor array by Carey et al. (2013): (a) system architecture; (b) illustration of in-pixel processing capabilities, clockwise from top-left: raw image, edge detection, Region-of-Interest (ROI) detection based on motion, extraction of interest points.

The operating speed can be also traded-off for power consumption, so if lower frame rates are acceptable then ultra-low power operation is possible.

In their ability to extract information on the focal plane, vision chips can even surpass biology, being capable of producing sparse, highly informative "events," for example describing locations of points of interest computed using elaborate feature extractors. These could be used in vision algorithms carrying out tasks such as object recognition and visual navigation, or as a peripheral system to determine regions-of-interest that should be further examined with a more elaborate, high-resolution (i.e. foveal) vision system.

Future directions

The systems described in this chapter exploit the advantages of modern microelectronic fabrication technologies that allow the integration of large numbers of transistors in a small silicon area. This enables the placement of sophisticated processing circuitry in each pixel of an image sensor. The space restrictions, however, still limit the achievable pixel densities and today's state-of-the-art vision chips have modest resolutions, in the range of 256×256 pixels. Biological systems place the processing circuitry in a tightly packed, multi-layered network, adjacent to a layer of photoreceptors. The advances in wafer-stacking silicon integration technologies might allow a more similar arrangement, and indeed some 3D integrated vision chips have been proposed (see for example Dudek et al. 2009). Another advance may come from flexible circuit substrates, to replace the stiff and brittle silicon wafer. This would allow the fabrication of more compact, and more tightly integrated, insect-like compound eyes.

Superior light sensitivities of biological photoreceptor cells can be matched by photodetectors based on Single Photon Avalanche Diodes (SPADs). These, as well as other sensor types (e.g. colour, infrared, polarization-sensitive, etc.) have not yet been fully explored in the context of vision chips. Animals recover depth information from binocular disparity, or estimating depth from motion (in addition to using other senses, e.g. auditory cues or echolocation). While brain-inspired systems pursuing these strategies have been investigated, engineered systems can also use methods relying on active illumination such as time of flight measurements or using structured light. These capabilities could be integrated into the vision sensors. Development of methods to fuse information from various sensor types is also a challenge.

Traditional computer vision algorithms are designed to execute efficiently on conventional sequential CPUs or GPU-type parallel hardware, and assume the availability of a sequence of high-resolution video frames. New algorithms and approaches, and better understanding of both the advantages and the limitations of carrying out some of the processing at the sensor level, and the desirable balance between the on-sensor and off-sensor computation, are needed to fully realize the potential of bio-inspired vision sensors. This can only be achieved by future research into the application of these devices in the context of complete vision systems.

Learning more

For more in-depth information on the biology of the eye and the retina, the "Webvision" website (http://webvision.med.utah.edu/) is an excellent resource (Kolb et al. 2013). Gollisch and Meister (2010) provide a good review of recent findings on neural computations in retinal circuits. *Flying insects and robots*, edited by Floreano et al. (2010), contains many contributions on state-of-the-art in insect-inspired vision. *Vision chips* by Moini (1999) charts many early attempts at building artificial retinas and bio-inspired vision chips, while several more recent vision chips with pixel-parallel processing capabilities are presented in *Focal-plane sensor-processor chips*, edited by Zarandy (2011). The chapters in this handbook by Bensmaia (Chapter 53), Lehman and van Schaik (Chapter 54), and Pearce (Chapter 17), discuss neuromorphic technologies for touch, audition, and chemosensation, for applications in areas including prosthetics. The chapter by Metta and Cingolani (Chapter 47) discusses the challenge of combining multiple neuromorphic systems, including neuromorphic vision, for the development of humanoid robots.

References

Barrows, G. L., Chahl, J. S., and Srinivasan, M. V. (2003). Biomimetic visual sensing and flight control. *Aeronautical Journal*,107, 159–68.

Brandli, C., Berner, R., Yang, M., Liu, S.-C., and Delbruck T. (2014). A 240×180 130dB 3us latency global shutter spatiotemporal vision sensor. *IEEE J. Solid State Circuits*, 49(10). doi: 10.1109/ JSSC.2014.2342715

Carey, S. J., Barr, D. R. W., Lopich, A., and Dudek, P. (2013). A 100,000 fps vision sensor with embedded 535 GOPS/W 256x256 SIMD processor array. In: *VLSI Circuits Symposium 2013*, Kyoto, June 2013, pp. 182–3.

Dudek, P., Lopich, A., and Gruev, V. (2009). A pixel-parallel cellular processor array in a stacked three-layer 3D silicon-on-insulator technology.In: *European Conference on Circuit Theory and Design, ECCTD 2009*, August 2009, pp.193–97.

Floreano, D., Zufferey, J.-C., Srinivasan, M.V., and Ellington, C. (eds.) (2010). *Flying insects and robots.* Berlin: Springer.

Floreano, D., et al. (2013). Miniature curved artificial compound eyes. *Proc. Natl Acad. Sci. USA*, 110(23), 9267–72.

Gollisch, T., and Meister, M. (2010). Eye smarter than scientists believed: neural computations in circuits of the retina. *Neuron*, 65, 150–64.

Kolb, H., Nelson, R., Fernandez, E., and Jones, B. (eds) (2013). *The organization of the retina and visual system*. Webvision, http://webvision.med.utah.edu/ [Internet].

Lichtsteiner, P., Posh, C., and Delbruck, T. (2008). A 128×128 120dB 15 us asynchronous temporal contrast vision sensor. *IEEE Journal of Solid-State Circuits*, 43(2), 566–76.

Moini, A. (1999). *Vision chips*. Dordrecht: Kluwer Academic Publishers.

Münch, T. A., Azeredo da Silveira, R., Siegert, S., Viney, T. J., Awatramani, G. B., and Roska, B. (2009). Approach sensitivity in the retina processed by a multifunctional neural circuit. *Nature Neuroscience*, 12(10), 1308–16.

Song, Y. M., Xie, Y., Malyarchuk, V., Xiao, J., Jung, I., Choi, K.-J., Liu, Z., Park, H., Lu, C., Kim, R.-H., Li, R., Crozier, K. B., Huang, Y., and Rogers, J. A. (2013). Digital cameras with designs inspired by the arthropod eye. *Nature*, 497, 95–99.

Zaghloul, K. A., and Boahen, K. (2006). A silicon retina that reproduces signals in the optic nerve. *Journal of Neural Engineering*, 3(4), 257–67.

Zarandy, A. (2011). *Focal plane sensor-processor chips*. Berlin: Springer.

Chapter 15

Audition

Leslie S. Smith

Department of Computing, Science and Mathematics, University of Stirling, UK

Audition is the ability to sense and interpret pressure waves within a certain range of frequencies. One can consider this as trying to solve the *what* and *where* tasks: what is the source of the sound (pressure waves), and where is it? Auditory environments are very varied, in the number and location of sound sources, their level and in the degree of reverberation, yet biological systems seem to have robust techniques that work over a large range of (though not all) conditions. We provide a brief review of the auditory system, attempting to connect its structure with the problems that it must solve: specifically locating some of the sound sources, and interpreting those that are important for the animal. Systems that use some of what is understood about animal auditory processing are discussed, including both CPU-based and neuromorphic approaches. It is, however, clear that the level of performance associated with animal auditory systems has not been achieved, and we discuss possible ways forward.

Biological principles of the auditory system

The biological auditory system comprises the outer, middle, and inner ears (where the sound is still a pressure wave), transduction to the neural code at the organ of Corti in the cochlea, processing and (re-)representation of this code in the brainstem and mid-brain nuclei, and finally cortical processing of sound. While this description seems to define a unidirectional system, from ear to cortex, there are many connections in the opposite direction as well. Further, auditory processing is often modified both by other sensory modalities and levels of alertness.

The pressure wave from a sound source arrives at the outer ears (pinnae) both directly and after reflecting from and diffracting around surfaces. After arriving there, the pressure wave is reflected down the auditory canal to the eardrum. With the exception of bone conduction, this is the sole source of signal to the rest of the auditory system. The (linear) path from sound source to eardrum not only introduces a delay, but also modifies the sound spectrum partly because the path has different attenuations at different frequencies, and partly because of resonances. The head-related transfer function (HRTF) (Blauert 1996) summarizes the response at each ear. Animals have two ears, and the differences in signal timing and spectrum at the ears are the primary cues for sound source location.

After arriving at the eardrum, the signal is impedance matched and transferred to the fluid-filled cochlea through the three bones of the middle ear. The effectiveness of this transfer can be reduced through the acoustic reflex, attenuating transfer in loud environments, and hence compressing the signal. The cochlea and organ of Corti (described in more detail in Chapter 54) are responsible for the separation of the sound into different frequencies, and for the transduction of this sound into auditory nerve spikes (action potentials). They enable further compression, adding to the compressive capabilities of the middle ear.

The auditory nerve (which has about 30 000 fibres in humans) mostly (90–95%) consists of type 1 fibres which transfer the signal transduced by the inner hair cells to the cochlear nucleus (CN) in the brainstem. These have a clear tonotopic organization, and this continues in the ascending nuclei. There are also type 2 fibres which transfer signal from the outer hair cells to the CN. In addition there are descending fibres (originating from the LSO and MSO in the superior olivary complex), which modulate the action of the outer and the inner hair cells. The brainstem contains many nuclei that are concerned with auditory processing: see Figure 15.1. These are generally in pairs (one per ear), and the ascending pathways sometimes stay on one side and sometimes cross. The lowest level crossing occurs between the anteroventral cochlear nucleus (AVCN) and the MSO and MNTB in the olivary complex (OC) (see Figure 15.1): this is implicated in detecting the timing and intensity differences that are the cues for sound source direction finding.

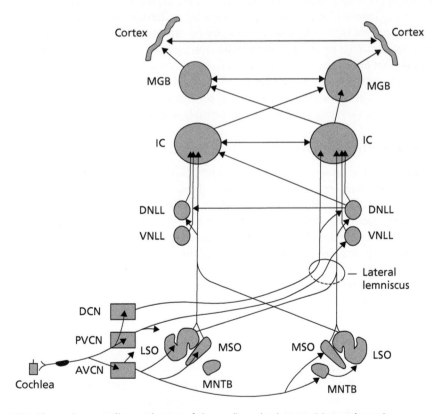

Figure 15.1 The main ascending pathways of the auditory brainstem. Many other minor pathways (as well as descending pathways) are not shown. Branching or joining of arrows does not imply that the fibres branch or join. Acronyms: DCN, PVCN, AVCN: Dorsal/Posteroventral/Anteroventral Cochlear Nucleus; LSO, MSO: Lateral/Medial Superior Olive; MNTB: Medial Nucleus of the Trapeziodal Body; VNLL, DNLL: Ventral/Dorsal Nucleus of the Lateral Lemniscus; IC: Inferior Colliculus; MGB: Medial Geniculate Body.

One component of the OC, the lateral superior olive, appears sensitive to intensity differences, and another, the medial superior olive, to timing differences. These timing differences are very small, being of the order of tens of microseconds, and require very specialized neural circuitry such as the giant synapse known as the Calyx of Held, a very large and fast inhibitory synapse in the Medial Nucleus of the Trapezoidal Body (MNTB), also in the olivary complex.

Another pair of brainstem nuclei, the lateral lemnisci, are innervated by the cochlear nucleus and by the olivary complex, both ipsi- and contralaterally. This suggests that they are important in binaural hearing, but exactly how is not clear. Their neurons' spectral receptive fields are more complex, and may relate to harmonic relations (see p 182 of Pickles (2008)). The inferior colliculus (IC) receives inputs from most of the brainstem auditory nuclei (ascending projections), and itself projects to the medial geniculate body (MGB). It is also the target of descending projections from both the cortex and the MGB. The IC is relatively large and highly connected, suggesting a critical role. It seems to be the first precortical nucleus that codes spectrotemporally, rather than purely spectrally, with sets of neurons which appear to code both envelope and frequency modulation.

The MGB is a thalamic nucleus, parts of which are tonotopically organized. It seems to function both as an area where modulation by other sensory and cortical areas first occurs, and it has major projections to the primary auditory cortex (PAC: marked cortex in Figure 15.1). The PAC has, like the rest of the cortex, a laminar structure, with six layers. Although it has a tonotopic organization, it has multiple other maps overlaid as well, including sharpness of tuning and frequency modulation. The spectrotemporal response functions (STRFs) are often non-linear (Theunissen et al. 2000), which means that they are difficult to investigate using simple engineered sounds. They tend to be tuned in some sense to the environmental sounds that matter to the animal. In addition, the PAC presumably has a cognitive aspect as well, discussed in chapter 18 of Rees and Palmer (2010).

Biomimetic and biohybrid systems

The best-known technological translations of the auditory system are hearing aids, and in particular cochlear implants. Hearing aids need to re-present the signal to the ear, so that while knowledge of auditory processing can be used in the design of algorithms, less advantage is gained. Cochlear implants present the signal more directly to the auditory nerve, and are discussed in detail in Lehmann and van Schaik (Chapter 54, this volume). For hearing aids, compression, specifically in those parts of the auditory spectrum where there is reduced sensitivity, is important. In addition, some binaural hearing aids attempt to maintain the precise (relative) timing at the two ears in order not to damage sound direction cues.

Another important technological translation is systems for recognizing speech: these are in daily use in many fields. While they all employ the (approximately) logarithmic frequency coding used in the cochlea, their general architecture otherwise owes little to early auditory processing. Most use a hidden Markov model applied to regularly produced (e.g. every 25 ms) mel-frequency cepstral coefficient (MFCC) vectors. However, the MFCC approach is inherently unable to ignore parts of the sound: it has to process the whole signal as if it was a single source whereas real sound fields contain energy from many concurrent sounds. Further, either the speaker/microphone configuration needs to be fixed, or else the system must be trained on a variety of plausible speaker/microphone configurations.

There is growing interest in landmark (also called event or feature) based approaches in which short sections of the utterance are identified as particular events (Stevens 2002; Hasegawa-Johnson et al. 2005; Lin and Wang 2011). This bottom-up approach has a connection to the responses of specific brainstem nucleus neurons, many of which appear to have an event-detecting capability. Such an approach can be used in conjunction with MFCC approaches (Lin and Wang 2011), or it can be used to attempt to stream the sound, that is, firstly to ignore parts of the sound not involved in the landmark, and secondly to allow some landmarks to be ignored (or to consider that the sequence of landmarks comes from more than one sound source). Further, landmarks are often chosen in such a way that they are relatively resilient to changes in the spatial configuration of the sound sources and microphones.

Audition is, however, more than speech recognition. Auditory scene analysis (Bregman 1990) aims to discover the sound sources and their locations. Attempting to find the location of a sound source (for example for a robot) must assume both multiple sources and unknown reverberation. MFCC approaches are entirely unsuitable here, because they lose all the fine time structure of the sound. There has therefore been an upsurge in interest in both hardware and software techniques that provide information about the spectral energy distribution of the signal, but without losing the fine time structure: these are often based on multiple concurrently running filters, usually placed approximately logarithmically, as in the cochlea.

Software based techniques are easier to develop, and many researchers have developed auditory front ends that start with filterbanks (most commonly the gammatone filterbank)(Lyon et al. 2010), then continue with models of hair cells, and of neurons within the auditory brainstem. However, they rarely run in real-time (at least on normal computers), and although biomimetic (and useful for developing techniques), they are not biomorphic. They have been invaluable as test-beds for development, for example in finding the direction of the foreground sound source in a robotic system by determining when to measure the time differences using sound onsets (Huang et al. 1997; Smith and Fraser 2004). This implements a version of the precedence effect, where it is the initial onset of a sound that animals use to determine the direction of the sound source: later parts of the sound arrive after reflection, and would give an erroneous estimate of the sound source direction.

To achieve sufficient real-time performance, cochlear models need to be implemented in hardware, whether analog or digital. To achieve appropriate performance, cochlear models need to be active (i.e. containing amplification). Early discrete component models were large and unwieldy. Integrated circuit biomorphic cochleae ("silicon cochleae") have been implemented since 1988 (see Lyon and Mead's chapter in Mead (1989)), and they have been applied to specific problems, such as localization and pitch perception. The real cochlea is active, and attempts have been made to implement this (Fragniére et al. 1997; Hamilton et al. 2008; Liu and Delbruck 2010). The van Schaik group's implementations go beyond modeling the organ of Corti, and model some of the neurons in the brainstem nuclei as well. Others (Liu et al. 2010) have developed silicon cochleae using digital technology. Although this area has not yet been incorporated into the mainstream, there is growing interest (indeed, in such neuromorphic systems in general) particularly because autonomous robots really need an auditory sense for location and interpretation of sound.

Future directions

Auditory scene analysis by machine is not yet approaching the capabilities of humans. Whether in the ability to identify concurrent sounds, or the ability to understand speech in the presence of many competing sources (often from other speakers, and therefore with similar statistical characteristics), humans currently thoroughly outperform machines. As noted above, there is a lot of processing applied to the signal before it gets to the cortex—considerably more than appears to be applied to visual signals. Trying to understand this early auditory processing so that a form of it can be applied is therefore an area of current research interest.

Currently the speech interpretation community is very much tied to the MFCC/HMM technologies: where the speech can be supplied as a very clean, non-reverberated signal, this may be appropriate, but for robotics something more is needed.

There is also considerable interest in music technologies: can one determine (for example) the title or the genre of music, or can one transcribe some music directly from the sound? Where biomorphic approaches may be appropriate is in the general case (rather than in the case of identifying a specific digital recording, which is a different problem that humans are generally unable to do). But biomorphic techniques, for example using onsets, amplitude, and envelope modulation, may be able to identify the start of each note or chord.

Sound is often not processed alone: often we have information from other modalities present as well. Examples such as the McGurk effect, or the well-known effect that visual direction will often take precedence over auditory direction show the strength of this in humans. There is therefore interest in joining these modalities together.

Learning more

Pickles (2008) has produced an excellent introduction to the animal hearing system. A more detailed description may be found in chapters 2 and 3 of Rees and Palmer (2010). There is a whole book on the Inferior Colliculus (Winer and Schreiner 2005), and the first chapter of that book reviews the whole of the central auditory system. Beyond that, there is a huge literature on the auditory neuroanatomy of many different animals, to be found often in the journal *Hearing Research*, or else in neurophysiology journals.

Lyon's review (Lyon et al. 2010) provides an excellent starting point for further investigation into software cochlear models. Many of these are freely available. Bregman's (1990) book on *Auditory Scene Analysis* remains a classic; Wang and Brown's (2006) book brings it more up-to-date. Blauert's classic on spatial hearing has a 1996 edition (Blauert 1996), the standard work in this area. Landmark based speech interpretation is thoroughly reviewed in Hasegawa-Johnson et al. (2005): it is also discussed in chapter 3 of Wang and Brown (2006). Although now ageing, Mead's (1989) book remains a useful introduction to Neuromorphic Systems: Liu and Delbruck (2010) provide an up-to-date review of sensory neuromorphic systems.

References

Blauert J. (1996). *Spatial Hearing*, Revised edition. Cambridge, MA: MIT Press.

Bregman, A. S. (1990), *Auditory scene analysis*. Cambridge, MA: MIT Press.

Fragnière, E., van Schaik, A., and Vittoz, E. (1997). Design of an Analogue VLSI Model of an Active Cochlea. *Analog Integrated Circuits and Signal Processing*, **12**, 19–35.

Hamilton, T.J., Jin, C., van Schaik, A., and Tapson, J. (2008). An Active 2-D Silicon Cochlea. *IEEE Transactions on Biomedical Circuits and Systems*, **2**(1), 30–43.

Hasegawa-Johnson, M., Baker, J., Borys, S., Chen, K., Coogan, E., Greenberg, S., et al. (2005). Landmark-Based Speech Recognition: Report of the 2004 Johns Hopkins Summer Workshop (Vol. 1, pp. 213–216). *Proceedings IEEE International Conference on Acoustics Speech and Signal Processing*. IEEE.

Huang, J. Oshnishi, N. and Sugie, N. (1997). Sound localization in reverberant environment based on a model of the precedence effect, *IEEE Trans. Instrum. Meas.* **46**, 842–6.

Lin, C.-Y., and **Wang, H.-C.** (2011). Automatic estimation of voice onset time for word-initial stops by applying random forest to onset detection. *Journal of the Acoustical Society of America*, **130**(1), 514–25.

Liu, J., **Perez-Gonzalez, D., Rees, A., Erwin, H.,** and **Wermter, S.** (2010). A biologically inspired spiking neural network model of the auditory midbrain for sound source localisation. *Neurocomputing*, **74**, 129–39.

Liu, S.-C., and **Delbruck, T.** (2010). Neuromorphic sensory systems. *Current Opinion in Neurobiology*, **20**(3), 288–95.

Lyon, R.F., **Katsiamis, A.G.,** and **Drakakis, E.M.** (2010). History and future of auditory filter models (pp. 3809–3812). *Proceedings of 2010 IEEE International Symposium on Circuits and Systems (ISCAS)*, May 30–June 2, 2010, Paris, France.

Mead C. (1989). *Analog VLSI and Neural Systems*. Boston, MA: Addison-Wesley.

Pickles, J.O. (2008). *An Introduction to the Physiology of Hearing*, 3rd edn. Bingley, UK: Emerald.

Rees, A., and **Palmer, A.R.** (2010). *The Oxford Handbook of Auditory Science: The auditory brain*. Oxford: Oxford University Press.

Smith, L.S., and **Fraser, D.** (2004). Robust sound onset detection using leaky integrate-and-fire neurons with depressing synapses. *IEEE Transactions on Neural Networks*, **15**(5), 1125–1134.

Stevens, K.N. (2002). Toward a model for lexical access based on acoustic landmarks and distinctive features. *The Journal of the Acoustical Society of America*, **111**(4), 1872–91.

Theunissen, F. E., **Sen, K.,** and **Doupe, A. J.** (2000). Spectral-temporal receptive fields of nonlinear auditory neurons obtained using natural sounds. *The Journal of Neuroscience*, **20**(6), 2315–31.

Wang, D., and **Brown, G.L.** (2006). *Computational Auditory Scene Analysis*. Hoboken, NJ: Wiley Interscience.

Winer, J.A., and **Schreiner, C.E.** (2005). *The Inferior Colliculus*. New York: Springer-Verlag,.

Chapter 16

Touch

Nathan F. Lepora

Department of Engineering Mathematics and Bristol Robotics Laboratory,
University of Bristol, UK

This article describes some principles underlying biological touch sensing in humans and non-human animals, and how these principles can result in biomimetic devices with improved capabilities compared with the state-of-the art in artificial touch. Three biological principles are considered. First, that cutaneous touch is superresolved, in that the acuity (accuracy) of perceiving fine stimulus detail is finer than the spacing between individual sensory mechano-receptors. Second, that touch is active, in that animals actively select and refine sensations in a purposive manner, rather than passively waiting for percepts to merely fall on our sensory organs. Third, that touch is exploratory, in that animals deploy purposive action patterns to encode properties of objects via a lexicon of exploratory procedures, such as tracing around or molding our fingers to an object. Biomimetic tactile systems have utilized these three principles to result in superior sensing capabilities, including systems that mimic the human fingertip and hand (cutaneous touch) and the rodent whisker system (vibrissal touch). Examples include using biomimetic tactile fingertips to attain superresolution sensing an order of magnitude better than the sensor resolution; actively controlling the contact between a fingertip or whisker array to optimize tactile perception; and deploying artificial hands to explore unknown objects. Future biomimetic touch could rival human capabilities, enabling tactile sensors to have widespread technological applications spanning across prosthetics, telehaptics, surgical robotics, biometric clothing, wearable computing, medical probes, and manufacturing

Biological principles

Touch is the ability to perceive and understand the world through physical contact (Prescott et al. 2016). The application of touch is also described as tactile (derived from the Latin *tactilis*) or as *haptic* (derived from the Greek *haptos*). As humans, we are intimately familiar with touch in the form of the sensory-muscular-skeletal system that has given success to our species: the human hand. From an appendage primarily for locomotion, in humans it has evolved into a specialized tactile system for grasping and manipulating objects and tools. Touch is also essential for other mammals, notably in the form of tactile whiskers in rodents that are used as a proximal sense around the snout. Rodents actively move their whiskers in a palpating "whisking" motion in a dynamic manner as they explore and interact with their surroundings (Figure 16.1).

In this article, we focus on three principles of biological touch that can underlie biomimetic tactile sensing and perception. The utility of these principles demonstrates how biological touch is a solid foundation on which to base robot touch.

Figure 16.1 (a) Depiction of superresolution, where the localization of stimulus extended over multiple sensor elements can be less than the sensor resolution. (b) Rats actively sweep their whiskers back and forth to sense their nearby environment with touch. (c) Enclosing an object with the hand to feel its shape is an example of haptic exploration, as is squeezing an object to feel its compliance.
(b) istock.com/scooperdigital. (c) istock.com/skodonnelly.

Touch is superresolved

Hyperacuity is the biological term for superresolution, and an aspect of perception in which the acuity (accuracy) of perceiving the fine detail of a stimulus is finer than the resolution (spacing) between individual sensory receptors. Hyperacuity is ubiquitous across the biological senses, including human touch, audition, and vision. Historically, hyperacuity was first studied in human visual localization. However, human touch is also known to attain hyperacuity and is the focus of this article. For example, subjects making brief static touches against embossed spatial patterns can estimate relative pattern size to 0.3 mm, an order of magnitude better than the 1–2mm spacing between (SA-I) mechanoreceptors in the fingertip (Loomis 1979). No physical laws are broken because perception can involve spatial averages over an array of activated sensory receptors, which can transcend the resolution limit. Thus, nature has discovered principles by which perceptual systems can function at a finer acuity than might be expected from the spatial arrangement of their sensory receptors.

The attainment of biological superresolution depends on the operation of both the peripheral sensory organs and central sensory processing. Our senses take their inputs via sensory receptors that transduce stimuli from the physical world into signals appropriate for neural processing. Each receptor has a receptive field (a region of space to which it responds), such as the region of skin that activates one tactile mechanoreceptor embedded within it. The spacing between receptors then defines the sensor resolution, which is surpassed when the sensing attains hyperacuity. Computationally, perceptual acuity is aided by coarse coding mechanisms over multiple overlapping receptive fields, enabling parallel encoding of the stimulus over a distributed population of sensory neurons. The somatosensory cortex functions to decode this population activity into internally coded evidence for distinct percepts, of use for planning and deciding what to do next. In consequence of sensory averaging during the perceptual process, these percepts can be at a finer detail than the spacing between sensory receptors, resulting in perceptual hyperacuity.

Touch is active

We not only touch, we feel. We do not just see, we look (Bajcsy 1988). We do more than hear, we listen. It is apparent from the most basic introspection that our senses are used in an active manner, by which we actively select and refine sensations rather than passively waiting for percepts to merely fall on our sensory organs. In this sense, perceptual activity is guided by our past sensations to search and probe for the perceptual information that we need to complete the tasks at hand.

Several definitions of active touch have been considered over the last 50 years (Prescott et al. 2011). However, all can be seen as stemming from influential work by the psychologist James J. Gibson in his 1962 article on *Observations on Active Touch*: "Active touch refers to what is ordinarily called touching. This ought to be distinguished from passive touch, or being touched. In one case the impression on the skin is brought about by the perceiver himself and in the other case by some outside agency." Here Gibson distinguishes between whether the agent is itself controlling its body to sense or, instead, whether a tactile sensation is exerted on the agent. This *purposive* aspect is central to his definition of active touch, as clarified later in the same article when he says: "The act of touching or feeling is a search for stimulation… When one explores anything with his hand, the movements of the fingers are purposive. An organ of the body is being adjusted for the registering of information."

Over the intervening half century, aspects of active touch with the human hand and fingertips have been documented across a large body of literature. One example of experimental work on active touch is that the sensation of surface roughness has been found to be altered by the manner that the surface is felt, in particular between active and passive modes of touch (Loomis and Lederman 1986). Much has been learnt also from touch in non-human animals, such as studies of how rodents deploy their whisker system, particularly on the role of feedback during active tactile sensing (Kleinfeld et al. 2006). Studies of active touch in other whiskered mammals and with insect antennae have also been made, including cockroach antennae, stick insect antennae, and tactile sensing in the naked mole rat, Etruscan shrew, and seals.

Touch is exploratory

The concept of *haptic exploration* was introduced by psychologists Susan Lederman and Roberta Klatzky in 1987, referring to purposive action patterns that perceivers execute to encode properties of surfaces and objects. In their words: "the hand (more accurately, the hand and brain) is an intelligent device, in that it uses motor capabilities to greatly extend its sensory functions" (Lederman and Klatzky 1987). Like Gibson, they emphasize the purposive nature of the movements guiding the hand and fingers to recognize stimuli, but they go further in proposing the motor mechanisms—a set of exploratory procedures—that constitute those purposive movements.

Haptic exploration seeks to be more concrete than an abstract notion of active touch by describing a taxonomy of *exploratory procedures*, each associated with an object property that is to be apprehended. The exploratory procedures associated with an object property are executed spontaneously when information about that property is desired. Examples include: lateral motion, to perceive surface texture, by passing the skin laterally across a surface to produce shear force; pressure, to perceive compliance or hardness, by exerting force on an object to bend or twist it; hand enclosure, to perceive volume, by molding the fingers around an object surface; and contour following, to perceive shape, by tracing the skin contact around the gradient of an object's surface.

Non-human animals also use exploratory procedures for tactile perception. For example, the back-and-forth whisking motion of rodent vibrissae is an exploratory procedure that can be used to perceive the location and identity of nearby objects.

Biomimetic systems

We now describe how some biomimetic tactile systems have utilized these three principals of biological touch to result in superior sensing capabilities, including systems that mimic the human fingertip and hand (cutaneous touch) and the rodent whisker system (vibrissal touch).

Biomimetic tactile superresolution

The potential impact of superresolution methods for tactile robotics is indicated by the advancements in technologies based on visual superresolution that earned the 2014 Nobel Prize in Chemistry (for superresolved fluorescence microscopy). Superresolution research in visual imaging has impacted the life sciences in ways unthinkable in the mid-90s and can be seen as analogous to biological hyperacuity of vision and touch where the discrimination is finer than the spacing between sensory receptors. While visual superresolution has revolutionized the life sciences by enabling the imaging of nano-scale features within cells, tactile superresolution has the potential to drive a step-change in tactile robotics, with applications from quality control and autonomous manipulators in manufacturing to sensorized prosthetics and probes in healthcare.

A recent study of tactile superresolution as biomimetic hyperacuity (Lepora et al. 2015) demonstrated superresolution with a biomimetic tactile sensor constructed for the iCub humanoid robot (Figure 16.2). A following study then demonstrated superresolution with another biomimetic tactile sensor, the TacTip, based instead on an optical tactile sensing technology (Lepora and Ward-Cherrier 2015). Both tactile sensors have a rounded shape with a compliant outer surface that enable contact sensing via compression. Their transduction methods differ, however, in that the iCub fingertip detects local pressure over distinct regions called taxels (tactile elements), whereas the TacTip detects local shear of the sensing surface that is apparent through sideways movement of internal pins (tracked visually via an internal camera). In both cases, tactile sensing is via discrete overlapping tactile elements mimicking aspects of mechanoreception in human fingertips;

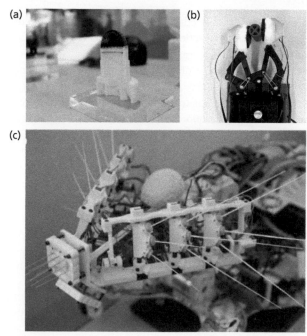

Figure 16.2 (a) Cut-through of the TacTip tactile fingertip, a 3d-printed optical tactile sensor. Discrete taxels are visible as pins embedded in the compliant sensing surface. Tactile superresolution can localize objects to finer than the taxel spacing. (b) A robot performing haptic exploration of an object by rolling it between two tactile sensors. (c) A tactile sensor (the BIOTACT whisker array) based on the rodent vibrissal system. This system can actively sweep its whisker sensors back-and-forth analogous to rodent whisking.

(a) Credit: Science Museum, London. (b) Tactile Robotics Group, Bristol Robotics Laboratory. (c) Bristol Robotics Laboratory.

for example, their receptive fields (areas sensitive to contact) are broader than the tactile element spacing with a sensitivity that peaks in the centre and decreases gradually away from that peak. As a consequence of this biomimetic design of the fingertip, the position of a cylinder relative to the fingertip can be localized to about one-tenth of a millimetre, giving an order of magnitude improvement over the taxel spacing (resolution) of about 4mm. To attain this superresolution, it is necessary to use statistical averaging over the population of taxels to extract evidence that can be used to form perceptual decisions (see Lepora, Chapter 28, this volume).

It is interesting that the degree of superresolution readily attainable in artificial systems is typically an order of magnitude, which is characteristic also of hyperacuity in human and animal vision, touch, and audition. One could interpret this coincidence as evidence that the methods for biomimetic superresolution capture the salient principles underlying biological hyperacuity.

Biomimetic active touch

The descriptions of active touch by psychologists such as J. J. Gibson motivated the engineering scientist Ruzena Bajcsy to give a more formal definition in her 1988 paper on *active perception* that could be applied directly to engineered systems. She defined active perception as "purposefully changing the sensors state parameters according to sensing strategies," such that these controlling strategies are "applied to the data acquisition process which will depend on the current state of the data interpretation and the goal or the task of the process." Her definition of active perception can be seen as a biomimetic version of Gibson's psychological treatment with her terminology and nomenclature adapted to engineering. From a biomimetic perspective, her phrase "changing the sensor's state parameters" generalizes "moving" or "adjusting" a sensory organ, while "data acquisition" means "sensing" and "data interpretation" is the engineering term for "perceiving."

Demonstrations of active perception with artificial tactile fingertips came soon after Bajcsy's original paper. For example, one study showed that controlling the speed, and hence spatial filtering, of a tactile fingertip can improve the measurement of surface roughness (Shimojo and Ishikawa 1993), which can be seen as analogous to Loomis and Lederman's (1986) experimental work on the tactile cutaneous perception of texture. Recent work on active touch with robot fingertips has considered how the active movements should be controlled during perception, for example to attain optimal perceptual decisions over the tactile stimuli. Examples include perceiving a tactile feature such as shape or texture while controlling contact location (Lepora 2016a) and accurately perceiving texture while controlling the contact force or movement speed of a biomimetic fingertip (Fishel and Loeb 2012). Biomimetic methods for superresolution have also been combined with active touch, with the benefit of attaining robust superresolution that is insensitive to how the object is initially contacted (Lepora et al. 2015).

Biomimetic active perception has also been demonstrated on mammal-like whiskered robots (see Prescott, Chapter 45, this volume). Early work used "active" to mean the sensor is moving but did not consider a feedback loop to modulate the whisking. Later whiskered robots implemented sensorimotor feedback between the tactile sensing and motor control, enabling demonstration that texture perception can be improved under real-time active control of the tactile sensors during sensing (Sullivan et al. 2012).

Biomimetic haptic exploration

There were several initial proposals that robot exploratory touch could be based around principles of haptic exploration and active perception, for example by the originators of those bodies of work (Bajcsy et al. 1987). Practical implementations of haptic exploration of unknown stimuli with dexterous robot hands followed a few years later in the early 1990s. The first attempts used haptic exploration to recognize an object by using exploratory moves that trace continually along its surface and by relating the exploratory procedures to the shape. These approaches were developed further to exploit haptic exploration where some fingers grasp and manipulate the object while others roll and slide over the object surface (Okamura and Cutkosky 2001). More recently, the field has expanded to include progress in tactually exploring objects so that they may be grasped more effectively, or even manipulated; however, such work is often not directly linked to the original biological origins of haptic exploration, but is rather seen as stemming from robot control or machine learning.

The original work on haptic exploration has continued to have a more direct influence on robotics when specific exploratory procedures have been implemented with single tactile sensors mounted on an actuator rather than a robot hand. Early work included motion control of a tactile fingertip for profile delineation of an unseen object (Maekawa et al. 1992), which is the artificial analog of contour following to feel shape (see "Touch is exploratory," above). Contour following has been considered since, not only as a pure robotic application to control a tactile sensor in contact with a surface, known as tactile servoing, but also as a biomimetic approach based on human and animal perceptual decision making (Lepora et al. 2017). The exploratory procedure of lateral motion to feel texture has also been treated from a biomimetic perspective, for example with tactile fingertips mounted on a sliding actuator (Fishel and Loeb 2012), and has reached the state of development that it now exceeds human capabilities.

Future directions

Whereas artificial vision is a mature technology that has many commercial applications, from security to hand-writing recognition, artificial touch is still relatively unexploited as a sensing modality. In part, this could be because artificial vision can be utilized effectively without using active perception by having a camera with a large field of view; conversely, tactile sensors can necessarily only contact an area the size of the sensor, and hence to be deployed effectively they must be controlled in an active manner. Thus we would claim that the lack of artificial active touch that can rival human capabilities is a challenge that must be overcome for tactile sensors to have widespread technological applications spanning across prosthetics, telehaptics, surgical robotics, biometric clothing, wearable computing, medical probes, and quality control in manufacturing.

Another challenge in the field of artificial tactile sensing is that there is no current agreement on the best manner in which to design and utilize artificial tactile sensors. Moreover, the resolution of tactile sensors is limited by the scale of engineered mechanoreceptors to typically a millimetre or larger. While miniaturization will result in finer arrays of tactile sensing elements, one must also contend with a need for touch sensors to be conformable to different shapes and sizes, and that the sensing elements will be embedded in a compliant medium that necessarily distributes the contact force across the sensor surface. Biomimetics has the potential to solve many of these problems, since these challenges are exactly those faced in biological systems.

Learning more

The Scholarpedia of Touch (Prescott et al. 2016) is a recent volume that collects many articles on biological touch systems and their emulation in artifacts via biomimetics. The review "tactile sensing—from humans to humanoids" is seen as a classic review relating the principals of biological tactile sensing to artificial touch (Dahiya et al. 2010). Another excellent survey of work in artificial tactile sensing can be found in Cutkosky and Provancher 2016), which in addition to covering work in tactile sensors and sensing covers a comprehensive history of progress on artificial active tactile perception and haptic exploration. The role of touch in grasping and manipulating objects is explored further in Cutkosky's contribution to this handbook (Chapter 30, this volume). The original paper by Ruzena Bajcsy on "Active perception" in 1988 is well worth studying, not least for the elegant manner in which she describes how biological sensing is inherently active and for her prescience in seeing how the field will develop in the future. Finally, overviews of modern research in the area of active touch and haptic exploration are given in the foreword by Prescott and colleagues to the Royal Society special issue on "Active touch sensing" (Prescott et al. 2011) and the article on "Active tactile perception" (Lepora 2016b) in *Scholarpedia of Touch*, which both cover biological and biomimetic studies from the perspective of human and non-human touch sensing.

References

Bajcsy, R., Lederman, S.J., and Klatzky, R.L. (1987). Object exploration in one and two fingered robots. *Proceedings of the 1987 IEEE International Conference on Robotics and Automation*, 3, 1806–10.

Bajcsy, R. (1988). Active perception. *Proceedings of the IEEE*, 76(8), 966–1005.

Cutkosky, M. R., and Provancher, W. R. (2016). Force and tactile sensing. In: B. Siciliano and O. Khatib (eds), *Springer Handbook of Robotics* (2nd edn). Berlin/Heidelberg: Springer, pp. 717–36.

Dahiya, R. S., Metta, G., Valle, M., and Sandini, G. (2010). Tactile sensing—from humans to humanoids. *Robotics, IEEE Transactions on*, 26(1), 1–20.

Fishel, J.A., and Loeb, G.E. (2012). Bayesian exploration for intelligent identification of textures. *Frontiers in Neurorobotics*, 6, 4. doi: 10.3389/fnbot.2012.00004

Gibson, J.J. (1962). Observations on active touch. *Psychological Review*, 69(6), 477.

Kleinfeld, D., Ahissar, E., and Diamond, M.E. (2006). Active sensation: insights from the rodent vibrissa sensorimotor system. *Current Opinion in Neurobiology*, 16, 435–44.

Lederman, S.J., and Klatzky, R.L. (1987). Hand movements: a window into haptic object recognition. *Cognitive Psychology*, 19(3), 342–68.

Lepora, N. F. (2016a). Biomimetic active touch with tactile fingertips and whiskers. *IEEE Transactions on Haptics*, 9(2), 170–83.

Lepora, N.F. (2016b). Active tactile perception. In: T. J. Prescott, E. Ahissar, and E. Izhikevich (eds), *Scholarpedia of Touch*. Amsterdam: Atlantis Press, pp. 151–9.

Lepora, N.F., Aquilina, K., and Cramphorn, L. (2017). Exploratory tactile servoing with active touch. *IEEE Robotics and Automation Letters*, 2(2), 1156–63.

Lepora, N. F., Martinez-Hernandez, U., Evans, M. H., Natale, L., Metta, G., and Prescott, T. J. (2015). Tactile superresolution and biomimetic hyperacuity. *IEEE Transactions on Robotics*, 31(3), 605–18.

Lepora, N. F., and Ward-Cherrier, B. (2015). Superresolution with an optical tactile sensor. In: *Intelligent Robots and Systems (IROS), 2015 IEEE/RSJ International Conference on*, pp. 2686–91.

Loomis, J.M. (1979). An investigation of tactile hyperacuity, *Sensory Processes*, 3, 289–302.

Loomis, J.M., and Lederman, S.J. (1986). Tactual perception. In: K. R. Boff, L. Kaufman, and J. P. Thomas (eds), Handbook of perception and human performance, Volume II Cognitive processes and performance. New York: John Wiley, chapter 31, pp.1–41.

Maekawa, H., Tanie, K., Komoriya, K., Kaneko, M., Horiguchi, C., and Sugawara, T. (1992). Development of a finger-shaped tactile sensor and its evaluation by active touch. In: *Robotics and Automation (ICRA), 1992 IEEE International Conference on*, pp. 1327–34.

Okamura, A. M., and Cutkosky, M. R. (2001). Feature detection for haptic exploration with robotic fingers. *The International Journal of Robotics Research*, 20(12), 925–38.

Prescott, T.J., Diamond, M.E., and Wing, A.M. (2011). Active touch sensing. *Philosophical Transactions of the Royal Society B: Biological Sciences*, 366(1581), 2989–95.

Prescott, T. J., Ahissar, E., and Izhikevich, E. (2016). *Scholarpedia of Touch*. Amsterdam: Atlantis Press.

Shimojo, M., and Ishikawa, M. (1993). An active touch sensing method using a spatial filtering tactile sensor. In: *Robotics and Automation (ICRA), 1993 IEEE International Conference on*, pp. 948–54.

Sullivan, J. C., Mitchinson, B., Pearson, M. J., Evans, M., Lepora, N. F, and Prescott, T. J. (2012). Tactile discrimination using active whisker sensors. *IEEE Sensors*, 12(2), 350–62.

Chapter 17

Chemosensation

Tim C. Pearce

Department of Engineering, University of Leicester, UK

Although we often take for granted our chemical senses (smell—olfactory, taste—gustatory, and common chemical sense—trigeminal) or are not even aware on their impact on our behaviour (vomeranasal), they must routinely solve an impressive set of detection and signal processing challenges. The diversity of molecular information that is detectable in the world is staggering: each molecule has an almost limitless variety of properties including different bonds, functional groups, and atomic configurations. The olfactory sense that we focus on here must solve a particularly difficult variant of this challenge—over 10 000 separate olfactory perceptions have been estimated to be elicited by an almost infinite variety of molecules. Any compound with molecular mass below ca. 300 Daltons has the potential to be detected, and new odorous molecules are being created every day and may be presented in concentrations ranging at least 10 orders of magnitude, yet each percept relies upon a stable and unique neural representation in the brain.

Certain compounds of behavioural importance are detectable at astonishingly low concentrations (such as mercaptan, with a detection threshold in humans below 1 part per billion). The detection task is made yet more challenging because most naturally occurring odours exist as complex multicomponent blends. Coffee, for instance, contains upwards of 1000 distinct chemical compounds of which approximately 40 are so-called key flavour volatiles, the relative concentrations of which are crucial to determining the overall sensory experience. Together these compounds generate a unified gestalt-like percept in the brain that is readily identified with a source "odour object." Recent evidence suggests that these olfactory representations can be segmented in similar ways to vision, as well as being the subject of attention, leading to ideas as to how some of the chemosensing capabilities of animals could be reproduced in Living Machine technologies.

Biological principles

In humans, olfaction begins by molecules entering the nose, where they are transported via the nasal turbinates across local pressure gradients, such that a fraction contact a specialized region, the olfactory mucosa (Figure 17.1). The olfactory mucosa has a surface area close to 1 cm^2 that is host to some ca. 10 million bipolar olfactory sensory neurons (OSNs) protected by only a thin mucous layer (approximately 200 µm in thickness). These cells are remarkable for at least two reasons: they require continual replenishment as the only neurons to be in contact with the outside world and are thought to each express a single olfactory receptor (OR) from the largest family of specialized G-protein coupled receptors (GPCRs) found in nature (human: 388, chimpanzee: 450, mouse: 1200, rat: 1430, zebrafish: 98). A given GPCR has a unique binding and signaling profile to a wide diversity of molecules (large molecular receptive range), making

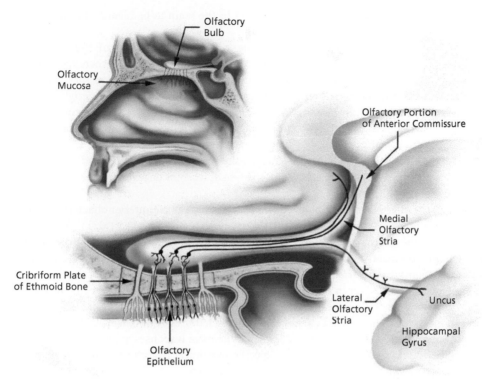

Figure 17.1 The human olfactory pathway.

Reproduced from Susan S. Schiffman and Tim C. Pearce, 'Introduction to Olfaction: Perception, Anatomy, Physiology, and Molecular Biology', in Tim C. Pearce, Susan S. Schiffman, H. Troy Nagle, and Julian W. Gardner (eds), *Handbook of Machine Olfaction: Electronic Nose Technology*, Figure 1.1, p. 2, doi: 10.1002/3527601597.ch1, Copyright © 2003 Wiley-VCH Verlag GmbH & Co. KGaA.

them of great interest for biosensor development. Stochastic gene regulation processes ensure that a large complement of the hundreds of OR types are expressed across the OSN population, together producing a patterned neural representation in the form of a "population code" (Sánchez-Montañés et al. 2002), that is unique to specific subsets of pure and complex odours.

Although population coding through diverse OR-mediated OSN responses is certainly crucial to olfactory sensing, particularly in mammals, a less studied aspect of olfactory coding is the chromatographic effect of the thin mucous layer surrounding the OSNs. This is thought to lead to segregation of compounds depending upon their sorption profiles, thereby imposing complex spatio-temporal patterns on the OSN population during the sniff cycle, that has potential to provide additional rich stimulus-dependent sensory information.

Specialized odorant binding proteins (OBPs) also add further specificity to OSN responses, as varieties of OBPs have selective binding properties to diverse groups of odour ligands and have been long known to be released into the mucous layer of the olfactory epithelium to traffic molecules to the binding site. These proteins are important for shifting sorption properties of predominantly hydrophobic odour molecules, which enhances concentration of odorant in the liquid phase, thereby selectively boosting sensitivity to certain key compounds.

What is less clear currently is how the neural representation of *natural* odours (as complex mixtures) relates to neural representations of their odour constituents driving the subsequent neural processing to create a unified odour percept in the brain to be identified as a distinct "odour object."

OSNs in the epithelium project axons through the cribriform plate of the skull (Figure 17.1) to the first stage of olfactory processing in vertebrates, the olfactory bulb (OB), where they specifically target sites of convergence known as glomeruli. Through a continual process of chemically guided axonal targeting and activity mediated synaptic competition (in which periglomerular PG cells are thought to play an important role), glomeruli act as sites of massive convergence for OSNs expressing identical GPCRs, which is known to significantly boost the sensitivity of the system, presumably through suppression of decorrelated receptor, neuronal, and stimulus noise. This axonal targeting and synaptic refinement continues into adulthood and ensures that a stable olfactory representation exists even in the face of receptor degeneration and neurogenesis of OSNs at the epithelium.

Read-out of the olfactory signal by higher brain centres such as the anterior olfactory nucleus, olfactory tubercle, and prepyriform cortex is mediated via mitral/tufted (M/T) cells in the deeper layers of the OB that each innervate an individual glomerulus and so receive an excitatory drive predominantly from a single OSN type. Lateral inhibition in the OB occurs via a complex arrangement of interglomerular connectivity and inhibitory granule cells (GR) that together creates decorrelation in the OB odour responses over time. This sharpens the odour representation through "on-centre off-surround" processing, similar to that found in the visual system as a mechanism for contrast enhancement. Mass action and the balanced excitation and lateral inhibition within the OB lead to complex dynamical responses amongst the M/T cell population, involving oscillations, stimulus specific dynamical trajectories, and odour codes multiplexed in time that are at the forefront of current research.

Connectivity to the higher centres, via the lateral olfactory tract in particular, to the prepyriform (olfactory) cortex forms sparse representations of both simple and complex odours and has been implicated in olfactory learning. Such sparse representations are easier to subsequently decode and reroute to the many higher brain centres that the olfactory pathway innervates.

Biomimetic and biohybrid systems

The motivation for looking to biology for building chemical sensing systems has a number of reasons.

- It has been very challenging to make highly specific chemosensors for certain analytes that can operate over wide concentration ranges, even if we know these a priori. Instead, taking a biological approach of deploying arrays of chemosensors with overlapping specificities can provide alternative and more efficient solutions to creating new technologies capable of detecting compounds and their mixtures not even yet invented.

- There is a need to create instrumentation able to predicting human olfactory responses in a variety contexts. Existing instrumentation such as gas chromatography and mass spectroscopy are largely analytic in nature, separating complex mixtures into lists of molecular species, whereas in mammals olfactory perception is synthetic and highly non-linear.

- There are undesirable properties of technological chemosensors shared by biological olfactory receptors, such as sensor noise processes, sensor degradation/instability that can lead to systematic drift in responses and non-linearities of ligand interactions via binding site competition for which the olfactory system architecture is well adapted.

- Solving challenging natural odour sensing problems performed routinely by animals, such as odour object identification, segmentation and attentional processing.

By following the architecture of olfactory systems as a blueprint to build so-called electronic noses (often abbreviated to e-noses), arrays of broadly tuned chemosensor arrays have been constructed for at least 30 years, bringing together a wide variety of sensor (micro-/nano-)technologies. These include chemically sensitive field effect transistors (ChemFETs), surface acoustic wave resonators (SAWRs), quartz crystal microbalances (QCMs), electrochemical cells, optical detectors, and photoionization detectors (PIDs), combined with all manner of chemically sensitive materials such as metal oxide semiconductor, heterocyclic polymers, electrochemically active materials, carbon black polymers, solvatochromic dyes, and biological factors such as olfactory receptors and specialized binding proteins. By using combinations of such sensor technologies and varieties of chemically sensitive materials it is in principle possible to create arrays with relatively decorrelated responses. However, many e-nose arrays in practice suffer from high correlation in their responses, which is a continuing topic of research.

For sensor arrays to be used as a front-end for biomimetic olfactory systems chemosensors should be independently addressable and deployed in large numbers from a wide diversity of materials to create sufficient decorrelation in the array responses. From the plethora of chemosensor technologies, high-density solvatochromic fluorescent dye microbeads (Figure 17.2a) and conducting polymer arrays (Figure 17.2c) are well suited to this since they can be microdeposited with high density and a wide variety of materials can be used to create diversity in response profiles (Albert et al. 2002; Bernabei et al. 2012).

The large set of vertebrate olfactory proteins (OBPs and ORs) provides a deep reservoir of naturally evolved solutions to chemical sensing with a huge variety of sensitivities and specificities. Recent advances in bio-sensing measurement and receptor deorphaning makes the direct real-time measurement of protein–ligand binding a practical possibility. A wide range of techniques have been applied to the OR measurement problem of measuring OR activation including optical (e.g. surface plasmon resonance), resonant (e.g. quartz-crystal microbalance), and electrochemical (patch and voltage clamping). Goldsmith et al. (2011) developed a relatively stable carbon nanotube transistor interface demonstrating transduction of binding events over tens of seconds for a variety of ORs (Figure 17.2d). These techniques are likely to pave the way for extremely high-density OR/OBP arrays for integration with future biomimetic chemical sensing systems.

In early electronic noses only time-invariant signal properties were used as a response measure for each chemosensor within an array (for instance fractional change in conductance before and after stimulus onset). More recently, the time dynamics of chemosensor responses have been exploited to enhance pattern separation and decorrelation. Whilst a number of chemosensor technologies demonstrate a certain degree of stimulus-specific temporal information for different chemical compounds this can be further enhanced through an "artificial mucosa" (Figure 17.3a) that mimics chromatographic selective segregation of ligands within a stationary phase similar to that observed in mammals. By imposing additional temporal dynamics on the stimulus delivery that depends on chemical identity, this biomimetic strategy for chemical segregation was shown to enhance the pattern separation of the overall system to boost classification performance (Covington et al. 2007).

Other biomimetic strategies have been demonstrated to improve e-nose performance. For example a simple model of competition based on wiring using Hebbian weight adaptation that models the process of axon refinement during OSN regeneration was shown to be able to automatically resolve beads types to target separate glomeruli (G1, G2 ... in Figure 17.2b) as they responded diversely to a battery of different odours (Albert et al. 2002). The resulting convergence of signals from identical chemosensor types has also been demonstrated to boost sensitivity and dynamic range by following the front-end architecture common to all olfactory systems

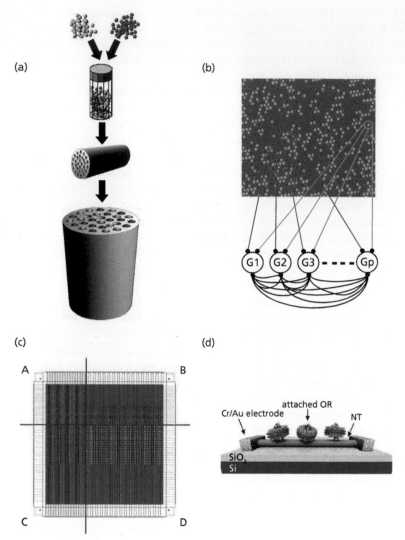

Figure 17.2 A variety of chemosensor technologies suited for biomimetic olfactory system development. (a) Solvatochromic microbead technology. (b) Glomeruli convergence and self-organization for automatic decoding of microbead type. (c) Large-scale conducting polymer array for biomimetic chemical sensing. (d) Transistor arrangement for olfactory GPCR measurement. Reprinted with permission from (Goldsmith et al. 2011).

(a) Reproduced from Keith J. Albert, Daljeet S. Gill, Tim C. Pearce, and David R. Walt, Automatic decoding of sensor types within randomly ordered, high-density optical sensor arrays, *Analytical and Bioanalytical Chemistry*, 373 (8), p. 792, doi:10.1007/s00216-002-1406-8, © Springer-Verlag 2002. With permission of Springer. (c) © 2012 IEEE. Reprinted, with permission, from Mara Bernabei, Krishna C. Persaud, Simone Pantalei, Emiliano Zampetti, and Romeo Beccherelli, Large-scale chemical sensor array testing biological olfaction concepts, *IEE Sensors Journal*, 12 (11), pp. 3174–83, doi: 10.1109/JSEN.2012.2207887. (d) Reprinted with permission from Brett R. Goldsmith, Joseph J. Mitala Jr, Jesusa Josue, Ana Castro, Mitchell B. Lerner, Timothy H. Bayburt, Samuel M. Khamis, Ryan A. Jones, Joseph G. Brand, Stephen G. Sligar, Charles W. Luetje, Alan Gelperin, Paul A. Rhodes, Bohdana M. Discher, and A. T. Charlie Johnson, Biomimetic Chemical Sensors Using Nanoelectronic Readout of Olfactory Receptor Proteins, *ACS Nano*, 5 (7), pp. 5408–16, doi: 10.1021/nn200489j. Copyright © 2011 American Chemical Society.

(Pearce et al. 2001). Another example has been to apply oscillating olfactory bulb models directly to chemosensor array data that has demonstrated some ability to segment odour mixtures, which has proven to be a difficult engineering challenge.

The ultimate level of biomimicry is to build a real-time neuromorphic implementation of the olfaction pathway. Figure 17.3b shows the architecture for an analog-VLSI chip based upon the mammalian olfactory bulb comprising integrate-and-fire spiking neurons, adaptive (exponential-decay) synapses, and lateral inhibition connectivity (Koickal et al. 2007). The model used was shown to be able to perform a degree of background supression to target learnt odours.

Future directions

While there are many examples of biological principles and even replica neuromorphic architectures that have been directly applied to developing real-world chemosensory instruments the field is still rich in outstanding opportunities. In terms of sensor technologies there are still many challenges. For example building biological-scale chemosensor arrays that are highly integrated, low-noise with decorrelated, repeatable, reversible, and stable responses. Tapping into the huge family of olfactory receptors as a resource for high dimensional biohybrid olfactory systems is also a huge technological challenge since many science and technology barriers stand in the way: optimizing measurement principles, immobilizing GPCR proteins for precise measurement, maintaining stability in GPCRs outside of the cellular environment a just a few of these issues. Another outstanding challenge that is becoming increasingly a focus is single molecule measurement as we move to sensor technologies capable of detecting discrete molecular events with data that is sensitive to molecular structure or other properties.

In terms of chemosensory processing there are also outstanding challenges. The reality of a neuromorphic olfactory processing architecture that shares the impressive performance of its biological counterpart has still some way to go: some of the principles of the biological system have been demonstrated in real-world implementations and full-blown neuromorphic olfactory systems with promising properties, yet we have not reached the point of large-scale integration of high-density chemosensor arrays (either biological or engineered chemosensors). A key challenge will be for these systems to be tested thoroughly in more naturalistic real-world contexts since the vast majority of studies in machine olfaction rely on precisely controlled pulsed stimuli. Neuromorphic solutions are well placed to show robust performance in the face of naturalistic stimuli as found, for instance, in a chemical plume that derives from unsteady, turbulent flow. This will allow such systems to be tested in robotic platforms suited for a variety of contexts and environments, of which there are already some promising examples.

Learning more

This overview has not covered invertebrate olfaction, which in itself has been a rich source of inspiration for engineered chemical sensing systems and chemosensing robots. For an open-access general olfaction book see Menini (2009), covering both vertebrate and invertebrate olfaction from a behavioural and neurobiological perspective. Glatz and Bailey-Hill (2011) provide a thorough review of measurement techniques for ORs.

For a comprehensive overview of machine olfaction see Pearce et al. (2003), covering topics of sensor technologies, biology and psychophysics, pattern recognition, olfactory guided robots, and signal processing strategies. Ayers (Chapter 51, this volume) discusses some of the challenges involved in building biohybrid sensing systems into a working robot. For some more

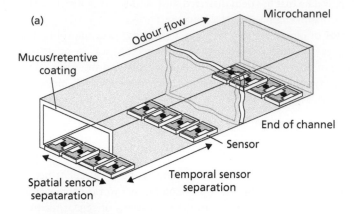

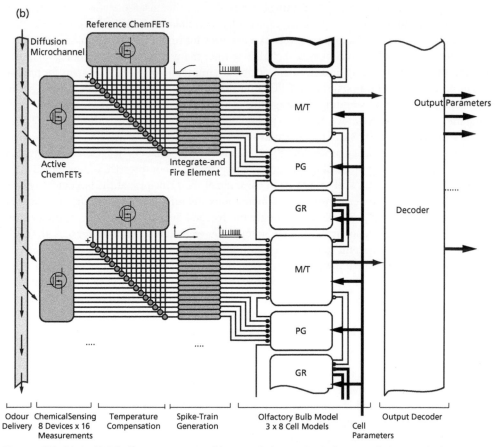

Figure 17.3 (a) Artificial olfactory mucosa. (b) AVLSI (advanced very large-scale integration) neuromorphic olfaction chip.

(a) Reproduced by permission of the Institution of Engineering & Technology. From Covington, J.A., Gardner, J.W., Hamilton, A., Pearce, T.C., Tan, S.L., Towards a truly biomimetic olfactory microsystem: an artificial olfactory mucosa, *IET Nanobiotechnology*, 1 (2), 15–21, © 2007, The Institution of Engineering & Technology.

recent reviews on neuromorphic approaches to olfaction see Persaud et al. (2013), Huerta and Nowotny (2012), and Raman et al. (2011).

References

Albert, K. J., Gill, D. S., Pearce, T. C., and **Walt, D. R.** (2002). Automatic decoding of sensor types within randomly ordered, high-density optical sensor arrays. *Analytical and Bioanalytical Chemistry*, 373(8), 792–802.

Bernabei, M., Persaud, K., Pantalei, S., Zampetti, E., and Beccherrelli, R. (2012). A large scale chemical sensor array testing biological olfaction concepts. *IEEE Sensors Journal*, 12(11), 3174–83.

Covington, J.A., Gardner, J.W., Hamilton, A., Pearce, T.C., and Tan, S.L. (2007). Towards a truly biomimetic olfactory microsystem: an artificial olfactory mucosa. *IET Nanobiotechnology*, 1(2), 15–21.

Glatz, R., and Bailey-Hill, K. (2011). Mimicking nature's noses: From receptor deorphaning to olfactory biosensing. *Progress in Neurobiol- ogy*, 93, 270–96.

Goldsmith, B.R., Mitala, J.J., Josue, J., Castro, A., Lerner, M.B., Bayburt, T.H., Luetje, C.W., Khamis, S.M., Gelperin, A., Jones, R.A., Rhodes, P.A., Brand, J.G., Discher, B.M., Sligar, S.G., and Johnson, A.T.C. (2011). Biomimetic chemical sensors using nanoelectronic readout of olfactory receptor proteins. *ACS Nano*, 5(7) 5408–16.

Koickal, T.J., Hamilton, A., Tan, S.L., Covington, J.A., Gardner, J.W., and Pearce, T.C. (2007). Analog VLSI circuit implementation of an adaptive neuromorphic olfaction chip. *IEEE Transactions on Circuits and Systems I*, 54 (1), 60–73.

Menini, A. (ed.). (2009). *The neurobiology of olfaction*. CRC Press: Boca Raton, USA.

Huerta, R., and Nowotny, T. (eds.) (2012). *Bioinspired solutions to the challenges of chemical sensing*. Frontiers E-books.

Pearce, T.C., Schiffman, S.S., Nagle, H.T., and Gardner, J.W. (eds.) (2003). *Handbook of Machine Olfaction: Electronic Nose Technology*. Wiley-VCH Verlag: Weinheim, Germany.

Pearce, T., Verschure, P., White, J., and Kauer, J. (2001). Robust stim-ulus encoding in olfactory processing: hyperacuity and efficient signal transmission. In *Emergent neural computational architectures based on neuroscience* Wermter, S., Austin, J., and Willshaw, D. (eds.), Springer-Verlag: Heidelberg, pp. 461–79.

Persaud, K.C., Marco, S., Gutierrez-Galvez, A. (2013). *Neuromorphic olfaction*. CRC Press: Boca Raton, USA.

Raman, B., Stopfer, M., and Semancik, S. (2011). Mimicking biological design and computing principles in artificial olfaction. *ACS Chemical Neuroscience*, 2(9), 487–99.

Sánchez- Montañés, M. A., and Pearce, T. C. (2002). Why do olfactory neurons have unspecific receptive fields? *Biosystems*, 67 (1) 229–38.

Chapter 18

Proprioception and body schema

Minoru Asada

Graduate School of Engineering, Osaka University, Japan

Proprioception is defined as our ability to sense the position of our own limbs and other body parts in space, and body schema is a body representation that allows both biological and artificial agents to execute their actions based on proprioception. The proprioceptive information used by existing artificial agents (robots) mainly relates to posture (and its change) and consists of joint angles (joint velocities) given a linked structure. Bongard (Chapter 11, this volume) discusses how a robot can use these signals to learn the structure of its body and how to control it. However, the counterpart in biological agents (humans and other animals) includes much more complicated components with associated controversies concerning the relationship between the body schema and the body image. The neural structures underlying both of these systems are far from well understood due to the limits of current imaging technology. A new trend of constructive approaches has been attacking this topic using computational models and robots. This chapter provides a brief overview of the biology of proprioception and body representation with a view to exploring how we can develop artificial analogs. Next, following a summary of the classical use of body schema in robotics, a series of constructive approaches addressing some of the mysteries of body representation are introduced. Finally, future directions and pointers to additional reading are given.

Biological principles

The original meaning of proprioception stems from the Latin *propius* meaning "one's own," or "individual" and *capio* meaning to grasp, hence its consideration raises not only neuroscientific issues but also psychological, and sometimes, philosophical ones, such as the distinction between conscious and unconscious awareness of the body (see Seth, Chapter 37, this volume). Therefore, it is difficult to clearly define "proprioception," and a similar situation is observed in defining body representation, especially the discrepancy between body image and body schema, since both are closely related to proprioception. Head and Holmes (1911) defined "body schema" as an unconscious neural map in which multi-modal sensory data are unified, and "body image" as an explicit mental representation of the body and its functions. Another issue is that the former is often viewed as for action and the latter for perception. There is a general consensus that body representations in biological systems are flexible and acquired by spatio-temporal integration of information from different sensory modalities, but the details of their structure and mechanism are far from uncovered.

Plasticity is one of the most significant properties of the body representation. Its origin may come from the period of fetal development in the womb where repetitive movements of touching self-body are often observed (Rochat 1998), and the fetus is thought to learn the relationship between its motions and consequent sensations. This early representation is thought to link to the important concepts of body ownership and agency, and to be the amalgam which is expected to differentiate into body schema and body image in later stages of development.

Flexibility and adaptability in body representation are desirable features caused by neural plasticity and can be observed in the case of tool use. Maravita and Iriki (2004) studied body schema extension during tool use by a macaque monkey that was retrieving food with a rake. Activity of neurons called bimodal neurons, which react to both somatosensory and visual stimulation, was recorded from the intraparietal cortex. They found two types of neurons: "distal type" which responded to somatosensory stimuli at the hand and visual stimuli near the hand, and "proximal type" whose visual receptive field was not centered around the hand, but spanned the whole space within reach (Figure 18.1). Both types of neurons adapt to the experience of tool use and also to the monkey's own motivation to use a tool (Figure 18.1 (d)).

The study of neuropsychological disorders can help in understanding the underlying mechanisms for the construction of body schema and how these may be affected by damage. One of the most interesting examples is the so-called "phantom-limb" phenomenon described by

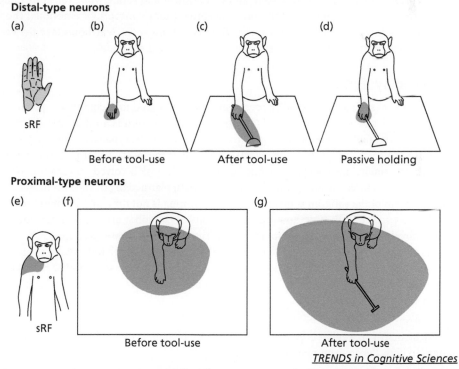

TRENDS in Cognitive Sciences

Figure 18.1 Changes in bimodal receptive field properties following tool use. The somatosensory receptive fields (sRF) of cells in this region were identified by light touches, passive manipulation of joints, or active hand use. The visual receptive field (vRF) was defined as the area in which cellular responses were evoked by visual probes (the most effective ones being those moving towards the sRF). (a) sRF (blue area) of the "distal type" bimodal neurons and their vRF (pink areas) (b) before tool use, (c) immediately after tool use, and (d) when just passively grabbing the rake. (e) sRF (blue area) of "proximal type" bimodal neurons, and their vRF (pink areas), (f) before, and (g) immediately after tool use.

Reprinted from *Trends in Cognitive Sciences*, 8 (2), Angelo Maravita and Atsushi Iriki, Tools for the body (schema), pp. 79–86, http://dx.doi.org/10.1016/j.tics.2003.12.008, Copyright © 2004 Elsevier Ltd., with permission from Elsevier.

Ramachandran and Blakeslee (1998). These authors showed that patients suffering from phantom pain, due to a missing limb, could alleviate their pain through visual feedback by observing their intact opposite limb in a mirror box, suggesting that the cortical representation of the body might have been restructured through this experience. More arguments on the definition and roles of body image and body schema, disorders, and other important concepts such as body ownership and agency, and forward models are given in Hoffman et al. (2010).

In a recent review, Naito et al. (2016) show that kinesthetic illusions can reveal the human body representations that contribute to motor control and corporeal awareness. They argue for the importance of three brain systems based on recent neuroimaging studies of proprioceptive bodily illusions: (1) a motor network, (2) specialized parietal systems, and (3) a right inferior fronto-parietal network. The first network contributes to fast online feedback control based on the processing of afferent inputs. The second one functions as transformation/integration of information from different coordinate systems for adaptive and flexible body representation. The third one seems to work as a monitor of this dynamic representation that can lead to corporeal awareness.

As a summary of the biological evidence, the body schema is generally considered to be a sensorimotor representation of the body used to guide movement and action while the body image is used to form our perceptual (body percept), conceptual (body concept), or emotional (body affect) judgments towards our body. However, there is not always a clear boundary between these two systems. Further investigations are needed to resolve these issues.

Biomimetic systems

For artificial agents such as robots, body representation is essential to accomplish a given task by executing a series of actions. A typical (and traditional) situation is a robot arm driven by a number of electric motors corresponding to the number of joints, typically between 3 and 6, that is required to reach and grasp some objects on a table. If the locations for pick and place of the target object, and its size (weight), are fixed and given, the task involved moving the robot's shoulder, upper arm, forearm, and gripper[1] following a path planned in advance. In most cases, an analytical form of the solution is given, and path planning is not difficult. The body schema is then a link structure indicating how the components of the robot arm are connected, with length and movable range information for each link. These parameters are given in advance, and joint angles are measured (usually, encoders are used) to monitor the path followed as a function of proprioception. This approach is called an explicit model; in constrast, implicit models are often used in uncalibrated systems where the desired parameters are estimated through interaction with the environment (Hoffman et al. 2010).

The biological case is quite different in the following ways.

- The link structure in a robot is generally fixed while a biological one is flexible and adapts to changes in both the environment and the self-body.

- In robotics, knowledge is given externally with high cost to estimate the parameters for the designer while a biological system is self-learning (the implicit model can partially correspond to this).

- No cross-modal association is typically included in robotics, while integration of multimodal sensory information is fundamental in the biological case.

---

[1] Generally, this means an end effector like a hand, but here we use the term "gripper" to avoid the complexity owing to the greater degrees of freedoms of a biomimetic robot "hand" that would typically include fingers.

While the explicit and implicit models discussed so far are based on the traditional robotics paradigm, a new trend of approaches known as cognitive developmental robotics (hereafter, CDR; Asada et al. 2009) has emerged that aims to understand human cognitive developmental processes using synthetic or constructive approaches. In CDR, computer simulations and real robot experiments are used to develop and test computational models that seek to explain findings from disciplines such as neuroscience and developmental psychology, and to verify hypotheses based on them.

There are a number of biological findings on body representation that can be explored from the viewpoint of CDR, for example related to: neural correlation and cross-modal associations in brain areas such as ventral and lateral interparietal areas (VIP and LIP), spatial and temporal contingencies (expectations) and invariances, and coordinate transformations among eye, head, and neck. Readers may find reviews of these studies in Asada et al. (2009) and Hoffman et al. (2010), two of which are briefly summarized next.

Hikita et al. (2008) presented a method that constructs a cross-modal body representation from vision, touch, and proprioception. When the robot touches something, the activation of tactile sensors triggers the construction process of the visual receptive field for body parts that can be found by visual attention based on a saliency map and can consequently be regarded as the end effector. Simultaneously, proprioceptive information is associated with this visual receptive field to achieve the cross-modal body representation. The computer simulation and the real robot results are comparable to the activities of parietal neurons found in the Japanese macaques (Maravita and Iriki 2004). Figure 18.2 shows the acquired visual receptive fields with ((c) and (d)) and without ((a) and (b)) a tool.

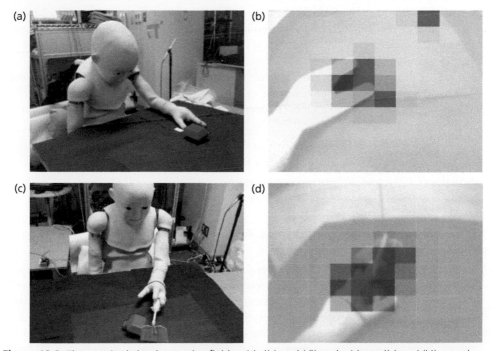

Figure 18.2 The acquired visual receptive fields with ((c) and (d)) and without ((a) and (b)) a tool.

Kuniyoshi and Sangawa (2006) have performed a striking simulation of fetal development in the womb and the emergence of neonatal behaviors (Figure 18.3). This is the simulation study to show how the fetal brain and body interact with each other in the womb. This work was recently extended by Yamada et al. (2016) as an embodied brain model of a human foetus based on much more detailed anatomical and physiological data. They showed that intrauterine sensorimotor experiences enable cortical learning of body representations and subsequent visual-somatosensory integration. Further, they showed that extrauterine sensorimotor experiences affect these processes. They developed a model involving 2.6 million spiking neurons and 5.3 billion synaptic connections based on a spike-timing-dependent plasticity (STDP) rule. Recent advances in computer technologies have enabled such a large-scale brain and body simulation. However, our brain has much more, about one hundred billion neurons and two trillion synaptic connections. Therefore, scaling is still a major problem.

Future directions

Body representation is one of the most fundamental issues in both natural and artificial systems which work in real worlds. So far, they often work separately, but in the future they can be expected to work together in biohybrid systems such as prostheses. In this chapter, we have briefly reviewed studies of body representation in both biology and robotics. However, the structure and underlying mechanisms by which body representation works are still far from uncovered. Cognitive developmental robotics may contribute to unravelling this mystery in two ways which are actually two sides of the same coin. First, to embody hypotheses, in the form of computational models, concerning the mechanisms underlying body representation, and to evaluate them by using computer simulations and real robot platforms. Second, to offer robots or any artificial equipment for psychological/behavioral and neuroscientific experiments as tools for systematic procedures to explore the mystery of body representation.

However, CDR is still nascent, and the achievements are at the preliminary level, therefore many more interdisciplinary approaches are essential and indispensable, not only to verify existing principles using robots, but more importantly to shed new light on the issue. To do this will require:

- Robot platforms with more biologically plausible body structure, such as artificial muscles (see Anderson and O'Brien, Chapter 20, this volume) and skeleton systems.

- More direct collaboration between engineers and researchers in other disciplines such as neuroscience, psychology, and cognitive science (rather than only referring to each other's studies).

With the efforts above, we can attack the following issues closely related to the body representation.

- The mirror neuron system (MNS) has been a focus point in both scientific and engineering approaches since it connects action observation and action execution. Understanding the actions of others through observation and execution is supposedly a kind of embodied cognition, another hot topic in philosophy, psychology, and AI. There are many controversial issues related to MNS (see Asada (2011) for debates), and these issues are often related to body representation (generation, manipulation, and modification). Oztop et al. (2013) have argued that computational models for MNS show the need for additional circuitry to lift the basic mirror neuron function found in monkeys to the higher cognitive functions seen in the human MNS.

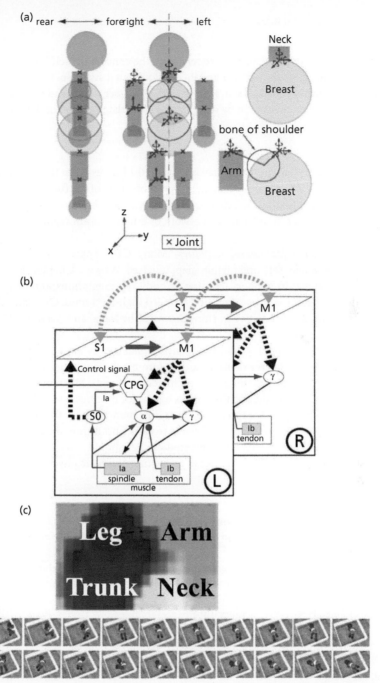

Figure 18.3 Fetal sensorimotor mapping and neonatal movements: (a) a fetus body model that consists of cylindrical or spherical body segments, connected to each other with constrained joints, (b) a brain model, lateral organization of the nervous system consisting of CPG (BVP neurons), S1 (primary somatosensory area), M1 (primary motor area), and so on, (c) the self-organized map from M1 to α motor neurons exhibiting separation into areas corresponding to different body parts, (d) the "neonate" model exhibits emergent motor behaviors such as rolling over and crawling-like motion.

Reproduced from *Biological Cybernetics*, 95 (6), Early motor development from partially ordered neural-body dynamics: experiments with a cortico-spinal-musculo-skeletal model, pp. 589–605, Figures 20 (adapted), 7, 21 (adapted), and 14 (adapted), doi:10.1007/s00422-006-0127-z, Yasuo Kuniyoshi and Shinji Sangawa, © Springer-Verlag 2006. With permission of Springer.

- ◆ The senses of ownership and agency are related to the concept of "self" which may have several levels depending on the context and its developmental stage, such as the ecological self, interpersonal self, and temporal self, as proposed by Neisser (1988) and others. One big question is how the development of body representation is related to that of the self.

Learning more

Readers who wish to explore this topic may learn aboutbe interested in the following:

Neuroscientific approaches. Purves et al. (2012) presents some of the fundamental principles of neuroscience and recent findings relevant to body schema (and body image) viewed as a multimodal representation distributed across a network of different brain regions. The latest edition is recommended.

Computational models for neural activities. Many CDR approaches adopt some form of Hebbian learning and self-organizing mapping (see Wilson, Chapter 5, and Herreros, Chapter 26, this volume) to support learning of body representations and/or motor control. These methods are still at the lower level, and we need high-level ones. One candidate could be the Bayesian framework of predictive learning discussed by Nagai and Asada (2015) and by Seth (Chapter 37, this volume).

Embodiment and soft robotics. Almost all conventional robots are equipped with electrical motors since it is easy to control them, and therefore many people use them and know-how has been accumulated. However, this motor architecture is quite different from biological systems, including the human body, and this makes it hard for robotics researchers to replicate the underlying mechanisms of body representation and its manipulation (see discussion in Metta and Cingolani, Chapter 47, this volume). Even if readers find it difficult to access biologically plausible soft materials and actuators for their studies, they should consider this point, and learn about approaches that use artificial muscles to create more realistic physical models.

Bibliography

Asada, M. (2011). Can cognitive developmental robotics cause a paradigm shift? In: J. L. Krichmar and H. Wagatsuma (eds), *Neuromorphic and Brain-Based Robots*, Cambridge, UK: Cambridge University Press, pp. 251–73.

Asada, M., Hosoda, K., Kuniyoshi, Y., Ishiguro, H., Inui, T., Yoshikawa, Y., Ogino, M., and Chisato Yoshida (2009). Cognitive developmental robotics: a survey. *IEEE Transactions on Autonomous Mental Development*, 1(1), 12–34.

Head, H. and Holmes, H. G. (1911). Sensory disturbances from cerebral lesions. *Brain*, 34(2–3),102–254.

Hikita, M., Fuke, S., Ogino, M., Minato, T., and Asada, M. (2008). Visual attention by saliency leads cross-modal body representation. In: *The 7th International Conference on Development and Learning (ICDL'08)*. doi: 10.1109/DEVLRN.2008.4640822

Hoffmann, M., Marques, H. G., Hernandez Arieta, A., Sumioka, H., Lungarella, M., and Pfeifer, R. (2010). Body schema in robotics: A review. *IEEE Transactions on Autonomous Mental Development*, 2(4), 304–324.

Kuniyoshi, Y., and Sangawa, S. (2006). Early motor development from partially ordered neural-body dynamics: experiments with a cortico-spinal-musculoskeletal model. *Biol. Cybern*, 95, 589–605.

Maravita, A., and Iriki, A. (2004). Tools for the body (schema). *Trends Cogn. Sci.*, 8(2), 79–86.

Nagai, Y., and Asada, M. (2015). Predictive learning of sensorimotor information as a key for cognitive development. In: *Proceedings of the IROS Workshop on Sensorimotor Contingencies for Robotics*, Vol. USB.

Naito, E., Morita, T., and **Amemiya, K.** (2016). Body representations in the human brain revealed by kinesthetic illusions and their essential contributions to motor control and corporeal awareness. *Neuroscience Research*, **104**, 16–30.

Neisser, U. (1988). Five kinds of self-knowledge. *Philosophical Psychology*, **1**, 35–59. doi:10.1080/09515088808572924

Oztop, E., Kawato, M., and **Arbib, M. A.** (2013). Mirror neurons: Functions, mechanisms and models. *Neuroscience Letters*, **540**, 43– 55.

Purves, D., Augustine, G. A., Fitzpatrick, D., Hall, W. C., LaMantia, A.-S., McNamara, J. O., and **White, L. E.** (eds) (2012). *Neuroscience* (5th edn). Sunderland, MA: Sinauer Associates, Inc.

Ramachandran, V. S., and **Blakeslee, S.** (1998). *Phantoms in the brain: probing the mysteries of the human mind*. New York: Harper Perennial.

Rochat, P. (1998). Self-perception and action in infancy. *Experimental Brain Research*, **123**, 102–109.

Yamada, Y., Kanazawa, H., Iwasaki, S., Tsukahara, Y., Iwata, O., Yamada, S., and **Kuniyoshi, Y.** (2016). An embodied brain model of the human foetus. *Scientific Reports*, **6**, 1–10 (27893).

Chapter 19

Electric sensing for underwater navigation

Frédéric Boyer and Vincent Lebastard

Automation, Production and Computer Sciences Department, IMT Atlantique (former Ecole des Mines de Nantes), France

Underwater navigation in turbid water for exploration in catastrophic conditions or navigation in confined unstructured environments is still a challenge for robotics. In these conditions, neither vision nor sonar can be used. In the case of sonar, for instance, echolocation has strong difficulties due to the signal scattering by particles along with the multiple interfering reflections by obstacles. Pursuing a bio-inspired approach in robotics, one can seek in nature which solution could be implemented to solve this difficult problem. In fact, several hundreds of fish species in families Gymnotidae and Mormyridae, which have evolved on both African and South American continents, have developed an original sense well adapted to this situation: the electric sense. In the mormyrid *Gnathonemus petersii*, the fish first polarizes its body with respect to an electric organ discharge (EOD) located at the base of its tail and generates a dipolar electric field in its near surroundings. Then, thanks to many transcutaneous electro-receptors distributed along its body, the fish "measures" the distortion of the electric field and infers an image of its surroundings. Known under the name of electrolocation, this omnidirectional active mode of perception is ideally adapted to navigation in confined spaces bathed by turbid waters. Thus, understanding and implementing this bio-inspired sense on our technologies would offer the opportunity to enhance the navigation abilities of our underwater robots. In this chapter, we will report some designs and implementations of a new sensor inspired by electric fish on an underwater robot. In particular, we shall see how we can design a sensor inspired by the fish and how we can use it by taking inspiration from the electrolocation strategies discovered by the fish and studied by biologists.

Biological principles

In the 50s, researchers such as Lissmann and Machin (1958) discovered that some fish living in freshwater can sense electric fields around them. These fish have electro-receptive organs with which they can detect the very weak currents which exist in their natural surroundings. Named electrolocation, this sensorial ability can be classified into passive and active electrolocation. Saltwater sharks and rays use passive electrolocation in order to navigate in their surroundings and find their prey, or to orient their body in telluric electric fields. These fish thus depend on exogenous fields that fortunately any animal produces with its muscular activity (Bullock and Heiligenberg 1986, Moller 1995). Unlike the passive electrolocators, the fish of Mormyridae and Gymnotidae families as well as the Apteronotes use active electrolocation to perceive their surroundings. In this case, these fish can measure the distortions of secondary

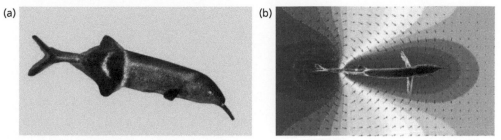

Figure 19.1 (a) The African mormyrid fish *Gnathonemus petersii* or elephant fish. (b) Top view of the fish basal electric field.

electric fields reflected by the objects that they polarize by the emission of a primary field named the "basal field." Thanks to active electrolocation, these fish (see Figure 19.1(a)) can explore their environment, and detects and analyze objects or prey that do not produce any electric field (see Figure 19.1(b)) (von der Emde et al. 2008, Engelmann et al. 2008).

Among the most efficient electric fish, the African fish *Gnathonemus petersii* (see Figure 19.1(a)) is probably the most extensively studied by biologists. As it is practiced by this fish, active electrolocation—which has a range of about the fish's length—is based on the emission of an electric field in the close surrounding of the fish by polarization of an electric organ discharge (EOD) located in its tail with respect to the rest of its body. Once this dipolar field is emitted, the fish senses its perturbations by the objects through an array of electro-receptors distributed all over its skin (see Figure 19.1(b)). Based on this property, Brian Rasnow in 1996 first demonstrated the relationship between some aspects of the environment and the electric intensity distribution on the skin of the fish, that we call the electric image (Rasnow 1996). He established a model that he derived from simple electromagnetism conditions to study the effect of the distance and the dimension of a sphere placed in the vicinity of the fish. His simple model helped him to show the relationship between the shape of the electric image and the distance and dimensions of the spheres. Though it is only applicable with restrictive conditions the model of Rasnow helped the robotic community to start the project of building a bio-inspired electric fish robot.

Biomimetic systems

Based on this principle, several designs of electric sensors were proposed (Solberg et al. 2008, Mayton et al. 2010, Bouvier et al. 2013, Servagent et al. 2013). In their principle, all designs consist of a set of conducting electrodes displayed on the boundaries of the insulating shell of a sensor or a robot. These electrodes are electrically connected between them by a circuit embedded within the sensor and such that once polarized by a voltage (or current generator) these electrodes produce, by virtue of Coulomb's law, a loop of currents in the conductive surrounding of the sensor. This loop emanates from the emitting electrodes (playing the role of the EOD) to meet the receiving electrodes and is closed by the internal circuit of the sensor which measures the electric variables perturbed on the receivers. To date, there exist two great types of technologies inspired by active electrolocation. The first one is named U-U (Solberg et al. 2008, Bouvier et al. 2013), and consists of generating an electric field (basal field) by imposing a voltage (U) between the emitters and the receivers (see Figure 19.2) and measuring the perturbed voltages induced by the surroundings (U) on the receivers. The second, named U-I (Bouvier et al. 2013, Servagent et al. 2013) produces its basal field in the same manner (U).

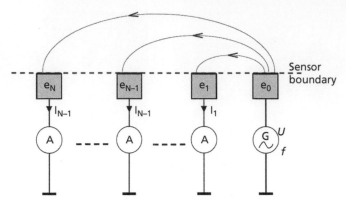

Figure 19.2 Principle of the U-I measurement mode for a probe with an arbitrary number of receptor electrodes. The electrode e_0 (emitter) is polarized with respect to the others (receivers) e_1 to e_N.

However, the measures of the perturbations produced by the objects on the receivers are currents (I). Remarkably, the U-I technology allowed one to obtain the range of the most efficient fish (the elephant fish previously evoked) while the range of U-U sensors does not exceed 1/3 of the total length of the fish. This feature is probably one of the successes of the birth of the first autonomous robot navigating with electric sense (Demonstrations of the Angels platform 2013). Beyond active electrolocation, other measurement modes have also developed allowing passive measurements, and which are consequently named 0-U and 0-I, the zero here indicating no emission, the letter being related to the measurement (Bouvier et al. 2013, Servagent et al. 2013).

These different realizations have given birth to several more fundamental results in particular in terms of modeling, perception, reconstruction of the environment, and navigation control. In the field of modeling, a novel approach has been developed. In its principle, the approach aims at replacing the simultaneous electric interactions between objects and the sensor by an expansion of successive reflections traveling between ones and others (Boyer et al. 2012). Remarkably, the approach allows one to obtain very concise analytic models. Moreover, it gives access to an intuitive hierarchy of electric interactions in complex scenes (several objects and sensors). Furthermore, it allows one to manage the order of approximation of the resulting model which can be compared with reference numerical codes as those based on the Boundary Elements Methods (Liu 2009, Porez et al. 2011).

Continuing on from these results on modeling, other works were pursued in the field of perception and control. In particular, a classical approach based on the use of a model integrated in a Kalman filter has been applied with some success to the problem of navigation in a tank. The approach is based on the reconstruction of the environment where the objects are modeled by spherical primitives (Lebastard et al. 2013). They show the efficiency of the approach especially for the 3D localization of small objects. Let us remark that other approaches respectively based on the particle filter (Solberg et al. 2008) and Bayesian filters (Silverman et al. 2012) have also been proposed.

More recently, it has been discovered that it was possible to address the problem of navigation in encumbered environments without reconstructing it, i.e. without using any model. The approach is based on bio-inspiration and the concept of embodiment. More precisely, by exploiting the morphology of the sensors (slender shape, bi-lateral symmetry), it is possible to address the problem of navigation as that of the interaction of the sensor's body with its electric field

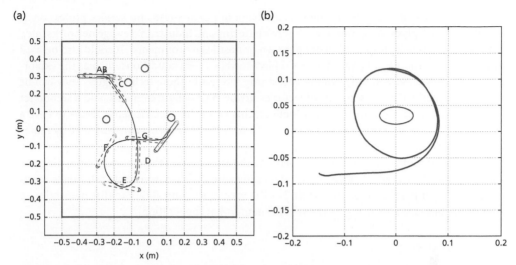

(a)

y (m)

x (m)

(b)

Figure 19.3 Application of a reactive control law on a slender probe: (a) (Insulating) obstacle avoidance and conductive object seeking. (b) Exploration of the shape of an object by revolving around it.

distorted by its physical environment. In this context, a set of reactive control laws have been proposed; its principle is based on the alignment of the body of the robot-sensor on the electric lines emitted by the polarized object. Remarkably, these strategies have been discovered in fish; the fish hunt by following the electric lines that lead them to their prey. Once implemented and tested on experimental tests, these bio-inspired (and embodied) strategies allow avoiding the use of any model, while giving relevant and robust behaviors to the sensor (robot) (see Figure 19.3). Among these behaviors, we find "obstacle avoidance" or the active seeking of conducting objects (see Figure 19.3) (Lebastard et al. 2012, Boyer and Lebastard 2012). Furthermore, the approach is a very nice example of the concept of morphological computation in the meaning of embodied intelligence (Pfeifer et al. 2007).

Based on this fundamental and experimental framework, the ANGELS project has proposed the first autonomous underwater robot equipped with electric sense using the reactive control approach previously evoked. The experiments (Demonstrations of the Angels platform 2013) have shown the feasibility of exploiting the electric sense in underwater robotics in constrained environments (encumbered by many obstacles, turbid water, etc.).

Future directions

Beside the impacts brought by the electric sense in itself, it has initiated a new generation of autonomous modular and reconfigurable underwater robots devoted to agile swimming in confined environments immersed by fluids which can be sufficiently dirty and encumbered to prevent the use of vision and sonar. The concept of adaptive morphology to a kind of electric body which self-adapts to its environment, combined with the mechanical possibilities and modularity, greatly multiplies the possibilities of adaptation (up to the most complex ones involved by sociologic interactions between small autonomous sub-agents sharing the same collective electric field as that of a single entity). Because of its high adaptability, a new underwater robot generation could be used for a number of different robotic applications. In particular, in the long term, modular robots with electric sense could be of use as assistants for service professionals (underwater and pipes inspection for offshore platforms and industries, sewage networks, etc.) and agents for security (exploration of unknown or hostile environments). It could also be used for autonomous robotics in environments subject to very severe conditions, such as the exploration of the abyss. Several industries have a strong interest in these technologies. They can be essentially classified in two sets: that of factories involved in water treatments, and a second set more involved in nuclear energy. Let us note on this point that electric sense seems to have a strong potential for the exploration of immersed pools with dirty waters contaminated by radioactivity. In this case, factories are interested in shape recognition and mapping. Going further, in the perspective of generalizing electric sense to the air, we could then imagine equipping our future robots with a kind of electric body interacting with their environment with simple reactive strategies. In particular, electric sense has the very interesting ability to distinguish living from dead objects, a feature highly relevant for man–robot cooperation.

Learning more

For the reader wishing to learn more on biological principles and experiments, see Bullock and Heiligenberg (1986), Moller (1995), Caputi et al. (1998), Hopkins (2009), von der Emde et al. (2008), Engelmann et al. (2008), and Pereira et al. (2012). A new approach of the modeling has been developed in Boyer et al. (2012) and this approach allows concise analytic models of the environment of the sensor. The classical control approach based on the reconstruction of the environment based on the use of a model integrated in a Kalman filter (Lebastard et al. 2013), in a particle filter (Solberg et al. 2008), in a Bayesian filter (Silverman et al. 2012), and in the MUSIC algorithm (Lanneau et al. 2017) have been applied with some success to the problem of reconstruction of environment and navigation. A more recent control based on the concept of morphological computation in the meaning of embodied intelligence is developed in Lebastard et al. (2012), Boyer and Lebastard (2012), and Lebastard et al. (2016). The broader challenge of designing biomimetic robots that can operate in underwater environments in discussed in Kruusmaa (Chapter 44, this volume).

References

Bouvier, S., Boyer, F., Girin, A., Gossiaux, P., Lebastard, V., and Servagent, N. (2013). Procédé et dispositif de controle du déplacement d'un système mobile dans un milieu conducteur d'électricité. Patent WO2013014392A1.

Boyer, F., Gossiaux, P., Jawad, B., Lebastard, V., and Porez, M. (2012) Model for a sensor bio-inspired from electric fish. *IEEE Transactions on Robotics*, **28**(2), 492–505.

Boyer, F. and Lebastard, V. (2012). Exploration of objects by an underwater robot with electric sense. In T. Prescott, N. Lepora, A. Mura, and P. Verschure (eds), *Biomimetic and Biohybrid Systems*, Vol. 7375 of *Lecture Notes in Computer Science*, Berlin/Heidelberg: Springer, pp. 50–61.

Bullock, T., and Heiligenberg, W. (1986). *Electroreception*. Hoboken, NJ: John Wiley and Sons.

Caputi, A., Budelli, R., and Bell, C. (1998). The electric image in weakly electric fish: physical images of resistive objects in *Gnathonemus petersii. Journal of Experimental Biology*, **201**(14), 2115–28.

Engelmann, J., Bacelo, J., Metzen, M., Pusch, R., Bouton, B., Migliaro, M., Caputi, A., Budelli, R., Grant, K., and von der Emde, G. (2008). Electric imaging through active electrolocation: Implication for the analysis of complex scenes, *Biol. Cybern.* **98**, 519–39.

Hopkins, C. D. (2009). Electrical perception and communication. *Encyclopedia of Neuroscience*, Vol. **3**. New York: Academic Press, pp. 813–31.

Lanneau, S., Boyer, F., Lebastard, V., and Bazeille, S. (2017). Model based estimation of ellipsoidal object using artificial electric sense. *The International Journal of Robotics Research*, **36**(9),1022–41.

Lebastard, V., Boyer, F., Chevallereau, C., and Servagent, N. (2012). Underwater electro-navigation in the dark. In: *IEEE Conference on Robotics and Automation*, ICRA 2012. St. Paul, Minnesota, USA: IEEE. ISBN 978-1-4673-1403-9, pp. 1155–60.

Lebastard, V., Chevallereau, C., Girin, A., Servagent, N., Gossiaux, P.-B., and Boyer, F. (2013). Environment reconstruction and navigation with electric sense based on Kalman filter. *International Journal of Robotics Research*, **32**(2), 172–88.

Lebastard, V., Boyer, F., and Lanneau, S. (2016). Reactive underwater object inspection based on artificial electric sense. *Bioinspiration and Biomimetics*, **11**(4), 045003.

Lissmann, H., and Machin, K. (1958). The mechanism of object location in *Gymnarchus niloticus* and similar fish. *The Journal of Experimental Biology* **35**, 451–86.

Liu, Y. (2009). *Fast multipole boundary element method*. Cambridge, UK: Cambridge University Press.

Mayton, B., LeGrand, L., and Smith, J. R. (2010). An electric field pretouch system for grasping and co-manipulation. In *IEEE International Conference on Robotics and Automation (ICRA 2010)*, pp. 831–8.

Moller, P. (1995). *Electric Fishes: History and Behavior*. New York: Chapman & Hall.

Pereira, A. C., Aguilera, P., and Caputi, A. A. (2012). The active electrosensory range of *Gymnotus omarorum*. *The Journal of Experimental Biology*, **215**, 3266–80.

Pfeifer, R., Lungarella, M., and Lida, F. (2007). Self-organization, embodiment, and biologically inspired robotics. *Science*, **318**(5853), 1088–93.

Porez, M., Lebastard, V., Ijspeert, A. J., and Boyer, F. (2011). Multi-physics model of an electric fish-like robot: Numerical aspects and application to obstacle avoidance. *IEEE/RSJ Int. Conf. on Intelligent Robots and Systems*, pp. 1901–06.

Rasnow, B. (1996). The effects of simple objects on the electric field of *Apteronotus*. *Journal of Comparative Physiology A*, **3**(178), 397–411.

Servagent, N., Jawad, B., Bouvier, S., Boyer, F., Girin, A., Gomez, F., Lebastard, V., and Gossiaux, P.-B. (2013). Electrolocation sensors in conducting water bio-inspired by electric fish. *IEEE Sensor Journal*, **13**(5), 1865–82.

Silverman, Y., Snyder, J., Bai, Y., and MacIver, M. A. (2012). Location and orientation estimation with an electrosense robot, *IEEE/RSJ Int. Conf. on Intelligent Robots and Systems*. IEEE, pp. 4218–23.

Solberg, J., Lynch, K., and MacIver, M. (2008). Active electrolocation for underwater target localization. *The International Journal of Robotics Research*, **27**(5), 529–48.

Demonstrations of the Angels platform (2013). http://www.youtube.com/watch?v=HoJu0OLyW4o.

von der Emde, G., Amey, M., Engelmann, J., Fetz, S., Folde, C., Hollmann, M., Metzen, M., and Pusch, R. (2008). Active electrolocation in *Gnathonemus petersii*: Behaviour, sensory performance, and receptor systems. *J. Physiol* **102**, 279–90.

Chapter 20

Muscles

Iain A. Anderson and Benjamin M. O'Brien

Auckland Bioengineering Institute, The University of Auckland, New Zealand

Mechanical devices that include home appliances, automobiles, and airplanes are typically driven by electric motors or combustion engines through gearboxes and other linkages. Airplane wings, for example, have hinged control surfaces such as ailerons. Now imagine a wing that has no hinged control surfaces or linkages but that instead bends or warps to assume an appropriate shape, like the wing of a bird. Such a device could be enabled using an electroactive polymer technology based on electronic artificial muscles. Artificial muscles act directly on a structure, like our leg muscles that are attached by tendon to our bones and that through phased contraction enable us to walk. Sensory feedback from our muscles enables proprioceptive control. So, for artificial muscles to be used appropriately we need to pay attention not only to mechanisms for muscle actuation but also to how we can incorporate self-sensing feedback for the control of position.

Biological principles

Natural muscle is a soft lightweight material whose speed, strain, and pressure for low density is unmatched by any man-made alternative. Mammalian skeletal muscles are usually configured antagonistically with other muscles across a joint. Given that the stiffness of muscle is very strongly influenced by its state of contraction when held at fixed length, the co-contraction of antagonistic muscles across a joint can be used to control position and stiffness simultaneously. For example, the control of stiffness in the arm is essential to correctly bow a violin string. Poor stiffness control that makes the arm too stiff, the result of "stage fright," can result in an embarrassing bouncing-bow vibration that will put an end to a concert.

Muscles can work in soft living bodies that lack rigid skeletal support, provided they work against an elastic structure or fluid pressure. For example, the neutrally buoyant bells of cnidarian jellyfish, supported against gravity by surrounding water, rely on muscle to provide the swimming contractions of the bell. But muscle contraction works antagonistically against the elastic bell. At the end of contraction and on relaxation the bell passively springs open under its stored elastic energy. Nerves and multiple pacemaker organs are arranged and connected so that contraction around the circumference of the bell is controlled for normal swimming, escape, or to intercept food.

In a soft organ such as the heart, the relaxed myocardium expands under the pressure of the blood. But when it contracts the overall soft tissue architecture ensures that the conducted signal from the pacemaker that is passed from gap junction to gap junction produces an orderly phasic contraction (LeGrice et al. 1995).

That we can touch our nose with our eyes closed and walk also demonstrates the efficacy of integrated muscle strain sensing. This is made possible through nerve cells that are embedded within muscle, known as muscle spindles, that give us the ability to know limb position through the sensing of length change. They also provide position feedback to the central nervous system (Martini et al. 2008) enabling fine motor control for maintaining correct balance and posture. Should we attempt too many kilos during a bench-press we will be warned by the pain feedback through nociceptors, sensory neurons that lie within muscle to warn us of potential tissue damage, or remind us to protect damaged tissue while it heals.

In conclusion, muscles control stiffness and position while sensing strain and pain. To emulate natural muscle in a soft robotic system we need to:

1. Identify a suitable material with a similar modulus and density to natural muscle.

2. Configure it in such a way that we can mimic natural muscle's use of antagonistic passive elements or active elements.

3. Mimic proprioception so that we obtain direct feedback for fine control. Dielectric Elastomer (DE) artificial muscles are good candidates for emulating real muscle. In this section we show how they operate, what they can do and the many interesting challenges that we face in adapting them to soft robotic applications.

Biomimetic or biohybrid systems

DE are typically built from soft polymers such as silicone (Modulus~ 0.1–1 MPa), acrylic (3M VHB, Modulus~ 2–3 MPa), and polyurethane (Modulus 17 MPa) (Brochu and Pei 2010) that have a modulus and specific density similar to muscle tissue (~ 1 MPa) (Kovanen et al. 1984). Their basic configuration (Figure 20.1) consists of a thin layer of one of the above materials sandwiched between flexible electrodes such as a carbon powder, grease, or nanotube layer. When an electric field is applied charges move onto the electroded surfaces leading to an electrostatic pressure, producing in-plane expansion and out-of-plane contraction. The electrostatic "Maxwell Pressure" P is given in equation 1 where, ε_0 and ε_r are the absolute and relative dielectric permittivities respectively, V is the voltage, and t is the thickness of the membrane (Pelrine et al. 1998):

$$P = \varepsilon_0 \varepsilon_r \left(\frac{V}{t}\right)^2 \tag{1}$$

Two basic ways of configuring DE actuators are illustrated in Figure 20.1: as a membrane that expands in-plane on actuation (b) or as a multi-layer stack that contracts out of plane on actuation (c). An example of an actuator that utilizes in-plane strain is the spring roll developed by the Pei group at UCLA (Pei et al. 2004). The spring roll is composed of layers of carbon electroded acrylic elastomer film rolled over a pre-compressed helical spring that provides support for the pre-stretched film as well as a passive antagonistic structure for the film to work against. Actuation of the DE film will cause the entire roll to lengthen. Electrical partitioning the DE into two 180° or four 90° separately actuated sectors can enable the roll to also bend about one or two axes, and using this feature they have produced robots with spring roll legs that could walk like the tube feet of starfish (Pei et al. 2004). A group at the Swiss Laboratory EMPA used multiple spring rolls to produce an arm-wrestling robot (Kovacs et al. 2007; Figure 20.2); one of three entered in the first human-robotic arm-wrestling competition for the 2005 SPIE Electroactive Polymers and Devices Conference (Bar-Cohen 2007). Although the device did not resemble a human arm, its internal actuators mimicked the agonist–antagonist arrangement of muscle around a joint using two opposing banks linked to each other across a pulley, enabling the robot

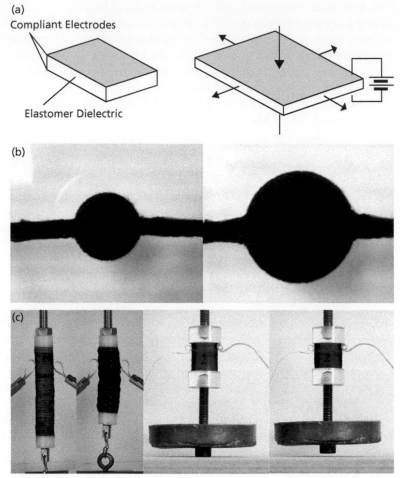

Figure 20.1 (a) DE actuation involves the response of a dielectric membrane material to having charge deposited on its free surfaces. Charges of opposite polarity draw the opposing surfaces together. Repulsion of like charges expands the dielectric surface area. (b) An in-plane actuator: on the left is a pre-actuation "expanding dot" DE membrane actuator, composed of 3M VHB4905. The circular blackened zone is the Nyogel carbon grease electrode. Application of charge results in more than doubling of area. (c) Stack actuators in action. Many layers of DE are stacked one atop to the other. On actuation the stack becomes shorter (right-hand views).

Reprinted from *Sensors and Actuators A: Physical*, 155 (2), G. Kovacs, L. Düring, S. Michel, and G. Terrasi, Stacked dielectric elastomer actuator for tensile force transmission, pp. 299–307, doi: 10.1016/j.sna.2009.08.027, Copyright © 2009 Elsevier Ltd., with permission from Elsevier.

to both wrestle and return to the starting point, a requirement of the competition (Kovacs et al. 2007). In another biomimetic example where muscles are used in an antagonistic way, a group based in Korea has developed a multi-segmented annelid worm robot with individual segments activated using DE (Jung et al. 2011); see Trimmer (Chapter 41, this volume) for further examples of soft robotic systems that exploit muscle-like actuators.

Earlier we described the jellyfish as a soft system with muscle acting against an elastic spring-like bell. A soft DE motor mimics this: radially arrayed DE actuators have been used antagonistically to control a soft elastic gear in a rotary motor. The muscles grip the shaft and roll

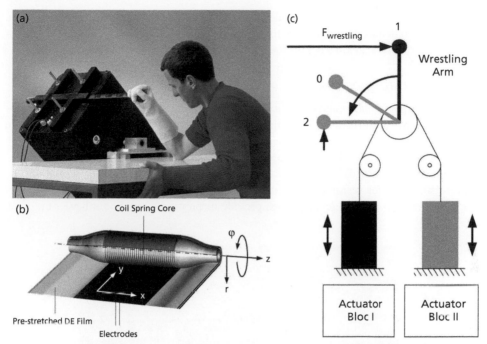

Figure 20.2 (a) The EMPA arm wrestling robot robot described by Kovacs et al. (2007). (b) Schematic of a spring roll actuator. These were arranged into two antagonistic groups and placed in the torso-sized volume (c).

Republished from *Smart Materials and Structures*, 16 (2), p. S306, An arm wrestling robot driven by dielectric elastomer actuators, Gabor Kovacs, Patrick Lochmatter and Michael Wissler, doi: 10.1088/0964-1726/16/2/S16 © 2007, IOP Publishing. Reproduced with permission. All rights reserved.

it like you would roll a pencil between thumb and forefinger. When two of these DE motors support opposing ends of the same shaft a 5 degree of freedom actuator is possible (Anderson et al. 2011). In another example of the coupling of DE muscles to an elastic structure, workers at EMPA have built a helium-filled blimp which swims like a fish through the air (Jordi et al. 2010). The skin of the blimp provides support for the DE muscles on part of its surface. The skin is also under tension from the pressure of the gas within so that actuation of the muscles produces an extension of the skin. This extension can be used to bend the body to promote a swimming motion.

Auxiliary sensors are not required for DE muscles. This is because they are able to self-sense. When stretched either through mechanical or electrical action their surface area increases and they also become thinner, a consequence of the near incompressibility of the dielectric material. Electrically DE resemble a simple parallel-plate capacitor. Their capacitance is given by:

$$C = \frac{Q}{V} = \varepsilon_0 \frac{A}{t} \qquad (2)$$

where Q is the charge on the DE, V is the voltage across it, A is the in-plane surface area of the DE, and t is its thickness. If we can measure a DE's electrical state we can infer something about how much it has been stretched. Algorithms for doing this have been developed by several workers (Anderson et al. 2012). Soft DE sensors are now commercially available.

DE are susceptible to several failure modes, the most notable of which is dielectric breakdown. Gisby et al. (2010) demonstrated that imminent DE failure can be monitored through the measurement of current leakage through the dielectric. This is synonymous with a pain signal.

Future directions

In the future the performance and reliability of DE materials needs to be continually improved—there are a number of workers exploring this (Brochu and Pei 2010). One key requirement is to reduce the voltage needed for muscle actuation (currently > 1kV). This can come about through improvements to electrode reliability, improved breakdown strength, and higher dielectric constants (Brochu and Pei 2010).

A promising area for research involves the exploration of control strategies for arrays of DE actuators. Comb jellyfish (ctenophores) provide inspiration: they do not have a brain yet

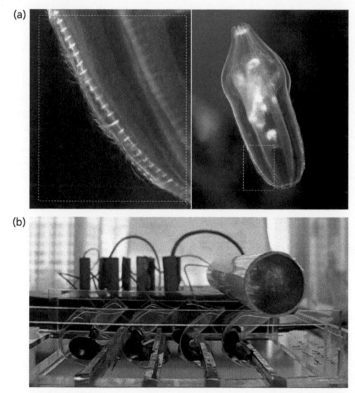

Figure 20.3 (a) A comb-jelly (*Lampea* sp.) depicting waves of cilia comb plate actuation passing in the forward direction (upwards) along a ctenophore comb row. (b) Inspired by this control strategy O'Brien et al. (2009) produced four touch-sensitive DE actuators based on the self-organized structure of Kofod et al. (2006). Each is activated on touch by the previous one and this results in the cylinder being pushed towards the left.

are capable of coordinated swimming control, facilitated through the sequential triggering of mechano-sensitive swimming paddles, each composed of numerous extra-long cilia. When activated a comb paddle pushes water backwards in a power stroke and this triggers the one in front to do the same. This simple control strategy has been mimicked using an array of mechano-sensitive DE actuators (Figure 20.3) that operate a conveyor mechanism (O'Brien et al. 2009).

The Dielectric Elastomer Switch (DES) (O'Brien et al. 2010) is opening new vistas for control. A DES consists of a surface printed electrode that undergoes very large resistance changes with strain and can switch power to a DE. Thus it is possible to control an arrangement of muscles with DES that are in turn influenced by direct mechanical interaction with the muscles. This has recently been used to control a self-commutating DE motor (O'Brien et al. 2012). For control of muscle arrays we look to emulate a pacemaker such as the one in the heart or the pacemaker mechanisms that cnidarian jellyfish use to produce coordinated rhythmic contractions of the bell. A ring oscillator could act as a pacemaker and such a device can be built using DE actuators with DES (O'Brien and Anderson 2011). The switch technology can also be used for higher-order control and this has led to the development of a Turing Machine based on DE actuators and DES (O'Brien and Anderson 2013). These examples demonstrate the capability of DE devices as a test bed for developing soft and smart robots.

Learning more

To learn more about DE transducers we refer the reader to the book *Dielectric Elastomers as Electromechanical Transducers* (Carpi et al. 2011). To read more about material aspects of DE actuators please see the review by Brochu and Pei (2010). A key issue with DE is electrode design. This is covered in the review by Rosset and Shea (2013). The functionality and control aspects of DE are reviewed by Anderson et al. (2012).

References

Anderson, I. (2007). *The Surface of the Sea, Encounters with New Zealand's Upper Ocean Life*. Auckland: Penguin Random House.

Anderson, I., Tse, T. C. H., Inamura, T., O'Brien, B. M., McKay, T., and Gisby, T. (2011). A soft and dexterous motor. *Applied Physics Letters*, **98**, 123704. doi: http://dx.doi.org/10.1063/1.3565195

Anderson, I. A., Gisby, T. A., McKay, T. G., O'Brien, B. M., and Calius, E. P. (2012). Multi-functional dielectric elastomer artificial muscles for soft and smart machines. *Journal of Applied Physics*, **112**(4), 041101.

Bar-Cohen, Y. (2007). Electroactive polymers as an enabling materials technology. *Proceedings of the Institution of Mechanical Engineers, Part G: Journal of Aerospace Engineering*, 221(4), 553–64.

Brochu, P. and Pei, Q. (2010). Advances in Dielectric Elastomers for Actuators and Artificial Muscles. *Macromolecular Rapid Communications*, 31(1), 10–36.

Carpi, F., et al. (eds) (2011). *Dielectric Elastomers as Electromechanical Transducers: Fundamentals*. Oxford: Elsevier.

Gisby, T.A., Xie, S. Q., Calius, E. P., and Anderson, I. A. (2010). Leakage current as a predictor of failure in dielectric elastomer actuators. *Proc. SPIE*, **7642**. doi: 10.1117/12.847835

Jordi, C., Michel, S., and Fink, E. (2010). Fish-like propulsion of an airship with planar membrane dielectric elastomer actuators. *Bioinspiration & Biomimetics*, 5(2), 026007.

Jung, K., Koo, J. C., Nam, J., Lee, Y. K., and Choi, H. R. (2011) Artificial annelid robot driven by soft actuators. *Bioinspiration & Biomimetics*, 2, S42–S49.

Kofod, G., Paajanen, M., and Bauer, S. (2006). Self-organized minimum-energy structures for dielectric elastomer actuators. *Applied Physics A: Materials Science & Processing*, 85(2), 141–3.

Kovacs, G., During, L., Michel, S., and Terrasi, G. (2009). Stacked dielectric elastomer actuator for tensile force transmission. *Sensors and Actuators A: Physical*, **155**(2), 299–307.

Kovacs, G., Lochmatter, P., and Wissler, M. (2007). An arm wrestling robot driven by dielectric elastomer actuators. *Smart Materials and Structures*, **16**(2), S306.

Kovanen, V., Suominen, H., and Heikkinen, E. (1984). Mechanical properties of fast and slow skeletal muscle with special reference to collagen and endurance training. *Journal of Biomechanics*, **17**(10), 725–35.

LeGrice, I. J., Smaill, B. H., Chai, L. Z., Edgar, S. G., Gavin, J. B., and Hunter, P. J. (1995). Laminar structure of the heart: ventricular myocyte arrangement and connective tissue architecture in the dog. *American Journal of Physiology—Heart and Circulatory Physiology*, **269**(2), H571–H582.

Martini, F. H., Timmons, M.J., and Tallitsch, R.B. (2008). *Human Anatomy*, 6th edn. San Francisco: Pearson Education.

O'Brien, B., and Anderson, I. A. (2011). An artificial muscle ring oscillator. *IEEE/ASME Transactions on Mechatronics*, doi: 10.1109/TMECH.2011.2165553.

O'Brien, B., and Anderson, I. (2013). An artificial muscle computer. *Applied Physics Letters*, **102**, 104102. doi: http://dx.doi.org/10.1063/1.4793648

O'Brien, B., Gisby, T., Calius, E., Xie, S., and Anderson, I. A. (2009). FEA of dielectric elastomer minimum energy structures as a tool for biomimetic design. *Proc. SPIE*, 2009, 7287.

O'Brien, B. M., Calius, E. P., Inamura, T., Xie, S. Q., and Anderson, I. A. (2010). Dielectric elastomer switches for smart artificial muscles. *Applied Physics A: Materials Science and Processing*, **100**(2), 385–9.

O'Brien, B. M., McKay, T. G., Gisby, T. A., and Anderson, I.A. (2012). Rotating turkeys and self-commutating artificial muscle motors. *Applied Physics Letters*, **100**(7), 074108.

Pei, Q., Rosenthal, M., Stanford, S., Prahlad, H., and Pelrine, R. (2004). Multiple-degrees-of-freedom electroelastomer roll actuators. *Smart Materials and Structures*, **13**(5), N86.

Pelrine, R. E., Kornbluh, R.D., and Joseph, J.P. (1998). Electrostriction of polymer dielectrics with compliant electrodes as a means of actuation. *Sensors and Actuators A: Physical*, **64**(1), 77–85.

Rosset, S., and Shea, H. R. (2013). Flexible and stretchable electrodes for dielectric elastomer actuators. *Applied Physics A*, **110**(2), 281–307.

Chapter 21

Rhythms and oscillations

Allen Selverston

Division of Biological Science, University of California, San Diego, USA

Oscillatory rhythms are a ubiquitous form of neuronal activity. They underlie such motor behaviors as locomotion, chewing, and breathing. Neural oscillations at frequencies ranging from 0.02 Hz to 600 Hz are found throughout the central nervous system (CNS) and have been shown to be involved in sensory and higher level information processing. To mimic this form of activity with neurotechnology, it is vital to understand the biological mechanisms that underlie rhythmic behaviors (rhythmogenesis). It is also important to understand how multi-phase patterns of rhythmic activity are produced and maintained. One of the best models for studying rhythmogenesis is the central pattern generator (CPG), a collection of nerve cells in the CNS that can produce rhythmic behaviors autonomously. Although CPGs are the main engine for rhythmic systems, to be functional they must be able to adapt to varying environmental conditions. To accomplish this, CPGs are controlled by descending commands which integrate higher-level visual, acoustic, and tactile stimuli and are able to modify the motor pattern on a cycle-by-cycle basis. In addition, almost all CPGs receive closed-loop sensory feedback that is functionally involved in rhythmogenesis and patterning. This chapter will highlight the principal features of oscillatory behavior that have been elucidated by the study of invertebrate CPGs. In particular, the two patterns produced by the lobster stomatogastric ganglion. In this simpler system, the basis of the rhythmicity and the logic behind the formation of the patterns is now well understood. This section will also include information about how the two CPGs are controlled by sensory feedback and central commands, and how chemical neuromodulators can functionally reconfigure them to produce different patterns. The last section will discuss how the biological principles can be incorporated into biomimetic and hybrid devices, such as those described elsewhere in this handbook (e.g. Ayers, Chapter 51, this volume), and the general challenges presented in reverse engineering neural systems.

Biological principles

The stomatogastric ganglion

The 30-celled stomatogastric ganglion (STG) is located on the upper surface of the crustacean stomach. Nerves exit the STG to supply striated (voluntary) muscles that control rhythmic movements of the stomach. One set of muscles controls three teeth, the gastric mill, that chews food and a second set, the pyloric, that performs a pumping and peristaltic function at the rear of the stomach. The role of the STG is to generate and synchronize these two separate rhythmic movements utilizing an eleven-cell gastric circuit and a fourteen-cell pyloric circuit. The cells are large (70–80 microns in diameter) and reidentifiable. By recording and stimulating all pairs of cells in each circuit and imposing severe tests for monosynaptacity, a detailed connectivity map for both CPGs was determined (Selverston and Moulins 1987). In terms of actual cell-to-cell connections, the stomatogastric circuits are the most detailed neuronal circuits available.

Neuronal identity in the STG

While neurons in vertebrate circuits can only be identified by type, invertebrate CPG neurons can be identified individually on the basis of their biophysical properties and their connections to other identified neurons. This makes it possible to label each neuron and return to it in preparations from different animals. The importance of cell identity for understanding how neural circuits work cannot be underestimated. The ability to return to the same neuron in a circuit is key to understanding design principles whether at the molecular, cellular, or network level. The membrane channels present in each cell confer distinct electrical properties and determine how each neuron functions within the network. In a sense, each neuron behaves as a hardware component in a complex mechanical or digital system, but reverse engineering the biological properties is not trivial (Marom et al. 2009). The connections between STG neurons are remarkably consistent and are comprised primarily of chemical inhibitory synapses and a few excitatory synapses. There are also many electrical synapses, especially between neurons of the same type that serve to synchronize their activity.

General properties of CPGs

The analysis of small CPG circuits has revealed both similarities and differences between biological and engineering solutions to equivalent problems. The topologies of CPG circuits are rigidly determined, with component neurons and their connections invariant from animal to animal. In one sense the biological circuits are as hard wired as if they were made by following a schematic diagram. The biophysical properties of each neuron are also reliably consistent, although the number of component channel proteins (the biological parts) for each neuron can vary up to fourfold as long as certain fixed ratios are maintained. For the STG, the determination of which channels are expressed (to give each neuron its identity) is highly degenerative. A description of all the ionic currents found in the STG as well as their function can be found in reviews (e.g. Harris-Warrick et al. 1992). All of the membrane currents are influenced by neuromodulators acting selectively on ion channels to alter their properties. The result of this action confers a broad range of biophysical properties to the network components—individual neurons and synapses. By changing the properties of channels, the circuit is functionally reconfigured over an extended period of time—minutes to hours (Nusbaum and Beenhakker 2002). Neural inputs from higher centers and sensory receptors are also directed to specific cells in the circuit to increase or decrease their activity in order to transiently direct motor output adjustments. In this case, the cell properties are altered only during the time the inputs are active. A CPG would only produce highly stereotyped "robot"-like rhythms, inadequate for autonomous behavior, without both neuromodulatory and synaptic inputs.

The logic of the pyloric rhythm

The pyloric stomach region acts as a pump and filter for food that has been macerated by the gastric mill. Essentially a tube, it has extrinsic muscles (PDs) on its upper and lower surface connected to the carapace so that when contracted the pylorus is dilated. It also contains intrinsic muscles (LP and PYs) along the rear surface that when contracted, from front to rear, produce a peristaltic-like pumping movement. The actual behavior consists of a rhythmic three-phase cycle, dilation followed by two phases of contraction at a frequency of about 1 Hz (Figure 21.1). The three phases are driven by three electrically coupled oscillatory neurons, the AB interneuron and the two PD motor neurons. These three, as well as all the other neurons in the pyloric circuit, are termed conditional bursters, i.e. they require neuromodulatory input from "higher" ganglia to burst. The AB and PDs burst (phase one) and strongly inhibit the other

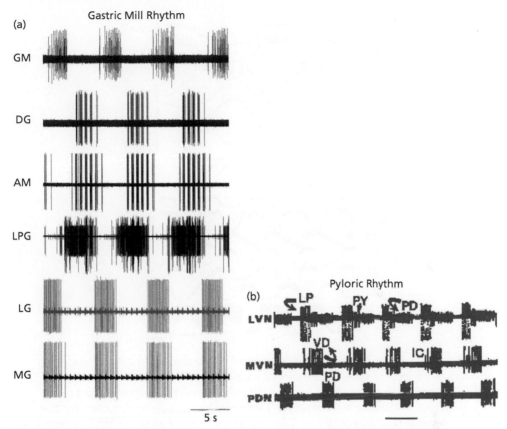

Figure 21.1 Canonical rhythmic bursts produced by the gastric and pyloric CPGs *in vitro* in a preparation receiving neuromodulatory input from higher centers. Axonal traces are identified according to the muscles they innervate. Note the strong pyloric modulation in the DG and AM neurons.

pyloric neurons (Figure 21.2). When they stop bursting, I_H currents cause LP to resume bursting first (phase two) followed by the eight PYs (phase three).

The logic of the gastric mill rhythm

The maceration of food occurs in a part of the stomach known as the gastric mill. It contains three serrated ossicles, one medial and two lateral, that serve as teeth. The teeth grind in a four-phase rhythm of ca. 0.1Hz, which consists of opening and closing the lateral teeth and a forward and backward movement of the medial tooth. Producing these four phases requires a CPG network far more complicated than the pyloric CPG. The gastric mill network utilizes more types of cellular and synaptic mechanisms and incorporates many more complicated network arrangements than the coupled oscillator–inhibitory mechanisms found in the pyloric CPG (Selverston et al. 2009). One can speculate that the additional complexity is needed because the behavior is not only intermittent, it is also more flexible and subject to more sensory feedback than is the pyloric rhythm. It is similar to the difference between the continuously beating heart and the intermittent movement of the legs that have to both walk and dance. There are only eleven neurons in the gastric CPG, two that open the lateral teeth (LPGs) and two that close it (LG and MG) (see Fig. 21.3 for entire

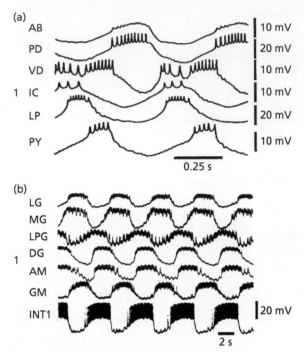

Figure 21.2 Simultaneous intracellular recordings from the neuronal machine that produces the axon bursts to the muscles. (a) Pyloric rhythm. (b) Gastric rhythm. The spikes riding on the slow waves correspond to the axonal spikes and the subthreshold activity from other neurons can be seen in most cells. Intracellular recordings enable any cell to be manipulated by current injection to increase or decrease its activity or fill it with dye to examine its morphology or photoinactivate it. Genetic material coding for specific channel proteins can also be inserted into single cells.

circuit). Four neurons move the medial tooth forward in a power sroke (GMs) and two neurons generate the return stroke movement (DG and AM). The lateral teeth and the medial tooth subset of neurons are driven and coordinated by the single Int 1 neuron which has strong intrinsic bursting properties. The medial tooth basic mechanism consists of two reciprocal inhibitory networks, between Int 1 and the LG/MG pair and the LG/MG pair and the LPGs. The medial tooth operates simply by excitation and feedforward inhibition with Int 1 exciting the DG/MG synergists and DG, MG, and Int 1 all inhibiting the tonically firing GMs. All of this activity generates the rhythm and the correct phase relationships between the four bursts leading to coordinated movements and effective "chewing" of food. The properties of the gastric mill neurons range from strong bursters to tonic firing to complete silence. Most synapses are inhibitory and the few excitatory synapses often have delays built into them. There is extensive electrical coupling between synergists. Like the pyloric interneuron AB, the Int 1 interneuron also sends a copy of the rhythm to neurons in higher ganglia as the feedback excitation to some of the gastric neurons (not shown in Figure 21.3).

To summarize, the gastric mill CPG produces rhythmic behavior in individual neurons by several different mechanisms:

◆ Conditional burster neurons produce rhythmic bursting in follower cells by utilizing synaptic excitation and inhibition.

(a)

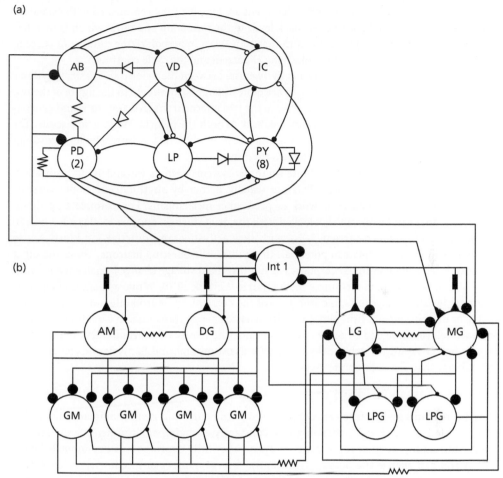

(b)

Figure 21.3 A simplified cartoon of the STG living machine circuits. The upper circuit (a) represents the pyloric CPG and the lower circuit (b) the gastric CPG. Black dots are chemical inhibitory synapses, black triangles are excitatory. Resistors represent electrotonic coupling as do the diodes which are rectifying. The size of the dot gives some indication of its strength. The black rectangles represent delay lines to their respective synapses. Each neuron except Int 1 and the AB also sends an axon out of the ganglion to a muscle.

- If a follower neuron is firing tonically it is periodically inhibited by the burster and therefore bursts out of phase with it.
- If a follower cell is silent, it can be periodically excited by the burster and fire in phase with it.
- The periodic excitation is delayed by various amounts, probably by an I_A current that causes follower neurons to fire with variable phase delays.
- Pairs of reciprocally inhibited neurons produce rhythmic bursting using a half-center network mechanism.

While combinations of these mechanisms can be found in other invertebrate systems, additional methods have been described for rhythmogenesis in vertebrate systems. In particular, recurrent excitation has been suggested for lamprey (Grillner et al. 2005) and frog tadpole swimming (Roberts and Perrins 1995). Here, cells are massively interconnected with excitatory synapses in a way so that positive feedback leads to the formation of a burst which is then terminated by a buildup of internal calcium that activates a hyperpolarizing potassium current. The hyperpolarization triggers a depolarizing H current and the cycle continues.

The STG circuit diagram shows connectivity exists between the pyloric and gastric rhythms, notable because the frequencies of the two are quite different. Not shown in the figure of the two circuits (Figure 21.3) are connections between the two interneurons (AB and Int 1) and neurons in higher ganglia that provide excitatory feedback to both the pyloric and gastric circuits. The postsynaptic potentials from these inputs can easily be seen in the intracellular recordings (Figure 21.2) and act to synchronize the two rhythms.

As with invertebrate CPGs, vertebrate rhythmic systems such as those driving respiration, locomotion, or mastication, the CPGs are usually driven by rhythmogenic ionic currents regardless of whether they use network or pacemaker mechanisms. For example a persistent Na current, that can be selectively blocked with the drug riluzole, appears to play a key role in rhythmogenesis. Another current, I_{CAN}, with slow activation and deactivation kinetics, also plays a role in producing plateau potentials in rhythmically bursting neurons. These and other currents, by causing non-linear responses to sub-threshold voltage changes, enable the neurons to sustain bursts and then terminate them (Harris-Warrick 2010). While a balance of currents is critical for rhythmogenesis, modification of the currents by neuromodulators can strongly enable or disable network function. In addition, fast synaptic transmission can also increase or decrease cell function and in some cases also cause the release of neuromodulators that affect the rhythm. Generally, vertebrate rhythmic systems use multiple mechanisms and contain more redundancy than do invertebrate systems.

Control of the STG rhythms

The STG rhythms and patterns are controlled by both neural and neuromodulatory input that adapts cycle frequency and patterning to a constantly changing environment. Some sensory input controls the ganglion on a cycle-by-cycle basis while neuromodulatory input is longer lasting and can change the output significantly. A large number of receptor types have been described morphologically (Dando and Maynard 1973):

1. Receptors to monitor movements of the mouth and lower esophagus.
2. Chemoreceptors that monitor the esophagus and foregut.
3. Stretch receptors near the STG that monitor the movements of the gastric mill.
4. Stretch receptors in the nerve that innervates the stomach wall that also monitors the gastric mill.
5. Proprioceptors in muscles near the pyloric region.
6. Stretch receptors near the digestive glands and initial parts of the midgut.

As already mentioned, in addition to neuronal inputs, STG circuits respond to chemical neuromodulatory substances delivered as hormones in the blood, from specialized neurons in higher centers and from some sensory neurons. They are mostly amines and small peptides that can be specifically stained and identified. They act via second messenger systems to alter the conductance properties of specific cells and synapses to which they are targeted and as a result alter the output of the entire circuit. The changes are robust and repeatable but last only in the order of minutes to hours. In the stomatogastric system, individual circuits can fundamentally change their output pattern, different circuits can be combined, entirely new circuit configurations formed, or neurons can simply switch from one circuit to another. So, a circuit can be hard wired anatomically but its function is heavily dependent on its surrounding chemical mileau. The basic rhythms shown in Figure 21.1 are the canonical lobster rhythms in a combined preparation that is receiving neuromodulatory input from higher ganglia. Remove this input and both rhythms are terminated. Add new neuromodulatory input and different patterns are formed.

Biomimetic and biohybrid systems

The reductionist approach to a particular biological system such as the STG has sought to uncover the underlying design principles for the production of rhythmic behavior. Other invertebrate CPGs use similar molecular components, but whose design principles at the cellular and circuit level are quite different. This despite the fact that the basic output patterns can be very much alike. This probably has less to do with a deep degeneracy, as has been suggested (Marom et al. 2009), than it has with evolution. Cell and synaptic properties evolve to suit selective pressures that are different for every species. These different CPG circuits cannot be determined by inductive reasoning but instead must be determined empirically if there is no previous knowledge. This does not mean having a detailed understanding of a biological circuit can lead to successful reverse engineering but it is a start. A circuit with known design principles may give satisfactory results but still may not be a representation of the biological design. Computer models of oscillatory networks can successfully mimic biological CPGs such as the STG and use them to better understand the properties of biological systems and to serve as the basis for physical models. Computer models of the STG networks, as well as others, can be considered at three levels: single neuron models, synaptic models, and whole network models. In general, models of single neurons fall into two categories, conductance-based and phenomenological. The conductance-based Hodgkin–Huxley (H–H) equations were originally proposed to describe the voltage-dependent conductances for ions generating the action potential but other conductances can be added. The model represents only a point process, namely the spike initiation site of the neuron. To consider the morphology of the cell in the model, other compartments must be added that compute the effects of passive and excitable regions of the neuron but a large number of compartments used greatly impacts the computation time. There are many other models that have been used to mimic the activity of single neurons. One is a very simplified three dimensional phenomenological model, the Hindmarsh–Rose (Torres and Varona 2012) consisting of three polynomial equations to approximate the H–H equations:

$$\frac{dV(t)}{dt} = W(t) + aV(t)^2 - bV(t)^3 - Z(t) + I$$

$$\frac{dW(t)}{dt} = C - dV(t)^2 - W(t)$$

$$\frac{dZ(t)}{dt} = r\left[s(V(t) - V_0) - Z(t))\right]$$

where *V(t)* is the membrane potential, *W(t)* describes the fast currents, *Z(t)* describes slow currents, and I is an external current.

This model can produce spiking and bursting regimes similar to the output of the STG and have been used to build artificial electronic neurons that can be connected together into networks. The electronic neurons can also interact bidirectionally with living neurons and have been used to rescue the pyloric rhythm after deletion of key cells or removal of neuromodulation (Figure 21.4). Despite the fact that the properties of electronic neurons are far removed from those of real neurons, they can, to a limited extent, substitute for their biological counterparts.

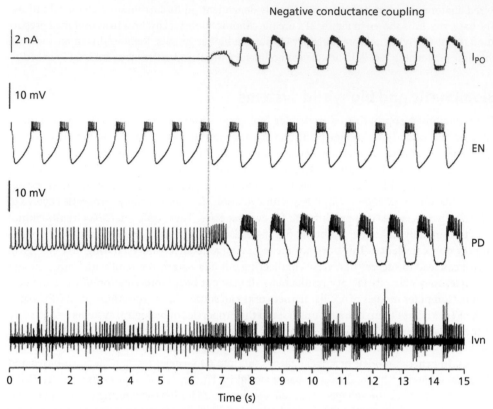

Figure 21.4 An artificial electronic neuron with Hindmarsh–Rose kinetics interfaced with a biological pyloric CPG. Here the recording shows the lack of bursting in one neuron (PD, third trace) and two other neurons recorded extracellularly (LP and PY, bottom trace) when the STG is deprived of neuromodulatory inputs. The top trace shows the artificial synaptic current and the second trace the oscillating potential of the artificial neuron. The artificial neuron is connected to a PD cell at the vertical line (6.5 seconds) with a simulated inhibitory synapse. This drives the PD out of phase with the electronic neuron and causes a resumption of rhythmic activity in the entire biological network at a frequency imposed by the artificial neuron.

Biomimetic synapses

The synapses used to couple the electronic neurons can range from simple to those with time-dependent non-linear properties such as accommodation and facilitation. In the simpler version, the summed synaptic currents can be injected into the model cell (Denker et al. 2005). More complete synaptic properties can be produced by using the dynamic clamp system (Szücs 2007; Pinto et al. 2001). The dynamic clamp is triggered by voltages recorded from presynaptic cells and converts them into programmed currents that mimic synaptic currents and that can be injected into postsynaptic cells via microelectrodes.

Biomimetic networks

By incorporating experimental cellular and synaptic data, the STG circuits have been modeled successfully (Harris-Warrick et al. 1992). Modeling small circuits with defined connectivity is useful in determining the role of particular cells in a network and the effects of manipulating the cellular activity can be then tested in the biological preparation. Furthermore, some fundamental questions regarding circuit operation can be examined with models far easier than with live preparations. For example the quantitative relationship between cell-driven and network bursting can be examined in terms of stability and flexibility, two opposing features of CPG networks (Ivanchenko et al. 2008). It had also been shown that individual biological neurons, when synaptically isolated, demonstrated chaotic behavior but that this behavior became regularized when the neurons were in a network and receiving synaptic currents. (Elson et al. 1999). This phenomenon could be modeled and explained using a simulation of the pyloric network (Falke et al. 2000). Biomimetic CPG networks have also been used to drive robotic systems (Ayers et al. 2010; Ayers, Chapter 51, this volume) but it should be noted that their usefulness is somewhat limited by the nonbiological properties of available actuator systems.

Hybrid networks

The dynamic clamp (Pinto et al. 2000; Figure 21.5) has been used to incorporate electronic neurons into both intact living circuits and circuits in which one or more biological neurons has been deleted by photoinactivation. A hybrid network consisting of two pyloric CPGs that were coupled with a dynamic clamp was used to study the mehanisms for coupling two independent CPGs. Such coupling of unit CPGs is a common phenomena in nature where for example the unit

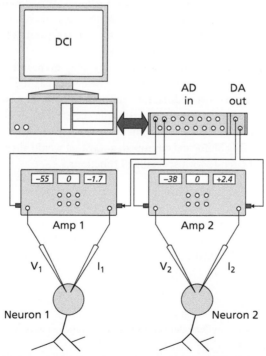

Figure 21.5 Connecting two neurons from the same or different STGs using the dynamic clamp system to emulate synapses. Two intracellular amplifiers are used (Amp 1 and Amp 2) each capable of recording membrane potentials and injecting current through two separate microelectrodes. The membrane voltages of the neurons goes to the AD input of the data acquisition interfaces. Current commands are sent to the cells via the DA output. The computer runs the dynamic clamp output to simulate the desired synaptic current.

oscillators for leg CPGs in quadruped locomotion can be coupled in phase, out of phase, or at some intermediate phase. It could be shown that excitatory or inhibitory coupling between specific neurons in the pyloric network could produce these different kinds of coupling functions.

Future directions

Hybrid circuits that couple electronic neurons and synapses to biological networks can be further refined to drive and control CPG and other types of networks. The opposite is also possible—biological motor patterns are already being used to drive artificial muscles and prosthetic devices. The basic principles of invertebrate circuit design have already been used for the construction of biomimetic CPGs that drive locomotion in lobster walking and lamprey swimming robots (Ayers et al. 2010).

Two new research methodologies are likely to have a major effect on the study of oscillatory circuits: (a) genetics based techniques for cell identification, monitoring, and manipulation; and (b) synthetic biology which aims to engineer cells and circuits to perform specific functions. The generation of oscillatory patterns in mammals is not yet understood in terms of cell-to-cell connectivity because an analysis at this level is hampered by the sheer numbers of cells involved, their lack of identifiability, and the inability to stimulate and record from pairs of neurons repeatedly. Types of neurons can now be identified based on their endogenous gene expression. Enhancers and promoters can be used to drive batteries of transgenes to encode labeled proteins. Transgenic animals expressing fluorescent proteins label particular cell types and can be used to visually guide recordings or to image neuronal and synaptic activity. Combined with optical methods (optogenetics), the activities of groups of neurons can be up- or down-regulated, or removed from a network entirely.

Synthetic biology, still mainly utilizing prokaryotes, may eventually be useful in creating eukaryotic circuits *de novo* by engineering neurons to behave in particular ways. By incorporating genetic material from a parts list of different channels, receptors, and other cell components, new cell properties can be created. The STG neurons may be ideal for this purpose since the cells are large enough to be directly injected with genetic material. The engineered neurons could be incorporated into functioning biological circuits in a way that would alter their activity patterns or be used to form hybrid circuits with specialized functions.

Learning more

Selverston (2010) provides a review of invertebrate central pattern generator circuits, Szücs (2007) discussing the challenge of modeling synapses in artificial networks, and Torres and Varona (2012) provide a useful introduction to the broader area of modeling biological neural networks.

References

Ayers, J, et al. (2010). Controlling underwater robots with electronic nervous systems. *Applied Bionics and Biomechanics*, 7, 57–67.

Dando, M. R., and Maynard, D. M. (1973). The sensory innervation of the foregut of *Panulirus argus* (Decapoda Crustacea). *Marine Behaviour and Physiology*, 2, 283–305.

Denker, M., Szucs, A., Pinto, R. D., Abarbanel, H. D. I., and Selverston, A. I. (2005). A network of electronic neural oscillators reproduces the dynamics of the periodically forced pyloric pacemaker group. *IEEE Trans Biomed Engineering*, 52, 792–9.

Elson, R. C., Huerta, R., Abarbanel, H., Rabinovich, M., and Selverston, A. (1999). Dynamic control of irregular bursting in an identified neuron of an oscillatory circuit. *J. Neurophysiol.*, **82**, 115–22.

Falke, M., et al. (2000). Modeling observed chaotic oscillations in bursting neurons: the role of calcium dynamics and IP_3. *Biol. Cyber.*, **82**, 517–27.

Grillner, S., Markram, H., De Schutter, E., Silberberg, G., and LeBeau, F. E. N. (2005). Microcircuits in action—from CPGs to neocortex. *Trends in Neurosciences*, **28**(10), 525–33.

Harris-Warrick, R. M., Marder, E. A., Selverston, A., Moulin, M. (eds) (1992). *Dynamic biological networks: the stomatogastric nervous system.* Cambridge, MA: MIT Press.

Harris-Warrick, R. M. (2010). General principles of rhythmogenesis in central pattern networks. *Progress in Brain Research*, **187**, 213–22.

Ivanchenko, M., Nowotny, T., Selverston, A., and Rabinovich, M. (2008). Pacemaker and network mechanisms of rhythm generation: cooperation and competition. *J. Theor. Biol.*, **253**(3), 452–61.

Marom, S, et al. (2009). On the precarious path of reverse neuro-engineering. *Frontiers in Computational Neuroscience*, **3**, 5.

Nusbaum, M. P., and Beenhakker, M. P. (2002). A small-systems approach to motor pattern generation. *Nature*, **417**(6886), 343–50.

Pinto, R. D., et al. (2000). Synchronous behavior of two coupled electronic neurons. *Physiol Rev E.*, **62**, 2644–56.

Pinto, R. D., et al. (2001). Extended dynamic clamp: controlling up to four neurons using a single desktop computer and interface. *J. Neurosci Methods*, **108**, 39–48.

Roberts, A., and Perrins, R. (1995). Positive feedback as a general mechanism for sustaining rhythmic and non-rhythmic activity. *Journal of Physiology (Paris)*, **89**, 241–8.

Selverston, A. (2010). Invertebrate central pattern generator circuits. *Phil. Trans. Roy. Soc. B*, **365**, 2329–45.

Selverston, A. I., and Moulins, M. (eds) (1987). *The Crustacean Stomatogastric System.* Berlin: Springer-Verlag.

Selverston, A. I., Szucs, A., Huerta, R., Pinto, R., and Reyes, M. (2009). Neural mechanisms underlying the generation of the lobster gastric mill motor pattern. *Front Neural Circuits*, **3**, 12.

Szücs, A. (2007). Artificial synapses in neuronal networks. In: J. A. Lassau (ed.), *Neural Synapse Research Trends*. New York, NY: Nova Science Publishers, Inc., pp. 47–96.

Torres, J. J., and Varona, P. (2012). Modeling biological neural networks. In: G. Rozenberg (ed.), *Handbook of Natural Computing*. Berlin: Springer-Verlag, pp. 533–64.

Chapter 22

Skin and dry adhesion

Changhyun Pang[1], Chanseok Lee[2], Hoon Eui Jeong[3], and Kahp-Yang Suh[2]

[1] School of Chemical Engineering, SKKU Advanced Institute of Nanotechnology, Sungkyunkwan University, South Korea
[2] School of Mechanical and Aerospace Engineering, Seoul National University, South Korea
[3] School of Mechanical and Advanced Materials Engineering, Ulsan National Institute of Science and Technology, South Korea

Explorations of the remarkable adhesive properties of animal skin have revealed various exquisite nanoscale architectures and their physical interactions. For example, for more than a decade, researchers from various disciplines have been fascinated by the unique structural features (e.g. high-aspect ratio, tilted angle, hierarchy, and spatulate head) of gecko foot hairs. More recently, beetles' attachment systems have been investigated extensively, revealing various functions such as mating/attachment to wax or wet surfaces, capillary adhesion, and wing fixation. In this chapter, gecko- and beetle-inspired adhesive properties are classified into and explained as three types: inclined, hierarchical, spatula-shaped pads of gecko foot hairs; mushroom-shaped pads of beetle feet; and interlocking of beetle wing-locking devices. After introducing the structural features responsible for the reversible adhesion mechanisms and functions of these systems, we describe how current nanofabrication methods can be applied to mimic or exploit the systems. Furthermore, potential applications for clean transportation devices, biomedical patches, and electric interlocking devices are described to widen the application range of artificial dry adhesives. Some outlooks and future challenges are briefly discussed in relation to next-generation, biomimetic dry-adhesive systems.

Biological principles

Various animal skins, including lizards, spiders, and insects have remarkable attachment abilities that enable them to move on a variety of surfaces encountered in nature (Arzt et al. 2003). The intriguing capability of adhesion is mostly caused by micro- or nano-scale hairy structures on the feet. With these fine hairy structures, numerous species exhibit strong yet reversible, repeatable, and directional attachment abilities against surfaces with varying roughness and orientation. In fact, these natural adhesive capabilities surpass the performance of current man-made adhesives that use acrylic materials or pressure-sensitive tapes, which are seldom reusable and easily contaminated. Recent advances in nanotechnology have made it possible to investigate the underlying mechanisms and structures of natural attachment systems in more detail.

Geckos' hierarchical foot hairs

The attachment pads of gecko lizards represent one of the most versatile and effective adhesive structures in nature, reflected in the exceptional ability to walk freely on various surfaces including vertical walls and even ceilings. This unusual adhesion capability is attributed to arrays of millions of microscopic foot hairs (setae) that split into hundreds of smaller nano-scale ends called spatulas, as shown in Figure 22.1a (Autumn et al. 2006). The spatular tips have a size of approximately 200 nm and generate van der Waals forces with strong adhesion (approximately 10 N/cm^2). The microsized setae provide sufficient structural height and compliance while preventing lateral collapse of the hairs. On the other hand, the nano-scale spatular tips enhance the adhesion strength by increasing the contact area as compared to that of a hemispherical or a simple flat tip. Also, the extremely thin (5–10 nm) dimensions of the pads facilitate intimate contact with surfaces even

(a) Hierarchical Spatular Pads

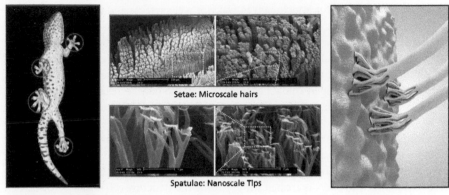

Setae: Microscale hairs

Spatulae: Nanoscale Tips

(b) Mushroom–Shaped Pads

(c) Wing Locking Device

Anterior Field on Hind Wing   Anterior Field on Beetle's Body

Tilted Microtrichias   Reversibility Directionality

Figure 22.1 A variety of attachment systems can be found on the surfaces of gecko lizards and beetles. (a) Slanted and hierarchical hairy structures of gecko feet, (b) mushroom-shaped pads of beetles for attachment during mating, and (c) photograph and SEM images of wing-locking device in beetle with conceptual illustration of its mechanism of operation.

(a) Reprinted from *Journal of Experimental Biology*, 209 (18), K. Autumn, A. Dittmore, D. Santos, M. Spenko, and M. Cutkosky, Frictional adhesion: a new angle on gecko attachment, pp. 3569–3579, Figure 1d, doi: 10.1242/jeb.02486, Copyright © 2006 The Company of Biologists. *NanoToday*, 4 (4), Hoon Eui Jeonga and Kahp Y. Suha, Nanohairs and nanotubes: Efficient structural elements for gecko-inspired artificial dry adhesives, pp. 335–346, doi: 10.1016/j.nantod.2009.06.004, Copyright © 2009 Elsevier Ltd., with permission from Elsevier, and *Advanced Functional Materials*, 21 (9), Moon Kyu Kwak, Changhyun Pang, Hoon-Eui Jeong, Hong-Nam Kim, Hyunsik Yoon, Ho-Sup Jung, and Kahp-Yang Suh, Towards the Next Level of Bioinspired Dry Adhesives: New Designs and Applications, pp. 3606–3616, doi: 10.1002/adfm.201100982 © 2011 WILEY-VCH Verlag GmbH & Co. KGaA, Weinheim.
(b and c) Reprinted from *NanoToday*, 7 (6), Changhyun Pang, Moon Kyu Kwak, Chanseok Lee, Hoon Eui Jeong, Won-Gyu Bae, and Kahp Y. Suh, Nano meets beetles from wing to tiptoe: Versatile tools for smart and reversible adhesions, pp. 496–513, doi: 10.1016/j.nantod.2012.10.009, Copyright © 2012 Elsevier Ltd., with permission from Elsevier.

under low preload. Interestingly, the setae arrays are slanted with a directional angle, which allows for exceptional anisotropic adhesion (strong attachment and easy detachment) because a surface with an angled structure is only adhesive when loaded in a particular direction. Furthermore, the geckos' hierarchical hairs exhibit superhydrophobic and self-cleaning properties, which enable the reversible and repeatable adhesion of their pads without surface contamination or fouling issues.

Beetles' mushroom-shaped pads

A tremendous number of beetle species also have remarkable capabilities of adhesion for locomotion on various complex surfaces (Pang et al. 2012b). Adhesive structures in beetles are present on the entire body, from wing to toe. Representative adhesion structures in beetles include mushroom-shaped contact elements for attachment to waxy or wet surfaces and the reversible interlocking device for wing fixation. In contrast to the slanted structures with asymmetric spatulate tips in geckos, optimized for temporary and directional adhesion, the mushroom-shaped features of beetles are configured vertically with symmetrical bulged tips. Therefore, these mushroom-shaped pads provide large normal adhesion and enhanced peeling-off strength for long-term attachment. The mushroom-shaped contact elements come into play, for example, when a male beetle is adhered on to the rough dorsal surface of a female beetle, or to various plant surfaces as shown in Figure 22.1b. It should be noted that the symmetrical mushroom-shaped tips induce intimate contact with various substrates, resulting in strong adhesion compared to that of other tip geometries, such as planar, spherical, asymmetric spatula, tubular, and concave shapes. In addition, the adhesion strength of mushroom-shaped structures can be further enhanced in wet conditions because a suction effect is created while pulling the structure off.

Beetles' interlocking hairs

The wing-locking device of beetles is also a versatile adhesive structure, the role of which is to fix and protect the delicate wings. Figure 22.1c shows a photograph of the wing-locking device with corresponding SEM images, and an illustration of its mechanism of operation. As shown, densely ordered, high aspect ratio (AR) structures of an angled corn shape (termed microtrichia) are present on the cuticular surface of the hind wing and on the beetle's body. The microtrichia on the hind wing are approximately 2.5 μm in diameter and 17 μm in height, and the cooperative microstructures on the body are approximately 1.5 μm in diameter and 15 μm in height (Pang et al. 2012a) Both structures are regularly ordered with a hexagonal packing layout with a spacing ratio (SR) of ~3. When beetles fold their wings, the microstructures on each surface are interlocked with high shear adhesion and directionality. As with the dry adhesion of gecko lizards, the underlying mechanism of the wing-locking device is van der Waals attraction between the interlocked microstructures. In contrast to other reversible binding systems (e.g. hooks or loops in commercial Velcro®), the van der Waals force-mediated interlocking mechanism in beetles' wings does not require physical grasping among complex structures. Based on these unique properties, the interlocking system can be adapted to a variety of devices, such as reversible fasteners, electric connectors, and pressure sensors for biomedical and robotics applications.

Biomimetic systems

Inspired by the fascinating adhesion properties of nature, extensive studies have been conducted to develop biomimetic, artificial dry adhesives by mimicking unique multi-scale structures of natural systems. A number of fabrication methods have been developed to make complex biomimetic structures with controlled leaning angle, hierarchy, aspect ratio, and tip-shapes

(Kwak et al. 2011b). These methods include multi-step capillary molding, angled etching, e-beam irradiation, metal deposition, and other techniques. In this section, three bio-inspired, artificial dry adhesive systems will be introduced: a gecko-inspired adhesive, a beetle-inspired mushroom-shaped adhesive, and a beetle-inspired interlocking system.

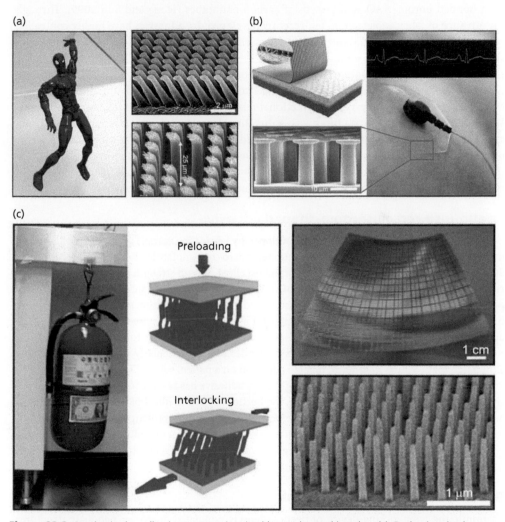

Figure 22.2 Synthetic dry adhesive systems inspired by gecko and beetles. (a) Gecko-inspired slanted and hierarchical adhesives, (b) biomedical skin patch consisting of beetle-like mushroom-shaped microstructures for affixing an ECG measurement module, and (c) photographs of beetle-inspired artificial nano-interlocker.

(a) Reprinted from *Nature Materials*, 2 (7), A. K. Geim, S. V. Dubonos, I. V. Grigorieva, K. S. Novoselov, A. A. Zhukov, and S. Yu. Shapoval, Microfabricated adhesive mimicking gecko foot-hair, pp. 461–463, doi: 10.1038/nmat917, Copyright © 2003, Macmillan Publishers Limited and *Proceedings of the National Academy of Sciences of the United States of America*, 106 (14), Hoon Eui Jeong, Jin-Kwan Lee, Hong Nam Kim, Sang Heup Moon, and Kahp Y. Suh, A nontransferring dry adhesive with hierarchical polymer nanohairs, pp. 5639–5644, Figures 2c and 5g, doi: 10.1073/pnas.0900323106, Copyright © 2009, National Academy of Sciences. (b) Reproduced from *Advanced Materials*, 23 (34), Moon Kyu Kwak, Hoon-Eui Jeong, and Kahp-Yang Suh, Rational Design and Enhanced Biocompatibility of a Dry Adhesive Medical Skin Patch, pp. 3949–3953, doi: 10.1002/adma.201101694, Copyright © 2011 WILEY-VCH Verlag GmbH & Co. KGaA, Weinheim. (c) Reproduced from *Advanced Materials*, 24 (4), Changhyun Pang, Tae-il Kim, Won Gyu Bae, Daeshik Kang, Sang Moon Kim and Kahp-Yang Suh, Bioinspired Reversible Interlocker Using Regularly Arrayed High Aspect-Ratio Polymer Fibers, pp. 475–479, doi: 10.1002/adma.201103022 Copyright © 2011 WILEY-VCH Verlag GmbH & Co. KGaA, Weinheim.

Gecko-inspired dry adhesive

To date, a number of methods have been proposed to develop gecko-inspired dry adhesives, including multi-step UV molding, nanodrawing, angled etching, multi-step photolithography, soft-lithography, microelectromechanical systems (MEMS) processes, replica molding with bonded porous AAO, and growth of carbon nanotubes (Jeong and Suh 2009). Through the advance of fabrication techniques, recently developed dry adhesives have highly controlled geometry (diameter, height, slanted angle, hierarchy, and tip shape), and show excellent adhesion performance by better mimicking of gecko foot hairs (Figure 22.2a; Jeong et al. 2009). For example, the adhesion strength of gecko-inspired dry adhesives is now up to approximately 100 N/cm^2, which is 10 times higher than that of gecko foot hairs (Jeong and Suh 2009). Synthetic dry adhesives also exhibit excellent directional-adhesion ability with slanted hairy structures. When a surface has hairy structures with a directional angle, such a surface represents strong attachment in the gripping direction but weak adhesion in the releasing direction. In terms of robustness, synthetic dry adhesives have demonstrated superior repeatability and durability through experiments with thousands of cycles of attachment and detachment, shedding light on practical and industrial applications of gecko-inspired dry adhesive systems.

Beetle-inspired mushroom-shaped adhesive

Beetle-inspired adhesives have also been developed by fabricating vertical micro-scale pillars with mushroom-shaped tips based on aforementioned micro- or nanofabrication methods (Pang et al. 2012b). The beetle-inspired adhesives represent excellent pull-off strength because their symmetrical bulged tips lead to easy generation of conformal contact with a substrate even under low preload. Furthermore, the adhesives also show moderately high adhesion strength even on rough or wet surfaces, broadening the potential applications of the beetle-inspired adhesive system. One application is as a biomedical skin patch. Since the mushroom-shaped adhesive provides strong normal adhesion without using wet or toxic chemicals, it is particularly useful as a biomedical skin patch that requires long-term, reliable adhesion on a patient's skin (Figure 22.2b; Kwak et al. 2011a). The biomedical skin patch utilizing the dry adhesive has several advantages over conventional wet medical bandages. For example, the beetle-inspired medical patch showed repeatable and restorable adhesion, providing better biocompatibility during prolonged exposure. Based on its unique advantages, the dry adhesive patch was integrated into a ubiquitous electrocardiogram (ECG) monitoring device. The monitoring device was firmly attached to a patient's chest and made successful measurements for 48 h without any inflammation issue.

Beetle-inspired interlocking system

To mimic the wing-locking device of beetles, densely populated micro- or nanohair structures were fabricated on a flexible substrate (Pang et al. 2012a). This artificial interlocking system was designed to produce high shear strength by bringing into contact two substrates having high-density hairy structures while other dry adhesives were designed for attachment on various surfaces. By hair-to-hair interlocking, attractive van der Waals forces are amplified and they exert very strong shear strength, while peeling off along a normal direction is effortless. Hairy structures with small diameter (~100 nm) and high AR (~10) are beneficial for increasing the contact area between the two adjacent structures and thus maximizing shear strength. A large-area film (9 cm × 13 cm) with nanopillar arrays (50-nm radius and 1-μm height) demonstrated remarkable shear strength (40 N/cm^2), which was much higher than that of conventional Velcro® systems (average 15 N/cm^2). A 5.25-kg fire extinguisher can be simply suspended by the aid of a small synthetic interlocker patch (1.5 cm^2) as shown in Figure 22.2c (Pang et al. 2012a).

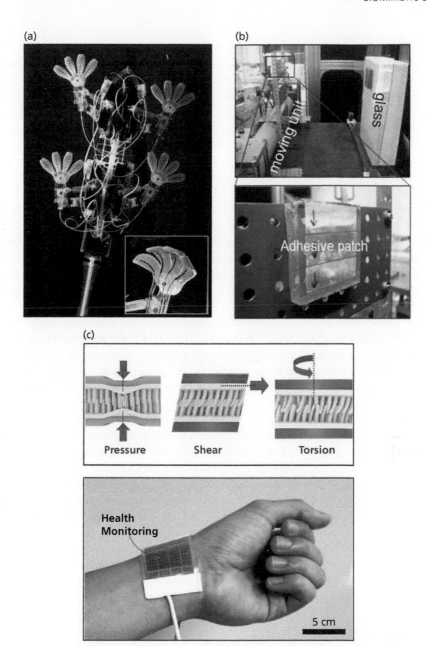

Figure 22.3 Applications of bio-inspired dry adhesive systems. (a) Wall climbing robot with directional, gecko-inspired footpad, (b) clean transportation system based on gecko-inspired synthetic adhesives, and (c) flexible strain-gauge sensor based on beetle-inspired interlocking device.

Future directions

Biomimetic dry adhesives with superior and smart adhesion properties have been developed in recent years with the aid of nanofabrication techniques. Compared to earlier dry adhesives, recently developed dry adhesives show excellent adhesion strength, smart directional adhesion, and structural robustness. Now, it is time to drive fundamental research on bioinspired dry adhesives into a new epoch by developing unique and practical applications for these materials. In fact, several applications of the dry adhesives have been demonstrated. For example, wall-climbing robots (e.g. StickyBot) were developed by application of gecko-inspired adhesives (Figure 22.3a; Sangbae et al. 2008). This robot can climb over various smooth vertical surfaces with the aid of directional adhesive pads. Another field to which dry adhesives can be applied is that of precision industries such as semiconductor or liquid crystal display (LCD) factories, where clean and residue-free transportation of fragile silicon or glass substrates is required during the product-assembly process. A prototype for a clean transportation system was demonstrated by exploiting dry adhesives with angled and hierarchical structures (Figure 22.3b; Jeong et al. 2009). A large-area glass substrate (47.5×37.5 cm^2) was successfully transported by utilizing the dry adhesive patch (3×3 cm^2) without surface contamination or damage even after many cycles of attachment and detachment.

A precise sensory device is also an important area of application of dry adhesive technology. A novel strain-gauge sensor was recently developed, inspired by the beetle's wing-locking device (Pang et al. 2012c). In general, such sensors contain a number of circuits or complex, layered matrix arrays. However, the strain sensor mimicking the wing-locking device found in beetles could detect pressure, shear, and torsion without using complicated electronic circuits. Instead, this device is based on two interlocked arrays of high-AR platinum-coated polymeric nanohairs that are supported on thin PDMS layers. When external pressure or stimuli were applied to the device assembly, the degree of interconnection and the electrical resistance of the sensor changed in a reversible, directional manner with specific, discernible strain-gauge factors. This kind of flexible biomimetic strain-gauge sensor is an essential component for the development of artificial systems that mimic the complex characteristics of the human skin.

Adhesives are used in many aspects of daily life and industry. Recent advances in nanofabrication via top-down and bottom-up approaches have made it possible to develop biomimetic synthetic dry adhesives with high performance. Several laboratory-level studies have shown potential applications of biomimetic synthetic dry adhesives. It is envisioned that a new kind of smart adhesive system with fascinating adhesion properties and structural robustness will be developed in the near future, and will be put to use in many practical applications including daily supplies, biomimetic robots, clean and precise manufacturing, and smart biomedical patches.

Learning more

Details of the underlying mechanism of geckos' attachment system can be found in Autumn et al. (2006). For readers who are interested in various attachment systems of insects, Arzt et al. (2003), Gorb (2008), and Pang et al. (2012b) will be helpful for understanding biological principles. For fabrication methods of synthetic dry adhesives, Jeong and Suh (2009) and Kwak et al. (2011b) provide knowledge about recent progresses and examples. Further applications of bio-inspired adhesive system are demonstrated in Pang et al. (2012c) and Kwak et al. (2011a). Vincent (Chapter 10, this volume) provides a broad introduction to the design and manufacture of biomimetic materials, while Fukuda and colleagues (Chapter 52, this volume) outline the state-of-the-art in material fabrication at the micro- and nano-scale. Bensmaia (Chapter 53, this volume) provides an overview of some of the mechanical and sensory properties of human skin and their possible emulation for prosthetic systems, while Metta and Cingolani (Chapter 47, this volume) briefly discusses biomimetic skin for humanoid robotics.

Dedication

This chapter is dedicated to the late Professor Kahp-Yang Suh, for his intense ardor and devotion, and for his endless support for exploring science and teaching; his selflessness will always be remembered.

Acknowledgment

We gratefully acknowledge support from the National Research Foundation of Korea (NRF-2017R1D1A1B03033272). This work was supported by the Ulsan National Institute of Science and Technology (UNIST) through the Creativity and Innovation Project Program (Grant UMI 1.130031.01).

References

Arzt, E., Gorb, S., and Spolenak, R. (2003). From micro to nano contacts in biological attachment devices. *Proc. Natl Acad. Sci. USA*, **100**, 10603–6.

Autumn, K., Dittmore, A., Santos, D., Spenko, M., and Cutkosky, M. (2006). Frictional adhesion: a new angle on gecko attachment. *Journal of Experimental Biology*, **209**, 3569–79.

Geim, A. K., Dubonos, S. V., Grigorieva, I. V., Novoselov, K. S., Zhukov, A. A. , and Shapoval, S. Y. 2003. Microfabricated adhesive mimicking gecko foot-hair. *Nature Materials*, **2**, 461–63.

Gorb, S. N. (2008). Biological attachment devices: exploring nature's diversity for biomimetics. *Philosophical Transactions of the Royal Society A: Mathematical, Physical and Engineering Sciences*, **366**, 1557–74.

Jeong, H. E., Lee, J.-K., Kim, H. N., Moon, S. H., and Suh, K. Y. (2009). A nontransferring dry adhesive with hierarchical polymer nanohairs. *Proc. Natl Acad. Sci. USA*, **106**, 5639–44.

Jeong, H. E., and Suh, K. Y. (2009). Nanohairs and nanotubes: Efficient structural elements for gecko-inspired artificial dry adhesives. *Nano Today*, **4**, 335–46.

Kwak, M. K., Jeong, H. E., and Suh, K. Y. (2011a). Rational design and enhanced biocompatibility of a dry adhesive medical skin patch. *Advanced Materials*, **23**, 3949–53.

Kwak, M. K., Pang, C., Jeong, H.-E., Kim, H.-N., Yoon, H., Jung, H.-S., and Suh, K.-Y. (2011b). Towards the next level of bioinspired dry adhesives: new designs and applications. *Advanced Functional Materials*, **21**, 3606–16.

Pang, C., Kim, T.-I., Bae, W. G., Kang, D., Kim, S. M., and Suh, K.-Y. (2012a). Bioinspired reversible interlocker using regularly arrayed high aspect-ratio polymer fibers. *Advanced Materials*, **24**, 475–9.

Pang, C., Kwak, M. K., Lee, C., Jeong, H. E., Bae, W.-G., and Suh, K. Y. (2012b). Nano meets beetles from wing to tiptoe: Versatile tools for smart and reversible adhesions. *Nano Today*, **7**, 496–513.

Pang, C., Lee, G.-Y., Kim, T.-I., Kim, S. M., Kim, H. N., Ahn, S.-H., and Suh, K.-Y. (2012c). A flexible and highly sensitive strain-gauge sensor using reversible interlocking of nanofibres. *Nature Materials*, **11**, 795–801.

Sangbae, K., Spenko, M., Trujillo, S., Heyneman, B., Santos, D., and Cutkosky, M. R. (2008). Smooth vertical surface climbing with directional adhesion. *IEEE Transactions on Robotics*, **24**, 65–74.

Section IV

Capabilities

Chapter 23

Capabilities

Paul F. M. J. Verschure

SPECS, Institute for Bioengineering of Catalonia (IBEC), the Barcelona Institute of Science and Technology (BIST), and Catalan Institute of Advanced Studies (ICREA), Spain

For as far as we can speak of a standard model of science it, in general, follows an atomistic perspective. Atomism and its associated method of reductionism goes back at least to the fifth century BC. Its pervasiveness in human culture can be interpreted as a suggestion for its status as a cognitive bias. Atoms, components and or modules do not automatically make systems. Hence, in this handbook, we have chosen to include an emphasis on the capabilities of living systems and its emulation in artifacts. In this way, we can be more specific in the benchmarks that we can set for current and future living machines. Capabilities generally arise from the integration of multiple components and thus sensitize us for the need to develop a system-level perspective on living machines. Here we summarize and consider the 14 contributions in this section which cover perception, action, cognition, communication, and emotion, and the integration of these through cognitive architectures into systems that can emulate the full gamut of integrated behaviors seen in animals including, potentially, our own capacity for consciousness.

Cruse and Schilling (Chapter 24) look at the phenomenon of pattern generation in the service of morphogenesis and behavior from pattern formation to locomotion and cognition. They distinguish between quasi-rhythmic and non-rhythmic patterns which are analyzed with respect to their specific substrate from homogeneous to discretely structured, with and without sensory feedback. Based on this taxonomy they subsequently show how specific neuronal controllers, using recurrent neural networks, can give rise to these spatiotemporal patterns. An interesting side-effect of the gate generating controllers described is that they do not need to explicitly store all information in an internal memory but rather capitalize on their direct coupling with the environment. Building on these phenomena and their explanation in terms of embodied neuronal controllers, the notion of "mind-patterns" is introduced as a substrate of subjective experience and complex social structures. Lastly, the problem is addressed of how motor control systems can be structured beyond the control of single or small sets of degrees of freedom, taking into account the problem of kinematic constraints and environmental perturbations. This analysis is of great importance to contribute to the realization of a new class of biomimetic control systems for advanced robot systems that are highly articulated with many degrees of freedom and/or continuous such as the hydrostatic octopus arm. Beyond this challenge lies the notion of the dynamic regulation of the degrees of freedom of an agent being dependent on task requirements; the confirmation of the agent and its load bearing capabilities will depend on the freezing, releasing, or creation of its degrees of freedom.

Leibo and Poggio (Chapter 25) analyze biological perceptual systems and their underlying computational principles. Starting from the sensory sheet of the retina and cochlea, which are described in traditional terms as a pixel array and frequency decomposition respectively, a model of perception is advanced that can conceptually be traced back to the hierarchical pandemonium model of Selfridge from the 1950s, i.e. complex feature detection emerges from combining simple feature detectors in a hierarchical fashion. The authors propose that the "canonical microcircuits" of the neocortex implement such schemes that are consistent with the classical framework of David Marr. The chapter also anticipates the rapid developments in the field of machine vision driven by so-called deep learning (Lecun et al. 2015; Schmidhuber 2015), indeed, strong similarities exist between the models described in this chapter and deep learning architectures.

The hierarchical neural models described by Leibo and Poggio and others currently define the state of the art and constitute the current standard against we will measure future attempts to understand perceptual processing in living machines. The functional hypothesis these models are based on, however, does have limitations that future work can overcome. For instance, with respect to accounting for the speed of processing displayed by biological perceptual systems (Kirchner and Thorpe 2006), their capability to extract invariances such as rotation, position, and scale using physical connections rather than algorithmic shortcuts (Wyss et al. 2003), the specific known topology of cortical networks (Markov et al. 2011), the role of top-down processing in perceptual processing (Bastos et al. 2015), the role of active inference (Friston 2010), and the sensitivity to biases due to input sampling or behavioral feedback (Verschure et al. 2003).

Herreros (Chapter 26) looks at the complexity of adaptive behavior from the perspective of control theory. This is a view that is often overlooked as we usually analyze the brain in terms of computation or information processing. However, as Claude Bernard and Ivan Pavlov already observed in the nineteenth century: brains control action. Moreover, the consideration that the brain is embodied and that this embodiment also constrains its functions gives further credence to this approach. Herreros first introduces basic notions of control theory and machine learning from which he advances a taxonomy that allows us to look in a more specific way at the control problems that living machines face. These are analyzed both at the level of innate reactive principles and in the form of anticipatory and learned control. Subsequently, these principles are linked to specific brain structures such as the basal ganglia and the cerebellum, thus providing a basis for the realization of biologically grounded control systems for living machines.

Mitchinson (Chapter 27) addresses the principles of attention and orienting starting with the description of key biological principles distinguishing the internal orienting associated with the notion of attention from overt orienting. The key to orienting is the notion of action selection and decision making, since the organism needs to select one target over another one. Subsequently, models of attention are described, distinguishing between bottom-up and top-down models and their combination in counter stream architectures. The psychological and biological principles of attention and orienting are then brought to the domain of living machines where Mitchinson analyzes specific examples of their application to different robots interacting with both the physical and social world. Central to the models described is the notion of excitatory biasing in attention where salient features and objects are amplified. Future models of attentional phenomena will also need to explain the selective suppression of sensory information such as observed in attentional blindness and change blindness (Cohen et al. 2012). Progress in this direction includes the validation gate hypothesis (Mathews et al. 2011), which considers both positive and negative feedback, and makes predictions concerning the attentional role of frontoparietal networks of the neocortex that have been confirmed in human psychophysics (Malekshahi et al. 2016).

Lepora (Chapter 28) addresses the topic of decision making, starting with the notion of the "race model" (Logan and Cowan 1984) which proposes that decision making is essentially defined through the integration of evidence until a decision threshold favoring an action is reached. This influential sequential analysis account of decision making has its origins in the code-breaking operations of the second world war (Kahn 1991). In addressing some of the biological principles of decision making, Lepora makes the point that decision making is at the core of survival for living systems, from single cells to individuals and collectives. Each of these levels is analyzed and their application to living machines described. Here the point is advanced that decision making in the real world faces an additional optimization problem where multiple cost functions must be satisfied such as energy and reaction time. The chapter concludes by advancing the common view in neuroscience that decision making essentially requires integration of evidence, a threshold, and competition between alternatives.

Much of the evidence for the sequential analysis account described by Lepora derives from studies of the correlation between neuronal firing in pre-motor areas of the non-human primate and decision making in simple perceptual tasks. Building on these findings, some researchers have gone further by claiming that average firing rate might be all we need to understand how the brain operates (Churchland et al. 2010). However, recent evidence has shown that this may not hold when tasks become more realistic, for instance when exceptions must be handled (Marcos et al. 2013). Hence, some secrets of the neuronal code behind decision making still need to be deciphered and may depend on a deeper understanding of temporal dynamics and spiking patterns in decision-making circuits.

Erdem et al. (Chapter 29) address the topic of spatial and episodic memory, starting with the suggestion that the so-called Grid Cells (GC) found in the medial entorhinal cortex (MEC) could enhance available methods for robot Simultaneous Localization and Mapping (SLAM). They elaborate on this point by first providing a detailed description of the GCs together with head direction cells and the hippocampal place cells. The key principles driving these different types of cells that are central in rodent navigation are subsequently compared to selected biomimetic models. The grid cell model used in this case is the so-called phase interference model, which proposes that grid cells are driven by distinct oscillatory inputs that through their dynamics realize the specific grid-like response pattern of the GCs. The first models of GCs, however, were based on hypothesized local attractor dynamics in MEC (Guanella et al. 2007) and subsequent empirical studies have shown that GCs indeed display response dynamics consistent with this hypothesis (Rowland et al. 2016). In addition, the model described poses the important question of how much functionality, for instance with respect to goal and reward processing, one should assume to occur in the hippocampus, especially in light of the recent discovery of dense interactions between this structure and the prefrontal cortex (Navawongse and Eichenbaum 2013). Indeed, alternative biologically grounded models have shown optimal robot foraging following this alternative perspective including attractor dynamics based GCs (Maffei et al. 2015).

Cutkosky (Chapter 30) describes the study of reaching, grasping, and manipulation, highlighting the large variety of solutions that have emerged during evolution both in terms of their morphology and behavior. However, a number of common design principles can be identified which are described in some detail including branching kinematics, compliance/conformability, non-linear actuation, and sensitivity. Furthermore, the control of a wide range of manipulators seems to follow common principles such as the hierarchical structuring of planning and the use of synergies. In addition, manipulation is based on a layered control structure which can exploit rapid sensor driven reflexes together with more advanced planning. The chapter closes by describing a number of advanced grasping living machines together with directions for future research.

Witte and colleagues (Chapter 31) address the challenge of autonomous displacement or locomotion. After an initial zoology based classification scheme, the chapter focuses on terrestrial and arboreal locomotion, noting the conservative nature of the evolution of neural control and contrasting this with the degree of adaptive change in the locomotor apparatus.

In this evolutionary analysis, Witte makes the point that, from an engineering or "Technical Biology" perspective, we would be able to field more robust machines by prioritizing amphibian over anthropomorphic systems: the former being the template on which also the latter was constructed during evolution. The concept of "muscle-tendon complexes" as adaptable springs is introduced as a primitive for the analysis of locomotion. This notion is formally introduced and it is shown that the experimentally observed linear spring behavior of a human leg is due to the linear character of gravitational potential energy. Subsequently, the concept of gait is introduced and analyzed. The chapter ends with a reflection of future trends in research and the current state of the art.

Hedenström (Chapter 32) describes the biology and technology of flight or avionics (which derives from the Latin for "bird", "avis"). Starting with the evolution of wings from forelimbs, the mechanics of biological flight is analyzed, and is proposed to be more efficient than human engineered flying machines. Subsequently, aerodynamics is analyzed, demonstrating that a distinct relationship exists between the power to stay airborne and the speed obtained. Several problems are identified that are still at the edge of empirical scrutiny such as our ability to measure energy expenditure and the exact aerodynamic forces that act upon the wings of flying animals. However, in some cases where this was accomplished, measured values for lift have been obtained that exceed theoretical limits. Flight behavior itself is serving various purposes and a number of these are described, in particular, long-range migratory flights. This analysis shows that there is an upper limit to self-powered flight, where the energy required to stay airborne can still be supplied through biological energy sources. Exceeding this limit implies one has to remain terrestrial or develop additional sources of power and lift. The chapter ends with an outlook and some useful suggestions for further reading.

Wortham and Bryson (Chapter 33) address communication from a comparative perspective including that of living machines. Communication is anchored in the combination of a signal and a receiver with "knowledge" to interpret this signal. This is a reduced form of Pearce's semiotics which is built on the triad of the object, sign, and agent-generated meaning (Atkin 2013). The authors question whether communication serves collaboration or whether it is a form of coercion imposed by Dawkins' hypothesis of "selfish genes." In answering this question, the notion of Evolutionary Stable Strategies (ESS) is applied to the phenomenon of communication. Subsequently a taxonomy of communication is introduced, derived from Tinbergen, which is linked to an information theoretic perspective advanced by Shannon and Weaver.

The study of communication raises the question of how a language could arise that would support and channel it. The authors introduce this important question by placing the origins of language in the domain of memetics—also stemming from Dawkins—the cultural replicator system structured around its "meme" atoms that operates in parallel to the genome. Wortham and Bryson also provide a description of some recent artificial biomimetic communication systems from swarm robotics and research on human–robot collaboration and human–robot interaction, analyzing some specific attempts to use collective robotics as an experimental tool to study animal communication and noting the minimal nature of some of the signaling systems used (for further on this see Krause et al. 2011). The chapter ends with a description of current and future challenges and pointers for further study.

A frequent target of research in robot communications, and the related field of artificial life, has been to explore synthetic analogs of chemical communication through pheromone trails in social insects (see, e.g. Brambilla et al. 2013). Indeed, from a wider living machines perspective we might consider chemical communication to be the precursor of all other forms of communication as it evolved to mediate intercellular signaling in communities of single-celled animals (see e.g. Prescott, Chapter 38, this volume; Keller and Surette 2006).

As Wortham and Bryson note, in human–robot collaboration the goal is to provide effective information exchange between machines and humans. Indeed, in the recent DARPA robot challenge one of the biggest problems was to obtain situational awareness for human remote operators (Yanco et al. 2015). In human–robot interaction, the bar is raised further because now we wish the machine to autonomously engage with the physical and social world. However, this also requires machines with much more advanced capabilities for modeling this world, including the capability to read the mind of other agents (see Lallée et al. 2015, and Verschure, Chapter 36, this volume). Indeed, machines exist today which can learn to communicate with humans using natural language, gesture, and implicit social cues (Moulin-Frier et al., in press).

Vouloutsi and Verschure (Chapter 34) address the emotions of living machines and propose that they can be seen from the perspective of self-regulation and appraisal. The authors first look at the pragmatic needs to endow machines with emotions and subsequently describe some of the historical background of the science of emotions and its different interpretations, links to affective neuroscience, and the specifics of the neuronal substrate of emotions. Based on this analysis, they argue that emotions can be cast in terms of self-regulation, where emotions provide for a descriptor of the state of the homeostatic processes that maintain the relationship between the agent and its internal and external environment. They augment the notion of homeostasis with that of allostasis, which signifies a change from stability through a fixed equilibrium to stability through continuous change. The chapter shows how this view can be used to create complex living machines where emotions are anchored in the need fulfilment of the agent: in this case considering both utilitarian and epistemic needs.

The components of a living machine must in turn be integrated into a functioning whole and Verschure (Chapter 35) argues that this requires a detailed understanding of the architecture of living machines. The chapter starts with a conceptual and historical analysis which, beginning from Plato, brings us to nineteenth-century neuroscience and early concepts of the layered structure of nervous systems. These concepts were further captured in the cognitive behaviourism of Tolman and came to full fruition in the cognitive revolution of the second half of the twentieth century. Verschure (Chapter 36) describes "Distributed Adaptive Control" (DAC), a specific proposal for a theory of cognitive architecture stemming from modern neuroscientific investigations of the systems architecture of the brain. An important challenge to a theory of cognitive architecture is how to benchmark it. Verschure proposes the Unified Theories of Embodied Minds (UTEM) benchmark which is an advance from the classic Unified Theories of Cognition benchmark proposed by Allen Newell.

Seth (Chapter 37) sketches a biomimetic approach towards consciousness emphasizing issues of integration and differentiation of conscious content, prediction, and the self. The chapter starts with working definitions which distinguish a number of core dimensions such as level and content and world and self. These are complemented with biological principles focusing on complexity and information theory, although one could argue that these are at best descriptions of biological principles and that there is a risk of a category error, i.e. mistaking the method for the phenomenon. The centrepiece of the chapter deals with the role of inference in consciousness, a concept that has emerged over recent decades (Merker 2007; Hesslow, 2012) and is currently gaining increasing traction in the field of consciousness science. Seth subsequently addresses the questions around body ownership and self-consciousness, again advancing a prediction-based perspective. Seth concludes with a sceptical view on the possibility that phenomenal consciousness can be synthesized—rather than being vehicles for artificial consciousness, he argues that, in the near-term at least, biomimetic systems could serve as a method for mechanistically understanding some of its components. The chapter ends with a sketch of future directions of research and suggestions for further reading.

While prediction may be a necessary condition for consciousness to arise, proponents of the prediction theory of consciousness will need to elaborate these ideas further by defining which forms of prediction, and their neural substrates, matter for conscious content (Verschure 2016). For instance, the majority of the neuronal hardware of the central nervous system comprises the cerebellum, which implements a prediction engine (multiple distinct forward models—Herreros and Verschure 2013) while operating outside of direct experience (see Herreros, Chapter 26, this volume).

The contributions in this section necessarily provide only partial coverage of the capabilities we might wish to develop for living machines; nevertheless they do show a breadth of research, and progress on multiple fronts, that suggests we are not too far off from building integrated systems that capture multiple characteristics of living things. In sections V and VI of this handbook we will peer into the menagerie of state-of-the-art biomimetic and biohybrid systems and see where this ambition is taking us.

Acknowledgments

Preparation of this chapter was supported by grant from the European Research Council under the European Union's Seventh Framework Programme (FP7/2007–2013)/ERC grant agreement no. 341196.

References

Atkin, A. (2013). Peirce's Theory of Signs. In Edward N. Zalta (ed.), *The Stanford Encyclopedia of Philosophy (Summer 2013 Edition)*. Stanford University. Retrieved from https://plato.stanford.edu/archives/sum2013/entries/peirce-semiotics

Bastos, A. M., Vezoli, J., Bosman, C. A., Schoffelen, J.-M., Oostenveld, R., Dowdall, J. R., ... Fries, P. (2015). Visual areas exert feedforward and feedback influences through distinct frequency channels. *Neuron*, 85(2), 390–401. http://doi.org/10.1016/j.neuron.2014.12.018

Brambilla, M., Ferrante, E., Birattari, M., and Dorigo, M. (2013). Swarm robotics: a review from the swarm engineering perspective. *Swarm Intelligence*, 7(1), 1–41. doi:10.1007/s11721-012-0075-2

Churchland, M. M., Byron, M. Y., Cunningham, J. P., Sugrue, L. P., Cohen, M. R., Corrado, G. S., ... others. (2010). Stimulus onset quenches neural variability: a widespread cortical phenomenon. *Nature Neuroscience*, 13(3), 369–378.

Cohen, M. A., Cavanagh, P., Chun, M. M., and Nakayama, K. (2012). The attentional requirements of consciousness. *Trends in Cognitive Sciences*, 16(8), 1–7. http://doi.org/10.1016/j.tics.2012.06.013

Friston, K. (2010). The free-energy principle: a unified brain theory? *Nature Reviews Neuroscience*, 11(2), 127–138. http://doi.org/10.1038/nrn2787

Guanella, A., Kiper, D., and Verschure, P. F. M. J. (2007). A model of grid cells based on a twisted torus topology. *International Journal of Neural Systems*, 17(4), 231–40. http://doi.org/10.1142/S0129065707001093

Herreros, I., and Verschure, P. F. M. J. (2013). Nucleo-olivary inhibition balances the interaction between the reactive and adaptive layers in motor control. *Neural Networks*, 47, 64–71. http://doi.org/10.1016/j.neunet.2013.01.026

Hesslow, G. (2012). The current status of the simulation theory of cognition. *Brain Research*, 1428, 71–9. http://doi.org/10.1016/j.brainres.2011.06.026

Kahn, D. (1991). *Seizing the enigma: the race to break the U-boat codes*. Boston, MA: Houghton Mifflin.

Keller, L., and Surette, M. G. (2006). Communication in bacteria: an ecological and evolutionary perspective. *Nature Reviews Microbiology*, 4(4), 249–258. http://doi.org/10.1038/nrmicro1383

Kirchner, H., and Thorpe, S. (2006). Ultra-rapid object detection with saccadic eye movements: visual processing speed revisited. *Vision Research*, 46(11), 1762–76. http://doi.org/10.1016/j.visres.2005.10.002

Krause, J., Winfield, A. F. T., and Deneubourg, J.-L. (2011). Interactive robots in experimental biology. *Trends in Ecology & Evolution*, **26**(7), 369–375. http://doi.org/10.1016/j.tree.2011.03.015

Lallée, S., Vouloutsi, V., Blancas, M., Grechuta, K., Puigbo, J., Sarda, M., and Verschure, P. F. M. J. (2015). Towards the synthetic self: making others perceive me as an other. *Paladyn: Journal of Behavioral Robotics*, **6**(1). http://doi.org/10.1515/pjbr-2015-0010

Lecun, Y., Bengio, Y., and Hinton, G. (2015). Deep learning. *Nature*, **521**, 436–44. http://doi.org/10.1038/nature14539

Logan, G. D., and Cowan, W. B. (1984). On the ability to inhibit thought and action: A theory of an act of control. *Psychological Review*, **91**(3), 295–327. http://doi.org/10.1037/0033-295X.91.3.295

Maffei, G., Santos-Pata, D., Marcos, E., Sánchez-Fibla, M., and Verschure, P. F. M. J. (2015). An embodied biologically constrained model of foraging: from classical and operant conditioning to adaptive real-world behavior in DAC-X. *Neural Networks*. http://doi.org/10.1016/j.neunet.2015.10.004

Malekshahi, R., Seth, A., Papanikolaou, A., Mathews, Z., Birbaumer, N., Verschure, P. F. M. J., and Caria, A. (2016). Differential neural mechanisms for early and late prediction error detection. *Scientific Reports*, **6**.

Marcos, E., Pani, P., Brunamonti, E., Deco, G., Ferraina, S., and Verschure, P. F. M. J. (2013). Neural variability in premotor cortex is modulated by trial history and predicts behavioral performance. *Neuron*, **78**(2), 249–55. http://doi.org/10.1016/j.neuron.2013.02.006

Markov, N. T., Ercsey-Ravasz, M. M., Gariel, M. A., Dehay, C., Knoblauch, K., Toroczkai, Z., and Kennedy, H. (2011). The tribal networks of the cerebral cortex. In: Chalupa, L. M., Caleo, M., Galli-Resta, L., and Pizzorusso, T. (eds), *Cerebral plasticity: new perspectives*. Cambridge, MA: MIT Press.

Mathews, Z., Bermúdez Badia, S., and Verschure, P. F. M. J. (2011). PASAR: An integrated model of prediction, anticipation, sensation, attention and response for artificial sensorimotor systems. *Information Sciences*. http://doi.org/10.1016/j.ins.2011.09.042

Merker, B. (2007). Consciousness without a cerebral cortex: a challenge for neuroscience and medicine. *The Behavioral and Brain Sciences*, **30**(1), 63–81–134. http://doi.org/10.1017/S0140525X07000891

Moulin-Frier, J. C., Fischer, T., Petit, M., Pointeau, G., Puigbo, J.-Y., Pattacini, U., Low, S. C., Camilleri, D., Nguyen, P., Hoffmann, M., Chang, H. J., Zambelli, M., Mealier, A. L., Damianou, A., Metta, G., Prescott, T. J., Demiris, Y., Dominey, P. F., and Verschure, P. F. M. J. (in press). DAC-h3: A proactive robot cognitive architecture to acquire and express knowledge about the world and the self. *IEEE Transactions on Cognitive and Developmental Systems*.

Navawongse, R., and Eichenbaum, H. (2013). Distinct pathways for rule-based retrieval and spatial mapping of memory representations in hippocampal neurons. *Journal of Neuroscience*, **33**(3), 1002–13.

Rowland, D. C., Roudi, Y., Moser, M.-B., and Moser, E. I. (2016). Ten years of grid cells. *Annual Review of Neuroscience*, **39**, 19–40.

Schmidhuber, J. (2015). Deep learning in neural networks: An overview. *Neural Networks*, **61**, 85–117.

Verschure, P. F. M. J. (2016). Synthetic consciousness: the distributed adaptive control perspective. *Philosophical Transactions of the Royal Society of London. Series B, Biological Sciences*, **371**(1701), 263–75. http://doi.org/10.1098/rstb.2015.0448

Verschure, P. F., Voegtlin, T., and Douglas, R. J. (2003). Environmentally mediated synergy between perception and behaviour in mobile robots. *Nature*, **425**, 620–4.

Wyss, R., Konig, P., and Verschure, P. F. M. J. (2003). Invariant representations of visual patterns in a temporal population code. *Proc. Natl Acad. Sci. USA*, **100**(1), 324–29. Retrieved from 10.1073/pnas.0136977100

Yanco, H. A., Norton, A., Ober, W., Shane, D., Skinner, A., and Vice, J. (2015). Analysis of human–robot interaction at the DARPA robotics challenge trials. *Journal of Field Robotics*, **32**(3), 420–44. http://doi.org/10.1002/rob.21568

Chapter 24

Pattern generation

Holk Cruse and Malte Schilling

Universität Bielefeld, Germany

The faculty to generate patterns is a basic feature of living systems. Such patterns may be generated as fixed spatial patterns mainly used for the development of morphological structures, which will be mentioned here only briefly, or as spatio-temporal patterns as used in the context of controlling behaviors (see Herreros, Chapter 26, this volume). Here we address two types of patterns: (1) quasi-rhythmic patterns, typically applied in locomotion, and (2) non-rhythmic patterns used to control specific behaviors and to allow for switching between various behaviors.

1. Quasi-rhythmic patterns are typically found in locomotory behaviors which can be ordered in a series being characterized by increasing unpredictability of the environment, as are swimming, flying, running, walking on flat surfaces, and walking and climbing on cluttered, i.e. unpredictable surfaces (see chapters by Kruusmaa (44), Hedenström (32), Witte et al. (31), and Quinn and Ritzmann (42) in this volume). The corresponding controllers range along a continuum from those which produce patterns endogenously based on circular attractors, to pattern generators where the loop through the world plays a crucial role, i.e. where body and environment are part of the computational system. Patterns may result from explicit implementation or result as emergent properties.

2. A brain, when controlling various sequential behaviors, has to switch between different internal states characterized by their ability to select motor output (action selection) and to select sensory input (top-down attention). These states can be represented by the activation of various combinations of procedural memory elements, whereby various coalitions of individual elements typically form discrete (i.e. point-) attractors and may be ordered in such a way that the complete system comprises a heterarchical structure.

Biological principles

The basic principles and examples shown are mainly based on insect studies and the models have been verified by application to simulated or physical robots. Seven different types of pattern generation will be addressed. After having started with morphological patterns, two cases will be presented that concern patterns describing locomotion as examples of quasi-rhythmic behaviors. The following case deals with how non-rhythmic temporal series of behaviors may be controlled, whereby all approaches mentioned are based on various types of recurrent neural networks (RNN). The next section discusses "mind patterns" that may emerge from neural architectures as discussed in the former section. The last two sections deal with cases of pattern generation where time is not the basic relevant factor. First, a system is explained that allows for formation of social structures in an insect colony. These patterns arise from application of simple local rules based on excitatory and inhibitory influences, and from influences of environmental structures. Second, an RNN is presented that allows an infinite number of attractors which form a smooth continuum like is needed in representations of the body as used in motor control.

Spatio-temporal patterns on homogenous substrate

Spatio-temporal patterns may emerge on a *homogenous substrate*. A classic example is given by the Belousov–Zhabotinsky reaction (Zhabotinsky 2007), producing patterns that appear to form rotating spirals, or the classical Rayleigh–Bènard patterns (Getling 2012). Mathematical description based on reaction-diffusion differential equations goes back to Turing (1952) (see Wilson, Chapter 5, this volume). It is important to note that boundary conditions (i.e. properties of the environment) can play a crucial role in shaping the structure of the pattern. This means that, even in these simple cases, the "loop through the world" cannot be neglected. Such mechanisms are assumed to be responsible for many morphological structures occurring during the development of a body. Impressive examples are given by Meinhardt (1995), who shows how complex patterns can be simulated that are found on the shells of snails and mussels. Although these patterns develop over time, usually the focus is on their static spatial properties.

In the remainder, this chapter concentrates on patterns where time plays an important role. To this end, we focus on spatio-temporal patterns characterizing animal behavior.

Locomotory patterns based on homogenous substrate

Although brains in general do not form a homogenous substrate, but are highly structured, there are cases where brain activity controlling locomotion can be approximated by a homogenous network. A well studied example concerns swimming in lamprey. Kozlov et al. (2009) developed a model of the spinal cord with excitatory and inhibitory connections that coarsely stand for diffusion and reaction mechanisms, respectively. The complete network contains about 10,000 model neurons. This structure allows for waves of activation and deactivation of neurons travelling along the spinal cord and being responsible for controlling rhythmic swimming movements over a wide range of frequencies. Swimming speed, turns, and direction (forward, backward) can be controlled showing behaviors very similar to the ones observed in real lampreys (videos can be found in www.pnas.org/cgi/content/full/0906722106/DCSupplemental).

Spatio-temporal patterns on discretely structured substrate, no sensory feedback

Discretely structured systems can be found in locomotory systems that have to control only a small number of limbs, for example six legs in the case of hexapods, four legs in the case of tetrapods, or four wings in many insects.

In the simplest case, each limb performs two alternating types of movements, power stroke and return stroke, in repetition, whereby these rhythmic movements are controlled by a central pattern generator (CPG) being characterized by the property to provide rhythmic output without need of rhythmic sensory input (Ijspeert 2008). These oscillators have to be coupled to produce ordered movements of the different limbs. Coupling may occur only neuronally, only mechanically via the substrate (Owaki et al. 2012; Dallmann et al. 2017), or, to take advantage of redundant solutions, by application of both pathways. When regarding pure neuronal coupling of the CPGs, the complete system consisting of (e.g. six) coupled local CPGs is sometimes considered a CPG because a clear separation is often not possible (Ijspeert 2008).

CPG-systems have the advantage that they can easily be constructed and that their behavioral properties can be exactly defined and predicted, including the mathematical investigation of stability limits. The drawback is that such (clock-like) systems cannot well adapt to changes in the environment. Therefore they may best be applied for swimming or flying in an homogenous environment, or walking on flat surfaces.

Things get more complex when limbs are equipped with more than one joint to be controlled. Legged movement in 3D space requires legs with at least three hinge joints, but further redundancy is helpful (some insects and decapods, for example lobsters, have more than three joints per leg). If each joint is controlled by its own local CPG, a highly structured architecture is required (for an interesting example see Ijspeert 2008, his Figure 1).

Spatio-temporal patterns on discretely structured substrate, with sensory feedback

To improve adaptability the world, i.e. the body and the environment, can be used as part of the computational system (often addressed by "embodiment" and "situatedness"). Due to the unpredictability of the world this approach requires an architecture that relies heavily on sensory feedback and an internal architecture that is able to produce stable behavior in spite of the, in general, unpredictable situations. A possible answer is the development of a neural control structure that by itself consists of many decentralized, i.e. loosely coupled, procedural elements together forming the procedural memory. An example is given by Walknet, a free gait controller that produces quasi-rhythmic spatio-temporal stepping patterns as found in insects (for a review see Schilling et al. 2013a). Although these patterns are usually termed as "tripod gait," "tetrapod gait," and "wave gait," there are no separate, "true" gaits. Rather these terms describe specific patterns out of a continuum depending essentially on one parameter, velocity. In simulation, these "gaits" appear as emergent properties of the whole system and are visible only in fairly undisturbed environments. Under disturbances very different patterns may emerge (e.g. when starting from "difficult" leg configurations, see Figure 24.1, or when negotiating tight curves or climbing over very large gaps).

Although Walknet simulations show that control of quasi-rhythmical patterns is possible without using CPGs, nature most probably exploits both solutions in parallel—another example of taking advantage of redundant structures. Approaches mixing both elements to investigate the tight coupling between CPGs and sensory feedback have been studied in simulation and on robots (Beer and Gallagher 1992; Kimura et al. 2007; Ijspeert 2008; Daun-Gruhn 2011).

Control of non-rhythmic behavior, temporal sequences

On a higher level, a brain must be able to select between different behaviors, as for example foraging, feeding, mating etc., and order them in a sensible temporal sequence. Such sequences

Figure 24.1 Footfall patterns of the simulated robot Hector (Schneider et al. 2014) controlled by Walknet when starting with an uncomfortable leg configuration. Contralateral legs show the same position, which, at the beginning, leads to a gallop-like coordination. After some steps the walker adopts a regular "wave-gait"-like pattern. Black bars indicate swing movement of the respective leg: left front (L1), middle (L2), and hind leg (L3), right front (R1), middle (R2), and hind leg (R3), from top to bottom. Abscissa is simulation time. The lower bar indicates 500 iterations corresponding to 5 s real time. See Figure 3.6 for a picture of the physical Hector robot.

may be stored by connecting the relevant behaviors via dedicated connections that implement an explicit temporal sequence. However, an alternative solution can be observed already on a lower level, for example the control of walking. In the case of the walking controller introduced above, Walknet does not contain such dedicated connections. Nonetheless, in an undisturbed environment and under steady-state conditions this network can show quasi-rhythmic behavior with a defined temporal sequence of behavioral elements. In principle, however, many different sequences are possible that depend on the current environmental conditions. An example is given for a hexapod starting with a "difficult" leg configuration, see Figure 24.1. The behavioral sequences depend on the actual physical situation, represented by the body and environment, as well as specific "neuronal" structures. Both influences together control the activation of, as such, non-rhythmic behavioral procedures (e.g. swing, stance).

Figure 24.2 depicts a control structure called MUBCA (Schilling et al. 2013b) that generalizes this approach. This structure allows selection of different procedures in a way that is on the one hand as flexible as possible, but on the other hand precludes combinations of behavioral elements that are not sensible. The upper part of Figure 24.2 (marked red) shows a specific control layer, the so-called motivation unit network. Some motivation units directly gait the output of specific procedures (e.g. controlling swing movement, or stance movement of a leg—in black or bluc—on the lower part of Figure 24.2). Motivation units may also influence other motivation units via excitatory or inhibitory connections as illustrated in Figure 24.2. This, at first sight,

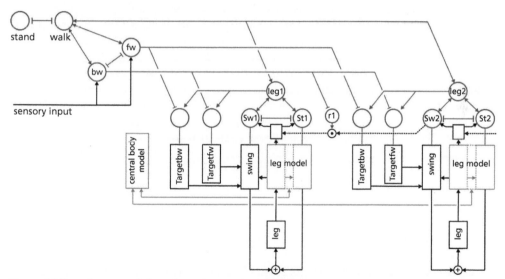

Figure 24.2 Walknet consisting of two layers, the motivation unit network (above, red) and the procedures (black, blue). Only two of the six leg controllers are depicted. Each consists of a Stance-net ("leg model," blue) and a Swing-net, the latter being connected with Target-nets ("Targetfw" for forward walking, "Targetbw" for backward walking). The motor output acts on the legs (boxes "leg" representing the body). The motivation units ("stand," "walk," "fw" for forward, "bw" for backward, "leg1," "leg2," "Sw1" for swing leg 1, "Sw2" for swing leg 2, "St1" for stance leg 1, "St2" for stance leg 2) are recurrently coupled via inhibitory connections (T-shaped connections, forming winner-take-all networks) and excitatory connections (arrow heads, allowing for forming coalitions). Sensory feedback is used by the motor procedures as well as to switch between the states (e.g. sensory input to motivation units "fw" and "bw"). r1 represents coordination rule 1 (for details see Schilling et al. 2013a).

hierarchical structure does not in general form a simple, tree-like arborization. As indicated by the bi-directional connections, motivation units form a recurrent neural network coupled by positive, or excitatory, and negative, or inhibitory, connections. Excitatory connections between motivation units allow for building coalitions. Motivation units coupled by mutual inhibition form local winner-take-all systems. In this way, the network allows for many, but dedicated, "internal" states. These states characterize the selection of various procedures, sequentially and/or in parallel, and protect the system from responding to inappropriate sensory input. In other words, such an internal state represents a specific context allowing for top-down attention. In the example given in Figure 24.2, the system may choose between stand and walk, and further between forward walking and backward walking.

In this network, sequences of behaviors are not stored as fixed chains in the memory, but emerge through the coupling via the environment while being limited by the internal boundary conditions. However, in specific cases unidirectional dedicated excitatory connections between motivation units appear to be sensible, for example when time is limited to allow for sensory feedback, i.e. for the control of fast behavioral sequences, or if adaptivity to changes in the environment is not required. The Distributed Adaptive Control (DAC, for review see Verschure 2012) architecture allows for learning of such behavioral sequences (see Verschure, Chapter 36, this volume).

Three other solutions to exploit the sensory input for the selection of different behavioral elements will be mentioned briefly (for more detailed discussion see Schilling et al. 2013a). Arena et al. (2009) used a two-layer network consisting of reaction diffusion Circular Neural Networks (CNN), which allow for the representation of Turing patterns. After a stable Turing pattern has emerged, it activates the appropriate behaviors via an additional "selector" layer. Instead of a Turing pattern network, Steingrube et al. (2010) used a network showing chaotic properties. This network is embedded in a preprocessor network and a postprocessor network, the latter transforming the attractor states adopted by the chaotic net into activation of specific behavioral elements. Tani (2007) proposed a two-layer RNN, where the upper layer deals with much longer time constants, thus being able to focus on more general aspects of the actual situation, in this way being able to represent different contextual aspects.

Higher level internal states (mind patterns)

A different class of patterns, which are however also relevant for the control of behavior, concerns mental, or mind, patterns such as, for example, specific internal states selected by (top-down) attention or various emotional states. Other mind patterns refer to switching between reactive and cognitive states, or between unconscious and conscious states.

At least in humans, these mind patterns are characterized as being accompanied by phenomenal experience, i.e. by the occurrence of subjective experience. To date neither the exact neuronal correlates nor possible functions of the subjective aspect, i.e. of being aware of the content of specific neuronal states, are known. The observation that mind patterns are equipped with a phenomenal aspect, at a first glance, may lead to the assumption that these patterns may belong to a domain fundamentally different from the internal states which were discussed in earlier sections and are not equipped with subjective experience. There are, however, arguments (e.g. Cruse and Schilling 2013), that understanding the functional aspect of these higher-level states might be possible while leaving the phenomenal aspect aside.

Following this argument, mind patterns have already been mentioned in the former section, when discussing states of the motivation unit network that represent different attentional states. A network of similar type might be able to control emotional states. However, instead of being connected to dedicated behavioral elements as is the case of controlling attentional states, a network controlling emotional states might be connected with specific functions related with the corresponding emotions.

The cognitive state as opposed to a reactive state has been defined as allowing the system for "internal trial-and-error," in order to test newly invented behaviors for solving an actual problem. This capability has been introduced in such a network, called reaCog, that requires specific neuronal structures for organizing the temporal sequence of detecting a problem, searching for a new behavior not given in the current context, testing the feasibility of that newly selected behavior by internal simulation, and, if successful, executing this new behavior (Schilling and Cruse 2017).

A very special, and for human beings fundamentally important, mind pattern is given by consciousness (see Seth, Chapter 37, this volume). There are many internal states a brain can adopt but, for a given moment in time, only a specific subset is accessible by (conscious) awareness. It is an open question what structures are responsible for selecting the elements contributing to the contents of consciousness and, in particular, how it is imaginable, after all, that neuronal, i.e. physical, states can lead to phenomenal experience. One current idea is that there is no special, separate structure required to represent consciousness, but that this mind pattern results as an emergent property of the network (Cruse and Schilling 2015).

Patterns forming social structures

Patterns can also be observed when studying social structures. An impressive example is given by Therolaz et al. (1991), who show how, via simple internal rules plus the boundary conditions of the environment, a three-layer social structure can emerge from a homogenous, genetically identical collection of animals in a wasp colony. The three groups consist of one queen and an about equal number of foragers and nest workers. The queen emerges from a winner-take-all coupling between the individuals realized by fights between any two individuals. The winner of the fight receives an increased "confidence value" which in turn increases the probability of winning the next fight. This mechanism results in a high confidence value for only one individual, thereby becoming the queen, and low confidence values in all other individuals. The distinction between the foragers and the nest workers depends on the structure of the nest. Larvae are found in the center, and food is stored near the outer margin of the nest. Contact with larvae stimulates non-queen individuals to forage for food. When outside the nest, these wasps have less contact with the queen, thus maintaining a somewhat higher confidence value compared to the other wasps, the nest workers. This division is stabilized because the queen, due to ongoing fights, moves the remaining individuals, becoming the nest workers, away from the centre of the nest and therefore away from the larvae. In this way, the nest workers receive less stimulation from the larvae and, thus, remain in the nest. Therolaz et al. (1991) have studied and verified the properties of this system by simulation. Interestingly, the social structure adapts to changes in the number of larvae as well as to changes in the number of foragers or nest workers.

Patterns forming a smooth subspace to distinguish geometrically possible states from geometrically impossible ones

The earlier cases concentrated on the specific case in which patterns are generated as a sequence of switching between discrete states, locomotion being the typical example. In these cases the underlying neuronal control structure is usually assumed to be characterized by distinct attractor states representing the different movements (see Figure 24.2, e.g. swing and stance). In contrast, many movements are better characterized as spanning a whole attractor subspace. An example is reaching movements. To reach a target, the control system has to actuate joints and in general has to deal with kinematics characterized by extra degrees of freedom. Furthermore, there are additional constraints on different levels (joint limits or blocking objects). Instead of enumerating all possible movements, an underlying representation is assumed which encodes the relations of the body with respect to the environment. Such a structure, usually called body schema, is assumed to be utilized in motor control (Wolpert et al. 1998). To control reaching or grasping movements the body schema has to deal with the so-called inverse problem of motor control. For prediction of movements the body schema is used as a forward model. Third, a body schema, or body model, integrates the multitude of sensory input systems into a coherent state. A body schema, in this respect, acts as a pattern generating system as it is encoding the body structure and can act as a pattern completion system which, as an autoassociator, provides corresponding patterns when only partial information is given.

An example of such a model is given by a MMC (Mean of Multiple Computation) network (Schilling 2011). The network consists of simple local relations between limbs, but contains redundant descriptions of these relations. Such a network can represent an infinite number of states corresponding to all geometrically possible configurations of this body, and only those. The network cannot adopt a pattern that would represent a geometrically impossible body configuration. To demonstrate the principle, Figure 24.3 depicts a network that is able to represent all geometrical positions adoptable by a planar arm equipped with three joints. For a given end position of the arm tip, due to the extra degree of freedom, in general an infinite number of arm configurations is geometrically possible. This body model is able to address all three problems mentioned above. The network finds a solution when a new target position is given and it can predict movement outcomes and thereby be used for planning ahead by means of

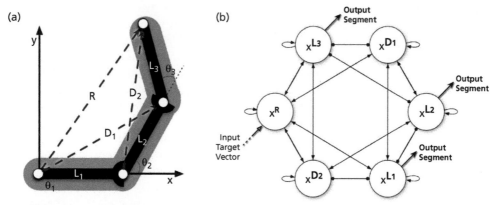

Figure 24.3 (a) A three degree, planar arm with the arm segments L_1, L_2, and L_3. R represents the Target vector pointing to the position of the tip of the arm. D_1 and D_2 do not represent segments of the arm, but are required for computation. The redundant geometry can be exploited by, for example, computing vector L_1 in two ways: $L_1 = D_1-L_2$, and $L_1 = R-D_2$. The mean value of these two calculations is then used to represent L1 in the next iteration. (b) depicts a recurrent neural network that implements all calculations used to represent the three-joint arm. (Only that part is shown that determines the x components of the vectors. A corresponding network is used for the y components.) This network shows the linear version (for the non-linear case see Schilling 2011). For computing the inverse kinematics, R is given as input, and L_1, L_2, and L_3 are used as output, as depicted. For computing the forward kinematics, L_1, L_2, and L_3 are used as input and R represents the output of the network.

internal simulation. In all the cases, the constraints of the network steer the pattern completion process leading to a coherent state that matches the (partial) input and is in agreement with the constraints.

Future directions

With respect to the networks using motivation units, an important open question concerns the problem of how these networks may be scaled up without loosing their abilities. More generally, this question addresses the problem as to how large memories should be organized while still allowing for effective storage and retrieval. Another aspect to be further studied concerns the question whether distributed representations (Tani 2007) instead of dedicated motivation units might be biologically more plausible.

Similarly, a body model may be realized by not only just one, but by several body models, each focusing on different aspects, for example, integration of different sensory modalities. Decoupling of different body models may, on the one hand, allow for a Theory of Mind (ToM), and, on the other hand, may lead to out-of-body experiences. Therefore, understanding of how these states may arise in a brain, what types of body models may exist, and how they may be coupled, is a research topic of general interest.

A further open problem concerns the contribution of learning the structures (see Arena et al. 2009, Verschure 2012), which relates to learning and storing of new memory elements, but also to stabilizing already existing elements.

Generally, all models mentioned are, compared to biological reality, formulated on a quite abstract level, which is in part due to lack of knowledge concerning the biological details. Therefore, study of brain functions, in particular the organization of memory, that are relevant for the construction of artificial systems, is required.

Learning more

A wonderful collection of examples as to how complex morphological patterns may emerge from simple local rules can be found in Meinhardt (1995), comparing biological patterns of sea shells and their simulation by simple local rules. A comprehensive introduction to the control of—in particular quadruped—locomotion with CPGs can be found in Ijspeert (2008). This article contains biological examples and mathematical models of CPGs plus an extensive literature review. A review of the control of hexapod walking without relying on CPGs is given by Schilling et al. (2013a).

References

Arena, P., De Fiore, S., and Patené, L. (2009). Cellular nonlinear networks for the emergence of perceptual states: application to robot navigation control. *Neural Networks*, **22**, 801–11.

Beer, R. D., and Gallagher, J. C. (1992). Evolving dynamical neural networks for adaptive behavior. *Adaptive Behavior*, **1**, 92–122.

Cruse, H., and Schilling, M. (2015), Mental states as emergent properties. From walking to consciousness. In: T. Metzinger and J. Windt (eds), *Open Mind.* , Frankfurt/M.: MIND Group Frankfurt/M.

Cruse, H., and Schilling, M. (2013). How and to what end may consciousness contribute to action? Attributing properties of consciousness to an embodied, minimally cognitive artificial neural network. *Front. Psychol.*, **4**, 324. doi: 10.3389/fpsyg.2013.00324

Dallmann, C. J., Hoinville, T., Dürr, V., and Schmitz, J. (2017), A load-based mechanism for inter-leg coordination in insects. *Proc. Royal Society B: Biological Sciences*, **284**, 1755. http://dx.doi.org/10.1098/rspb.2017.1755.

Daun-Gruhn, S. (2011). A mathematical modeling study of inter-segmental coordination during stick insect walking. *J. Comp. Neurosci.*, **30**, 255–78.

Getling, A.V. (2012). Rayleigh–Bénard convection. *Scholarpedia*, 7(7),7702.

Ijspeert, A. (2008). Central pattern generators for locomotion control in animals and robots: a review. *Neural Networks*, **21**, 642–53.

Kimura, H., Fukuoka, Y., and Cohen, A.H. (2007). Realization of dynamic walking and running of the quadruped using neural oscillators. *Autonomous Robots*, **7**, 247–58.

Kozlov, A., Huss, M., Lansner, A., Hellgren Kotaleski, J., and Grillner, S. (2009). Simple cellular and network control principles govern complex patterns of motor behaviour. *Proc. Natl Acad. Sci. USA*, **106**, 20027–32.

Meinhardt, H. (1995). *The algorithmic beauty of sea shells*. Berlin/Heidelberg: Springer.

Owaki, D., Kano, T., Nagasawa, K., Tero, A., and Ishiguro, A. (2012). Simple robot suggests physical interlimb communication is essential for quadruped walking. *J. R. Soc. Interface*, **10**. doi: 10.1098/rsif.2012.0669

Schilling, M. (2011). Universally manipulable body models—dual quaternion representations in layered and dynamic MMCs. *Autonomous Robots*, **30**(4), 399–425.

Schilling, M., and Cruse, H. (2017). ReaCog, a minimal cognitive controller based on recruitment of reactive systems. *Front. Neurorobot.*, **11**(3). doi: 10.3389/fnbot.2017.00003

Schilling, M., Hoinville, T., Schmitz, J., Cruse, H. (2013a). Walknet, a bio-inspired controller for hexapod walking. *Biol. Cybern.*, **107**, (4) 397–419. doi: 10.1007/s00422-013-0563-5

Schilling, M., Paskarbeit, J., Hüffmeier, A., Schneider, A., Schmitz, J., and Cruse, H. (2013b). A hexapod walker using a heterarchical architecture for action selection. *Frontiers in Computational Neuroscience*, **7**. doi: 10.3389/fncom.2013.00126

Schneider, A., Paskarbeit, J., Schilling, M., and Schmitz, J. (2014). HECTOR, a bio-inspired and compliant hexapod robot. In: A. Duff, T. Prescott, P. Verschure, and N. Lepora (eds): *Living Machines 2014*, LNAI 8608. New York: Springer, pp. 427–429.

Steingrube, A., Timme, M., Wörgötter, F., and Manoonpong, P. (2010). Self-organized adaptation of a simple neural circuit enables complex robot behaviour. *Nature Physics*, **6**, 224–30.

Tani, J. (2007). On the interactions between top-down anticipation and bottom-up regression. *Frontiers in Neurorobotics*, **1**, 2. doi: 10.3389/neuro.12/002.2007

Therolaz, G., Goss, S., Gervet, J., and Deneubourg, J. L. (1991). Task differentiation in *Polistes* wasp colonies: a model for self-organizing groups of robots. In: J. A. Meyer and S. W. Wilson (eds), *From Animals to Animats*. Cambridge, MA: MIT Press, pp. 346–55.

Turing, A. M. (1952). The chemical basis of morphogenesis. *Phil. Trans. Roy Soc. London B*, **237**, 37–72.

Verschure, P. F. M. J. (2012). Distributed adaptive control: a theory of the mind, brain, body nexus. *Biologically Inspired Cognitive Architectures*, **1**, 55–72.

Wolpert, D., Miall, R., and Kawato, M. (1998). Internal models in the cerebellum. *Trends in Cogn. Sc.*, **2**(9), 338–47.

Zhabotinsky, A.M. (2007). Belousov–Zhabotinsky reaction. *Scholarpedia*, 2(9),1435.

Chapter 25

Perception

Joel Z. Leibo[1,2] and Tomaso Poggio[1]

[1] McGovern Institute for Brain Research, Massachusetts Institute of
Technology, USA
[2] Google DeepMind

The brain's perceptual systems routinely perform tasks that remain spectacularly beyond the capabilities of engineered systems. Most children can describe all the objects in a scene and understand a voice speaking from a crowd. But no one knows how to produce the same behavior by hooking up a camera and microphone to a computer. At the same time, primary sensory physiology is one of the most well-studied areas of neuroscience, and the biologically inspired approach to building computer perception systems is now yielding results. In fact, there are signs that engineered systems are catching up with the brain. Vision-based pedestrian detection systems are now accurate enough to be installed as safety devices in (for now) human-driven vehicles (Markoff 2013) and the (accidental –?) politics of mobile-phone voice-recognition systems provoke popular uproar (Rushe 2011). While not being entirely biologically based, *computational neuroscience* makes up a considerable portion of both systems' intellectual pedigree.

   In keeping with its importance to human life, perception is also the topic of a massive—and rapidly growing—scientific literature. Any attempt to summarize will inevitably overlook many important areas. In writing this chapter we have endeavored to choose examples that support a certain view of how the brain implements perception which is directly related to recent engineering applications. All our examples will be drawn from primate auditory and visual systems. However, the principles they illustrate, both computational and biological, are applicable to other organisms and other modalities.

Biological and computational principles

The inputs to perception

In many ways, the retina really is like a camera (though see Dudek, Chapter 14, this volume for ideas on building a more life-like neuromorphic camera). It consists of an array of light-sensing devices called photoreceptor cells. Just as the CCD pixels of a digital camera each sense light at a particular spatial location on the 2D imaging surface, each photoreceptor responds to light at a particular location in the visual field. A specific cell may respond to flashes of light occurring 3 degrees to the left of the center of gaze. Perceptual scientists would say that location is in the cell's *receptive field* (RF).

   For most of our purposes, we can think of the collective activity of all the eye's photo-receptor cells as identical to a bit-map image with one pixel per photoreceptor RF. However, it has a few properties that distinguish it from the images you capture with your iPhone. First, there are two different major subtypes of photoreceptors: *cones*—which function best in bright light and underlie color vision—and *rods*—which work in dim light but are not sensitive to color. Unlike most cameras, the retina's sensor distribution is not isotropic. At its peak in the central 4 degrees of visual angle (called the *fovea*) the average cone density is 199,000 cones/mm^2, but it declines to 50% at 1.75 degrees from the center (still within the fovea!). It continues to drop from there—by 20 degrees the cone density is down to 5% of its peak. Rod density also varies with eccentricity; in particular, in the central 1.25 degrees there are no rods at all. After considering the retina's extreme anisotropy, the fact that we perceive a unified visual world without a constant blur extending into our peripheral vision becomes remarkable in itself.

The primary sensory organ of audition, the cochlea, can be thought of as a bank of band-pass filters (see Smith, Chapter 15, and Lehmann and van Schaik, Chapter 54, this volume, on progress towards building an artificial cochlea). A sound signal is a pressure wave which causes the vibration of an inner ear structure called the basilar membrane. *Inner hair cells* sense movement of the basilar membrane and relay auditory information to the brain via the auditory nerve. The thickness and width of the basilar membrane vary along its length causing subsections of it to vibrate at different frequencies. As a consequence of the basilar membrane's varying mechanical properties, the output of the cochlea is organized *tonotopically*. That is, on one end of its spatial axis the inner hair cells respond selectively to low frequency tones, on the other end, they are selective for high frequency sounds. The frequency for which a given cell is selective is called its *characteristic frequency* (CF) (see Figure 25.1B).

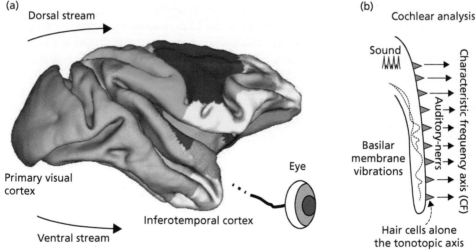

Figure 25.1 (a) The brain of a macaque monkey—a commonly studied animal in perceptual neuroscience. Colors indicate regions responsible for different sensory (and motor) modalities. Vision is blue and audition is red. The brain model and coloring is from SumsDB (Van Essen and Dierker 2007). Arrows indicate the flow of visual information. The dorsal stream—the "where" pathway—is concerned with object motion and localization. The ventral stream—the "what" pathway—is concerned with object identity (Mishkin et al. 1983). (b) The tonotopic organization of the cochlea and auditory nerve.

For tone frequencies < 4 kHz, the responses of inner hair cells and afferent auditory nerve fibers are synchronized (phase-locked) with the vibration of the basilar membrane. Above 4 kHz, responses are not phase-locked, however, the tonotopic organization of the cochlea is valid up to 20 kHz, the high-frequency limit of human hearing, so tones above 4 kHz are still represented, just without temporally precise spiking. In this chapter we emphasize the computational and algorithmic properties that follow from the cochlea's tonotopic organization. However, there is an alternative perspective on auditory processing that emphasizes phase-locking and temporal precision over spatial tonotopic organization. We note that much of these "place" and "time" theories of auditory perception can be reconciled with one another (see Shamma 2001), but that discussion is beyond the scope of this chapter.

In this section we highlighted the notions of a visual cell's receptive field and an auditory cell's characteristic frequency. Both primary sensory organs are organized topographically. That is, retinal cells with receptive fields near one another in the visual field are also located near one another on the 2D retina (this is obvious if you think of the retina as a camera with a sensor for each pixel location). Similarly, neighboring cochlear cells have similar characteristic frequencies. In the next sections we will consider perceptual processing beyond the primary sensory organs. We will see that the central nervous system maintains the topographic organization of its sensory inputs and uses it in the computation of later perceptual representations.

Early perception

It is useful to think of early perception as an inverse problem, e.g. early vision as inverse optics or early audition as inverse acoustics. The forward problems are well-posed; they have a unique solution. For example, the problem of determining the pressure wave sensed by a microphone (or an ear) given some set of sound sources in the world has a unique solution. However, the same is not true for the corresponding inverse problem. All the sounds in an auditory scene sum together to generate the signal measured at the sensor. Thus, many different scenes are physically consistent with the same sensory data.

Consider a simple two-tone sound stimulus (as in Figure 25.2). The plots on the left are a representation of the responses of auditory-nerve fibers (the cells that carry information from inner hair cells to the brain) as a function of time. Each tone evokes a traveling wave along the basilar membrane's tonotopic axis. In this representation, a traveling wave appears as an oscillation in amplitude at a range of frequencies, phase-locked to the stimulus. The traveling wave evoked by one tone has its strongest amplitude at 300 Hz while the other tone's characteristic frequency is 600 Hz. Since the two waves are not in phase with one another, their collision causes sharp boundaries to appear in the auditory-nerve representation between regions that are phase-locked to different tones. The CF location of these *edges* depends on the amplitude and frequency of the two tones. Thus, any mechanism to detect the locations of the edges could be used to extract the sound spectrum (as in Figure 25.2). In particular, Shihab Shamma suggested the use of a lateral inhibitory network for this purpose (Shamma 1985). Lateral inhibition has the effect of detecting and enhancing discontinuities in the input pattern.

The visual system is met with essentially the same computational problem. As in the auditory system, the task is to detect and localize edges, which, in this case, are sharp changes in luminance caused by the boundaries between objects. In both cases, the underlying problem is that of numerical differentiation with respect to a topographically organized feature map. Since early visual areas inherit a retinotopic organization from the eye, the derivative may be taken with respect to the spatial coordinates of the feature map since they correspond directly

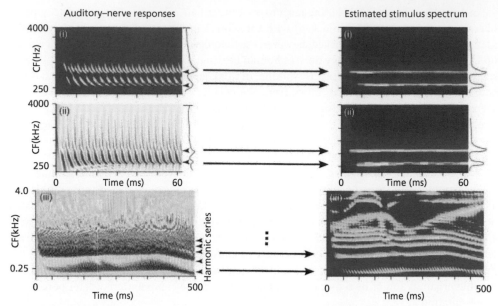

Auditory–nerve responses

Estimated stimulus spectrum

Figure 25.2 Schematic model of the early stages of auditory processing The abscissa is time. On the left the ordinate indicates the CFs of simulated auditory nerve fibers. On the right the ordinate indicates the CFs of cells in a downstream region, possibly the anteroventral cochlear nucleus. Left side: The spatiotemporal response patterns of auditory-nerve fibers evoked by (i), a quiet two-tone stimulus of 300 Hz and 600 Hz with neither tone saturating the nerve responses. (ii) A louder two-tone stimulus that does saturate the auditory nerve fiber responses for a wide range of CFs. As a result, the time-averaged response (i.e. a function of frequency) is barely modulated. In contrast, the time-averaged response of the quiet stimulus in (i) has two prominent peaks at 300 and 600 Hz. (iii) Speech contains many frequency components that are harmonically related to a "fundamental" frequency which corresponds to the voice's pitch. Right side: The estimated stimulus spectrum, computed by processing (i, ii, and iii) with a lateral inhibitory network. Lateral inhibition has the effect of detecting and enhancing the discontinuities in the input pattern between regions that are phase-locked to different tones. (i and ii) Regardless of loudness, the estimated stimulus spectrum is similar. (iii) Harmonics are clearly visible in the network's output.

to coordinates on the image. In the auditory case, the derivative is still taken with respect to the map's spatial coordinate. But since the auditory nerve is tonotopically organized, this corresponds to differentiating with respect to frequency.

It turns out that, like most problems of early perception, edge detection is an ill-posed problem (Poggio et al. 1985). The definition of ill-posedness was introduced by Hadamard (1923). A problem is well-posed when its solution (1) exists, (2) is unique, and (3) depends continuously on the initial data. Ill-posed problems fail to satisfy one or more of these criteria. Edge detection is ill-posed because the solution does not depend continuously on the data. That is, differentiation is not robust against noise. To see this, consider a function $f(x)$, and its perturbation $g(x) = f(x) + \varepsilon \sin(wx)$. For small ε, the two functions may be very close (e.g. close in L2 norm), and how close they are does not depend much on w. On the other hand, the difference between their derivatives $f'(x)$ and $g'(x)$ may be very large if w is large.

Ill-posed problems can be made well-posed by restricting the class of admissible solutions through the introduction of suitable a priori assumptions. This process of using a priori knowledge to convert an ill-posed problem into a well-posed problem is called *regularization*. Edge detection can be regularized by assuming the underlying image is smooth.

Consider the inverse problem of discovering z from data y:

$$Az = y$$

In standard (Tikhonov) regularization theory, A is a linear operator and P is linear (Tikhonov et al. 1977). The regularized problem is to find z that minimizes

$$\|Az - y\|^2 + \lambda \|Pz\|^2$$

The first term measures the difference between the solution and the data while the second term penalizes solutions that violate the a priori assumption encoded in P. The weighting between the two terms is controlled by a scalar λ, called the regularization parameter. Through the choice of λ it is possible to control the compromise between the solution's degree of regularization and its closeness to the data.

For edge detection, the suitable a priori assumption is that the image is smooth. Thus P should be chosen so that it penalizes solutions that are not smooth. One way to do this is to choose $P = d^2/dx^2$, the second derivative operator (many other choices are possible too (Poggio et al. 1988).

It turns out that the regularized solution to the problem of edge detection is equivalent to convolving the image with the derivative of a Gaussian-like filter (Poggio et al. 1985). Edges are zeros of the second derivative or critical points (maximum, minimum, or inflection points) of a first derivative of the image. Thus, an algorithm for edge detection consists of two steps: (1) convolution with the derivative of a Gaussian, and (2) computing the maximum over local regions. This is how essentially all edge detection algorithms work—e.g. Canny (1986).

In general, for any ill-posed inverse problem, whenever the data is given on a regular grid and A is space-invariant, the (Tikhonov) regularized solution can be computed by a convolution (Poggio et al. 1988). Many of the brain's inverse problems are space-invariant in this sense. Returning to the example of extracting the profile of a sound spectrum (Figure 25.2), Shamma suggested that a lateral inhibition network operating on the responses of the auditory nerve would have the effect of localizing the discontinuities that appear between regions that are phase-locked to different tones (Shamma 1985). In his network, each afferent fiber is treated the same way: inhibited according to a function of the responses of its local tonotopic neighborhood. This operation can be seen as the convolution of the auditory nerve representation with a particular filter followed by a non-linear pooling of the responses of nearby cells.

So far this chapter has been concerned with early perceptual problems. In the next section we will consider a problem of high-level perception: object recognition. We will find that convolution, non-linear pooling, and topographic organization remain fundamental concepts in that setting as well.

High-level perception: recognition and invariance

If early vision is like inverse optics, then high-level vision is like inverse graphics. Likewise, early audition is like inverse acoustics and, for the case of speech, high-level audition can be thought of as inverse speaking. At this stage, the central problem facing natural or artificial perceptual systems is the variability of the sensory world. Many stimuli must be considered equivalent—e.g. images of the same object from different viewpoints—but the patterns of sensory stimulation evoked by different views of equivalent scenes may bear little resemblance to one another. This problem of invariance arises in all sensory modalities (though we will mostly focus on vision here).

Visual information passes from the retina along the optic nerve to the thalamus, from there it goes to primary visual cortex (V1): a large region in extreme posterior part of the brain (occipital lobe) (Figure 25.1A). V1 cells are commonly divided into two categories: *simple* and *complex*. Hubel and Wiesel, who first made the distinction, found that some V1 cells—which they named simple cells—were optimally tuned to oriented bars. That is, one simple cell may respond when a horizontal line appears in its receptive field while another simple cell may respond when a diagonal line appears in its receptive field. Simple cells are sensitive to the exact location of their preferred stimulus within the receptive field. They have oriented "on" and "off" regions; the appearance of a stimulus in the former increases the cell's firing, while in the latter it suppresses it. Complex cells are also tuned to oriented lines but tolerate shifts in the line's exact position within their receptive field. That is, they have no "off" regions. Most complex cells also have somewhat larger receptive fields than simple cells (Hubel and Wiesel 1962).

Simple cells represent stimuli by an orientation at each position (the coordinates are x, y, and angle). Complex cells also represent stimuli by those three coordinates but their spatial sensitivity is diminished. In this sense, the pattern of activity over the population of complex cells can be thought of as a blurred version of the representation carried by the simple cells. V1 inherits the retinotopic organization of its inputs. Thus, cells with receptive fields in nearby regions of the visual field are also located nearby one another in cortex.

Hubel and Wiesel conjectured that complex cells are driven by simple cells (Hubel and Wiesel 1962). A complex cell tuned to an orientation θ tolerates shifts because it receives its inputs from a set of simple cells optimally tuned to θ at different (neighboring) positions. In the popular *energy model*, the response of a complex cell is modeled as the sum of squares of a set of neighboring simple cell responses (Adelson and Bergen 1985).

In downstream higher visual areas, cells respond to increasingly large regions of space. At the end of this processing hierarchy, in the most anterior parts of the ventral visual system—particularly in the anterior parts of inferotemporal cortex (IT)—there are cells that respond invariantly despite significant shifts (up to several degrees of visual angle). A small object could be translated so that it no longer falls in the receptive fields of any of the photoreceptors that originally imaged it. Even then, the representation in IT could remain largely unchanged. Hubel and Wiesel proposed that a similar organization to the one they discovered in V1 may be repeated beyond early vision; the cells in downstream visual areas could be optimally tuned to more complex image fragments, e.g. corners and Ts may come after V1's lines—all the way up to whole object representations in IT (see Figure 25.3).

Fukushima's neocognitron is an early example of a computational model of object recognition built around Hubel and Wiesel's proposal (Fukushima 1980). In that model, alternating layers of "simple" (S) and "complex" (C) cells compute either a "tuning" or a "pooling" operation; other more recent models maintain this organization (LeCun and Bengio 1995; Riesenhuber and Poggio 1999).

S cells can be thought of as template detectors; they respond when their preferred image (their template) appears in their receptive field. Since visual areas are approximately retinotopically organized, the pattern of activity of all the cells tuned to a particular template, each at a different location in the visual field—the feature map—can be understood as a map of where in the visual field the template appears. That is, the feature map is the convolution of the inputs with the template. A layer that represents many different features, like V1 with a template for each orientation, is said to contain many feature maps.

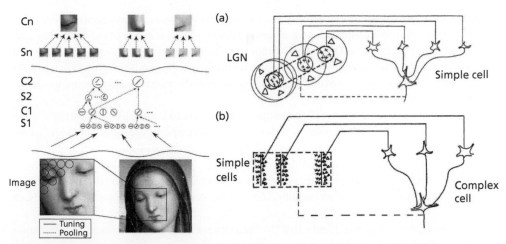

Figure 25.3 Left: Illustration of a model with a convolutional architecture. In early layers, units represent simple stimuli like oriented bars. In later levels, units are tuned to complex stimuli like patches of natural images. Right: Hubel and Wiesel's illustration of their proposal that complex cells are driven by simple cells. It also shows that simple cell tuning can be produced by a combination of "center-surround" receptive fields like those in the lateral geniculate nucleus (LGN) of the thalamus.

Reprinted from *The Journal of Physiology*, 160 (1), D.H. Hubel and T.N. Wiesel, Receptive fields, binocular interaction and functional architecture in the cat's visual cortex, pp. 106–54, Figures 19 and 20, doi: 10.1113/jphysiol.1962. sp006837, Copyright © 1962 The Physiological Society.

C cells pool together the responses of S cells at different locations. Thus C cells are tuned to the same templates as their input S cells, but they are invariant to stimulus translations within their receptive field. Pooling is typically modeled with a non-linear operation—as in the energy model: sum of squares of a set of S cell responses—or the HMAX model: the max of a set of S cell responses. Since the cortex is retinotopically organized, the S feature maps are computed by convolving the template with the inputs. C feature maps are computed by applying the nonlinear pooling function to local neighborhoods of S cells.

A good question to ask at this point is: how could the proposed organization be wired up through either visual experience or evolution? There are two separate questions here, one is how to learn the templates (the wiring into the S cells), the other is how to learn the pooling (the wiring into the C cells). The former question is answered by the statistics of commonly viewed image fragments (e.g. Olshausen and Field 1996). On small spatial scales, low complexity features like the oriented edges of V1 are represented. On larger spatial scales, the cells represent more complex features.

The second problem, that of developing invariance via this scheme, is equivalent to the problem of associating units representing the same template at different positions (and scales) in the visual field. While it is straightforward to achieve this in a computational model, for visual cortex it is one of a class of notoriously difficult correspondence problems. One way the brain might solve it is via temporal-association-based (TAB) methods. The most famous TAB method is Földiák's trace rule (Földiák 1991). The trace rule explains how many cells having the same selectivity at different spatial positions could be wired to the same downstream cell by exploiting continuity of motion: cells that fire to the same stimulus in close temporal contiguity are all presumably selective to the same moving stimulus.

TAB methods are based on the assumption that objects normally move smoothly over time. In the natural world, it is common for an object to appear first on one side of the visual field and then travel to the other side as the organism moves its head or eyes. A learning mechanism that takes advantage of this property would associate temporally contiguous patterns of activity. As a system employing such an algorithm gains experience in the visual world it would gradually acquire invariant template detectors.

Psychophysical studies have tested the temporal association hypothesis by exposing human subjects to altered visual environments in which the usual temporal contiguity of object transformation is violated. Exposure to rotating faces that change identity as they turn around leads to false associations between faces of different individuals (Wallis and Bülthoff 2001). Similarly, exposure to objects that change identity during saccades leads to increased confusion between distinct objects when asked to discriminate at the retinal location where the swap occurred (Cox et al. 2005).

There is also physiological evidence that the brain uses a TAB method to acquire invariance. Li and DiCarlo showed that exposure to an altered visual environment in which highly dissimilar objects swap identity across saccades causes anterior IT neurons to change their stimulus preference at the swapped location (Li and DiCarlo 2008). They went on to obtain similar results for scale invariance (Li and DiCarlo 2010); objects grew or shrank in size and changed identity at a particular scale. After an exposure period, AIT neurons changed their stimulus preference at the manipulated scale. A control unswapped location and scale was unaffected in both experiments.

The canonical microcircuit hypothesis

We have now seen the same two mathematical operations appear in the solution to an early vision problem (edge detection), an early auditory problem (sound spectrum estimation), and a high-level perception problem: visual object recognition. If we had considered other perceptual problems, e.g. motion detection, we would have found the same operations appearing there as well (Adelson and Bergen 1985). It is tempting to propose that circuits for computing these operations are part of a "canonical microcircuit": a structural motif reappearing at all levels of perceptual (or other) processing—in all parts of cortex. While not conclusively established, there is now converging evidence for this *canonical microcircuit hypothesis* from several fields. The neuroanatomical evidence is reviewed in Douglas and Martin (2004), and the algorithmic properties of canonical microcircuits are considered in (Kouh and Poggio 2008).

Biomimetic systems

In the review above, we have already alluded to several biomimetic systems in the context of their role as models of biology. Here we discuss the application of a particular family of these models to engineering problems. Convolutional neural nets (CNNs) have recently been used successfully, surpassing the previous state-of-the-art, on a variety of problems.

Convolutional neural nets have two alternating types of layers, analogous to the S and C layers of the Neocognitron and HMAX. Convolutional layers compute the convolution of their input with a set of stored filters. The convolution is typically followed by a non-linear activation function (such as a sigmoid). Pooling layers typically compute the max (or another function) over local regions of the output of the previous convolution layer. Pooling layers are said to sub-sample the convolution layers because they have less units, effectively lowering the network's resolution while making it invariant to local translations.

In the domain of speech recognition, convolutional networks have recently shown strong performance on the TIMIT phoneme recognition benchmark (Abdel-Hamid et al. 2012). Convolutional networks for phoneme recognition work by pooling over local frequency ranges. It is believed that this works because it effectively normalized several common acoustic variations, e.g. different speakers pronouncing the same vowel and various kinds of channel noise.

Recently, Krizhevsky et al. used a convolutional neural network to obtain image categorization results that significantly improved the previous state-of-the-art (Krizhevsky et al. 2012; Figure 25.4). They used graphics processing units (GPUs)—the development of which was mostly driven by videogame application—to drastically improve the speed at which their network could be trained, thus enabling them to use a far larger network, and considerably more training examples than previous networks.

Future directions

Lately, the field has been moving in two complementary directions. One direction, exemplified by Krizhevsky et al.'s approach, has been toward using bigger networks and accommodating more and more training data. This is an appropriate approach for the "age of big data;" Internet companies can easily gather terabytes of data hourly.

The other direction focuses on what can be done with much smaller amounts of data. Humans learn visual concepts from many fewer examples than Krizhevsky's network requires. One avenue of research along these lines is to use TAB methods to associate neural representations of stimuli across other transformations besides translations. The networks resulting from such processes are still convolutional, but now with respect to other dimensions of stimulus variation, e.g. 3D rotation in depth (Stringer and Rolls 2002; Poggio et al. 2012). The goals of this approach are still unrealized, but the hope is that it will yield networks that are capable of mimicking the human ability to learn from very small numbers of examples—since, for any new concept, all natural stimulus transformations could be handled automatically.

Learning more

The famous book, *Vision* by David Marr, describes the theoretical approach that we have attempted to follow in this review (Marr 1982). Marr distinguishes between three different levels on which perceptual systems can be studied: (1) computational theory, (2) algorithms, and (3) implementation in hardware (or wetware). The book also contains a good overview of many of the vision problems that we neglected here, e.g. stereopsis and motion processing. The main methodological point has recently been discussed and updated in Poggio (2012).

More information on early vision, from the perspective of inverse problems, can be found in Poggio et al. (1985). The high-level perception problem of visual object recognition was recently reviewed by DiCarlo et al. (2012).

Shihab Shamma reviews the connections between vision and audition (in Shamma 2001) and the anatomical evidence for a canonical microcircuit is reviewed in Douglas and Martin (2004). Many other papers discuss the algorithms that could be implemented by canonical microcircuits, see, for example, Kouh and Poggio (2008).

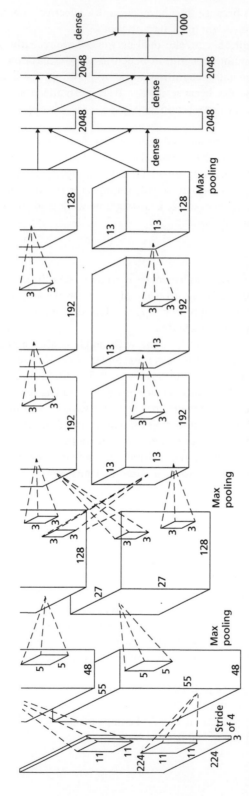

Figure 25.4 Illustration of the architecture of the convolutional network used by Krizhevsky et al. to obtain significant improvements over the previous state-of-the-art on a large image classification task (Krizhevsky et al. 2012). This diagram highlights the interconnections between the two GPUs on which their system was implemented.

References

Abdel-Hamid, O., et al. (2012). Applying convolutional neural networks concepts to hybrid NN-HMM model for speech recognition. In *IEEE International Conference on Acoustics, Speech and Signal Processing (ICASSP)*, pp. 4277–80.

Adelson, E.H., and Bergen, J.R. (1985). Spatiotemporal energy models for the perception of motion. *Journal of the Optical Society of America A*, **2**(2), 284–99.

Canny, J. (1986). A computational approach to edge detection. *IEEE Transaction on Pattern Analysis and Machine Intelligence*, **6**, 679–698.

Cox, D., et al. (2005). "Breaking" position-invariant object recognition. *Nature Neuroscience*, **8**(9), 1145–7.

DiCarlo, J., Zoccolan, D., and Rust, N. (2012). How does the brain solve visual object recognition? *Neuron*, **73**(3), 415–34.

Douglas, R., and Martin, K. (2004). Neuronal circuits of the neocortex. *Annu. Rev. Neurosci.*, **27**, 419–51.

Van Essen, D., and Dierker, D. (2007). Surface-based and probabilistic atlases of primate cerebral cortex. *Neuron*, **56**(2), 209–225.

Földiák, P. (1991). Learning invariance from transformation sequences. *Neural Computation*, **3**(2), 194–200.

Fukushima, K. (1980). Neocognitron: A self-organizing neural network model for a mechanism of pattern recognition unaffected by shift in position. *Biological Cybernetics*, **36**(4), 193–202.

Hadamard, J. (1923). *Lectures on the Cauchy Problem in Linear Partial Differential Equations*. Yale: Yale University Press.

Hubel, D., and Wiesel, T. (1962). Receptive fields, binocular interaction and functional architecture in the cat's visual cortex. *The Journal of Physiology*, **160**(1), 106.

Kouh, M., and Poggio, T. (2008). A canonical neural circuit for cortical nonlinear operations. *Neural Computation*, **20**(6), 1427–51.

Krizhevsky, A., Sutskever, I., and Hinton, G. (2012). ImageNet classification with deep convolutional neural networks. In *Advances in neural information processing systems*. Lake Tahoe, CA, pp. 1106–14.

LeCun, Y., and Bengio, Y. (1995). Convolutional networks for images, speech, and time series. In: M. A. Arbib (ed.), *The handbook of brain theory and neural networks*. Cambridge, MA: MIT Press, pp.255–8.

Li, N., and DiCarlo, J.J. (2008). Unsupervised natural experience rapidly alters invariant object representation in visual cortex. *Science*, **321**(5895), 1502–7.

Li, N., and DiCarlo, J.J. (2010). Unsupervised natural visual experience rapidly reshapes size-invariant object representation in inferior temporal cortex. *Neuron*, **67**(6), 1062–75.

Markoff, J. (2013). At high speed, on the road to a driverless future. *The New York Times*, May 27, 2013, p.2.

Marr, D. (1982). *Vision: A computational investigation into the human representation and processing of visual information*, New York, NY: Henry Holt and Co., Inc.

Mishkin, M., Ungerleider, L., and Macko, K. (1983). Object vision and spatial vision: two cortical pathways. *Trends in Neurosciences*, **6**, 414–17.

Olshausen, B.A., and Field, D.J. (1996). Emergence of simple-cell receptive field properties by learning a sparse code for natural images. *Nature*, **381**(6583),607–9.

Poggio, T. (2012). The Levels of Understanding framework, revised. *Perception*, **41**, 1017–23.

Poggio, T, Torre, V., and Koch, C. (1985). Computational vision and regularization theory. *Nature*, **317**(26), 314–19.

Poggio, T, Voorhees, H., and Yuille, A. (1988). A regularized solution to edge detection. *Journal of Complexity*, **4**(2), 106–23.

Poggio, T., et al. (2012). The computational magic of the ventral stream: sketch of a theory (and why some deep architectures work). *MIT-CSAIL-TR-2012-035*.

Riesenhuber, M., and Poggio, T. (1999). Hierarchical models of object recognition in cortex. *Nature Neuroscience*, 2(11), 1019–25.

Rushe, D. (2011). iPhone uproar: Is Siri anti-abortion? *The Guardian*, December 1, 2011, p.1.

Shamma, S. (2001). On the role of space and time in auditory processing. *Trends in Cognitive Sciences*, 5(8), 340–8.

Shamma, S. (1985). Speech processing in the auditory system II: Lateral inhibition and the central processing of speech evoked activity in the auditory nerve. *The Journal of the Acoustical Society of America*, 78(5), 1622–32.

Stringer, S.M., and Rolls, E.T. (2002). Invariant object recognition in the visual system with novel views of 3D objects. *Neural Computation*, 14(11), 2585–96.

Wallis, G., and Bülthoff, H.H. (2001). Effects of temporal association on recognition memory. *Proc. Natl Acad. Sci. USA*, 98(8), 4800–4.

Tikhonov, A. N., Arsenin, V. I., and John, F. (1977). *Solutions of ill-posed problems* (Vol. 14). Washington, DC: Winston.

Chapter 26

Learning and control

Ivan Herreros

SPECS, Institute for Bioengineering of Catalonia (IBEC), the Barcelona Institute of Science and Technology (BIST), Barcelona, Spain

Artificial systems have come to outperform humans in a series of tasks, most of them taking place in highly structured and abstract domains, such as playing chess or, more recently, Go (Silver et al. 2016). However, the robots' performance still pales in comparison to that of animals in many fields. Low level behaviors such as motor control is one of them. For instance, the dexterity and balance of a two-year-old child as it runs and bumps with other children in a playground is clearly far more advanced than that of state-of-the-art of humanoid robots, despite enormous resources having been invested in this field of research both in the public and private sectors.

In this chapter, we put forward the idea that in order to close this gap we first have to analyze animal behavior using the mathematical tools of the engineering branches of control theory and machine learning. Only then could we clearly pinpoint the computational components of the tasks faced and their solutions achieved in the animal realm. Then, secondly, equipped with this conceptual language, we can expect to enrich the original engineering disciplines with these solutions that had evolved in animals. Indeed, the thrill of research in the biomimetic field lies not only in explaining animal behavior in terms of control theory or machine learning but also in advancing control theory or machine learning with insights coming from biology. However, answering the first question, which belongs to the interest of neuroscience, is a prerequisite to performing research for the second.

This chapter is structured as follows. First, we introduce basic concepts of control theory and machine learning in order to provide the reader with an intuitive understanding of the kind of problems that are addressed within each discipline and the nature of the solutions they provide. Next, within the biological domain, we will classify animal actions both from a control theory perspective and according to whether they are innate or acquired. Next, we will relate the mechanisms of action acquisition that have been considered in animal psychology with different classes of machine learning problems introduced. Next, we will address the question of the neural substrate of those mechanisms. Finally, we will mention a few examples of biomimetic systems that have been designed along these lines.

Working definitions

The control problem

Control theory is an engineering discipline that develops methods with the goal of obtaining the desired performance from systems. The field is also known as *automatic control*, as it originated during the industrial revolution with the rationale of removing human operators from the production process. In general, the desired performance of a system is defined in terms of maintaining or reaching a desired state or following a determined *target signal* (possibly taking into account the cost of obtaining that desired performance; e.g. amount of energy consumed). For instance: an air-conditioning system has to reach and maintain the state of having a room at a given temperature; an automatic pilot has to guide a plane's descent before landing, decreasing the altitude according to a predetermined pattern (e.g, the target signal).

In control theory parlance, the desired state (or output) is known as the *reference (r)*. The difference between the actual state (or output) and such a reference is the *error (e)*. To define a controlled system we also have to determine the object controlled, known as the *plant (P)*, its measurable *output (y)*, and the input or inputs by which the system can be actuated to modify its state. When such input proceeds from a controller (C) it is referred to as the control signal (u).

There are two fundamental control strategies: feedback and feed-forward control (Figure 26.1). This divide arises from the role played by the current state in computing the control signal.

Feed-forward control

Knowing the system's initial and desired states, we can pre-compute an entire control signal, e.g. a sequence of commands, and apply it to the system. In order to achieve this we need a good model of the controlled system, such that we can precisely predict how inputs affect its state. Otherwise, even slightly inaccurate predictions and small variability on the effector side will accumulate with time and the longer the sequence of commands, the more the system may end up deviating from the desired state. In other words, the feed-forward control strategy is by necessity *model-based* and requires very precise mathematical models and accurate physical systems (i.e. sensors and actuators).

Feedback control

The feedback control strategy consists of constantly updating the control signal according to the ongoing output or state information. In its simpler version, all that needs to be defined for implementing a feedback control strategy is the direction in which to act in order to reduce a

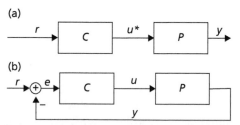

Figure 26.1 Feedback and feed-forward control schemes. (a) In a feed-forward control scenario the controller (C) receives the desired output (r). Taking into account an inverse model of the plant, and knowing its initial state, it computes an output signal (u*) that is sent to the plant (P) and that results in it producing the final output y. If the control program was successful the final value of y must be equal to r. C and P form an open-loop system because the system's output (y) does not enter it again. (b) In a feedback scenario, the controller (C) continually receives an error signal (e) measuring the distance between the current output (y) and the desired one (r). According to this information, C outputs a control signal (u) that changes the state of P and, through this, y. Because of y entering C again via the error computation and affecting the generation of a new u, C and P form a closed-loop system.

given error, which is defined as the difference between the current output and the desired one. In other words, whether increasing the control signal increases or decreases the output. As an example, consider that turning up a heating device augments the room temperature but turning up an air-conditioning unit decreases it. Thus, the same error in the desired output (a room temperature two degrees below the desired one) is corrected with opposite control signals (turning up versus turning down) depending on the system controlled.

Feedback controllers can either compute their signal directly from the error in the output (*output feedback*) or estimate the internal state of the system from its output and compute the control signal according to the complete system state (*state feedback*). Output feedback is *model-free*. That is, it does not require a model of the controlled system for implementing a controller (even though, from an engineering perspective, it will be beneficial to have access to such a model during the controller design). The air-conditioning unit example above can be interpreted as output feedback. State feedback is model-based, as in order to infer the states of a system given its output it needs to model, at least, the relation between outputs and the system states. However, this doesn't mean that state feedback suffers from the same sensitivity to modeling inaccuracies as feed-forward control. To the contrary, the feedback mechanism inherent to state feedback will correct deviations from the desired state stemming from model inaccuracies. In short, whereas in feed-forward control model inaccuracies can cause divergence between desired and actual states, in state feedback they only limit the maximum performance.

To conclude, we have considered one type of problem in control theory, that is achieving that the plant follows (or reaches) a particular reference, and have introduced the two families of methods for doing so. On the one hand, feedback control is robust but reactive, as it needs for errors to occur before they can be corrected. On the other, feed-forward control is efficient, as it can avoid errors before they occur, but remains very sensitive to modeling inaccuracies, noise, or disturbances. If one were to have a perfect model of the controlled object, absolutely accurate (noiseless) sensors and actuators, and the guarantee that no external elements could interfere with the task (no disturbances), the choice should be to apply only feed-forward control, computing the control signal beforehand. In real applications and especially in the case of agents acting in open environments, the previous conditions are not easily met and thus it will often be required to rely upon feedback control mechanisms. That being said, usually the best strategy will be to balance both feedback and feed-forward control according to task- and/or agent-related demands.

We can exemplify all the concepts introduced above in the context of controlling an arm. In that case the plant is the arm and the central nervous system acts as the controller. The state of the system can be defined according to the angular positions and velocities of the joints or as the degree of contraction of the arm muscles. The output is provided by the sensory systems carrying information about the position of the arm, namely, the visual and proprioceptive systems (and indirectly the tactile and nociceptive systems). The control signal would be coded by the discharge of the motor neurons causing muscle movement. Finally, whether a feedback or a feed-forward strategy will generate the control signal will depend upon the context: very rapid voluntary movements will mostly be executed according to a feed-forward plan whereas feedback control will play a role in executing slower actions, or fast reactions to painful stimuli.

Machine learning schemes

Machine learning is a discipline at the intersection between computer science, mathematics, and statistics. Its goal is to define algorithms that allow machines to extract knowledge from data. Historically, machine learning has been oriented towards a type of processing where the learning algorithm has no role in the data generation. That is, the data was given to the algorithm a priori, and then the trained algorithm was exploited with previously unseen data. For instance, a statistical speech recognizer has little or no control over the generation of the speech that it is recognizing. This type of approach might be dominant even till today, especially in the domain of deep learning, where machine learning algorithms are usually used together with very large amounts of data (or Big Data). However, the reader of this manual will probably be more interested in a situation where the learning algorithm is embedded in an autonomous agent, such as a robot. In that case, the so-called behavioral loop establishes a closed-loop between the process of learning and the stimuli that drive that learning (Verschure et al. 2003). In other words, as learning will shape action it will also shape the perceptions that result from these actions, which are the stimuli that provide the learning algorithm with information in the first place. That being said, this doesn't imply that the machine learning techniques are not applicable in real-world behaving agents, but it does imply that the statistics of the data generation might violate assumptions usually made in the classical machine learning community (such as that the data points are identical and independently sampled).

According to whether the behavioral loop is present, we could classify the problem into *passive* and *active* machine learning approaches. Indeed, one heritage of the initial research in machine learning having taken place mostly in passive setups is the classical division of the operation of learning algorithms into training and application phases. This division usually does not apply to active setups, as in that case the agent will most likely be constantly updating its model; that is, there will be just one phase, where the learning algorithm will be both exploited and trained.

Classes of machine learning algorithms and problems

There is a fundamental division of machine learning problems and algorithms into three classes: supervised, unsupervised, and reinforcement learning (Figure 26.2). This division is relevant from the standpoint of biological and biohybrid systems, because it allows identifying the basic computational aspects inherent to a particular learning problem or task. Indeed, the process of acquiring a complex behavior in a biological or an artificial system comprises of, and can be described by, sub-problems belonging to these different classes, which will be further described in the sections that follow.

A second source of interest in introducing this division stems from the possibility of associating different brain structures with different classes of learning. Indeed, regions of the brain have been attributed with different computations in the terms that we will further elaborate in the next section.

At this point, one should note that in control theory, when considering the division between feedback and feed-forward strategies, both could in principle be applied to solve the same problem. This is not the case for classification of machine learning, where the three families of problems reflect not a methodological distinction but one related to the computational nature of the problem.

Supervised learning

In a supervised learning problem a system learns an input–output transformation given data examples. In mathematical terms, the goal of the algorithm is to approximate a function, or a set of functions, from paired data samples. Depending on whether the outputs are defined over a continuous or a discrete domain, we are faced with a function approximation or a classification problem.

The automatic diagnosis of medical images (e.g. detecting tumors in X-ray images) and speech-to-text translation are classical classification problems that have long since been addressed in

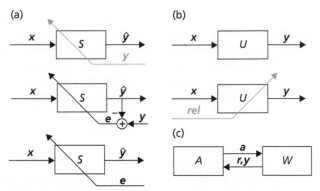

Figure 26.2 Basic schemes in machine learning. (a) In supervised learning, the algorithm (S) receives data (x) and produces an output ($\hat{y}$), which in general is an approximation of a desired value (y). In the most classical scheme (top), S receives the desired output only for an initial set of data (the training set) and afterwards S is employed to map unseen inputs into outputs according to the mapping function acquired. Additionally (middle), S may simply receive an error signal (e) computed as the difference between y and $\hat{y}$ or (bottom) receive error information, which might be an unknown function of the difference between y and $\hat{y}$. (b) In unsupervised learning, the algorithm (U) receives input data x from which it produces a new representation y, either compressed or expanded. During processing, all inputs might be considered equally important (above) or weighted according to an input-dependent relevance signal (*rel*, below). (c) Minimal representation of the information exchange between the agent (A) and the world (W) in reinforcement learning: A executes an action (a) from which it receives an immediate reward (r) and a perception (or observation, y) from W. Given y, A infers the state (internal or external) of W; updates the value of the preceding state, a set of states, or the state evaluation function, according to r; possibly updates its action-selection policy; and finally, chooses a next action.

the supervised learning domain. Stock market prediction is probably the most exploited of the supervised learning problems, along with weather forecasting (and more recently, the prediction of whether user will click or not on ads appearing in webpages).

From a methodological perspective, supervised learning problems can be either *off-line* or *on-line*. In the first case, all classified data samples are available for the training period and afterwards the algorithm is only applied to unlabeled data. In the case of on-line learning labeled data samples become continuously available to the system such that it can constantly update its model. The prediction of time signals provides an example of on-line scenarios. For instance, if we forecast the value of a stock market asset on a day-to-day basis, even though the target (correct) value is not available at the time of the prediction it will be so one day later. Hence, the predictive model can be constantly updated during operation.

Additionally, the learning process can rely upon *iterative* or *batch* updates. In the first case, the model is updated after the presentation of each data sample, whereas in the second case, updates occur only after computing the error for a set of samples, i.e. a batch. In the case of off-line learning, the batch can comprise the entire training set.

In the biological domain, the on-line iterative approach is the most relevant. For instance, motor behavior will be constantly updated during performance and memories formed as soon as stimuli are experienced. However, it is proposed that off-line learning occurs during sleep, driven by the replay of daily experiences. That is, while we sleep we would process a batch of data that was provided by our experiences during that day.

What is learned in the supervised scheme? Note that since data pairs are labeled, learning could consist of simply storing the input–output associations in a look-up table. However, even though with this method we could retrieve the labels of previously seen data, a simple look-up table would not allow the labeling of unseen data. Indeed, from a supervised learning algorithm we expect that it should *generalize over the training set*, and thus be able to infer the underlying features of the data that determine the correct input-output associations, i.e. the mapping function.

Finally, in the most general interpretation, the term *supervision* implies that during learning, for each output generated, the algorithm is provided with error information allowing it to locally correct the mapping function such that the error would decrease. On the one hand, this implies that supervised learning is strictly error-driven: if there is no error, there is no update of the model and therefore no learning. On the other hand, this also implies that the algorithm does not necessarily require access to the desired output, it solely needs information on how to change its output to reduce the error. To provide an intuitive account of this, say that we have to control a mechanism throwing a ball such that it stops at a particular distance and that we can control the velocity at which the ball is released. Even if we are not given the target velocity (which is our desired output value), knowing whether the ball stopped before or after the desired position will inform us about whether to increase or decrease the velocity of release. In other words, the error in the position will *supervise* the velocity. Of course, in the general case, knowing the desired output value allows exact computation of the error. In that case, and assuming that the goal is minimizing the error squared, the gradient of the error function can be obtained and the mapping function updated following a gradient descent.

Neural networks provide very popular implementations of supervised learning algorithms, with the multi-layer perceptron being the most widely employed general-purpose supervised neural network. Multi-layer perceptrons are non-linear function approximators that generate an intermediate level of representation between the input and output domains. Supervised learning algorithms are implemented as neural networks also in the deep-learning field; in that case with multiple intermediate hidden levels.

Unsupervised learning

Unsupervised learning is related to statistical modeling of a data domain. An unsupervised learning algorithm aims at extracting patterns from a set of unlabeled data. The objective of this type of algorithm can also be understood in terms of data compression. That is, given a set of data, provide a compressed representation that preserves all or sufficient information. Thus, unsupervised learning processes can be related to data transmission or storage. For instance, the minimum description length algorithm, which searches for the compression code that for a particular data set minimizes the joint size of the code plus the data compressed, provides an example of an unsupervised learning algorithm (Goldsmith 2001) (specifically an example of the lossless-compression approach).

Data clustering is a typical unsupervised learning problem. In this case, the compression comes from replacing each data item by a class identifier, in the deterministic case, or by a list of probabilities of belonging to a particular class, in the probabilistic one. For instance, consider a data set with multi-electrode, extracellular recordings of neural activity. Each data item contains electrode traces during identified action potentials. An electro-physiologist will be interested in sorting all these spike events according to the neurons that fired. However, the physiologist himself will not be able to assign spikes to particular neurons or even know the number of neurons present in a given recording. But given the relative positions between neurons and electrodes, each neuron's spikes will leave a distinct fingerprint in the recording. The spike-sorting algorithm will infer such a number of distinct fingerprints present in the data, thereby associating each spike to a particular neuron. Note that in terms of data compression, a spike-sorting algorithm reduces the original data size by several orders of magnitude, from electro-physiological signals sampled at high time resolution to a list of spike events associated to different neurons.

Probabilistic methods are particularly successful in unsupervised tasks as they allow the defining of the problem in terms of finding the parameters for a statistical model that maximizes the likelihood of obtaining the data. The approach consists of assuming that the observations are generated by an underlying probabilistic model, then finding the parameters making the data more likely according to that model. Expectation-maximization is a widely used and powerful algorithm for estimating the parameter values when we know a priori the number of causes generating the data (e.g. in the previous example the different neurons are the different *causes* of the recorded signals).

Two final remarks on the unsupervised learning algorithms: first, even though most frequently unsupervised learning algorithms produce compressed representations of the data, they might at times be used to expand the data representation. For instance, assuming that data is represented in an *n*-dimensional space, we might use an unsupervised learning algorithm to project it into an *m*-dimensional one, with $m>n$. Generally, the rationale of such a process is to increase (linear) separability. To bring in an example from the next section, cerebellar granule cells, a type of small cells in the input layer of the cerebellum, are supposed to implement an unsupervised learning algorithm. Since by themselves, cerebellar granule cells account for more than half of the neurons in the human brain, by necessity, they have to provide an expanded representation of the input that they receive from other parts of the nervous system.

Secondly, either way, whether the goal of the unsupervised learning algorithm is to project data into a higher or a lower dimensional space, until this point we have assumed that all data samples were equally relevant. However, data may additionally be tagged according to relevance. In that case, if we consider that under a given unsupervised learning algorithm data samples may be competing for representational space, more relevant data would be assigned more such space. Put differently, from the data compression approach, the algorithm might receive not only the data to compress, but also an indication regarding which data should be stored more faithfully.

In animals, we could posit that unsupervised learning occurs mostly in the perceptual domains, where the stimuli are the raw data and the causes of the stimuli (a tree, a face) are the patterns to be discovered.

Reinforcement learning

In reinforcement learning problems, an artificial agent must learn how to act in order to maximize its experienced reward (and consequently minimize the amount of punishment or negative reward). In contrast with the previous two machine learning schemes, reinforcement learning is usually an active learning paradigm, i.e. the agent learns while acting. That is, accordingly to the behavioral feedback previously, the agent determines through its actions the stimuli that it receives.

The minimal formulation of a reinforcement learning problem comprises a set of *states, rewards*, and *actions*. Actions define transitions between states, and rewards are associated with those transitions by measuring the immediate reward for the action execution. In reinforcement learning parlance we differentiate between the agent and the environment. Thus, once the agent chooses an action, the environment provides it with an immediate reward (e.g. a scalar) and information regarding the new state reached.

A crucial concept in reinforcement learning is the *policy*, which is a function that maps states into actions and directs the agent on which action to select, either deterministically or probabilistically, at any given state. The objective of a reinforcement learning algorithm is to determine an *optimal policy* where departing from a certain state will result in selecting a sequence of actions that maximize a function of the current plus future rewards. That function can be either the cumulative reward for finite sequences of actions or, for infinite ones, the discounted cumulative reward, where the further in the future the less the contribution of a reward to the choosing of the current action.

At a particular state, the total (discounted) cumulative reward expected under the current policy determines the *value* of a state. In simpler words, the value of a state measures how much reward the agent expects to gather in the future acting according to the current policy. Policies prescribe actions to an agent according to their expected value. Hence, there is a clear circular dependence between values and policies. Moreover, an agent does not know a priori the real value of the states it encounters. In general, an agent can only assign a value to a state after visiting it and experiencing the subsequent rewards (or, more generally, infer the value of that state according to its similarity with previously experienced states).

The basic reinforcement learning scheme allows for many additional complications. For instance, actions themselves can be probabilistic, which is tantamount to having inaccurate effectors, and therefore, their consequences cannot be fully determined a priori; states might be non-observable, for which the agent will have to estimate its current state based on its history of observations and actions; or, in real-world problems, it will not be possible to experience all the states because their number might be too large or infinite. In the latter case, instead of having the value associated to each state stored in a lookup-table there will be a value function that for a certain state will return its predicted value. In other words, reinforcement learning problems imply solving a supervised learning problem, namely the learning of how to predict the states' values for a given policy.

Despite what may appear as a convoluted mathematical problem definition, reinforcement learning basically addresses the problem of deciding how to act based on incomplete knowledge of the environment, for instance, based solely in individual experience.

Conceptually, where supervised learning can be understood as a guided process, in which at each point of time the algorithm is given explicit information as to how to update its prediction, reinforcement learning implicitly entails a search-like process. This is to say, after each action the agent might receive a reward but it will not know whether taking another action would have resulted in receiving a higher or a lower one. An agent can only approximate an answer to this question by taking alternative actions whenever a same situation repeats, i.e. by performing a search in the action/state space. It is usual to describe a particular reinforcement learning solution in terms of its exploration/exploitation trade-off. There, exploiting refers to an agent choosing actions optimally according to its current (and possibly incomplete) knowledge of the world, and exploration implies behaving in a locally suboptimal manner to increase that knowledge.

Biological principles

Reactive/feedback actions

The separation between reactive and anticipatory behavior is partially reflected on the one between feedback and feed-forward control, respectively. In those terms, reactive actions are triggered by perceived errors that are decreased or corrected by the action execution. There is, between the action and the error stimulus, a so-called closed-loop (or feedback) relationship such that the perception that triggers the action is modified as a result of the action. Thus, "reactive" can be interpreted in terms of "negative error feedback control." The most basic example of this type of motor behavior is a spinal reflex, like the spinal withdrawal reflex (Schouenborg and Weng 1994). In the spinal withdrawal reflex, a noxious stimulation of the cutaneous receptors on a limb results in a limb movement (e.g. a retraction) that terminates contact with the cause of the noxious stimulus (e.g. a stinging plant, a stimulating electrode, etc.).

Noxious stimuli signal errors unambiguously; pain implies a deviation from the desired state of avoiding being hurt and maintaining the body integrity. However, in general, other stimuli are only errors as long as they signal deviations from a desired state. For instance, seeing a cursor on a particular position on a screen can only be interpreted an error if that was not the desired position. This also means that reactive actions are not limited to either reflexes or protective behavior. For instance, a high level task such as moving a cursor to a particular position on a screen is reactive. Moreover, in addition to avoidance, also appetitive behavior can be described appealing to negative feedback. For instance, when an experimenter delivers an appetitive stimulus, such as sucrose solution, into the mouth of an animal, for that animal the desired state becomes ingesting the liquid, and perceiving it in the mouth becomes the error driving swallowing. In summary, the term "error," borrowed from control theory, doesn't imply having made a mistake or the experience of a negative event, it simply indicates a distance between the actual and the desired state.

Anticipatory/feed-forward actions

Anticipatory actions are *predictive* by nature. They imply issuing motor commands in accordance not to the current state and/or perceptions, but to presumed future ones. Anticipatory actions can also occur in reaction to external stimuli, but the crucial difference with reactive actions is that the anticipatory action will not affect the stimulus that triggered it. In control theory terms, anticipatory actions define *open-loop* systems. In the case of anticipatory reflexes, the triggering cue stimulus will only predict a sensory error of the ones we introduced above. With this, anticipatory reflexes become reactive actions to predicted errors.

One should note that if we could extend this formulation of anticipatory reflexes to all types of anticipatory actions, it would introduce the radical notion that, in biological systems, all actions ultimately lead to the activation of a feedback mechanism or, put differently, that feed-forward actions do not exist in isolation but only in relation to feedback primitives. This same principle has been expressed within the action oriented predictive coding framework (Clark 2013) or the active inference scheme (Friston et al. 2010).

Innate and acquired actions

By definition, innate actions should be predefined and do not depend upon the agent's experience. An inborn action should manifest, after maturation, independently from previous sensory experience and/or should not be affected or modified by a later one.

Analogous to artificial systems, innate actions are a repertoire of pre-programmed behaviors. However, whereas in the case of artificial systems it might seem straightforward to define or identify this type of actions, to what extent there exist purely inborn actions in animals is unclear. Elementary actions that are shared by all healthy individuals of a species could be interpreted as innate actions. However they could still result from an innate *action acquisition mechanism* that shapes a sensorimotor response according to a contingency that all individuals experience during development. Moreover, actions that we tend to classify as innate are flexible. For instance, the muscular stretch reflex, by which the activation of motor neurons, in response to a perceived elongation of the muscle, contracts it to its previous length, can be rapidly retrained, i.e. suppressed, in a few trials in healthy subjects (Nashner 1976).

Thus, it is more meaningful to define innate actions as sensory motor transformations for which there exists a basic specification prefiguring the stimulus-action response that does not need to be acquired through experience. Proving this statement experimentally requires complex manipulations depriving animals from their normal sensory experience during maturation. Nonetheless, there are examples in the literature. Ocular tracking responses, which stabilize images in the retina, are found in diurnal animals reared in complete darkness. Even in darkness, when such animals turned their head to one side their eyes turned in the opposite direction (a behavior known as the vestibulo-ocular reflex [VOR]), and thus they were acting to minimize an error (the displacement of the visual field on the retina) that they have never experienced.

Acquired responses

Acquired responses are learned through experience. In general terms, whereas innate responses are the product of evolution maximizing fitness, acquired or learned responses are developed through experience to increase reward and/or diminish punishment. An acquired behavior that achieves this goal is said to be *adaptive*, in the opposite case a detrimental acquired action is *maladaptive*.

Adaptive actions are then the joint result of innate learning mechanisms and the agent's experience. Experimental psychology has proposed two main sources for the learning of actions: Thorndike's law of effect and the Rescorla–Wagner (RW) model.

Thorndike's law of effect and operant conditioning

The law of effect states that *in a particular situation, animals are more likely to repeat actions that led to satisfactory outcomes and less likely to repeat actions that led to unsatisfactory ones*. This type of learning is studied with the operant conditioning protocol of experimental psychology, in which animals may be rewarded for the execution of given actions and/or punished for the execution of others. For instance, a rat introduced into a chamber with two levers where one delivers sucrose and the other nothing will at first push any of the levers with equal probability. However, very rapidly, the rat will acquire the action–reward association and then whenever it is placed again in the same chamber, it will directly actuate the reward lever.

At this point it is easy to observe that reinforcement learning is the computational translation of an operant conditioning scheme. For instance, a simple setup where an animal has to chose between two levers, one delivering reward and the other nothing, could be minimally translated into a reinforcement-learning problem with two actions (choosing left or right) and a reward value associated to each action. In this case, the obvious optimal policy that could be known after trying both levers would consist in choosing always the same rewarded lever.

The RW model and classical or Pavlovian conditioning

According to the RW model *animals only learn when events violate their predictions*. It is therefore an error-driven type of learning that was at first developed as a means to account for the acquisition of conditioned reflexes in the classical (or Pavlovian) conditioning paradigm.

Described in behavioral terms, classical conditioning results from the pairing of a so-called unconditioned stimulus (US) that naturally (and innately) elicits an unconditioned response (UR) with an initially neutral conditioning stimulus (CS). Through the repeated paired exposure of CS and US, the animal develops a reaction to the CS, the so-called conditioned response (CR), that resembles UR. For instance, in eyeblink conditioning, an airpuff (US) is preceded by a tone (CS), such that, after training, the protective blink (UR) that initially followed the US is partially replaced by an anticipatory response (CR) triggered by the CS. In psychological terms, the RW model accounts for this type of learning as the building of a sensory-sensory association between the CS and the US, such that the former comes to predict the latter (Rescorla 1988). The ensuing CR represents the animal reaction to the predicted US instead of being the result of a direct transformation of the CS into a behavioral response. Thus, we can consider classical conditioning as a sensory-prediction task.

The RW model accounts for the phenomenology of learning in classical conditioning by the failure or success of that sensory prediction. That is, once an animal successfully anticipates the occurrence of a US (or its absence) no further learning occurs and its response (or lack of response) is maintained, reaching an asymptote in the case of acquisition. On the other hand, errors in the prediction strengthen or weaken the CR, depending upon whether there was a failure to predict an incoming US or a predicted US did not occur, respectively. In the RW model animals do not learn according to what occurs (*perceptual* or *sensory errors*), but only according to the difference between what occurs and what was predicted (*sensory prediction errors*). In these terms, the perception of a US conveys unambiguously a sensory error, but internally, it will become a sensory prediction error as far as it was an unexpected US.

Pavlov defined classical conditioning to be *not* instrumental. In other words, the expression of the CR should not affect (or ameliorate) the perception of the US. For instance, eye-blink conditioning might be performed using a periorbital mild electric shock as a US. In that case, the anticipatory blink has no protective effect upon the noxious stimulus, and thus the resulting CRs observed in this preparation are not adaptive. Even though this issue fired much debate during the sixties and seventies of the nineteenth century one should not be too concerned from a biomimetics perspective. Indeed, classical conditioning exploits an innate mechanism for the acquisition of anticipatory reflexes that assumes that the feedback reflexes were useful in the first term. In other words, in a setup where URs have no adaptive value, the question should not be why CRs are acquired, but why URs persist. Why does the animal keep blinking to the periorbital electric shock even though that blink achieves no purpose?

Hence, if we are interested in applying the learning mechanisms uncovered in classical conditioning in a robot, we should not be troubled by the non-instrumentality inherent to the Pavlovian definition. Instead, we should worry that the reactive behavior onto which we are grounding our anticipatory behavior is beneficial to the robot. However, one must be particularly aware that once the anticipatory actions become instrumental, they activate a behavioral feedback mechanism. In short, the sensory perception that initiated learning may not the same sensory perception at the end of learning.

To conclude, classical conditioning is related with supervised learning. First, if one describes classical conditioning as the acquisition of a sensory prediction then it becomes a supervised learning task by definition. More subtly, in classical conditioning of musculoskeletal reflexes, the perceived sensory error acts as an error signal informing the organism about an action that should have been taken previously. In this case, the goal of the system is not to predict the sensory error, but to change its behavior in subsequent trials according to the sensory error such that it is suppressed (Figure 26.3).

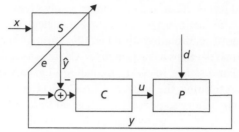

Figure 26.3 Adaptive control architecture for conditioned avoidance (anticipatory reflex). From a control theory perspective, this controller includes a closed-loop system, comprising a negative feedback controller (*C*) and a plant (*P*), supplemented by a feed-forward input (*ŷ*). The feed-forward input is acquired by the supervised learning process (*S*). The state of *P* is affected by a disturbance (*d*), e.g. a noxious stimulus, and thus the output *y* carries a pain signal. The pain signal is implicitly compared to a reference equal to zero (see the change in sign) and fed to *S* and *C* as an error signal (*e*). *S* attempts to associate *e* with a preceding context (*x*), such that whenever a same *x* is experienced in the future it will issue an anticipatory *ŷ*, thus triggering the execution of a protective action in prediction of *d*. Note that whenever *d* is totally avoided, *y* is equal to zero and no further learning is induced in *S*. In nature, *P* would be a part of the musculoskeletal system, *C* a brainstem nuclei or a spinal reflex arc, and *S* a cerebellar microcircuit.

Perceptual learning and unsupervised learning

When presenting reinforcement learning and classical conditioning we obviated the problem of stimulus identification. For instance, in an operant scheme two stimuli may signal different action–reward contingencies (e.g. choose left for reward after stimulus A and right after stimulus B). But what happens if the difference between both stimuli is too slight for the subject to detect? In that case the agent will succeed only if it is previously able to develop an acuter perception of the stimuli. Perceptual discrimination tasks show that subjects can improve their perceptual acuity with prolonged exposure to stimuli. In a perceptual discrimination task subjects are instructed to make judgments according to particular stimulus properties; e.g. assessing whether two consecutive tones differ in pitch. Note that even if the subject is given the correct answer after each trial, the task is not a supervised learning one. The problem faced is not about generalizing the decision criterion that one is explicitly given, but in acquiring an internal response to the stimuli (or a perception) sufficiently detailed to be able to apply that criterion. It is, thus, a *representational* problem. For a wide range of tasks, discrimination performance improves after training, evidencing perceptual learning.

In nature, unsupervised learning problems are embedded in higher-level problems. For instance, in learning a second language the production problem is subsidiary to the perception one. For example, a non-native French speaker will not be able to produce the different French vocalic sounds until he/she can separately perceive them (admittedly, learning to differentiate new phonemic sounds is particularly difficult for adults, even when explicitly taught). Indeed, native speakers acquired their phonemic repertoire by the mere repeated exposure to the target sounds during maturation, mostly by engaging in conversation with their caregivers in infancy (who not necessarily correct the children's mispronunciations). Again, one can appreciate the analogy between unsupervised learning and data compression: a huge variety of utterances produced by different speakers, added to background noise, articulated at different speeds, volumes, and so forth, are summarized in the perception of a particular phoneme.

A tractable approach to quantify the substrate of perceptual learning in animals is measuring the changes in the receptive fields of neurons in the sensory cortices. For instance, assuming that the tonotopic arrangement of the cells in the primary auditory cortex comes about jointly as a product of evolution and unsupervised learning over an animal's lifetime, such an arrangement can be rapidly altered in a fear conditioning experiment. There, a particular sound frequency, the CS, becomes very salient because of its association with a highly aversive US, e.g. an electric shock. As a result, the number of cells *coding*, or being activated by, the CS in the auditory cortex increases. Put differently, the CS becomes better (or more) represented in the animal's brain after being experienced coupled to the US. That change corresponds to relevance-modulated unsupervised learning (as we introduced earlier) where, in particular, the US dictates the relevance of the CS.

Biological and biohybrid systems

How do the machine learning strategies introduced above relate with the actual computations in the central nervous system? An influential answer was put forward by Doya (1999). He proposed that one should look at different brain structures from a computational perspective rather than from a functional one. He assigned unsupervised learning to the cortex, supervised learning to the cerebellum, and reinforcement learning to the basal ganglia. In that view, asking whether the cerebellum is involved in motor control or cognition is mistaken. Instead, the cerebellum, with its regular structure, provides the rest of the brain with a computational facility, namely a supervised learning algorithm, that can be applied in all the suitable functional contexts. Doya's useful categorization should not be taken word for word though. For instance, even if error-driven learning is the distinctive feature of the cerebellar cortex, that doesn't exclude unsupervised learning processes occurring within the cerebellar structure (e.g. Schweighofer et al. 2001) or to have learning modulated by reward signals in the cerebellum.

Conversely, the necessary role of basal ganglia in the action selection process of reinforcement learning (Gurney et al. 2001) doesn't imply either that neither cortex nor cerebellum have a role in reinforcement learning. Instead, for complex environments, learning a representation of the state space will call upon unsupervised learning from the cortex, and predicting the value of the available actions may call upon the cerebellum.

When it comes to biomimetic systems, thus far brain-based controllers have usually addressed the problem of reproducing a single type of learning with a controller built around a particular brain area. For instance, action selection in robots in a foraging task has been achieved with a controller implementing learning in a model of the basal ganglia (Prescott et al. 2006), and conditioned avoidance with a computational model where learning occurred in the cerebellum (Herreros et al. 2013; McKinstry et al. 2006). Brain-based robot controllers including a comprehensive set of brain areas and able to face a diversity of tasks have not yet been developed. The notable exception of the Spaun computational model (Eliasmith et al. 2012) has the drawback that even though it includes multiple areas and learns to perform a variety of tasks, it lacks embodiment (Maffei et al. 2015).

Future directions

The building of brain-based controlled robots is still in its infancy. Past an initial stage where artifacts were aimed at demonstrating single learning principles or hinged on the computational power of a particular brain structure, now it is time for integrated models where system-level interactions are studied.

Several principles extracted from the brain can guide these new developments. On the one hand, from a top-down perspective, one can originate new systems in cognitive theories that give an integrated account on how the different brain areas interact, such as the Distributed Adaptive Control architecture (see Verschure, Chapter 36, this volume; and Verschure et al. 2014) or the Distributed Processing Modules (Houk et al. 2007). On the other hand, from a bottom-up perspective, following the biological design as close as possible can pay off with unattended ideas even when attempting to model simple behaviors. For instance, brain-based controllers inspired on the cerebellum have since long existed (McKinstry et al. 2006). However, it hasn't been until the implementation in a computation model of the well-documented cerebellar nucleo-olivary inhibition, a negative feedback connection whose function was difficult to interpret from a motor control perspective (Lepora et al. 2010), that it has become clear that rather than replacing reactive control by anticipatory control, the spino-cerebellar circuitry has the potential for blending both types of control (Herreros and Verschure, 2013). In summary, when implementing brain-based controllers one should go beyond cherry-picking features from the biology just to make the model work, that is, ignoring a priori problematic knowledge. There is more mileage in understanding the benefits hiding in the apparent restrictions that the biological design imposes than in just putting together loosely bio-inspired components to produce a behavior that could have been more easily met with standard engineering methods in the robotics field.

Moreover, even though in the last few years we have begun to witness sci-fi-like robot performances, recent developments in the field of robotics will necessitate a change of current control paradigms. Indeed, robots such as Big Dog (Boston Dynamics) and its descendants display heavy and rigid bodies, whose control, though complex and flexible, still results from pure feedback strategies. However, to enable safe human interaction and energy efficiency, the next generation of robots will need to be made of soft and light compliant materials. In that case, achieving accurate behavior while maintaining low impedance will only be possible via incorporating feedfoward controllers (Della Santina et al. 2017) whose complexity will be such that it would make their a priori pre-specification unfeasible. Practitioners should then very likely incorporate (machine) learning in the process of the control design. Robotics will soon find itself in a situation, both in terms of compliancy properties and complexity of the control task, that resembles the one of vertebrate motor control systems. Hence, in the near future one should expect brain-based schemes (where both learning and control are interwined) to be of great value for solving practical problems in robotics.

General abstract theories of brain function can also inspire new developments. For instance, the Free Energy Principle (FEP) account of brain function (Friston et al. 2010) and the action-oriented predictive coding (AOPC) approach (Clark 2013) refer to Bayesian inference, a technique that can be applied to optimal control and reinforcement learning. However, both FEP and AOPC still have to bridge the gap between simulations and the control of real devices. At this point, and given the difficulty of implementing the whole Bayesian inference machinery in a brain-based controller, a good strategy for testing the utility of the predictive coding approach is to focus on specific assumptions of that theory. In this vein, recent work has shown that implementing an anticipatory postural adjustment as the reaction to a chain of sensory predictions (from distal to proximal to proprioceptive) yields more robust a behavior than setting it up as a combination of parallel sensory–motor associations driven by the distinct type of stimuli (Maffei et al. 2014, 2017). Note that in this case, the insight we get from nature is not how to implement individual learning algorithms, but how to wire them together in adaptive systems.

Finally, returning to the Bayesian approach, two of the big questions in computational neuroscience are how neurons code uncertainty and how it affects computations. In the domain of living machines this translates into how brain-based controllers can contextually weight perception or action execution according to their variable expectations. This is largely uncharted territory and will be a research direction that will change qualitatively the adaptability and robustness of the behavior of synthetic living machines.

Learning more

This chapter avoided using mathematical notation to introduce both the control and the learning problem. The intention was showing that a set of basic notions in both disciples can be intuitively understood without resorting to mathematical formalisms; and that such an understanding can be readily applied to identify the parallels between the issues studied in both disciplines and the basic elements underlying animal learning and motor control. However this is probably as far as one can reach without mathematics in this field. A deeper understanding of the control theory and machine learning methods, opening the possibility of applying these concepts in real-world artifacts, requires a background in algebra, statistics, and basic calculus.

Assuming such a background, a good way to approach control theory is by the study of linear dynamical systems. For this, an excellent resource is the series of video lectures by Professor Boyd (Boyd 2008). For control theory itself, Astrom and Murray (2012) is a freely available textbook with a modern approach, mostly focused in the state-space formulation of dynamical systems.

Regarding machine learning, two excellent resources are Bishop (2006) and Mackay (2003). Both follow a Bayesian and information theoretic approach to the field and cover basic topics, such as regression techniques, as well as advanced methods, like the expectation maximization algorithm and Kalman filtering. Both textbooks exist as hardcover, but there are free electronic versions on their respective authors' homepages. The best introduction to reinforcement learning is still the textbook by Sutton and Barto (1998).

Finally, Wolpert et al. (2011) offer an excellent review of the control and learning principles underlying motor behavior, with a focus on experimental data from human motor learning studies.

Acknowledgments

Preparation of this chapter was supported by grant from the European Research Council under the European Union's Seventh Framework Programme (FP7/2007–2013) / ERC grant agreement no. 341196.

References

Astrom, K. J., and Murray, R. M. (2012). *Feedback Systems: An Introduction for Scientists and Engineers.* Princeton: Princeton University Press.

Bishop, C. (2006). *Pattern recognition and machine learning.* New York: Springer. Retrieved from http://soic.iupui.edu/syllabi/semesters/4142/INFO_B529_Liu_s.pdf

Boyd, S. (2008). Introduction to linear dynamical systems. Retrieved from https://see.stanford.edu/Course/EE263

Clark, A. (2013). Whatever next? Predictive brains, situated agents, and the future of cognitive science. *Behavioral and Brain Sciences,* 36(3), 181–204. doi:10.1017/S0140525X12000477

Della Santina, C., Bianchi, M., Grioli, G., Angelini, F., Catalano, M., Garabini, M., and Bicchi, A. (2017). Controlling soft robots. *IEEE Robotics & Automation Magazine,* 24(3). doi: 10.1109/MRA.2016.2636360

Doya, K. (1999). What are the computations of the cerebellum, the basal ganglia and the cerebral cortex? *Neural Networks*, **12**(7–8), 961–74. doi:10.1016/S0893-6080(99)00046-5

Eliasmith, C., Stewart, T. C., Choo, X., Bekolay, T., Dewolf, T., Tang, Y., and Rasmussen, D. (2012). A large-scale model of the functioning brain. *Science*, **338**, 1202–05.

Friston, K. J., Daunizeau, J., Kilner, J., and Kiebel, S. J. (2010). Action and behavior: A free-energy formulation. *Biological Cybernetics*, **102**(3), 227–60. doi:10.1007/s00422-010-0364-z

Goldsmith, J. (2001). Unsupervised learning of the morphology of a natural language. *Computational Linguistics*, **27**(2), 153–98. doi:10.1162/089120101750300490

Gurney, K., Prescott, T. J., and Redgrave, P. (2001). A computational model of action selection in the basal ganglia. I. A new functional anatomy. *Biological Cybernetics*, **84**(6), 401–10. doi:10.1007/PL00007984

Herreros, I., Maffei, G., Brandi, S., Sanchez-Fibla, M., and Verschure, P. F. M. J. (2013). Speed generalization capabilities of a cerebellar model on a rapid navigation task. In: N. Amato et al. (eds), *2013 IEEE/RSJ International Conference on Intelligent Robots and Systems*. IEEE, pp. 363–8. doi:10.1109/IROS.2013.6696377

Herreros, I., and Verschure, P. F. M. J. (2013). Nucleo-olivary inhibition balances the interaction between the reactive and adaptive layers in motor control. *Neural Networks*, **47**, 64–71. doi:10.1016/j.neunet.2013.01.026

Houk, J. C., Bastianen, C., Fansler, D., Fishbach, A., Fraser, D., Reber, P. J., … Simo, L. S. (2007). Action selection and refinement in subcortical loops through basal ganglia and cerebellum. *Philosophical Transactions of the Royal Society of London. Series B, Biological Sciences*, **362**(1485), 1573–83. doi:10.1098/rstb.2007.2063

Lepora, N. F., Porrill, J., Yeo, C. H., and Dean, P. (2010). Sensory prediction or motor control? Application of marr-albus type models of cerebellar function to classical conditioning. *Frontiers in Computational Neuroscience*, **4**(October), 140. doi:10.3389/fncom.2010.00140

Mackay, D. J. C. (2003). Information theory, inference, and learning algorithms. *Learning*, **22**(3), 348–49. doi:10.1017/S026357470426043X

Maffei, G., Sánchez-Fibla, M., Herreros, I., and Verschure, P. F. M. J. (2014). The role of a cerebellum-driven perceptual prediction within a robotic postural task. In: A.P. del Pobil, E. Chinellato, E. Martínez-Martín, J. Hallam, E. Cervera, and A. Morales (eds), *From Animals to Animats 13*, 13th International Conference on Simulation of Adaptive Behavior, SAB 2014, Castellón, Spain, July 22-25, 2014, Proceedings. (pp. 76–87). Basel: Springer.

Maffei, G., Santos-Pata, D., Marcos, E., Sánchez-Fibla, M., and Verschure, P. F. (2015). An embodied biologically constrained model of foraging: from classical and operant conditioning to adaptive real-world behavior in DAC-X. *Neural Networks*, **72**, 88–108.

Maffei, G., Herreros, I., Sanchez-Fibla, M., Friston, K.J., and Verschure, P. F. M. J. (2017). The perceptual shaping of anticipatory actions. *Proc. R. Soc. B*, **1780**.

McKinstry, J. L., Edelman, G. M., and Krichmar, J. L. (2006). A cerebellar model for predictive motor control tested in a brain-based device. *Proc. Natl Acad. Sci. USA*, **103**(9), 3387–92. doi:10.1073/pnas.0511281103

Nashner, L. M. (1976). Adapting reflexes controlling the human posture. *Experimental Brain Research. Experimentelle Hirnforschung. Experimentation Cerebrale*, **26**(1), 59–72. doi:10.1007/BF00235249

Prescott, T. J., Montes González, F. M., Gurney, K., Humphries, M. D., and Redgrave, P. (2006). A robot model of the basal ganglia: behavior and intrinsic processing. *Neural Networks*, **19**(1), 31–61. doi:10.1016/j.neunet.2005.06.049

Rescorla, R. A. (1988). Pavlovian conditioning. It's not what you think it is. *The American Psychologist*, **43**(3), 151–60. doi:10.1037/0003-066X.43.3.151

Schouenborg, J., and Weng, H. R. (1994). Sensorimotor transformation in a spinal motor system. *Experimental Brain Research. Experimentelle Hirnforschung. Experimentation Cerebrale*, **100**(1), 170–4. doi:10.1007/BF00227291

Schweighofer, N., Doya, K., and Lay, F. (2001). Unsupervised learning of granule cell sparse codes enhances cerebellar adaptive control. *Neuroscience*, **103**(1), 35–50.

Silver, D., Huang, A., Maddison, C. J., Guez, A., Sifre, L., Van Den Driessche, G., ... and Dieleman, S. (2016). Mastering the game of Go with deep neural networks and tree search. *Nature*, **529**(7587), 484–9.

Sutton, R. S., and Barto, A. G. (1998). *Introduction to reinforcement learning* (Vol. **135**). Cambridge, MA: MIT Press.

Verschure, P. F. M. J., Pennartz, C., and Pezzulo, G. (2014). The why, what, where, when and how of goal-directed choice: neuronal and computational principles. *Philosophical Transactions of the Royal Society*, **369**(September). Retrieved from http://rstb.royalsocietypublishing.org/content/369/1655/20130483.short

Verschure, P. F. M. J., Voegtlin, T., and Douglas, R. J. (2003). Environmentally mediated synergy between perception and behaviour in mobile robots. *Nature*, **425**(6958), 620–24. doi:10.1038/nature02024

Wolpert, D. M., Diedrichsen, J., and Flanagan, J. R. (2011). Principles of sensorimotor learning. *Nature Reviews. Neuroscience*, **12**(12), 739–51. doi:10.1038/nrn3112

Chapter 27

Attention and orienting

Ben Mitchinson

Department of Psychology, University of Sheffield, UK

"Everyone knows what attention is. It is the taking possession by the mind, in clear and vivid form, of one out of what seem several simultaneously possible objects or trains of thought."
William James, Principles of Psychology, 1890

These words of William James serve to confirm our lay understanding of the term "attention": that it is a "focusing" function of the mind. This understanding is exemplified by a teacher's order to "pay attention" in class, the intention being that the students focus their minds on the lesson. However, the teacher does not need to be a mind-reader to identify the children who are paying attention—simply checking which ones are looking at the board will do. To give another example: a golfer aiming to improve their game might be advised to "keep their eye on the ball". It goes without saying that they must focus their brain and their golf club on the ball as well. These everyday examples suggest a simplified, pragmatic model of biological attention: that it is that function of the brain that focuses sensory, motor, and processing resources *in concert* and *sequentially* onto *spatial* targets through *physical* orienting. These principles are intuitive—any act that any agent performs upon a physical object will require the focusing of multiple resources upon the location in space occupied by the object. Thus, attention-driven orienting both reveals the mind's focus of attention and underpins much of animal behavior. This article looks at the scope and implications of these biological principles, discusses a range of biomimetic systems that take advantage of them, and considers how future artificial systems will be shaped by them. In the final section, we direct the interested reader to sources of more detailed information.

Biological principles

Definitions of attention and orienting

William James's remark that "Everyone knows what attention is" belies the fact that the study of attention today is vivacious, diverse, cross-disciplinary, and covers a very broad set of ideas so that there is no single universally agreed definition of attention (Tsotsos et al. 2005). The remainder of James's words point to another bias in our understanding of attention: that it is a function of the mind. This understanding is reflected in the thousands of publications each year in psychology that use manipulation of attention as a window into mental processing. The majority of psychology research papers study *covert* orienting, which refers to internal routing of information so that some things are processed more than others (it is this to which James's words refer).

However, like the teacher and the golfer in the examples above, in everyday life we mostly observe the expression of attention through *overt* orienting (in animals, as well as humans). Discrete targets of attention are usually delineable in space, and attention is "paid" most concretely by the "orienting" of sensory resources to a target location. More than any internal mechanism, the "physical switch" of overt sensory orienting controls what is available to be processed from moment to moment in the brain, so that "neural orienting" usually follows sensory orienting. Furthermore, since interactions are usually made with the object of attention (allowing sensorimotor loop closure), motor resources follow, so that all classes of resource are oriented in concert. This principle is expressed also morphologically: for example, the tactile whiskers relied upon by many mammals are arrayed around the mouth, a key motor manipulator, so that both are always directed towards the same region of space.

Overt orienting, then, is absolutely central to behavior, and—correspondingly—it is observed across all classes of vertebrate, as well as insects (Bernays and Wcislo 1994). In simpler animals that lack mobile sensory organs it can be expressed by turning the whole body (Collett 1988). For some senses—touch is one example—orienting actually requires locomotion and approach (and even remote sensors perform better at shorter range). Thus, orienting—in some form—is a core component of the behavior of even the simplest of animals. For the current generation of Living Machines, it is this physically expressed orienting, and its control through the management of attention, that is the most relevant. Therefore, in this article, we use a simplified definition of attention and orienting: attention is that function of the brain, *ubiquitous* amongst behaviors and across the animal kingdom, that focuses all resources (sensory, motor, and processing) *in concert* and *sequentially* onto *spatial* targets through *physical* orienting.

Orienting is acting

Its role in resource assignment means that attention is closely tied to action, beyond orienting. This relationship is brought out most clearly in animals that rely on short-range sensors, such as rats. Rats explore a "tactile scene" by moving their snout from place to place, supporting these movements, as necessary, by locomotion of the body. Taken together, such a series of orients describes what we might call the medium-scale behavior of the animal, with local behaviors (such as sniffing, tasting, or consuming) taking place against this backdrop. During such exploration, a list of spatial targets of attention would be enough to define quite well the complete behavior of the animal at a medium scale.

Action selection is a basic requirement for an animal interacting with its environment (see Lepora, Chapter 28, this volume). It requires that an object in the environment be chosen to act upon, and an action be chosen to perform upon it. Owing to the way resources follow the focus of attention, attention management is very closely related to the selection of an object of any action. However, the relationship between attention and action selection is closer still because most objects offer only one or a very limited set of affordances (there is really only one thing that a giraffe can do to an acacia tree).

Thus, attention and action are intertwined by the twin facts that (i) much of medium-scale behavior can be understood as "approach to attend" and (ii) because the object of attention is the object of action. We can summarize this principle as *orienting is acting*.

Orienting in social interaction

The centrality of attention to behavior means that the focus of the mind is very useful in predicting the behavior of the animal. But the converse holds also, a principle which we might call *reciprocity*: one of our clearest insights into an animal's mind is its orienting behavior. We can judge the quality of orienting behavior—at the simplest, its presence or absence—but we can gain more information from the target of orienting. As the teacher can judge where the attention of the pupils lies by watching how they orient, so orienting plays a key social role—for instance, in conversation. Overt orienting is a strong signal of attention, and plays a key role in managing social interactions.

Models of attention

But how are spatial targets chosen? Again, the psychology literature suggests considerable complexity. However, simplified models have been developed that offer robust components of the answer. A particularly mechanistic (and, thus, particularly relevant to this discussion) class of models emphasize the concept of spatial "saliency," that quality of a stimulus in its environment that marks it as worthy of attention. The computation of saliency integrates "bottom-up" influences (from external sensory stimuli) and "top-down" influences (from internal state). These models usually focus on the visual system, but are more widely applicable.

In a concrete model of visual search within an image (Itti et al. 1998; see Figure 27.1), regions that are different from their surroundings (in features such as intensity, color, edge orientation) have elevated saliency. Bottom-up contributions from each of these features are integrated into a single "saliency map," from which the highest peak is chosen as the next attended location by a "winner-take-all" function that corresponds to there being only a single focus of attention at any one time. A function known as "inhibition-of-return" then modulates the map so that the next most salient location is attended next, and so on, the result being an ordered sequential search through the salient locations. An important feature of this type of model is that the bottom-up computations are simple and fixed and can be performed across all locations in parallel.

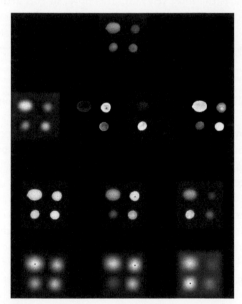

Figure 27.1 Simplified illustration of bottom-up top-down visual attention model (Frintrop et al. 2005). (Top) Four bottom-up "features" are extracted from the image: "fruit" (round luminance pattern), "red," "green," "yellow". (Bottom) Three different sets of weights, dependent on an imagined top-down modulation, combine these features into an additive saliency map, which is then low-pass filtered, and the peak (black dot) chosen as the focus of attention. (Left) No modulation—all features equally weighted; lemon is chosen as the largest/brightest object. (Middle/right) The tomato/lime is selected for attention when only the red/green color feature has non-zero modulated weight.

Reproduced from Simone Frintrop, Gerriet Backer, and Erich Rome, 'Goal-Directed Search with a Top-Down Modulated Computational Attention System', in Walter G. Kropatsch, Robert Sablatnig, and Allan Hanbury (ed.), *Pattern Recognition: 27th DAGM Symposium, Vienna, Austria, August 31—September 2, 2005. Proceedings*, pp. 117–24, Copyright © 2005, Springer-Verlag Berlin Heidelberg.

Models of top-down modulation affect the computation of the overall saliency map, enhancing the saliency attached to certain features, combination of features, or spatial regions, allowing what is known as "guided search". One way to implement this is to adjust the weights associated with different feature maps of a bottom-up model (Frintrop et al. 2005). For instance, during a search for a particular fruit, the weighting of the feature map for the appropriate color might be elevated (see Figure 27.1).

Attention can also be directed solely by top-down influences ("goal-driven" attention) whereby the attention is directed to a location known to correspond to a goal. In a complete system, all of these functions may be active somewhat in parallel. For instance, if a bird flies in the window whilst we are focusing on the television weather forecast, the bird will nonetheless grab our immediate attention owing to its high bottom-up salience. Thus, these simple models allow that an agent focus on a particular task, whilst remaining alert to unexpected salient events.

Biomimetic systems

What can these biological models offer to designers of artificial systems? First, the principles distilled above apply to robots as well as animals: to a first approximation, an agent can only do one thing at once, and most behaviors begin with attending to a spatial region. Thus, a system that focuses all of a robot's resources on a single spatial target both concretizes an abstract design constraint and offers a useful component of the behavioral control loop. This architecture remains relevant and useful however abstract target selection becomes. Meanwhile, the corollaries that orienting constitutes behavior and that orienting is a key social signal hint at the breadth of functional roles that attention management can play. Finally, biological models (particularly from the visual system) offer processing architectures that can be realized usefully and efficiently. Thus, the centrality of attention to animal behavior is rapidly becoming reflected in the increasing use of attention management and orienting as central components of autonomous system designs (as well as in the hundreds of patents that have been filed describing techniques for generating saliency maps).

Attention management

In purely algorithmic terms, saliency-based models such as those due to Itti et al. (1998) and Frintrop et al. (2005) can greatly improve the computational efficiency of image search versus a brute force approach by allowing computational effort to be focused on promising image regions. Accordingly, a variety of systems are now starting to employ models of this type to add a degree of autonomy. One example is the adaptive attention-driven pan/zoom security camera model of Davis et al. (2007), which chooses human-like motion characteristics to drive saliency in a bottom-up fashion, and incorporates an inhibition-of-return-like mechanism, maximizing the opportunity for observing future human activity, and making best use of limited recording bandwidth by storing only salient data. (Such an application can equally be viewed as performing data compression.) Orienting (both of the sensors and, thereby, of computational resources) to salient stimuli is also being used to support the operation of, and also the unsupervised training of, object recognition systems that hinge on object localization, and are so central to interactive robotic systems (for example, iCub, Figure 27.2). In a similar vein, landmark identification for robot localization (including as part of Simultaneous Localization And Mapping or SLAM) becomes much more tractable if a simple attention system picks candidates for further processing. Other recent applications proposed for attention-driven focused processing include safety-critical tasks such as emergency management in hazardous industrial environments and search and rescue. Several groups (for example, Hülse et al. 2011) have presented systems that follow

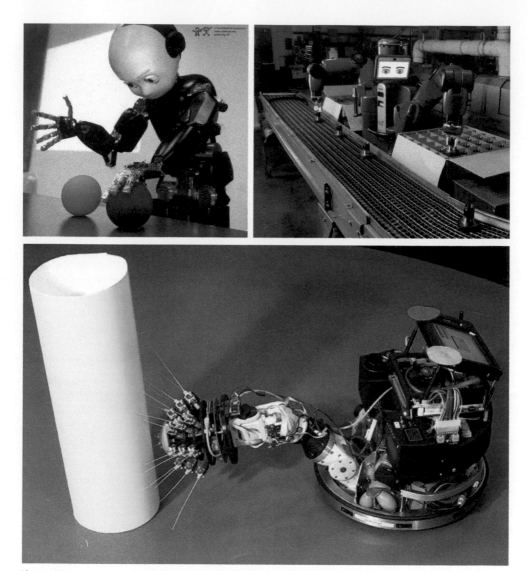

Figure 27.2 Three robots that use orienting behavior. (Top left) iCub orienting eyes, processing, and hand in concert. (Below) Shrewbot performing tactile orienting to an object. (Top right) Baxter industrial robot uses visual orienting to indicate its focus of spatial attention, helping its interactions with human workers.

(a) © iCub Facility, IIT, 2017. (b) and (c) © Rethink Robotics™, 2013.

visual attention with reaching, thus orienting all resources in concert to a single spatial location and underlining the relationship between orienting and action selection.

At the implementation level, the inherent parallelism of saliency-based models has inspired dedicated analog, mixed-mode, digital, and event-based hardware designs (in both VLSI and FPGA), achieving significantly improved computational efficiency versus general purpose serial hardware, and offering power-efficient and real-time saliency map computation. Such hardware systems can be crafted even more efficiently by focusing processing effort on a fovea-like region in the central visual region if it can be assumed that the camera will be pointing directly at an object of interest. Offloading full-field visual computations to low-power dedicated hardware in this way is helping to make attention management accessible even on power-limited mobile platforms.

Orienting as acting

Integrating the orienting of mobile sensors with locomotion is an emerging field of research. Shubina and Tsotsos (2010) created a parallel system of attention at the map level (to complement the visual field level) to drive efficient exploration in a mobile robot object search task. In our own recent work with mammal-like mobile robots with tactile whiskers (see Prescott, Chapter 45, this volume) we have transposed visual models of attention into the tactile realm. In these models, saliency is excited by tactile "intensity," whilst top-down modulation tends to inhibit distant locations as well as implementing inhibition-of-return (Mitchinson and Prescott 2013; Prescott et al. 2015). Given the short range of tactile sensors, orienting behavior in these robots brings the tip of the snout to each selected focus of attention. Thus, medium-scale behavior is specified as a series of orients, and these models generate rodent-like exploratory behavior as the robot (Shrewbot, see Figure 27.2) investigates an experimental arena (Pearson et al. 2013, and references therein), offering a first example of a mobile system completely driven by (tactile) attention. It is easy to imagine that the integration of local behavior (object recognition, manipulation, say) onto such platforms could offer complete and practically useful systems (resource harvesting, for instance). These studies, driven as they were by the necessities of mobile robotics, provide alternative—though not exclusive—hypotheses as to how orienting and locomotion may be integrated in biological systems. Not only that, but they highlight that from an attention system left to itself, with some function that inhibits the saliency of already-visited locations, the particularly useful behavior of exploration (or search) emerges naturally.

Orienting in social interaction

One of the earliest examples of an attentive social robot was Kismet, developed at MIT in the late 1990s, which displayed surprisingly effective interactions with humans, underpinned by physical orienting of its head and eyes to indicate its attention either to an object or to a human. When two agents attend the same location it is known as "joint attention"; this behavior is both functionally, and to an extent socially, required for some communication or cooperation tasks. Human–robot interaction (HRI) is now a rapidly expanding field focusing on the design of robots to facilitate interaction. A swathe of studies has shown the very significant impact that the indication (or, at least, simulation) of attention—usually through visual orienting—has on the quality of human–robot interactions. One example is the report of Mutlu et al. (2009) that a robot involved in a conversation group was able to effectively manipulate, using only gaze control, the roles human participants played in the conversation, their feelings of being involved in a group, and their feelings towards the robot participant, including generating discomfort in human participants through appearing to ignore them. Accordingly, production systems that are intended to interact with humans are increasingly being equipped with mobile eyes, heads, or both. One example is the industrial robot Baxter (Rethink Robotics™, Figure 27.2), which uses physical orienting movements of its "head" as well as digitally generated "eyes" on a screen to indicate where its attention is focused. Another is the service/pedagogical robot Tico (Adele Robots), which indicates attention with mobile head and eyes, encouraging interaction from customers and children alike. The Huggable™ robot under development at MIT, that orients to touch, will join an expanding set of "companion robots" (such as the Paro robot seal) that are already being deployed in patient emotional care.

Future directions

Much effort in robotics research focuses on mapping and navigation—essentially, "how do we get to where we're going?"—and as a result fairly robust practical solutions to these large-scale problems already exist. By contrast, robots remain relatively poor at medium- and local-scale behaviors, and it is in this area that biomimetic solutions may be able to push technology forward most effectively. As we have seen, it is relatively straightforward to take lessons from biology and construct attention systems that guide medium-scale behavior. In future, as exemplified by current work with reaching and manipulating in research robots, the role of attention in local behaviors such as interaction and manipulation will become increasingly important. A further challenge will be the smooth and elegant integration of behavior across these scales. The requirement that practical systems perform integration of this sort means that robot designers cannot compartmentalize problems in the way that biologists can, and it seems likely at this stage that models of integration will come first from biomimetics.

Within that integrated context, and as in the example above, truly flexible and effective biomimetic agents need to be able to prioritize tasks as and when they arise. Determining how to switch effectively between goal-directed behavior and reactive behavior (and when not to) is something that animals do robustly and elegantly. Attention management systems are likely to play a key role in these more flexible designs, though more sophisticated management of attention will be required. Furthermore, and beyond the somewhat simplified model discussed here, some tasks within themselves require switching and/or sharing of attention between multiple targets, with perhaps some resources being directed to one and some to another. One example is legged locomotion, where attention may need to be constantly switched between the placement of the feet and planning the path ahead, not to mention a companion. For such a task, an attention system is perhaps the key part of any control system, governing both the collection of sensory data and the deployment of computing resources.

Learning more

This article has only skimmed the surface of the literature on attention and orienting. Whilst much of this literature is likely to be of limited relevance to contemporary autonomous systems, Ward's excellent article at Scholarpedia (2008) provides an interesting and succinct roadmap of the broader field. Several of the other Scholarpedia articles referenced therein are of more direct relevance, discussing in detail matters we have only touched on. After that, I highly recommend the article by Frintrop et al. (2010) which provides a comprehensive and highly readable survey of contemporary models of visual attention. For a great deal more detail, Tsotsos et al. (2005) provide a brief history of thinking on the subject of attention as part of the introductory material to the collection *Neurobiology of Attention* (ed. Itti, Rees, Tsotsos 2003), which consists of a large number of brief articles acting as jumping-off points for a wide range of topics in the field (including computational topics). Also useful are the collection *Attention and Orienting* (Lang et al. 1997) which focuses on processing of sensory stimuli and the modulatory role on attention of motivation, and *Orienting of Attention* (Wright and Ward 2008) which provides a detailed review of the literature on covert orienting. The reader interested in history may also like to dip into Pavlov's reports from 1927 of his famous experiments with dogs which includes one of the first descriptions of overt sensory orienting (which he described as the "investigatory reflex"); the text of his book is currently available on the web courtesy of Christopher D. Green, York University, Toronto, at http://psychclassics.yorku.ca/Pavlov. Several books on the subject of biomimetic robotics are now available; however, this field is undergoing rapid development, so the most up-to-date information will be found at robotics conferences (such as ICRA, IROS, and RSS) and at biomimetics conferences (such as Living Machines and ISAB). The recently inaugurated *Journal of Human–Robot Interaction* will prove a useful starting point for the field of HRI.

Acknowledgments

This work was supported by the EU Framework 7 project EFAA (ICT-270490).

References

Bernays, E. A., and Wcislo, W. T. (1994). Sensory capabilities, information processing, and resource specialization. *Quarterly Review of Biology*, **69**(2), 187–204.

Collett, T. S. (1988). How ladybirds approach nearby stalks: a study of visual selectivity and attention. *Journal of Comparative Physiology A*, **163**(3), 355–63.

Davis, J. W., Morison, A. M., and Woods, D. D. (2007). An adaptive focus-of-attention model for video surveillance and monitoring. *Machine Vision and Applications*, **18**(1), 41–64.

Frintrop, S., Backer, G., and Rome, E. (2005). Goal-directed search with a top-down modulated computational attention system. In: W. G. Kropatsch, R. Sablatnig, and A. Hanbury (eds), *Pattern Recognition: 27th DAGM Symposium, Vienna, Austria, August 31—September 2, 2005. Proceedings.* Berlin/Heidelberg: Springer-Verlag, pp. 117–24.

Frintrop, S., Rome, E., and Christensen, H. I. (2010). Computational visual attention systems and their cognitive foundations: A survey. *ACM Transactions on Applied Perception (TAP)*, **7**(1), 6.

Hülse, M., McBride, S., and Lee, M. (2011). Developmental robotics architecture for active vision and reaching. In: *Development and Learning (ICDL), 2011 IEEE International Conference on* (Vol. 2). IEEE, pp. 1–6.

Itti, L., Koch, C., and Niebur, E. A. (1998). Model of saliency-based visual attention for rapid scene analysis. *IEEE Trans. on PAMI*, **20**(11), 1254–9.

Lang, P. J., Simons, R. F., and Balaban, M. T. (eds) (1997). *Attention and orienting: Sensory and motivational processes.* Hillsdale, NJ: Lawrence Erlbaum Associates.

Mitchinson, B., and Prescott, T. J. (2013). Whisker movements reveal spatial attention: a unified computational model of active sensing control in the rat. *PLoS Comput. Biol.*, **9**(9), e1003236. doi:10.1371/journal.pcbi.1003236

Mutlu, B., Shiwa, T., Kanda, T., Ishiguro, H., and Hagita, N. (2009). Footing in human-robot conversations: how robots might shape participant roles using gaze cues. In: *Proceedings of the 4th ACM/IEEE international conference on human robot interaction.* IEEE, pp. 61–68.

Pearson, M. J., Fox, C., Sullivan, J. C., Prescott, T. J., Pipe, T., and Mitchinson, B. (2013). Simultaneous localisation and mapping on a multi-degree of freedom biomimetic whiskered robot. In: *IEEE International Conference on Robotics and Automation (ICRA), Karlsruhe, 6–10th May.* IEEE.

Prescott, T. J., Mitchinson, B., Lepora, N. F., Wilson, S. P., Anderson, S. R., Porrill, J., Dean, P., Fox, C. W., Pearson, M. J., Sullivan, J. C., and Pipe, A. G. (2015). The robot vibrissal system: understanding mammalian sensorimotor co-ordination through biomimetics. In: P. Krieger and A. Groh (eds), *Sensorimotor Integration in the Whisker System.* New York: Springer, pp. 213–240.

Shubina, K., and Tsotsos, J. K. (2010). Visual search for an object in a 3D environment using a mobile robot. *Computer Vision and Image Understanding*, **114**(5), 535–47.

Tsotsos, J. K., Itti, L., and Rees, G. (2005). A brief and selective history of attention. In: L. Itti, G. Rees, and J. K. Tsotsos (eds), Neurobiology of attention. San Diego, CA: Elsevier Academic Press, p. i.

Ward, L. M. (2008). Attention. *Scholarpedia*, **3**(10), 1538.

Wright, R. D., and Ward, L. M. (2008). *Orienting of attention.* New York: Oxford University Press.

Chapter 28

Decision making

Nathan F. Lepora

Department of Engineering Mathematics and Bristol Robotics Laboratory,
University of Bristol, UK

In this chapter, we describe recent progress in understanding how living organisms make decisions and the implications for engineering artificial systems with autonomous decision-making capabilities. Nature appears to re-use common design principles for decision making across a hierarchy of organizational levels, from microscopic single cells to the brains of animals to entire populations. One common principle is that decision formation is realized by accumulating sensory evidence up to a preset decision threshold. An explanation for this mechanism lies with the statistical technique of sequential analysis, from which this principle follows as a mathematical consequence of optimal decision making under uncertainty. Sequential analysis has applications spanning from cryptography to clinical drug testing that, given the similarity with biological decision making, may be considered as biomimetic methods. Artificial perception based on sequential analysis has also advanced the state of the art in robot capabilities, such as enabling perceptual hyperacuity and robust sensing under uncertainty. Swarms of robots can make collective decisions to solve tasks ranging from exploration and foraging to aggregation and group coordination. Future applications could lead to individual robots or artificial swarms that perceive and interact upon complex environments with the remarkable ease and robustness now achievable only by living organisms.

Biological principles

Decision making is the process whereby alternatives are considered and then chosen based on the values and goals of the decision maker. Therefore, decision making underlies the operation of any biological organism, since all organisms interact with their environment and any particular course of interaction is but one of many possible actions. In animals, the process of decision making is enabled by perception, in which sensory information about an animal's environment is represented within its nervous system. Perception then results in action: information processing in the nervous system drives motor commands to muscles that produce movement. The outcomes of these perceptual decisions are apparent in the overt actions that an animal makes in conjunction with its environment, constituting the animal's behaviour.

There are many categories of decision and ways of making choices. The simplest types of decisions are binary choices: yes/no, either/or, stop/go, do/don't; for these choices, the decisions are based on weighing the pros and cons of the two possible outcomes. More complex decisions involve multiple alternatives, with the choice based on how well each alternative compares with criteria related to the value and goals of the decision maker. In principle, a random choice between alternatives is a form of decision making, but is generally not a useful method when the outcomes have differing consequences. Therefore, most decisions are made on the basis of evidence. In animals, perception can thus be considered as a process for gathering evidence, with the drives and needs of the animal determining how this perceptual evidence is used to make a decision.

Living organisms form a hierarchy of decision makers (see Figure 28.1). The most basic functional unit for a living decision maker is a single cell, whether as a unicellular entity or as part of a multicellular organism. Neurons of the central nervous system are a notable example of cells involved in decision making. At the next level of the hierarchy are multicellular organisms, including animals. Human decision making is of particular interest, as studied by the discipline of psychology. At the top of the hierarchy, decision making can also be considered at a social level, by which populations or groups of living organisms collectively make decisions.

We now give a more detailed description of decision making at these three levels of hierarchy: cellular, individual, and collective decision making.

Cellular decision making is ubiquitous at the level of the single cell. Cells sense external stimuli via receptors in the cell membrane to make regulatory changes to their internal environment that affect how they interact with their surroundings. Neurons are cells that have honed these mechanisms

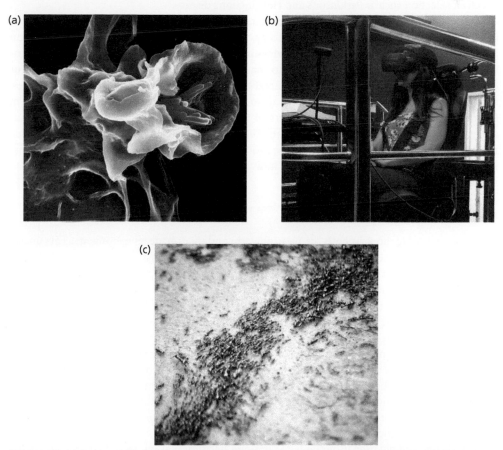

Figure 28.1 (a) Decision making by an immune cell that identifies and engulfs a threat. (b) Human participating in a psychological investigation of decision making. (c) Collective decision making in ants.

(a) © Science Photo Library. (b) Reproduced with permission of Adam Zaidel (personal copyright). See https://zaidel.wixsite.com/zaidel-lab for details of the experiment. (c) Mehmet Karatay/Wikimedia Commons (CC BY-SA 3.0).

to become dedicated signalling and decision-making devices. Receptors called synapses in their branching dendrites trigger electrical signals that converge on their cell body to trigger spikes of electrical activity along output axons that synapse onto other cells. Decision making in animals is evident at the level of a single neuron. For example, the majority of sensory neurons synapse in the spinal cord onto motor neurons to form reflex arcs that control action reflexes. These reflexes can give simple decisions with the choice hardwired by the animal's evolutionary history.

Individual decision making is found in the behavior of animals and other multicellular organisms. Indeed, the central nervous system could be viewed as having evolved for animals to perform complex decisions over multiple alternatives while taking account of past experience and the needs of the individual. Some reflexive decisions are akin to action reflexes but are learnt from repeated exposure to sensory contingencies, known as conditioning. For example, freezing or startling due to conditioned fear is a reflexive reaction to neutral stimuli previously paired with aversive events, and depends critically on a brain structure called the amygdala. Other decisions are more deliberative, involving comparing evidence for multiple alternatives and selecting the most appropriate choice. Many brain regions are involved in forming and mediating such decisions, for example the basal ganglia whose dysfunction is associated with neurological disorders of action selection such as Parkinson's and Huntingdon's disease.

Collective decision making is apparent in humans and other social animals when a group of individuals make a choice together. Humans have developed formal systems for group decision making, including consensus agreement and voting. Social animals also utilize hierarchical dominance relationships between individuals to manage group behaviour, such as alpha males or females that lead the group. Some species, notably insects of the order Hymenoptera (ants, bees, and wasps), have evolved an extreme form of social behaviour called eusociality. This results in a social organization where sterile members of the species carry out specialized tasks, such as defence or resource gathering, to care for the reproductive members of that society. Dedicated methods for collective decision making have evolved in these animals, for example quorum sensing that utilizes the population densities of individuals that each sense competing alternatives to make a decision.

Biomimetic systems

Uncertainty is ubiquitous in both the natural world and man-made environments. This uncertainty can be because sensorial information is generally stochastic or contaminated with undesired signals, but also environments change and can be unstructured in their form and dynamics. Hence, decision making may be viewed as the process of sufficiently reducing uncertainty about the alternatives to allow the appropriate choice to be made. But costs are incurred in any process of reducing uncertainty, since the process of gathering evidence uses motor, sensorial, or computational abilities that could be better utilized for other tasks. Moreover, time delays during decision making may also lead to losing the resource that is the subject of the decision, for example to competitors or because the outcome is time limited. Therefore, the process of reaching a decision should seek to balance these competing objectives: decisions should take adequate time to reduce uncertainty and hence mistakes, but should also be sufficiently quick to not lose or waste precious resources. Formally, the problem is to minimize the costs of the expected outcome, known as optimal decision making.

Historically, computational methods for optimal decision making were first applied during World War II by British code-breakers. Alan Turing and his colleagues at Bletchley Park, an estate in the countryside north of London, developed the first completely programmable digital computer (Colossus) and a crypto-analytic method (Banburismus) to break enciphered German naval messages (see Gold and Shadlen 2002; Figure 28.2(a)). Because this work remained secret

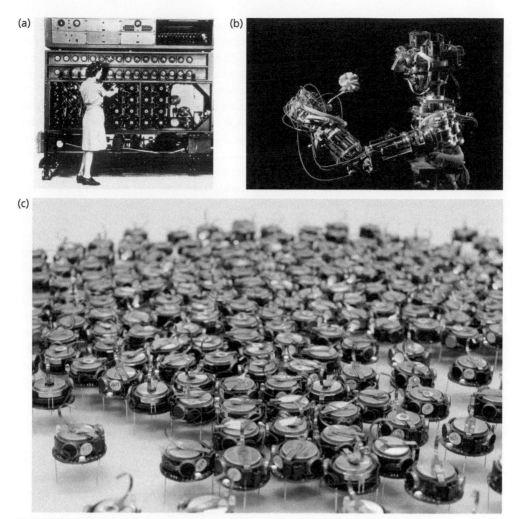

Figure 28.2 (a) Original electronic computer used to break German cyphers in World War II, using methods now recognized as relating to how neurons make decisions. (b) A humanoid robot making a decision about which object to pick up. (c) Robot swarm aggregating into a single collective entity.

(a) © US Air Force Photo. (b) Massimo Brega, The Lighthouse/Science Photo Library. (c) Sabine Hauert, Bristol Robotics Laboratory.

until the 1980s, the earliest published work is by Abraham Wald in 1947 on a statistical technique that he called *sequential analysis*.

A diverse range of studies of biological systems ranging from single cells to individuals to collectives indicates three shared computational principles underlying decision formation in the natural world (e.g. Perkins and Swain 2009, Gold and Shadlen 2007, Sumpter 2006). The first common principle is that sensory information is represented as *evidence* signals for the decision alternatives, for example the rate of protein binding in a signalling pathway of a bacterium, the neuronal activity in the brain of an animal, and the population densities of bees returning from potential nest sites in a swarm. The second principle is that evidence accumulates until reaching a *threshold* that triggers a decision, such as the concentration to switch a gene in a cell, the activation level for a motor neuron, or the quorum number to reach swarm consensus. The last common principle is *competition* between alternatives, in that evidence for one outcome disfavours all others; examples include competition for protein binding between distinct signalling pathways in cells, inhibition between neurons in the brain, and recruitment of ants from one potential nest site to another in a swarm. Remarkably, these three computational principles also underlie the statistical technique of sequential analysis for making optimal decisions under uncertainty.

Similarly to how Banburismus could, in retrospect, be considered as implementing biological principles for decision making, there are a wide range of statistical applications of sequential analysis that could now also be considered biomimetic. Since the 1980s, clinical trials have been designed around a procedure of testing new individuals until reaching a threshold level of evidence to decide whether a drug is effective. Another application of sequential analysis is for industrial quality control on production lines to decide upon a change point in fault detection, with decision threshold based on the desired quality and risk associated with that defect. Financial trading can also employ sequential analysis methods to decide whether to buy or sell stock options, again with the threshold based on the risk of the transaction.

Recently, this approach for optimal decision making has been applied to enabling robots to perceive their environments with animal-like capabilities. A strength of this approach to robot perception is that it connects closely with leading work in neuroscience, allowing insights from animal perception to be transferred to robot perception. For example, these biomimetic methods have enabled the demonstration of hyperacuity in robot touch (Lepora et al. 2015), giving perceptual acuity finer than the sensor resolution, as is common in animal perception (see also Lepora, Chapter 16, and Metta and Cingolani, Chapter 47, this volume). The decision making also appears far more lifelike, such as having broad reaction time distributions resembling those from human psychophysics, rather than the fixed decision times that are conventional in robotics.

In another application of biomimetic decision making, the self-organized decision processes in insect colonies have encouraged the development of swarm robotics (Şahin 2005), a branch of collective robotics that uses biomimetic principles to coordinate groups of robots. The decision formation can involve recruiting individual robots within a group from one course of action to another, such as identifying the location of resources, to proceed analogously to evidence accumulation methods in natural decision making. These biologically based principles have enabled robot swarms to perform various useful tasks, including dispersing individuals over a landscape to forage for resources, aggregating the swarm onto a single optimal site analogously to nesting behaviour, and coordinating the motion of the entire swarm to collectively achieve a goal (Figure 28.2(c); see also Figure 3.3, this volume, and Nolfi, Chapter 43, this volume)).

Future directions

Biomimetic systems provide both a test-bed for theoretical ideas in biology and a means for generating solutions to challenges in science and technology. Both of these aspects are important for the future direction of decision making with artificial devices.

Many key problems in robotics and engineering have the potential to be solved by decision-making capabilities that mimic those of animals. For example, while robots are successful at tasks where they can be rigidly controlled for predictable environments like factory assembly lines, they have failed to make significant impact in unstructured environments like our homes, hospitals, and workplaces. Yet animals are able to perceive and act with apparent ease in the complexity of the natural world, suggesting a key ingredient is missing from the decision-making capabilities of robots. Solving this problem will revolutionize the use of robotics in society, with radical implications for automating where we live and work. Moreover, swarm robotics (see Nolfi, Chapter 43, this volume) also offers an entirely new range of robot capabilities beyond those achievable with present technology. For example, swarms could enable distributed perception across natural and artificial environments, with applications such as monitoring environmental health and deciding upon corrective actions when necessary—in effect allowing an entire environment to form a biohybrid robot system.

By virtue of their mimicry of living systems, biomimetic artifacts also provide excellent models of their natural counterparts, giving complementary ways to formulate and test hypotheses about biological function. As Richard Feynman famously said: "What I cannot create, I do not understand." In this sense, the ultimate test of understanding the brain would be to implement an artificial one in a robot and probe whether it can perceive and interact with complex environments with the remarkable ease and robustness now achievable only by animals.

Learning more

There are many excellent reviews of decision making at the cellular, individual, or collective level, of which for reasons of space we can only offer a limited selection. For cellular decision making, Perkins and Swain's (2009) excellent article on "Strategies for cellular decision making" reviews recent progress in the information-processing strategies that cells use for decision making under uncertainty. For neural and individual decision making, we refer to Gold and Shadlen's (2007) classic review of "The neural basis of decision making" and Bogacz et al.'s (2006) treatment of "The physics of optimal decision making," both of which survey advances in understanding how the basic elements of decision formation are implemented in the brain. A pedagogical and interesting survey of collective decision making behaviour in animals can be found in Sumpter's (2006) review of "The principles of collective animal behaviour" published in *Philosophical Transactions of The Royal Society*.

To learn more about the relationship between theories of decision making in neuroscience and code-breaking from World War II, we refer to another article by Gold and Shadlen (2002) on "Banburismus and the brain: decoding the relationship between sensory stimuli, decisions, and reward," published in the neuroscience journal *Neuron*. This article also makes an influential proposal that the brain controls decision making through neural circuits that calculate the rate of reward.

Finally, information about robot decision making from biological principles is distributed across many research papers, from which we select a few here by their relation to examples in this article. An example of how implementing biologically based decision making in robots can inform about biological perception is covered by the author and colleagues in "Optimal decision making in mammals: insights from a robot study of rodent texture discrimination," (Lepora et al. 2012), with a related paper on "Tactile superresolution and biomimetic hyperacuity" (Lepora et al. 2015) describing how biological principles can be useful for robot perception. Collective decision making in robots based on swarm cognition in insects is treated in "Self-organized aggregation triggers collective decision making in a group of cockroach-like robots" by Garnier and colleagues (Garnier et al. 2009), and the implications for biological experimentation in Krause and colleagues "Interactive robots in experimental biology" (Krause et al. 2011), published in *Trends in Ecology and Evolution*.

References

Bogacz, R., Brown, E., Moehlis, J., Holmes, P., and Cohen, J.D. (2006). The physics of optimal decision making: a formal analysis of models of performance in two-alternative forced-choice tasks. *Psychological Review*, 113(4), 700.

Garnier S., Gautrais J., Asadpour, M., Jost, C., and Theraulaz, G. (2009). Self-organized aggregation triggers collective decision making in a group of cockroach-like robots. *Adaptive Behavior*, 17(2), 109–33.

Gold, J. I., and Shadlen, M. N. (2002). Banburismus and the brain: decoding the relationship between sensory stimuli, decisions, and reward. *Neuron*, 36(2), 299–308.

Gold, J. I., and Shadlen, M. N. (2007). The neural basis of decision making. *Annual Reviews in Neuroscience*, 30, 535–74.

Krause, J., Winfield, A. F., and Deneubourg, J. L. (2011). Interactive robots in experimental biology. *Trends in Ecology & Evolution*, 26(7), 369–75.

Lepora, N. F., Fox, C. W., Evans, M. H., Diamond, M. E., Gurney, K., and Prescott, T. J. (2012). Optimal decision-making in mammals: insights from a robot study of rodent texture discrimination. *Journal of the Royal Society Interface*, 9(72), 1517–28.

Lepora, N. F., Martinez-Hernandez, U., Evans, M. H., Natale, L., Metta, G., and Prescott, T. J. (2015). Tactile superresolution and biomimetic hyperacuity. *IEEE Transactions on Robotics*, 31(3), 605–18.

Perkins, T. J., and Swain, P. S. (2009). Strategies for cellular decision-making. *Molecular Systems Biology*, 5(1), 326. doi: 10.1038/msb.2009.83

Sumpter, D. J. (2006). The principles of collective animal behaviour. *Philosophical Transactions of the Royal Society B: Biological Sciences*, 361(1465), 5–22.

Şahin, E. (2005). Swarm robotics: From sources of inspiration to domains of application. In: Şahin, E., and Spears, W.M. (eds), *Swarm Robotics*. SR 2004. Lecture Notes in Computer Science, vol 3342. Berlin/Heidelberg: Springer, pp. 10–20.

Wald, A. (1947). *Sequential analysis*. Reprinted 1973. Mineola, NY: Dover Books.

Chapter 29

Spatial and episodic memory

Uğur Murat Erdem[1], Nicholas Roy[2], John J. Leonard[2], and Michael E. Hasselmo[3]

[1] Department of Mathematics, North Dakota State University, USA
[2] Computer Science and Artificial Intelligence Laboratory, Massachusetts Institute of Technology, USA
[3] Department of Psychological and Brain Sciences, Center for Systems Neuroscience, Boston University, USA

Biological data inspires the development of navigation mechanisms for autonomous agents assisting humans in the exploration and mapping of hazardous environments. These biologically inspired mechanisms can be used to enhance the simultaneous localization and mapping in cluttered and dynamic environments, transmission of the generated maps to a human operator and the operator's capacity to guide exploration. This review will present two computational models using a network of head direction cells, grid cells, and place cells for autonomous goal-directed navigation. In the first model the agent creates a cognitive map of a novel environment by random exploration. During navigation the agent decides on its next movement direction towards a goal by probing linear look-ahead trajectories in several candidate directions while stationary and picking the one activating place cells representing the goal location. The second model improves the range of linear look-ahead probes significantly by imposing a hierarchical structure on the cognitive map consistent with the experimental findings of differences in the firing field size and spacing of grid cells recorded at different positions along the dorsal to ventral axis of the entorhinal cortex. The new model represents the environment at different scales by populations of simulated hippocampal place cells with changing firing field sizes.

Biological principles

One of the crucial features of many living organisms capable of locomotion is their ability to navigate their daily environment performing life-critical tasks. For instance, squirrels are surprisingly good at rediscovering locations of food they previously buried and rats can learn to revisit or to avoid previously visited food locations. Many animals retreat to a previously visited shelter in the presence of an immediate threat or of a long-term threat like a bear retreating to a cave for hibernation to conserve energy during harsh seasons. It is a plausible assumption that for the living organisms to perform such navigation tasks they should possess a cognitive mechanism to represent their environment as a collection of critical regions, e.g. nest locations, food locations, etc., and to recall these regions when the need arises, and the means to exploit relations between such regions (O'Keefe and Nadel 1978). The entorhinal cortex and hippocampus play a role in goal-directed behavior towards recently learned spatial locations in an environment. Rats show impairments in finding the spatial location of a hidden platform in the Morris water-maze after lesions of the hippocampus, postsubiculum, or entorhinal cortex. Recordings from several brain areas in behaving rats show neural spiking activity relevant to goal-directed spatial behavior. These include grid cells in the entorhinal cortex that fire when the rat is in a repeating regular array of locations in the environment falling on the vertices of tightly packed equilateral triangles (Hafting et al. 2005), place cells in the hippocampus that respond to mostly unique spatial locations (O'Keefe and Nadel 1978), head direction cells in the postsubiculum that respond to narrow ranges of allocentric head direction (Taube 2007), and cells that respond to translational speed of running (O'Keefe et al. 1998).

Head direction cells

A head direction cell is a neuron that significantly increases its firing rate when the rat's allocentric head orientation in the world horizontal plane, i.e. the head azimuth angle, approaches a specific angle which is referred to as its *preferred direction*. Since the preferred direction of a head direction cell is bound to its environment's global coordinate system rather than the rat's local coordinate system it is considered as allocentric. Head direction cells preserve their tuned firing properties even in the dark. The head direction cell's preferred direction is independent of the rat's actual location as long as its environment does not change. Although the head direction cell's signal might be similar to what a compass would measure, their underlying mechanisms differ significantly. While a compass relies on Earth's electromagnetic field to show a fixed orientation, the head direction cell depends on the environmental cues and proprioceptive inputs. Previous experimental data show that the tuned directions of all head direction cells of a single rat tend to be locked up to an offset to a specific main orientation, usually towards a salient visual cue in the environment (Taube 2007). Head direction cells also differ in their tuning curves, i.e. their firing rate versus polar angle profiles. While most head direction cells respond to a tight range of polar angles, some show a broader tuning curve; some might even have multiple peaks in their tuning curves. Extensive experimental data describe head direction cells in the deep layers of the entorhinal cortex, the postsubiculum, and the anterior thalamus (Taube 2007).

Grid cells

Another type of spatially tuned neuron is the grid cell. A grid cell is a neuron type which increases its firing rate significantly when the animal traverses a regular array of periodic locations in the environment. The collection of locations where an individual grid cell fires, i.e. the grid cell's firing fields, forms a two-dimensional periodic pattern with regular inter-field intervals and similar field areas. More specifically, the firing fields of a single grid cell tile the infinite two-dimensional plane as the vertices of equilateral triangles tessellating the plane. Four parameters uniquely define the organization of a single grid cell's firing fields: the firing field size, the inter-field spacing, the overall orientation of their hexagonal pattern, and the spatial phase. Extensive experimental data show the existence of grid cells with different inter-field spacing and field areas along the dorsal to ventral axis of the medial entorhinal cortex (Sargolini et al. 2006). Recordings from neurons along the dorsal–ventral axis of the entorhinal cortex show grid cell firing fields gradually increasing in size and separation. A discrete organization of scale has recently been reported (Stensola et al. 2012). In a single rat, grid cells in the medial entorhinal cortex are organized in anatomically overlapping modules with distinct scale and orientation. The clustering of grid cells is of hierarchical nature and the relative increase of firing field sizes between neighboring modules varies between individual rats, but has a mean value across all data of about $\sqrt{2}$.

Place cells

One of the most interesting types of spatially tuned neurons is the place cell which is a type of neuron found throughout subfields of the hippocampus. A place cell increases its firing rate when the animal crosses a compact region of the environment (O'Keefe and Nadel 1978). The firing field of a place cell is called its *place field*. The main difference between a place cell and a grid cell is that while each grid cell exhibits multiple firing fields covering the space with a regular and periodic pattern, place cells are tuned to fire exclusively when the animal is inside a single spatially unique location (though some place cells show multiple firing fields). Hence, place cells are appropriate as building blocks of a cognitive spatial map. The place fields usually stabilize within minutes when the animal is first exposed to a new environment and remain stabilized between exposures to the same environment. A phenomenon known as *remapping*, i.e. random change in place field locations, occurs when the animal enters a new unknown environment after its place fields stabilize in a previous known environment. While the salient visual and other sensory cues play an important role in the formation of place fields, evidence supports the potential effect of proprioceptive inputs for place cell firing as place fields tend to maintain their stabilization for extended durations in the dark.

Biomimetic systems

The ability of an autonomous agent to internally represent its environment as a spatial map and to accurately estimate its position at the same time has been studied extensively in the robotics community. The problem is known as SLAM (Simultaneous Localization and Mapping) and several approaches have been proposed that perform extremely well but mostly in relatively static indoor areas and for limited durations. Most autonomous systems rely on two types of sensor information to perform the SLAM parameter estimates. The first type of sensing information is gathered from external cues via active sensors (e.g. range finders and obstacle detectors based on radar, sonar, or laser) and passive sensors (e.g. cameras, microphones, etc.). The second type of information is provided by proprioceptive cues such as inertia, wheel odometry, compass readings, etc. The external cues work relatively well for localization estimation as long as the environment stays relatively constant over time. However, they are usually susceptible to failure as the environment changes either due to natural phenomena such as changes in lighting conditions, weather, etc., or due to man-made changes such as relocating furniture, denying access to a previously available route, etc. Conversely, proprioceptive sensors are indifferent to external cues by definition and in theory they should provide supporting information when external cues are ambiguous. Unfortunately, sensors based on internal cues suffer from noise accumulation which makes their readings unreliable after a relatively short period of time in the absence of some corrective input from external cues. Hence, the SLAM problem becomes a challenge to maintain over extended periods of time in a dynamic environment.

Biological systems seem not to suffer as much from the before-mentioned shortcomings of artificial navigation. Rats are, for instance, able to forage, explore, and navigate successfully in their relatively large and dynamic environment for extended periods of time. They can adapt to changes relatively quickly and deduce solutions to immediate problems, i.e. finding novel routes when denied access to a previous route or exploiting potential short-cuts when new access points become available. Hence, several computational goal-directed navigation mechanisms have been proposed aiming to provide partial explanations as to how the brain might be representing the space and using such representations to perform navigation tasks.

One of the computational models proposed is RatSLAM (Milford 2008). In this approach the environment is represented as a collection of *pose cells* and each pose cell gets associated with a *view cell*. Each pose cell encapsulates pose information, i.e. the location and the orientation of the agent, at some time instance during navigation. The pose cell network maintains an activity bump via both excitatory and inhibitory connections which involves using attractor dynamics fed by self-motion cues. The network topology is locally sensitive, i.e. preserves metric local neighborhoods. Each pose cell is also associated with a view cell which selectively responds to a specific view from the location and orientation of its associated pose cell. During exploration, each pose cell that is activated gets a new view cell associated with it. If a familiar scene is re-encountered, the view cell encoding the familiar scene guides the activity bump in the pose cell network towards the previously established associations, hence providing a simple case of loop closure. RatSLAM has been used to successfully map both small and large areas. Although RatSLAM provides an elegant biologically inspired model for mapping and spatial representing, it does not specifically address the goal-directed navigation problem in the same biologically motivated framework.

A second biologically inspired computational model for goal-directed navigation has been proposed by Erdem and Hasselmo (2012). In this model, head direction cells modulated by proprioceptive velocity data provide inputs to downstream grid cells driven by a *phase interference model*. Several grid cells with different scale and field spacing converge to form a single place cell. When the agent encounters salient locations during exploration, it recruits new place cells to represent (encode) its current location in its cognitive spatial map. Each place cell is also associated by a *reward cell*. The lateral connection weight among two reward cells is proportional to the time between the agent's consecutive visits to place cells associated by the two reward cells. Maintaining the constant speed assumption, the reward cell network topology is quasi-metric. When the agent picks one of the visited locations as its current goal, an exponentially decaying reward signal propagates starting from the reward cell representing the chosen goal location. The propagation generates a reward signal gradient field with its peak at the goal location. The agent then selects the direction towards the chosen goal location by generating several vicarious linear forward trajectories originating from its current location and picking the one activating the reward cell with maximum reward signal. The crucial point here is the fact that the place fields in the spatial map do not necessarily need to be dense (they can be non-overlapping and farther apart from each other). The phase space of grid cells provides the underlying continuity allowing for each vicarious linear trajectory to traverse areas of the environment not represented by any place cell. The inspiration for this model comes from experiments showing hippocampal spiking sweep events encoding forward and backward trajectories of behaving rats during goal directed tasks (Johnson and Redish 2007) and sharp wave ripple events observed during sleep (Jadhav et al. 2012). Furthermore, this model allows the discovery and exploitation of novel shortcuts in the environment as observed with behaving rats in a hairpin maze task.

An extended version of the linear look-ahead model (Erdem et al. 2015) relaxes the linear probe range restriction of the previous model by using a hierarchy of place cell maps each representing the environment at different scales. This multi-scale approach allows arbitrary extension of the maximum probe range while keeping the duration of a single probe constant and equivalently guaranteeing a predefined maximum level of noise accumulation regardless of the probe range. In this extended model, the agent constructs the spatial map of the environment in a similar way to the previous model but this time each salient location is represented by place fields at different scales. This approach creates a scale-space of the environment. The place cell hierarchy allows propagation of separate forward linear probes simultaneously at each scale level hence covering longer and longer ranges guaranteeing a noise accumulation level limited from above regardless of the probe range. Recordings from neurons along the dorsal–ventral axis of the entorhinal cortex show grid cell firing fields gradually increasing in size and separation (Hafting et al. 2005). Place cells have not been shown to have discrete spatial scales, but they clearly vary their scale at different dorsal to ventral positions within the hippocampus.

Future directions

State-of-the-art goal-directed robotic navigation systems perform extremely well for limited durations and within relatively static environments. Higher-level living organisms, however, appear not to suffer from the degrading effects of persistent navigation for extended periods of time and in dynamic environments. The technical challenge is bridging the spatial representation that autonomous systems use and the spatial representation created by grid cells in the entorhinal cortex and place cells in the hippocampus. Grid cells show stable firing over long time periods (10 min) even in darkness, indicating robust path integration despite the noise inherent in neural systems, which is an extremely challenging feature for the state-of-the-art robotic navigation. If the robust biological mechanisms of grid cells could be implemented in robots they would provide a dramatic advance over their current capabilities.

An open question in the biological representation of space is the trigger to associate hippocampal cells to certain spatial locations. There is compelling evidence that the association trigger might not only depend on spatial cues but on context as well (Komorowski et al. 2009). Further understanding of how the brain prioritizes contextual and spatial associations could have significant impacts on selection of sensory cues to encode locations and their organization in a persistent database in robotic SLAM systems. Another interesting biological phenomenon not very well understood so far is the *remapping* of the place cells. It is not yet very clear why or how the remapping happens. More insight into this phenomenon might result in more efficient encoding of space in robotic navigation.

Learning more

An excellent review of head direction cells can be found in Taube (2007). Readers interested in different computational models describing grid cell mechanisms may find Zilli (2012) extremely informative. O'Keefe and Nadel (1978) provide good coverage of how the brain might be representing its surrounding environment and using this representation for survival tasks. A recent treatment of biologically inspired robotic navigation is given in Milford (2008).

References

Erdem, U. M., and Hasselmo, M. E. (2012). A goal-directed spatial navigation model using forward trajectory planning based on grid cells. *The European Journal of Neuroscience*, 35(6), 916–31.

Erdem, U. M., Milford, M. J., and Hasselmo, M. E. (2015). A hierarchical model of goal directed navigation selects trajectories in a visual environment. *Neurobiology of Learning and Memory*, 117, 109–21. doi: 10.1016/j.nlm.2014.07.003

Hafting, T. et al. (2005). Microstructure of a spatial map in the entorhinal cortex. *Nature*, 436(7052), 801–6. Available at: http://dx.doi.org/10.1038/nature03721 [Accessed March 8, 2012].

Jadhav, S. P., et al. (2012). Awake hippocampal sharp-wave ripples support spatial memory. *Science*, 336(6087), 1454–8. Available at: http://www.ncbi.nlm.nih.gov/pubmed/22555434 [Accessed March 4, 2013].

Johnson, A., and Redish, A. D. (2007). Neural ensembles in CA3 transiently encode paths forward of the animal at a decision point. *The Journal of Neuroscience*, 27(45), 12176–89. Available at: http://www.ncbi.nlm.nih.gov/pubmed/17989284 [Accessed July 12, 2012].

Komorowski, R. W., Manns, J. R., and Eichenbaum, H. (2009). Robust conjunctive item-place coding by hippocampal neurons parallels learning what happens where. *The Journal of Neuroscience*, 29(31), 9918–29. Available at: http://www.pubmedcentral.nih.gov/articlerender.fcgi?artid=2746931&tool=pmcentrez&rendertype=abstract [Accessed June 12, 2013].

Milford, M. J. (2008). *Robot navigation from nature: simultaneous localisation, mapping, and path planning based on hippocampal models* 1st ed., Springer Verlag. Available at: http://www.amazon.com/dp/3540775196 [Accessed November 4, 2011].

O'Keefe, J. et al. (1998). Place cells, navigational accuracy, and the human hippocampus. *Philosophical Transactions of the Royal Society of London. Series B: Biological Sciences*, 353(1373), 1333–40. Available at: http://rstb.royalsocietypublishing.org/content/353/1373/1333.abstract [Accessed November 4, 2011].

O'Keefe, J., and Nadel, L. (1978). The hippocampus as a cognitive map. *Philosophical Studies*, 2(04), 487–533. Available at: http://www.pdcnet.org/collection/show?id=philstudies_1980_0027_0263_0267&pdfname=philstudies_1980_0027_0263_0267.pdf&file_type=pdf [Accessed July 19, 2011].

Sargolini, F. et al. (2006). Conjunctive representation of position, direction, and velocity in entorhinal cortex. *Science*, 312(5774), 758–62. Available at: http://www.sciencemag.org/content/312/5774/758.short [Accessed June 10, 2011].

Stensola, H. et al. (2012). The entorhinal grid map is discretized. *Nature*, 492(7427), 72–8. Available at: http://www.ncbi.nlm.nih.gov/pubmed/23222610 [Accessed February 28, 2013].

Taube, J. S. (2007). The head direction signal: origins and sensory-motor integration. *Annual Review of Neuroscience*, 30(1), 181–207. Available at: http://www.annualreviews.org/doi/abs/10.1146/annurev.neuro.29.051605.112854 [Accessed July 14, 2012].

Zilli, E. A. (2012). Models of grid cell spatial firing published 2005–2011. *Frontiers in neural circuits*, 6(April), p.16. Available at: http://www.pubmedcentral.nih.gov/articlerender.fcgi?artid=3328924&tool=pmcentrez&rendertype=abstract [Accessed October 9, 2012].

Chapter 30

Reach, grasp, and manipulate

Mark R. Cutkosky

School of Engineering, Stanford University, USA

" . . . an instrument for instruments"

Aristotle, Parts of Animals

The ability to interact physically with one's environment is a basic need shared by animals and robots. The instrument of this interaction can be a hand or gripper but may also be a mouth, as when a dog carries a stick, or even tentacles, as in the case of the octopus. Hence, when considering the problem of designing and controlling robotic hands we have a wide range of biological examples for inspiration and instruction.

This chapter seeks to identify principles that we can glean from nature regarding the design and operation of hands, and to show how they influence robotic hands and can improve their performance, and simplify their control. The chapter does not attempt a comprehensive survey of animal solutions to grasping and manipulation, nor does it cover the full range of robotic grippers used in manufacturing and other specialized operations. Instead, it focuses on the emerging area of *mobile manipulation*: hands designed for mobile robotic platforms which, like animals, must interact with objects and people in the world at large. In this emerging field, the robustness, sensitivity, and versatility found in the hands, pincers, mouths, and other "end effectors" of animals have particular lessons for hand design and operation.

The biological literature on grasping and manipulation is dominated by studies of the human hand, and most multi-fingered robotic hands are clearly anthropomorphic. However, as noted in the next section, some particular ideas from insects, amphibians, and other animals are also valuable when designing hands for mobile manipulation, affording solutions that are much simpler than those found in humans and primates. The following section examines three hands designed for mobile manipulation, showing how they have incorporated specific design principles from nature to increase robustness and versatility and simplify control.

The chapter concludes with a discussion of future directions. Robotic hands are at a tipping point—the cost and complexity are finally dropping to the point that bio-inspired compliant, multi-fingered hands can be used on mobile robots used for search and rescue, exploration, and interactions with people in business, home, and health care settings. The same technologies can also be applied to prosthetics and even bionics as we look into the future.

Biological principles

Lessons from human and animal examples

Regarding hand function and operation in nature, the greatest literature by far is centered on the human hand, which has been a subject of systematic inquiry since at least ancient Greek times.

The current literature dates mainly from 1900 onward, informed especially by hand surgery (e.g. to restore functionality after an injury) and in recent years increasingly by neurography to understand the functioning of mechanoreceptors and hand control (Tubiana et al. 1998; Castiello 2005; Johansson and Flanagan 2009). Applications beyond surgery range from physical therapy to design of prostheses. Starting in the 1980s, the study of human grasping and tactile sensing has also been motivated by a desire to inform robotic solutions.

However, the human hand is remarkably complex, consuming a substantial fraction of the human motor cortex and requiring years of experience before adult levels of skill are obtained in everyday manipulation (Poole et al. 2005). Human hands also serve various additional functions including communication and thermoregulation. Therefore, some of the complexity of the human hand is likely to be irrelevant from a robotics perspective. In light of these observations, we should not be surprised that even the most sophisticated robotic hands come nowhere close to human hands in terms of their capabilities. Fortunately, there are also valuable simplifying strategies that can be identified when humans grasp and manipulate objects. In search of simplifying principles it is also useful to examine the hands and grasping strategies adopted by animals such as insects, frogs, and even the octopus, which is remarkably dexterous despite having nothing like the vertebrate system of fingers with tendons and bones.

In the following subsection we examine some of the common design principles found in human and animal hands that are particularly relevant for the design of hands for mobile manipulation.

Design principles

The design principles described in this section have chiefly been identified in the context of human hands; however, many of them are also relevant to other multi-fingered hands in frogs, primates, etc. Hence they are applicable to a wide range of multi-fingered robotic hands whether anthropomorphic or not.

Branching kinematics

The common kinematic structure for hands in vertebrates, ranging from frogs to primates, is to have multiple fingers branching from a common palm and wrist. Even the octopus has a variation on this theme, with eight tentacles emanating from a central body. Each finger constitutes a serial kinematic chain. The configuration space represents the space of unique poses that the hand can adopt, with each finger varying from fully extended to fully flexed and with additional motions for adduction and abduction (Melchiorri and Kaneko 2016). A group of fingers represents a parallel set of series chains, and presents a very large configuration space whose dimension is the total number of degrees of freedom, except in cases where the motions of some joints are always coupled. For example, the human hand is commonly said to have 27 degrees of freedom (El Koura and Singh 2003). Moreover, the hand is located at the end of a wrist and arm, giving it an enormous workspace for grasping and manipulating. Other vertebrates with hands may have have fewer fingers, fewer joints per finger, and more coupling between adjacent joints or fingers; nonetheless they too have a large configuration space. Even larger is the number of muscle and tendons used to control the hand, with large muscles complemented by small muscles for a combination of strength and sensitivity (Tubiana et al. 1998). The control of all these muscles is a challenge and consumes a substantial portion of the brain's motor cortex. However, as discussed later in this chapter, there are also strategies for simplifying grasp planning and control.

The advantage of the hand's large configuration space can be seen in the wide range of grasps used when picking up and working with objects. For example, humans often switch among diverse grasps with a single object to accommodate changing task requirements (Figure 30.1, left). The fingers can pinch objects, using internal forces to hold the objects with friction, or they can envelop them in a wrap grasp for greater security. Other grasps such as the *lateral pinch* and the *index finger extension* grasp have the hand oriented to exploit wrist rotations to increase the mobility of the grasped object.

Compliance and conformability

Most hands are compliant both at the contact surfaces and in terms of actuation. For example, the contact surfaces of human hands are composed of soft, dense tissue (subcutus) covered by the dermis and outer epidermis, which can easily conform to irregular surfaces, such as those of rock, and even wrap around corners to provide a firm grip with little grasp force. The epidermis additionally has fingerprint ridges which provide much more reliable friction on smooth and moist surfaces and enhance tactile sensing (Johansson and Flanagan 2009). The compliance of the finger tissues also helps to prevent damage during collisions and whenever interaction forces between the hand an object become unexpectedly high. Robustness is an important

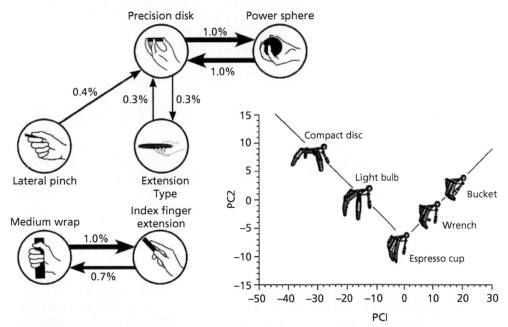

Figure 30.1 Left: Sequence of common grasps in everyday manipulation shows how readily humans switch among grasp types in response to changing task requirements. Right: Despite significant individual variations in joint angles, many common grasps can be approximated with just a few principal components (PC1, PC2), reducing the size of the grasp search space.

Left: © 2013 IEEE. Reprinted, with permission, from Ian M. Bullock, Joshua Z. Zheng, Sara De La Rosa, Charlotte Guertler, and Aaron M. Dollar, Grasp Frequency and Usage in Daily Household and Machine Shop Tasks, *IEEE Transactions on Haptics*, 6(3), pp. 296–308, doi: 10.1109/TOH.2013.6. Right: Reproduced from Marco Santello, Martha Flanders, and John F. Soechting, Postural Hand Synergies for Tool Use, *The Journal of Neuroscience*, 18(23), pp. 10105–10115, Figure 8 © 1998, The Society for Neuroscience.

consideration because hands and feet are the most frequent sites of energetic interactions with the world. A notorious difficulty in designing and maintaining prosthetic hands is to avoid damage from such daily interactions (Belter et al. 2013).

Additional compliance exists in the musculoskeletal structure that connects actuators (muscles) to finger forces and motions. Large forces applied in opposition to the muscles will cause the fingers to move backward. This back-driveability is a safety feature, and prevents damage in the case of unexpected loads and collisions. Even the hard claws and pincers of crustaceans have compliance in their muscle and tendon apparatus. This feature is often lacking in conventional robot grippers which use inherently non-backdrivable worm gears or very high transmission ratios, and makes them more vulnerable to damage in uncontrolled environments.

Non-linear actuation

One reason that traditional robot hands often use a non-backdrivable drivetrain is to avoid having to keep the motor active for a prolonged time at stall, producing heat and consuming energy, while grasping objects. Muscles are quite different, they have non-linear force/displacement curves and can resist forces pulling against their intended direction of motion with high forces and low effort, in comparison to the energy they require to impart a motion (Hogan 1984). This property substantially increases the maximum force required to pull objects out of a grasp and reduces hand fatigue when carrying a heavy object.

Moreover, unlike robot actuators, muscles often do not have a one-to-one mapping with joints—there may be multiple muscles operating on a single joint or a single muscle that actuates multiple joints. In the human hand, large muscles are located in the forearm and drive the fingers via tendons through the wrist; small muscles permit fine control. In insects, multiple muscles on a single joint may be specialized, for example, for fast or slow motions. When a single muscle operates across multiple joints, the system may be underactuated meaning that the joint motions are not independently controllable. For example, in human hands, we cannot vary the distal joint angle independently of the middle or metacarpophalangeal joint. In insects, we see more dramatic simplifications. For example, a single muscle pulling on a tendon or apodeme may be routed down the entire length of the leg for actuating the tarsal claw, as shown in the illustration in Figure 30.2 (left). This muscle has no opposing muscle for extension; it acts against the passive elasticity of the exoskeleton (Busshart et al. 2011). Moreover, it is activated in such a way that the opening and closing of the distal claw is largely independent of the motions of the limb (Radnikow and Bassler 1991). In Figure 30.3 (right) we see an analogous arrangement for an under-actuated hand. A single tendon is used to flex the fingers and acts in opposition to elasticity in the finger joints.

Sensitivity

The hand's frequent interactions with the world are also an important source of information. Not surprisingly, hands are endowed with an especially rich collection of mechanorceptors for measuring pressures, local skin deformations, vibrations, temperatures, etc. In humans, the density of mechanoreceptors on the glabrous (non-hairy) skin of the hands is several times that of most other parts of the body. In addition, tactile sensing is inherently multi-modal. There are specific classes of mechanoreceptors for responding to different kinds of local events, including variations in pressure, skin deformation, vibrations, temperature, etc. The mechanoreceptors can also be divided into "fast acting" and "slow acting" categories specialized for responding respectively to continued deformations (e.g. the sensation imparted by the corner of an object held steadily against the fingertips) and to transient phenomena, such as those that arise from scanning our fingertips over a textured surface (Johansson and Flanagan 2009). Even arthropods with hard exoskeletons have numerous hairs or spines on their legs and pincers and

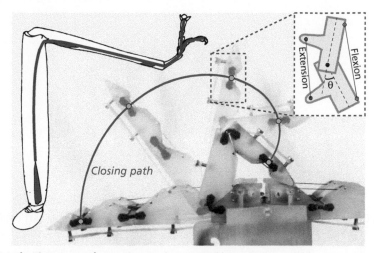

Figure 30.2 Left: The tarsus of many insects is actuated by a single, variable geometry apodeme (tendon) that acts against the elasticity of the exoskeleton. Right: a hand for underwater grasping (Stuart et al.2014, 2017) uses a similar scheme with a variable geometry flexor tendon and extensor springs at each flexural joint.

Left: Adapted from G. Radnikow and U. Bässler, Function of a Muscle Whose Apodeme Travels Through a Joint Moved by other Muscles: Why the Retractor Unguis Muscle in Stick Insects is Tripartite and Has No Agonist, *The Journal of Experimental Biology*, 157(1), p. 90, Figure 1 © The Company of Biologists Limited. Right: © 2014 IEEE. Reprinted, with permission, from Hannah S. Stuart, Shiquan Wang, Bayard Gardineer, David L. Christensen, Daniel M. Aukes, and Mark Cutkosky, A compliant underactuated hand with suction flow for underwater mobile manipulation, *IEEE International Conference on Robotics and Automation*, doi: 10.1109/ICRA.2014.6907847.

specialized campaniform or slit sensilla that measure the strains produced in their exoskeletal segments by contact forces (Barth 1985). To provide a comparable sensitivity remains a major challenge in robotics. Even the most sophisticated robot hands have tens or perhaps 100 sensors in their hands, but animals have thousands.

Special features

The need to perform special tasks or work with particular objects gives rise to special features such as claws and fingernails (many vertebrates), sticky pads (insects), and suckers (octopus).

Robotic hands can also benefit from such features. For example, it is difficult for most multi-fingered robotic hands to pick up objects such as small coins from a tabletop unless they have fingernails or suction. Other specialized features may include locking joints to save energy when grasping heavy objects (Figure 30.5, right) and specialized mechanisms for changing the arrangement of fingers about the palm (e.g. to switch from a radially opposed grip similar to the *precision disk* in Figure 30.1 to a grip more akin to the *lateral pinch*).

Manipulation and control principles

Simplifying grasp planning

Early work established that human grasps can be organized into a hierarchical taxonomy and found that grasp preshaping is evident while the hand is in motion toward an object. Subsequent work has examined the neural processes behind grasp and manipulation planning in some detail (Castiello 2005) and has led to systems for categorizing grasps according to object shape and task requirements (Bullock et al. 2013). Initially a general grasp shape is adopted according to object shape attributes (e.g. whether flat, long and thin, or round). Then the grasp is specialized based on what we intend to do with the object. Hence the grasp for picking up a hammer is

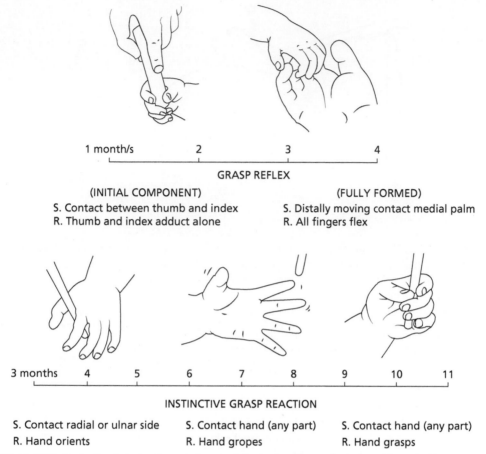

Figure 30.3 Tactually elicited reflexes in human infants simplify manipulation although they are overlaid with more sophisticated responses and behaviors in adults.

Reprinted from *Neuropsychologia*, 3, Thomas E. Twitchell, The automatic grasping responses of infants, pp. 247–59, http://dx.doi.org/10.1016/0028-3932(65)90027-8, Copyright © 1965. Published by Elsevier Ltd., with permission from Elsevier.

different from the grasp used for hammering. The importance of task requirements is also seen in Figure 30.1 (left) which illustrates some of the grasps, and transitions between them, with the highest frequency in video observations of people involved in manual work. People switch easily and frequently between diverse hand configurations.

As noted earlier, human grasp choice and hand control are highly complex. They consume a substantial fraction of the motor cortex and require years for children to approach adult proficiency (Poole et al. 2005). Hence, it should be no surprise that robotic grasp planning and hand control remain open challenges. Fortunately, there are also some important simplifying principles at work. For example, while there are many idiosyncratic grasps that people adopt to suit variations in object properties, task-related forces and sensing requirements (as exemplified by the transitions in the left diagram in Figure 30.1), it is also the case that most common grasps can be approximated with just a few principal components, also termed "grasp synergies." For example Figure 30.1 (right) shows the results of a principal components analysis of grasp configurations, defined by the joint angles of the fingers, for a range of grasps applied by people to a wide range of common objects (Santello et al. 1998). Despite the many degrees of freedom and the many individual variations in joint angles, just two principal components, represented as

PC1 and PC2 in the plot, capture the most significant variations in hand postures. The significance of this finding is that by designing and controlling hands in a space defined by just a few principal components, one can readily obtain grasps for most common objects and tasks that closely approximate the many idiosyncratic grasps that humans use. This approach informs the design of hands such as the Dexmart hand (Palli et al. 2014) shown in Figure 30.4, left.

Another, related, simplification is to recognize that the third, fourth, and fifth fingers often operate essentially as a single "virtual finger" somewhat analogous to grasping with mittens (Venkatamaran and Iberall 1990). Hence, one can achieve satisfactory results for grasping many objects by designing robotic hands to approximate the two or three "virtual fingers" used by humans in handling most objects. In addition, the human propensity for fashioning and using tools allows us to duplicate many of the specialized strategies seen in animals. Thus an ant picking up and carrying a seed with its pincers reminds us of grasping and manipulating an object with pliers. Conversely, studies of human manipulation with pliers can provide insight into the observed behaviors of ants, and both of these sets of observations can provide insight for the developers of robotic hands.

Detailed hand design and control

The material considered thus far addresses grasping and hand design from the standpoint of "type selection:" how many fingers, how many degrees of freedom per finger, how to arrange the fingers, etc. Once a general type of hand has been chosen (e.g. three opposed fingers with three phalanges each, actuated by a single motor) there are many tools available for parametric design and optimization. Examples include Grasp It! (Miller and Allen 2000), SynGrasp (Malvezzi et al. 2015), and SimGrasp (Wang 2017). Useful starting points for such analyses are Melchiorri and Kaneko (2016) and Prattichizzo and Trinkle (2016).

Exploiting automatic sensor-based responses

In animals, many stereotypical behaviors are sensor-initiated. For example, frogs perform stereotypical behaviors in response to visual and tactile stimuli that are commonly associated with prey in the vicinity. The responses include orienting the head, striking with the mouth and tongue, and, when tactile stimuli on the forelimbs are also involved, performing enveloping grasps with the hands to funnel food toward the mouth (Comber and Grobstein 1981). Other examples of reflexive grasping and object acquisition include the triggered responses of the mantid to secure prey (Corrette 1990). In this case, the morphology of the limbs, with inward pointing spines, helps to ensure object capture and stabilization. In humans, although primitive actions are much harder to detect amidst the very complex voluntary and skilled behaviors that we perform every day, they remain present and are important for monitoring normal neural development, as illustrated in the examples in Figure 30.3 (Twitchell 1965).

More generally, automatic responses help us to handle objects gently and securely. As the fingers begin to make contact, sensory information is used in specializing the grasp based on detailed task requirements (e.g. whether it is more important to hold securely, or to hold gently so that we can more easily monitor vibrations from the object) and object attributes (e.g. whether rough or slippery). We also exploit sensory reflexes to adjust grasp forces continuously and unconsciously based on the perception of small incipient slips that precede gross sliding (Johansson and Flanagan 2009). This capability allows us to handle objects with a consistent safety factor despite variations in object weight and coefficient of friction; when our fingers become numb due to cold we lose this sensitivity and are correspondingly clumsier.

At a more advanced level, people adopt stereotypical exploratory procedures to elicit certain kinds of information about objects they are handling (Lederman and Klatzky 1993). For example, to perceive texture we rapidly move our fingertips back and forth, laterally, over a surface; to gage firmness we squeeze an object between the fingers and thumb.

Biomimetic systems

For as long as robots have been working with objects in the world, they have needed grippers or hands. The formal study of robotic grasping dates to the late 1970s. Among the best known early multi-fingered hands are the Okada (Okada 1982), Stanford/JPL hand (Salisbury 1985), and Utah/MIT hands (Jacobsen et al. 1984), which remain impressively sophisticated by modern standards. The ability of robots to choose grasps and perform useful manipulations with multi-fingered hands has, however, developed much more slowly. This is not surprising when we consider the complexity of grasping and manipulation in humans and primates. Consequently, alongside the research on dexterous hands with many controllable degrees of freedom, there has been a steady interest in "under-actuated" hands. These hands have fewer actuators or motors than degrees of freedom and rely on compliance to grasp a wide range of objects. Although they may be less conspicuously anthropomorphic than fully actuated hands, they also take lessons from biology, drawing ideas from the hands of lower vertebrates (e.g. frogs) and arthropods.

Figure 30.4 illustrates three recently developed hands that demonstrate different approaches to achieving versatility and robustness in grasping and manipulation, taking advantage of current materials, sensors, and rapid prototyping technologies.

At left in Figure 30.4 is the Dexmart hand (Palli et al. 2014), which aims to closely approximate most human grasps while significantly simplifying the hand mechanism in comparison to human hands. Like the human hand, it has its primary actuators in the forearm, connected by tendons routed through the wrist in such a way that wrist angle and finger flexion are largely decoupled.

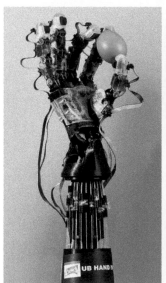 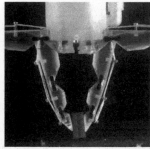  

Figure 30.4 Three robotic hands that incorporate diverse principles from biology. Left: the Dexmart hand aims to approximate most human grasps, drawing upon the principal components or "grasp synergies" observed (Figure 30.2) to simplify the mechanics (courtesy of the University of Bologna, Laboratory of Automation and Robotics). Middle: The KAUST hand is under-actuated with a single tendon per finger, with joint stiffnesses designed to permit wrap objects on large objects and fingertip pinch grasps on small objects. Right: the SRI ARM-H hand is also under-actuated with one tendon per finger but uses electrostatic brakes to selectively lock joints, enabling grasps like that shown for holding a flashlight and switching it on.

The hand aims specifically to achieve the principal components associated with human grasps of everyday objects, based on an extension of the principal components analysis shown in Figure 30.1. The complex geometries of the finger phalanges, palm, and wrist are created using a 3D printing process, with tendons routed through the fingers as illustrated in the detail at left in Figure 30.5.

Each finger of the Dexmart hand is controlled by four motors and tendons. This "n+1" arrangement hearkens back to the Stanford/JPL hand (Salisbury 1985) and enables independent torque control at each joint, despite the limitation that tendons can pull but not push. A fifth tendon couples the motions of the middle and distal joints in a manner somewhat similar to the coupling found in human fingers and in the apodeme (tendon) routing for some insects.

The middle images show a prototype of a hand developed by a team from Stanford University and KAUST, later used with a two-armed underwater robot working in marine archaeology (Stuart et al. 2014, 2017). The hand needs to grasp delicate corals and other small objects but also to grasp large pipes and frames securely. The need to operate at tens of meters below the sea surface prompted an insect-inspired solution in which a single tendon acts against elastic flexures and extension springs, without pivots that would require waterproof shaft seals at each joint. As seen in Figure 30.2, the kinematics as designed provide varying amounts of leverage to the flexion tendon and the extension springs, so that the the finger curls initially at the base and ultimately at the tip. This variable tendon geometry, in combination with the various stiffnesses at the joints, automatically provides fingertip pinch grasps for small objects and encircling wrap grasps for large objects. The fingers are made from two grades of polyurethane, stiff for the finger phalanges and compliant at the joints. This kind of variation in material properties is common in animals. For example, consider insects and crustaceans which have stiff exoskeletons, except at the joints. As in the case of insects and crustaceans, the monolithic structure with flexures in lieu of pivots with shafts and bearings is also robust, accommodating large unexpected loads without damage.

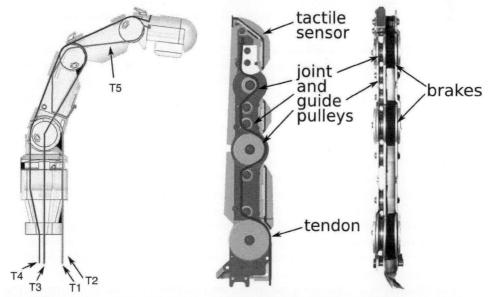

Figure 30.5 Tendon routing details for the Dexmart hand (left) and the SRI ARM-H hand (middle). While the Dexmart hand uses four actuated tendons (T1–T4) to achieve full control of three joints, the SRI ARM-H hand uses a single tendon and electrostatic brakes (right) to lock specific joints on demand.
Left: Courtesy of the University of Bologna, Laboratory of Automation and Robotics.

A special difficulty encountered in underwater grasping is that motions of the hand, as it approaches an object, will create localized fluid dynamics that tend to push the object away. To overcome this effect, the KAUST hand employs a gentle suction flow at the fingertips. While there doesn't seem to be a direct precedent for suction flow in biological fingertips, suction is widely used to enhance prey acquisition in oral feeding by fish and frogs (Comber and Grobstein 1981).

The right image in Figure 30.4 shows one of three hands developed for the DARPA ARM-H project, aimed at developing new multi-fingered hands for mobile manipulation. The pictured hand, designed by a team from SRI, Meka Robotics Inc., and Stanford, is under-actuated with a single tendon for each finger acting in opposition to passive elastic elements (Aukes et al. 2014). In addition, there are electrostatic brakes at each joint, as detailed in the right view in Figure 30.5. When locked, the brakes allow the hand to carry heavy objects with low energy expenditure, roughly analogous to the ability of muscles to resist large forces with low effort when those forces act against the muscle's direction of pull. By selectively locking joints, the brakes also permit particular grasp configurations that would otherwise be difficult to achieve with a single tendon, as exemplified in Figure 30.4 by holding a flashlight and independently depressing the trigger to turn it on. The hand also has a reconfigurable palm allowing it to switch between opposed pinch grasps for small objects and interlaced wrap grasps for heavy objects.

Future directions

A goal of the DARPA ARM-H and Dexmart projects was to produce hands that are relatively inexpensive as well as sufficiently robust, lightweight, and versatile for mobile manipulation. Common themes among these hands include tendon-driven fingers and the exploitation of rapid prototyping techniques to create complex three-dimensional geometries at modest cost. Other enabling technologies include compact, powerful motors and motor drivers and, perhaps most important, the availability of robust, low-cost sensors that can communicate over a network.

The main driver behind the last of these technological advances has been the smartphone industry. A smartphone integrates dozens of sensors for touch, vibration, sound, etc., as well as microprocessors for sampling and filtering signals and communicating with other devices. Such devices reduce the cost of sensing and the amount of wiring required to place sensors at the joints and in the fingertips. Nonetheless, whether wires or flexible printed circuits are used, it remains a major challenge to create paths for the electrical connections that will not be vulnerable to damage and will not suffer fatigue failures as the fingers are repeatedly flexed. As a consequence, even the most sophisticated robotic hands are covered with only a small fraction of the sensors found in animals ranging from frogs to humans. For example, the hands in Figure 30.4 have just a few tactile sensors, and only on their gripping surfaces. Consequently, they cannot respond to stimuli like a light brushing contact against the back of the hand or a lateral contact, as illustrated in the lower left image in Figure 30.3. As yet there is nothing akin to the distributed neural networks that permeate biological hands.

With continued advances in the fabrication of complex multi-material components, in actuation and in sensing, robots will increasingly be applied to situations involving human interaction, in applications ranging from health care and home assistance to small-scale manufacturing. The same technology is also relevant to prosthetics and has already started to produce prosthetic hands that can be controlled using nerve impulses and that are endowed with a sense of touch (Gilja et al. 2011). As these applications mature, we can expect the cost of hands to drop and the availability and sophistication of software for controlling them to increase. Indeed, the advancement of software for grasp planning and manipulation control has been hampered by a lack of suitable hands for experimentation and software development. With increasing numbers of hands we may see advances in software akin to those that have already taken place for speech recognition and computer vision.

Learning more

There is a rich literature on the human hand and its functional capabilities. A good example of the medical literature on hand morphology and functionality is provided in Tubiana et al. (1998). Other reviews that examine human grasping performance specifically from the standpoint of providing insight for robotics include the collection of chapters in Venkataraman and Iberall (1990).

As noted in this chapter, tactile sensing is indispensable for grasping and manipulation and is perhaps the least fully developed area of hand technology at present. To better understand tactile sensing and its role in human manipulation, a good starting point is the review by Johansson and Flanagan (2009); Castiello (2005) provides a complementary review, focusing on the neuroscience behind hand control. Another interesting point of departure is the study of exploratory procedures or EPs, that humans typically use for tactile exploration of objects and surfaces. For this and additional information on human tactile sensing a useful reference is Lederman and Klatzky (1993). A summary of robotic tactile sensing with a discussion of current challenges and references is provided in Cutkosky et al. (2016); see also Lepora (Chapter 16, this volume) .

Numerous articles have appeared on robotic hands and algorithms for grasp choice and manipulation planning. A couple of useful starting points are the chapters on Robot Hands (Melchiorri and Kaneko 2016) and Grasping (Prattichizzo and Trinkle 2016) in the *Springer Handbook of Robotics*, the latter providing a thorough introduction to the mathematical analysis of grasps and hand control.

Finally, for a snapshot of the current state of the art in robotic hands aimed especially at mobile manipulation applications a useful starting point is the special issue of the *International Journal of Robotics Research* on the Mechanics and Design of Robotic Hands that includes references Aukes et al. (2014) and Palli et al. (2014) cited earlier in this chapter, along with several other noteworthy hands.

References

Aukes, D., Heyneman, B., Ulmen, J., Stuart, H., Cutkosky, M. R., Kim, S., Garcia, P., and Edsinger, A. (2014). Design and testing of a selectively compliant underactuated hand. *The International Journal of Robotics Research*, 33(5), 721–735.

Barth, F. G. (1985). Slit sensilla and the measurement of cuticular strains. In: F. G. Barth (ed.), Neurobiology of arachnids. Berlin/Heidelberg: Springer, pp. 162–88.

Belter, J. T., et al. (2013). Mechanical design and performance specifications of anthropomorphic prosthetic hands: a review. *Journal of Rehabilitation Research and Development*, 50(5), 599–618.

Bullock, I. M., Zheng, J. Z., De La Rosa, S., Guertler, C., and Dollar, A. M. (2013). Grasp frequency and usage in daily household and machine shop tasks. *IEEE Transactions on Haptics*, 6(3), 296–308.

Busshart, P., Gorb, S. N., and Wolf, H. (2011). Activity of the claw retractor muscle in stick insects in wall and ceiling situations. *The Journal of Experimental Biology*, 214, 1676–84.

Castiello, U. (2005). The neuroscience of grasping. *Nature Reviews*, 6, 726–36.

Comber, C., and Grobstein, P. (1981). Tactually elicited prey acquisition behavior in the frog, *Rana pipiens*, and a comparison with visually elicited behavior. *J. Comp. Physiol.*, 142, 141–50.

Corrette, B. J. (1990). Prey capture in the praying mantis *Tenodera aridifolia sinensis*: coordination of the capture sequence and strike movements. *Journal of Experimental Biology*, 148(1), 147–80.

Cutkosky, M. R., Howe, R. D., and Provancher, W. R. (2016). Force and tactile sensing. In: B. Siciliano and O. Khatib (eds.), Springer Handbook of Robotics, 2nd edition. Berlin/Heidelberg: Springer-Verlag, Chapter 28.

ElKoura, G., and Singh, K. (2003). Handrix: animating the human hand. *Proceedings of the 2003 ACM SIGGRAPH/Eurographics symposium on Computer animation*, San Diego, CA, July 26–27, 2003. Aire-la-Ville, Switzerland: Eurographics Association.

Gilja, V., Chestek, C. A., Diester, I., Henderson, J. M., Deisseroth, K., and Shenoy, K. V. (2011). Challenges and opportunities for next-generation intracortically based neural prostheses. *Biomedical Engineering, IEEE Transactions on*, **58**(7), 1891–9.

Hogan, N. (1984). Adaptive control of mechanical impedance by coactivation of antagonist muscles. *IEEE Transactions on Automatic Control*, **29**(8), 681–90.

Jacobsen, S. C., Wood, J. E., Knutti, D. F., and Biggers, K. B. (1984). The UTAH/M.I.T. dextrous hand: work in progress. *The International Journal of Robotics Research*, **3**(4), 21–50.

Johansson, R. S., and Flanagan, J. R. (2009). Coding and use of tactile signals from the fingertips in object manipulation tasks. *Nature Reviews Neuroscience*, **10**, 345–59.

Okada, T. (1982). Computer control of multijointed finger system for precise object-handling. *Systems, Man and Cybernetics, IEEE Transactions on*, **12**(3), 289–99.

Lederman, S. J., and Klatzky, R. L. (1993). Extracting object properties through haptic exploration, *Acta Psychologia*, **84**(1), 29–40.

Malvezzi, M., Gioioso, G., Salvietti, G., and Prattichizzo, D. (2015). Syngrasp: a matlab toolbox for underactuated and compliant hands. *IEEE Robotics & Automation Magazine*, **22**(4), 52–68.

Melchiorri, C., and Kaneko, M. (2016). Robot hands. In: B. Siciliano and O. Khatib (eds.), *Springer Handbook of Robotics*, 2nd edition. Berlin/Heidelberg: Springer-Verlag, Chapter 19.

Miller, A. T., and Allen, P. K. (2000). Graspit!: a versatile simulator for grasp analysis. In: *Proc. of the ASME Dynamic Systems and Control Division*, Orlando, FL, **2**, 1251–8.

Palli, G., Melchiorri, C., Vassura, G., Scarcia, U., Moriello, L., Berselli, G., Cavallo, A., De Maria, G., Natale, C., Pirozzi, S., May, C., Ficuciello, F., and Siciliano, B. (2014). The DEXMART hand: mechatronic design and experimental evaluation of synergy-based control for human-like grasping. *The International Journal of Robotics Research*, **33**(5), 799–824. doi: 10.1177/0278364913519897

Poole, J. L., Burtner, P., Torres, T., McMullen, C., Markham, A., Marcum, M., Anderson, J., and Qualls, C. (2005). Measuring dexterity in children using the nine-hole peg test. *Journal of Hand Therapy*, **18**(3), 348–51.

Prattichizzo, D., and Trinkle, J.C. (2016). Grasping. In: B. Siciliano and O. Khatib (eds.), *Springer Handbook of Robotics*, 2nd edition. Berlin/Heidelberg: Springer-Verlag, Chapter 38.

Radnikow, G., and Bassler, U. (1991). Function of a muscle whose apodeme travels through a joint moved by other muscles: why the retractor unguis muscle in stick insects is tripartite and has no agonist. *J. Exp. Biol.*, **157**, 87–99.

Salisbury, J. K. (1985). Kinematic and force analysis of articulated hands. In: M. T. Mason and J. K. Salisbury (eds.), *Robot hands and the mechanics of manipulation.*Cambridge, MA: MIT Press, pp. 1–167.

Santello, M., Flanders, M., and Soechting, J.F. (1998). Postural hand synergies for tool use. *Journal of Neuroscience*, **18**(23), 10105.

Stuart, H. S., Wang, S., Gardineer, B. G., Christensen, D. L., Aukes, D. M., and Cutkosky, M. R. (2014). A compliant underactuated hand for underwater mobile manipulation. In: *IEEE International Conference on Robotics and Automation, Hong Kong, China, 2014*. IEEE.

Stuart, H.S., et al. (2017). The Ocean One hands: an adaptive design for robust marine manipulation. *The International Journal of Robotics Research*, **36**(2), 150–66.

Tubiana, R., Thomine, J.-M., and Mackin, E. (1998). *Examination of the hand and wrist*. London: Martin Dunitz. ISBN 1-85317-544-7

Twitchell, T.E. (1965). The automatic grasping responses of infants. *Neuropsychologia*, **3**, 247–59.

Venkataraman, S., and Iberall, T. (1990). *Dextrous robot hands*. New York, NY: Springer-Verlag. ISBN-10: 1461389763

Wang, S. (2017). SimGrasp: Grasp Design Simulation Package. https://bitbucket.org/shiquan/sim-grasp

Chapter 31

Quadruped locomotion

Hartmut Witte[1], Martin S. Fischer[2], Holger Preuschoft[3], Danja Voges[1], Cornelius Schilling[1], and Auke Jan Ijspeert[4]

[1] Biomechatronics Group, Technische Universität Ilmenau, Germany
[2] Institute of Systematic Zoology and Evolutionary Biology with Phyletic Museum, Friedrich-Schiller-Universität Jena, Germany
[3] Institute of Anatomy, Ruhr-Universität Bochum, Germany
[4] Biorobotics Laboratory, EPFL, Switzerland

Locomotion (latin "*locus*" = place) is the movement of an organism or a machine from one place to the other. It is helpful to describe motions in terms of mechanics. For locomotion a simply applicable criterion is that the Center of Mass (CoM) of the organism in predefined time leaves a predefined volume, e.g. an ellipsoid, and does not re-enter it within the x-fold time. The term "predefined" mirrors the acknowledgement that due to body shape (and this means, ultimately, taxon) one scalable general set of parameters (yet) may not be defined.

In organisms, locomotion is usually driven by a central element (called a "body") and/or appendices. This may cover the range from hydraulic/pneumatic propulsion in fluids to legged locomotion (Figure 31.1). Commonly, periodically deformed elements transfer the energy provided by actors (e.g. muscles) within the structure (transmission) to the environment (effectors). Seemingly simple, but remarkable: due to phylogeny bodies without appendages are "automotive" (e.g. snakes), but appendages without a body do not even exist. Concentration of metabolic functions in a central body provides short diffusion distances, thus the uneven mass distribution (in relation to a globe as an ideal) is not necessarily for mechanical reasons, while the concentration of masses on the proximal end ("near to the trunk") of extremities could be made plausible just by mechanical effects (cf. Hildebrand 1985, Witte et al. 1991, and corresponding work). This points to one restriction of biomimetic approaches: evolution provides systems that function, not optima—organisms are multifunctional and never completely specialized to one function, as may be realized in technical design.

Biological principles

For practical means, zoologists classify locomotion by the substrates the organisms interact with:

◆ Most organisms still live and move in water—in multicellular organisms this is called "aquatic locomotion" (latin "*aqua*" = "water").

◆ The other fluid of relevance is air: flying is called "aviatic locomotion" (zoological terminology: "*avis*" = bird; Latin "*via*" = "way"—"a-via-tic" means "where no fixed way is given").

◆ For reasons of completeness, the locomotion in quasi-fluidic media like sand is called "subterranean locomotion" (Latin "*sub terram*" resp. "*sub terra*" means "under ground").

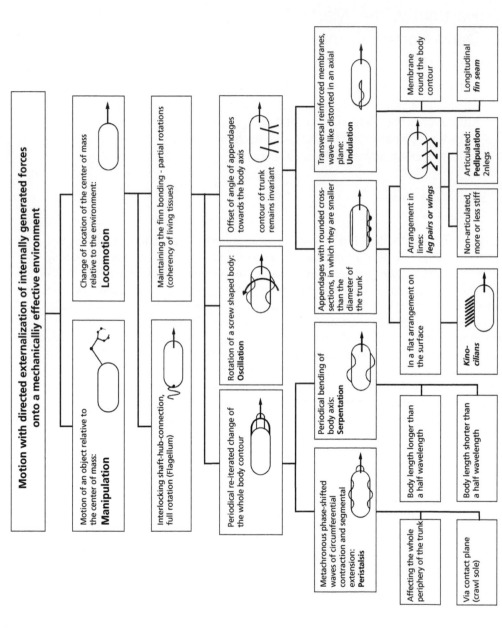

Figure 31.1 Some principles of organismic locomotion. Aviatic locomotion does not really fit into the scheme and thus is neglected here.

◆ Locomotion on the ground is called "terrestrial locomotion" (Latin "*terra*" = earth, here in the sense of "ground").

◆ Often but not always plants as substrates of locomotion are used in a comparable manner to the ground, thus the category "arboreal locomotion" ("*arbor*" = "tree") exists. If specialization to living in trees is stated, the term "arboricol" instead of "arboreal" is applied (cf. ᴘʀᴇᴜsᴄʜᴏꜰᴛ 2002). Since arboreal locomotion contains elements of climbing but is not identical to it, on the other hand, climbing does occur on steep substrates on the ground—even in biorobotics climbing is called just "climbing" (cf. CLAWAR: "Climbing and Walking Robots").

Vertebrate locomotion

Vertebrates are characterized by the existence of a spine ("*Columna vertebralis*"). A spine or its phylogenetic precursor "*Chorda dorsalis*" implies the mechanics of an endoskeletal system ("stiff in the center, compliant in the periphery"), especially in tetrapods with four (resp. two: e.g. kiwis) extremities, called legs (resp. arms). Following ꜰɪsᴄʜᴇʀ (1998a), for the purposes of robotics, terrestrial locomotion of quadrupedal vertebrates roughly may be categorized into an "amphibio-reptile type" and a "mammal type" (Figure 31.2; cf. ꜰɪsᴄʜᴇʀ and Blickhan 2006, ᴋᴀʀᴀᴋᴀsɪʟɪᴏᴛɪs et al. 2013).

The amphibio-reptile type shows oscillations of the spine (and thus the body stem = head + neck + trunk + tail) mainly in the horizontal plane, which are coupled to the ground by legs with two long segments. This locomotion type is some 400 million years old, while the mammal type arose 140 million years ago as a derivative of the first amphibio-reptile one, complementing but not replacing it. In terms of purely functional aspects like transportation, amphibio-reptile locomotion with two segmented, sprawled legs coupled to a horizontally oscillating trunk is a clever biomimetic solution: effective, high stability due to large track width, spare number of parts, and as paragons (i.e. as models for biomimetics) show high locomotor performance. Nevertheless, the number of robots with real reptile-like locomotion is sparse. It appears that, roboticists being mammals themselves show a strong orientation towards mammals as their nearest relatives, especially pets, thus choosing a much more complicated locomotion type as their paragon.

In the remainder of this chapter we will focus therefore on terrestrial, and to a lesser extent arboreal, locomotion of mammals.

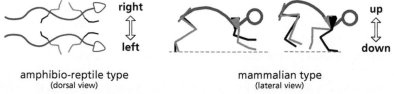

amphibio-reptile type
(dorsal view) mammalian type
(lateral view)

Figure 31.2 Two main principles of vertebrate quadrupedal locomotion. Left: amphibio-reptile type. The spine is oscillating horizontally, its motion is coupled to the substrate by extremities with two long segments. Vertebrae and proximal segments primarily rotate around the vertical axis, the distal segments around horizontal (transversal) axes. Right: mammal type. The spine is oscillating vertically in the sagittal plane, its motion is coupled to the substrate by extremities with three long segments. Vertebrae and all segments primarily rotate around horizontal (transversal) axes.

Data from Martin S. Fischer, Crouched posture and high fulcrum, a principle in the locomotion of small mammals: the example of the rock hyrax (*Procavia capensis*)(Mammalia: Hyracoidea), *Journal of Human Evolution*, 26 (5–6), pp. 501–52, 1998.

Mammalian locomotion

The vertical oscillations of the mammalianspine and trunk are coupled to the ground by legs with three long segments. In some species, the movements of the trunk provide half of the spatial gain of a stride. The extremities are derived from the amphibio-reptile leg by elongation of pre-existing skeletal elements: on the forelimbs, the scapula is extended and used as a central (on the trunk) kinematic element in the vertical plane; on the hindlimbs the bones of the middle foot (metatarsals) form an extended element in the periphery, near the ground. All leg segments more or less move in a vertical plane ("para-median-sagittal plane," parallel to the symmetry plane of the body), allowing reduction of the mechanical description to a planar model. The basic control structure remains about the same as in reptiles—thus a renewed locomotor apparatus is driven by an old control system: in terms of evolutionary biology the neural control is conservative (cf. Reilly et al. 2006), the locomotor apparatus adaptive. For sure, short-time adaptations to the environment are ruled by neural control—including life-long learning, but the life of an individual in measures of evolutionary time is ultra-short. This evolutionary adaptation process is mirrored by terms like "Intelligent mechanics" (introduced by M. S. Fischer and H. Witte in workgroup discussions at SAB '98 in Zurich) or "Morphological computation" (Pfeifer and Bongard 2006)—the best control is not always necessary. The trigger for the transformation of the leg structure is unclear—M. S. Fischer hypothesizes that the transfer from allothermy to homoiothermy (cold- to warm-blooded) was decisive. The thermal insulation for homoiothermy is realized by hairs (fur), and "who has a fur owns parasites" (cf. Fischer 1998b). This leads to the need of scratching even on the back, impossible with the two-segmented reptile legs. The transformation of legs and consecutively the usage of the trunk from amphibio-reptile to mammal type thus may be due to non-locomotor reasons, and it may be senseless to search for mechanical advantages. The decision of roboticists to adopt mammals as paragons instead of reptiles is a typical human attitude—"What is newer, has to be better." For more details see Fischer and Witte (2007).

Pendula and springs—mechanical models of mammalian locomotion

Since human ambition tends to realize complex things, and in light of the mainstream goal in robotics to build mammalian-type quadrupedal machines (oriented to experiences with typically mammalian pets) we will explore this item in more detail, and later due to man's tendency to mimic the creator we will also discuss some issues in bipedal locomotion for humanoid robotics. Starting with Weber and Weber (1836), and extending through to the late nineteenth century (Braune and Fischer 1895), the view on biomechanics of human and, due to anthropocentric attitudes thus mammalian, locomotion was ruled by rigid body mechanics. For periodic movements, the physical model was a suspended pendulum—in terms of Full and Koditschek (1999) a template for several "anchors" ("embodiments," cf. Brooks 1991, Brooks et al. 1998). This paradigm was extended by the concept of inverted pendula in combination with elasto-mechanics, leading to the concept of "ballistic walking"(cf. Mochon and McMahon 1980a, b). Consecutive work was based on the spring-mass model introduced by Blickhan in 1986 (cf. Blickhan 1989), as derivatives of the SLIP—Single Loaded Inverted Pendulum (cf. Geyer et al. 2006), in a substrate-oriented form already starting with analyses of Dawson andTaylor (1973) on kangaroo tendons, compiled as state-of-the-art by Alexander (1988), and extended to the concept of the muscle as a tunable spring (Mussa-Ivaldi et al. 1988) by the discovery and structural and functional identification of elastic giant proteins in vertebrate muscles. Figure 31.3 illustrates the current concepts of muscle-tendon complexes as tunable springs. For details cf. Hill (1938), Labeit et al. (1997), and consecutive work.

This development was accompanied by a discussion on the geometric or elastic allometric scaling of vertebrates (McMahon 1973, Alexander 1977), based on Taylor et al. (1970), repeatedly

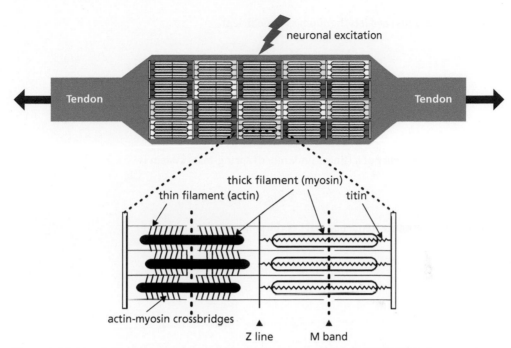

Figure 31.3 Muscle-tendon springs, embodiment of Hill's muscle model (Hill 1938). Top: muscle belly with tendons—sarcomeres, different activation level is illustrated by colors from white to red. No muscle physiologically is ever activated homogenously and completely, excitation waves running from proximal to distal change the pattern—collagen (blue) in several structures, serial and parallel to sarcomeres: tendons, endo-, epi-, perymysium). Bottom: sarcomere with active apparatus (left) and passive elastic elements (right: titin).

triggered by the identification of elastic effects on macro to molecular levels. The difference of allometry coefficients between geometric scaling (0.333—mass, gravity, pendula) and elastic scaling (0.375—mass, springs) is small but led to discussions which, in view of the measurement errors underlying allometry, appear to be speculative. But additionally, they seem to be obsolete. Blickhan and Full (1983) identified a relative descriptor for the springy behavior of moving animals to be equal for a wide range of species, from six-legged cockroaches via four-legged mammals to two-legged human beings. This seemed to be a strong indicator of the dominance of compliance in legged locomotion.

If we start with the term introduced by Blickhan and Full (1993) to be a constant:

$$K_{rel} = \frac{F_{max}/G}{\Delta L/L} = const$$

where K_{rel} is normalized stiffness, F_{max} is peak ground reaction force during cyclic locomotion, G is body weight, ΔL is the (springy) length change of a virtual leg (ground contact—CoM) with length L, and assume the peak force to be in an allometric relation (scaling) to body mass:

$$F_{max} \sim m^a$$

where m is body mass and a is the allometry coefficient, then

$$\Delta \frac{L}{L} \sim \frac{m^a}{m} = m^{a-1}$$

and

$$\Delta L - m^{a-1} \cdot L$$

The resonance frequency of a (linear undamped) spring-mass system is

$$f_c - \sqrt{\frac{c}{m}}$$

where f is the resonance frequency and c is the stiffness of the spring, leading to:

$$c = \frac{F_{max}}{\Delta L} \sim \frac{m^a}{\Delta L} \sim \frac{m^a}{m^{a-1} \cdot L} = \frac{m}{L}$$

$$f_c \sim \sqrt{\frac{m/L}{m}} = \sqrt{\frac{1}{L}}$$

while the resonance frequency of a mathematical pendulum is

$$f_p \sim \sqrt{\frac{g}{L}} \sim \sqrt{\frac{1}{L}}$$

Analyzing human gait and a theoretical limit on human height

Thus, for any size of the animal, and any allometric relation between mass and ground reaction force, the resonance mechanisms of gravitational and spring-mass pendula may be tuned to one another. Figure 31.4 demonstrates this aspect as an explanation for the limitation of human body height. For animals with long feet (like human beings) for the stance phase the SLIP model is extended by a rotational spring (like in a metronome—"MSLIP") (cf. the results of Weiss et al. (1988) on the rotational stiffness of the human ankle joint). The resonance properties of this mass-spring-damper system are tuned to those of the suspended pendula in swing phase. In the case of upright locomotion of human beings, the forelimbs (arms) are freed from the cyclic coordinated kinematic coupling with the ground; the open kinematic chain allows purely dynamic interaction of the arms with trunk and legs. At the same time, the upright human trunk's dominating and most regular periodic movement is axial torsion of the trunk. In walking, the amplitudes depending on velocity show a typical resonance graph (cf. Witte 2002), indicating that the energetically optimal velocity is also defined by the resonant swing of suspended leg and arm pendula with the torsional spring-mass (moment of inertia) system of the trunk. "Diagonal" gait resembles the contra-lateral swing of hand and feet, 180° phase difference. 90° of this phase difference is contributed by the torsion of the trunk. Andrada (2008) showed that this value could change within a wide range without major mechanical disadvantage—but at 90° the lateral motion of the thorax is compensated by axial rotation of the head in such a manner that the gaze fixation realized is purely mechanical—"intelligent mechanics". As an aside, due to the

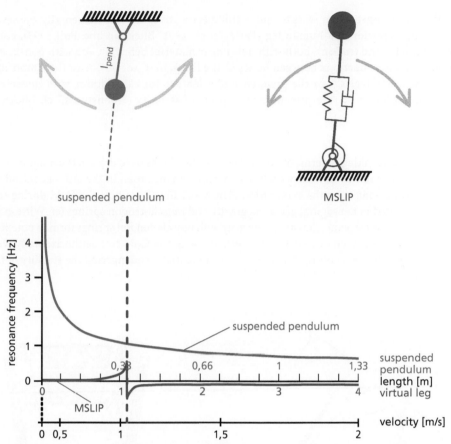

Figure 31.4 Explanation for the observable limitation of human body height. Interaction of suspended pendula (blue: l_{pend} = length of leg and arm pendula) as models for the extremities (cp. Weber &Weber 1836, Witte 1992), and an inverted pendulum with linear spring and damper in the virtual leg, plus a rotational spring around the pivot (red: "MSLIP": Metronome-like Single Loaded Inverted Pendulum) as model for the stance phase, lead to a common resonance frequency of both mechanisms. According to anthropometry, the ratio of length of suspended pendulum to virtual leg length is taken to be 0.37 m/1.1 m. Top left: template (term according to Full and Koditschek (1999)) of swing phase. Top right: template for stance phase "MSLIP". Bottom: tuning of parameters leads to intersection of the two resonance functions at suspended pendula lengths of 0.37 m and virtual leg lengths of about 1.1 m, with corresponding walking velocities of about 1.1 m/s (the energetically optimal velocity following Cavagna et al. (1976)).

use of axial torsional spring effects by the paragons, in humanoid robots which do not show this effect the trunk is covered by something looking like a spacesuit—in astronauts, we are used to not seeing the trunk motion, thus our alarm signal for seeing the wrong motion pattern is turned off due to learning about other technical objects.

The interplay between kinetic energy and potential energy in the gravitational field (both bound to mass) allows control of motion even in systems without springs, by lifting and lowering and accelerating and decelerating mass components of the body in a coordinated manner—coupling a special (quadratic) non-linear behavior with a linear one (cf. Cavagna et al. 1977 and consecutive work). Under this perspective, the experimentally observed linear spring behavior of a human leg (Farley et al. 1991, Blickhan and Full 1993), composed of muscles and tendons, both with non-linear material behavior, and with non-linear kinematic transfer functions between bones in the links, may not be due to the simplification of control strategies (nor the preference of scientists for calculations with linearized equations), but has to be interpreted under aspects of stability ("robustness," cf. Blickhan et al. 2007).

Achieving stability

This approach of thinking in terms of potential fields also allows us to explain the usage of non-linear spring mechanisms in tasks with lower control demands than in the task of controlling vertical substrate contact by the extremities. Figure 31.5 illustrates how the CoM during trot of a horse is guided in the sagittal plane via gravity and muscle-tendon springs (cf. Witte et al. 1995). In the transverse plane, gravity in interplay with muscle-tendon springs forms a potential groove with progressive characteristics of walls, allowing the CoM to float horizontally with small excursions, but preventing higher excursions potentially endangering the stability of the overall system.

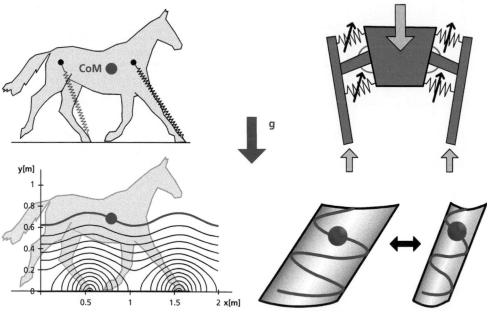

Figure 31.5 Guidance of Center of Mass (CoM) by potential fields. Gravity: g. Left: Guidance of CoM in a trotting horse, sagittal plane (2D) (experimental results, cf. Witte et al. (1995)). Top left: One instant in a trot cycle, red dot represents CoM. Bottom left: Guidance of CoM (red dot, trajectory in red) in the potential field formed by gravity and leg springs. Right: Guidance of CoM in a trotting horse or a walking human being (3D) (concept, abstracted from experimental results on horses and human beings). Top right: tunable elastic bearing of the upper body in the frontal plane, around the hip joints. Grey arrows represent loads. Bottom right: 3D-guidance of CoM during diagonal gaits in a potential rim defined by gravity and tunable stiffness of muscles (around hip or scapula). In adaptation to substrates, the lateral swing of CoM may be changed. The steepness of the rim's "walls" progressively limits the excursions, and may serve as a safeguard mechanism.

Within the boundary conditions derived before, a three-segmented mammalian leg undergoes several constraints. To achieve stable kinematic behavior under axial compression, the proportion of the lengths of the 1st plus 3rd segments in relation to the 2nd segment has to be kept within limits (Seyfarth et al. 2002, Fischer and Blickhan 2006). A pantograph coupling between segment "n" and segment "n + 2" by biarticular muscles (crossing two joints) provides energetic advantages (Jacobs et al. 1996). In mammalian legs, the pantograph mechanism described by M. S. Fischer occurs twofold (Figure 31.6). For small mammals, the kinematics and dynamics of such a leg are described in Fischer et al. (2002) and Witte et al. (2002).

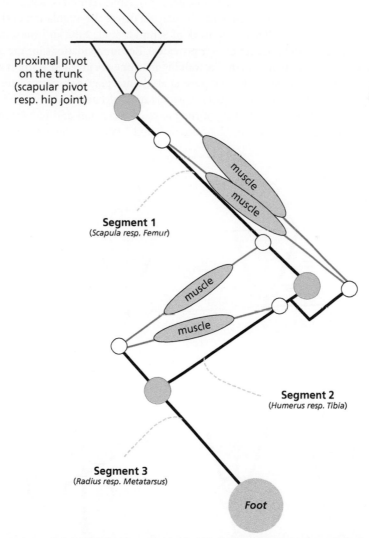

Figure 31.6 Pantograph leg of a three-segmented mammalian leg. (cf. Fischer (1998a), Fischer and Blickhan (2006), Fischer and Witte (2007)) Two four joint–four bars gears are formed by three bony bars (constant length) and one coupling muscle each. The geometry under compressive load is defined by muscles in the diagonal. All muscles anatomically are classified to be extensor muscles.

Adapting to different substrates and body sizes

In climbing, the feet at the end of the 3rd segment longitudinally (e.g. around the branch axis) have to transfer higher torques than in walking and running on flat ground (Figure 31.7). This need for grasping paves the way for the development of manipulation (see Cutkosky, Chapter 30, this volume), but it leads to higher masses for feet (and hands in primates; for details on mass distribution cf. Preuschoft et al. (1998), and on primate climbing Schmidt and Fischer (2000)), and thus influences the proportions of extremities and central kinematic mechanisms like the usage of the scapula. Due to the allometric gap between $1/L$ and $1/L^2$ (where L is a length representative for the animal's size) between load (mass prop. L^3, mass moment of inertia prop. L^5) and drive (muscle force prop. L^2, muscle torque prop. L^3), large mammals have more extended legs than smaller ones. But even in elephants the scapula drives the angulation of the forelimb "column" (Hutchinson et al. 2006; cf. the work of Schmidt et al. 2002 and consecutive work, especially what concerns the implications for primates). Recommendations for the mechanical structure of a quadrupedal mammalian-like walking machine are given in Witte et al. (2000).

Following Hildebrand (1985), the temporo-spatial interplay between extremities and trunk in terrestrial locomotion may cover a wide range of combinations. The formation of a nervous plexus allowed the shifting of the phase of excitation and movement between legs, extending the "bound gaits" to "half-bound" and "gallop," where a temporo-spatial

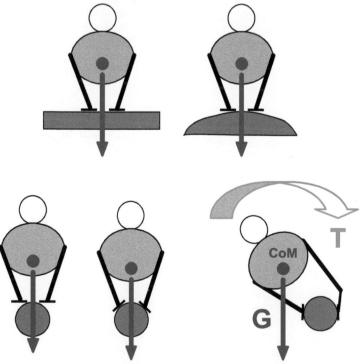

Figure 31.7 Interdependencies between locomotion on flat ground and on branches. The transversal size of the animal relative to the diameter of the branch is decisive for the attribute "arboreal locomotion" in contrast to just walking or running If the vector of the weight force G passes outside the branch, a compensating torque T is necessary, grasping defines pivots (cf. Jenkins (1974), Preuschoft (2002), Lammers and Gauntner (2008)).

difference between the ground contact of the fore limbs (which allows the animal to explore the substrate contact at first) increases adaptivity to substrates. The tendency to develop more lateral footfall sequences in a walk is complemented by diagonal gaits (e.g. trot) with more evenly distributed phase differences between fore and hind limbs at medium speeds, adopted from reptile precursors. In the extreme, diagonal gaits are used by endurant trotters like horses or camels, which reduced muscle-tendon complexes (especially the superficial digital flexors) to rather pure collagenous springs (cf. Preuschoft & Günther (1994) and corresponding work of Preuschoft for functio-morphological consequences). To abandon the tuning option allows maximization of elastic energy storage and thus optimization of power consumption for locomotion. Such kind of specialization usually is a feature of large mammals (small mammals have a restricted overview of their environment)—the decrease of acceleration by usage of elasticity may allow predators to finish each of those steady evolutionary experiments by selection of the prey as an individual. Thus, small mammals, disregarding their fur, more or less all look alike: ellipsoid trunk with ellipsoid head, eventually plus tail (cf. Fischer 1998a,b).

Up to 60° of incline, rats simply use their normal terrestrial half-bound and especially symmetrical (diagonal) patterns. On steeper substrates, they project their weight upwards "relapsing" into the phylogenetically older synchronous gait patterns, in extreme bounds (cf. Andrada et al. 2013). Surprisingly, the energetic contribution of the trunk to rat climbing is small in relation to the limbs: less than 20% of conservative mechanical energy components. Since the overall DOFs (degrees of freedom) of the trunk are high and steadily used, we always have to keep in mind the difference between arguments derived from kinematics and those from dynamics/kinetics.

Biomimetic systems

How far are these insights of technical biology (mainly of functional morphology) biomimetically transferable into machines? As yet, only to a minor extent. In humanoid robots, where humans are paragons (being tetrapods with an evolutionary heritage common with the other mammals, but not quadrupeds), close biomimesis is needed due to human pattern recognition and psychology and the need to avoid the "uncanny valley", Mori (1970). In contrast, quadrupedal robots may be designed much more abstractly and adapted to specific tasks.

Should quadrupedal robots make use of the principles just described? Not necessarily: Kimura`s *Tekken* robot (Kimura et al. 2001, Fukuoka et al. 2003) and Boston Dynamics' *BigDog* (Raibert et al. 2008) illustrate the high performance ability of machines with rigid bodies and even two-segmented legs, without provoking an experience of uncanniness in human observers, looking at a copy of a pet. But in terms of energy efficiency, either for high velocity and/or high endurance, it could be helpful to follow biomimetic principles. The high-ambition M.I.T. *Cheetah* (Seok et al. 2013) makes use of up to three-segmented legs (see Figure 45.1), to allow the usage of "pronk," a special gait in mammals shown, for example, by springboks. Presently, the only robotics group seeking to faithfully implement bionic the principles of quadrupedal mammalian locomotion mechanics, including the use of the trunk for locomotion, and in combination with biomimetics of nervous systems, is the Biorobotics Laboratory (BioRob) at Ecole Polytechnique Fédérale de Lausanne (Figure 31.8; cf. Rutishauser et al. (2008), Spröwitz et al. (2013), Ijspeert (2014), Eckert et al. (2015)).

In conclusion, it seems likely that a need to improve energy autonomy may provide the strongest drive to prefer biology-driven approaches over technological ones in quadruped robots.

Figure 31.8 Examples for quadrupedal robots from Biorobotics Laboratory BioRob at EPFL. Left: Cheetah Cub. Right: Turning performance of Cheetah Cub S ("S" for "spine).

Learning more

Ijspeert (2014) has provided a useful review of recent attempts to develop biomimetic locomotion systems, focused towards quadruped robots. A good number of other chapters in this handbook consider the challenge of locomotion in different animal phyla and in biomimetic living machines. Cruse and Schilling (Chapter 24) describe some the central pattern generating circuits underlying the locomotion gaits of animals and how this design can be co-opted to support robust locomotion for robots. Trimmer (Chapter 41) considers worm-like locomotion in soft invertebrates, while Quinn and Ritzmann (Chapter 42) focus on locomotion in invertebrates with hard exoskeletons and how robotics has learned from insect biology to design robust hexapod walking and running machines. Prescott (Chapter 45) discusses some mammal-like quadrupedal robots including MIT's *Cheetah*, while Metta and Cingolani (Chapter 47) discuss the challenge of creating stable bipedal locomotion in humanoid robots. Elsewhere, the chapter by Kruusmaa (Chapter 44) describes the design of efficient swimming for aquatic robots based on fish morphology, and those by Hedenström (Chapter 32) and Send (Chapter 45) consider flight in insects and birds and its emulation in aviatic living machines. Animals have evolved different kinds of external coverings (integuments) adapted to locomotion in specific environments, e.g. fish scales, insect wings, bird feathers. Pang et al. (Chapter 22) discuss and exemplar adaptation—how vertebrate skin in reptiles such as geckos has adapted at a micro-geometric scale to provide better grip for climbing on steeply angled surfaces.

References

Alexander, R. M. (1977). Mechanics and scaling of terrestrial locomotion. In: T. J. Pedley (ed.), *Scale effects in locomotion*. London: Academic Press, pp. 93–110.

Alexander, R. M. (1988). *Elastic mechanisms in animal movement*. Cambridge, UK: Cambridge University Press.

Andrada, E. (2008) *A new model of the human trunk mechanics in walking*. Berichte aus der Biomechatronik 1, Ilmenau: Univ.-Verl.

Andrada, E., Mämpel, J., Schmidt, A., Fischer, M. S., Karguth, A., and Witte, H. (2013). From biomechanics of rats' inclined locomotion to a climbing robot. *Int. J. Des. Nat. Ecodyn.*, 8(3), 192–212.

Blickhan, R. (1989). The spring-mass model for running and hopping. *J. Biomech.*, 22, 1217–27.

Blickhan, R., and Full, R. J. (1993). Similarity in multilegged locomotion: bouncing like a monopod. *J. Comp. Physiol. A Neuroethol. Sens. Neural. Behav. Physiol.*, 173(5), 509–17.

Blickhan, R., Seyfarth, A., Geyer, H., Grimmer, S., and Wagner, H. (2007). Intelligence by mechanics. *Philos. Transact. A: Math. Phys. Eng. Sci.*, **365**, 199–220.

Braune, W., and Fischer, O. (1895). Der Gang des Menschen. I. Teil. *Abh. math.-phys. Kl. kgl.-sächsischer Wiss.*, **21/4**, 151–322.

Brooks, R. A. (1991). Intelligence without representation. *Artificial Intelligence*, 1–3(47), 139–159.

Brooks, R. A., Breazeal, C., Irie, R., Kemp, C. C., Marjanovic, M., Scassellati, B., and Williamson, M. M. (1998). Alternative essences of intelligence. *Proc. AAAI/IAAI*, **1998**, 961–68.

Cavagna, G. A., Heglund, N. C., and Taylor, C. R. (1977). Mechanical work in terrestrial locomotion: two basic mechanisms for minimizing energy expenditure. *Am. J. Physiol.*, **233**, 243–61.

Cavagna, G. A., Thys, H., and Zamboni, A. (1976). The sources of external work in level walking and running. *J. Physiol.*, **262**, 639–57.

Dawson, T. J., and Taylor, C. R. (1973). Energetic cost of locomotion in kangaroos. *Nature*, **246**, 313–14.

Eckert, P., Spröwitz, A., Witte, H., and Ijspeert, A. J. (2015). *Comparing the effect of different spine and leg designs for a small bounding quadruped robot.* In: Robotics and Automation (ICRA), 2015 IEEE International Conference, pp. 3128–33.

Farley, C. T., Blickhan, R., Saito, J., and Taylor, C. R. (1991). Hopping frequency in humans: a test of how springs set stride frequency in bouncing gaits. *J. Appl. Physiol.*, **71**(6), 2127–32.

Fischer, M.S. (1998a). Crouched posture and high fulcrum, a principle in the locomotion of small mammals: the example of the rock hyrax (*Procavia capensis*)(Mammalia: Hyracoidea). *J. Hum. Evol.*, **26**(5–6), 501–52.

Fischer, M. S. (1998b). *Die Lokomotion von* Procavia capensis *(Mammalia, Hyracoidea): Zur Evolution des Bewegungssystems bei Säugetieren.* Keltern: Goecke & Evers.

Fischer, M. S., and Blickhan, R. (2006). The tri-segmented limbs of therian mammals: kinematics, dynamics, and self-stabilization—a review. *J. Exp. Zool. Part A: Ecological Genetics and Physiology*, **305**(11), 935–52.

Fischer, M. S., Schilling, N., Schmidt, M., Haarhaus, D., and Witte, H. (2002). Basic limb kinematics of small therian mammals. *J. Exp. Biol.*, **205**(9), 1315–38.

Fischer, M.S., and Witte, H. (2007). Legs evolved only at the end! *Philos. Transact. A: Math. Phys. Eng. Sci.*, **365**(2007), 185–98.

Full, R., and Koditschek, R.J. (1999). Templates and anchors: neuromechanical hypotheses of legged locomotion on land. *J. Exp. Biol.*, **202**(23), 3325–32.

Fukuoka, Y., Kimura, H., and Cohen, A.H. (2003). Adaptive dynamic walking of a quadruped robot on irregular terrain based on biological concepts. *Int. J. Rob. Res.*, **22**(3–4), 187–202.

Geyer, H., Seyfarth, A., and Blickhan, R. (2006). Compliant leg behaviour explains basic dynamics of walking and running. *Proc. R. Soc. London B: Biol. Sci.*, **273**(1603), 2861–67.

Hildebrand, M. (1985). Walking and running. In: M. Hildebrand, D. M. Bramble, K. F. Liem, and D. B. Wake (eds), *Functional vertebrate morphology*. Cambridge, MA: Harvard University Press, pp. 38–57.

Hill, A.V. (1938). The heat of shortening and dynamics constants of muscles. *Proc. R. Soc. Lond. B: Biol. Sci.*, **126** (843), 136–95.

Hutchinson, J. R., Schwerda, D., Famini, D. J., Dale, R. H., Fischer, M. S., and Kram, R. (2006). The locomotor kinematics of Asian and African elephants: changes with speed and size. *J. Exp. Biol.*, **209**(19), 3812–27.

Ijspeert, A. J. (2014). Biorobotics: Using robots to emulate and investigate agile locomotion. *Science*, **346**(6206), 196–203.

Jacobs, R., Bobbert, M. F., and van Ingen Schenau, G. J. (1996). Mechanical output from individual muscles during explosive leg extensions: the role of biarticular muscles. *J. Biomech.*, **29**(4), 513–23.

Jenkins Jr., F.A. (1974). Tree shrew locomotion and the origins of primate arborealism. In: F.A. Jenkins, Jr (ed.), *Primate Locomotion*. New York: Academic Press, pp. 85–115.

Karakasiliotis, K., Schilling, N., Cabelguen, J. M., and Ijspeert, A. J. (2013). Where are we in understanding salamander locomotion: Biological and Robotic Perspectives on Kinematics. Biol. Cybern., **107**(5), 529–44.

Kimura, H., Fukuoka, Y., and Konaga, K. (2001). *Towards 3D adaptive dynamic walking of a quadruped robot on irregular terrain by using neural system model.* Proceedings. 2001 IEEE/RSJ International Conference on Intelligent Robots and Systems, 2001. IEEE.

Labeit, S., Kolmerer, B., and Linke, W.A. (1997). The giant protein titin. Emerging roles in physiology and pathophysiology. *Circ. Res.*, **80**(2), 290–94.

Lammers, A. R., and Gauntner, T. (2008). Mechanics of torque generation during quadrupedal arboreal locomotion. *J. Biomech.*, **41**(11), 2388–95.

McMahon, T. A. (1973). Size and shape in biology. *Science*, **179**, 1201–4.

Mochon, S., and McMahon, T. A. (1980a). Ballistic walking. *J. Biomech.*, **13**, 49–57.

Mochon, S., and McMahon, T. A. (1980b). Ballistic walking: an improved model. *Math. Biosc.*, **52**, 241–60.

Mori, M. (1970). Bukimi no tani [the uncanny valley]. *Energy*, **7**, 33–35.

Mussa-Ivaldi, F. A., Morasso, P., and Zaccaria, R. (1988). A distributed model for representing and regularizing motor redundancy. *Biol. Cybern.*, **60**, 1–16.

Pfeifer, R., and Bongard, J. (2006). *How the body shapes the way we think: a new view of intelligence.* Cambridge, MA: MIT Press.

Preuschoft, H. (2002). What does "arboreal locomotion" mean exactly and what are the relationships between "climbing", environment, and morphology? *Z. Morphol. Anthropol.*, **83**(2–3), 171–88.

Preuschoft, H., and Günther, M.M. (1994). Biomechanics and body shape in primates compared with horses. *Z. Morphol. Anthropol.*, **80**, 149–65.

Preuschoft, H., Christian, A., and Günther, M. M. (1998). Size dependence in prosimian locomotion and their implications for the distribution of body mass. *Folia Primatol.*, **69** (Supplement 1), 69–81.

Raibert, M., Blankespoor, K., Nelson, G., Playter, R., and the BigDog Team (2008). *BigDog, the rough-terrain quadruped robot.* Proceedings of the 17th World Congress of the International Federation of Automatic Control, Seoul, Korea, July 6–11, 10822–25.

Reilly, S. M., McElroy, E. J., Odum, R. A., and Hornyak, V. A. (2006). Tuataras and salamanders show that walking and running mechanics are ancient features of tetrapod locomotion. *Proc. R. Soc. London B: Biol. Sci.*, **273**(1593), 1563–8.

Rutishauser, S., Sproewitz, A., Righetti, L., and Ijspeert, A. J. (2008). *Passive compliant quadruped robot using central pattern generators for locomotion control.* In: Biomedical Robotics and Biomechatronics. 2nd IEEE RAS & EMBS International Conference on Biomedical Robotics and Biomechatronics BioRob, Scottsdale, 2008. IEEE, pp. 710–15.

Schmidt, M., and Fischer, M. S. (2000). Cineradiographic study of forelimb movements during quadrupedal walking in the brown lemur (*Eulemur fulvus*, Primates: Lemuridae). *Am. J. Phys. Anthropol.*, **111**(2), 245–62.

Schmidt, M., Voges, D., and Fischer, M. S. (2002). Shoulder movements during quadrupedal locomotion in arboreal primates. *Z. Morphol. Anthropol.*, **83**(2–3), 235–42.

Seok, S., Wang, A., Chuah, M. Y., Otten, D., Lang, J., and Kim, S. (2013). *Design principles for highly efficient quadrupeds and implementation on the MIT Cheetah robot.* Robotics and Automation (ICRA), 2013 IEEE International Conference, 3307–12.

Seyfarth, A., Geyer, H. Günther, M., and Blickhan, R. (2002). A movement criterion for running. *J. Biomech.*, **35**(5), 649–55.

Spröwitz, A., Tuleu, A., Vespignani, M., Ajallooeian, M., Badri, E., and Ijspeert, A. J. (2013). Towards dynamic trot gait locomotion: Design, control, and experiments with Cheetah-cub, a compliant quadruped robot. *Int. J. Rob. Res.*, **32**(8), 932–50.

Taylor, C. R., Schmidt-Nielsen, K., and Raab, J. L. (1970). Scaling of energetic cost of running to body size in mammals. *Am. J. Physiol.*, **219**, 1104–107.

Weber, E., and Weber, W. (1836). *Die Mechanik der menschlichen Gehwerkzeuge.* Göttingen: Dietrich.

Weiss, P. L., Hunter, I. W., and Kearney, R. E. (1988). Human ankle joint stiffness over the full range of muscle activation levels. *J. Biomech.*, **21**(7), 539–44.

Witte, H. (1992). *Über mechanische Einflüsse auf die Gestalt des menschlichen Körpers.* Ph.D. thesis, Ruhr-Universität Bochum, Germany.

Witte, H. (2002) Hints for the construction of anthropomorphic robots based on the functional morphology of human walking. *J. Rob. Soc. Jpn.*, **20**(3), 247–54.

Witte, H., Biltzinger, J., Hackert, R., Schilling, N., Schmidt, M., Reich, C., and Fischer, M. S. (2002). Torque patterns of the limbs of small therian mammals during locomotion on flat ground. *J. Exp. Biol.*, **205**(9), 1339–53.

Witte, H., Fremerey, M., Weyrich, S., Mämpel, J., Fischheiter, L., Voges, D., Zimmermann, K., and Schilling, C. (2013). *Biomechatronics is not just biomimetics.* In: Proceedings of the 9th International Workshop on Robot Motion and Control, Wasowo Palace, Wasowo, Poland, July 3–5, 2013, pp. 74–79.

Witte, H., Ilg, W., Eckert, M., Hackert, R., Schilling, N., Wittenburg, J., Dillmann, R., and Fischer, M. S. (2000). Konstruktion vierbeiniger Laufmaschinen. *VDI-Konstruktion*, **9**(2000), 46–50.

Witte, H., Lesch, C., Preuschoft, H., and Loitsch, C. (1995). Die Gangarten der Pferde: sind Schwingungsmechanismen entscheidend? Teil II: Federschwingungen bestimmen den Trab und den Galopp. *Pferdeheilkunde*, **11**(4), 265–72.

Witte, H., Preuschoft, H., and Recknagel, S. (1991). Human body proportions explained on the basis of biomechanical principles. *Z. Morphol. Anthropol.*, **78**(3), 407–23.

Chapter 32

Flight

Anders Hedenström

Department of Biology, Lund University, Sweden

Powered flight is perhaps the ultimate biomimetic challenge that confronts engineers, where nature provides illustrious material in the form of birds, bats, insects, and the extinct pterosaurs. With the independent evolutionary tinkering leading to flight among these four groups, there are manifold illustrations on functional variation and the suite of adaptations necessary for aerial locomotion, including aerodynamic lift generators (wings), flight muscle adaptation (engine), energy deposits for endurance flight (fuel), and sensory systems for control of actuators and navigation. In many respects biological systems show superior performance to that of any man-made mimics, which makes animal flight the ideal model for biomimetic design. In this chapter I describe basic features of aerodynamics and flight performance in animal flyers.

Biological principles

Flight morphology

The wings of vertebrates are modified forelimbs, while the wings of insects are probably derived from dorsal extensions of the legs. In birds the majority of the wing surface is formed by flight feathers, which are attached along the ulna (secondaries) and the hand skeleton (primaries). Feathers deform aero-elastically under aerodynamic load. The shaft of each flight feather is attached to the wing skeleton via reinforced follicles, and the combined arrangement of primaries and secondaries form a "distributed spar" that absorbs the aerodynamic force from the wing (Pennycuick 2008). Birds have a cantilever wing, in which a stiff structure delivers all the bending and torsional loads to the shoulder joint.

In contrast to birds, bats have a flexible wing membrane with little resistance to bending or torsion. The membranous skin is stretched between elongated finger bones. The wing can be flexed at the elbow and wrist joints, while the first digit (thumb) can be moved to lower an anterior leading edge flap to increase the camber. The bat wing membrane has muscles running fore–aft that are not attached to the skeleton, which when shortened flatten the wing surface and when relaxed increase the camber.

The insect wing consists of a thin membrane supported by a system of veins containing fluid. There are no joints in the wing, preventing it from being flexed as in birds or bats, but through the axillary sclerites (anatomical features of the body at the insertion at the hinge of the wing) the veins can be affected to adjust the wing shape.

Aerodynamics

To fly, an animal must generate forces to counteract the pull of gravity and aerodynamic drag. In steady gliding flight the wings are held in a fixed position to generate aerodynamic lift to balance the weight, while potential energy is converted to do work against the drag. This is the reason why a glider loses height against the surrounding air. In active (powered) flight an aerodynamic force is generated by the flapping wings that counteracts both weight and aerodynamic drag. The continuous motion and shape deformation of flapping wings reflect the complex variation of the aerodynamic force during the wingbeat cycle, which often calls for simplification of the analysis (Pennycuick 2008).

The aerodynamic power required to fly can be divided into three main components: (1) *induced power* due to the generation of downwash (lift), (2) *profile power* due to the drag of the flapping wings, and (3) *parasite power* due to the drag of the non-lifting body. In some systems also inertial power, due to the angular acceleration of the wing's inertia, is added to the total, but in cruising forward flight this component is small or negligible. The sum of the power components yield the "power curve," which is the function of power required to fly (P) in relation to speed through the air (U_a). A generic power curve is shown in Figure 32.1, exhibiting the characteristic U-shape. This power curve describes the required mechanical power output that the animal has to generate by the periodic contractions by the flight muscles. Muscle work requires the mobilization of increased metabolism when carbohydrates or fatty acids are converted into muscle contractions, which generates excessive heat as a by-product. In vertebrates the energy conversion efficiency is somewhere near 20–30% (Pennycuick 2008), but mechanical power output has proven notoriously difficult to measure (Askew and Ellerby 2007). However, the total empirical literature strongly supports the existence of a general U-shape for the power–speed relationship in vertebrates (Engel et al. 2010). It has recently been shown that insects, such as hawkmoths (*Manduca sexta*) flying freely in a wind tunnel, also exhibit a U-shaped power–speed relationship, by measuring the rate of kinetic energy added to the wake by the flapping wings (Warfvinge et al. 2017).

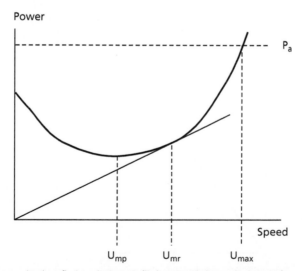

Figure 32.1 Power required to fly in relation to flight speed through the air for an animal. The horizontal dotted line denotes the power available from the flight muscles (P_a), and the illustrated characteristic speeds are: minimum power speed (U_{mp}), maximum range speed (U_{mr}), and maximum sustainable speed (U_{max}).

The power curve indicates that it is relatively expensive to fly slowly and to hover ($U_a = 0$ m/s), while especially many insects are adapted for this type of flight mode. True hoverers flap their wings back and forth in a nearly horizontal stroke plane, where the wings beat in a propeller-like way around their root. At the turn of each half-stroke the wings flip over by being rotated around a span-wise axis before reversing direction during the next half-stroke, so that during the back-stroke (morphologically the upstroke) the underside of the wing acts as the aerodynamic upper side. Seen from the side the wing tip follows a tilted figure-8 in true hovering. Hummingbirds (Trochilidae) have a similar hovering kinematics as insects, but in contrast to insects that have a symmetric force generation between up- and downstroke, the downstroke generates about 75% of the required lift in hummingbirds (Warrick et al. 2005). In other birds and bats the wings are usually flexed during the upstroke to various degrees, and in slow flight the upstroke is often aerodynamically inactive.

Aerodynamic lift of a fixed wing is

$$L = \frac{1}{2}\rho U^2 S C_L,$$

(1)

where ρ is air density, U is airspeed, S is wing area, and C_L is a dimensionless lift coefficient that describes the efficacy of the wing to generate lift. The lift coefficient includes parameters such as shape, surface structure, and camber. An alternate way of expressing lift is to invoke circulation, Γ, which is the integrated vorticity over a fluid area and in the case of an airfoil it measures the bound vorticity (dimension length squared divided with time), as

$$L = \rho \Gamma b U,$$

(2)

where b is wing span. By noting that $S = bc$, where c is the mean wing chord, and combining equations (1) and (2) we obtain $C_L = 2\Gamma/Uc$. Under steady flow conditions a wing can attain a certain maximum value of C_L at a certain angle of attack (Figure 32.2). If the angle of attack is increased beyond the point of maximum C_L the flow separates into a turbulent wake and C_L drops dramatically, which is when the wing has stalled. Empirical measurements show

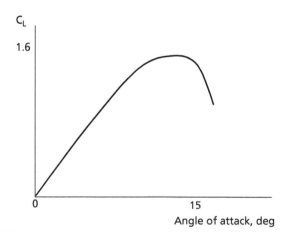

Figure 32.2 Lift coefficient (C_L) in relation to angle of attack under steady conditions. At some angle of attack the flow separates and the lift coefficient drops dramatically, which is when stall occurs.

that maximum values of C_L are about 1.6 or less, depending on wing shape and Reynolds number[1].

To calculate the aerodynamic force on a flapping wing according to eqn (1) is complicated because the local speed and angle of attack changes continually throughout the stroke cycle. By dividing the wing in thin strips and calculating the force on each such strip for different time-steps, the total force is obtained by adding the calculated force for each strip and time-instance throughout the wingbeat. This procedure implicitly assumes that the force at each wing location and time step can be represented by the steady force obtained from eqn (1), i.e. that the quasi-steady assumption applies. When applying this approach to the kinematics of certain insects, perhaps most famously to the bumblebee, the required C_L was found to be larger than $C_{L,max}$. The same result was found for hovering flycatchers (Norberg 1975). This paradox could be resolved by the discovery of several unsteady aerodynamic mechanisms that generate additional lift to that given by steady-state aerodynamics (Weis-Fogh 1975). Common to these mechanisms is the formation of a Leading Edge Vortex (LEV), which forms due to flow separation at high angles of attack. If the LEV remains attached to the wing surface, which it does during the short translation of a wingbeat, the circulation of the LEV enhances the lift. The LEV will be present due to delayed stall or the clap-and-fling mechanisms, but the time history of its growth will differ. In a delayed stall it takes some time, and translation of the wing, before the LEV assumes full strength, while in a clap-and-fling the LEV is formed immediately when the wings separate. Transiently enhanced lift can also arise due to the fast rotation along the spanwise axis at supination and pronation when the wing changes direction.

According to Kelvin's circulation theorem a vortex is shed into the wake if lift changes on a wing, and the circulation of that vortex is equal to the associated change in bound circulation but has opposite spin. Flying animals, gliding or flapping, leave behind a vortex wake that can be interpreted as an aerodynamic "footprint." At slow speed and hovering, when the upstroke is unloaded, the wake consists of elliptic vortex loops that encircle the induced downwash. When the speed increases the upstroke also generates force and the wake will consist of a pair of undulating wing-tip vortices with transverse vortices shed when the lift changes during the wingbeat cycle (Figure 32.3). At slow flight the wing may move through a vortex that was shed on the previous half-stroke, which under certain circumstances can be recaptured and help increase the lift. This wake capture is yet another unsteady aerodynamic mechanism "discovered" by evolution (Lehmann 2008).

By means of flow visualization, such as Particle Image Velocimetry (PIV), vortex wakes can be depicted and quantities such as vorticity and circulation measured (Spedding et al. 2003). In slowly flying bats measured values for $2\Gamma/Uc$, i.e. C_L, were higher than allowed by steady aerodynamics, and subsequent flow visualization near the wing showed that small bats also develop LEVs that may contribute up to 40% of the weight support (Muijres et al. 2008). This mechanism is also found in hummingbirds and flycatchers, which are adapted for slow/hovering flight.

Flight performance

Flight is used in many ways among animals, ranging from display flight for mate attraction, escape maneuvers, and foraging and food search, to long-distance migration. Depending on how flight is used, animals have evolved wing morphology and size according to their ecology, while the resulting design is a compromise between conflicting selection pressures. Migrants

---

[1] Reynolds number is defined as the ratio of inertial to viscous forces, equal to Ul/ν, where U is velocity, l is a linear dimension (typically the chord c), and ν is kinematic viscosity.

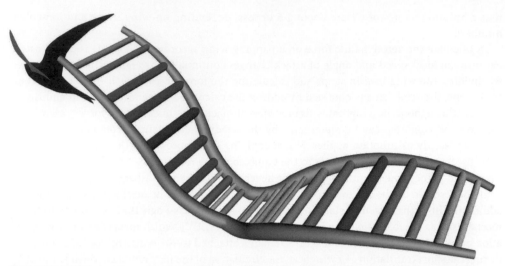

Figure 32.3 Wake vortices shed behind by a bird in cruising flight. The main structures are the two undulating wing-tip vortices, and the transverse vortices reflect changes in lift during the wingbeat cycle.

Reprinted from *Journal of Experimental Biology*, 211 (5), P. Henningsson, G. R. Spedding, and A. Hedenström, Vortex wake and flight kinematics of a swift in cruising flight in a wind tunnel, pp. 717–730, doi: 10.1242/jeb.012146, Copyright © 2008 The Company of Biologists.

have often evolved relatively long and slender wings (high aspect ratio) as an adaptation to reduce flight costs, while maneuvering flight requires relatively short wings and low wing loading (weight divided by wing area).

For sustained flight the power available from the flight muscles must equal the power required according to the power curve (Figure 32.1). In most cases power available is much larger than required for medium flight speeds, allowing for acceleration, maneuvers, or climbing flight. The power margin also allows for fuel accumulation before long migratory flights. Added fuel will increase the intercept of the power curve because added mass increases flight costs, and therefore the power available sets the limit for how much fuel (mainly fat in migrants) can be stored, thereby also limiting the flight range. In long-distance migrating birds the weight can be more than doubled due to fuel stores before long flights across barriers such as oceans and deserts, an increase that often is accompanied with increased size of the flight muscles. The record long-haul flight is performed by the Alaskan bar-tailed godwit (*Limosa lapponica baueri*), which makes an 11 000 km non-stop flight from Alaska to New Zealand in the autumn (Gill et al. 2009). This epic fight takes 7–9 days to complete depending on winds en route. The spring migration is split into two flights with a stopover in the Yellow Sea region, but these flights are also impressive. It was recently demonstrated, using accelerometers, that alpine swifts (*Tachymarptis melba*) and common swifts (*Apus apus*) may be airborne for 6–10 months during the non-breeding season (Liechti et al. 2013, Hedenström et al. 2016)—while these species normally feed on insects in the open air this means they also have to sleep while on the wing. That it is possible for birds to sleep while flying was recently shown for frigatebirds (Rattenborg et al. 2016).

The shape of the power curve defines characteristic speeds associated with minimum power and maximum range (U_{mp} and U_{mr}; Figure 32.1). The former is the best alternative to stay airborne for as long time as possible given a certain amount of energy, while U_{mr} should be favored when commuting or migrating at minimum energy cost per unit distance. Observations show that birds and bats do select an appropriate flight speed according to ecological context. The maximum range speed is modified if there is a head or tail wind so that U_{mr} increases into head winds and decreases with tail winds. Also this prediction has been observed in migrating birds. The power available from flight muscles also defines if a bird can hover (U = 0 m/s) and determines the maximum sustainable flight speed, U_{max} (Figure 32.1).

The atmosphere is never at rest and because of features in the landscape and weather, vertical winds occur as slope-lift or lee waves when horizontal winds are deflected by mountain ridges or pass over topographical features. On sunny days air near the ground is heated differentially depending on the heat reflectance characteristics of the ground. Because warm air expands its density is reduced and the air will rise as convective thermals. Flying animals have discovered this "free energy" and use it for soaring. For example raptors can reach high altitudes by soaring within thermals, and then glide in the desired direction until another thermal is encountered for a new climb phase. Such a flight strategy may be favorable, since gliding flight is energetically much cheaper than flapping flight, especially in large birds (Pennycuick 2008).

In a series of geometrically similar birds flight mechanical theory predicts that power required for flight scales as $M^{7/6}$ and flight speed is expected to scale as $M^{1/6}$ (Pennycuick 2008). The power available from the flight muscles increases in direct proportion to the body mass, which means that the curves for available power and that for required power cross at a certain body mass. This limit for maximum body mass while still being able to fly by powered flight is about 12–15 kg, which is where the largest flying extant bird species are found. Flightless species such as ostriches and cassowaries have relaxed this requirement and have evolved much larger sizes. Historically there existed larger birds (e.g. *Argentavis*, 70 kg), but these most likely relied on soaring flight that requires less muscle power than flapping flight. Measurements of flight speeds in birds show that speed increases less steeply with body mass than the expected 1/6 exponent (Pennycuick et al. 2013), which can partially be explained by the fact that real birds deviate from isometric scaling with larger birds having progressively longer wings than small birds.

Biomimetic systems

The ultimate goal in biomimetic micro air vehicle (MAV) design is to create an object that has the same size as its biological model. It should be propelled by flapping wings and armed with sensory systems providing continuous input, with feedback to the motor output for appropriate response when encountering obstacles. Take-off and landing should be possible when wind gusts occur. Between-platform communication about position and direction should allow movement in groups (flocks). Millions of years of evolutionary tinkering have solved all these problems among thousands of animal species, which serve as inspiration for engineering efforts. Even though engineering designs lag behind their biological models on all fronts, scientists have recently made significant progress both regarding miniaturization and control of flight (Ma et al. 2013, Fuller et al. 2014).

Learning more

To understand all aspects of animal flight a genuinely cross-disciplinary approach is needed, drawing from different fields such as fluid dynamics, zoology, biomechanics, neurobiology, physiology, and engineering. Updated reviews about bird, bat, and insect flight can be found in Pennycuick (2008), Swartz et al. (2012), and Dickinson (2006), respectively. Papers about biomimetic flight have recently been summarized by Lentink and Biewener (2010). Work towards a bird-like flapping wing MAV includes the SmartBird, supported by Festo, and described by

Send (Chapter 46, this volume); see Gerdes et al. (2012) for a review of other efforts in this direction. Bat flight has also inspired MAV design such as the robot developed by Ramezani et al. (2017) and shown in Figure 64.1 (this volume). Liu et al. (2016) provide a review of the biomechanics of insect flight and their emulation in small MAVs. A bibliometric review of progress in the development of MAVs, including biomimetic systems, is provided by Ward (2017).

References

Askew, G.N., and Ellerby, D.J. (2007). The mechanical power requirements of avian flight. *Biol. Lett.*, 3, 445–8.

Engel, S., Bowlin, M.S., and Hedenström, A. (2010). The role of wind-tunnel studies in integrative research on migration biology. *Integrative & Comparative Biology*, 50, 323–35.

Fuller, S. B., Karpelson, M., Censi, A., Ma, K. Y., and Wood, R. J. (2014). Controlling free flight of a robotic fly using an onboard vision sensor inpsired by insect ocelli. *J. R. Soc. Interface*, 11, 20140281.

Gerdes, J. W., Gupta, S. K., and Wilkerson, S. A. (2013). A review of bird-inspired flapping wing miniature air vehicle designs. *J. Mech. Robot.*, 4, 021003 (2012).

Gill, R.E., Jr., Tibbitts, T.L., Douglas, D.C., Handel, C.M., Mulcahy, D.M., Gottschalck, J.C., Warnock, N., McCaffery, B.J., Battley, P.F., and Piersma, T. (2009). Extreme endurance flights by landbirds crossing the Pacific Ocean: ecological corridor rather than barrier? *Proc. R. Soc. B*, 276, 447–57.

Hedenström, A., Norevik, G., Warfvinge, K., Andersson, A., Bäckman, J., and Åkesson, S. (2016). Annual 10-month aerial life phase in common swift *Apus apus*. *Current Biology*, 26, 3066–70.

Henningsson, P., Spedding, G.R., and Hedenström, A. (2008). Vortex wake and flight kinematics of a swift in cruising flight in a wind tunnel. *J. Exp. Biol.*, 211, 717–30.

Lehmann, F.-O. (2008). When wing touch wakes: understanding locomotor force control by wake-wing interference in insect wings. *J. Exp. Biol.*, 211, 224–33.

Lentink, D., and Biewener, A.A. (2010). Nature-inspired flight—beyond the leap. *Bioinspiration and Biomimetics*, 5, doi:10.1088/1748-3182/5/4/040201.

Liechti, F., Wivliet, W., Weber, R., and Bächler, E. (2013). First evidence of a 200-day non-stop flight in a bird. *Nature Communications*, 4, 2554.

Liu, H., Ravi, S., Kolomenskiy, D., and Tanaka, H. (2016). Biomechanics and biomimetics in insect-inspired flight systems. *Philosophical Transactions of the Royal Society B: Biological Sciences*, 371(1704), 20150390. doi:10.1098/rstb.2015.0390

Ma, K.Y., Chirarattananon, P., Fuller, S.B., and Wood, R.J. (2013). Controlled flight of a biologically inspired, insect-scale robot. *Science*, 340, 603–7.

Muijres, F.T., Johansson, L.C., Barfield, R., Wolf, M., Spedding, G.R., and Hedenström, A. (2008). Leading-edge vortex improves lift in slow-flying bats. *Science*, 319, 1250–3.

Norberg, U.M. (1975). Hovering flight in the pied flycatcher (*Ficedula hypoleuca*). In: *Swimming and Flying in Nature*, Vol. 2 (Wu, T.Y.T., Brokaw, C.J., and Brennen, C., eds.), pp 869–81. Plenum Press, New York.

Pennycuick, C.J. (2008). *Modelling the flying bird*. Academic Press: London.

Pennycuick, C.J., Åkesson, S., and Hedenström, A. (2013). Air speeds of migrating birds observed by ornithodolite and compared with predictions from flight theory. *J Roy. Soc. Interface*, 10, 20130419.

Ramezani, A., Chung, S.-J., and Hutchinson, S. (2017). A biomimetic robotic platform to study flight specializations of bats. *Science Robotics*, 2(3). doi:10.1126/scirobotics.aal2505

Rattenborg, N. C., Voirin, B., Cruz, S. M., Tisdale, R., Dell'Omo, G., Lipp, H.-P., Wikelski, M., and Vyssotski, A. L. (2016). Evidence that birds sleep in mid-flight. *Nat. Commun.*, 7, 12468.

Spedding, G. R., Rosén, M., and Hedenström, A. (2003). A family of vortex wakes generated by a thrush nightingale in free flight in a wind tunnel over its entire natural range of flight speeds. *J. Exp. Biol.*, 206, 2313–44.

Swartz, S.M., Iriarte-Diaz, J., Riskin, D.K., and Breuer, K.S. (2012). A bird? A plane? No, it's a bat: an introduction to the biomechanics of bat flight. In: *Evolutionary History of Bats: Fossils, Molecules and Morphology* (Gunnell, G. and Simmons, N.B., eds.), pp 317–52. Cambridge University Press, Cambridge, UK.

Ward, T. A., Fearday, C. J., Salami, E., and Soin, N. B. (2017). A bibliometric review of progress in micro air vehicle research. *International Journal of Micro Air Vehicles*, **9**(2), 146–165. doi:10.1177/1756829316670671

Warfvinge, K., KleinHeerenbrink, M., and Hedenström, A. (2017). The power-speed relationship is U-shapd in two free-flying hawkmoths (*Manduca sexta*). *J. R. Soc. Interface*, **14**, 20170372.

Warrick, D.R., Tobalske, B.W., and Powers, D.L. (2005). Aerodynamics of the hovering hummingbird. *Nature*, **435**, 1094–97.

Weis-Fogh, T. (1975). Flapping flight and power in birds, conventional and novel mechanisms. In: *Swimming and Flying in Nature*, Vol. 2 (Wu, T.Y.T., Brokaw, C.J., and Brennen, C., eds.), pp 729–62. Plenum Press, New York.

Chapter 33

Communication

Robert H. Wortham and Joanna J. Bryson

Department of Computer Science, University of Bath, UK

From a traditional engineering perspective, communication is about effecting control over a distance, and its primary concern is the reliability of transmission. As with much in biomimetics, we find that such traditional problems and practices are not entirely separable from the situation in nature, but that a natural perspective better informs the true requirements of even engineered autonomous systems. In particular, nature allows for collaborative control with minimal signals, generating robust, distributed systems. Human communication also employs the exceptional capacity of language, but robot–human interaction requires an understanding of implicit mechanisms of human communication as well.

Here we review communication in nature, describing the evolution of communication from the perspective of the selfish gene. We explore whether communication arose from collaboration or manipulation, and the reasons why communication in nature is ubiquitous and generally honest, given the apparent evolutionary benefits of free-riding. We consider the content necessary for communication, and show that context and relevance allow a message to be effectively communicated with very little information transfer. Humans possess the unique ability to communicate with language, and we explore how this differs from the non-verbal communication we share with other animals, and the challenges that robots face when using language for communication.

We then review communication in contemporary biomimetic systems. From the perspective of communication, these fall into several categories, depending on whether the communication is between robots, or between robot and human, and also whether the robotic system is fully autonomous (has its own goals), or is in some sense collaborating with a human to achieve a goal. Swarm robotics, inspired by social insects such as bees and ants, requires non-centralized, distributed communication mechanisms and we consider some of the work in this area to date (see Nolfi, Chapter 43, this volume). Designing autonomous human–robot interaction predicates an understanding of human social interaction, and we review relevant background psychology. We also review work to date in the generation of synthetic emotion and the ability of robots to sense and classify human emotion. Finally, we provide some pointers for future directions and further reading.

Biological principles

Collaboration or manipulation?

When thinking of the fundamentals of natural communication—in fact, all communication—we need to define what communication essentially entails. First, there must be some kind of signal. Second, there must be a receiver that has beliefs or behavior which are modified by reception of the signal. The degree to which receipt of a signal benefits either sender or receiver necessarily depends on what the receiver would have chosen to do without the signal. For intentional communication, the sender must knowingly want the signal to achieve the definite purpose of modifying the beliefs or behavior of another. But the term *communication* is used widely in the natural sciences, including to describe behavior in plants and microbia that are not ordinarily considered to have intentions. Even linguistic communication in humans can include unintended, unconscious signaling, sometimes even counter to the known goals of the sender. For example, a speaker's tone may reveal an emotional state the speaker wishes to keep hidden.

Intuitively, one might think of communication as facilitating cooperation, both in animals and humans. We might imagine that the capacity to communicate arose from a desire by natural agents to interact with each other to achieve shared objectives. Communication may indeed appear cooperative when it facilitates the achievement of a shared goals, such as reproduction, food collection in colonies, and so on. Even antagonistic communication can be seen as cooperative if it facilitates finding lower-cost resolution to conflicting goals, as in competition over a particular mate or access to another scarce resource. We might therefore suppose that, historically, organisms that cooperated via communication were better able to achieve their goals, and so prospered from an evolutionary perspective. Unfortunately this kind of common sense thinking arises from what Oxford ethologist David McFarland called the "incurable disease" of anthropomorphism (McFarland and Bösser 1993, p. 1). That is, we might apply human characteristics (or *assumed* human characteristics) to non-human entities and thus infer human thought patterns where none exist.

In stark contrast, Dawkins (1976) forwards the perspective of Darwinian evolution. All living organisms are driven by a set of "selfish genes" that succeed only through replicating themselves. However, genes do not fight the necessary battles for survival and reliable replication on their own. Rather, they collaborate, fighting inside "survival machines" or "Vehicles" composed also of other genes—the organisms we observe in nature. These organisms are themselves driven to compete in order to advance the reproductive goals of their genes, and any communication that evolves between organisms will only evolve if it benefits the organism sending the signal (or at least the gene triggering it to do so) relative to others. Thus Dawkins (1982) frames communication as arising as coercion, not cooperation. Dawkins argues that all such inter-organism communication, whether between individuals (e.g. sexes) of the same species, or between differing species can be viewed as one organism manipulating another with the ultimate goal of replicating its genes. Further, Scott-Phillips et al. (2012a) show that a state of non-interaction is an evolutionarily stable strategy, and so communication will not necessarily emerge even when it is in both parties' interest.

However, these observations do not mean that cooperative, mutually beneficial, or even altruistic communication cannot evolve. Because genes are shared within and even across species, they can also motivate behavior that is costly to an individual Vehicle, so long as that behavior is beneficial to securing copies of those genes existing in the future (Hamilton 1964; Gardner and West 2014). Nevertheless, it is worthwhile for a robotics developer to remember that biologists always think of communication in terms of its costs, risks, and benefits for all individuals involved, and that coercion may be a more useful metaphor, particularly for systems that are self-learning, or involve components owned by multiple parties, each with their own goals and constraints.

Origins and stability of communication

Communication incurs measurable cost arising both from the energy expended and time used in order to generate signal, and from the risk that the signal may transmit information to competitors rather than collaborators, for example a predator drawn by bird song. Additionally, the presence of a signaling mechanism may otherwise affect the performance of an animal, for example the weight of the peacock's tail. Biology now views this cost as part of the signal's mechanism—the cost itself communicates the value(s) of the signaler, for example the quality of the individual that can support such a handicap (Zahavi and Zahavi 1997). The essential argument of the handicap principle is that observable handicaps serve as necessarily honest signals—signals that can be trusted by the recipient—because of the increased cost to the signaler incurred by the handicap. Honest signals are important for recipients given the evolutionary pressure for coercion, for example a peahen prefers to find the best mate to inseminate her eggs. Honesty may also be assured via two mechanisms: Indices and Deterrents (Scott-Phillips 2011). Indices are mechanisms where the signal form is tied to signal meaning. For example, the frequency of an animal's roar being directly tied to its size. Deterrents are mechanisms that punish dishonest signals with excessive cost, such as weight, visibility, or duration (e.g. Tibbetts and Izzo 2010).

Viewing communication through this evolutionary lens raises two questions: Firstly, how does communication come to emerge at all, and secondly, why is it an evolutionarily stable strategy (ESS), when there are selfish individual benefits to both non-participation and dishonesty?

The first question raises the chicken and egg problem of communication: how did communication first arise, since there is always a cost (opportunity, risk, or metabolic) to creating a signal, and may be some similar cost to having the apparatus to receive the signal (Smith and Harper 1995). Who would bear this cost first? There is no benefit to creating a signal if there is no capable receiver. Similarly, why would there be an adaptation to receive signals, when then are no signals to receive? This question is neatly addressed by Hauser (1996) and Scott-Phillips et al. (2012a), who show how communication evolves from precursor interactions that do not fully meet our definition of *communication*. These interactions can be categorized as either ritualization or sensory manipulation. For example, an animal may use urine or faeces to mark territory. How does this arise? Initially the animal may experience fear when at the boundary of its territory, and so may relieve itself there, without communicative intent. The presence of this material may act as a cue for others and thus become associated with territorial boundaries. Through ritualization of this behavior a communication channel is established. Hauser makes the useful distinction between *cues* and *signs* as precursors to communicative intent: cues vary over time, whereas signs are temporally invariant.

Sensory manipulation involves one individual stimulating the senses of another in order to affect the receiver's behavior to the benefit of the signaler, but without intending to communicate that intention to the receiver. The example often given here is of a male insect wishing to mate with a female. He brings the female food, coercing her to stay and eat the food, allowing mating. Since this behavior is beneficial to the species (or more precisely, beneficial to the replication of genes shared by two individuals capable of breeding), females also evolve to recognize the bringing of food as a precursor to mating, and hence communication arises. Only if recognizing the signal is generally of benefit to the receiver will the receiver evolve to do so. However, sometimes evolution may exploit a normally useful signal in a way that does not benefit the receiver, as in the case of mimicry (Wickler 1968; Johnstone 2002). If recognizing the signal is not of more general benefit than cost, then receivers will evolve to ignore the signal, and the signaler will incur its costs without benefit. This begins to show us why animal communication is generally "honest" rather than "deceptive;" for further discussion see Hauser (1996). Scott-Phillips et al. (2012a) postulate that ritualization is the more common route to communication, since the prerequisites are less restrictive, and also the initial cues are implicitly honest.

Content necessary for communication

Much research in artificial multi-agent systems has focused on the expressive power of communicative languages (Wooldridge 2009). In contrast, natural communication systems tend to exploit relatively simple and minimal signals, the meaning of which essentially derives from extensive models. In other words, evolution, or a shared phylogenetic history, provides adequate priors such that minimal data is required to communicate context. For example, mating calls primarily discriminate which species is calling, and secondarily the quality and possibly other attributes of the caller. Calls between offspring and mothers signal identity and emotional state, and must be hard for a predator to imitate coercively (de Oliveira Calleia et al. 2009; in robotics, identity and associated forgery concerns are ordinarily handled with encryption.) The pronking of gazelles not only signals to herdmates the presence of a predator, but also to the predator that that the individual at least is fit enough to have detected them and as such is probably a bad target for an attempted kill (Caro 1994).

Tinbergen (1952) offers four useful perspectives through which natural communication can be understood:

- Mechanistic: the physical communication mechanisms—neural, physiological, psychological.
- Ontogenetic: the genetic and environmental factors that guide development of communication.
- Functional: the effect of communication on survival and reproduction.
- Phylogenetic: the historic origins of the species and its communicative traits, which determine available mechanisms.

Tinbergen's fascinating research with gulls set the precedent for demonstrating that only very basic signaling is required for effective communication to facilitate the entire lifecycle (Tinbergen and Falkus 1970). The power of this sort of communication is still present in human communicative systems, and may still often implicitly determine our behavior. Humans are communication machines, having not only language, but an extrordinary array of pheremones for an ape (Stoddart 1990). Even in language, we are influenced by implicit information such as affect and dialect. Newborn babies are more likely to attend to (and therefore learn from) people sharing their mother's dialect (Fitch 2004). Of course, human language also contains explicit content, discussed further below.

Shannon and Weaver (1971) provide us with a quantitative methodology for the evaluation of the information content involved in any communication. They start from the assumption that we can measure information transmission by investigating the reduction in uncertainty that follows from the successful transmission of a message. The general architecture employed in the analysis is

$$source \rightarrow message \rightarrow signal \rightarrow +noise \rightarrow receiver \rightarrow destination \qquad (1)$$

The theory shows that information content is proportional to the logarithm of the number of choices in the message.

$$H(X) = -\sum_i P(x_i)\log_n P(x_i) \qquad (2)$$

where X is a sequence of symbols forming the message, i is the number of symbols in the message, and n is the number of choices for each symbol in the message (for a message consisting of binary 1's and 0's then $n = 2$).

If there are few choices (i.e. the received signal is from a small set of possible choices) then this implies little actual information transfer. If we then add to this context the interpretation of the signal by the receiver, i.e. given the internal state of the receiver, what valid subset of messages it is expecting to receive, we see that often the actual information required to transfer a message is very low. The addition of *noise* during signal transmission reduces the rate at which a communication channel of given capacity can accurately transmit the desired message.

$$C = W \log_2 \frac{P + N}{N}$$

(3)

Where C is the capacity of the channel given the noise, W is the bandwidth of the channel, P is the power of the signal, and N is the power of the noise. We can see that the higher the noise level, the lower the rate of information transmission, determined by the relative power of the noise, compared with the power of the signal. Shannon's theory can be very usefully applied to natural, human, and artificial communication (see further Allen and Hauser 1993).

Language as an exception

Humans have long assumed that other animals have languages we just haven't learned, yet despite extensive research there is in fact no evidence that any other extant species shares a communication system with anything like the power for expression and innovation displayed in all human languages. Given its core importance not only for communication but for cognition, explanations for this unique trait are a holy grail for much of biological anthropology and cognitive science. Possible answers range from the theological to the technical, the latter including an also-mysteriously-unique capacity for representations supporting indefinite recursion (Hauser et al. 2002) or at least being able to reason about minds (Stiller and Dunbar 2007; Moll and Tomasello 2007). A simpler explanation is that hominids share with other apes adaptations such as long lives and big brains that facilitate exploiting *culture*—the communicating of novel behavior between local conspecifics—but are unique among simians in the otherwise rather common adaptation for vocal imitation (Bryson 2009). However, since there is no consensus on explaining human language in particular or uniqueness more generally, we leave this question to the section on further reading.

One radical theory related to language origins that is becoming increasingly accepted (though not in its details) is the idea of *memetics*, again first postulated by Dawkins (1976). This is the theory that humans have a dual replicator system, that is, that we are subject to two orthogonal strands of evolution (Richerson and Boyd 2005). The first is genetic evolution, common to the rest of biology, but the second is the evolution of concepts and ideas—memetic material. Although no one has identified exactly what a meme might be, the same is true of genes. Although DNA, the genetic material and substrate by which biological information is transmitted, has now been identified (long after it was first hypothesized), the discrete units thought to be genes are still not well understood (Dennett 1995). Language is currently the main substrate by which human ideas are transmitted and innovated, but it is by no means the only one: we also learn from gestures, models, and other artifacts. In addition, language itself is generally presumed to be a result of memetic cultural evolution.

What is critical to understand for a roboticist is that language evolves continuously. Contrary to much schooling, there is no "correct" way to speak English or any other language, only ways that signal particular educational or social backgrounds, or ways that are more or less effective for communicating in a particular community. Also, the words in languages represent a set of concepts found historically to be of value for communication and thinking. Just because a concept has its own word in one language does not guarantee that there is a single-word translation for that concept in any other language, or indeed that that concept will make sense in another culture. As our concepts and communities constantly change, so do our languages. Nevertheless, people have historically managed to transact complex commercial and individual negotiations with people with whom they do not share a language. This is again partly because we can assume a set of likely desires and capacities for any other human, so can guess the meaning of some essential phrases and gestures with only a limited amount of context. We also tend to create simplified languages, called *pigeons*, when two language groups come together. These concepts may be useful for communicating with robots.

When human language is involved, there is little apparent cost to employing deception to gain advantage that would be evolutionarily advantageous. The problem of creating an evolutionarily stable strategy (ESS) for human language is therefore more complex. Scott-Phillips (2011) has proposed a mechanism by which human communication could become evolutionarily stable, and this relies heavily on the idea of reputation as a social governor of honesty. Uniquely in nature, human communication also involves epistemic vigilance; the ability to evaluate reliably received communications as to whether they are true or false before acting on them. Human language is also famously symbolic, with one term able to reference multiple real-world objects, but as implied by our earlier discussion of information theory, this capacity for generalization may be a fairly common attribute of communicative systems.

Biomimetic systems

Typically in robotics we broaden the idea of *communication* to include all types of social interaction between actors. Robot social interaction can thus be divided into two broad categories: robot–robot interaction, and human–robot interaction (HRI). For biomimetic robots, the major current field of study for multiple autonomous interacting robots is termed "swarm robotics." HRI can be further subdivided into human-directed collaboration with robots versus interaction with autonomous robots. In the former, the robot and human work collaboratively to achieve a given task. The robot, whilst intelligent, does not have sufficient cognitive capacity, or more simply the motivational agency, to choose high-level goals or perform tasks alone. Human collaborators may be required to assist with planning, expert knowledge, dexterous manipulation, or other aspects of the task. More autonomous HRI provides the greatest challenge, and here various techniques have been adopted to enable the robot to communicate effectively with humans.

Swarm robotics

Parker (2008) describes three commonly used paradigms for building distributed intelligent systems:

◆ bio-inspired, emergent swarms paradigm;

◆ organizational and social paradigms; and

◆ knowledge-based, ontological, and semantic paradigms.

Swarm robotics takes its inspiration from social insects, and involves the distributed self-coordination of significant numbers of relatively simple robots. Robot swarms may be homogeneous, where all robots have the same physical morphology and AI capabilities, or heterogeneous, where robots within the swarm may be specialized for different tasks, or have differing capabilities. Rather than some centralized or remote controlling system, swarm robotics predicates local sensing and communication abilities (Şahin 2005). The advantages of such a decentralized approach are that it provides both scalability (number of robots) and robustness (resilience to failure; Winfield 2000; Rubenstein et al. 2014).

One of the notable characteristics of swarm robotics research is the (relatively) complex behaviors that can be coordinated with very simple signaling. Parker (1998) developed the ALLIANCE heterogeneous swarm control architecture, where all communication was broadcast, rather than directed to individuals within the swarm. Each robot broadcast its location and current behavior, avoiding the need for others to gather this information via sensors. The robot's location and activity were broadcast at a known frequency (rate). If others did not receive this information after a pre-determined timeout, then they assumed the robot was no longer active.

Howard et al. (2006) demonstrate a large heterogeneous mobile robot team, with the shared task of mapping an unknown territory. Each robot is identified with a unique fiducial marker rather like a "bar code," simply and effectively communicating its identity. RFID tags may also be used for this task. Each robot measures the identity, range, and heading of adjacent robots. The estimated position and pose data of each robot is combined using UDP (broadcast) messaging over a wireless network to create a shared global map. Interesting communication features here include:

- The use of "mapper/leader" robots that have better sensors, leading "sensor" robots with much cheaper and less functional sensor capabilities.
- The use of broadcast messaging to achieve scalable information transfer, and the ability of the overall system to tolerate lossy communications.

Fredslund and Matarić (2002) examine the problem of achieving and maintaining robot formations. In keeping with the norms of artificial swarm research more generally (Reynolds 1987), robot formations were found that could be maintained by a simple shared algorithm that focuses solely on a neighbor in the formation. Minimal communication is needed to establish the desired formation and the role of each robot within the formation, e.g. who is the leader. Werfel et al. (2014) have extended such work, presenting an algorithm that, taking a three-dimensional shape as input, produces distributed instructions such that a robot swarm can construct that shape. Rubenstein et al. (2014) demonstrate a similar algorithm that works in two dimensions for a swarm of 1000 robots. The lesson here is that once again, very minimal broadcast communication is needed, in addition to sensor input, to achieve sophisticated and coordinated behavior, although how to generate the individual heuristics or "plans" for the robots is an ongoing and promising area of research.

Swarm robotics can be sufficiently biomimetic that they may be used for scientific studies in the the evolution of communication. Floreano et al. (2007) investigated whether altruistic communication could evolve in robot swarms. Robots were able to emit and perceive a blue light, and also a red light that was colocated with a food source (charging station). Each robot was equipped with an omnidirectional, color-sensitive camera. In this experiment the learning was not individual (ontogenetic), but evolutionary (phylogenetic), using genetic algorithms over a neural controller. A physics-based robot simulation environment was used for many of the generations, but at regular intervals generations occurred on real robots. As predicted by the theory of inclusive fitness described above, Floreano et al. (2007) found that communication readily evolves when colonies consist of genetically similar individuals and when selection acts at the colony level. Further, Mitri et al. (2011) found that when associated with unrelated individuals, the robots evolved "deceptive" signals, leading to worldwide headlines that scientists had evolved lying robots.

More often though, science demands the simplest model necessary for parsimony, validation, and analytic tractability. There is a recent trend in using biomimetic "robots" to provide controlled input for examining animal social cognition. However, in many cases the robots are only biomimetic in shape or other appearance, with no appreciable cognitive skills (Faria et al. 2010).

Human–robot collaboration

Robots may be thought of as tools, facilitating the achievement of goals for humans (Bryson 2010). Some tasks are currently too complex, or involve too much risk, to allow a fully autonomous robot to attempt them unaided. Human–robot collaboration (HRC) provides a methodology enabling humans and robots to work as partners to accomplish these tasks. Fong et al. (2002) explore human–robot dialogue to facilitate mutual assistance for the achievement of common goals. Interestingly, in this study the human functions as a resource for the robot, providing information and processing like other system modules. Using a very structured textual communication system, the robot can ask questions of the human as it works. This approach provides the robot with more freedom in execution and is more likely to find good solutions to its problems. Effective collaboration requires information sharing between parties, so collaborative control considers both user (human) and robot needs (Sheridan 1997). The need to make robots comprehensible to human partners is in fact sometimes used as a justification for bio-inspired approaches to robotics (Brooks and Stein 1994; Sengers 1998; Novikova and Watts 2015).

One effective means to share information with human collaborators is via Augmented Reality (AR), the overlaying of computer-generated graphics onto the real worldview. Green et al. (2008) introduce AR techniques by first discussing the communication necessary for human–human collaborative activity. In order to achieve useful work together, a common ground of shared understanding about the world must be achieved. Within a human–robot collaboration scenario, this grounding of symbols and their meanings is similarly vitally important (Steels 2008). Green et al. (2008) review work carried out at NASA and elsewhere to implement AR. Within an AR environment, in addition to speech, humans can communicate using a wide variety of non-verbal cues such as gaze, nodding, and other deictic gestures, with AR providing the grounding context.

More recent work takes a probabilistic approach to collaborative decision making (Kaupp et al. 2010). Human "operators" can be regarded as remotely located, valuable information sources which need to be managed carefully. Robots then decide when to query operators using Value-Of-Information theory, i.e. humans are only queried if the expected benefit of their observation exceeds the cost of obtaining it. In this study a navigation task is executed jointly by robot and human. The robot navigates a maze with local sensory input (a laser scanner), but transmits its visual camera signal to the remote human operator for interpretation. When the robot needs high-level visual information in order to make a decision, it simply queries the human, rather than having to interpret the raw visual information from the camera. This sort of "human in the loop" approach is also often proposed for actions that may have ethical consequences, such as assaults by military robots (Vallor 2014; Hellström 2013).

Autonomous human robot interaction

In order to use a biomimetic approach for autonomous HRI, we first need to consider human–human social interaction. When we communicate, we claim the attention of one or more others. This implies that the information communicated is relevant to the receiver. Therefore relevance may be seen as the key to human communication and cognition (Sperber and Wilson 1986). In order to communicate in a relevant manner, we may need to understand the mental state of others. Understanding this mental state is known as Theory of Mind (ToM). There is strong empirical evidence for ToM in adult humans, including fMRI studies of contexts where ToM is exploited (Saxe et al. 2006). However, Gallagher (2006) proposes a more direct understanding of the mental state and intent of others from directly observable phenomena. For example, one can read emotion and infer intention directly from the faces of others, without having to conduct any "mind reading." This background psychology is important for understanding and designing effective human–robot communications, not least because all human–robot interactions inevitably involve some anthropomorphization on the part of the human.

Kanda et al. (2002) describe studies using a humanoid robot able to generate many human-like communication behaviors, and equipped with sensors enabling human non-verbal and verbal communication to be sensed. This work develops the idea of communicative relationships between human and robot, based on relevance theory. For example, humans more easily understand a robot's verbal utterances once they have built a prior relationship with the robot. This idea of communication requiring perceived relevance in order to be successful is reinforced by Kuchenbrandt et al. (2014), who found that the efficacy of human–robot interaction is significantly affected by the perceived gender of the robot, the gender of the human, and the stereotypic gender association of the task.

One of the major breakthroughs in autonomous robotics has been the design and development of robotic imitation learning. This enables robots to be taught new tasks by human demonstration (supervised learning), and then to perfect these tasks through practice (unsupervised learning). This essentially biomimetic—and memetic—approach of social learning through communication (Meltzoff and Moore 1995; Schaal 1999) continues to deliver increasingly outstanding results, and is a major success in human–robot (primarily) nonverbal communication. The work of Billard and Hayes (1997), Klingspor et al. (1997), and Breazeal and Scassellati (2002) foresaw that this approach would produce machines that are useful, flexible, and easy to use. Movement primitives can be learned from demonstration and then combined generatively by the robot to produce novel behaviors for new tasks as they arise. Grasping and manipulation are good examples of such behaviors. Recently this work has developed using advanced probabilistic techniques to allow previously unseen objects to be effectively grasped and manipulated by interpolating known behaviors (Huang et al. 2013; Kopicki et al. 2014).

Non-verbal communication by robots can be used to great effect to significantly improve the productivity of human–robot teams, both in terms of speed of communication, and robustness to errors in communication (Breazeal et al. 2005). Humans maintain a mental model of the robot and non-verbal communication has been found to help with inference of the internal "mental" state of the robot. The design principles for non-verbal communication can again be taken from psychology, e.g. feedback, affordances, causality, and natural mappings (Sengers 1998; Breazeal et al. 2005). Humans typically display emotions non-verbally as part of normal communication, and whilst this may be synthesized readily in humanoid robots (Zecca et al. 2009), body expressions alone have also been successfully used to indicate emotional responses in non-humanoid robots (Novikova and Watts 2014). It should not be forgotten that only a small percentage of human communication involves words. Non-verbal, and particularly synthetic emotional, communication can significantly enhance the persuasive power of an autonomous robot (André et al. 2011).

Future directions

There is clear market pressure for further biomemetic communication in domestic robotics. Users find natural communication intuitive and often attractive. This approach is not without hazards or at least critics, however. The imitation of life may cause humans to over-respond to robots or misattribute value or fragility, leading consumers to waste time or money on attending to robots (Ashrafian et al. 2015). However, such concerns only apply to transmission *by* the robot, and may be ameliorated by sufficiently transparent signals (Bryson and Kime 2011; Boden et al. 2011). As mentioned in our review of nature, *reception* and comprehension of signals will generally take precedence in any communication system involving learning or evolving actors, as recognizing and categorizing stimuli and associating these with appropriate beliefs or actions are fundamental to cognitive capacity.

Substantial advances are currently being made not only in vision and speech recognition, but also in sensing human emotion. By fusing Bayesian classification of speech audio, facial expression, and gesture, higher recognition rates have been achieved than by analysing speech alone. Kessous et al. (2009) already report high recognition rates for anger, irritation, joy, pride, and sadness. Even deception can be detected (Schuller et al. 2008). However, there is still much work to do in this area. Schuller (2012) lists "confidence, deception, frustration, interest, intimacy, pain, politeness, pride, sarcasm, shame, stress, and uncertainty," as reasonable targets for comprehension, and the same may be needed for production.

Returning to consider nature, we are only just beginning to understand the importance of altruistic communication—the sharing of knowledge—and information networks as an adaptive strategy for animals. At the same time, we are eliminating most of the non-human biomass on our planet (Barnosky 2008) creating gaps in critical ecosystem webs, including information networks. Williams (2013) suggests that biomemetic robots might serve not only to replace animals, but to guide or instruct those animals that remain. However, the tractability and economic feasibility of large-scale application of technologies of this type have yet to be demonstrated. In contrast, the use of robotics in experimental biology and psychology is a promising trend likely to accelerate (Krause et al. 2011). Such applications produce immediate value in scientific research, and may also serve as testbeds for more ambitious ecological projects.

Finally, the widespread use of robotics promises a growing need for the improved capacity for distributed autonomous robot control and communication. From robot mining (Bonchis et al. 2014; hopefully a means to return to the less ecologically damaging but more individually hazardous strategy of pit mining) to urban disaster intervention (Sahingoz, 2013; dos Santos and Bazzan, 2011) to bridge and sewer maintenance, the deployment of self-organizing, self-healing robot-only or human-and-robot systems holds promise for a better future with more robust and responsive infrastructure (Abelson et al. 2000). Realizing that future in a way that benefits everyone involved requires—among many other advances—excellent communication.

Learning more

For further reading we suggest the following. For those wishing to get to grips with Darwinian evolution and evolutionarily stable strategies, Dawkins (1982) is still the obvious starting point. The emergence of communication is neatly covered by Scott-Phillips et al. (2012b). Those with a further interest in animal communication should read Bradbury and Vehrencamp (2011) or Hauser (1996). For further details on ecological and evolutionary constraints that affect the design of signals, see Davies et al. (2012). Hauser and Konishi (1999) supplement this with additional detail on the design of animal communication. McFarland (1986) is an excellent book covering animal behavior from the perspective of psychobiology, ethology, and evolution. Finally, Scott-Phillips (2014) provides a recent book-length introduction to human language from a comparative biological perspective.

McFarland and Bösser (1993) is an early but seminal book, concerned with the application of animal behavior (including communication) to robots. Parker (2008) provides a useful overview of the field of distributed intelligence and its application in multi-robot systems. Breazeal et al. (2005) provide a useful introduction to nonverbal human–robot interaction. Pobil et al. (2014) is an example of the wide range of communication-related topics included within the study of autonomous agents, and these proceedings contain several papers covering topics such as perceptual prediction and collective and social behavior achieved using swarm robotics techniques. Nourbakhsh (2013) predicts how we might interact with robots in the near and far future.

References

Abelson, H., Allen, D., Coore, D., Hanson, C., Homsy, G., Knight, T. F. Jr., Nagpal, R., Rauch, E., Sussman, G. J., and Weiss, R. (2000). Amorphous computing. *Commun. ACM*, 43(5), 74–82.

Allen, C., and Hauser, M. D. (1993). Communication and cognition: is information the connection? *PSA: Proceedings of the Biennial Meeting of the Philosophy of Science Association*, 1992(2), Symposia and Invited Papers. Chicago: University of Chicago Press/Philosophy of Science Association, pp. 81–91.

André, E., Bevacqua, E., Heylen, D., Niewiadomski, R., Pelachaud, C., Peters, C., Poggi, I., and Rehm, M. (2011). Non-verbal persuasion and communication in an affective agent. In: R. Cowie, C. Pelachaud, and P. Petta (eds), *Emotion-oriented systems*, Cognitive Technologies series. Berlin/Heidelberg: Springer, pp. 585–608.

Ashrafian, H., Darzi, A., and Athanasiou, T. (2015). A novel modification of the Turing test for artificial intelligence and robotics in healthcare. *The International Journal of Medical Robotics and Computer Assisted Surgery*, 11(1): 38–43. doi: 10.1002/rcs.1570

Barnosky, A. D. (2008). Megafauna biomass tradeoff as a driver of quaternary and future extinctions. *Proc. Natl Acad. Sci. U SA*, 105(Supplement 1), 11543–8.

Billard, A., and Hayes, G. (1997). Learning to communicate through imitation in autonomous robots. In: W. Gerstner, A. Germond, M. Hasler, and J.-D. Nicoud (eds), *Artificial Neural Networks (ICANN'97)*, volume 1327 of *Lecture Notes in Computer Science*. Heidelberg: Springer, pp. 763–8.

Boden, M., Bryson, J., Caldwell, D., Dautenhahn, K., Edwards, L., Kember, S., Newman, P., Parry, V., Pegman, G., Rodden, T., Sorell, T., Wallis, M., Whitby, B., and Winfield, A. (2011). *Principles of robotics*. The United Kingdom's Engineering and Physical Sciences Research Council (EPSRC). http://www.epsrc.ac.uk/research/ourportfolio/themes/engineering/activities/principlesofrobotics/

Bonchis, A., Duff, E., Roberts, J., and Bosse, M. (2014). Robotic explosive charging in mining and construction applications. *Automation Science and Engineering, IEEE Transactions on*, 11(1), 245–50.

Bradbury, J. W., and Vehrencamp, S. L. (2011). *Principles of animal communication*. Sunderland, MA: Sinauer Associates.

Breazeal, C., Kidd, C., Thomaz, A., Hoffman, G., and Berlin, M. (2005). Effects of nonverbal communication on efficiency and robustness in human–robot teamwork. In: *2005 IEEE/RSJ International Conference on Intelligent Robots and Systems*, Alberta, Canada. New York: IEEE, pp. 708–13.

Breazeal, C., and Scassellati, B. (2002). Robots that imitate humans. *Trends in Cognitive Sciences*, 6(11), 481–87.

Brooks, R. A., and Stein, L. A. (1994). Building brains for bodies. *Autonomous Robots*, 1(1), 7–25.

Bryson, J. J. (2009). Representations underlying social learning and cultural evolution. *Interaction Studies*, 10(1), 77–100.

Bryson, J. J. (2010). Robots should be slaves. In: Y. Wilks (ed.), *Close engagements with artificial companions: key social, psychological, ethical and design issues*. Amsterdam: John Benjamins, pp. 63–74.

Bryson, J. J., and Kime, P. P. (2011). Just an artifact: Why machines are perceived as moral agents. In: *Proceedings of the 22nd International Joint Conference on Artificial Intelligence*, Barcelona. Burlington, MA: Morgan Kaufmann, pp. 1641–6.

Caro, T. M. (1994). Ungulate antipredator behaviour: Preliminary and comparative data from african bovids. *Behaviour*, 128(3/4), 189–228.

Davies, N. B., Krebs, J. R., and West, S. A. (2012). *An introduction to behavioural ecology*. Chichester, UK: Wiley-Blackwell.

Dawkins, R. (1976). *The selfish gene*. Oxford: Oxford University Press.

Dawkins, R. (1982). *The extended phenotype: the gene as the unit of selection*. Oxford: W.H. Freeman & Company.

de Oliveira Calleia, F., Rohe, F., and **Gordo, M.** (2009). Hunting strategy of the margay (*Leopardus wiedii*) to attract the wild pied tamarin (*Saguinus bicolor*). *Neotropical Primates*, **16**(1), 32–4.

Dennett, D. C. (1995). *Darwin's dangerous idea*. New York: Penguin.

dos Santos, F., and Bazzan, A. (2011). Towards efficient multiagent task allocation in the RoboCup Rescue: a biologically-inspired approach. *Autonomous Agents and Multi-Agent Systems*, **22**(3), 465–86.

Faria, J., Dyer, J., Clément, R., Couzin, I., Holt, N., Ward, A., Waters, D., and Krause, J. (2010). A novel method for investigating the collective behaviour of fish: introducing 'Robofish'. *Behavioral Ecology and Sociobiology*, **64**, 1211–18. 10.1007/s00265-010-0988-y.

Fitch, W. T. (2004). Kin selection and 'mother tongues': A neglected component in language evolution. In D. K. Oller and U. Griebel (eds), *Evolution of communication systems: a comparative approach*. Cambridge, MA: MIT Press, pp. 275–96.

Floreano, D., Mitri, S., Magnenat, S., and Keller, L. (2007). Evolutionary conditions for the emergence of communication in robots. *Current Biology*, **17**(6), 514–9.

Fong, T., Thorpe, C., and Baur, C. (2002). Collaboration, dialogue, and human–robot interaction. In: R. Jarvis and A. Zelinsky (eds), *Springer Tracts in Advanced Robotics 6*. Berlin/Heidelberg: Springer, pp. 255–66.

Fredslund, J., and Matarić, M. (2002). A general algorithm for robot formations using local sensing and minimal communication. *IEEE Transactions on Robotics and Automation*, **18**(5), 837–46.

Gallagher, S. (2006). The narrative alternative to theory of mind. In: R. Menary (ed.), *Radical enactivism: intentionality, phenomenology, and narrative*. Amsterdam: John Benjamins, pp. 223–29.

Gardner, A., and West, S. A. (eds) (2014). Inclusive fitness: 50 years on. Special issue. *Philosophical Transactions of the Royal Society B: Biological Sciences*, **369**(1642).

Green, S., Billinghurst, M., Chen, X., and Chase, G. (2008). Human–robot collaboration: a literature review and augmented reality approach in design. *International Journal of Advanced Robotic Systems*, **5**(1), 1–18.

Hamilton, W. D. (1964). The genetical evolution of social behaviour. *Journal of Theoretical Biology*, **7**, 1–52.

Hauser, M. D. (1996). *The Evolution of Communication*. A Bradford book. Cambridge, MA: MIT Press.

Hauser, M. D., Chomsky, N., and Fitch, W. T. (2002). The faculty of language: what is it, who has it, and how did it evolve? *Science*, **298**, 1569–79.

Hauser, M. D., and Konishi, M. (1999). *The design of animal communication*. A Bradford book. Cambridge, MA: MIT Press.

Hellström, T. (2013). On the moral responsibility of military robots. *Ethics and Information Technology*, **15**(2), 99–107.

Howard, A., Parker, L., and Sukhatme, G. (2006). Experiments with a large heterogeneous mobile robot team: exploration, mapping, deployment and detection. *The International Journal of Robotics Research*, **25**(5–6), 431–47.

Huang, B., Bryson, J., and Inamura, T. (2013). Learning motion primitives of object manipulation using Mimesis Model. In: *IEEE International Conference on Robotics and Biomimetics, ROBIO 2013*. IEEE, pp. 1144–50.

Johnstone, R. A. (2002). The evolution of inaccurate mimics. *Nature*, **418**, 524–6.

Kanda, T., Ishiguro, H., and Ono, T. (2002). Development and evaluation of an interactive humanoid robot: Robovie. In: *IEEE International Conference on Robotics and Automation*, Washington, DC, Volume 2. New York: IEEE, pp. 1848–55.

Kaupp, T., Makarenko, A., and Durrant-Whyte, H. (2010). Human–robot communication for collaborative decision making: a probabilistic approach. *Robotics and Autonomous Systems*, **58**(5), 444–56.

Kessous, L., Castellano, G., and Caridakis, G. (2009). Multimodal emotion recognition in speech-based interaction using facial expression, body gesture and acoustic analysis. *Journal on Multimodal User Interfaces*, **3**(1–2), 33–48.

Klingspor, V., Demiris, J., and Kaiser, M. (1997). Human–robot communication and machine learning. *Applied Artificial Intelligence*, 11, 719–46.

Kopicki, M., Detry, R., Schmidt, F., Borst, C., Stolkin, R., and Wyatt, J. L. (2014). Learning dexterous grasps that generalise to novel objects by combining hand and contact models. In: *IEEE International Conference on Robotics and Automation*, number Sec II, Hong Kong, China. New York: IEEE, pp. 5358–65.

Krause, J., Winfield, A. F., and Deneubourg, J.-L. (2011). Interactive robots in experimental biology. *Trends in Ecology & Evolution*, 26(7), 369–75.

Kuchenbrandt, D., Häring, M., Eichberg, J., Eyssel, F., and André, E. (2014). Keep an eye on the task! How gender typicality of tasks influence human–robot interactions. *International Journal of Social Robotics*, 6(3), 417–27.

McFarland, D. (1986). *Animal behaviour: psychobiology, ethology and evolution*. London: Longman.

McFarland, D., and Bösser, T. (1993). *Intelligent Behavior in Animals and Robots*. Bradford book. Cambridge, MA: MIT Press.

Meltzoff, A. N., and Moore, M. K. (1995). *Infants' understanding of people and things: From body imitation to folk psychology*. Cambridge, MA: MIT Press.

Mitri, S., Floreano, D., and Keller, L. (2011). Relatedness influences signal reliability in evolving robots. *Proceedings of the Royal Society B: Biological Sciences*, 278(1704), 378–83.

Moll, H., and Tomasello, M. (2007). Cooperation and human cognition: the Vygotskian intelligence hypothesis. *Philosophical Transactions of the Royal Society B: Biological Sciences*, 362(1480), 639–48.

Nourbakhsh, I. R. (2013). *Robot futures*. Cambridge, MA: MIT Press.

Novikova, J., and Watts, L. (2014). A design model of emotional body expressions in non-humanoid robots. In: *The Second International Conference on Human-Agent Interaction*, Tsukuba, Japan. New York: ACM, pp. 353–60. doi:10.1145/2658861.2658892

Novikova, J., and Watts, L. (2015). Towards artificial emotions to assist social coordination in HRI. *International Journal of Social Robotics*, 7(1), 77–88.

Parker, L. (1998). ALLIANCE: an architecture for fault tolerant multirobot cooperation. *IEEE Transactions on Robotics and Automation*, 14(2), 220–40.

Parker, L. (2008). Distributed intelligence: overview of the field and its application in multi-robot systems. *Journal of Physical Agents*, 2(1), 5–14.

Pobil, A. P., Chinellato, E., Martínez-Martín, E., Hallam, J., Cervera, E., and Morales, A. (eds) (2014). *From Animals to Animats 13: 13th International Conference on Simulation of Adaptive Behavior*. Castellón, Spain: Springer International Publishing.

Reynolds, C. W. (1987). Flocks, herds, and schools: a distributed behavioral model. *Computer Graphics*, 21(4), 25–34.

Richerson, P. J., and Boyd, R. (2005). *Not by genes alone: how culture transformed human evolution*. Chicago: University Of Chicago Press.

Rubenstein, M., Cornejo, A., and Nagpal, R. (2014). Programmable self-assembly in a thousand-robot swarm. *Science*, 345(6198), 795–99.

Sahingoz, O. K. (2013). Mobile networking with UAVs: opportunities and challenges. In: *International Conference on Unmanned Aircraft Systems (ICUAS '13)*, 28–31 May 2013, Atlanta, GA. IEEE, pp. 933–41.

Saxe, R., Schulz, L. E., and Jiang, Y. V. (2006). Reading minds versus following rules: dissociating theory of mind and executive control in the brain. *Social Neuroscience*, 1(3–4), 284–98.

Schaal, S. (1999). Is imitation learning the route to humanoid robots? *Trends in Cognitive Sciences*, 3(6), 233–42.

Schuller, B., Eyben, F., and Rigoll, G. (2008). Static and dynamic modelling for the recognition of non-verbal vocalisations in conversational speech. In: E. André, L. Dybkjær, H. Neumann, and R.

Pieraccini (eds), *Perception in Multimodal Dialogue Systems*. Berlin/Heidelberg: Springer, pp. 99–110.

Schuller, B. W. (2012). The computational paralinguistics challenge [social sciences]. *Signal Processing Magazine, IEEE*, **29**(4), 97–101.

Scott-Phillips, T. (2014). *Speaking Our Minds: Why human communication is different, and how language evolved to make it special*. London: Palgrave Macmillan.

Scott-Phillips, T. C. (2011). Evolutionarily stable communication and pragmatics. In: A. Benz, C. Ebert, G. Jäger, and R. van Rooij (eds), *Language, Games, and Evolution*. Berlin/Heidelberg: Springer, pp. 117–33.

Scott-Phillips, T. C., Blythe, R. A., Gardner, A., and West, S. A. (2012a). How do communication systems emerge? *Proc. R Soc. B: Biol. Sci.*, **279**(1735), 1943–9.

Scott-Phillips, T. C., Kirby, S., and Ritchie, G. (2012b). The origins of human communication. *Science*, **25**(11), 2008–9.

Sengers, P. (1998). Do the thing right: an architecture for action expression. In: K. P. Sycara and M. Wooldridge (eds), *Proceedings of the Second International Conference on Autonomous Agents*. Rio de Janeiro: ACM Press, pp. 24–31.

Shannon, C. E., and Weaver, W. (1971). *The Mathematical Theory of Communication*. Champaign, IL: University of Illinois Press.

Sheridan, T. (1997). Eight ultimate challenges of human–robot communication. In: *6th IEEE International Workshop on Robot and Human Communication*, pages 9–14.

Smith, M. J., and Harper, D. (1995). Animal signals: models and terminology. *Journal of Theoretical Biology*, **177**(3), 305–11.

Sperber, D., and Wilson, D. (1986). *Relevance: communication and cognition*. Oxford: Blackwell.

Steels, L. (2008). The symbol grounding problem has been solved, so what's next? In: M. de Vega, A. Glenberg, and A. Graesser (eds), *Symbols and Embodiment: Debates on meaning and cognition*. Oxford: Oxford University Press, pp. 223–44.

Stiller, J., and Dunbar, R. I. M. (2007). Perspective-taking and memory capacity predict social network size. *Social Networks*, **29**(1), 93–104.

Stoddart, D. M. (1990). *The scented ape: the biology and culture of human odour*. Cambridge, UK: Cambridge University Press.

Tibbetts, E. A., and Izzo, A. (2010). Social punishment of dishonest signalers caused by mismatch between signal and behavior. *Current Biology*, **20**(18), 1637–40.

Tinbergen, N. (1952). "Derived" Activities; Their Causation, Biological Significance, Origin, and Emancipation During Evolution. *Quarterly Review of Biology*, **27**(1), 1–32.

Tinbergen, N., and Falkus, H. (1970). *Signals for Survival*. Oxford: Clarendon Press.

Vallor, S. (2014). Armed robots and military virtue. In: L. Floridi and M. Taddeo (eds), *The Ethics of Information Warfare*, volume 14 of *Law, Governance and Technology Series*. Cham, Switzerland: Springer,pp. 169–85.

Werfel, J., Petersen, K., and Nagpal, R. (2014). Designing collective behavior in a termite-inspired robot construction team. *Science*, **343**(6172), 754–8.

Wickler, W. (1968). *Mimicry in plants and animals*. World University Library. London: McGraw-Hill.

Williams, C. (2013). Summon the bee bots: can flying robots save our crops? *New Scientist*, **220**(2943), 42–5.

Winfield, A. F. T. (2000). Distributed sensing and data collection via broken ad hoc wireless connected networks of mobile robots. In: L. E. Parker, G. Bekey, and J. Barhen (eds), *Distributed autonomous robotic systems 4*. Tokyo: Springer Japan, pp. 273–82.

Wooldridge, M. (2009). *An introduction to multiagent systems*. Chichester, UK: John Wiley & Sons.

Zahavi, A. and Zahavi, A. (1997). *The Handicap Principle: A missing piece of Darwin's puzzle.* Oxford: Oxford University Press.

Zecca, M., Mizoguchi, Y., Endo, K., Iida, F., Kawabata, Y., Endo, N., Itoh, K., and Takanishi, A. (2009). Whole body emotion expressions for KOBIAN humanoid robot: preliminary experiments with different emotional patterns. In: *18th IEEE International Symposium on Robot and Human Interactive Communication*, Toyama, Japan. IEEE, pp. 381–6.

Şahin, E. (2005). Swarm robotics: from sources of inspiration to domains of application. In: E. Şahin, W. M. Spears, and A. F. Winfield (eds), *Swarm Robotics*. Berlin/Heidelberg: Springer, pp. 10–20.

Chapter 34

Emotions and self-regulation

Vasiliki Vouloutsi[1] and Paul F. M. J. Verschure[1,2]

[1] SPECS, Institute for Bioengineering of Catalonia (IBEC), the Barcelona Institute of Science and Technology (BIST), Barcelona, Spain
[2] Catalan Institute of Advanced Studies (ICREA), Spain

There is no consensus on what constitutes an emotion, what the relation is between emotions and perception and action, nor on what their functional roles are. This chapter takes a pragmatic perspective by taking the synthesis of emotions in machines as its starting point. From this perspective, emotions will be linked to notions of motivation and self-regulation as serving a key role in the regulation of control systems underlying behavior and communication.

Our world is constantly advancing, both technologically and scientifically. New, complex tools allow us to construct sophisticated machines whose usage goes beyond direct utilitarian purposes. According to the International Federation of Robotics (IFR), in 2015, 5.4 million service robots for personal and domestic use were sold, accounting for an increase of 16% from the previous year (Haegele 2016). These robots vary from domestic tasks (like vacuum cleaning or lawn-mowing), medical or healthcare, robots for education, to companions, assistants, or toys (see Figure 34.1). In fact, sales are expected to increase to almost 30 million units by the end of 2019. The future where robots will be fully integrated into our daily lives is very near. However, this raises certain challenges, as now robots will need to adopt social roles in dynamic social environments that may not always involve well-defined tasks.

Robots that interact with humans need to display rich social behaviors, following the social rules that are attached to the roles they assume; they need to be understood and accepted by humans also as communication partners. This raises a fundamental question: what kind of behavioral characteristics should robots have in order to be socially accepted by their human peers? One approach is to follow an anthropomorphic heuristic and impose that robots should exhibit behaviors and characteristics that are similar to those of humans, such as the ability to communicate via natural language or the usage of gestures and other natural cues, as well as to express and perceive emotions (Breazeal 2003).

However, is it necessary for a robot to display human-like behavior and emotion, and if so, do humans interact with robots in a similar way as to how they would interact with other humans? Early influential studies in human–computer interaction have shown that humans tend to treat smart technology as a peer using the same social rules they apply to humans (Reeves and Nass 1998).

When designing robots that interact socially with humans, almost all aspects of the robot have been found to affect that interaction, from their physical appearance to the speed and smoothness of their movements. For this reason, the tendency is to design anthropomorphic social robots, since they provide an intuitive interface and fulfill social expectations (Duffy 2003). Appearance is not the sole factor that affects social interaction; the behavior of the robot also systematically influences people's perception and expectations. Research in this domain suggests that, in the case of social robots, their behavior must be autonomous and transparent

Figure 34.1 An example of a social robot that has the role of a city guide and a robotic dog as a companion.
Artwork by Alvaro Martinez. Copyright © 2017, SPECS_lab.

and similar to that of humans in order to be effective. When a robot's behavior is modeled after a human, it encourages people to interact with it intuitively, without the need of special training. However, one does have to avoid the so-called uncanny valley (Mori 1970) which lurks when the robot violates the expectations of the highly specialized social perception systems of humans.

Regardless of whether the primary role of the robot is to interact socially with humans or to assist them in accomplishing a specific task, it needs to employ similar communication and behavioral mechanisms to those deployed in human–human interaction. For instance, humans learn to use paralinguistic cues and social rules that they intuitively follow, thus social robots will need to do the same; they need to not only understand what goes on but also exploit similar communication channels and socially contextualize the information they receive. To be more precise, in a social context with humans, the robot should be able to establish and/or maintain social relationships, learn and recognize states of other agents and humans, as well as express and perceive emotions. This raises the question of what emotions are and whether they should play any role in the realization of living machines. On the one hand, we look at the evaluative role of emotions or appraisal (Frijda 1987), which is predicated on motivation. Here the latter sets the context of the former (Verschure 2012), i.e. whether food will trigger happiness depends on whether the consumer is satiated or hungry. On the other, the outcome of emotional appraisal can inform internal processing such as learning and memory and/or define communicative signals. In this way, we can distinguish between epistemic and utilitarian emotions. Hence, in this perspective emotions play a much deeper role in the organization of behavior than solely as a cue system.

Biological principles

Darwin proposed that humans and a number of animal species use similar behaviors to express emotions (Darwin 1876). This would suggest that a common set of biological principles underlie this ability. In this chapter, in particular, the close relationship between motivation and emotion is emphasized.

Motivation, goals, and drives

Anyone who is interested in understanding, influencing, or even mimicking biological behavior (and therefore implementing it in a robot) has to start with understanding motivation. Why do animals behave the way they do under different (or the same) conditions? What is the "primal force" that guides behavior? Motivation is the *"process that energizes, directs and maintains goal-oriented behavior"* (Huitt 2001) or that which causes us to act generating the so-called "why" of behavior (Verschure 2012; see also Verschure, Chapter 35, this volume). In the framework presented here, emotion is seen as predicated on the state of the motivational system.

The study of motivation is a rich field and a number of different theories have been proposed (Graham and Weiner 1996). In the early twentieth century, the psychologist William McDougall coined the notion of *instinct theory*, which has its roots in evolution theory and postulates that organisms behave in certain ways because they are biologically determined to do so. This notion also influenced the psychoanalysis of Sigmund Freud and the ethology of Lorenz. Although this theory can certainly describe animal behavior, such as the famous imprinting experiments of Lorenz, it fails to explain them in terms of underlying processes. The *arousal theory* of motivation (Lindsley 1951) suggests that organisms behave in a certain way to maintain an optimal level of arousal that varies depending on the properties of the individual or the situation. The *incentive theory* hypothesizes that animals act in certain ways because of external rewards or punishments, or *incentives*. Incentives can be primary (not learned) and secondary reinforcers (they become rewards after being associated with other primary incentives). According to this theory, animals act because they strive toward goals driven by reward-seeking or hedonism, a perspective that has also informed many models of machine learning. This theory mainly focuses on association and learning to control motivation and through that behavior. The *humanistic theory* of motivation, mainly represented by Maslow's hierarchy of needs (Maslow 1943), suggests that needs can be categorized in a hierarchical way, ranging from basic survival and biological needs to self-actualization, where higher needs cannot be pursued if lower ones are not satisfied. Finally, *drive theory* (also known as drive reduction theory) posits that an organism's unsatisfied need is the source for motivation: actions satisfy needs (Hull 1943). Based on this theory, all organisms need to be in a state of balance. Changes in the environment can cause imbalance, which in turn creates a state of arousal and unpleasant feeling or tension called *drive*. Once out of balance, the organism will try to engage in behaviors that reduce this drive (hence the name drive reduction). According to this theory, there are two main types of drive: *primary*, which reflect biological needs, and *secondary*, which are learned drives. We can see that in general motivation theories allude to the ability of the organism to maintain a "steady state." Already Hippocrates (350 BC) equated health with the harmony between mind and body, while the nineteenth-century French physiologist Claude Bernard spoke of the organism maintaining its internal environment, or "milieu," in balance while facing a fluctuating external environment. The Russian physiologist Ivan Pavlov generalized this notion of stasis to the relation between an organism and the external environment.

Homeostasis and allostasis

According to Walter Cannon, the coordinated physiological process that maintains a steady state of the organism can be called *homeostasis* (Cannon 1932). Homeostasis refers to the control of physiological processes with the aim of keeping them within certain bounds using negative feedback: a sensor detects a state of the system which is compared with a reference value and a control signal is generated proportional to the difference, which in turn drives cells, tissue, organs, or the whole organism to reduce the detected discrepancy. Hence, homeostasis (or "identical state") is the self-regulation of a dynamical system towards constancy. Cannon focused on five homeostatic processes critical to the biochemistry of life involving the essential variables of pH, temperature, plasma osmolality, glucose, and calcium. However, other physiological processes are believed to follow similar principles.

Homeostasis is essentially based on a predefined reactive negative feedback system which precludes the inclusion of anticipation and learning. In addition, it raises the question of scalability when many partially conflicting essential variables must be regulated. As a result, the complementary notion of *allostasis* has been advanced as achieving stability through change, in particular by changing the boundaries within which essential variables are held through learning and anticipation (Sterling and Eyer 1988). For instance, a glucose deficit might be tolerated in order to evade a predator. The cost to the organism of maintaining an essential variable from its set point is called the allostatic load. Whereas homeostatic processes are independent and autonomous, in the case of allostasis auto-regulation depends on a central control system, i.e. the brain.

Emotions and affect

Emotions are associated with feelings and moods and have phenomenal aspects; they are experiences with distinct intensities and qualities. *Emotions* are seen as being transient and directed towards someone or something, while *moods* are feelings that last longer, are less intense, and often lack immediate triggering stimuli. Emotions can both activate and direct behavior as in fear and anger. Although emotions mostly accompany motivated behaviors, they are fundamentally different in the way they are triggered: while motivations are dependent on internal needs, emotions can be elicited by a variety of external stimuli in the absence of pre-existing needs and goals. An emotion is a complex episode that creates a readiness to act and has several different components (Frijda 1988). An emotion usually begins with an *appraisal* of a given situation or stimulus, i.e. the interpretation of the situation or stimulus relevant to needs, goals, and/or well-being. This appraisal differentiates various emotions and leads to its specific quality. Other components of emotions include associated *thoughts* and *action* tendencies as well as *bodily reactions*, usually accompanied by *facial expressions*, and finally responses that aim at coping with or reacting to the emotional state. None of these components in itself can be seen as an emotion; rather, an emotion is complex and the result of the interplay of all of them.

To this day, emotion is a controversial subject, as there is no general agreement on its definition or its underlying processes. In fact, there are so many different definitions regarding emotions and their properties that a broad definition is needed to include their most significant aspects. Despite the lack of general consensus, emotions are usually responses to events that are relevant to the individual (Frijda 1988). For some, emotions are the result of somatic responses to affective stimuli as postulated in the classical *James–Lange theory*. This theory was heavily criticized by Walter Cannon and Philip Bard, as the experience of emotions seems to precede the occurrence of bodily changes and can be seen as the result of a simultaneous activation of physiological responses and identification of emotional cues from sensory information. Damasio in his popular book *Descartes Error* revived the James–Lange body-centered idea of emotion seeing that emotions are anchored in somatic markers, but recently the author has changed his mind in the face of neuroscientific evidence and the current status of the James–Lange theory is again under debate. Recent studies do confirm the link between bodily reactions and emotions, as the first seem to affect the latter; however, bodily reactions do not seem to be the *cause* of emotions. While emotions do not directly derive from somatic responses, they are seen as being linked to them. Other theorists support the notion that emotions are an experience, subject to motivational situations placed in an approach–avoid continuum: behavior is oriented towards or away from a stimulus and emotion is, therefore, any experience with high intensity and hedonicity (Cabanac 2002) that will make an animal work towards or away from a stimulus (Rolls 2000). Another fundamental question on emotions is whether they are discrete or continuous and universal or contingent on local contexts. Paul Ekman defined six basic discrete emotions that can be found in most cultures and can be considered primitive and universal: anger, disgust, fear, happiness, sadness, and surprise (Ekman 1992). These discrete emotions are believed to be innate and fundamentally different from other more complex emotions since they can be uniquely expressed, and exhibit different behavioral, physiological, and neural reactions (Colombetti 2009). In contrast, continuous emotions are defined by one or more dimensions such as valence/arousal, where the former defines the quality on a positive (happy) to negative (sad) dimension and the latter defines the intensity (Russell 1980).

Recently, the neural mechanisms underlying emotions have gained increasing attention in the scientific community, emphasizing the role of the amygdala found in the medial temporal lobe (LeDoux 2000, 2012; Scherer 1993). Indeed, it has been proposed that the amygdala can be seen as a generic valence assignment system, which mediates between primitive behavioral control systems of the brainstem and midbrain and the perceptual and cognitive systems of the neocortex. An alternative influential proposal is that by Jaak Panksepp that defines seven basic emotional systems found in the midbrain/brain stem: CARE, FEAR, LUST, RAGE, PANIC, PLAY, and SEEKING (Panksepp and Biven 2012). These systems are seen as underlying the full spectrum of emotions and are linked to the regulation of adaptive behavior. Conversely, Craig has placed emphasis on the anterior insular cortex as the structure where a broad range of subjective states are represented including the feelings associated with simple and complex emotions (Craig 2009). Hence, this suggests that emotions are dependent on a broad hierarchy of systems from the brainstem to the frontal cortex and should be considered in terms of the architecture of the brain as opposed to a singular module (Verschure 2012).

Despite the heterogeneity of the definitions, some invariants stand out where emotion involves some kind of appraisal, regardless of whether something is good or bad, rewarding or punishing, or even something one would work for or avoid. This has also been proposed as a form of emotional learning (LeDoux 2012). Invertebrates and vertebrates need to learn from their environment, in real time, to survive. For instance, in classical conditioning, the behavioral signature of emotional learning has been observed in *C. elegans*, sea slugs, and moths. Survival not only requires the identification of which of the stimuli in the environment are relevant for behavior, i.e. appraisal, but also how to modify behavior accordingly—action preparation and shaping. The way the brain develops representations of such stimuli and their associated actions has been the subject of the study of classical and operant conditioning (LeDoux 2012). In general, it is found that learning in both cases depends on motivating stimuli, e.g. food or shocks, that trigger mechanisms gating learning and memory. Gating utilizes neuromodulatory systems originating in subcortical structures such as the ventral tegmental area, the nucleus basalis of Meynert, or the Locus Coereleus. Computationally one can interpret these systems as issuing a "print now" signal that regulates synaptic plasticity, allowing local learning rules to be controlled by global mechanisms (Sánchez-Montañés et al. 2002). Indeed, this principle is mirrored in many machine-learning approaches. For example, in a model of auditory classical conditioning, the amygdala provides emotional appraisal which drives the nucleus basalis of Meynert (NBM), which facilitates learning in the primary auditory cortex, remodeling the receptive fields to better detect the tone that predicts a shock. This model is consistent with the physiology of learning in A1 and demonstrates robust tonotopic map formation and adjustment, even in the presence of noise or inhomogeneities in stimulus sampling (Sánchez-Montañés et al. 2002). This example demonstrates an epistemic impact, i.e. remodeling of A1 representations of tones, dependent on stimulus appraisal realized by the amygdala driven by a motivating stimulus. This illustrates the latest trend in research where emotions are considered from a system's perspective, as they can alter perception, motivational priorities, learning, attention, memory, and decision-making (Dalgleish 2004; Verschure 2012). Thus, emotions can be instrumental in assisting communication, by expressing one's internal state (external/utilitarian emotions) and for organizing perception, cognition, and behaviors (internal/epistemic).

Biomimetic systems

Now we return to the question of the role of motivational and emotional systems for a robot. So far in research emphasis has been placed on the display and/or recognition of utilitarian emotions. However, as the analysis above shows, a good reason to endow artificial agents with affective mechanisms is that they can modify perception, motivational priorities, and action selection and help overcome uncertainties and biases in dynamic environments (Fellous and Arbib 2005; Verschure et al. 1992).

Inspired by the drive motivation theory or by ethology, many robotic systems are endowed with homeostatic mechanisms in which each drive needs to be maintained in balance and their action selection mechanisms aim at achieving that (Breazeal and Brooks 2005). For instance, it has been argued that motivational systems not only allow the robot to successfully complete a task but also focus on a predefined goal, facilitating interaction with humans (Stoytchev and Arkin 2004). The first robot control model that explicitly brought together associative learning of sensorimotor mappings with motivations and emotions mapped a model of classical conditioning to foraging robots (Verschure et al. 1992). Here, simple stimuli such as collisions and rewards triggered internal states of negative and positive valence respectively, which in turn triggered avoidance or approach actions. This appraisal, in turn, gated the epistemic learning process, such that neutral stimuli predictive of these simple ones could be associated with the same behaviors. This model has been explicitly linked to utilitarian emotions in an avant-garde human accessible artificial organism "ADA: the sentient space" (Eng et al. 2005; Wassermann et al. 2003). ADA was developed for the Swiss national exhibition Expo.02 and had, in total, over 500,000 visitors. ADA's main goal was to realize an intense and consistent interaction with its visitors; ADA's behavioral system comprised of three main drives (survival, recognition, and interaction) and emotional states now reflected the perturbations to this homeostatic system. In every cycle, ADA was calculating overall happiness and used this value to determine if the action selected contributed to its drive reduction. The emotional state of ADA reflected the error in drive reduction and was expressed through a composition of sounds and lights in the exhibition. Therefore, this artificial organism was maximizing its own goal functions (or maximizing "happiness") by keeping drives in homeostasis while communicating this process through externalizing its utilitarian emotions, influencing the behavior of the visitors in a way to reduce its drives.

The interaction of homeostasis and allostasis has been extended to behavioral control in robot models of foraging as well as in Human–Robot Interaction (HRI) scenarios. By combining homeostatic and allostatic levels of control, animals can perform complex real-world tasks like foraging, regulating their internal states, and maintaining a dynamic stability with their environment. Such systems have been implemented in social agents (Lallee et al. 2014; Vouloutsi et al. 2013) as well as robots performing foraging tasks (Sánchez-Fibla et al. 2010). In this scenario, each drive is influenced by its homeostatic state; however, adaptation is achieved through *allostasis*, as now the homeostatic limits are adjusted dependent on overall demands on the system (Figure 34.2).

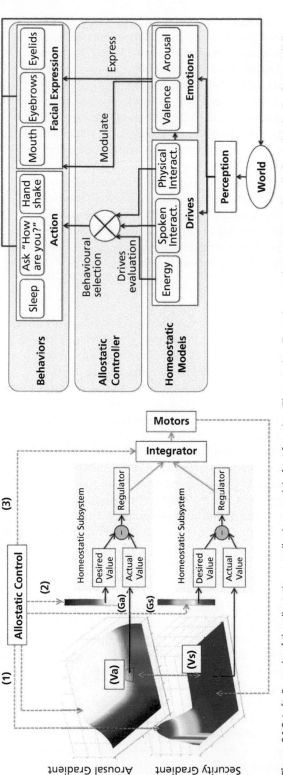

Figure 34.2 Left: Example of the allostatic controller in a model of rat foraging. The reactive allostatic control system comprises dedicated Behavior Systems (BS) that combine predefined sensorimotor mappings with homeostatic drive reduction mechanisms that are predicated on the needs of the organism, serving Self Essential Functions (SEF). In order to map needs into behaviors, the essential variables served by the BSs have a specific distribution in space called an "affordance gradient." Here we consider the (internally represented) "attractive force" of the home position supporting the *Security* SEF or of open space defining an *Exploration* SEF. The values of the respective SEFs are defined by the difference between the sensed value of the affordance gradient (V) and its desired value given the prevailing needs (G). The regulator of each BS defines the next action as to perform a gradient ascent on the SEF. An integration and action selection process across the different BSs forces a strict winner-take-all decision that defines the specific behavior emitted. The allostatic controller regulates the internal homeostatic dynamics of the BSs to set priorities defined by needs and environmental opportunities through the modulation of the affordance gradients (1), desired values of SEFs (2), and/or the integration process (3). Right: Detailed diagram of the behavior generation system with a humanoid robot. As the world is perceived, it affects the robot's drives (homeostatic subsystems) and emotions. The homeostatic value of each drive is evaluated by the allostatic controller which selects which drive needs to be satisfied and the corresponding behavior from the set of available actions, and executes it. The behaviors shown in the diagram are just a subset example to illustrate the underlying principles. In parallel, emotions are constantly updated based on the content of the environment and the global satisfaction of the drives. In turn, emotions are expressed through facial expression and they modulate the execution of actions.

Left: Reprinted from *Advances in Complex Systems*, 13(3), Marti Sanchez-Fibla, Ulysses Bernadet, Erez Wasserman, Tatiana Pelc, Matti Mintz, Jadin C. Jackson, Carien Lansink, Cyriel Pennartz, and Paul F. M. J. Verschure, Allostatic Control For Robot Behavior Regulation: A Comparative Rodent–Robot Study, pp. 377–403, doi: 10.1142/S0219525910002621, Copyright © 2010 The Company of Biologists. Right: Adapted from Stephane Lallee, Vasiliki Vouloutsi, Maria Blancas Munoz, Klaudia Grechuta, Jordi-Ysard Puigbo Llobet, Marina Sarda, and Paul F. M. J. Verschure, Towards the synthetic self: making others perceive me as another, Paladyn, *Journal of Behavioral Robotics*, 6(1), pp. 136–164, doi: 10.1515/pjbr-2015-0010, Copyright © 2015. The Authors. Reprinted under the terms of the Creative Commons Attribution License, which permits unrestricted use, distribution, and reproduction in any medium, provided the original author and source are credited.

Given our analysis, the reasons for using emotional mechanisms in robots are twofold. On the one hand, the robot can communicate its internal states in a human readable way (through facial expressions, body posture, or prosodic features) using utilitarian emotions. When the robot smiles or displays a sad expression, people can readily read these expressions, assess the situation, understand what was the cause of this emotional display, and increase the transparency of the interaction. This transparency, however, is a coherent binding of these utilitarian emotions to the motivational state of the robot as, for instance, used in the ADA system. On the other hand, emotional systems influence current and future internal states, appraisal and action tendencies, as they may provide varying reward and punishment signals with their associated epistemic impact. The examples we have considered here show that the implementation of a system of drives and appraisal mechanisms, combined with utilitarian and epistemic emotions, serves well many of the capabilities robots need. Drives and goals are necessary to motivate behavior, emotions are needed to appraise situations and organize behaviors as well as define communicative signals, both of which act as effective mechanisms to promote interaction and adaptation. However, we also see that the notion of emotion pertains to properties of an integrated architecture (see Verschure, Chapter 35, this volume) rather than an isolated emotion module.

Future directions

The potential benefit of having socially fit robots integrated into our daily lives has created several opportunities not only in industry but also in science where robots embody models of mind and brain. Such applications create interesting challenges to the domain of Living Machines and artificial intelligence, as robots that socially interact with humans will be required, among other things, to learn from their environment, understand complex social interaction, and be able to reason about human intentions and actions. Psychologists and social scientists can also benefit, as robots can become the testbed for psychological, emotional, and communication models aimed at helping us understand ourselves in a future hybrid human–robot society. The latest trend in realizing advanced functionalities in robots (or have the robot perform a specific task) is to go back to biology, and study and understand the brain and the behavioral mechanisms of biological counterparts known to excel in the task under study or exhibit relevant functionalities. Thus, scientists can assess behavioral models grounded in biology by implementing them on robotic replicas. The field of social robotics should be the outcome of close collaboration between psychologists, social scientists, neuroscientists, and engineers; only then can we get closer to fully functional social living machines.

Learning more

A very insightful analysis incorporating neuroscientific studies in humans as well as animal research to understand emotions is the book by Panksepp and Biven, *The Archaeology of Mind: Neuroevolutionary Origins of Human Emotions* (2012). Here, the authors identify common emotional circuits among mammals, suggesting that humans are not the only beings capable of emotions. A complementary analysis of the neuronal correlates of feelings and emotions is provided by Bud Craig in *How Do You Feel? An Interoceptive Moment with Your Neurobiological Self* (2014). A detailed overview of the relationship between emotions and robots can be found in Fellous and Arbib's *Who needs emotions? The brain meets the robot* (Fellous and Arbib 2005). A general resource related to emotion research and affective computing can be found at http://emotion-research.net/. Finally, an interesting review on social robots can be found in Fong et al. (2003). Here the authors provide extensive information and resources related to the topics of robot design, appearance, and behavior, as well as the role of emotional systems and personality in human–robot interactions.

Acknowledgments

Preparation of this chapter was supported by the EC FP7 project WYSIWYD (FP7-ICT-612139), EASEL (FP7-ICT-611971), and EC H2020 ERC project cDAC (341196).

References

Breazeal, C. (2003). Emotion and sociable humanoid robots. *International Journal of Human-Computer Studies*, **59**(1), 119–55.

Breazeal, C. and Brooks, R. (2005). Robot emotion: a functional perspective. In: J.-M. Fellous and M. A. Arbib (eds), *Who needs emotions? The brain meets the robot*. Oxford: Oxford University Press, pp. 271–310. Available at: http://excedrin.media.mit.edu/wp-content/uploads/sites/14/2013/07/Breazeal-Brooks-03.pdf [Accessed October 21, 2014].

Cabanac, M. (2002). What is emotion? *Behavioural Processes*, **60**(2), 69–83. Available at: http://www.ncbi.nlm.nih.gov/pubmed/12426062.

Cannon, W. B. (1932). *The wisdom of the body*. New York: W.W. Norton & Company, Inc. Available at: http://books.google.gr/books?id=zdkEAQAAIAAJ.

Colombetti, G. (2009). From affect programs to dynamical discrete emotions. *Philosophical Psychology*, **22**(4), 407–25.

Craig, A. D. (2009). How do you feel—now? The anterior insula and human awareness. *Nature Reviews Neuroscience*, **10**(1), 59–70. Available at: papers3://publication/uuid/2CB3E0C6-D12A-4CD4-976C-4EB906C77CF1%5Cnhttp://www.nature.com/doifinder/10.1038/nrn2555.

Craig, A. D. (2014). *How do you feel? An interoceptive moment with your neurobiological self*. Princeton: Princeton University Press.

Dalgleish, T. (2004). The emotional brain. *Nat. Rev. Neurosci.*, **5**(7), 583–9.

Damásio, A. R. (1994). *Descartes' error: emotion, reason and the human brain*. New York: Putnam. ISBN 0-399-13894-3

Darwin, C. (1876). *The Expression of the Emotions in Man and Animals*. London: Penguin Classics.

Duffy, B.R. (2003). Anthropomorphism and the social robot. *Robotics and Autonomous Systems*, **42** (3–4), 177–90. Available at: http://linkinghub.elsevier.com/retrieve/pii/S0921889002003743 [Accessed August 28, 2014].

Ekman, P. (1992). An argument for basic emotions. *Cognition & Emotion*, **6**(3–4), 169–200.

Eng, K. E. K., Douglas, R. J., and Verschure, P. F. M. J. (2005). An interactive space that learns to influence human behavior. *IEEE Transactions on Systems Man and Cybernetics Part A Systems and Humans*, **35**(1), 66–77. Available at: http://ieeexplore.ieee.org/lpdocs/epic03/wrapper.htm?arnumber=1369346.

Fellous, J. M., and Arbib, M. A. (2005). *Who needs emotions?: The brain meets the robot*. Oxford: Oxford University Press. Available at: http://books.google.com/books?hl=fr&lr=&id=TvDi5V03b4IC&oi=fnd&pg=PR11&dq=Emotions:+from+brain+to+robot&ots=7kJ2_bXXqe&sig=IHdFhtmBKnHn5xFzJDq_C9aSUs4.

Fong, T., Nourbakhsh, I., and Dautenhahn, K. (2003). A survey of socially interactive robots. *Robotics and Autonomous Systems*, **42**(3–4), 143–66. Available at: http://linkinghub.elsevier.com/retrieve/pii/S092188900200372X [Accessed March 1, 2012].

Frijda, N. H. (1987). *The emotions*. Cambridge, UK: Cambridge University Press. Available at: http://books.google.com/books?hl=fr&lr=&id=QkNuuVf-pBMC&oi=fnd&pg=PR11&dq=The+Emotions+frijda&ots=BJK9l47qRv&sig=Vo9cOIwN_xy5MROqREJI_vtjafg.

Frijda, N. H. (1988). The laws of emotion. *American Psychologist*, **43**(5), 349–58. Available at: http://doi. apa.org/getdoi.cfm?doi=10.1037/0003-066X.43.5.349.

Graham, S., and Weiner, B. (1996). Theories and principles of motivation. In: P. A. Alexander and P. H. Winnhe (eds), *Handbook of Educational Psychology*. Abingdon, UK: Routledge, pp. 63–84.

Haegele, M. (2016). *World Robotics Service Robots 2016*. Frankfurt am Main: International Federation of Robotics.

Huitt, W. (2001). Motivation to learn: An overview. *Educational psychology interactive*, 12. Valdosta, GA: Valdosta State University. Available at: http://chiron.valdosta.edu/whuitt/col/motivation/ motivate.html

Hull, C. (1943). *Principles of behavior*. New York: Appleton-Century-Crofts. Available at: http://www. citeulike.org/group/2050/article/1282878.

Lallee, S., Vouloutsi, V., Munoz, M. B., Grechuta, K., Puigbo Llobet, J.-Y., Sarda, M., and Verschure, P. F. M. J. (2014). Towards the synthetic self: making others perceive me as another. *Paladyn, Journal of Behavioral Robotics*, **6**(1), 136–164. doi: 10.1515/pjbr-2015-0010

LeDoux, J. (2012). Rethinking the emotional brain. *Neuron*, **73**(4), 653–76. Available at: http:// linkinghub.elsevier.com/retrieve/pii/S0896627312001298 [Accessed February 29, 2012].

LeDoux, J. E. (2000). Emotion circuits in the brain. *Annu. Rev. Neurosci.*, **23**, 155–84.

Lindsley, D. B. (1951). Emotion. In: S. S. Stevens (ed.), *Handbook of experimental psychology*. New York: John Wiley & Sons, pp. 473–516.

Maslow, A. H. (1943). A theory of human motivation. *Psychological Review*, **50**(4), 370. Available at: http://scholar.google.com/scholar?q=A+theory+of+human+motivation&btnG=&hl=fr&as_ sdt=0#0.

Mori, M. (1970). Bukimi no tani [The uncanny valley], *Energy*[Online], **7**(4), 33–5.

Panksepp, J., and Biven, L. (2012). *The Archaeology of Mind: Neuroevolutionary Origins of Human Emotions*. Norton Series on Interpersonal Neurobiology. New York: WW Norton & Company.

Reeves, B., and Nass, C. (1998). The media equation. In: B. Reeves and C. Nass, *The Media Equation: How People Treat Computers, Television, and New Media Like Real People and Places*. Stanford, CA: The Center for the Study of Language and Information Publications, pp. 19–36.

Rolls, E. T. (2000). Précis of The brain and emotion. *The Behavioral and Brain Sciences*, **23**(2), 177–91; discussion 192–233.

Russell, J. A. (1980). A circumplex model of affect.pdf. *Journal of Personality And Social Psychology*, **39**(6), 1161–78.

Sánchez-Fibla, M., Bernadet, U., Wasserman, E., Pelc, T., Mintz, M., Jackson, J. C., Lansink, C., Pennartz, C., and Verschure, P. F. M. J. (2010). Allostatic control for robot behavior regulation: a comparative rodent-robot study. *Advances in Complex Systems*, **13**(3), 377–403. Available at: http:// www.worldscientific.com/doi/abs/10.1142/S0219525910002621.

Sánchez-Montañés, M. A., König, P., and Verschure, P. F. M. J. (2002). Learning sensory maps with real-world stimuli in real time using a biophysically realistic learning rule. *IEEE Transactions on Neural Networks*, **13**(3), 619–32.

Scherer, K. R. (1993). Neuroscience projections to current debates in emotion psychology. *Cognition & Emotion*, **7**(1), 1–41. Available at: http://www.tandfonline.com/doi/abs/10.1080/02699939308409174.

Sterling, P., and Eyer, J. (1988). Allostasis: a new paradigm to explain arousal pathology. In: S. Fisher and J. Reason (eds), Handbook of life stress, cognition and health. New York: John Wiley & Sons, pp. 629–49.

Stoytchev, A., and Arkin, R. C. (2004). Incorporating motivation in a hybrid robot architecture. *Journal of Advanced Computational Intelligence and Intelligent Informatics*, **8**, 290–5. Available at: http:// www.ece.iastate.edu/~alexs/papers/JACI_2004/JACI_2004.pdf.

Verschure, P. (2012). Distributed adaptive control: a theory of the mind, brain, body nexus. *Biologically Inspired Cognitive Architectures*, 1(1), 55–72. Available at: http://www.sciencedirect.com/science/article/pii/S2212683X12000102 [Accessed October 21, 2014].

Verschure, P., Kröse, B. J. A., and Pfeifer, R. (1992). Distributed adaptive control: The self-organization of structured behavior. *Robotics and Autonomous Systems*, 9(3), 181–96. Available at: http://www.sciencedirect.com/science/article/pii/0921889092900543.

Vouloutsi, V., Lallée, S., and Verschure, P. F. M. J. (2013). Modulating behaviors using allostatic control. In: N. F. Lepora, A. Mura, H. G. Krapp, P. F. M. J. Verschure, and T. J. Prescott (eds), *Biomimetic and Biohybrid Systems*. Living Machines 2013. Lecture Notes in Computer Science, vol 8064. Berlin/Heidelberg: Springer, pp. 287–98.

Wassermann, K., Eng, K., and Verschure, P. F. M. J. (2003). Live soundscape composition based on synthetic emotions. In: *IEEE Computer Society*. IEEE, pp. 82–90.

Chapter 35

The architecture of mind and brain

Paul F. M. J. Verschure

SPECS, Institute for Bioengineering of Catalonia (IBEC), the Barcelona Institute of Science and Technology (BIST), and Catalan Institute of Advanced Studies (ICREA), Spain

The etymology of the word architecture goes back to the Greek notion ἀρχιτέκτων (arkhitéktōn, "architect") or ἀρχι- (arkhi-, "chief") and τέκτων (téktōn, "builder"). Architecture thus denotes the product that a chief builder realizes based on an implicit or explicit master plan or blueprint. With this, the question of "architecture" thus deals with understanding the blueprint that organizes components and processes into a complete structure or system. The architecture of a building sculpts matter into human compatible affordances and as such virtualizes its function away from physical instantiation. The question of architecture becomes relevant when we aim at describing, emulating, or synthesizing multi-component biological systems: synthetic bodies and brains. The human body comprises 206 bones, over 640 muscles, and about 78 organs[1]. The human brain, one of the vital organs of the body, comprises about 90 billion neurons (Herculano-Houzel 2009) arranged into nine main areas[2] that give rise to the mind comprising at least seven distinct processes[3]. Understanding the organization of these structures and their associated processes is the realm of the architecture of mind and brain. This illustrates that in the study of architecture we encounter two main approaches that fall along the lines of Cartesian dualism: the functional architectures of res cogitas of the mind, e.g. cognitive architectures, and the material architecture of res extensa of the body and brain, e.g. brain architecture. This immediately points us towards the overarching challenge of how to bridge these two domains and to identify the virtualization they perform.

Architectures organize components or modules into a functioning whole whose system-level properties transcend that of the conglomerate. Similarly, architecture constitutes a distinct level of organization and description that adds specific operating capabilities to the resultant system. We can think of processes of resource allocation and management, regulation of the (in)activation of components, information exchange, the protocols and systems required to monitor component (mal)function, etc. Similarly, components of an architecture must both be functionally encapsulated in order to support virtualization and localize specific operations, and display system-level transparency to facilitate their integration and the exchange of operationally

---

[1] These enumerations depend on the specific definitions used.

[2] From caudal (tail) to rostral (nose) we can distinguish the following main developmental divisions and main structures: *Myelencephalon*: Medulla; *Metencephalon*: Pons, Cerebellum; *Mesencephalon*: Superior colliculus, Central gray matter, Substantia nigra; *Diencephalon*: Hypothalamus, Thalamus; *Telencephalon*: Cerebral cortex, Basal ganglia, Limbic system.

[3] A simple distinction can be made between Motivation, Emotion, Perception, Memory, Cognition, Action, and Consciousness.

relevant states. Above all, an architecture must equilibrate its physical realization with its multi-scale functions. The notion of "architecture" is essential and historically consistent when advancing an integrated system-level perspective on both the functional and structural organization of mind and brain.

Historical antecedents

The multi-faculty view of mind goes back to the classics. For instance, Plato's *Phaedo* (360 BC) and *Republic* (380 BC) sketch a nativist view of mind (or soul) where reason is in opposition of the passions of appetite and the spirit of the bodily non-rational soul. The linking of such mental faculties to the brain can be found in Hippocrates (~400 BC) and elaborated in Galen (~150 AD), who proposed a particular organization of function along the ventricles of the brain, which would stay in vogue for the following 1500 years. One of the first attempts to define an integrated comparative perspective on behavior came from George Romanes who, in the nineteenth century, inspired by the Darwinian revolution, saw behavior from amoeba to humans as a mixture of distinct functions: Reflex, Instinct, and Reason (Romanes 1888). The first explicit statement of a specific multi-layered brain organization came from John Heulings Jackson, who in his Croonian lecture of 1884 proposed that higher motor centres re-represent the properties of lower ones. In the same period the notion of "brain architecture" was already in use as in Mickle's president's address to the British medical association of 1895, while the leading US brain anatomist of the time, Edward Spitzka, described his study of brain structure as the science of brain architecture. Ivan Pavlov, half a world away in Russia, conceptualized the functional organization of the brain responsible for classical conditioning as comprising subcortical areas that provide stimulus–response reflexes, which in turn are substituted by associations formed at the level of the "cortical analyzer." In the same period, the European school of structuralism and its associated psychophysical methods advanced functional models of perception, cognition, and action. For instance, Donders used mental chronometry to show that information processing proceeded along discrete stages of detection, recognition, and decision (Jensen 2006).

During the beginning of the twentieth century, no elaborate models of the organization of mind and/or brain were advanced under the dogmatic influence of the behaviorist search for the operational laws of adaptive behavior. Welcome breaks from this tradition were the cognitive behaviorists Hull and Tolman, who proposed elaborate models of the functional organization of mind. The former defined an intricate parallel scheme of the processes that map a stimulus into a response, comprising 17 distinct components (Hull 1952). Tolman's so-called Sowbug constituted an early example of an embodied model of goal-oriented action using a hypothetical synthetic organism (Tolman 1939). Hence, as the study of mind and brain advanced beyond simple models, the need emerged for concepts that would facilitate a coherent description of the relationship between structure and function. The concept "architecture" provides this bridge.

Architecture of mind

Until the mid twentieth century theories on the organization of mind and brain were not cast in terms of architecture. For this, the key transition was the advent of the computer and its associated hardware and software organization. With the development of programmable electronic computing machinery during the Second World War, such as Colossus and ACE (strongly influenced by Turing) in the UK and the Harvard Mark I and ENIAC in the US, a fundamental question became how these machines should be designed. The view that became most popular and is still behind contemporary computer design is now known as the von Neumann architecture.

Von Neumann's breakthrough notion of the stored program was developed in collaboration with Eckert and Mauchly and was inspired by Turing's early work (von Neumann 1945/1993). The first stored program computer EDVAC had its processing unit access a memory unit over a communication interface or bus both for accessing and storing data as well as retrieving instructions. Note that von Neumann does not speak of an architecture but rather compares the operation of the EDVAC in terms of the nervous system.

The concept "architecture" entered the computer literature in a description of IBM's Stretch supercomputer to denote the combined functional and structural description of data formats, instruction sets, hardware properties, and optimizations (Brooks 1962). As such the notion of architecture thus describes the different abstraction layers at which computers are designed and built, from the application and programming language to the operating system and the instruction set, and the underlying devices and their physics virtualizing function from implementation.

The computer revolution brought about a new metaphor to explain the mind and brain, which combined with the dissatisfaction felt with the dogmas of behaviorism created the conditions for a paradigm shift: the cognitive revolution driven by the computer metaphor. This transition implied that the mind could also now be decomposed in abstraction layers. Early models of cognition, which take an architectural perspective in terms of identifying a number of specific processes underlying mind, are the General Problem Solver proposed by Newell, Shaw, and Simon, EPAM by Feigenbaum and Simon, and Anderson's Adaptive Control of Thought model (see Langley et al. 2009 for a review). Newell and Simon summarized the underlying idea of these models as a Physical Symbol System that comprises transduction of states of the world, memory for symbols and expressions, and operators that could change these expressions. The fundamental proposal of this approach was that problem-solving constitutes search. All models of cognition developed during this era of Good Old Fashioned AI (GOFAI) followed the GPS model. This tradition culminates in the influential work of Allen Newell on Unified Theories of Cognition (UTC) and his candidate theory SOAR (Newell 1990), where the challenge is to answer the "great psychological data puzzle" through a single set of mechanisms for all cognitive behavior. Newell also devised a list of criteria that any Cognitive Architecture which aspired to be a UTC had to satisfy:

1. Behave flexibly as a function of the environment.
2. Exhibit adaptive (rational, goal-oriented) behavior.
3. Operate in real-time.
4. Operate in rich, complex, detailed environments.
5. Use symbols and abstractions.
6. Use language.
7. Learn from the environment and from experience.
8. Acquire capabilities through development.
9. Operate autonomously, but within a social community.
10. Be self-ware and have a sense of self.
11. Be realizable as a neural system.
12. Be constructible by an embryological growth process.
13. Arise through evolution (Newell 1990, p.19).

This is a tall challenge and indeed so far nobody has succeeded, or even knows how to measure exactly these properties in biological or synthetic living machines.

An alternative view on how to define the architecture of cognition was proposed by David Marr who, also with an eye toward what later would be called computational neuroscience, suggested that in order to understand information processing, in this case focusing on vision, we should distinguish three distinct levels or questions: the computational question of what the system does, the questions on the representations and algorithms it uses, and the way these are implemented in hardware (Marr 1982).

The GOFAI computer metaphor and its associated definition of cognitive architecture is an example of *functionalism*: explanations of mind build on the functional role of basic elements and not on their intrinsic organization and/or implementation. We can think of explaining the operations of a computer in terms of the rules of its program and the symbols they operate on and not the electrons moving about in its hardware circuits. Hence, the GOFAI notion of architecture stipulates a top-down specification process where ultimate implementation imposes no relevant constraints on high-level function. This implies that implementation can follow multiple instantiations, as in the case of a calculator that can be implemented on contemporary computing machinery from desktops, to laptops, to phones, and microcontrollers to the Pascaline, a slide ruler, or an abacus: different hardware, same function. The powerful concept of architecture as defining abstraction layers from structure to function thus comes with a strong bias with respect to the explanatory dominance of these layers, i.e. abstract computation over physical implementation.

GOFAI was superseded by approaches that professed inspiration by biology such as the connectionism of parallel distributed processing, neural networks, behavior-based robotics, and artificial life. This was due to the inability of GOFAI to overcome some fundamental challenges such as the symbol grounding problem and the frame of reference problem (see Verschure (1996) for a summary). The notion of architecture, however, stuck and models were now advertised as architectures such as the robotics-oriented subsumption architecture and a range of connectionist models (for a review see Pfeifer and Scheier 1999). It was largely an expression of aspiration than of realization—none of the proposals made in this period dominated by biological metaphors has survived as a scientific theory of brains and bodies. This was largely due to the blanket negation of GOFAI that was used to define this new movement. However, they have placed the issue of embodiment and physical instantiation of architectures in bodies and brains firmly on the agenda (Chiel and Beer 1997). This transition reaffirms the core tenets of the Newell test and opens the door to a biomimetic perspective on the explanation of biological cognition in terms of synthetic architectures of mind.

The transition from a mind/reason to a body/behavior centered perspective has important implications for the notion of architecture as introduced by GOFAI and the von Neumann framework: strict top-down specification and multi-instantiation must now be augmented by bottom-up considerations, in particular how these two counter-currents of operation are coordinated. For instance, if one includes the morphology of the body itself in the operations that a cognitive architecture performs, then instantiation constrains computation and we have to consider a bi-directional exchange of constraints among the abstraction layers of the architecture. One can also consider the specific requirements of the hardware to maintain its components, e.g. metabolic demands, which should be virtualized with respect to overall function such as reaching the finish in an Ironman or writing a book chapter. In other words, bodies and brains also matter in order to explain the mental operations that minds can perform. Another consideration is that both bodies and brains are in turn organized at multiple scales: from the atomic interactions between CaM kinase enzymes and microtubules regulating biological memory and the intracellular processes supporting neuronal signal transduction, to the interaction of multiple brain systems and the demands of the physical, social, and cultural environment. In this perspective, we have to consider a system where the cognitive level, or the knowledge level (Newell 1990), is just one level of organization and description among many. This multi-scale organization of bodies, brains, and mind shows that we have to move from a strict top-down computational view, which indeed so far has failed to explain mind or brain, to a hybrid view of an architecture where body, brain, and environment are tightly coupled and dynamically regulated at a multitude of scales, rather than following a static hierarchical relationship. We can consider "Mind" as a virtualization layer of this architecture (Verschure 2012).

Brain architecture

Only a few coherent proposals exist on the architectural organization of the brain (Edelman 1989; Merker 2007; Vanderwolf 2007; Fuster 2008; Swanson 2011) and their mapping to both the microscopic organization of neuronal circuits or macroscopic behavior is the great challenge of modern neuroscience. An early proposal for overall brain organization originated in the work of Penfield and Jasper who, on the basis of brain stimulation during neurosurgery of epilepsy patients, advanced the centrencephalic theory of brain organization where core brain systems of the mid brain and the basal ganglia were seen as the pivot on which all other brain operations depend (Penfield and Jasper 1954).

Computational neuroscience, which emerged in the late 1980s, has largely stayed away from the architecture challenge despite a liberal use of the notion "architecture." This is largely due to a dominant reductionist physics-oriented view in models of explanation in neuroscience. Examples of architecture models that take into account questions of multi-scale organization and virtualization are the Distributed Adaptive Control theory of mind and brain (DAC, see below), Leabra (O'Reilly et al. 2012) and ACTr (Stocco et al. 2010).

The DAC architecture is the only model that has so far succeeded in bridging the gap between symbolic architectures, rational and social behavior, and the embodied and situated brain (for a review see Verschure, Chapter 36, this volume, and Verschure 2012). DAC states a clear explanandum, explaining the goal-oriented action in physical and social worlds, and shows how it can result from a control architecture which is organized in four layers with tight coupling within and between these distinct layers: the Soma (the body), Reactive, Adaptive, and Contextual layers. Across these layers, a columnar organization exists that at every level of the hierarchy deals with the processing of states of the world grounded in exteroception, the self, derived from interoception and action sensed through proprioception with an increased dependence on memory. The latter mediates between the former two via the environment. The reactive layer can be seen as a model of the evolutionary ancient Core Behavior Systems (CBS) of the brainstem which drives behavior and learning, the adaptive layer facilitates the learning of the perception and action state space and targets the limbic system and cerebellum, and the contextual layer affords acquisition of goal-oriented policies modeling the frontal areas of the neocortex and thalamus. The control flow in DAC is distributed following both bottom-up and top-down regulation. DAC has directly addressed the question of virtualization by progressing via a series of models at both the psychological level of behavior and function and that of the embodied brain, including pertinent anatomical and physiological constraints following a methodology of convergent validation. For instance, DAC showed that optimal decision-making requires conjunctive perception–action representations. Detailed biologically grounded models of the hippocampus have shown how this brain structure directly realizes this functional requirement, while associated models of the prefrontal/premotor cortex have shown how they are deployed in the neurodynamics of decision-making.

Core challenges for architectures of mind and brain

Newell made an important step in the advancement of UTCs. However, his challenge is anthropocentric, for instance by emphasizing language use, and cogitate-centric, i.e. how about emotion, motivation, or perception and action? Now, about 25 years later, we can thus recalibrate UTC by redefining it as Unified Theories of Embodied Minds (UTEM) that will take the form of instantiated architectures of the brains and bodies of living machines. Candidate UTEM has to answer both functional and structural constraints for the UTEM benchmark:

Functional constraints:
Level 1: Display autonomous adaptive[4] flexible real-time goal-oriented behavior in complex physical environments (Newell test: 1, 2, 3, 4, 7, 10-self).
Level 2: Display autonomous adaptive flexible real-time goal-oriented behavior in complex social environments including the use of symbols and language (Newell test: Level 1 + 5, 6, 9, 10-self-aware)

Structural constraints:
Level 1: Biological validity: plausibly be the product of biological evolution and be demonstrably constructible through neuro- and morphogenesis (Newell test: 11–13).
Level 2: Physical realizability: perform in real time, in the real world, using resources (e.g. energy, communication, and computation) comparable to biological systems.

The UTEM benchmark refines and elaborates Newell's original UTC benchmark by defining levels of performance (physical vs social), psychological and biological validity, and by including behavioral and epistemic autonomy together with real-world and real-time performance as specific constraints. This implies that UTEM argues that UTCs should be both scientific theories of natural minds and brains as well as models for synthesizing artificial ones. In this way the misleading effects of mimicry that the Turing test faces is avoided while biological inspiration is rejected as a method.

At the heart of the challenge is to find a scientific explanation of goal-oriented behavior, which has been anathema to the standard reductionist model pursued in our physics-based scientific paradigms. In other words, this means re-appreciating teleological explanations as Tolman advocated about 90 years ago. Goals are indeed a high-level form of virtualization comprising low-level needs, environmental opportunities, and cues and memory. Answering Newell's challenge will require us to develop situated and embodied architectures of mind and to understand their processes of virtualization and control.

With respect to the control architectures themselves, one largely untouched challenge, which looms on the horizon, is the question of how subsystems of the brain are controlled and organized through regulatory signals. For instance, how is the integration of sensory information with their modality-specific time constants (vision—fast transduction, chemical sensing—slow transduction) realized in the service of decision-making and action control, or how are processes of memory acquisition, retention, and expression managed? One insight from the study of evolution and cellular metabolism is that biological architectures follow bow-tie structures where transformations and control are achieved by relying on only a few critical elements (Csete and Doyle 2004), or constraints that de-constrain (Kirschner and Gerhart 2006). In the case of the brain we can think of the use of a few neurotransmitters such as dopamine to provide a global signal for reward and surprise (Redgrave et al. 2010). Hence, the constructed architecture not only allows us to place the pieces of the great psychological and biological data puzzle into a coherent framework, it also raises new questions and opens up new avenues for research which will allow us to identify the pieces that are still missing.

---

[4] Adaptation is derived from both learning and development.

Learning more

Historical context:
Gardner (1987) provides a great introduction to the cognitive revolution and the advances in Good Old Fashioned AI (GOFAI), while Pfeifer and Scheier (1999) does the same for what followed after its demise in the form of new AI, artificial life, and neural networks. Boden (2008) provides an encyclopedic overview of the search for understanding mind from the perspective of machines. Boden (2016) can be seen as an up-to-date overview of the evolving field of AI.

Cognitive architectures:
Allen Newell (1990) summarizes about four decades of work on advancing artificial intelligence towards a unified theory of cognition. In an influential article, Anderson and Lebiere (2014) analyze this tradition with special emphaszis on the question of how to benchmark such theories.

Brain architecture:
Swanson (2011) provides an excellent overview of the basic architecture of the mammalian brain while Kaas et al. (2007) deliver a masterly overview of brain evolution; Vanderwolf (2007) links brain evolution to behavior while Panksepp and Biven (2012) do the same from the perspective of emotion; Edelman (1989) provides an exceptional example of system level brain theory that still stands today as unique in its scope.

Evolution:
Kirschner and Gerhart (2006); Deacon (2011)
Journal: *Biologically Inspired Cognitive Architectures*, edited by Alexei Samsonovich
Book series: The Oxford Series on Cognitive Architectures, edited by Frank Ritter

Acknowledgments

Supported by European Commission H2020 project socSMCs (641321) and the European Research Council project cDAC (ERC 341196).

References

Anderson, J. R., and Lebiere, C. J. (2014). *The atomic components of thought*. Abingdon, UK: Psychology Press.

Boden, M. (2008). *Mind as machine: a history of cognitive science*. Oxford: Oxford University Press.

Boden, M. (2016). *AI: its nature and future*. Oxford: Oxford University Press.

Brooks, F. P. J. (1962). Architectural Philosophy. In: W. Buchholz (ed.), *Planning a computer system: Project Stretch*. New York: McGraw-Hill, pp. 5–16.

Chiel, H. J., and Beer, R. D. (1997). The brain has a body: adaptive behavior emerges from interactions of nervous system, body and environment. *Trends Neurosci.*, **20**(12), 553–7.

Csete, M., and Doyle, J. (2004). Bow ties, metabolism and disease. *Trends in Biotechnology*, **22**(9), 446–50.

Deacon, T. W. (2011). *Incomplete nature: how mind emerged from matter*. New York: WW Norton & Company.

Edelman, G. M. (1989). *The Remembered Present: A biological theory of consciousness*. New York: Basic Books.

Fuster, J. M. (2008). *The prefrontal cortex*. New York: Academic Press.

Gardner, H. (1987). *The Mind's New Science: A history of the cognitive revolution*. New York: Basic Books.

Herculano-Houzel, S. (2009). The human brain in numbers: a linearly scaled-up primate brain. *Frontiers in Human Neuroscience*, **3**, 1–11.

Hull, C. L. (1952). *A Behavior System: An introduction to behavior theory concerning the individual organism*. New Haven, CT: Yale University Press.

Jensen, E. R. (2006). *Clocking the mind: Mental chronometry and individual differences*. Amsterdam: Elsevier.

Kaas, J. H., Striedter, G. F., Rubenstein, J. L., Bullock, T. H., Krubitzer, L. A., and Preuss, T. M. (2007). *Evolution of nervous systems*. Amsterdam: Elsevier.

Kirschner, M. W., and Gerhart, J. C. (2006). *The plausibility of life: resolving Darwin's dilemma*. New Haven, CT: Yale University Press.

Langley, P., Laird, J. E., and Rogers, S. (2009). Cognitive architectures: research issues and challenges. *Cognitive Systems Research*, **10**, 141–60.

Marr, D. (1982). *Vision: a computational investigation into the human representation and processing of visual information*. New York: Freeman.

Merker, B. (2007). Consciousness without a cerebral cortex: a challenge for neuroscience and medicine. *Behavioral and Brain Sciences*, **30**(01), 63–81.

Newell, A. (1990). *Unified Theories of Cognition*. Cambridge, MA: Harvard University Press.

O'Reilly, R. C., Hazy, T. E., and Herd, S. A. (2012). *The leabra cognitive architecture: how to play 20 principles with nature and win!* Oxford: Oxford University Press.

Panksepp, J., and Biven, L. (2012). *The archaeology of mind: neuroevolutionary origins of human emotions*. Norton Series on Interpersonal Neurobiology. New York: WW Norton & Company.

Penfield, W., and Jasper, H. (1954). *Epilepsy and the functional anatomy of the human brain*. Oxford: Little, Brown and Co.

Pfeifer, R., and Scheier, C. (1999). *Understanding intelligence*. Cambridge, MA: MIT Press.

Redgrave, P., Rodriguez, M., ..., and Smith, Y. (2010). Goal-directed and habitual control in the basal ganglia: implications for Parkinson's disease. *Nat. Rev. Neurosci.*, **11**(11): 760–772. doi: 10.1038/nrn2915

Romanes, G. J. (1888). *Animal Intelligence*. New York: Appleton.

Stocco, A., Lebiere, C., and Anderson, J. R. (2010). Conditional routing of information to the cortex: a model of the basal ganglia's role in cognitive coordination. *Psychological Review*, **117**(2), 541.

Swanson, L. W. (2011). *Brain architecture: understanding the basic plan*. New York: Oxford University Press.

Tolman, E. C. (1939). Prediction of vicarious trial and error by means of the schematic sowbug. *Psychological Review*, **46**, 318–36.

Vanderwolf, C. H. (2007). *The evolving brain: the mind and the neural control of behavior*. Dordrecht: Springer Verlag.

Verschure, P. F. M. J. (1996). Connectionist explanation: taking positions in the mind–brain dilemma. In: G. Dorffner (ed.), *Neural networks and a new artificial intelligence*. London: Thompson, pp. 133–188.

Verschure, P. F. M. J. (2012). Distributed Adaptive Control: a theory of the mind, brain, body nexus. *Biologically Inspired Cognitive Architecture*, **1**(1), 55–72.

von Neumann, J. (1945/1993). First draft of a report on the EDVAC. *IEEE Annals of the History of Computing*, **4**, 27–75.

Chapter 36

A chronology of Distributed Adaptive Control

Paul F. M. J. Verschure

SPECS, Institute for Bioengineering of Catalonia (IBEC), the Barcelona Institute of Science and Technology (BIST), and Catalan Institute of Advanced Studies (ICREA), Spain

Distributed Adaptive Control (DAC) is a theory of mind and brain introduced in 1991 against the background of developments in the fields of artificial intelligence and cognitive science (see Verschure and Prescott, Chapter 2, and Verschure, Chapter 35, this volume)[1]. In the second phase of development, DAC was generalized towards neuroscience and robotics. DAC departs from the paradox between rationalism and empiricism: does knowledge originate in reason as advanced by Plato or in the senses as advanced by his student Aristotle? At the heart of the contradiction between these two schools of thought stands the obstacle of the inability of mapping sensory data into valid knowledge and reasoning, a criticism levelled by Chomsky against Skinner triggering the AI revolution of the 1950s and by the cognitivists against the early connectionism of the 1980s. The British empiricist David Hume had already articulated this dilemma and observed that there is no necessity to interpret physical events following each other, such as object A hitting object B which subsequently moves, as having a causal relation as in "A causes B to move" (Hume 1748). Hence, how can we be certain about "causation"? This sceptical analysis led Kant, after famously having been awakened from his dogmatic slumbers, to postulate that minds can have valid knowledge, but it will be limited to their own experience and require priors of space and time to structure this knowledge (De Pierris and Friedman 1997). There is no consensus on the solution of this fundamental problem of epistemology, and for this reason, I have argued for the constructive empiricism of convergent validation (Verschure and Prescott, Chapter 2, this volume). However, given that we can see the mind/brain as a knowledge organ, DAC has taken this fundamental challenge in epistemology as its point of departure. At the time that DAC was formulated this problem was reintroduced in the field of cognitive science as the symbol grounding problem, or "how can an artificial intelligence conceived as a computer assign meaning to its symbols"? (Harnad 1990; see Wortham and Bryson, Chapter 33, and Verschure, Chapter 35, this volume). In a broader perspective, we can speak of the problem of priors (Verschure 1998). The driving intuition behind DAC is that knowledge is grounded in the interaction between embodied agents and their physical and social environments, constrained by mechanisms of learning and memory, and bootstrapped from minimal priors that have been selected during evolution (see Prescott and Krubitzer, Chapter 8, and Deacon, Chapter 12, this

---

[1] A more detailed description and tutorials on DAC can be found under http://csnetwork.eu/CSN%20 Book%20Series

volume). DAC was of immediate relevance to the nascent field of "New AI" as witnessed by the many talks that early proponents of this field such as Luc Steels and Rolf Pfeifer gave on the first results that were obtained with DAC (Verschure et al. 1993). It was also one of the first neural models that was ported to both simulated and physical mobile robot, through the collaboration with Francesco Mondada and the Khepera microrobot (Mondada and Verschure, 1993).

The Distributed Adaptive Control (DAC) architecture

Let us start with some definitions. DAC takes "mind" to be the functional macroscopic properties of embodied brains that are directly or indirectly expressed in action. Mind is an amalgamation of processes including motivation, perception, cognition, attention, memory, learning, consciousness, etc. This mind is situated in physical and social environments. Because of the tight coupling of body, brain, mind, and environment, especially once we take into account the memory of the agent, we can speak of a nexus (Verschure 2012). "Behavior" is defined as autonomous changes in the position or shape (confirmation) of the body or soma of an agent. Once behavior serves internally generated goals we can speak of "action", i.e. it is intentional or conative. The "brain" is a distributed, wired control system that exploits the spatial organization of connectivity combined with the temporal response properties of its units to achieve transformations from sensory states, derived from both the internal (body) and external environment, into action. The core variable this mind/brain controller maintains in a dynamic equilibrium is the integrity of the organism in the face of the second law of thermodynamics (see also Prescott, Chapter 4, Deacon, Chapter 12, and Mast et al., Chapter 39, this volume). This drive for self-preservation and realization is fundamentally defined through the organism's needs that are continuously shaped by somatic and environmental change (see also Prescott and Verschure, Chapter 1, and Vouloutsi and Verschure, Chapter 34, this volume). DAC follows nineteenth-century physiologists Claude Bernard and Ivan Pavlov in conceptualizing the mind/brain as a control system that generates action to maintain a multi-stable equilibrium between the body and the environment, rather than an information processor or a computational system. In other words, information is processed, and processes realized, that might be describable in computational terms, but they all serve and are predicated on current and future action. A control system is usually considered in separation from what it controls, i.e. the plant (see also Witte et al., Chapter 31, this volume). This separation raises the question what top-level functions and essential variables the mind/brain maintains to generate the *How* of action? DAC (Figure 36.1) proposes that these are just five: needs, drives and motivation or "Why", states of the world such as objects or "What", spatial structure of the task or "Where", the temporal dynamics of the task and the agent or "When" and "Who", in case the agent deals with other agents. We can this say that the function of the brain is to solve the H4W optimization problem when a single agent confronts its physical world and H5W in the case where the world also contains other agents such as predators, prey, and conspecifics (Verschure 2016).

The *somatic level* designates the body and defines three fundamental sources of information: *sensation, needs*, and *actuation*. Sensation is driven by external and internal sources of stimulation (or exo- and endosensing, respectively). Needs are dictated by the essential variables that assure survival, while actuation is defined by the control of the musculoskeletal system[2].

---

[2] At this stage DAC does not include detailed analysis of the processes that comprise the core of the body such as the organs, smooth and cardiac muscle systems, and their control by the peripheral nervous system.

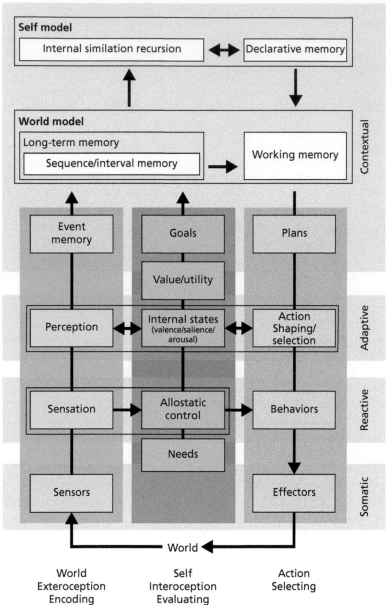

Figure 36.1 A highly abstracted representation of the Distributed Adaptive Control (DAC) theory showing its main processes (boxes) and dominant information flows (arrows). DAC is organised along four layers (Soma, Reactive, Adaptive, and Contextual) and three columns (World, Self, Action). Across these layers three functional columns of organisation exist: exosensing, the sensation and perception of the *external world* (left, blue); endosensing, detecting and signalling states derived from the physically instantiated *self* (middle, green); and action which establishes the interface between self and the world *action* (right, yellow). The arrows show the primary flow of information mapping exo- and endosensing into action, defining a continuous loop of interaction with the world. *Soma* designates the body and its sensors, organs, and actuators. It defines the needs, or Self Essential Functions (SEF), the organism must satisfy to survive. The *Reactive Layer* (RL) comprises dedicated Behavior Systems (BS) each implementing predefined sensorimotor mappings

The reactive layer

The *reactive layer* (Verschure et al. 1993) comprises fast predefined sensorimotor loops that support the behaviors serving Self Essential Functions (see also Vouloutsi and Verschure, Chapter 34, this volume). These reflexes are coupled via need and drive systems creating sense-affect-act triads. Behaviorally the performance is what one would expect from a so-called Braitenberg vehicle (Braitenberg 1984; Walter 1951) or the subsumption architecture (Brooks 1986). However, the distinguishing feature of the DAC reactive layer is that it is part of a larger architecture. The activation of a reflex carries essential information on the interaction between the agent and the world that is a key control signal for subsequent layers driving conflict resolution and epistemic needs, i.e. knowledge acquisition (for a detailed analysis, see Herreros, Chapter 26, this volume). A competitive relationship exists between the different internal states and to resolve conflicts between these states, such as between avoidance and approach, the Internal State Selector (ISS) is introduced. A further distinguishing feature of this layer is that it is modeled after an allostatic process that regulates the homeostatic behavior subsystems, which is both closer to the dynamics of physiological systems and scalable, as opposed to more phenomenological behavior found in behavior-based robotics (Arkin 1998). The Self Essential Functions of RL are both oriented towards direct survival as well as epistemic functions such as exploration and novelty seeking as in the notion of infotaxis already found in insects (Vergassola et al. 2007; see also Pearce, Chapter 17, this volume).

The adaptive layer

The *adaptive layer* extends the predefined sensorimotor loops of the reactive layer with acquired sensor and action states (Figure 36.2). Hence, it allows the agent to escape from strictly predefined reflexes through learning. The *adaptive layer* is interfaced to the full sensorium of the agent, its internal needs, and its effector systems receiving internal state information from the reactive layer and in turn generating action. This layer of DAC was first modeled in Verschure and Coolen (1991) and validated with simulated robots in Verschure et al. (1993). At this layer, adaptive mechanisms are deployed to deal with the fundamental unpredictability of both the internal and the external environment. Through learning, a state space of sensation and action is acquired together with their direct association. The adaptive

serving the SEFs. To allow for action selection, task switching and conflict resolution, all BSs are in turn regulated via an allostatic controller that sets their internal homeostatic dynamics relative to overall system demands and opportunities. The *Adaptive Layer* (AL) acquires representations of the states of the world and the agent and shapes action constrained by the value functions derived from the allostatic control of the RL. Learning by the AL minimizes perceptual and behavioral prediction error, building a model free action generation system. The *Contextual Layer* (CL) further expands the time horizon in which the agent can operate, realising model based policies, through the use of sequential short- and long-term memory systems (STM and LTM respectively). STM acquires conjunctive sensorimotor representations that are generated by the AL as the agent acts in the world. STM sequences are retained as goal-oriented models in LTM when positive value is encountered, as defined by the RL and AL. The contribution of these stored LTM policies to goal-oriented decision-making depends on four factors: perceptual evidence, memory chaining, valence, and the expected cost of reaching a given goal state. The content of working memory (WM) is defined by the memory dynamics that represents this four-factor decision-making model. The autobiographical memory system allows the restructuring of memory around the unifying notion of Self which DAC proposes is essential to engage with the social world and interpret the states of others regarding that of the self. See text for further explanation.

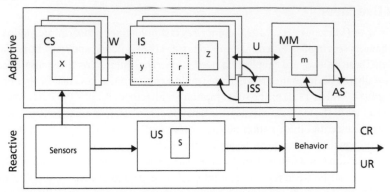

Figure 36.2 Schematic representation of the Adaptive Layer. Boxes are simulated neuronal populations labelled using the basic elements of classical conditioning: CS: Conditioned Stimulus defined through sensory modalities, US: Unconditioned Stimulus defined through internal states, UR: Unconditioned Response defined by predefined motor system, CR: Conditioned Response acquired through association with CS. In classical conditioning, the initially neutral stimulus (CS) substitutes the ability of the US to trigger a predefined response (UR) through their paired presentation. DAC specific labels are: IS: Internal State, MM: Motor Map, ISS: Internal State Selection, AS: Action Selection. State variables in solid boxes denote activity (CS: x; IS: z; US: s; MM: m) while those in dashed boxes denote inputs (IS: y and r). W, U, and V are weight matrices.

layer is originally a model of classical conditioning (see Herreros, Chapter 26, and Verschure, Chapter 35, this volume).

The learning dynamics of the adaptive layer is based on prediction-based Hebbian learning (Verschure et al. 1992) and has been phrased in a general formal framework called *correlative subspace learning* (CSL; Duff and Verschure 2010). We define the following abbreviations for the activities in the different simulated neuronal populations:

Activity:

$$s = \text{US} \in \mathbb{R}^{M}$$

$$x = \text{CS} \in \mathbb{R}^{N}$$

$$z = \text{IS} \in \mathbb{R}^{K}$$

$$m = \text{MM} \in \mathbb{R}^{L}$$

Weight matrices:

$$V = \text{US to IS} \in \mathbb{R}^{M \times K}$$

$$W = \text{CS to IS} \in \mathbb{R}^{N \times K}$$

$$U = \text{IS to MM} \in \mathbb{R}^{K \times L}$$

Inputs to *IS*:

$$r = \text{US to IS} \in \mathbb{R}^{K}$$

$$y = \text{CS to IS} \in \mathbb{R}^{K}$$

With these definitions the dynamics of the adaptive and reactive layer can be written as:

$$r = V^T s$$

$$y = W^T x$$

$$z = y + r$$

$$e = Wy$$

$$m = U^T z H \left(z - \Theta_A \right) \qquad (1)$$

The weights are updated following the CSL rule:

$$\Delta W = \eta (x - e)((1 - \zeta)y + (1 + \zeta)r)^T \qquad (2)$$

where η is a variable learning rate. In CSL the learning dynamics is determined by two product terms $xy^T \infty \, xx^T$ and xr^T, which we can interpret as the sensory prediction of x and the motor prediction of r, respectively. ζ varies between -1 and 1 and balances the relative impact of either perceptual or motor learning. With a ζ of -1 perceptual learning dominates and Eq. 1 is identical to the so-called Oja learning rule (Oja 1982). With a value of 1, the learning dynamics aims at minimizing the motor prediction error. In perceptual learning, the weight changes are driven by the discrepancy between actual sensory events, x, and those expected given the state of IS, e. DAC refers to e as the perceptual *prototype*.

The adaptive layer of DAC provides a solution to the problem of priors described earlier because it acquires the state space of the agent by its interaction with the environment. In DACI perceptual learning was achieved by changing W through Hebbian learning. The insistence on Hebbian learning was to build models that were based on minimal biologically plausible assumptions, i.e. locality of information, and as such a response to supervised learning schemes that require global signalling, structured input sampling, and human pre-labelling of input patterns. DACI also showed emergent behavioral patterns such as "wall following". In a further extension, DACI was applied to a robot foraging task based on block sorting where the model was augmented with a value-based learning system guiding plasticity (Verschure et al. 1995). However, DACI showed fundamental shortcomings such as an excessive dependence on the details of input sampling and the overgeneralization of actions. DACII was defined to solve these problems using a new prediction-based local learning rule of which CSL, defined above, is the most general formulation which is still in use.

CSL captures the law of associative competition formulated by Rescorla and Wagner that describes the impact of the paired presentation of a particular Conditioned Stimulus (CS) and Unconditioned Stimulus (US) on the probability of observing Conditioned Responses (CR). This law emphasizes that associative change will depend on how unexpected the CS is, given the US (Rescorla and Wagner 1972). The CSL model is also consistent with adaptive filter methods going back to the Kalman filter and other derived approaches (Kalman 1960). Indeed, DACII anticipated the "predictive brain" hypothesis which is currently much in vogue and is seen as an integrative framework to understand the brain (Friston 2010). CSL is a specific instantiation of this so-called free energy principle and its notion of active inference which we have further generalized towards hierarchical forward model based control (Maffei et al. 2017).

A unique application of DACII has been the 180 m^2 human-accessible exhibition space Ada that supported complex interactions with its visitors (Eng et al. 2003). Over half a million people visited the exhibit during the summer of 2002 as part of the Swiss National Expo.02. The reactive layer of

DAC was used to set the behavioral goal functions of Ada to optimize interaction, recognition, and distribution of visitors. Based on the changes in these goal functions and the ability to achieve need reduction the emotional states of Ada were defined following a valence and arousal model consistent with the notion of epistemic emotions (see Vouloutsi and Verschure, Chapter 34, this volume). DACII was successfully used to acquire guiding cues generated by the luminous floor tiles of Ada to purposefully influence the position and movement direction of visitors (Eng et al. 2005).

The AL allows the agent to overcome the predefined behavioral repertoire of the RL and to successfully engage with the unpredictable aspects of the world. The weak pre-specification of the RL combined with the learning mechanisms of the AL allows the DAC system to bootstrap itself to deal with novel and a priori unknown states of the world and the agent, i.e. resolving the symbol grounding problem. This adaptation, however, occurs in a restricted temporal window of immediate interaction and to escape from the "now" other learning systems must be engaged. The adaptive layer and the state space it acquires provide a foundation for these more advanced learning and memory intense learning systems: the contextual layer. The *contextual layer* of DAC integrates the state space acquired by the adaptive layer to create behavioral plans or policies. The atomic elements of these plans are formed by the state space of exo- and endo-sensing constructed by the adaptive layer or its sensorimotor contingencies (Figure 36.3). The

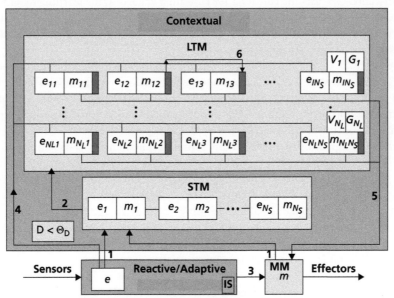

Figure 36.3 Contextual Layer of DAC: (1) The perceptual prototype *e* and the motor activity *m* generated by the adaptive layer are acquired and stored as a *segment* if the discrepancy between predicted and occurring sensory states, *D*, falls below its predefined threshold v_D . *D* is the time-averaged reconstruction error of the AL perceptual learning system: $x - e$. (2) If a goal state is reached, e.g. reward or punishment detected, the content of STM is acquired in LTM as a sequence conserving its order, and STM is reset. Every sequence is labelled with the specific goal and internal state it pertains to. (3) The motor population *MM* receives input from the *IS* populations per the rules of the adaptive layer. (4) If the input of the *IS* population to *MM* is sub-threshold, the values of the current *CS* prototypes, *e*, are matched against those stored in LTM. (5) The *MM* population receives the motor response calculated as a weighted sum over the memory segments of LTM as input. (6) The segments that contributed to evidence supporting the executed action will prospectively bias other LTM segments they are associated with.

contextual layer comprises systems for short- and long-term and working memory (STM, LTM, and WM, respectively). These memory systems allow for the formation of sequential representations of states of the environment (perception) and actions generated by the agent or its acquired sensory–motor contingencies. The acquisition and retention of these sequences are conditional on the goal achievement of the agent and the absence of reactive layer activation. These behavioral plans can be recalled through perceptual matching and internal chaining among the elements of the retained memory sequences. The dynamic variables that this process entails are DAC's working memory system.

Initially, DAC's behavior is dominated by the reactive layer from which the adaptive layer is bootstrapped. Once the average sensory reconstruction error falls below a specific threshold, Θ_d, the contextual layer is enabled, and ongoing sensorimotor contingencies are stored in its STM buffer comprising the sensory prototype e and the action m_{AL} respectively. Once a goal state is reached the STM sequence is retained in LTM. The contribution of LTM segment k of sequence n, d_{nk}, to decision making and action generation given a certain goal, is defined by four factors: *perceptual evidence*, or the state of the collector unit, c_{nk}, *memory bias* defined by the state of the trigger unit, t_{nk}, the expected distance to the *goal* state, g_{nk}, and *valence*, v_n:

$$d_{nk} = c_{nk}t_{nk}g_{nk}H\left(v_n, v\right)$$

$$c_{nk} = \alpha c_{nk}\left(t-1\right) + dist\left(e, e_{nk}\right)$$

$$t_{nk} = \beta t_{nk}\left(t-1\right) + \sum_{i \in N_{LTM} j \in N_{STM}} P_{ij}d_{ij}\left(t-1\right)\delta\left(m_{ij}, m_{mm}\right) \tag{3}$$

c_{nk} is the time-integrated perceptual similarity of the stored and current sensory prototype. This similarity is calculated using distance function *dist*, which is usually the normalized Euclidian distance. The trigger unit, t_{nk}, is the time-integrated memory bias defined by the LTM connectivity matrix P that sums over all LTM segments, N_{LTM}, and the sequence length, N_{STM}. All LTM segments that contributed to the generation of the executed action project their decision-making score, d_{ij}, to segment k of sequence n if they are connected. This arrangement allows for the chaining of the segments of sequences based on memory. t implements the working memory of the contextual layer since it stored the decision-making dynamics and LTM segments are contributing to its state when their stored motor state, m_{ij}, matches the motor state executed by the adaptive layer. $m_{mm} \cdot g_{nk}$ defines the inverse normalized distance of segment k of sequence n to the goal state of this sequence. This favors segments closer to goal states to influence decision-making. All segments whose decision-making score d exceed the decision-making threshold Θ_d project their motor state, m_{nk}, onto the motor population MM. After action selection through the winner-take-all process of MM, the selected action is executed, and segments that are associated with this action update the values of the trigger units t through spreading activation. Subsequently, the LTM connectivity matrix P can be updated to reflect specific rules (see below).

The contextual layer

The extension of DAC with the *contextual layer* was designated DACIII and was first proposed in Verschure (1993) and tested in foraging benchmarks (Verschure and Voegtlin 1998) while a recurrent neural network model of the STM-LTM systems was formulated in DACIV (Voegtlin and Verschure 1999). Further experiments with simulated and physical robots revealed that a unique feedback loop exists between action and perception that stabilizes the interaction between the AL and CL through behavior itself called behavioral feedback (Verschure et al. 2003). This demonstrated that the model-based policies or behavioral plans acquired by the contextual layer, through their mapping to action, carve out an effective behavioral space or niche rendering the world more predictable and therefore perceptual reconstruction error smaller. A further question was whether the solutions that DACIII found for foraging could be considered optimal. Lacking consensus benchmarks and comparable approaches (see also Verschure, Chapter 35, this volume) a more formal approach was taken based on work by Massaro on multi-modal speech perception, where Bayesian integration was proposed as a universal principle of perception and cognition (Massaro 1997). Indeed, DAC comprises constructs that are analogs of the central components of a Bayesian analysis: goals, actions, hypotheses, observations, experience, prior probabilities, and a score function. By phrasing the foraging tasks performed with the DAC architecture in Bayesian terms, it was shown that the contextual layer generates actions that are optimal in a Bayesian sense; this version was called DACV (Verschure and Althaus 2003). The contextual layer of DACV acquires hypotheses on how to achieve specific goals by the adaptive layer and the interaction with the environment. The matching and competition processes explore this knowledge base of LTM with the objective of selecting the optimal action, where optimality is defined as maximizing reward (e.g. targets in a foraging task) while minimizing punishment (e.g. collisions). In a Bayesian perspective the optimal action, a, is the one that optimizes the expected gain $\langle g \rangle_a$:

$$\langle g \rangle_a = \sum_{s_n \in S} p(s_n \mid r) G_g (s_n, a) \tag{4}$$

where $p(s_n|r)$ is the posterior probability defined by Bayes' rule and $G_g (s_n,a)$ is a score function that defines the gain obtained from executing action a if s_n is true. In case of DACV, s and r are the perceptual prototypes of the contextual and adaptive layer, respectively, and G is determined by the labelling of the LTM sequences in terms of the goal states they are associated with. These goal states are the top-level representations generated by the Self column of the architecture comprising needs, drives, and value and their associations to states of the world. DACV showed that the actions selected by the contextual layer are optimal with respect to G. This demonstration showed that DACV is an autonomous rational system that bootstraps its own knowledge through its interaction with the environment and subsequently uses it in a Bayes optimal fashion to reach its goals, a key step towards UTEM (see Verschure, Chapter 35, this volume).

The chaining mechanism of the contextual layer displays both sequence and goal fidelity. Sequence fidelity is defined by the modulation of t by other segments belonging to the same sequence. DACIII and DACV exploited a nearest neighbor forward chaining where segments propagate a biasing signal to their direct neighbor closer to the goal state of the sequence. To be more robust to variability in the sensor states and thus to noise in the perceptual prototypes that drive the memory matching process (Eq. 3), LTM chaining was expanded to so-called *memory smoothing*, where the propagation of the memory biasing signal along a sequence follows a Gaussian distribution (Ringwald and Verschure, 2007). Memory smoothing was introduced to allow the contextual layer to integrate across some modalities such as spatial information and sensory cues. Goal-based sequences contain further implicit information which was captured in the notion of *goal fidelity*. Goal fidelity describes a process that modulates the decision-making score of each segment by its inverse normalized distance to the goal state of its sequence. However, distance regarding memory segments and actual physical distance to be covered to physically reach a goal are not necessarily similar. Moreover, the goal state of a sequence could either be appetitive or aversive, which could also be expressed in the goal distance measure.

Hence, the goal fidelity measure was expanded to reflect these factors. This augmented model was applied to foraging benchmarks comparing it to an optimal Partially Observable Markov Decision Process (POMDP) solution that contains global information about the task (Thrun et al. 2005). DAC's performance was only ten percent lower than that of the MC-POMDP method despite that it used only information that could be directly sensed by the robot. This result is remarkable considering that DAC is fully self-contained and does not require a priori global information, i.e. it uses purely egocentric actions as opposed to the POMDP that relies on allocentric ones. Another step towards UTEM (see Verschure, Chapter 35, this volume).

A fundamental limitation of DAC was that all actions were encoded in egocentric terms. Hence, slight variations in the position of the agent would lead to a rapid divergence in the prediction-based matching mechanisms of the AL and CL, and thus the behavioral trajectories generated would increasingly diverge from those followed during learning. This illustrates another fundamental non-linearity that living machines must face and correct. To overcome this ergodicity problem, experiments were performed that replaced egocentric action encoding with an allocentric definition of a movement vector that points to the location of the next landmark given the current position (Marcos et al. 2010). This was achieved using a path integration mechanism that accumulates rotations and translations as the agent moves between landmarks. It was shown how this approach leads to robust maze navigation and learning even when significant noise is applied to the motor output. This result illustrates the relationship between egocentric taxon-based versus allocentric route-based navigation strategies (see Witte et al. Chapter 31, this volume), and shows how DAC bootstraps representations of routes and model-based allocentric policies from predefined reactive egocentric taxon-based strategies. Another variation on the acquisition of allocentric navigation strategies with DAC was recently proposed for mapless navigation. This variation explored the lower bounds on the elements of a contextual memory, exploring the notion that a contextual memory supporting navigation could be conceptualized as a graph where the nodes are defined by landmarks and their connections by acquired heading directions (Kubie and Fenton 2009). Indeed, the LTM of the CL can be seen as a graph structure linking episodes defined by integrated sensorimotor states with movement vectors. It has been shown that this approach towards mapless navigation is both robust on robots performing multi-modal chemical search tasks and is consistent with ant navigation behavior (Mathews et al. 2009). These DAC robots are capable of returning to their original home position after reaching a target location and recovering from kidnapping, i.e. sudden displacements within the task area. This has been further generalized to mammalian spatial cognition in Maffei et al. 2015).

Current status and future directions: Answering the UTEM challenge with DAC

DAC has been widely used to address the H4W challenge as in robot foraging tasks. However, to satisfy the Universal Theory of Embodied Minds benchmark (see Verschure, Chapter 35, this volume), it must both generalize to the social world, or H5W, and satisfy structural constraints derived from the brain.

DAC has been successfully mapped to the specifics of the neuroscience of mind and brain along two types of models to realize convergent validation (see Verschure, Chapter 35, this volume). On the one hand, a whole brain architecture approach was followed which facilitates the mapping to behavior, while components of the architecture and basic operating principles have been linked to the invertebrate and mammalian brain through anatomically and physiologically constrained models (Verschure et al. 2014). These two lines of the DAC project have been integrated into a first embodied whole brain model comprising detailed models of core brain structures including cerebellum, entorhinal cortex, hippocampus, and prefrontal/premotor cortex called DACX. DACX validates this model in the context of foraging including obstacle avoidance, hoarding, exploration, and homing (Maffei et al. 2015). Also, predictions of the DAC derived hypotheses have in turn been validated in the laboratories of experimental neuroscientists and core principles of DAC have been mapped to a highly effective neurorehabilitation technology (see Rubio Ballester, Chapter 59, this volume). These results show the commitment of DAC to convergent validation with the goal of obtaining an empirically adequate description of biological living machines.

To generalize from H4W to H5W, the DAC architecture has been mapped to the control of an anthropomorphic robot that engages in dyadic interactions with humans in a third phase of development of the theory (Lállée et al. 2015). This step has included the augmentation of the functions of the architecture to include drives to socially engage, to seek knowledge of the world through interaction with humans, the ability to acquire models of other agents and "read their minds", to use social cues to establish and maintain interaction, and to learn language. Underlying the successful deployment of DAC in H5W scenarios, the system was augmented with an autobiographical memory system that allows the robot to anchor its experiences in the H5W ontology (Lallee and Verschure 2015). This DACh, h for humanoid, architecture has been shown to successfully solve the H5W challenge in restricted task domains with a single human[3]. This is another key step towards UTEM. The current challenge is to scale this H5W property up to the ability to make anything a task, the requirement for Robot Artificial General Intelligence. The DAC theory predicts that this will require a form of robot consciousness (see Seth, Chapter 37, this volume).

How has DAC held up over the last 25 years? Currently, artificial intelligence is in the grips of a third wave of neural network modeling after the early models of Rashevsky and Rosenblatt and their reprise in the connectionism of the late 1980s (see Prescott and Verschure, Chapter 1, this volume). Two approaches stand out that again fall at opposite sides of the rationalism–empiricism divide. First, due to advances in computing technology and the availability of massive data sets, learning in multi-layered neural networks has made significant progress into complex task domains (Lecun et al. 2015; Schmidhuber 2015). Hence, deep learning assumes minimal priors compensated by extremely large data sets for training. These data sets, however, are still labelled by humans to drive supervised learning. Subsequently, deep learning has been combined with reinforcement-based learning with the goal to reach human or super-human performance in benchmark tasks such as Atari video games (Mnih et al. 2015) and complex games such as Go (Silver et al. 2016). These approaches still require a lot of trials to train the network, much higher than what humans need. In response, others have proposed that one-shot learning can be achieved when pre-existing core knowledge is provided in the form of physics and psychology simulation engines, typically Bayesian causal models (Lake et al. 2015; Lake et al. 2017). These approaches do express core tenets of the DAC theory such as the need for learning-dependent acquisition of the state space, its compression, and the combined realization of this state space with the development of action policies. However, these approaches are also still problematic when applied to real-world living machines. Deep learning approaches require large data sets made up of uncorrelated adjacent samples as well as significant computational resources, making them hardly compatible with the real-time constraints faced by living machines. Bayesian causal models, in turn, require extensive prior knowledge dependent

---

[3] See http://wysiwyd.upf.edu/research for illustrative videos.

on the physical and social environment, which still must be accounted for, is hardly generalizable across different applications, and will remain brittle in the face of real-world variability. Both approaches are critically dependent on human labelling of data and are thus failing on the problem of priors. DAC advances and has implemented an alternative approach considering that embodiment and situatedness provide a grounded source of priors for living machines based on the following principles. First, embodiment strongly constrains the viable set of sensorimotor states of the agent through the specific physical coupling of the body and the environment. These constraints are particular to the distinct morphology of a living machine, its internal needs (e.g. in term of energy consumption and damage minimization), and the environment in which it operates, precluding a full explicit predefinition as assumed regarding full pre-labelling of data and an intuitive physics/psychology engine. Second, situatedness places the robot in the social environment governed by specific norms and conventions and provides crucial cues about well-adapted behavioral policies (Lallée et al. 2015). Here again, this is specific to the context in which the living machine operates and cannot be fully predefined as assumed by an intuitive psychology engine. Also, DAC has taken benchmarks of greater ecological validity, such as foraging, language learning, and logical puzzles, than computer and board games. These more contemporary approaches are essentially providing alternative views of subsystems accounted for in the DAC theory and consequently are still far removed from achieving the UTEM challenge. In contrast, DAC has already made significant inroads towards this benchmark of Artificial General Intelligence by addressing both levels of functional constraints (H4W and H5W) and by having established strong links with the neuronal principles of perception, cognition, and action. However, to quote Alan Turing: "We can only see a short distance ahead, but we can see plenty there that needs to be done."

Learning more

The general outline of DAC as applied to the H4W problem can be found summarized in Verschure (2012) with its mapping to the brain described in Verschure et al. (2014). The summary of its solution of H5W and the role of consciousness therein is presented in Verschure (2016), while the implementation on humanoid robots can be found in Lallée et al. (2015) and Moulin-Frier et al. (2017a). The mapping of extent models of contemporary artificial intelligence such as deep learning and reinforcement learning to DAC can be found in (Moulin-Frier et al. 2017b). The mapping from DAC to neurorehabilitation is summarized in Verschure (2011) and Ballester et al. (2016), while its application in robot-based tutoring systems is described in Vouloutsi et al. (2016).

Acknowledgments

The development of the DAC theory has received support from a number of projects supported by the European Commission in particular in FP7 Synthetic Forager (SF, 217418), Experimental Functional Andriod Assistant (EFFA, 270490), GoalLeader (270108), eSMCs (270212), What You Say IsWhat You Did (WYSIWYD, 612139) and H2020: socSMCs (641321) and the European Research Council project cDAC (ERC 341196).

References

Arkin, R. C. (1998). *Behavior-based robotics*. Cambridge, MA: MIT Press.

Ballester, B. R., Maier, M., San Segundo Mozo, R. M., Castañeda, V., Duff, A., and Verschure, P. F. M. J. (2016). Counteracting learned non-use in chronic stroke patients with reinforcement-induced movement therapy. *Journal of NeuroEngineering and Rehabilitation*, **13**(1), 74. http://doi.org/10.1186/s12984-016-0178-x

Braitenberg, V. (1984). *Vehicles: experiments in synthetic psychology.* Cambridge, MA: MIT Press.

Brooks, R. A. (1986). A robust layered control system for a mobile Robot. *IEEE: Journal on Robotics and Automation*, **2**, 14–23.

De Pierris, G., and **Friedman, M.** (1997). Kant and Hume on Causality. In E. N. Zalta (ed.), *Stanford encyclopedia of philosophy* (Winter 2013). Stanford University. https://plato.stanford.edu/archives/win2013/entries/kant-hume-causality/.

Duff, A., and **Verschure, P. F. M. J.** (2010). Unifying perceptual and behavioral learning with a correlative subspace learning rule. *Neurocomputing*, **73**(10–12), 1818–1830. http://doi.org/10.1016/j.neucom.2009.11.048

Eng, K., Baebler, A., Bernardet, U., Blanchard, M., Costa, M., Delbruck, T., ... **Verschure, P. F. M. J.** (2003). Ada—Intelligent Space: an artificial creature for the Swiss Expo.02. In: *IEEE/RSJ International Conference on Robotics and Automation (ICRA 2003)*, Taipei, Taiwan. IEEE, pp. 4154–9.

Eng, K., Mintz, M., and **Verschure, P. F. M. J.** (2005). Collective human behavior in interactive spaces. In: *International Conference on Robotics and Automation (ICRA 2005)*, Barcelona, Spain. IEEE, pp. 2057–62.

Friston, K. (2010). The free-energy principle: a unified brain theory? *Nature Reviews Neuroscience*, **11**(2), 127–38. http://doi.org/10.1038/nrn2787

Harnad, S. (1990). The symbol grounding problem. *Physica D*, **42**(1990), 335–46. http://doi.org/10.1016/0167-2789(90)90087-6

Hume, D. (1748/2007). *An Inquiry concerning Human Understanding.* (Oxford World's Classics edition, P. Millican, ed.) Oxford: Oxford University Press.

Kalman, R. E. (1960). A new approach to linear filtering and prediction problems. *Transactions ASME Journal of Basic Engineering*, **82**, 35–45.

Kubie, J., and **Fenton, A.** (2009). Heading-vector navigation based on head-direction cells and path integration. *Hippocampus*, **19**(5), 456–79. Retrieved from http://onlinelibrary.wiley.com/doi/10.1002/hipo.20532/full

Lake, B. M., Salakhutdinov, R., and **Tenenbaum, J. B.** (2015). Human-level concept learning through probabilistic program induction. *Science*, **350**(6266), 1332–8. http://doi.org/10.1126/science.aab3050

Lake, B. M., Ullman, T. D., Tenenbaum, J. B., and **Gershman, S. J.** (2017). Building machines that learn and think like people. *Behavioral and Brain Sciences*, September 2017. http://doi.org/10.1017/S0140525X16001837

Lallee, S., and **Verschure, P.** (2015). How? Why? What? Where? When? Who? Grounding ontology in the actions of a situated social agent. *Robotics*, **4**(2), 169–193. http://doi.org/10.3390/robotics4020169

Lallée, S., Vouloutsi, V., Blancas, M., Grechuta, K., Puigbo, J., Sarda, M., and **Verschure, P. F. M. J.** (2015). Towards the synthetic self: making others perceive me as an other. *Paladyn Journal of Behavioral Robotics*, **6**(1). http://doi.org/10.1515/pjbr-2015-0010

Lecun, Y., Bengio, Y., and **Hinton, G.** (2015). Deep learning. *Nature*, **521**, 436–44. http://doi.org/10.1038/nature14539

Maffei, G., Herreros, I., Sanchez-Fibla, M. R., Friston, K. J., and **Verschure, P. F. M. J.** (2017). The perceptual shaping of anticipatory actions. *Proc. R. Soc. B*, **1780**. http://dx.doi.org/10.1098/rspb.2017.1780.

Maffei, G., Santos-Pata, D., Marcos, E., Sánchez-Fibla, M., and **Verschure, P. F. M. J.** (2015). An embodied biologically constrained model of foraging: from classical and operant conditioning to adaptive real-world behavior in DAC-X. *Neural Networks*, **72**, 88–108. http://doi.org/10.1016/j.neunet.2015.10.004

Marcos, E., Sánchez-Fibla, M., and **Verschure, P. F. M. J.** (2010). The complementary roles of allostatic and contextual control systems in foraging tasks. In: S. Doncieux, B. Girard, A. Guillot, J. Hallam,

J. A. Meyer, and J. B. Mouret (eds), *From Animals to Animats 11.* SAB 2010. Lecture Notes in Computer Science, vol **6226**. Berlin/Heidelberg: Springer, pp. 370–9.

Massaro, D. W. (1997). *Perceiving talking faces : from speech perception to a behavioral principle.* Cambridge, MA: MIT Press.

Mathews, Z., Lechón, M., Calvo, J. B., Dhir, A., Duff, A., Badia, S. B., … Verschure, P. F. M. J. (2009). Insect-like mapless navigation based on head direction cells and contextual learning using chemo-visual sensors. In: *The 2009 IEEE/RSJ International Conference on Intelligent RObots and Systems IROS.* IEEE, pp. 2243–50.

Mnih, V., Kavukcuoglu, K., Silver, D., Rusu, A. A., Veness, J., Bellemare, M. G., … Hassabis, D. (2015). Human-level control through deep reinforcement learning. *Nature,* **518**(7540), 529–33. http://doi.org/10.1038/nature14236

Mondada, F., and Verschure, P. F. M. J. (1993). Modeling system-environment interaction: The complementary roles of simulations and real world artifacts. In: J. L. Deneubourg, H. Bersini, S. Goss, G. Nicolis, and R. Dagonnier (eds), *Proceedings of the Second European Conference on Artificial Life, Brussels.* Brussels, Belgium: MIT Press, pp. 808–17.

Moulin-Frier, C., Fischer, T., Petit, M., Pointeau, G., Puigbo, J.-Y., Pattacini, U., … Verschure, P. (2017a). DAC-h3: a proactive robot cognitive architecture to acquire and express knowledge about the world and the self. Submitted to *IEEE Transactions on Cognitive and Developmental Systems.*

Moulin-Frier, C., Puigbò, J.-Y., Arsiwalla, X. D., Sánchez-Fibla, M., and Verschure, P. F. M. J. (2017b). Embodied artificial intelligence through distributed adaptive control: an integrated framework. Paper submitted to the ICDL-Epirob 2017 conference.

Oja, E. (1982). A simplified neuron model as a principal component analyzer. *Journal of Mathematical Biology,* **15**, 267–73.

Rescorla, R. A., and Wagner, A. R. (1972). A theory of Pavlovian conditioning: variations in the effectiveness of reinforcement and nonreinforcement. In: A. H. Black and W. F. Prokasy (eds), *Classical Conditioning II: Current Research and Theory.* New York: Appleton- Century-Crofts, pp. 64–99.

Ringwald, M., and Verschure, P. F. M. J. (2007). The fusion of multiple sources of information in the organization of goal-oriented behavior: spatial attention versus integration. In: *Proceedings for the European Conference on Mobile Robots—ECMR 2007,* Freiburg, Germany, pp. 1–6. Retrieved from http://ecmr07.informatik.uni-freiburg.de/proceedings/ECMR07_0049.pdf

Schmidhuber, J. (2015). Deep learning in neural networks: an overview. *Neural Networks,* **61**, 85–117.

Silver, D., Huang, A., Maddison, C. J., Guez, A., Sifre, L., van den Driessche, G., … Hassabis, D. (2016). Mastering the game of Go with deep neural networks and tree search. *Nature,* **529**(7587), 484–9. http://doi.org/10.1038/nature16961

Thrun, S., Burgard, W., and Fox, D. (2005). *Probabilistic robotics.* Cambridge, MA: MIT Press.

Vergassola, M., Villermaux, E., and Shraiman, B. I. (2007). "Infotaxis" as a strategy for searching without gradients. *Nature,* **445**, 406–9.

Verschure, P. F. M. J. (1993). The cognitive development of an autonomous behaving artifact: the self-organization of categorization, sequencing, and chunking. In: H. Cruze, H. Ritter, and J. Dean (eds), *Proceedings of Prerational Intelligence* (Studies in). Bielefeld: ZiF, pp. 95–117. Retrieved from http://tinyurl.com/yhz3ztk

Verschure, P. F. M. J. (1998). Synthetic Epistemology: The acquisition, retention, and expression of knowledge in natural and synthetic systems. In: *IEEE World Congress on Computational Intelligence* (Vol. **98**). Anchorage, AK: IEEE, pp. 147–53.

Verschure, P. F. M. J. (2011). Neuroscience, virtual reality and neurorehabilitation: brain repair as a validation of brain theory. In: *Conference proceedings for the Annual International Conference of the IEEE Engineering in Medicine and Biology Society. IEEE Engineering in Medicine and Biology Society.* (Vol. **2011**). IEEE, pp. 2254–7. http://doi.org/10.1109/IEMBS.2011.6090428

Verschure, P. F. M. J. (2012). Distributed Adaptive Control: a theory of the mind, brain, body nexus. *Biologically Inspired Cognitive Architectures*, **1**(1), 55–72. http://doi.org/10.1016/j.bica.2012.04.005

Verschure, P. F. M. J. (2016). Synthetic consciousness: the distributed adaptive control perspective. *Philosophical Transactions of the Royal Society of London. Series B, Biological Sciences*, **371**(1701), 263–75. http://doi.org/10.1098/rstb.2015.0448

Verschure, P. F. M. J., Verschure, P. R., and Pfeifer, R. (1992). Categorization, representations, and the dynamics of system-environment interaction: a case study in autonomous systems. In: J. A. Meyer, H. Roitblat, and S. Wilson (eds), *From Animals to Animats: Proceedings of the Second International Conference on Simulation of Adaptive behavior. Honolulu: Hawaii.* Cambridge, MA: MIT Press, pp. 210–17.

Verschure, P. F. M. J., and Coolen, A. C. C. (1991). Adaptive fields: distributed representations of classically conditioned associations. *Network: Computation in Neural Systems*, **2**(2), 189–206. http://doi.org/10.1088/0954-898X/2/2/004

Verschure, P. F. M. J., Kröse, B., and Pfeifer, R. (1993). Distributed adaptive control: The self-organization of structured behavior. *Robotics and Autonomous Systems*, **9**(3), 181–96. http://doi.org/http://dx.doi.org/10.1016/0921-8890(92)90054-3

Verschure, P. F. M. J., and Althaus, P. (2003). A real-world rational agent: unifying old and new AI. *Cognitive Science*, **27**(4), 561–590. Retrieved from http://dx.doi.org/10.1016/S0364-0213(03)00034-X

Verschure, P. F. M. J., Pennartz, C. M. A., and Pezzulo, G. (2014). The why, what, where, when and how of goal-directed choice : neuronal and computational principles. *Philos. Trans. R. Soc. Lond. B: Biol. Sci.*, **369**, 20130483. http://doi.org/10.1098/rstb.2013.0483

Verschure, P. F. M. J., and Voegtlin, T. (1998). A bottom up approach towards the acquisition and expression of sequential representations applied to a behaving real-world device: Distributed Adaptive Control III. *Neural Networks*, **11**, 1531–49.

Verschure, P. F. M. J., Wray, J., Sporns, O., Tononi, G., and Edelman, G. M. (1995). Multilevel analysis of classical conditioning in a behaving real world artifact. *Robotics and Autonomous Systems*, **16**, 247–65.

Verschure, P. F., Voegtlin, T., and Douglas, R. J. (2003). Environmentally mediated synergy between perception and behaviour in mobile robots. *Nature*, **425**, 620–4.

Voegtlin, T., and Verschure, P. F. M. J. (1999). What can robots tell us about brains? A synthetic approach towards the study of learning and problem solving. *Rev. Neurosci.*, **10**(3–4), 291–310.

Vouloutsi, V., Blancas, M., Zucca, R., Omedas, P., Reidsma, D., Davison, D., et al. (2016). Towards a synthetic tutor assistant: the EASEL Project and its architecture. In: N. F. Lepora, A. Mura, M. Mangan, P. F. M. J. Verschure, M. Desmulliez, and T. J. Prescott (eds), *Biomimetic and Biohybrid Systems.* Proceedings 5th International Conference, Living Machines 2016, Edinburgh, UK, July 19–22, 2016. Berlin/Heidelberg: Springer, pp. 353–64.

Walter, W. G. (1951). A machine that learns. *Scientific American*, **184**(8), 60–3.

Chapter 37

Consciousness

Anil K. Seth

Sackler Centre for Consciousness Science, University of Sussex, UK

Consciousness is perhaps the most familiar aspect of our existence, yet we still do not know its biological basis. The standard approach within neuroscience has been to look for "neural correlates" of consciousness, which are brain regions or processes that co-activate or co-occur with conscious (but not unconscious) phenomena. However, because correlations are not explanations, there is a need to describe testable causal neural mechanisms that could account for experiential (phenomenal) properties of consciousness. This chapter outlines such a "biomimetic" approach, identifying three principles linking properties of conscious experience to potential biological mechanisms. First, conscious experiences generate large quantities of information in virtue of being simultaneously integrated and differentiated. Second, the brain continuously generates predictions about the world and self, which account for the content of conscious scenes. Third, the conscious self depends on active inference of self-related signals at multiple levels, from interoceptive responses reflecting physiological integrity, to social signals determining the social self. Each principle relates to a distinct dimension of consciousness: *level* (being conscious at all), *content* (being conscious of this, rather than that), and *self* (as opposed to being conscious of the world).

What is consciousness anyway?

Consciousness has at times been at the centre of neuroscience and psychology, at other times it has been banished from scientific circles altogether. Over recent years the biological investigation of consciousness has gathered momentum and legitimacy so that it now stands as a major scientific challenge for the present century.

Before describing candidate biological principles underlying consciousness it is first necessary to offer some working definitions and distinctions. Put simply, for a conscious organism *there is something it is like to be that organism*. Even more simply, consciousness is what disappears when we fall into a dreamless sleep and what returns the next morning when we wake up. For conscious organisms there exists a continuous (though interruptable) stream of conscious scenes or experiences—a phenomenal world—which has the character of being subjective and private, in the sense of being experienced as particular to the experiencing organism.

Consciousness can be analyzed into separate dimensions of *level* and *content*. Conscious *level* defines a scale ranging from total unconsciousness (as in brain death, coma, general anesthesia, and probably dreamless sleep) to states of vivid alert wakefulness populated by continuously changing conscious scenes. It is notable that wakefulness and conscious level can dissociate, as happens in neuropathologies like the vegetative state. Conscious scenes are composed of conscious *contents* which refer to the distinguishable elements of a given conscious experience: colors, shapes, objects, thoughts, moods, smells, and the like. Conscious contents are often referred to collectively as *qualia* in the philosophical literature.

Conscious contents further divide into *world*-related and *self*-related. Most conscious scenes integrate elements of both, though via attention either may dominate. (Attention and consciousness, while closely related, are distinguishable.) Some conscious content is *higher-order* or metacognitive, reflecting that we (humans, at least) can be conscious *of being conscious* (of something). Self-related conscious content may include experiences of the body (from the inside and from the outside) as well as experiences of being a particular conscious self that is continuous over time (the metacognitive "I"). A further distinction is often attempted between *phenomenal* consciousness and *access* consciousness (Block 2005). Those who believe in this distinction believe that the conscious contents within any given conscious scene may exceed those which we have reportable access to, where reportable access refers to—for example—what one can *say* one experiences at a particular time. Whether or not this distinction is real or apparent, it is certainly true that the neural underpinnings of conscious contents per se may differ from those supporting processes like verbal (or other behavioral) report.

Finally, one can distinguish between those organisms capable of being conscious (i.e. capable of having non-zero conscious level entailing some conscious contents) and those that are not. There is little consensus on this issue. While many researchers are happy to include primates and perhaps most mammals, there is little agreement when it comes to birds, other vertebrates, and invertebrates (e.g. cephalopods), even though some of these creatures display highly sophisticated behaviors suggestive of consciousness. And there is a widespread (but not uniform) scepticism regarding the prospect of conscious (non-biological) machines. The lack of consensus on these issues underlines the need for an approach to consciousness based on explanatory principles linking properties of experience to properties of mechanism—what one might call the "real problem" of consciousness.

Biological principles

Starting from these working definitions and distinctions, three explanatory principles can be identified, each connecting an experiential property of consciousness with a potential neurobiological mechanism.

Dynamical complexity

The first principle is *dynamical complexity*. Phenomenologically, this reflects that conscious scenes are simultaneously highly *differentiated* (each conscious scene is different from a vast repertoire of alternative possibilities) and highly *integrated* (each conscious scene is experienced as a unity). Formally, this means that conscious scenes are highly informative for the experiencing organism, in the specific sense of reduction of uncertainty. This basic insight underpins a collection of related proposals about the mechanisms potentially underlying the level of consciousness. Prominent among these is *integrated information theory*, according to which conscious level is determined by the quantity of information generated by a system over and above that generated by its parts considered independently, indexed by the measure Φ ("phi;" see Tononi et al. (2016) and Figure 37.1A). An alternative measure, "causal density," quantifies the overall density of information flow within a system as operationalized using Granger causality, which is a robust time-series analysis method for detecting directed functional connectivity in the dynamics of complex systems (Seth et al. 2011). While specific evidence for these or similar measures is still mostly lacking, accumulating data underline that neurodynamic states consistent with consciousness indeed occupy a middle-ground characterized by the coexistence of functional integration and differentiation: dynamical states at either extreme result in unconsciousness, as shown by functional disintegration during anesthesia and hyper-synchrony during absence epilepsy (see, for example, Casali et al. (2013)). Thus, characterizing this dynamical middle-ground provides both an insight into the neurodynamics of consciousness and a potential design principle for biomimetic devices.

The principle of dynamical complexity is also evident in "global workspace" architectures. On this set of views conscious experience happens when some mental content (e.g. a perception, intention, or cognition) gains access to a neuronally implemented "global workspace" (GW) connecting disparate thalamo-cortical regions, making this content globally available for action selection and other cognitive and behavioral processes (Baars 2005; Dehaene and Changeux 2011). A wide range of evidence is consistent with GW theory, showing that stimuli eliciting reportable conscious contents evoke widespread modality-independent neural responses (as measured, e.g., by functional MRI) as compared to stimuli that remain below reportability thresholds. Recent studies, however, cast doubt on the generality of these findings by suggesting that widespread frontoparietal activity may have more to do with providing behavioral report or noticing changes in perception, than in supporting perception itself (Tsuchiya et al. 2015).

Much of the evidence supporting dynamical complexity theories is also consistent with GW theory (and vice versa). However, there are important differences. While GW theory is mostly a theory about access consciousness, dynamical complexity theories address the more difficult challenge of phenomenal consciousness. Also, GW theory is frequently associated with specific brain regions (e.g. the frontoparietal network, or the precuneus) while complexity theories focus more on dynamical activity patterns independently of their regional expression. Teasing apart the implications of these distinctions is one important area in which biomimetic approaches could usefully contribute.

Predictive processing

The second principle is that of *predictive processing* (or, in a more restricted interpretation, "predictive coding"). This has a long history, originating with the insights of Hermann von Helmholtz and reaching recent prominence in the "Bayesian Brain" hypothesis and the "free energy principle" (Clark 2016; Friston 2009; Hohwy 2013; Seth 2014). The idea is that, in order to support adaptive responses, the brain must discover information about the likely causes of sensory signals (i.e. perception) without direct access to these causes, using only information in the flux of sensory signals themselves (see Herreros, Chapter 26, and Verschure, Chapter 36, this volume, for a further analysis in the context of control). According to predictive processing, this is accomplished via probabilistic inference on the causes of sensory signals, computed according to Bayesian principles. This means estimating the probable causes of data (the posterior) given observed conditional probabilities (likelihoods) and prior "beliefs" about probable causes. This in turn means inducing predictive or "generative" models of the sensory data. Applied to cortical networks, the concept of predictive processing overturns classical notions of perception as a largely "bottom-up" process of evidence accumulation or feature detection. Instead, perceptual content is specified by *top-down* signals emerging from multi-level (hierarchical) generative models of the causes of sensory signals, which are continually modified by *bottom-up* prediction errors signaling mismatches between predicted and actual signals (see Figure 37.1B).

Importantly, prediction errors can be minimized either by optimizing inference or updating generative models (perceptual inference and learning; changing the model to fit the world) or by performing actions to bring about sensory states in line with predictions (active inference; changing the world to fit the model). In most incarnations these processes are assumed to unfold continuously and simultaneously, underlining a deep continuity between perception and action.

While predictive processing provides a rich framework for understanding perception, action, and cognition, and their neural underpinnings, some basic questions about consciousness

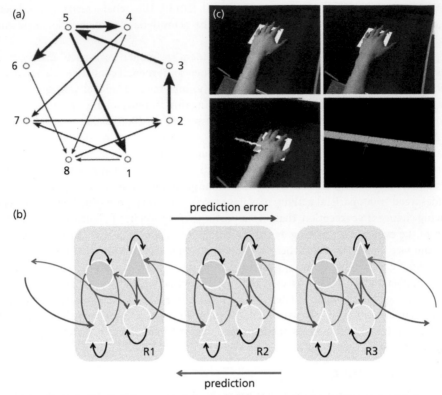

Figure 37.1 Biological principles potentially underlying aspects of consciousness. (a) An example of a simulated network optimized to generate high integrated information (Φ). (b) A functional architecture for hierarchical predictive processing across three cortical levels (R1, R2, and R3). Prediction errors are conveyed in a bottom-up direction (red arrows) and originate from superficial layers; predictions flow top-down (blue arrows) and originate in deep layers. Triangles represent projection neurons and circles interneurons. (c) A virtual-reality 'rubber hand' (top left) that can be stroked with a paintbrush (bottom left) or modulated by cardiac feedback (top right; the virtual hand pulses to red and back according to the cardiac cycle). The bottom-right panel shows a virtual 'ruler' used to assess the extent of experienced ownership of the virtual hand.

(a) Reproduced from Barrett AB, Seth AK, Practical Measures of Integrated Information for Time-Series Data, *PLoS Computational Biology*, 7(1), e1001052. doi:10.1371/journal.pcbi.1001052. Copyright © 2011 Barrett, Seth. Reprinted under the terms of the Creative Commons Attribution License, which permits unrestricted use, distribution, and reproduction in any medium, provided the original author and source are credited. (b) Adapted from *Trends in Cognitive Sciences*, 13 (7), Karl Friston, The free-energy principle: a rough guide to the brain?, pp. 293–301, doi: 10.1016/j.tics.2009.04.005, Copyright © 2009 Elsevier Ltd., with permission from Elsevier. (c) Reproduced from Anil K. Seth, Interoceptive inference, emotion, and the embodied self, *Trends in Cognitive Sciences*, 17(11), pp. 565–573 doi:10.1016/j.tics.2013.09.007, Copyright © 2013 The Author. Reprinted under the terms of the Creative Commons Attribution License, which permits unrestricted use, distribution, and reproduction in any medium, provided the original author and source are credited.

remain unanswered. In particular we would like to know whether conscious contents arise when predictive models are verified by incoming sensory input, or when surprising events falsify current perceptual predictions. Emerging data seem to favor the former (Melloni et al. 2011; Pinto et al. 2015), and this evidence is compatible with conscious phenomenology inasmuch as we often consciously see what we "expect" to see. This idea has been extended to account for psychotic hallucinations and delusions in terms of the reshaping of higher-level perceptual (or cognitive) priors to "explain away" abnormally persistent prediction errors (Fletcher and Frith 2009). Further research is needed to examine the possibility of more complex interactions between behavioral context and the computational components of Bayesian inference in specifying conscious contents.

The predictive processing framework is well suited to biomimetic approaches because it both describes a detailed neurocomputational architecture (Bastos et al. 2012) and incorporates agent–environment interactions, most notably via the concept of active inference where perceptual predictions are confirmed through behavior (see also Herreros, Chapter 26, this volume). The emphasis on top-down signaling conveying predictions is also well aligned with abundant evidence showing the importance of such top-down (or "re-entrant") signaling for conscious awareness (Lamme 2010). Moreover, predictive processing raises the important question of whether neural systems are "strongly" Bayesian (i.e. they actually implement Bayesian inference via explicit manipulation of representations of probability distributions) or only "weakly" so (i.e. Bayesian terms provide a useful shorthand for describing the operation of fundamentally non-Bayesian neural circuits).

Body-ownership and self-consciousness

Key to conscious experience is also the experience of being a *self*. This integrates multiple aspects including (i) a "background" experience of bodily or homeostatic integrity; (ii) an experience of identifying with or owning a particular body; (iii) the emergence of a first-person perspective on the world (including the body), and (iv) specific higher-order properties including the experience of continuity of the self over time, and of ownership over intentions, actions, thoughts, and perceptions. A biological principle potentially underlying many of these characteristics is *predictive multisensory integration*. Taking a strong cue from predictive processing just described, this says that experience of selfhood is shaped by active inference (i.e. the brain's "best guess") of the causes of *self-related* sensory signals. Multisensory correlations provide strong evidence for common causes and thus provide a useful principle for distinguishing self-related from non-self-related signals. A popular example of this is the "rubber hand illusion" where congruent visual and tactile information leads to the experience that a fake hand is part of one's own body (Botvinick and Cohen 1998), and which can be readily understood in terms of predictive inference (Apps and Tsakiris 2014). A similar framework can also be applied to understanding the origin of a first-person perspective, since this too can be modified experimentally by inducing multisensory correlations including viewing the world (and body) from a different place by clever use of virtual reality equipment (Lenggenhager et al. 2007).

A rather different tradition emphasizes the grounding of conscious selfhood in the processing of *interoceptive* signals, where interoception refers to the sense of the internal state of the body. Damasio's "somatic marker" hypothesis proposes that subjective emotional responses (or "feeling states") involve the mapping and regulation of continuously changing internal bodily states, mediated by interoceptive pathways (Damasio 2000). These feeling states help to guide behavioral decisions by representing the emotional significance of stimuli, or of potential future actions via a further "as-if-body loop" mechanism. More recently, the machinery of predictive processing has been applied to interoception in a model which sees emotional feeling states as resulting from active inference of the (internal and external) causes of interoceptive signals (Seth 2013; Seth and Friston 2016). This model integrates exteroceptive predictive perception (as described above) with an interoceptive equivalent, again providing a neural mechanism by which self and non-self can be disambiguated. In line with this model, recent data has shown that experienced ownership of a virtual "rubber hand" can be facilitated by congruency between visual input and visual representation of (interoceptive) cardiac signals (Figure 37.1C).

Predictive self-modeling provides a powerful biomimetic principle by specifying functional architectures by which self and non-self can be distinguished (see also Bongard, Chapter 11, and Verschure, Chapter 36, this volume), and by emphasizing the importance of homeostasis and active (i.e. predictive) regulation of internal bodily states in determining and maintaining selfhood with roots reaching back to early principles of cybernetics (Seth 2015). They also provide an avenue by which the functional implementation of emotional responses can be considered, both for guiding behavior and as underpinning important aspects of conscious selfhood (see Vouloutsi and Verschure, Chapter 35, this volume).

Biomimetic systems

Attempts to construct "conscious machines" or "artificial consciousness" are in their infancy and with scarce exceptions do not reflect the principles just described. As with artificial life and artificial intelligence, there is a debate about whether synthetic approaches should be considered *weak* (i.e. simulations) or *strong* (i.e. instantiations). This metaphysical question cannot be resolved by experiment; however, adopting a weak view in practice is likely to enhance scientific progress by avoiding the circularities that confound strong approaches (Seth 2009).

One exception which does reflect a principle just described is the (Learning) Intelligent Distributed Agent (L/IDA) of Franklin (see Baars and Franklin 2007) which is directly based on GW theory. This model uses a GW architecture to assign naval personnel to tasks via a natural language dialogue. Reflecting a strong view, this model has been called "functionally conscious" by Franklin, though many may disagree. Indeed most similar models focus on showing how GW dynamics can be related to neuronal mechanisms and fit nicely within a weak view of artificial consciousness (Dehaene and Changeux 2011).

Similar differences of interpretation can be applied to probably all other attempts under the rubric of artificial/machine consciousness and this is not the place for a comprehensive review (see instead Reggia 2013). Suffice it to say here that these attempts include models which associate consciousness with the internal control of attention (Taylor 2007), with a metacognitive "virtual machine" architecture (Sloman and Chrisley 2003), with self-modeling in embodied humanoid robotics (Holland 2007), and with much else besides. Although strong artificial consciousness remains a possibility in principle, the prospect remains at best very distant. Therefore, the greatest impact of the biomimetic approach will derive *not* from brute attempts to construct "conscious machines" de novo, but instead from synthetic models which attempt to mechanistically account for properties of conscious phenomenology.

Future directions

Much important work remains in elaborating the three principles described in this chapter. Synthetic models are needed to understand the complex relationships between neural architecture, embodiment, and agent–environment interactions in generating high levels of Φ or related measures. Such models could also address whether global workspaces architectures are necessarily associated with high dynamical complexity. In addition, biomimetic approaches could broach the fundamental question of the adaptive advantages (if any) accruing from control systems tuned for high Φ (Albantakis et al. 2014). In predictive processing, a key challenge is to understand which aspects of hierarchical Bayesian inference are important in specifying conscious contents and disambiguating them from unconscious processes. Biomimetic approaches could also shed light on the neural circuits potentially able to implement predictive perception and action (Bastos et al. 2012) and their relation to the deep convolutional neural networks that have radically advanced computer vision and reinforcement learning for control (LeCun et al. 2015; Mnih et al. 2015). Biomimetic approaches are perhaps especially suited to predictive accounts of self-consciousness in virtue of the centrality of embodiment to these approaches. By equipping simulated neural systems with (virtual or real) bodies, having closely regulated internal and proprioceptive states, the functional architecture of the predictive self can be usefully elaborated.

Beyond even these ambitious endeavors, perhaps the most exciting avenues ask how the above principles can be integrated. For example, does validation of predictions in a predictive processing architecture imply high dynamical complexity or integrated information? And even more speculatively, biomimetic approaches raise the possibility of new answers to old questions about the prevalence of conscious experience in non-human animals, infants, and following severe brain injury in vegetative and coma patients (Tononi and Koch 2015). In closing, the possibility in principle of strong artificial consciousness raises important ethical issues: For example, how could we ensure that "conscious machines" would not suffer, and would it be ethical to switch them off? The present implausibility of a conscious machine should not discourage such pre-emptive ethical research.

Learning more

The recent state of the art in machine or artificial consciousness is well reviewed by Reggia (2013); see also Gamez (2008). The *Journal of Machine Consciousness* (World Scientific) is also a useful resource, though this has now ceased publication. Information integration theory is described in detail by Tononi (Tononi 2012; Tononi et al. 2016), there are excellent presentations of predictive processing by both Clark and Howhy (Clark 2016; Hohwy 2013), and predictive accounts of selfhood can be found in recent review articles (Apps and Tsakiris 2014; Limanowski and Blankenburg 2013; Seth and Friston 2016). Good general resources for the neuroscience of consciousness include the textbook *Cognition, Brain, and Consciousness* by Baars and Gage (Baars and Gage 2010), and the academic journals *Frontiers in Consciousness Research* (Frontiers) and *Neuroscience of Consciousness* (Oxford University Press). The chapters of *Scholarpedia* are also useful (www.scholarpedia.org/article/Category:Consciousness), as are the webpages of the *Association for the Scientific Study of Consciousness*; their annual conferences are highly recommended (www.theassc.org). The author's perspectives on consciousness science are summarized for a general audience in this online article (Aeon, 106): https://aeon.co/essays/the-hard-problem-of-consciousness-is-a-distraction-from-the-real-one

Acknowledgments

Preparation of this chapter was supported by the Dr. Mortimer and Theresa Sackler Foundation which supports the work of the Sackler Centre for Consciousness Science. The author also acknowledges support from ERC project CEEDs (ERC-FP7-ICT-258749) and from the Canadian Institute for Advanced Research (Azrieli Programme in Mind, Brain, and Consciousness).

References

Albantakis, L., Hintze, A., Koch, C., Adami, C., and **Tononi**, G. (2014). Evolution of integrated causal structures in animats exposed to environments of increasing complexity. *PLoS Comput. Biol.*, **10**(12), e1003966. doi:10.1371/journal.pcbi.1003966

Apps, M. A., and Tsakiris, M. (2014). The free-energy self: A predictive coding account of self-recognition. *Neurosci. Biobehav. Rev.*, 41C, 85–97. doi:10.1016/j.neubiorev.2013.01.029

Baars, B. J. (2005). Global workspace theory of consciousness: toward a cognitive neuroscience of human experience. *Prog. Brain Res.*, 150, 45–53. doi:S0079-6123(05)50004-9

Baars, B. J., and Franklin, S. (2007). An architectural model of conscious and unconscious brain functions: Global Workspace Theory and IDA. *Neural Netw.*, 20(9), 955–61. doi:10.1016/j.neunet.2007.09.013

Baars, B. J. and Gage, N. M. (2010). *Cognition, brain, and consciousness: introduction to cognitive neuroscience.* St Louis, MO: Academic Press.

Barrett, A. B., and Seth, A. K. (2011). Practical measures of integrated information for time-series data. *PLoS Computational Biology*, 7(1), e1001052. doi:10.1371/journal.pcbi.1001052.

Bastos, A. M., Usrey, W. M., Adams, R. A., Mangun, G. R., Fries, P., and Friston, K. J. (2012). Canonical microcircuits for predictive coding. *Neuron*, 76(4), 695–711. doi:10.1016/j.neuron.2012.10.038

Block, N. (2005). Two neural correlates of consciousness. *Trends Cogn. Sci.*, 9(2), 46–52.

Botvinick, M., and Cohen, J. (1998). Rubber hands 'feel' touch that eyes see. *Nature*, 391(6669), 756. doi:10.1038/35784

Casali, A. G., Gosseries, O., Rosanova, M., Boly, M., Sarasso, S., Casali, K. R., … Massimini, M. (2013). A theoretically based index of consciousness independent of sensory processing and behavior. *Sci. Transl. Med.*, 5(198), 198ra105. doi:10.1126/scitranslmed.3006294

Clark, A. (2016). *Surfing uncertainty.* Oxford: Oxford University Press.

Damasio, A. (2000). *The feeling of what happens: Body and emotion in the making of consciousness.* Boston, MA: Harvest Books.

Dehaene, S., and Changeux, J. P. (2011). Experimental and theoretical approaches to conscious processing. *Neuron*, 70(2), 200–27. doi:S0896-6273(11)00258-3

Fletcher, P. C., and Frith, C. D. (2009). Perceiving is believing: a Bayesian approach to explaining the positive symptoms of schizophrenia. *Nat. Rev. Neurosci.*, 10(1), 48–58. doi:10.1038/nrn2536

Friston, K. J. (2009). The free-energy principle: a rough guide to the brain? *Trends Cogn. Sci.*, 13(7), 293–301.

Gamez, D. (2008). Progress in machine consciousness. *Conscious Cogn.*, 17(3), 887–910.

Hohwy, J. (2013). *The predictive mind.* Oxford: Oxford University Press.

Holland, O. (2007). A strongly embodied approach to machine consciousness. *Journal of Consciousness Studies*, 14(7), 97–110.

Lamme, V. A. (2010). How neuroscience will change our view on consciousness. *Cognitive Neuroscience*, 1(3), 204–40.

LeCun, Y., Bengio, Y., and Hinton, G. (2015). Deep learning. *Nature*, 521(7553), 436–44. doi:10.1038/nature14539

Lenggenhager, B., Tadi, T., Metzinger, T., and Blanke, O. (2007). Video ergo sum: manipulating bodily self-consciousness. *Science*, 317(5841), 1096–9.

Limanowski, J., and Blankenburg, F. (2013). Minimal self-models and the free energy principle. *Front. Hum. Neurosci.*, 7, 547. doi:10.3389/fnhum.2013.00547

Melloni, L., Schwiedrzik, C. M., Muller, N., Rodriguez, E., and Singer, W. (2011). Expectations change the signatures and timing of electrophysiological correlates of perceptual awareness. *J. Neurosci.*, 31(4), 1386–96. doi:31/4/1386

Mnih, V., Kavukcuoglu, K., Silver, D., Rusu, A. A., Veness, J., Bellemare, M. G., … Hassabis, D. (2015). Human-level control through deep reinforcement learning. *Nature*, 518(7540), 529–33. doi:10.1038/nature14236

Pinto, Y., van Gaal, S., de Lange, F. P., Lamme, V. A., and Seth, A. K. (2015). Expectations accelerate entry of visual stimuli into awareness. *J. Vis.*, 15(8), 13. doi:10.1167/15.8.13

Reggia, J. A. (2013). The rise of machine consciousness: studying consciousness with computational models. *Neural Netw.*, **44**, 112–31. doi:10.1016/j.neunet.2013.03.011

Seth, A. K. (2009). The strength of weak artificial consciousness. *Journal of Machine Consciousness*, 1(1), 71–82.

Seth, A. K. (2013). Interoceptive inference, emotion, and the embodied self. *Trends Cogn. Sci.*, **17**(11), 565–73. doi:10.1016/j.tics.2013.09.007

Seth, A. K. (2014). A predictive processing theory of sensorimotor contingencies: Explaining the puzzle of perceptual presence and its absence in synesthesia. *Cogn. Neurosci.*, **5**(2), 97–118. doi:10.1080/17588928.2013.877880

Seth, A. K. (2015). The cybernetic Bayesian brain: from interoceptive inference to sensorimotor contingencies. In: J. M. Windt and T. Metzinger (eds), *Open MIND* Frankfurt-am-Main: MIND Group, pp. 1–24.

Seth, A. K., Barrett, A. B., and Barnett, L. (2011). Causal density and integrated information as measures of conscious level. *Philosophical Transactions Series A: Mathematical, Physical, and Engineering Sciences*, **369**(1952), 3748–67. doi:10.1098/rsta.2011.0079

Seth, A. K., and Friston, K. J. (2016). Active interoceptive inference and the emotional brain. *Philosophical Transactions of the Royal Society B: Biological Sciences*, **371**(1708), 20160007. doi:10.1098/rstb.2016.0007

Sloman, A., and Chrisley, R. (2003). Virtual machines and consciousness. *Journal of Consciousness Studies*, 10(4–5), 133–72.

Taylor, J. G. (2007). CODAM: a neural network model of consciousness. *Neural Netw.*, **20**(9), 983–92. doi:10.1016/j.neunet.2007.09.005

Tononi, G. (2012). Integrated information theory of consciousness: an updated account. *Archives italiennes de biologie*, **150**(4), 293–329.

Tononi, G., Boly, M., Massimini, M., and Koch, C. (2016). Integrated information theory: from consciousness to its physical substrate. *Nat. Rev. Neurosci.*, **17**(7), 450–61. doi:10.1038/nrn.2016.44

Tononi, G., and Koch, C. (2015). Consciousness: here, there and everywhere? *Philosophical Transactions of the Royal Society B: Biological Sciences*, **370**(1668), 20140167. doi:10.1098/rstb.2014.0167

Tsuchiya, N., Wilke, M., Frassle, S., and Lamme, V. A. (2015). No-report paradigms: extracting the true neural correlates of consciousness. *Trends Cogn. Sci.*, **19**(12), 757–70. doi:10.1016/j.tics.2015.10.002

Section V

Living machines

Chapter 38

Biomimetic systems

Tony J. Prescott

Sheffield Robotics and Department of Computer Science, University of Sheffield, UK

So far in this volume we have considered the nature of living things and some of their key build-ing blocks and capabilities. This has set the stage for the current section and the next where we will describe some exemplar integrated biomimetic and biohybrid systems—living machines. To place these contributions in some additional context this introduction briefly reviews the history of life and of its variety, noting some of the critical branching points in the phylogen-etic tree, identifying some of the organisms that have been the focus of research on biomimetic systems, and exploring why they might be seen to be important or pivotal. Whilst the follow-ing chapters are placed in a roughly historical order, based on when organisms similar to those modeled first appeared in the fossil record, this is not to endorse a "great chain of being" view of evolution culminating in the arrival of humankind. Rather, with very few exceptions, bio-mimetics has taken its inspiration from species that are alive today and therefore have been evolving for just as long as we have. Moreover, evolution continues, and, with the age of the Anthropocene, now has a guiding hand in the form of human science and engineering. Our species is playing an increasingly critical role in determining the biota of tomorrow's Earth through the combined impacts of a wave of human-driven extinction, selective breeding and gene modification, new human-engineered ecosystems, and the merging of the biological with the artificial as described in this book.

Beginnings

The spark of life on Earth began to glow somewhere between 3.5 and 4 billion years ago. It trig-gered a blaze that radiated across the planet and through time, generating billions of different species. Like a wildfire burning out from the middle, nearly all of the different flames of life have become extinct during the long history of our planet and today there are thought to be around nine million distinct species (Mora et al. 2011)—less than 1% of those that have ever existed. But what ignited that first spark?

As we explained in section II (see Prescott, Chapter 4, and Deacon, Chapter 12), an essential attribute of living systems is that they exist far from equilibrium and must stave off disintegra-tion, through an inflow and outflow of energy. So the key question is, how did such self-sustain-ing systems first get going? In recent decades a community of physicists and chemists have taken up the challenge of understanding the conditions of early Earth that would have given rise to the first replicators. Their goal is also synthetic—to capture the essential properties of that environ-ment in the laboratory and to see if life can be made to emerge again; perhaps for the first time in four billion years. Chapter 39 by Mast and colleagues discusses this endeavor, highlighting the idea that the emergence of far-from-equilibrium systems requires an environment that is itself at disequilibrium. These authors discuss the kind of swirling chemical soup in which the first eddies of life could have emerged, and review the progress that has been made in the quest to build artificial replicators.

Once life got going, the first living things took the form of unicellular organisms[1] that we call prokaryotes, a name that means "before the nucleus." Tiny capsules of water-based gel, just micrometers across, every prokaryote contains a single chromosome of DNA alongside other complex molecules, surrounded by a double layer of non-soluble fatty acids that form its boundary with the world.[2] These cells are able to fulfill the basic functions of life, most importantly, to maintain themselves, grow, and reproduce (by splitting in two). Artificial life researchers are investigating, in simulation, how simple replicators could give rise to model "protocells" (Agmon et al. 2016); in the longer term, artificial cells based on chemical building-blocks may be realizable through nanofabrication methods (see Fukuda and colleagues, Chapter 52, this volume).

Some considerable time later—perhaps half a billion years after the life's initial flicker— colonies of photosynthesizing prokaryotes, the cyanobacteria, began to create large complex differentiated structures that would leave some of the first fossils (stromatolites). This breakthrough was made possible through the evolution of an important trick—the division of labour (Rossetti et al. 2010). Some cells used energy from photosynthesis to make copies of themselves, others, carrying the same genetic code, converted themselves into machines for fixing atmospheric nitrogen, helping the self-copiers to grow, whilst settling for a non-reproductive life themselves. Through chemical signalling and transfer of genetic material these simple creatures also discovered the means to rapidly propagate information from cell to cell creating the first "societies" and early glimmers of collective intelligence (Bloom 2000).

Around 1.5 billion years on from the first differentiated bacterial colonies, a further defining event in evolution happened. Some strains of prokaryotes found ways to live inside each other forming the first eukaryotic—"true nucleus"—organisms. Different microbial incomers discovered ways to make themselves useful inside the bigger creature, forming internal factories, known as organelles, that could build proteins, convert glucose to energy, or process waste materials. Others acted as transporters shifting material around inside the cell or moving the whole organism around by deforming its body to swim or crawl. The genetic material contributed by all of these life-time tenants became contained within a bounded nuclear structure, from where it coordinated activity throughout the rest of the cell using messenger molecules and reproduced as one. Eukaryotic evolution also saw the invention of sex, creating the new operator for genetic search, crossover, and providing the basic template for evolutionary algorithms that have inspired thousands of studies in artificial life (Downing 2015; Mitchell 1998).

Single-celled eukaryotes such as amoebas developed sophisticated chemical and mechanosensory systems and motor capabilities that could support active lifestyles such as predation.

---

[1] Viruses also emerged early in the history of Earth. Since these replicators cannot reproduce outside of a cellular host, whether they should be considered alive is a moot point that depends on your definition. Koonin and Starokadomskyy (2016) argue that the class of replicators forms a continuum from the cooperative through to the selfish. Living organisms, viewed as "communities of interacting, coevolving replicators of different classes" (ibid., p. 125), lie towards the cooperative end of this spectrum, viruses towards the more selfish extreme.

[2] Prokaryotes divide into two distinct kingdoms: bacteria and archaea; the latter live at high temperatures or in places with high methane concentrations, i.e. in circumstances that are extremely hostile to life for other organisms. The differences between bacteria and archaea could help us to understand the origins of early life (see Mast et al., Chapter 39).

Societies of some eukaryotic organisms evolved to form jelly-like aggregates allowing them to move around together making foraging more efficient when food was sparse and forming reproductive "stalks" that could release spores higher into the air. Chapter 40 by Ishiguro and Umedachi takes inspiration from these simple multicellular entities—the "slime molds"—to create a soft-bodied, deformable robot with decentralized control.

An explosion of multicellularity

Around 700 million years ago the evolution of life really picked up speed. Building on the possibility of differentiated germ cells and somatic (body) cells, first discovered by the cyanobacteria, living things began to form structured colonies of increasing intricacy. The age of complex multicellular life had begun (Knoll 2011). On top of the other achievements of advanced eukaryotes, this new kind of multicellularity depended on a number of additional key mechanisms—adhesion (ways of sticking cells together), intra-cellular communication and transport, and a developmental program. Critical for the latter were the new networks of regulatory genes. Bilateral symmetry emerged, along with a two-ended gut, sensory organs, and tentacles (Erwin and Davidson, 2002). Fossil evidence for such creatures began to accumulate, taking the form not of body fossils but tracks left by the behaviour of active, locomoting animals. Computer and robotic simulations (Figure 38.1, top left) suggest that these animals possessed an integrative centralized neural ganglion—the first brain (Prescott 2007; Prescott and Ibbotson 1997). Early bilaterians may have been similar to the modern polyclad flatworm that inspired Kazama et al. (2013) to build a sheet-rubber deformable swimming robot.

With the emergence of complex multicellularity evolution accelerated to such a pace that within a period of just tens of millions of years, known as the Cambrian explosion, virtually all of the major classes of bilateral animals made their first appearance (Budd 2003). Bilaterian animals split between the protostomes, which would form groups such as the arthropods and molluscs, and the deuterostomes, made up of echinoderms, such as sea urchins and starfish, and chordates, which includes the vertebrates (chordates with backbones). A fundamental protostome/deuterostome difference is inversion of the main body axis—deuterostomes are upside-down protostomes (or perhaps its the other way round). However, both types of animal share fundamental mechanisms for development and body patterning based on regulatory homeobox gene networks. Bilaterians radiated out to occupy a vast range of ecological niches within the Earth's marine ecosystem. Together with plants (see Mazzolai, Chapter 9, this volume) and fungi, soft- and hard-bodied invertebrates quickly colonized the land. The chapter by Trimmer (Chapter 41) describes how the structure and behavior of invertebrate animals such as worms and caterpillars is inspiring the design and control of new types of deformable soft-bodied robots (see also Figure 38.1 top right), whilst Quinn and Ritzmann (Chapter 42) show how careful study of locomotion in arthropods has provided many useful lessons for legged robotics. Biomimetics has also explored the arachnid class of arthropods, an interesting model for climbing (Waldron et al. 2013), and the cephalopod molluscs, particularly the octopus, for soft robotic actuation, sensing and control (see Figure 38.1 center right and Laschi et al. 2012). Marine crustaceans, which belong to the arthropod phylum, have also been the target for biomimetic engineering focusing on how their (relatively) simple and well-characterized nervous systems give rise to patterned behavior (see Ayers and Witting (2007) and Selverston, Chapter 21, this volume). The arthropods have also given rise to some of the best known super-organisms—the social insects—capable of collective behaviors such as the "terraforming" and foraging activities of bee, ant, and termite colonies. The behavior of animal swarms has inspired a whole field of bio-inspired collective robotics discussed here in the chapter by Nolfi (Chapter 43).

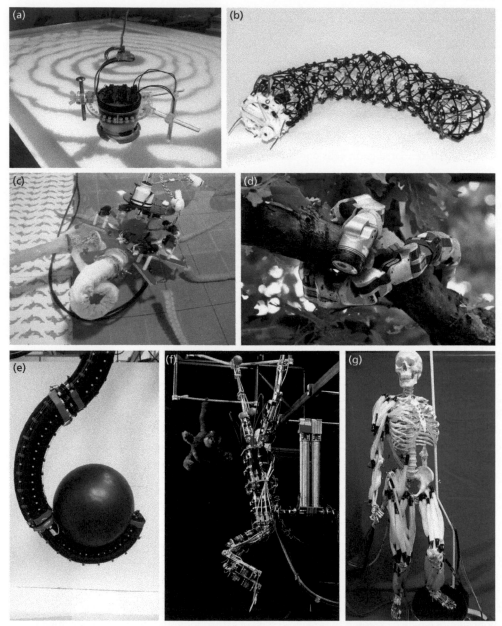

Figure 38.1 Biomimetic robots. (a) This trace-maker robot from the University of Sheffield creates a spiral pattern similar to some Precambrian trace-fossils made by early bilaterians. The spiral pattern emerges from the interaction of three simple reactive behaviors that respond to the presence or absence of the robot's own trail (a projected history of the robot's trajectory substitutes for an actual physical trail). The ability to generate spiral and meandering traces suggests the presence of a centralized ganglion, or brain, in early bilaterians that was able to exert control over peripheral reflexes. (b) Worm locomotion is a fascinating process that involves peristaltic waves that propagate along the animal's segmented body. This prototype worm robot from Case Western University is being used to understand how worms travel over irregular ground (see also Trimmer, Chapter 41, this volume). (c) The octopus robot developed by the Biorobotics Institute at the Scuola Superiore

Biomimetics with a backbone

During the Cambrian period there were already creatures similar to early fish, though vertebrate life would evolve in the oceans for over a hundred million more years before venturing significantly onto land. Chapter 44 by Kruusmaa explores fish swimming as a model for efficient aquatic locomotion in robots, and also describes how a morphology similar to that of a turtle can make for a highly maneuverable sub-sea platform. The ability to swim by propagating an oscillating pattern down the spine can morph into tetrapod walking by replacing fins with legs (see Witte et al., Chapter 31, this volume and Ijspeert et al. (2007)) or snake-like crawling by getting rid of fins and legs altogether. The snake has been a popular reptilian model for the design of robots that can access confined spaces including inside the human body (Webster et al. 2006) or safely manoeuvre on difficult surfaces such as a tree limb (e.g. Figure 38.1 center right), whilst adhesion mechanisms in the reptilian gecko toe (see Pang et al., Chapter 22, this volume) inspired *Stickybot*, a robot that can climb smooth vertical surfaces (Kim et al. 2008). One group of synapsid reptiles would go on to evolve thermoregulation, hair, specialized sensory systems, a bigger brain, and giving birth to live young; hence, the evolution of mammals. A number of characteristic mammalian innovations have been explored through biomimetics, as described by Prescott (Chapter 45). Examples include whiskered sensing systems modeled on rodent vibrissae, active sonar inspired by dolphins and bats, and soft actuators modeled on the elephant trunk (see Figure 38.1, bottom left, and Hannan and Walker 2003). Around the same time that the first mammals were appearing, vertebrates finally followed invertebrates into the air. The special character of bird flight is considered in the contribution by Send (Chapter 46; see also Ledenström, Chapter 32 on the principles of flight) using the example of Festo's *Smartbird*— a flapping-wing robot inspired by the herring gull.

The primate branch emerged from the mammalian family tree around 85 million years ago, developing bigger brains, forward-facing eyes for stereo, often color, vision, prehensile hands and tails adapted for foraging in arboreal environments, and a lifestyle that often revolved around large social groups. Biomimetic models have included brachiating robots (Nakanishi et al. 2000) that swing from branch to branch like an ape, and research on social learning based on models of the primate "mirror neuron" system (e.g. Demiris and Khadhouri 2006). Some of

Sant'Anna, Italy mimics some of the capabilities of octopus arms for reconfiguration and grasping by squeezing. (d) This modular robot snake developed by the Biorobotics Lab at Carnegie Mellon University can use its many degrees of freedom to explore difficult-to-access spaces such as this tree limb. (e) Like octopus arms, elephant trunks are hyper-redundant manipulators giving lots of options for movement and grasping but creating challenges for control. Similarities between the elephant trunk and the octopus arm are the consequence of convergent evolution to soft manipulator systems that substitute hydrostatic components (compressed volumes of liquid) for hard body parts. This pneumatically actuated "continuum" robot from Clemson University, USA, is one of several proboscidean-inspired robotic systems (Walker 2013). (f) The Brachiator III robot from the University of Tokyo mimics the ability seen in many primates to move quickly and safely through arboreal environment by swinging from tree to tree. (g) There are many different designs for humanoid robots. This one from the Suzumori-Endo laboratory at the Tokyo Institute of Technology models the musculoskeletal system of the human body using pneumatic actuators and is able to move in a very human-like way (Kurumaya et al. 2016).

(a) Photograph courtesy of Sarah Prescott; (b) Photograph courtesy of Alexander Kernbaum, Roger Quinn and Hillel Chiel; (d) Photograph courtesy of Howie Choset; (f) Photograph © Peter Menzel/menzelphoto.com. (g) Photograph courtesy of Shunichi Kurumaya.

the descendants of simian primates switched from life in the forest to begin walking upright and, around two million years ago, to making tools for use in hunting and foraging. Following several further expansions in brain size, and, in the last half a million years or less, our species, *Homo sapiens*, developed language and culture. The quest to create biomimetic artifacts that are human-like in form and function, known as humanoid robotics, is perhaps the most popular and headline-grabbing area of biomimetics, and the subject of our final chapter in this section, Chapter 47, by Metta and Cingolani. Research in humanoid robots is addressing many challenges from the opposable thumb to consciousness but with particular emphases on bipedal locomotion, object reach/grasp and manipulation, and social cognition (see Chapters 25, 30, 34, 35, and 37).

Conclusion

Whilst the variety of biomimetic systems is impressive (and the review here is far from comprehensive) there is still much of the natural world to explore. With few exceptions, notably some insect, mammal-like, and humanoid robots, biomimetics has focused on translating a single exceptional characteristic of a target animal into an engineered artifact, leaving scope for considerable future work on how different classes of animal achieve long-term autonomy by trading off different drives and needs to generate behavior that is integrated over both space and time (see Prescott 2007). Also of future interest is the capacity of animals to thrive in ecosystems containing thousands of species at different scales. Symbiosis between very different kinds of animals should be a particularly interesting target for biomimetics as we scale up to create ecologies of artificial systems that cooperate and collaborate in the real world.

References

Agmon, E., Gates, A. J., and Beer, R. D. (2016). The structure of ontogenies in a model protocell. *Artificial Life*, **22**(4), 499–517. doi:10.1162/ARTL_a_00215

Ayers, J., and Witting, J. (2007). Biomimetic approaches to the control of underwater walking machines. *Phil. Trans. R. Soc. Lond. A*, **365**, 273–95.

Bloom, H. (2000). *Global Brain: The Evolution of Mass Mind from the Big Bang to the 21st Century*. New York: John Wiley & Sons.

Budd, G. E. (2003). The Cambrian fossil record and the origin of the Phyla. *Integrative and Comparative Biology*, **43**(1), 157–65. doi:10.1093/icb/43.1.157

Demiris, Y., and Khadhouri, B. (2006). Hierarchical attentive multiple models for execution and recognition of actions. *Robotics And Autonomous Systems*, **54**(5), 361–9. doi:http://dx.doi.org/10.1016/j.robot.2006.02.003

Downing, K. L. (2015). *Intelligence Emerging: Adaptivity and Search in Evolving Neural Systems*. Cambridge, MA: MIT Press.

Erwin, D. H., and Davidson, E. H. (2002). The last common bilaterian ancestor. *Development*, **129**(13), 3021–32.

Hannan, M. W., and Walker, I. D. (2003). Kinematics and the implementation of an elephant's trunk manipulator and other continuum style robots. *Journal of Robotic Systems*, **20**(2), 45–63. doi:10.1002/rob.10070

Ijspeert, A. J., Crespi, A., Ryczko, D., and Cabelguen, J.-M. (2007). From swimming to walking with a salamander robot driven by a spinal cord model. *Science*, **315**(5817), 1416–20.

Kazama, T., Kuroiwa, K., Umedachi, T., Komatsu, Y., and Kobayashi, R. (2013). A swimming machine driven by the deformation of a sheet-like body inspired by polyclad flatworms. In: N. F. Lepora, A. Mura, H. G. Krapp, P. F. M. J. Verschure, and T. J. Prescott (eds), *Biomimetic and Biohybrid*

Systems: Second International Conference, Living Machines 2013, London, UK, July 29—August 2, 2013. Proceedings. Berlin/Heidelberg: Springer, (pp. 390–92).

Kim, S., Spenko, M., Trujillo, S., Heyneman, B., Santos, D., and Cutkosky, M. R. (2008). Smooth vertical surface climbing with directional adhesion. *IEEE Transactions on Robotics*, 24(1), 65–74. doi:10.1109/TRO.2007.909786

Knoll, A. H. (2011). The multiple origins of complex multicellularity. *Annual Review of Earth and Planetary Sciences*, 39(1), 217–39. doi:doi:10.1146/annurev.earth.031208.100209

Koonin, E. V., and Starokadomskyy, P. (2016). Are viruses alive? The replicator paradigm sheds decisive light on an old but misguided question. *Studies in History and Philosophy of Science Part C: Studies in History and Philosophy of Biological and Biomedical Sciences*, 59, 125–34. doi:http://dx.doi.org/10.1016/j.shpsc.2016.02.016

Kurumaya, S., Suzumori, K., Nabae, H., and Wakimoto, S. (2016). Musculoskeletal lower-limb robot driven by multifilament muscles. *ROBOMECH Journal*, 3(1), 18. doi:10.1186/s40648-016-0061-3

Laschi, C., Cianchetti, M., Mazzolai, B., Margheri, L., Follador, M., and Dario, P. (2012). Soft robot arm inspired by the octopus. *Advanced Robotics*, 26(7), 709–27. doi:10.1163/156855312X626343

Mitchell, M. (1998). *An introduction to genetic algorithms*. Cambridge, MA: MIT Press.

Mora, C., Tittensor, D. P., Adl, S., Simpson, A. G. B., and Worm, B. (2011). How many species are there on Earth and in the ocean? *Plos Biology*, 9(8), e1001127. doi:10.1371/journal.pbio.1001127

Nakanishi, J., Fukuda, T., and Koditschek, D. E. (2000). A brachiating robot controller. *IEEE Transactions on Robotics and Automation*, 16(2), 109–23. doi:10.1109/70.843166

Prescott, T. J. (2007). Forced moves or good tricks in design space? Landmarks in the evolution of neural mechanisms for action selection. *Adaptive Behavior*, 15(1), 9–31.

Prescott, T. J., and Ibbotson, C. (1997). A robot trace-maker: modeling the fossil evidence of early invertebrate behavior. *Artificial Life*, 3, 289–306.

Rossetti, V., Schirrmeister, B. E., Bernasconi, M. V., and Bagheri, H. C. (2010). The evolutionary path to terminal differentiation and division of labor in cyanobacteria. *Journal of Theoretical Biology*, 262(1), 23–34. doi:http://dx.doi.org/10.1016/j.jtbi.2009.09.009

Waldron, K. J., Tokhi, M. O., and Virk, G. S. (2013). *Nature-Inspired Mobile Robotics*. Singapore: World Scientific Publishing Company.

Walker, I. D. (2013). Continuous backbone "continuum" robot manipulators: a review. *ISRN Robotics*, 2013(July), 1–19. doi:http://dx.doi.org/10.5402/2013/726506

Webster, R. J., Okamura, A. M., and Cowan, N. J. (2006). *Toward Active Cannulas: Miniature Snake-Like Surgical Robots*. Paper presented at the 2006 IEEE/RSJ International Conference on Intelligent Robots and Systems, 9–15 October 2006.

Chapter 39

Toward living nanomachines

Christof Mast[1], Friederike Möller[1], Moritz Kreysing[2],
Severin Schink[3], Benedikt Obermayer[4],
Ulrich Gerland[3], and Dieter Braun[1]

[1] Systems Biophysics, Ludwig-Maximilians-Universität, München, Germany
[2] Max Planck Institute of Molecular Cell Biology and Genetics, Dresden, Germany
[3] Theory of Complex Biosystems, Technische Universität München, Garching, Germany
[4] Berlin Institute for Medical Systems Biology, Max Delbrück Center for Molecular Medicine, Berlin, Germany

While some time ago, physics was the science of inanimate matter only, today the distinction between physics, biology, biochemistry, and medicine has become blurry and fluid. The physics of living systems is studied with very diverse experimental and theoretical approaches. This competence extension for physics is also reflected in a re-emerging research field, which explores how the transformation of inanimate matter to animate matter was possible about 4 billion years ago.

Life is evolution and evolution is necessary for life (see Prescott, Chapter 4, this volume). To create living systems, an evolutionary mechanism needs to be created in a stable way such that more complex systems can be created by the magic of continuous replication and selection (see Deacon, Chapter 12, this volume). Recent experimental approaches between the disciplines have revealed new insights into how such a process could be started by geological gradients, and have shown a way by which non-living matter could become living matter.

These experiments have focused on non-equilibrium conditions and address the question of how nanomachines can copy themselves in far from equilibrium conditions. In this chapter we specifically address the question: how can natural non-equilibrium boundary conditions push simple molecular entities towards the ever more complex dance of Darwinian evolution? As we will show, the search for the beginning of evolution entails interesting biotechnological surprises and hosts a lot of unexpected physical chemistry.

Amazingly, and for a very long time, there was no research on the origin of life. The answer seemed too simple: life creates itself all the time. The continuous, spontaneous creation of life from humus appeared very compelling—hence the long-standing theory. The evidence is simple: everyone has probably already experienced this him/herself at home. After a power failure or from waiting too long, the contents of a refrigerator become alive spontaneously! It needed a long debate and very elegant and straightforward experiments by Pasteur in the nineteenth century to prove that we just do not see the microbes and seeds, and that living things do not simply arise by themselves from dirt.

Since then, science has had to dodge the question. At best, individual small sub-steps could be addressed, typically driven by synthetic chemistry. However, the development of modern biological techniques now allows us to explore diverse hypotheses about the origin of life experimentally in much more detail. Here, we will focus on a series of steps under non-equilibrium conditions, and discuss experiments that ultimately focus on the central question of the new field of synthetic biology: how could we create synthetic, autonomous life in the lab?

Biological principles

The puzzle

By comparing DNA sequences from different species and protein classes, we know a great deal about evolution from the time at which the first single-celled organisms arose and have since inhabited the Earth. Geological evidence suggests that the first life arose shortly after water became available on early Earth. While the first cells apparently became alive relatively quickly, the evolution of more complex cells and multicellular organisms like ourselves took an order of magnitude more time. This suggests—in spite of the complexity of the first cell—a surprisingly rapid development towards living matter early on. Evolution probably was fast in the beginning.

It is very hard to infer the structure and mechanisms of the first cells through evolutionary biological extrapolation. The otherwise so successful back-extrapolation of biology is no longer possible and faces a wall of lateral gene transfer. We must utilize chemical, geological, and physical arguments. We cannot say for sure whether the historical truth on the origin of life will ever come to light. But for sure we believe that we have not yet tried hard enough. At the core of the problem is the aim is to find plausible and experimentally verifiable ways that lead from the first organic molecules to complex life. This would answer a large scientific question. It would also close the cultural and theoretical gap to understanding our own origins (see Deacon, Chapter 12, this volume).

From the time of the first living cell, it is quite clear how life evolved progressively towards more complex organisms: cell division and mutation of the genetic material led to ongoing organismic change, optimized by selection pressure from other living cells. But with the first cellular system, we expect that the molecules must have been subjected to Darwinian evolution already for quite a while. What processes have driven the evolution of the first molecules? Evolution at this stage was not the struggle for life against other living species, but a struggle to sustain the evolutionary dynamics that keep the genetic information despite the degradation of individual molecules. If you asked scientists working in the field, the consensus on the important pieces of the puzzle would probably lead to the image shown in Figure 39.1.

First, one needs to create the fundamental biological molecules (nucleotides, amino acids, etc.). Their formation through geological processes has been investigated by chemistry for some time already, with new results still showing novel insights. Non-equilibrium processes are certainly helpful also for chemistry, but despite UV illumination these processes are complex and hard to study. For these molecules to interact and create stable complexes, they are needed in high local concentrations—an unstable state, since thermal diffusion, especially at small scales, will always dilute the species. It is assumed that this early life soon needed some sort of a shell, be it a cell membrane or just pores in rock, such that the molecular reactions would be ongoing with sufficiently high efficiency. All these are non-equilibrium requirements that need a continuous supply of energy.

The next step is the generation of polymers from single molecules, such as DNA and RNA—of which RNA is likely to have been the first carrier of genetic information—as well as proteins as catalysts for biochemical reactions. Probably it was only later when a division of tasks was established where DNA was used as information storage and stable proteins were established for far more efficient catalysis. Once Darwinian evolution was established from replication and

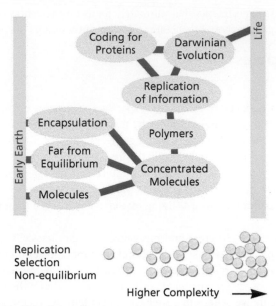

Figure 39.1 The puzzle of life. For replication and selection at the molecular level to come into existence, distinct properties are required to be combined. Understanding the origin of life requires the realistic combination of these puzzle pieces. The ultimate aim is to find autonomous, robust Darwinian evolution on the molecular level. From there the generation of the first living cells by evolution seems plausible.

selection between the molecules, the development of the first single-celled organisms and a complex metabolism is plausible.

We will discuss experiments to tackle the above steps in a model of the simple non-equilibrium condition of a temperature difference across a rock pore or cleft. It should be noted that pH and reduction–oxidation (redox) gradients were both present on the early Earth and possibly equally important non-equilibrium drivers of early molecular life.

Basic boundary conditions for life

The natural laws of physics are formulated as partial differential equations, which, by construction, depend very sensitively on spatial and temporal boundary constraints. Although we obtain detailed information only through experiments, a general consensus in the field states three basic conditions for the emergence and existence of life:

1. Life exists only far from equilibrium. For physicists, this insistence on the second law of thermodynamics is obvious: to generate information, local entropy must decrease. Only a system which is kept from equilibrium has the ability to locally generate a complex structure and sustain life against, for example, the degradation of constituent molecules.

2. The necessary molecules must be provided by the environment at least in a very rudimentary form. Their concentrations probably were very low initially. Many biological facts indicate that RNA was the first polymer of molecular evolution. It is involved in the most important processes of life. RNA translates into proteins and defines the genetic code, it forms the reaction center of the ribosome, and RNA monomers still play a major role as an energy store (ATP) and signal molecules (cGMP, cAMP). It was long unclear how RNA could be

synthesized prebiotically, until the group of Sutherland (Powner 2009) showed a new mechanism for the synthesis of RNA bases. For RNA polymerization, the group of Ernesto di Mauro showed that cGMP may be polymerized into long RNA polymers (Costanzo 2009), at best in the dry state.

3. The prebiotic genetic molecules must be included in a container so that they react with each other for a long time and protect themselves against rapid flow. However, the separation must be open enough to allow food molecules to enter and waste products to be expelled. A fully insulating protocell would fall into local thermal equilibrium. Porous volcanic rock membranes offer the possibility of molecular inclusion (Figure 39.2) with simultaneous transmission of smaller reaction products. The concentrations required in such rock pores can be accumulated by the non-equilibrium of a temperature gradient as shown below.

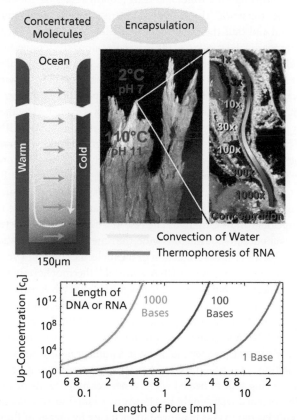

Figure 39.2 Rock sponges near warm sources are interesting niches for early life. The primordial ocean initially was a desert for molecules: there were only few and far apart. In order not to lose the first biomolecules by diffusion, natural barriers are interesting. Such rock pores are for example generated by underwater hot springs, but also quite common in volcanic basalt. These settings include temperature gradients which—by convection and thermophoresis—provide a strong accumulation mechanism. The orientation and exact geometry of the pore is not crucial. Interestingly, longer molecules are significantly better concentrated.

Thermal molecular traps to concentrate molecules

Often it is assumed that the first precursor of the modern cell was surrounded by a lipid membrane. However, the membrane prevents the directed, active transport of molecules and thus their active accumulation. In modern cells, this is made possible by highly evolved, complex membrane pump proteins. It is quite controversial as to when, during cellular evolution, vesicles and modern cells based on lipid membranes developed. The metabolism of lipids in archaea and bacteria (the two different kingdoms of prokaryotic organisms) developed independently, suggesting that membrane development was late. It also seems reasonable that early evolution benefited at the start from lateral gene transfer, i.e. it maintained a free exchange of genetic molecules. A grouping of molecules into genetic individuals was probably only needed later.

Which is the non-equilibrium that allows the accumulation of molecules? We have shown that a simple temperature difference across a water-filled pore is an elegant solution (Figure 39.2). Similar to a Carnot heat engine, which generates work from temperature differentials, the molecules are highly concentrated (Baaske et al. 2007; Weinert and Braun 2009; Mast 2010), and continue to diffuse freely through the water and thus remain accessible for chemistry.

How does the molecule trap work? The difference in temperature has two mutually reinforcing effects. It keeps the liquid and thus the molecules circulating by thermal convection in a circuit. Superimposed is the perpendicular motion of the molecules along the temperature difference. The latter mechanism is called thermophoresis or the Soret effect. Charged molecules have a tendency to wander to the cold side. There, they are very effectively transported by downwards convection and, as a result, molecules entering the long pore from the top become concentrated at its bottom.

We have also shown, with Stefan Duhr, that thermophoresis is explained by the entropy the molecules induce by their contact with salty water solutions. The motion of DNA/RNA molecules in a temperature gradient can be measured separately. In this experiment, the temperature is applied by infrared lasers, and movement of molecules is detected optically by means of fluorescence. The same approach was used by the startup company *NanoTemper*, created by Duhr and Baaske, to measure the affinity of biomolecules in free solution, showing an unexpected and elegant solution for biological and pharmaceutical fundamental research. This company has now more than 70 employees and demonstrates compellingly how Origins of Life research can lead to successful biotechnology.

If the elongated pore is about 100–200 microns wide, the timescales of the convection and the thermophoresis are comparable and the accumulation of the molecules becomes very efficient (Baaske et al. 2007). Based on the measurements of DNA and RNA thermophoresis we expect to find a virtually unlimited up-concentration of these molecules. RNA and DNA behave almost identically and are preferred by their strong charge over other molecules. The accumulation is largely independent of the orientation of the pore or the precise geometric shape. Interestingly, it is found that rock pores allow long DNA/RNA to accumulate much better than shorter molecules (Figure 39.2).

We confirmed experimentally these dependencies in a number of implementations of a thermal trap. In one implementation, an infrared laser generates the temperature gradients very precisely. At the same time, the moving spot of heated water generates movements of water inside a given geometry. High quality thermal traps could be shown in a radial or in a linear geometry (Mast 2010). The molecule concentration was measured in each case by means of fluorescence and successfully modeled using fluid dynamic calculations. Compared to alternative mechanisms such as concentration or dry freezing, the molecules continue to move freely in the liquid and to be able to interact and react with each other.

Interestingly, lipids, which are components of cell membranes, can also be concentrated locally. The Szostak group has shown that they then form vesicles (Budin et al. 2009). The concentrated molecules can thus be readily packaged by such a thermal trap at the same time.

Polymerization in the molecular trap

It becomes interesting when a chemical reaction—in our case, the polymerization of RNA—is combined with the thermal trap. One expects a mutual gain: RNA polymerizes longer due to the enhanced concentration of the RNA monomers. On the other hand, the trap accumulates long polymers significantly better. How does the combined system react?

We have shown, both by experiments using sticky DNA strands and by theoretical calculations by Severin Schink and Ulrich Gerland, that the polymerization is strongly supported by a thermal trap—even under the most adverse conditions of reversible "polymerization" by the sticky DNA strands. From a certain length of the trap, the DNA length increases rapidly (Mast 2013). Since the thermophoretic characteristics of DNA and RNA are the same, the trap allows the polymerization equilibrium to shift massively from short towards long RNA molecules. In the steady state of the reaction, one finds the almost paradoxical situation that the concentration of long polymers is significantly higher than the concentration of the monomers offered outside the trap. In short, if the geometry is fitting, a pore of rock becomes a polymerization machine.

Darwinian evolution

Living things evolve through replication and selection. Replication has been at the focus of Origin of Life research for a long time. Less attention has been focused on how selection may have occurred on an initially inanimate Earth. So far, the replication of interesting molecules in the laboratory has usually succeeded only through man-made intervention such as artificial replication, selection, and specific purification. But how did the inanimate Earth select for the first replicating molecules? An important milestone of molecular evolution was the production of the first proteins. Only with proteins can more complex metabolisms and first cells be imagined. The selection of the conditions under which this genetic apparatus has evolved to encode proteins is an interesting problem.

Replication and selection with a molecule trap?

As described above, a laminar thermal convection is able to replicate genetic molecules largely independent of its length. We can hope that the molecules are also accumulated in the same setting, allowing the creation of an open system which traps molecules and replicates them at the same time, enabling a microscopic autonomously running Darwin machine for long and complex molecules.

In our experiment we were able to show this combination fundamentally (Mast 2010). We replicated DNA by the melting and cooling steps implemented by the thermal convection. At the same time, we accumulated the replicated molecules with the temperature difference. Both processes are driven by the same temperature gradient. One can observe the process using a fluorescence microscope and by labeling the DNA with fluorescent dyes. We found that the DNA very rapidly doubles every 50 seconds, matching the time period of the thermal convection. For comparison, the "house animal" of biology, the stomach bacterium *E. coli*, doubles every 20 minutes. The demonstrated rapid replication in this case should allow for interesting evolutionary experiments.

Replication towards increasing complexity

Life is made from polymers. For evolution, it is important that longer polymers can prevail against shorter polymers. This is far from self-evident. On the contrary, as famously shown by Spiegelman and colleagues (Mill et al. 1967), a tendency towards ever shorter polymers is found if the replication is performed at the same temperature. The reason is that short RNA molecules can be replicated faster, outgrowing the longer polymers. Since the short RNA is made in an exponentially increasing concentration, they replace the longer molecules very fast. In Spiegelman's experiment, the RNA had lost all genetic information by shortening after only a few generations.

If the molecular evolution of complex genetic information is to be carried over many generations into the future, such a length degeneration must be prevented at any cost. In part, this can be achieved by performing the replication in a temperature oscillation. For example, the molecules could be circulated continuously by a laminar thermal convection. The double-stranded DNA is melted in the hot part of the convection into two single strands (Figure 39.3). Both strands can now be replicated in the cold again to double strands. As with Spiegelman, the short strands are again faster in this replication step. However, all double-stranded molecules must now wait together to be transported by convection in order that the molecules are moved back into the hot zone for re-melting. If diffusion can be neglected, short and long DNA molecules are duplicated at the same speed. In addition, thermal convection solves the second hard problem: it melts double-stranded structures into single strands so that they can be replicated exponentially.

It was possible, in this particular case, to feed the replication and selection process continuously by adding new monomers that are pushed by a slow flow of water through the pore. The flow must be slow enough that the thermal convection is not perturbed too much. At the site of

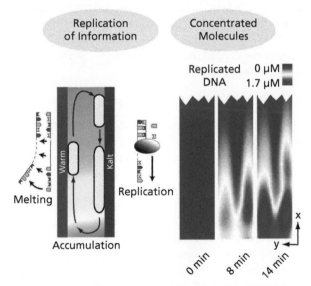

Figure 39.3 Replication and accumulation. To be able to probe boundary conditions by experiment, we harness the fast and simple base-ligation chemistry evolved by biology. We use DNA that is replicated by a polymerase chain reaction (PCR). The DNA melts at a high temperature, and is replicated in colder regions by a protein. Thermal convection provides the temperature oscillation required for this replication. Interestingly, this reaction can be combined with a thermal trap mechanism based on the thermophoresis of molecules. The replicated molecules are then concurrently concentrated in the lower part of the chamber. As shown in the experiment on the right side, accumulation and replication of DNA are driven by the same temperature difference across the pore.

the event, DNA is then permanently replicated and accumulated by length. We recently showed that replication and selection can be fed in a through-flow chamber (Kreysing 2015). The resulting replication–selection process allows longer genes (75 bases) to outcompete short genes (35 bases). The length selectivity can be finely tuned by the flow and opens many possible scenarios of autonomous evolution driven by thermal gradients.

Replication without proteins

In modern organisms, information is stored on the DNA/RNA strands by means of proteins, creating a very low error rate. However, to encode a protein in this way, more than one thousand bases are required, which again need very accurate replication. This is a chicken-and-egg problem: long RNA cannot be replicated without protein and proteins cannot be created without long RNA. Very simple molecules with only a few bases and activation groups were shown by Kiedrowski (1986) to replicate spontaneously. An interesting chemical approach is to evolve RNA molecules capable of catalyzing replication of the RNA itself. The best candidates can also replicate several bases. The problem is how to build the specific sequences required for this process and the high salt concentrations required, which can quickly degrade RNA.

One can also use the fact that, in some conditions, double-stranded RNA chemically decomposes slower than single stranded RNA; a computer analysis showed that sequence information survived significantly longer by this asymmetric degradation (Obermayer et al. 2011). This process is identical to a biological base-by-base replication, but with only a 30% efficiency. So it could very well be, that replication in the beginning was an emergent passive property of many molecules.

Replication and translation by transfer RNA

In modern organisms, the conversion of RNA into proteins is accomplished by a ribosome. This is a very large complex consisting mainly of RNA. Sequences of three bases, called codons, are detected by a pool of mediator molecules, the transfer RNAs, which are chemically associated with a specific amino acid. The amino acids are aligned and then polymerized by the ribosome. The result is a polymer of amino acids encoded by RNA: the protein. Central to the origin of life problem is to understand how early molecules could generate this fundamental, yet very complex, coding machinery.

Experiments show that transfer RNA can also take on the role of a replicator (Krammer et al. 2012). It will then not be replicating individual bases, but a sequence of base combinations and thus a predecessor of the later trimeric codons. Again, the temperature in the non-equilibrium system is a very important factor in the replication reaction. Initially, the transfer RNAs are cooled quickly and the conformational energy is saved in a metastable hairpin state (Figure 39.4). With only a moderate temperature oscillation, this conformational energy is discharged in the replication reaction. The simplest two-letter code replicator could be experimentally demonstrated by starting from transfer RNA sequences and this reaction could be modeled theoretically.

This replicator only rarely copies wrong base combinations, but has the tendency to spontaneously generate new sequences at higher temperatures. Interestingly, the replication is very fast with a doubling time of only 30 seconds. This copying process is clearly suggestive of the later association of codon sequences with amino acids and possibly allowed the encoding of the first protein-like reaction centers (Figure 39.4). How viable this new approach is for longer sequences and full tRNA structures will be explored in the future.

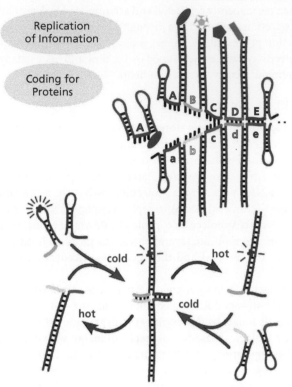

Figure 39.4 Combining replication and protein synthesis. Does the synthesis of proteins and the replication of their amino acids have a common origin? Using DNA-machine approaches, one can implement a replicator of sequence snippets from transfer RNA molecules. Transfer RNA is today used to implement the genetic code. Here, the same molecule is used as a thermally driven, exponential replicator. The final structure suggests that the sequence of amino acids is grouped by the genetic information encoded into the sequence of tRNA molecules.

Future directions

We have tried to show how simple experiments can shed new insights into the possible beginnings of life. This research—currently still dominated by chemical approaches—is now opening itself to interdisciplinary viewpoints. For example, here we have looked into the combination of biological molecules, non-equilibrium physics and modeling, and biophysical non-equilibrium experimentation in a geological context. The famous words of Feynman, "What I cannot create, I do not understand" (see also, Verschure and Prescott, Chapter 2, this volume), probably nowhere apply as strongly as in the Origins of Life field. Only once we are able to create nanomachines that can sustain molecular evolution in geological settings without human intervention, can we say that we understand how life came about. Our approaches show that this task, while being very bold, might still be simpler than expected.

Learning more

Good and accessible books to introduce yourself to the topic are *Life ascending* by Nick Lane, *Emergence of life on Earth* by Iris Fry, and *The emergence of life* by Pier Luigi Luisi. For seminars and talks on the topic, check the Origins of Life initiatives that are establishing themselves at selected locations or attend the biannual Gordon Conference on the Origins of Life.

References

Baaske, P., Weinert, F. M., Duhr, S., Lemke, K. H., Russell, M. J., and Braun, D. (2007). Extreme accumulation of nucleotides in simulated hydrothermal pore systems. *Proc. Natl Acad. Sci. USA*, **104**, 9346–51.

Budin, I., Bruckner, R. J., and Szostak, J. W. (2009). Formation of protocell-like vesicles in a thermal diffusion column. *Journal of the American Chemical Society*, **131**, 9628–9.

Costanzo, G., Pino, S., Ciciriello, F., and Di Mauro, E. (2009). Generation of long RNA chains in water. *Journal of Biological Chemistry*, **284**, 33206–16.

Fry, I. (2000). *Emergence of life on Earth: a historical and scientific overview*. New Brunswick, NJ: Rutgers University Press.

Obermayer, B., Krammer, H., Braun, D., and Gerland, U. (2011). Emergence of information transmission in a prebiotic RNA reactor. *Physical Review Letters*, **107**, 018101.

Krammer, H., Möller, F. M., and Braun, D. (2012). Thermal, autonomous replicator made from transfer RNA. *Physical Review Letters*, **108**, 238104.

Kreysing, K., Keil, L., Lanzmich, S., and Braun, S. (2015). Heat flux across an open pore enables the continuous replication and selection of oligonucleotides towards increasing length. *Nature Chemistry*, **7**, 203–8.

Kiedrowski, G. v. (1986). A self-replicating hexadeoxynucleotide. *Angwandte Chemie Int. Ed.*, **25**, 932–5.

Lane, N. (2010). *Life ascending: the ten great inventions of evolution*. London: Profile Books Ltd.

Luigi, P. (2016). *The emergence of life: from chemical origins to synthetic biology*. Cambridge, UK: Cambridge University Press.

Mast, C. B., and Braun, D. (2010). Thermal trap for DNA replication. *Physical Review Letters*, **104**, 188102.

Mast, C. B., Schink, S., Gerland, U., and Braun, D. (2013). Escalation of polymerization in a thermal gradient. *Proc. Natl Acad. Sci. USA*, **110**, 8030–5.

Mill, D. R., Peterson, R. L., and Spiegelman, S. (1967). The synthesis of a self-propagating and infectious nucleic acid with a purified enzyme. *Proc. Natl Acad. Sci. USA*, **58**, 217–24.

Powner, M. W., Gerland, B., and Sutherland, J. D. (2009). Synthesis of activated pyrimidine ribonucleotides in prebiotically plausible conditions. *Nature*, **459**, 239–42.

Weinert, F. M., and Braun, D. (2009). An optical conveyor for molecules. *Nano Letters*, **9/12**, 4264–7.

Chapter 40

From slime molds to soft deformable robots

Akio Ishiguro[1] and Takuya Umedachi[2]

[1] Research Institute of Electrical Communication, Tohoku University, Sendai, Japan
[2] Graduate School of Information Science and Technology, The University of Tokyo, Tokyo, Japan

An autonomous decentralized control mechanism, where the coordination of simple individual components yields non-trivial macroscopic behavior or functionalities, could be key to understanding how animals orchestrate the large degrees of freedom of their bodies in response to different situations. However, a systematic design methodology for autonomous decentralized control is still missing. To alleviate this problem, we focused on the plasmodium of a true slime mold (*Physarum polycephalum*), which is a primitive multinucleate single-cell organism. Despite its primitiveness, and lacking a brain and nervous system, the plasmodium exhibits surprisingly adaptive and versatile behavior (e.g. taxis, exploration). This ability has undoubtedly been honed by evolutionary selection pressure, and there likely exists an ingenious mechanism that underlies the animals' adaptive behavior. We successfully extracted a design scheme for decentralized control and implemented it in an amoeboid robot with many degrees of freedom. The experimental results showed that adaptive behaviors emerge even in the absence of any centralized control architecture. The obtained results shed new light on understanding how to orchestrate large degrees of freedom of animals' bodies to induce adaptive behavior.

Biological principles

Slime molds (Eumycetozoa) are eukaryotic organisms that can live separately as single-celled animals or can aggregate together to form multicellular structures. The plasmodium of a "true" slime mold, such as *Physarum polycephalum*, is an aggregate form created by the fusion of many individual cells and is mainly comprised of two parts: an outer skin in gel form and protoplasm inside in sol form (see Figure 40.1(A)). The plasmodium contains a network of vein-like protoplasmic structures and many cell nuclei. As a plasmodium, the aggregate creature can actively search for food, for instance in piles of leaves or decaying logs. Slime mold movement is created by biochemical oscillators that are distributed spatially in the plasmodium and generate a rhythmical mechanical contraction (cycle period of 1–2 min), which in turn induces protoplasmic streaming between the body parts. As a whole, spatiotemporal oscillation patterns can be observed as variations in thickness, and the plasmodium switches patterns to generate different locomotion/behaviors, such as a traveling wave for taxis and spiral wave for exploration behavior, in response to different situations. Despite the absence of a central nervous system or specialized organs, such global behaviors are induced by interaction between homogeneous components (i.e. biochemical oscillators) in the plasmodium. Therefore, an ingenious design principle for decentralized control systems likely exists in this organism. However, how the individual components interact with each other to generate such adaptive behaviors is not well understood.

There are two essential components that enable the plasmodium to exhibit adaptive behavior: a phase modification mechanism based on mechanosensory information and physical communication (i.e. morphological communication; Rieffel et al. 2010; Paul 2006) stemming from the protoplasm. The oscillators of the plasmodium are similar to a central pattern generator (CPG) (Takamatsu et al. 2001; Takamatsu 2006; see also Holk and Cruse, Chapter 24, this volume). The difference is that the oscillators of the plasmodium interact with each other using mechanical constraints, not a neural network. The oscillators contract to increase the pressure in the protoplasm, which leads to protoplasmic streaming inside the outer skin according to the pressure gradient (Kobayashi et al. 2006); this in turn allows distant oscillators to interact with each other physically. The biochemical oscillator is able to feel force from the protoplasm and modify its own contraction rhythm to reduce the pressure applied by the protoplasm (Yoshiyama et al. 2009).

Biomimetic systems

Inspired by this living organism, we developed a plasmodium model that reproduces amoeboid locomotion. Inspired by the body structure of the plasmodium, the robot comprises two main parts: outer skin and protoplasm (Figure 40.1(B)). The outer skin comprises multiple modules: each is equipped with a *real-time tunable spring* (RTS) (Figure 40.1(C)), friction control unit (Figure 40.1(D)), and local controller corresponding to the biochemical oscillator.

The RTS is a key actuator device we developed to mimic an actin–myosin filament. The significant feature of RTS is its capability of changing the resting length of spring at any time by forcibly winding/unwinding a coil spring. The resting length is controlled according to the phase of the oscillator, which reproduces the rhythmic mechanical contraction of the body part. Note that a discrepancy between the resting length and actual length of the spring always exists due to the mechanical passivity. The discrepancy can be measured as the tension on the spring by adding a pressure sensor to one terminal of the spring. The device can also change softness by changing its resting length. The friction control unit switches between anchor and anchor-free friction modes according to the phase of the oscillator.

A flat balloon is embedded inside the outer skin to act as protoplasm for the robot. Note that this structure allows distant modules to interact with each other physically: force exerted by one module is conveyed via the balloon to the other modules when the RTS in one module contracts and pushes the protoplasm due to conservation of the balloon volume.

To control the RTS and friction control unit of each module, we modeled the biochemical oscillator of the plasmodium using one of the simplest oscillator models, i.e. the phase oscillator model (Kuramoto 2003). The dynamics of the oscillator is described by:

$$\dot{\theta}_i = \omega + \Sigma_{j=i+1,i-1} \sin(\theta_j - \theta_i) - \frac{\partial I_i}{\partial \theta_i}. \tag{1}$$

The first term on the right-hand side is the intrinsic frequency of the oscillator, which is identical for all modules. The second term simulates the chemical diffusion interaction between neighbouring oscillators. The third term is the phase modification mechanism (i.e. local sensory feedback) based on mechanosensory information of the oscillators; it is designed to reduce the value of function θ_i. As noted in the previous section, the biochemical oscillator of the plasmodium can sense the force from the protoplasm and tends to reduce this force by modifying its

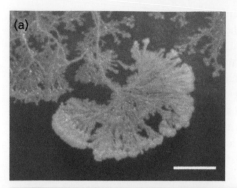

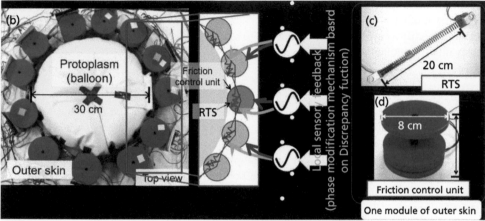

Figure 40.1 Plasmodium of true slime mold and real physical robot inspired by plasmodium.
(a) Plasmodium of true slime mold (white bar indicates 1 cm). (b) The real physical robot consists
of modules; each is controlled by a phase oscillator. One module consists of (c) a real-time tunable
spring (RTS) which can change the resting length and friction control unit (d) to switch between
anchor and anchor-free modes.

own phase. Based on this biological finding, we designed the function using tension T_i for this
model as follows:

$$I_i = \sigma \frac{T_i^2}{2}. \qquad (2)$$

σ is a parameter specifying the strength of the local sensory feedback. We named this function the "discrepancy function" because it quantifies the discrepancy taking place between the
control system, mechanical system, and environment.

To induce protoplasmic streaming effectively, we also introduced a symmetry-breaking mechanism. More specifically, we changed the stiffness distribution of the outer skin, as illustrated in
Figure 40.2 (Umadachi et al. 2010, 2011). This is done in a fully decentralized manner by increasing the nominal value of the resting length oscillation when a module detects an attractant. At
the beginning of the experiment, all oscillators started from the in-phase condition. This means
that all modules tried to contract by shorting their resting lengths.

However, this was impossible because of the mechanical constraint stemming from the protoplasm; the area surrounded by the outer skin was conserved. In this condition, each module
felt high tension, which led to a high value for the discrepancy function and in turn caused a
phase modification to reduce the value in each module. As a result of this decentralized phase

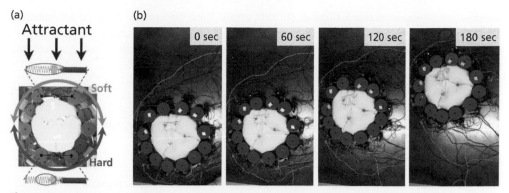

Figure 40.2 Experimental result with amoeba-like soft robot. (a) Example of softness distribution of outer skin as response to attractant. (b) Taxis locomotion toward attractant generated in fully decentralized manner.

modification using mechanical constraints, a phase gradient from the tail to head was generated, and the robot moved toward the attractant. Interestingly, distant modules that do not communicate via the second term in equation (1) seemed to cooperate with each other to generate the taxis locomotion. The distant modules were communicating via the third terms, i.e. a phase modification mechanism relying on mechanical constraints.

This result strongly suggests that soft bodies allow the control system to be simpler and less centralized (Rieffel et al. 2010; Paul 2006). Our research group has demonstrated that this design method of reproducing the adaptive behavior of animals is applicable to other morphologies, such as snake-like robots (Sato et al. 2010) and legged robots (Owaki et al. 2012).

Future directions

The plasmodium exhibits not only taxis locomotion but also a huge diversity of behaviors by generating variable oscillation patterns. For instance, a traveling wave can be observed when the plasmodium moves in one direction; however, a complex spiral wave can be observed before the well-ordered traveling wave is generated or after a new chemical stimulus. This behavioral diversity is a very important feature to realizing adaptive behavior against unpredictable environmental changes. Nevertheless, most robots and artificial systems lack behavioral diversity. We previously demonstrated that this behavioral diversity can be achieved with the same design (i.e. exploiting morphological computation and phase modification mechanism based on mechanosensory information; Umedachi et al. 2013). Hence, the design method can also help reproduce behavioral diversity for other robot morphologies.

Another interesting topic is designing multi-timescale adaptivity. At present, the robot only has relatively short-term adaptivity, e.g. the phase modification mechanism in this research. However, plasmodium is known to be capable of changing the topology of the body depending on the environment. The topology becomes a tree-like structure when the medium contains toxic chemicals; on the other hand, it becomes a mesh-like structure when the medium contains nutrients (Ito et al. 2011). Adding such long-term adaptivity to the short-term adaptivity can be an extremely interesting topic for understanding the primitive yet surprisingly sophisticated intelligence of this life form.

Learning more

Details on plasmodium oscillation are given in Kobayashi et al. (2006), which presents both the structure of the body and mathematical descriptions of the oscillation. Other famous and interesting plasmodium behaviors include solving a maze (Nakagaki et al. 2000) and constructing an effective transport network (Tero et al. 2007, 2010). These papers are also good for learning about the general biology of true slime molds. Readers may be interested to compare pattern generation and locomotion in soft-bodied invertebrates such as caterpillars, that have central nervous systems, as described by Trimmer (Chapter 41, this volume).

Acknowledgments

This work was funded in part by the Japan Science and Technology Agency through CREST.

References

Kobayashi, R. Tero, A. and Nakagaki, T. (2006). Mathematical model for rhythmic protoplasmic movement in the true slime mold. *Journal of Mathematical Biology*, **53**, 273–86. doi: 10.1007/s00285-006-0007-0

Rieffel, J. A., Valero-Cuevas, F. J., and Lipson, H. (2010). Morphological communication: exploiting coupled dynamics in a complex mechanical structure to achieve locomotion. *Journal of the Royal Society Interface*, 7(45), 613–21.

Takamatsu, A. (2006). Spontaneous switching among multiple spatio-temporal patterns in three-oscillator systems constructed with oscillatory cells of slime mold. *Physica D: Nonlinear Phenomena*, **223**, 180–8. doi: 10.1016/j.physd.2006.09.001

Takamatsu, A., Tanaka, R., Yamada, H., Nakagaki, T., Fujii, T., and Endo, I. (2001). Spatiotemporal symmetry in rings of coupled biological oscillators of physarum plasmodial slime mold. *Physical Review Letters*, **87**, 0781021. doi: 10.1103/PhysRevLett.87.078102, 2001

Yoshiyama, S. Ishigami, M. Nakamura, A. and Kohama, K. (2009). Calcium wave for cytoplasmic streaming of *Physarum polycephalum*. *Cell Biology International*, **34**, 35–40. doi: 10.1042/CBI20090158

Kuramoto, Y. (2003). *Chemical oscillations, waves, and turbulence*. Mineola, NY: Dover Books.

Paul, C. (2006). Morphological computation: A basis for the analysis of morphology and control requirements. *Robotics and Autonomous Systems*, Special Issue on Morphology, Control and Passive Dynamics, **54**, (8), 619–30.

Umedachi, T., Takeda, K., Nakagaki, T., Kobayashi, R., and Ishiguro, A. (2010). Fully decentralized control of a soft-bodied robot inspired by true slime mold. *Biological Cybernetics*, **102**, 261–9. doi: 10.1007/s00422-010-0367-9

Umedachi, T., Takeda, K., Nakagaki, T., Kobayashi R., and Ishiguro, A. (2011). A soft deformable amoeboid robot inspired by plasmodium of true slime mold. *International Journal of Unconventional Computing*, 7, 449–62.

Umedachi, T. Idei, R. Ito, K. and Ishiguro, A. (2013). A fluid-filled soft robot that exploits spontaneous switching among versatile spatio-temporal oscillatory patterns inspired by true slime mold. *Artificial Life*, **19**(1). doi: 10.1162/ARTL_a_00081

Sato, T., Kano, T., and Ishiguro, A. (2010). A decentralized control scheme for an effective coordination of phasic and tonic control in a snake-like robot. *Journal of the Acoustical Society of America*, 7, 016005.

Owaki, D., Kano, T., Nagasawa, K., Tero, A., and Ishiguro, A. (2012). Simple robot suggests physical interlimb communication is essential for quadruped walking. *Journal of the Royal Society Interface*, 23097501.

Ito, M., Okamoto, R., and Takamatsu, A. (2011). Characterization of adaptation by morphology in a planar biological network of plasmodial slime mold. *J. Phys. Soc. Jpn*, **80**, 074801.

Nakagaki, T., Yamada, H., and Toth, A. (2000). Maze-solving by an amoeboid organism. *Nature*, **407**(470). doi: 10.1038/35035159,.

Tero, A., Takagi, S., Saigusa, T., Ito, K., Bebber, D. P., et al. (2010). Rules for biologically-inspired adaptive network design. *Science*, **327**, 439–42. doi: 10.1126/science.1177894

Tero, A., Kobayashi, R., and Nakagaki, T. (2007). A mathematical model for adaptive transport network in path finding by true slime mold. *Journal of Theoretical Biology*, **244**, 553–64. doi: 10.1016/j.jtbi.2006.07.015

Chapter 41

Soft-bodied terrestrial invertebrates and robots

Barry Trimmer

Biology Department, Tufts University, Medford, USA

Animals are composed almost entirely of soft materials. This is even true for vertebrates, whose internal skeleton of hard materials constitutes less than 15% of their body weight. This important fact is largely ignored in studies of terrestrial locomotion where the primary focus has been on modeling the kinematics of limbs and joints as rigid bodies, notwithstanding the necessity of springs and dampers added to produce realistic movements. This tendency to overlook highly deformable materials is partly because their mechanical properties are so complex and poorly understood. However, ignoring soft materials has also led to serious problems in trying to emulate animal capabilities in biomimetic robots. This chapter will describe our current understanding of locomotion in terrestrial invertebrate species that are fundamentally soft structures. The focus will be on new findings from studies on caterpillars which show that soft animals can exploit stiff materials in the environment to remain flexible and conformable (an *environmental skeleton*). This will be contrasted with the more traditional model of soft animal locomotion that assumes such animals must become stiffer through pressurization and controlled hydrostatics. The final section will describe biomimetic devices that have been inspired by both animal models and how such biological solutions might be employed to build controllable, highly deformable mobile machines.

Biological principles

Animals that have stiff skeletons (such as vertebrates and adult insects) rely on constraints imposed by joints to reduce the degrees of freedom and simplify movement control. A stiff skeleton also permits high loading, fast locomotion, and easy inter-conversion of forces and displacements using levers. In contrast, soft animals can deform in more complex ways (e.g. twisting, crumpling, inflating) and have virtually unlimited degrees of freedom. Interestingly, there is no general theory of how soft animals control their movements in a computationally efficient manner. It would be very useful to understand this control so that it can be applied to more conformable and adaptable robots. Research on soft-bodied locomotion has focused on mollusks (e.g. cephalopods), annelids (e.g. leeches, earthworms), and insect larvae. Cephalopods (e.g. octopus) have astoundingly complex movements and elaborate musculature, making them ideal models for flexible structures (see Prescott, Chapter 38, this volume) but the complexity and size of their peripheral nervous system (*50 million peripheral neurons within each octopus arm*), make it difficult to interpret the relationship between neural signals and specific movements.

Locomotion in annelids. Annelids are much more accessible in this regard. Their primary locomotory muscles are organized into longitudinal and circumferential blocks with smaller muscles connecting the dorsal and ventral surfaces for flattening the animal (Quillin 1999). Each body segment is generally separated from the rest by a flexible septum and the worms as a whole can be considered a fixed-volume flexible cylinder (Figure 41.1A). Consequently, the two main muscle groups are antagonistic (Figure 41.1B)—contraction of longitudinal muscles results in short segments of increased radius whereas circumferential muscles squeeze segments into narrow long cylinders (Figure 41.1C). The degree and speed of lengthening (analogous to the mechanical "advantages" of levers) are determined by the aspect ratio of the cylinder (Kier and Smith 1985) (Figure 41.1D). As constant volume cylinders, worms have a classical "hydrostatic" skeleton; they can become stiffer by contracting muscles in the body wall which increases both internal pressure and body wall tensile loading (hoop stress). Consequently, a more pressurized worm will be able to resist higher bending forces. However the advantage of doing this

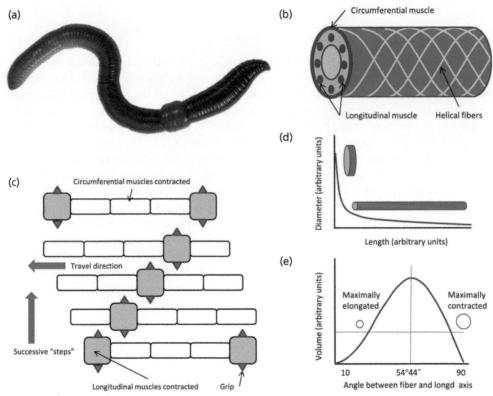

Figure 41.1 An overview of worm biomechanics and locomotion. (a) Annelids such as the earthworm are segmented pressurized cylinders. (b) Major muscles are arranged around the central cavity in bundles of longitudinal fibers and circumferential rings. The body wall is often reinforced with helical fibers. (c) A simplified scheme of crawling shows that contraction of longitudinal muscles shortens each segment and anchors the body. Contraction of circumferential muscles extends the segment. The sequence of segment movements can change in different contexts. (d) By maintaining a constant volume each segment of the worm can have a different aspect ratio. The starting aspect ratio affects the speed and magnitude of extension. (e) The crossing angle of helical reinforcing fibers affects the tendency to passively restore body shape.

does not scale particularly well since hoop stress is proportional to the body radius and disproportional to the body wall thickness. As animals increase in size they must increase their body wall thickness and internal pressure disproportionately to support their own body weight and the increased forces of locomotion. This probably explains why large hydrostatic animals live in environments where their body is supported by the surrounding medium, either swimming or burrowing.

The structure of the body wall has major consequences for movement control in hydrostatic animals. Typically, the flexible tissue matrix is reinforced with relatively stiffer elastic fibers organized in helical array cross patterns that help to distribute stresses. These fibers also make the body wall anisotropic and, depending upon the fiber crossing angle, produce quite different mechanical advantages. Elastic fibers with crossing angles greater than 54° 4″ will passively assist in restoring the body position after lateral bending, whereas angles below this value help to restore the resting body diameter after circumferential muscle contractions (Figure 41.1E). These general mechanical principles are well-matched to the neural commands that generate peristaltic locomotion or other bending movements, using reciprocal antagonism of muscle groups acting on the hydrostatic skeleton in much the same way as vertebrates' skeletons.

Locomotion in caterpillars. Relatively recent work has discovered a different strategy for soft-bodied terrestrial locomotion that does not rely upon hydrostatic pressure to interconvert forces and displacements. Instead, animals such as caterpillars remain relatively soft and conformable by using internal pressure to produce baseline turgor but not to regulate body and limb movements. Locomotion is accomplished by transmitting compressive forces to the substrate and generally keeping the body in tension. This strategy has been called the "*environmental skeleton*" (Lin and Trimmer 2010). The mechanism has been studied in detail in larval stages of the tobacco hornworm, *Manduca sexta* (Figure 41.2A).

Caterpillars like *Manduca* lack circular muscle (Figure 41.2B), do not maintain constant segment volume, and do not constrict body segments as worms do. Instead, they crawl through a wave-like series of segment contractions starting at the posterior and moving forward (Figure 41.2C). Most of this crawl cycle is dominated by the abdominal segments with up to three segments in swing phase simultaneously and the prolegs carried forward passively. During stance, the prolegs grip the substrate passively (Lin and Trimmer 2010) using cuticular hooks (crochets), which are actively retracted just after the start of the segment swing phase. Unlike in burrowing or swimming annelids and marine cephalopods, *Manduca* movements are entirely terrestrial, comparatively slow, and usually confined to a small workspace, making them relatively easy to study by three-dimensional motion capture. *Manduca* is also advantageous because its neural and biomechanical components are equally accessible. Every abdominal segment contains about 70 distinct muscles (Figure 41.2B), each innervated by just one (occasionally two) motor neuron(s), and there are no inhibitory motor units. Therefore, most of *Manduca*'s movements are controlled by only a few hundred *identified motor neurons* whose activity can be monitored using electrodes implanted in the muscles of freely moving animals (Figure 41.2D, E). The muscle attachment points have been mapped to external features, permitting kinematic muscle length measurements, and also providing an accurate site for electrode placement. In *Manduca* the observed kinematics, dynamics, and tissue properties can be related directly to neural activity.

These studies reveal that caterpillars do not control the timing of locomotory muscle contractions very precisely but instead produce overlapping waves of segmental muscle activation while pushing the gut and other internal tissues forward. The stepping pattern of a crawl is probably dictated by the activation timing of the proleg retractor motor neurons in each segment leading to crochet release. This form of locomotion is remarkably robust, requiring no

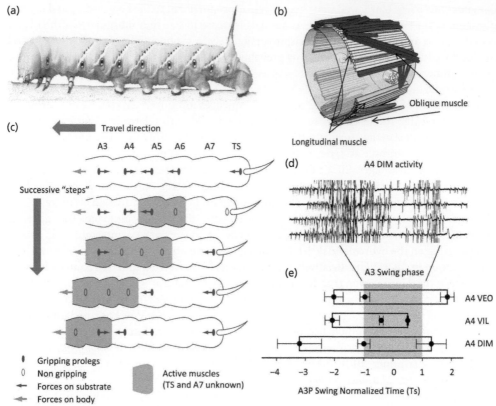

Figure 41.2 Caterpillar biomechanics and locomotion. (a) Caterpillars such as *Manduca sexta* are not constant volume cylinders and unlike worms their body does not alternate between compression and extension. (b) The muscles are not organized into antagonistic blocks but groups of dorsal, ventral, and oblique fibers, each controlled by one (occasionally two) motor neurons. (c) Waves of muscle contraction move forward accompanied by progressive release of the prolegs; compressive forces are largely carried by the substrate. (d) EMG recordings of the dorsal internal muscle (DIM) during four successive steps aligned to the proleg swing phase. (e) A summary of the timing of muscle activation in segment A4 showing broad muscle co-activity.

(a) and (b) Adapted from B. A. Trimmer and Huai-ti Lin, Bone-Free: Soft Mechanics for Adaptive Locomotion, *Integrative and Comparative Biology*, 54 (6), pp. 1122–1135, Figure 1a, doi: https://doi.org/10.1093/icb/icu076 © The Author, 2014. (d) Reprinted from *Journal of Neuroscience Methods*, 195 (2), Cinzia Metallo, Robert D. White, and Barry A. Trimmer, Flexible parylene-based microelectrode arrays for high resolution EMG recordings in freely moving small animals, pp. 176–84, doi: 10.1016/j.jneumeth.2010.12.005, Copyright © 2010 Elsevier B.V. All rights reserved, with permission from Elsevier. (c) and (e) Reproduced from *Journal of Experimental Biology*, 213 (13), Michael A. Simon, Steven J. Fusillo, Kara Colman, and Barry A. Trimmer, Motor patterns associated with crawling in a soft-bodied arthropod, pp. 2303–2309, Figures 6 and 7, doi: 10.1242/jeb.039206, © Company of Biologists, 2010.

significant changes in kinematics even when the animal transitions from horizontal crawling to a vertical climb.

Although caterpillars can increase their turgor to cantilever across large gaps in the substrate, they do not appear to use pressurization to control their movements. Presumably the extensive air-filled, and compressible, tracheal system prevents them accurately controlling pressure or maintaining a constant internal volume. One advantage of the environmental skeleton strategy is that the body can remain soft and able to conform to the substrate, a distinct advantage for animals living on their food source and usually trying to be cryptic. As soft animals caterpillars can also change their shape, even curling into a wheel to generate fast ballistic rolling as an escape reflex (Lin et al. 2011). One of the major limitations of the environmental skeleton is that it requires the substrate to be stiffer than the caterpillar itself; *Manduca* does not climb very effectively on hanging threads or materials that can buckle.

Biomimetic systems

Robotic applications of soft material technology

Compliant structures have frequently been incorporated into joints and actuators of biomimetic robots to increase their adaptability and robustness. Because these materials are "added" to a traditional design, they often create control problems requiring additional sensors or more powerful software. Attempts to build completely soft robots are rare but include designs based on peristalsis (Menciassi and Dario 2003) and "conformable wheels." However, even these machines include rigid components such as McKibben actuators with air valves, metal springs, and thermoplastic bearings. Recently, a variety of pneumatic robots have been constructed from silicon elastomers and made to move by controlling the inflation of different compartments (Shepherd et al. 2011). These devices use off-board compressed gases and electrically controlled solenoid valves to control pre-determined deformations. With the future development of onboard pressure storage and distribution systems these robots could have a promising future. These pneumatic robots are not based on a particular biological system (with the exception of spider leg joints there are very few animals that inflate their limbs to move around) and will not be discussed further here (but see Ishiguro and Umedachi, Chapter 40, this volume).

The ultimate solution to creating highly deformable moving machines is to build them entirely from soft materials. The primary challenges are in finding practical compliant actuators ("artificial muscles") and in devising effective control systems. Although all artificial muscles have significant problems, there are now a variety available (see Anderson, Chapter 20, this volume) including electroactive polymers (EAPs), ionic polymer metal composites (IPMCs), shape-memory alloys, and even muscles themselves (see Ayers, Chapter 51, this volume). Novel electroactive polymers have been applied to make gel robots resembling shrimp, lizards, seaweed, and starfish. Also, a continuous toroidal-drive actuator has been demonstrated using soft materials. Other examples might better be described as compliant robots, rather than soft-material robots, as they incorporate flexible components that undergo only small deformations or that are attached to an otherwise rigid structure.

Worm-like robots

A wide variety of worm-like robots have been developed based on peristaltic waves with a range of hard and soft actuators (Jung et al. 2007). Perhaps the most successful approach has been to use shape-memory alloy (SMA) actuators. This was pioneered in worm-like crawlers by shaping SMA wire into a coil to create forceful tensile linear actuators with large strain (over 100%) (Menciassi and Dario 2003) and the approach continues to be used in other crawling and swimming robots. Most recently the technology was developed into an impressively robust crawler called Meshworm (Seok et al. 2010). Interestingly the Meshworm uses a constant length design rather than a constant volume as would be expected in a worm; radial SMA contraction in one segment causes radial expansion of an adjacent segment. The whole structure is constrained by a mesh of crossed fibers to help direct deformations between segments. Propulsion is derived from waves of ground contacts and feedback is provided by linear potentiometers that detect the length of each segment. Using iterative learning control to maximize objective functions, the duration of each SMA actuation was adjusted to either maximize Meshworm's speed or traveling distance and energy consumption. Steering was achieved by replacing two of the passive tendons with longitudinal SMA coils. Activation of one coil shortens one side of the robot to bias its movements in that direction.

Caterpillar-like robots

Using insights gained from studies of *Manduca* locomotion, a family of caterpillar-like robots has been developed that also use SMA-coil actuators (Figure 41.3). Because these robots are not controlled by pressure, they can be constructed from almost any material that is soft enough to deform (Figure 41.3A, C). Moreover, these materials can be porous or penetrated by other components. The robots are not constrained to particular shapes and can be made in hollow or monolithic body forms. Typically they are either cast from silicone elastomer (Figure 41.3C, D) or 3D-printed directly in a flexible rubber polymer such as TangoPlus® (Objet Ltd.; Figure 41.3A, B) and require little or no assembly. SMA coils are attached to the body (Figure 41.3D) and arranged in specific orientations to bend different parts of the chassis. Locomotion is achieved by controlling frictional interactions with the ground by lifting different parts of the body or deploying retractable sticky pads. They can crawl, inch, or roll (Lin et al. 2011) and even climb

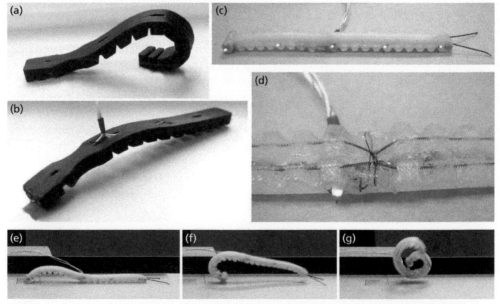

Figure 41.3 Examples of caterpillar-like robots developed to test material utility and actuation. (a) and (b) Softworm robots 3D-printed from soft elastomeric polymers can also be cast from biomaterials. (c) GoQBot is an example of a robot cast from several mixtures of silicone rubbers. (d) The Softworms are actuated with shape memory alloy (SMA) coils threaded through the body. (e–g) High speed images of GoQBot in a fast ballistic roll (frames at 40, 120, and 200 ms).

(d) Reprinted from *Bioinspiration and Biomimetics*, 6 (2), Huai-Ti Lin, Gary G Leisk, and Barry Trimmer, GoQBot: a caterpillar-inspired soft-bodied rolling robot, pp. 026007–21, doi: 10.1088/1748-3182/6/2/026007, © Copyright 2011 IOP Publishing.

steep inclines. These devices can be steered by differential activation of the left and right SMAs. The most inexpensive and robust versions are powered and controlled through fine wire tethers; however, untethered versions carrying their own lithium polymer battery and a small four channel radio receiver have also been constructed.

The development of this soft robot platform has helped to confirm that complex caterpillar gaits such as crawling and inching can be generated through simple changes in actuator timing and amplitude. Much of the movement coordination arises through coupled deformations within the body and through its interaction with the environment. In addition these caterpillar-like robots demonstrate another important attribute of highly deformable devices: they can morph to exploit other body shapes. As an example, the GoQBot (Figure 41.3C–D) has an elongated narrow body that can be deformed into a circle. When done quickly, this change releases enough stored elastic energy to produce ballistic rolling locomotion (Figure 41.3E–G). The GoQBot changes conformation within 100 ms, generating approximately 1G of acceleration and 200 rpm, enough to propel the 10-cm-long robot at a linear velocity of 200 cm/s (Lin et al. 2011).

Future directions

By understanding and exploiting soft materials in robotic applications, new avenues of research and development have started to develop. Perhaps most exciting and far reaching are the recent attempts to grow robots using a combination of synthetic materials and living cells. There have been several attempts to build hybrid devices in which living tissue is incorporated into a mechanical device. This includes dissected muscles attached to levers, electrodes implanted into intact animals, and "cyborg" machines in which animals "drive" a robot vehicle. However, most attempts to incorporate cellular systems into robots have used vertebrate tissues which have very exacting environmental requirements. For example, the long-term survival of vertebrate muscle requires a functional vascular system to deliver oxygen and glucose and to remove waste. Most vertebrate muscles are also very sensitive to changes in temperature (less than 10°C operating range) and because of their reliance on tightly regulated blood flow they are easily disrupted or killed by variations in their chemical environment.

Therefore, a key innovation is the use of insect cells, which have remarkable environmental tolerance, as building blocks for useful devices (Baryshyan et al. 2012). Insects can function at temperatures below 0°C to above 45°C and their tissues are typically at these ambient temperatures. Some insects (including *Manduca sexta*) can be frozen and still crawl and develop normally when thawed. Because they live in niche (often extreme) environments insects also have tremendous tolerance to toxic compounds, changes in pH, oxidative stress, anoxia, and radiation. Their open circulatory system (lacking blood vessels, oxygen carriers, or a lymphatic system) does not selectively regulate the local milieu of different tissues so it is relatively easy to mimic *in vitro*. Partly because of this organization, each insect tissue has a large degree of autonomy, functioning well in isolation. Other advantages relate to commercial, ethical, and pragmatic concerns. Insects are widely available as source materials and they are the most diverse group of organisms on the planet. Their use is minimally regulated (usually related to agricultural significance) and they are cheap to maintain and to harvest. Insects are among the most studied animals, particularly at the genomic level, so there are abundant tools for manipulating metabolic and developmental pathways. Finally, insect proteins are already used commercially on a massive scale (silk production) and are available as basic building blocks for the robot components.

Learning more

Details of worm crawling and modeling can be found in Quillin (1999), which describes earth-worm movements, studies of scaling issues, and references to mathematical descriptions of undulatory and peristaltic locomotion. For general biomechanics related to hydrostatic organisms the work of Kier and colleagues (Kier and Smith 1985) is highly recommended. The work on caterpillars, soft robots, and the environmental skeleton is described in a series of publications that are referenced in Lin et al. (2011) and Lin and Trimmer (2010). Readers may be interested to compare pattern generation and locomotion in slime molds, that lack any kind of nervous system, as described by Ishiguro and Umedachi (Chapter 40, this volume). Additional information is available on the websites of most of the authors cited in this article and new developments are published in journals such as *Bioinspiration & Biomimetics, Journal of Experimental Biology*, and *Soft Robotics*.

Acknowledgments

The author's work reported here was funded in part by National Science Foundation grants (IOS 1050908, DBI-1126382, IGERT 1144591) and DARPA Army W911NF-11-1-0079.

References

Baryshyan, A. L., Woods, W., Trimmer, B. A., and Kaplan, D. L. (2012). Isolation and maintenance-free culture of contractile myotubes from *Manduca sexta* embryos. *PLoS ONE*, 7, e31598.

Jung, K., Koo, J. C., Nam, J. D., Lee, Y. K., and Choi, H. R. (2007). Artificial annelid robot driven by soft actuators. *Bioinspiration and Biomimetics*, 2, S42–S49.

Kier, W. M.,and Smith, K. K. (1985). Tongues, tentacles and trunks - the biomechanics of movement in muscular-hydrostats. *Zoological Journal of the Linnean Society*, 83, 307–24.

Lin, H.-T., Leisk, G., and Trimmer, B. A. (2011). GoQBot: A caterpillar-inspired soft-bodied rolling robot. *Bioinspiration and Biomimetics*, 6, 026007-21.

Lin, H.-T., and Trimmer, B. A. (2010). The substrate as a skeleton: ground reaction forces from a soft-bodied legged animal. *Journal of Experimental Biology*, 213, 1133–42.

Menciassi, A., and Dario, P. (2003). Bio-inspired solutions for locomotion in the gastrointestinal tract: background and perspectives. *Philos Transact A Math Phys Eng Sci*, 361, 2287–98.

Metallo, C., White, R. D., and Trimmer, B. A. (2010). Flexible parylene-based microelectrode arrays for high resolution EMG recordings in freely moving small animals. *Journal of Neuroscience Methods*, 195 (2), 176–84. doi: 10.1016/j.jneumeth.2010.12.005

Quillin, K. J. (1999). Kinematic scaling of locomotion by hydrostatic animals: ontogeny of peristaltic crawling by the earthworm *Lumbricus terrestris*. *Journal of Experimental Biology*, 202, 661–74.

Seok, S., Onal, C. D., Wood, R., Rus, D., and Kim, S. (2010). Peristaltic locomotion with antagonistic actuators in soft robotics. *2010 IEEE International Conference on Robotics and Automation (Icra)*, 1228–33.

Shepherd, R. F., Ilievski, F., Choi, W., Morin, S. A., Stokes, A. A., Mazzeo, A. D., Chen, X., Wang, M., and Whitesides, G. M. (2011). Multigait soft robot. *Proc. Natl Acad. Sci. USA*, 108, 20400–3.

Simon, M. A., Fusillo, S. J., Colman, K., and Trimmer, B. A. (2010). Motor patterns associated with crawling in a soft-bodied arthropod. *Journal of Experimental Biology*, 213, 2303–9.

Trimmer, B. A., and Lin, H.-T. (2014). Bone-free: soft mechanics for adaptive locomotion. *Integrative and Comparative Biology*, 54(6), 1122–35. doi: https://doi.org/10.1093/icb/icu076

Trueman, E. R. (1975). *The locomotion of soft-bodied animals*. London: Edward Arnold.

Wainwright, S. A. (1988). *Axis and circumference*. Cambridge, MA: Harvard University Press.

Chapter 42

Principles and mechanisms learned from insects and applied to robotics

Roger D. Quinn[1] and Roy E. Ritzmann[2]

[1] Mechanical and Aerospace Engineering Department,
Case Western Reserve University, USA
[2] Biology Department, Case Western Reserve University, USA

Insects have been providing inspiration for robotic designs for decades as biologists continue to learn more about these complex invertebrate systems. Initial legged robots typically mimicked insects only in terms of their basic six-legged designs and walking gaits. Since then robots have been developed that take advantage of insect leg and wing designs, compliant structures, movement behaviors, reflexes, and even local neural control systems identified in their central nervous systems. Future robots may be controlled with models of entire insect nervous systems including their brains.

Insects are attractive models for robotic development. Many of their behaviors and capabilities are desirable to enable robots to perform a variety of missions in urban and natural environments. Insects that can fly, swim, or walk, or perform a combination of these behaviors, can move from a starting location to a goal while overcoming a variety of obstacles along the way. There are beetles that can climb walls and trees and walk across ceilings. Grasshoppers and crickets can jump up steps and across crevices many times their body lengths. Flies can hover in the air and fly in any direction. Moths can fly to track and locate an odor source tens of meters away through woodlands. Engineers are modeling the neuromechanical principles responsible for these insect capabilities and implementing them in robots.

In this chapter we limit the discussion to some advances in ground robots that were inspired by insects because doing justice to all of the insect robot literature would require an entire volume. In particular we describe some robots that have been designed taking advantage of cockroach and stick insect locomotion and behavioral mechanisms. Then we discuss some controllers that are based upon these insects' behaviors and neural pathways. Finally, we describe perceived future robot capabilities based upon the current direction of insect neuroscience.

Simplified robots implementing cockroach locomotion principles

Insects are complex neuromechanical systems far more intricate than any existing robot. For example, cockroach legs have seven active degrees of freedom and many more passively compliant joints in their tarsi (feet). There are more than 10^5 neurons in their central nervous systems including their thoracic ganglia (akin to a vertebrate spinal cord) and their brain. Software and hardware models of insects are increasing in fidelity as more is learned.

Because of this daunting complexity there is a class of robots that have been engineered to take advantage of fundamental insect locomotion principles and mechanisms rather than their detailed leg designs and neural systems. These principles include the basic six-leg design which can provide a large support polygon for robust static stability. A mechanism that is important for their dynamic stability is compliance in their joints and tarsi that results in mechanically compliant legs despite their relatively stiff exoskeleton leg segment structures. Cockroaches have been shown to use this compliance to reject dynamic disturbances (Jindrich and Full 2002). The fundamental leg cycle is another important principle of locomotion. The leg cycle consists of the stance phase, which supports the body, and the swing phase that returns the foot in preparation for the next stance phase. Insect gaits are well known and include the alternating tripod gait that is normally used by the cockroach where the front and rear legs on one side of the body move in synch with the middle leg on the contralateral side. However, this coordination pattern with the leg pairs out of phase may change when the animal moves over irregular terrain. For example, cockroach leg pairs have been shown to move in phase when they climb steps. Cockroaches also have a joint in their thorax between their middle and front legs that they use to prevent high centering when climbing obstacles. Also, when approaching a step in preparation for climbing, they rotate a joint in their middle legs that causes these legs to extend and pitch the body upward. Many of these mechanisms and principles of locomotion have been implemented in robots with relatively simple designs with good results.

RHex robots are remarkably agile six-legged vehicles because they benefit from several important locomotion principles observed in cockroaches (Saranli et al. 2001). They have just six motors, one per leg, that spin their feet in a circle. Each leg is rotated in the stance phase at the desired robot speed and then is rapidly rotated in the same direction through the swing phase to return it to the stance phase. This is an unusual but effective way of implementing the fundamental leg cycle. RHex's legs are also designed to be passively compliant in the radial direction to model compliance observed in cockroach legs with similar gains in stability. RHex's six motors are controlled to organize its legs into insect gaits including the alternating tripod gait and a gait where the leg pairs move in phase for climbing stairs. Using these cockroach locomotion principles, RHex robots have been shown to walk and run through irregular terrain, up and down stairs, and even jump (Johnson and Koditschek 2013).

Whegs robots follow from their RHex precursor and also benefit from a number of cockroach locomotion principles but with even fewer motors and a greater use of passive degrees of freedom (Quinn et al. 2003). To simplify their control systems and reduce the number of motors, Whegs use a device called a wheel-leg. A wheel-leg has a hub and multiple evenly spaced spokes with a foot at the end of each spoke. Each spoke can act as a leg and the fundamental leg cycle is fulfilled by simply rotating the wheel-leg at a constant speed. As one spoke completes its stance cycle another spoke can enter the stance phase. The standard Whegs robot has just one propulsion motor which drives all of its six wheel-legs through a system of sprockets and chains. The standard wheel-leg has three spokes. The wheel-legs are assembled on the robot so that the leg pairs are out of phase resulting in the robot walking in a tripod gait with just one propulsion motor.

Compliant mechanisms are used to enable a Whegs robot to change its gait rather than adding more motors to accomplish this task. The robot described above cannot change from its tripod gait. In order to do so, devices with torsional springs are installed in its hubs. The input axle driven by the motor applies torque to one side of the spring and the wheel-leg is driven by the other side of the spring. The rotation of the springs is mechanically limited and the springs are pre-tensioned so that the springs do not rotate until a particular threshold value of torque is applied between the input axle and the environment. Therefore, a Whegs robot walks in a tripod gait on flat horizontal surfaces but terrain disturbances perturb its gait so that it passively adapts its gait to the environment. When walking uphill, more than three legs will be in stance and when climbing steps its leg pairs move in phase as observed in cockroach climbing.

Whegs robots also have an actuated joint that allows them to bend their bodies, which requires a second motor. For design simplicity, the axle for this "body-joint" is co-located with the middle leg axle. The Whegs body joint is used to implement two cockroach locomotion principles. First, it can bend downward to help the robot to avoid high centering when it climbs. Second, it can bend upward to raise the front end of the robot so the front legs can reach on top of taller obstacles, which the cockroach accomplishes by pitching its entire body upward. A Whegs robot with its body joint frozen climbs rectangular obstacles that are 1.58 times its leg length, whereas the same robot using the body joint to its best advantage can climb obstacles that are greater than twice the leg length (Figure 42.1). We found that springy compliance in series with the body-joint motor both reduces mechanical failure in the transmission and effects a more even leg loading (Boxerbaum et al. 2008).

Simplified robots turn less elegantly than their animal models. Cockroaches' rear legs drive them forward while the front and middle legs rotate to direct the animal's motion. A front leg on the side toward the turn reaches out in swing and pulls inward in stance whereas the contralateral leg does the opposite. Because each of RHex's legs is driven by a separate motor, it can be skid-steered. With only the two motors described above, Whegs could not turn. To accomplish turning, Whegs use two small servo motors, electrically coupled, to rotate their front and rear wheel-leg pairs in opposite directions.

Figure 42.1 A still from a video in which DAGSI Whegs uses its body flexion joint to climb an obstacle. Note that its front legs are in-phase and both are supporting and lifting the body whereas the middle legs are not loaded and, therefore, are in their normal out-of-phase orientation.

The implementation of cockroach locomotion principles in Whegs robots provides them with surprising agility despite their simplicity. Whegs robots have climbed and run through rubble piles and through orchards with high weeds and fallen branches. They have climbed tall obstacles and run up and down stair cases. Palm size versions called "Mini-Whegs" with four wheel-legs have set speed records (in body lengths per second for a legged robot), jumped, flown, and climbed glass walls. These capabilities are a result of the incorporation of insect locomotion principles into their designs. Furthermore, their local control system is embedded into their mechanics, allowing them to passively adapt to their environments.

One criticism of RHex and Whegs are that because they spin their feet in a circle, they are "just wheeled vehicles." This is not true for two reasons. First, their legs can apply a normal force to climb a rectangular obstacle taller than their leg length, which is not possible with wheels, which must rely upon friction in this situation. Second, their gait matters for their stability and agility, whereas gait has no meaning for a wheeled vehicle and an ordinary wheeled vehicle cannot change its posture to improve its stability.

Robots with insect-inspired legs

Insect legs provide much greater agility than the simplified legs of RHex and Whegs. Insects can reach and place their feet on targeted locations so that they can walk through sparse terrain, turn efficiently, and change their posture to enhance stability and crawl through small crevices. Multi-segmented legs are necessary to afford robots these capabilities.

Dozens of insect-inspired robots with multi-segmented legs have been developed. For brevity we shall describe just a few. Most of these robots have six identical legs, which is not the case for any insects.

Robot II has six identical legs, each with three actuated degrees of freedom and a springably compliant tarsus. Its design was not based upon any particular insect. It was notable for its control system that combined a stick insect-inspired gait generation network and local leg reflexes. It could walk forward and backward across sparse terrain using its searching reflex, comply with and walk over irregular terrain, adapt its leg movement to swing over obstacles using its elevator reflex, turn, and walk reflexively when disturbed. A number of robots since Robot II have performed these movements with similar configurations but with much more refined and robust platforms using more modern technology.

Robot III (Figure 42.2) was one of the first robots to model a particular insect (the cockroach *Blaberus discoidalis*) at the level of its leg kinematics (Nelson and Quinn 1999). The leg segment lengths of a cockroach were scaled up by a factor of 17. The rear legs are the largest and the most powerful, the middle legs are smaller, and the front legs are the smallest. One degree of freedom at each of the thorax-coxa, coxa-femur, and femur-tibia joints on its rear legs are the same as those used by the cockroach to drive itself forward. The middle legs on the robot have those same joints and an additional degree of freedom at the thorax-coxa. These were the joints we then thought were most important for cockroach middle legs. However, since then we have discovered that a trochanter-femur joint is also important in these legs. The front legs on the robot, as in the insect, are the most agile and include the same joints as the middle leg plus a third degree of freedom at the thorax-coxa. All of the robot's joint ranges of motion were also modeled after those of the cockroach.

Robot III's control system was notable for two accomplishments. First, the robot could cycle its legs in a cockroach-like manner moving each of its joints through movements similar to those observed in the cockroach and forming the tripod gait. The leg pairs performed the same functional movements as in the insect. The rear legs' tarsi began stance behind their thorax-coxa joint, the middle legs' tarsi started in front of their thorax-coxa joint and cycled through to

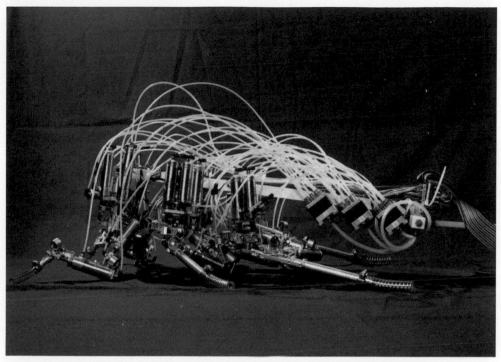

Figure 42.2 The legs of Robot III model those of *Blaberus discoidalis* cockroach. The front legs are the smallest and most dexterous while the rear legs are the largest and most powerful.

behind it, and the front legs' tarsi remained in front of their thorax-coxa joint throughout their leg cycle. Second, the robot was capable of postural stability despite large disturbances.

HECTOR (Figure 3.6, this volume) is a 20 times scaled up morphologically detailed model of the stick insect *Carausius morosus* (Schilling et al. 2013). In addition to modeling the segments, degrees of freedom, attachment positions, and orientations and functionality of the legs, HECTOR also copies the thorax structure of the insect. It has three thorax (pro-, meso-, and metathoracic) segments to which the front, middle, and rear leg pairs are attached, respectively. Actuated joints between each of these body segments allow for up/down and right/left rotations mimicking these degrees of freedom in the insect. Its leg joints are driven by electric motors through rotary springs (series elastic actuators) to mimic the insect's compliant joints and legs. The joints of its legs and its gait will be coordinated by an artificial neural network called Walknet. Walknet has been trained to incorporate decades of invertebrate locomotion data, and displays several adaptive behaviors, including directional walking, local reflexes, and elementary navigation (Schilling et al. 2013).

Local neural control systems

It is now possible to control robot legs using models of insect local neural systems that are strongly rooted in biology. Previous insect robots were typically controlled with finite state machine implementations of behaviors and reflexes observed in insects because the neurobiology was not known in sufficient detail. Neuroscientists have now determined the architecture of the neural system and many of the pathways responsible for the coordination of the joints in a leg. Models of local neural systems have been developed and been used to control individual legs and entire insect robots.

Each joint in an insect leg is controlled by its own central pattern generator (CPG) which is modulated by sensory signals (Ekeberg et al. 2004; see Cruse and Schilling, Chapter 24, this volume, for a description of CPGs). The joint CPGs in a leg are not directly coupled. Instead, the joints of a leg are coordinated through sensory inputs. For example, joint movements associated with stance are encouraged by loading and an angle threshold helps to trigger swing movements. A number of pathways have been identified for stance, swing, and turning movements.

Models of this neural system have been shown to be sufficient to cause a robotic leg to mimic insect leg behaviors. When it was first modeled as a finite state machine network with no CPGs and just discrete event sensory switching, it could become stuck in a state waiting for an event that might never happen. Biological neural network (BNN) implementations do not have this problem (Daun-Gruhn 2010; Szczecinski et al. 2014). The continuous time dynamics of the CPGs cycle the joints to avoid them getting stuck and have the additional benefit of smoothing motions. In fact, BNN models have shown how a small amount of information descending from higher centers can cause the local neuromechanical system to smoothly transition between walking and turning movements.

Autonomy

Robots with even the most streamlined and agile legs, body joints, sensors, local control systems, and efficient actuation and power systems are still of limited use without the capacity to perform movements autonomously. To perform a significant mission a robot should be able to move from one location to another while avoiding obstacles without a human operator specifying detailed motions. A higher-level system or brain must gather information about the robot's immediate surroundings and its own state, integrate and evaluate the quality of this information, and make sensible, context-dependent decisions. Insects are also excellent models for the development of these autonomous systems.

Within the insect brain, the central complex (CC) appears to play a role in matching behavioral choice to current internal and external conditions (Ritzmann et al. 2012; Strausfeld and Hirth 2013). Neurons within the central complex monitor massive amounts of sensory information, including visual and antennal cues, and alter their firing patterns in anticipation of changes in movement (Guo and Ritzmann 2013). Using a range of techniques on numerous insect models, neuroscientists are working to decipher the function of the CC. These include genetic tools and electrophysiological recordings performed as the animals move. In the future these data will be used to develop context-dependent models of the CC, which can then be used to control the autonomous movements of robots; these can be coupled to biomimetic sensors such as those inspired by the insect compound eye (see Dudek, Chapter 14, this volume).

In the meantime, behavioral studies and modeling provide insight into how insects make decisions. These techniques have been used to show that cockroach wall-following behavior can be modeled using a PID control system (Lee et al. 2008). In another study, videos of cockroaches walking in a lit rectangular arena with a small shaded shelter in one location were digitized and then modeled using a stochastic finite state machine. The resulting agent's decisions regarding when to follow a wall, turn, and explore, and its affinity for the shelter and time to reach the shelter are statistically similar to the animal (Daltorio et al. 2013).

Discussion

There are two major questions in the field of bio-inspired robotics. First, after more than a quarter century of work, has robotics benefited from animal data? We believe so. One could argue that RHex and Whegs could have been designed without cockroach inspiration. But, the point is they had *not* been. One can always say after the fact that animal inspiration was not necessary, but the fact is that, in these cases, observations of insect behavior pointed the way and these robots were the result. As another example, the advantages animals gain from their compliant mechanisms have compelled roboticists to develop muscle-like actuators such as series-elastic and various polymer types. This work may eventually lead to energy-efficient legged robot designs. Yet, it must be admitted that development of robots that capture the detail of insect leg design and neural control has been a challenge, with most of the mission-capable robots currently coming from more abstracted designs that only superficially resemble insect architecture. It is possible that in the near term properties of insect neural control, such as that being studied in the brain as well as goal-seeking behavior in complex situations (Daltorio et al. 2013), will play a more important role in vehicles that bear little resemblance to insects. These could include wheeled, treaded, or even flying drone devices where autonomous control would benefit from the context-dependent properties of insect behavior that in the animal are imparted by brain circuitry.

The second question is: Have scientists gained from working with engineers? Again, we believe so. Models of animal systems that engineers develop are being used in investigations to test scientific hypotheses. In this case software models are often more practical than hardware models because software models can be developed and manipulated relatively rapidly as compared to their hardware counterparts and they are not restricted by available components. However, demonstrations in hardware are valuable to verify simulation results.

Learning more

Beer et al. (1998) and Quinn et al. (2003) provide further reviews of biomimetic approaches to the design of insect-like robots. Morrey et al. (2003) discuss the extension of the Wheg design towards small robots capable of robust and fast locomotion. A broader discussion of the role of the insect brain in controlling walking is provided by Ritzmann et al. (2012) and Martin et al. (2015).

References

Beer, R. D., Chiel, H. J., Quinn, R. D., and Ritzmann, R. E. (1998). Biorobotic approaches to the study of motor systems. *Current Opinion in Neurobiology*, 8(6), 777–82. doi:https://doi.org/10.1016/S0959-4388(98)80121-9

Boxerbaum, A. S., Oro, J., Peterson, G., and Quinn, R. D. (2008). The latest generation Whegs™ robot features a passive-compliant body joint. In: *Proceedings of the IEEE/RSJ International Conference on Intelligent Robots and Systems (IROS'08)*, Nice, France.

Daltorio, K. A., Tietz, B. R., Bender, J. A., Webster, V. A., Szczecinski, N. S., Branicky, M. S., Ritzmann, R. E., and Quinn, R. D. (2013). A model of exploration and goal-searching in the cockroach, *Blaberus discoidalis*. *Adaptive Behavior*, 21(5), 404–20.

Daun-Gruhn, S. (2010). A mathematical modeling study of inter-segmental coordination during stick insect walking. *J. Comput. Neurosci.*, 30. 255–78.

Ekeberg, O., Blumel, M., and Bueschges, A. (2004). Dynamic simulation of insect walking. *Arthropod Structure and Development*, 33(3), 287300.

Guo, P., and Ritzmann, R. E. (2013). Neural activity in the central complex of the cockroach brain is linked to turning behaviors. *J. Exp. Biol.*, 216, 992–1002.

Jindrich, D. L., and Full, R. J. (2002). Dynamic stabilization of rapid hexapedal locomotion. *J. Exp. Biol.*, 205, 2803–23.

Johnson, A. M., and Koditschek, D. E. (2013). Toward a vocabulary of legged leaping. *Proceedings of the 2013 IEEE Intl. Conference on Robotics and Automation*, May 2013, pp. 2553–60.

Lee, J., Sponberg, S. N., Loh, O. Y., Lamperski, A. G., Full, R. J., and Cowan, N. J. (2008). Templates and anchors for antenna-based wall following in cockroaches and robots. *IEEE Transactions on Robotics* (Special Issue on Bio-Robotics), **24**(1), 130–43.

Martin, J. P., Guo, P., Mu, L., Harley, C. M., and Ritzmann, R. E. (2015). Central-complex control of movement in the freely walking cockroach. *Current Biology*, **25**(21), 2795–803. doi:https://doi.org/10.1016/j.cub.2015.09.044

Morrey, J. M., Lambrecht, B., Horchler, A. D., Ritzmann, R. E., and Quinn, R. D. (2003). Highly mobile and robust small quadruped robots. In: *Proceedings 2003 IEEE/RSJ International Conference on Intelligent Robots and Systems* (IROS 2003) (Cat. No.03CH37453), vol.1, pp. 82–87.

Nelson, G. M., and Quinn, R. D. (1999). Posture control of a cockroach-like robot. *IEEE Control Systems*, **19**(2), 9–14.

Quinn, R.D., Nelson, G.M., Ritzmann, R.E., Bachmann, R.J., Kingsley, D.A., Offi, J.T., and Allen, T.J. (2003). Parallel strategies for implementing biological principles into mobile robots. *Int. Journal of Robotics Research*, **22**(3), 169–86.

Ritzmann, R., Harley, C., Daltorio, K., Tietz, B., Pollack, A., Bender, J., Guo, P., Horomanski, A., Kathman, N., Nieuwoudt, C., Brown, A., and Quinn, R. (2012). Deciding which way to go: how do insects alter movements to negotiate barriers? *Front. Neurosci.*, **6**(97). doi:10.3389/fnins.2012.00097

Saranli, U., Buehler, M., and Koditschek, D. (2001). RHex a simple and highly mobile hexapod robot. *Int. J. Robotics Research*, **20**(7), 616–31.

Schilling, M., Paskarbeit, J., Hoinville, T., Hüffmeier, A., Schneider, A., Schmitz, J., and Cruse, H. (2013). A hexapod walker using a heterarchical architecture for action selection. *Frontiers in Computational Neuroscience*, 7. https://doi.org/10.3389/fncom.2013.00126

Strausfeld, N. J., and Hirth, F. (2013). Deep homology of arthropod central complex and vertebrate basal ganglia. *Science*, **340**, 157–61.

Szczecinski, N. S., Brown, A. E., Bender, J. A., Quinn, R. D., and Ritzmann, R.E. (2014). A neuromechanical simulation of insect walking and transition to turning of the cockroach *Blaberus discoidalis*. *Biol. Cyber.*, **108**(1), 1–21. doi: 10.1007/s00422-013-0573-3

Chapter 43

Cooperation in collective systems

Stefano Nolfi

Laboratory of Autonomous Robots and Artificial Life, Institute of Cognitive
Sciences and Technologies (CNR-ISTC), Rome, Italy

Living systems rarely operate in isolation. By interacting with other individual entities, molecules, cells, animals, and humans are able to achieve behaviors, functionalities, structures, and other properties that are not achievable by individuals acting alone. In this chapter we will illustrate the key principles and properties that characterize natural collective cooperating systems and methods through which artificial collective systems displaying analogous properties can be synthesized. In doing that we will focus in particular on collective systems formed by colonies of cooperating individuals such as social insects and collective robotics systems.

Biological principles

Cooperative collective systems, such as social insects, display amazing behaviors that combine efficiency, flexibility, and robustness. For example, leafcutter ants (*Atta* spp.) forage for leaves hundreds of meters away from their nest by organizing highways to and from their foraging sites. Similarly, fungus-growing termites (*Macrotermes* spp.) build large air-conditioned nest structures which include thick protecting walls and a labyrinth of ventilation ducts (Bonabeau et al. 1999). What is particularly fascinating is that such complex behaviors and structures are produced by colonies of individuals that are relatively simple. The whole system thus is much more than the sum of its parts.

These phenomena, that have puzzled naturalists and philosophers for years, can be explained as the result of few fundamental principles.

The first principle is *emergence*, namely the fact that the behavior of agents that are embodied and situated emerges from numerous interactions between the individuals and the environment. At any time-step the perceived environment co-determines the action produced by the individual that in turn modifies the physical environment and/or the individual–environment relation that determines the next perceived environmental state. A sequence of these bi-directional interactions gives rise to a dynamical process, a behavior, that cannot be traced back to the characteristics of the individual only and that consequently can exceed the complexity of the individual.

The emergent nature of behavior stated above characterizes either solitary or social individuals but assumes a stronger role in the latter case due to the fact that social individuals do not only interact with a physical but also with a social environment constituted by other cooperating, and eventually competing, individuals (Nolfi 2009). To illustrate this point, let me consider the case of actions that alter the physical environment. Individuals might modify their physical environment to self-simplify their own subsequent activities. For example, humans tend to annotate their planned activities on calendars to eliminate the need to mentally remember the corresponding information. In a social context, however, a similar action can be used also to coordinate the activity of a working team. Indeed, social insects widely exploit this kind of strategy to coordinate their behavior. For example, the coordination of individuals during nest construction is realized by exploiting the fact that construction actions, e.g. the release of oil pellets over a specific part of the building nest, produce modifications of the perceived environment of the other individuals that can be used to later trigger the appropriate consequent construction actions. Similarly, the coordination necessary to discover and maintain a navigation path between the nest and a distant food source is realized by releasing chemical substances over the terrain and by reacting to sensed chemicals in a way that enables the progressive formation and maintenance of a chemical trail between the two distant locations.

The second principle is *self-organization*, i.e. the spontaneous formation of spatial, temporal, or spatiotemporal structures or functions in systems composed of many components or individuals (see Camazine et al. 2001; Wilson, Chapter 5, this volume). This property refers to a broad range of pattern-formation processes occurring both in physical and in biological systems that occur on the basis of interactions internal to the system without intervention by external directing influences. Self-organization is thus a specific type of emergent process that results from the interactions occurring between the components or individuals forming the system. The rules that regulate the interactions are based on local information, without reference to the global pattern that, therefore, cannot be considered as a property imposed by external factors.

It is important to consider that any collective system operating on the basis of any possible given set of interacting rules will tend to display some form of organization since, as the result of the interactions between the system's elements, it will tend to display only a small subset of all the possible patterns. What is interesting from a biological (and engineering) point of view is the fact that, when the characteristics of the elements are appropriately shaped and/or the characteristics of the rules that regulate the interactions are appropriately set, the system can display *purposeful, self-regulatory*, and *robust* organizations, i.e. organizations that achieve useful functionality, are able to reconfigure in changing conditions, and are able to keep functioning in spite of perturbations. A well-studied case of self-organization exhibiting these properties is constituted by the pheromone trail pattern dynamically generated by colonies of interacting ants that enable them to effectively navigate toward food sources. The pheromone trail pattern tends to spontaneously restructure itself as a result of significant environmental variations, e.g. as a consequence of the depletion of a food source, as well as to rearrange as a result of minor perturbations, e.g. as a consequence of the occlusion of a small passage.

The third principle that characterizes collective systems is their *distributed nature* and their ability, as a whole, to achieve behaviors and functionalities that are not achievable by the individuals forming the system. This aspect provides several advantages with respect to centralized and monolithic systems. Indeed it enables the solution of complex problems through the coordination and cooperation of individuals with relatively simple sensory–motor systems and computational capabilities. It enables the system as a whole to adapt its morphology to different conditions through *self-assembling*, i.e. through the formation of temporary structures obtained by the physical assembling and disassembling of individuals forming the system (see e.g. Ishiguro and Umedachi, Chapter 40, this volume). It enables the realization of capacities that are robust with respect to failures of parts/individuals since other elements might play the required role. Finally, it might enable the realization of solutions that are scalable with respect to the number of individuals involved and the complexity of the problem.

Biomimetic systems

The possibility of synthesizing artificial collective systems displaying the properties discussed in the previous section is obviously very appealing from an engineering point of view. Although the first attempts to design systems of this kind are relatively recent, the first concrete demonstrations of the idea have been realized and progress has been made toward the elaboration of the theoretical and methodological foundation of this research area (Kernbach 2012). Indeed, according to Frei and Di Marzo Serugendo (2011), this line of research is leading to the development of a radical new engineering method that has been named Complexity Engineering.

In this section I will review two examples of artificial collective systems that display some of the properties described above. Each system is constituted by a swarm of autonomous robots designed through an evolutionary technique. I will briefly discuss methodological issues in the concluding section. Some other examples of collective behavior in artificial systems are described elsewhere in this handbook. For instance, Moses and Chirikjian (Chapter 7, this volume), focus on reproduction and self-assembly, and Wortham and Bryson (Chapter 33, this volume) focus on communication. Wilson (Chapter 5, this volume) also provides a general introduction to self-organization in distributed systems.

Self-organized path formation in a swarm of robots

The first example that I would like to illustrate concerns a swarm of mobile robots that have been evolved for the ability to forage in an unknown environment (Sperati et al. 2011). This experiment shows how a swarm of cooperating robots can exploit self-organization properties to effectively accomplish a navigation task. More specifically, in the experiment, a team of 10 ePuck robots (Figure 43.1a) has been evolved for the capacity to find two distant target areas (Figure 43.1d), situated at variable distance, and efficiently navigate between them.

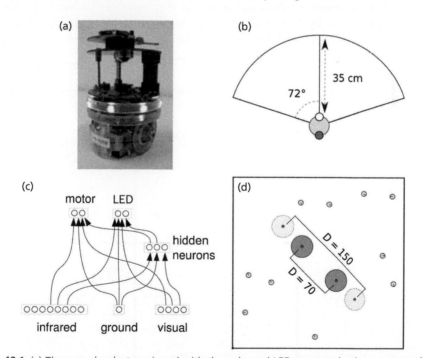

Figure 43.1 (a) The e-puck robot equipped with the coloured LED communication turret and the omni-directional camera. (b) A schematic representation of the robot body and camera sensors. The blue and red LED positions are indicated respectively as a white and a gray dot on the robot's body. (c) The neural network architecture. (d) Snapshot of the simulated environment. The two gray disks represent the circular target areas, with a red light in the center. The distance between the centers varies from trial to trial in the range [70, 150] cm.

Each robot is provided with a neural network controller (Figure 43.1c) that includes eight sensory neurons that encode the state of the corresponding infrared sensors, distributed uniformly around the robots' body, a ground sensory neuron that encodes the state of an infrared sensor located under the frontal portion of the robots' body, and four visual sensors that encode the intensity of red and blue color perceived on the frontal-left and frontal-right portion of the robot's visual field (Figure 43.1b). The network also includes three internal neurons, two motor neurons that encode the desired speed of the robot's two wheels, and two light actuator neurons that determine whether the blue and red LEDs situated on the frontal and rear side of the robot are turned on or off.

The circular target areas are marked by the gray color of the ground and by a red LED located at their center that is always turned on. The connection weights, biases, and time constants of the neural controllers are evolved (Trianni and Nolfi 2012). The genotype of each evolving individual is translated into N identical neural controllers (i.e. the group of robots is homogeneous) that are embodied into N corresponding robots. Robots are evaluated for their ability to move back and forth between the two target areas as quickly as possible. For more details see Sperati et al. (2011).

Since no explicit map of the environment is available, and since the robots' sensory range is limited (i.e. targets can be perceived only from a short distance), the robots have to explore the environment to find the target areas. Moreover, to navigate efficiently back and forth between the two areas they need to know their relative positions.

The analysis of the evolved robots indicates that they solve the problem by creating a dynamic path, constituted by the robots themselves, and by using the path to navigate (Figure 43.2). The robots organize in two lines and keep moving between the two areas in opposite

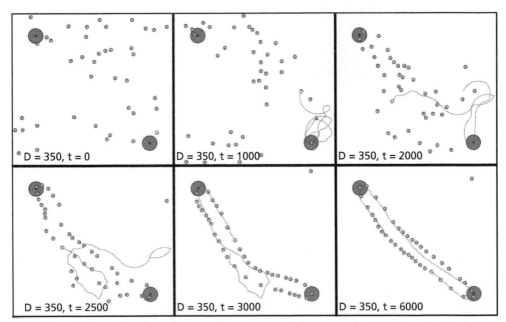

Figure 43.2 Six snapshots taken during a successful trial with 40 robots. It is possible to notice the chain formation and optimization through time. The trajectory of a single robot within the group is also shown as a gray line.

directions. This dynamic path forms as a result of the interactions among the robots, mediated by simple rules that regulate how each individual reacts to the local information provided by environmental and social cues. Once this collective structure is formed, it influences the local interactions among the robots in a way that ensures that the path is self-sustained, and that each robot moves directly toward the next target area. In other words, the formation of the dynamic path allows each individual robot to effectively travel back and forth between the two areas and allows the swarm, as a whole, to preserve the information on the location of the two areas.

We refer to this structured spatio-temporal pattern formed by the robots as a dynamic chain. The term "dynamic" well illustrates two interesting features of this structure. First, each robot within the chain is not static, but moves continuously along it, swinging between the target areas as requested by the fitness function. Second, the chain connecting the two targets adapts its shape according to the current distance D between areas: the chain direction is optimized, choosing the shortest path between the two areas (a straight line in our setup), and the inter-robot distance varies to fit all robots in the chain.

The robots produce large circular trajectories and move in a straight line toward target areas, as soon as they start to perceive them, then producing a tight U-turn once the areas are reached. Moreover, the robots avoid nearby obstacles, always maintaining their frontal blue LED on, and dodge on the left when they perceive a blue light ahead. This simple action set constitutes the basic mechanism for the formation of dynamic chains. Indeed, the repeated interactions regulated by the dodging behavior lead to the formation of chains of robots moving in line. The production of straight movements toward nearby areas and the production of a U-turn once the area is reached combined with the left-dodging behavior leads to the formation of double chains of robots that move to and from target areas. These multiple chains might become instable and fall apart, as a result of the interference caused by robot–robot avoidance behaviors, or join so as to form a single chain connecting the two target areas. The progression toward the formation of a single chain connecting two areas is ensured by the fact that chains of robots unconnected to target areas are more unstable than chains moving to and/or from a single target area and by the fact that the latter chains are more unstable than chains connecting two target areas. The instability of chains is also essential for enabling the robots to explore the environment until a single chain connecting the two targets is formed. Indeed, the partial instability of incomplete chains leads to a progressive deformation of their trajectory in space that enables sub-chains to enter into contact and to merge, and single chains to progressively explore new portions of the environment to reach still unconnected target areas.

Coordination and behavior generalization in swarm-bots

In this section I will review an experiment that shows how the behavior that emerges from robot–environment interactions can display a multi-level and multi-scale organization that supports generalization at the level of behavior (i.e. the spontaneous production of behavioral capabilities). The example also illustrates how self-assembling, i.e. the physical attachment of different individuals, can support a tight form of interaction and coordination.

In the experiment a swarm-bot, i.e. a group of physically assembled robots, have been evolved for the ability to locomote in a coordinated manner and to move toward the light (Baldassarre et al., 2006, 2007). Each robot (Figure 43.3) is a fully autonomous individual formed by two parts that can rotate one with respect to the other. The bottom part includes the wheels that allow the robot to move. The top part includes the gripper that allows the robots to attach to other individuals. The robots are also provided with infrared sensors that detect the presence

Figure 43.3 A swarm-bot formed by four robots assembled into a linear structure.

of nearby objects, light sensors that measure the intensity of light, and a traction sensor that measures the traction that the top part exerts on the bottom part.

Once the robots are self-assembled, they have to tightly coordinate their direction of movement to locomote. This means that they have to negotiate a common direction of movement when they are initially oriented in significantly different way, they have to compensate for misalignments that originate during motion, and they have to accommodate individuals' attempts to vary the direction of movement with respect to the overall direction of the group.

By evolving robots assembled into a linear structure for the ability to locomote toward a light target in an empty arena, we observed that they manage to successfully solve their task in a robust and effective way. Interestingly, by post-evaluating the evolved robots in different environmental conditions we observed how they are able to generalize their skills and to spontaneously display new behavioral capabilities that were never observed or rewarded during the evolutionary process. Indeed, when the evolved robots are assembled into a circular structure and situated in the maze environment shown in Figure 43.4, the swarm-bot displays a capacity to move in a coordinated manner, to avoid obstacles, to explore the arena, to adapt its shape so to overcome narrow passages, and to finally reach the light target.

The analysis of the evolved solutions indicates that such remarkable generalization capacities are an effect of the multi-level and multi-scale organization of the robots' behaviors. Indeed, when we analyze the evolved robots we can see how, first of all, they display the following four low-level behaviors that emerge from robot–environment interactions mediated by simple control rules:

1. A *move-forward behavior* that consists of the individuals' ability to move straight when the robot is coordinated with the rest of the group, is oriented toward the direction of the light gradient (if any), and does not collide with obstacles. This behavior results from the combination of: (a) a control rule that sets the desired speed of the two wheels to the maximum positive value when the perceived traction has a low intensity and when the intensity of the light perceived on the left and right side of the robot is comparable, and (b) the sensory effects of the execution of such action, mediated by the external environment, that does not alter the robots' sensory state until the conditions for the exhibition of this behavior hold.

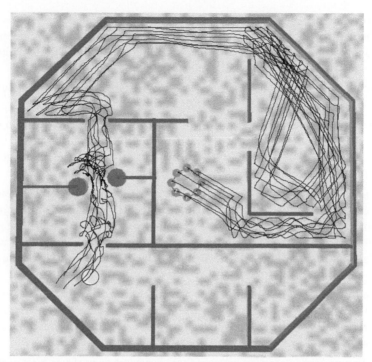

Figure 43.4 The behavior produced by eight robots assembled into a circular structure in a maze environment including walls and cylindrical objects (represented by gray lines and circles). The robots start in the central portion of the maze and reach the light target located in the bottom-left side of the environment (represented by an empty circle) by exhibiting a combination of coordinated-movement behaviors, collective obstacle-avoidance, and collective light-approaching behaviors. The irregular lines, that represent the trajectories of the individual robots, show how the shape of the assembled robots changes during motion by adapting to the local structure of the environment.

2. A *conformist behavior* that consists of the individuals' ability to conform its orientation with that of the rest of the team when the two orientations differ significantly. This behavior results from the combination of: (a) a control rule that makes the robot turns toward the direction of the traction when its intensity is significant, and (b) the sensory effects produced by the execution of this action, mediated by the external environment, that lead to a progressive reduction of the intensity of the traction until the orientation of the robot conforms with the orientation of the rest of the group.

3. A *phototaxis behavior* that consists of the individuals' ability to orient toward the direction of the light target. This behavior results from the combination of: (a) a control rule that makes the robot turn toward the direction in which the intensity of the light perceived is higher, and (b) the sensory effects produced by the execution of this action, mediated by the external environment, that lead to a progressive reduction of the difference in the light intensity detected on the two sides of the robot until the robot is oriented toward the light.

4. An *obstacle avoidance behavior* that consists of the individuals' ability to change direction of motion when the execution of a motor action produces a collision with an obstacle. This behavior results from the combination of: (a) the same control rule responsible for behavior

2 described above that makes the robot turns toward the direction of the perceived traction, and (b) the sensory effect of the actions that led to the collision with obstacles, mediated by the environment, that lead to the generation of a force oriented toward the opposite direction with respect to the robot's direction of motion.

The combination and the interaction between these four elementary behaviors produces the following higher-level behaviors that extend over a longer time span:

5. A *coordinated motion behavior* which consists of the ability of the swarm-bot to negotiate a common direction of movement and then to keep moving along this direction by compensating further misalignments originating during motion. This behavior emerges from the combination of the move-forward and conformist behaviors.

6. A *coordinated light-approaching behavior* that consists of the ability of the swarm-bot to move in a coordinated manner toward a light target. This behavior emerges from the combination of the conformist, move-forward, and phototaxis behaviors. When the intensity of the perceived traction is high, the conformist and move-forward behaviors play the main role, while, when the intensity of the traction is low, the main role is played by the phototaxis and move-forward behaviors.

7. A *coordinated obstacle avoidance behavior* that consists of the ability of the swarm-bot to modify in a coordinated way its direction of motion so as to avoid nearby obstacles. This behavior arises as the result of the combination of the obstacle avoidance, conformist and move-forward behaviors.

The combination and the interaction between these behaviors gives rise to a set of still higher-level behaviors that extend over still longer time spans:

8. A *collective exploration behavior* that consists of the ability of the team to visit different areas in the environment when the light target cannot be detected. This behavior emerges from the combination of the coordinated movement and the coordinated obstacle avoidance behavior that ensures that the swarm-bot can explore different parts of the environment without getting stuck and without entering into limit cycle trajectories.

9. A *shape re-arrangement behavior* that consists of the ability of the swarm-bot to dynamically adapts its shape to the current structure of the environment so as to pass through narrow passages. This behavior emerges from the combination of the coordinated motion, coordinated obstacle avoidance behavior, and eventually the coordinate light-approaching behavior, and from the effects produced by the forces originating during collisions that affect primarily the robots that collided with the obstacles.

Finally, the combination of these behaviors leads to the following higher-level behavior:

10. A *collective navigation behavior* which consists of the ability of the swarm-bot to navigate toward the light target by producing coordinated movements, exploring the environment, passing through narrow passages, and producing a coordinate light-approaching behavior. This overall behavior, which corresponds to the behavior shown in Figure 43.4, arises as the result of the combination of collective exploration, shape re-arrangement, and collective light-approaching behaviors.

What is important to notice is that the robots have been evolved for producing behaviors 5 and 6 only (i.e. the fitness function used to select evolving individuals only rates the extent to which the swarm-bot is able to locomote and the extent to which it is able to locomote toward the light). The development of behaviors 1–3 could be explained by considering that these behaviors (i.e. the move-forward, conformist, and phototaxis behaviors) are instrumental for the production of behaviors 5 and 6. The other behaviors, instead, are the result of generalization processes.

Indeed, behavior 4 (obstacle avoidance) arises as a result of the interaction between robots and physical obstacles mediated by the same control rule that is responsible for the production of behavior 2 (conformist). This leads to a form of generalization in which the robots react to new circumstances through the production of new behaviors. These new behaviors, which arise as a result of generalization processes, are not necessarily adaptive and might indeed produce counter-adaptive consequences. On the other hand, the production of new spontaneous behaviors that are similar to those exhibited by evolved robots in similar robot–environment circumstances enables the robots to face new situations appropriately in most cases.

Behaviors 7–10, on the other hand, are the result of the interaction and combination of behaviors 1–6. These behaviors therefore are the result of another kind of generalization process that originates from the fact that the interaction between the existing behaviors necessarily leads to the production of new higher-level behaviors that originate from the execution of several lower-level behaviors over time or from the concurrent execution of multiple lower-level behaviors. Once again, these behaviors that have not been shaped by evolution directly might produce counter-adaptive consequences in some cases but will tend to play useful functionalities in the majority of cases thanks to their relation with the existing adaptive capacities from which they originate.

Overall this implies that the behaviors of the adapting robots display multi-level and multi-scale organization characterized by behaviors that emerge from the interaction of the same control mechanisms with different environmental circumstances and by the interaction and combination of different lower-level behaviors. This process might lead to generalization processes at the level of behavior in which the robots tend to react to new circumstances through the production of new behaviors. By producing new behaviors that are similar to those exhibited by the robot in similar circumstances they might face new situations appropriately without the need for further training.

More generally, this experiment shows how the evolution of collective systems might lead to multi-level and multi-scale skills formation that might enable the synthesis of a progressively larger and more complex behavioral repertoire.

Synthesizing collective behaviors

The complex system nature of behavior (Nolfi 2009) makes the synthesis of behavioral systems through standard design techniques extremely hard. This difficulty is due to the indirect relationship between the characteristics of individuals and the behaviors that emerge from agent–environment interactions. The problem becomes even more complex in the case of collective systems in which behavior is the emergent result not only of the interactions between each individual and the environment but also of the interactions between individuals. For this reason the most widely used approaches take inspiration from nature and attempt to replicate either the solutions discovered by nature (bio-inspired approaches) or the adaptive process through which nature generates effective solutions (evolutionary approaches).

In bio-inspired approaches researchers attempt to replicate a capacity displayed by a specific species by analyzing in detail the rules that regulate the behavior of the natural individuals, by creating colonies of artificial agents that operate on the basis of similar rules embedded in similar environmental conditions, and by manually tuning critical selected parameters until the desired collective behavior is produced. This approach has been used, for example, to design cockroach-like robots that are able to self-organize into subgroups located under multiple sheltered areas (Garnier et al. 2005) or ant-like robots able to discover and navigate toward multiple foraging areas (Garnier et al. 2007).

In evolutionary approaches, on the other hand, the rules that regulate how the robots react to different environmental circumstances and eventually the characteristics of the robots' bodies are initially set randomly and are subjected to an artificial evolutionary process driven by a fitness function that estimates the extent to which the behaviors exhibited by the colonies of robots approach the desired target behaviors (Trianni and Nolfi 2012). Two examples of this approach have been reviewed in the previous section.

From the point of view of the synthesis of collective behavior in artificial systems, bio-inspired and evolutionary approaches both have advantages and disadvantages. Both approaches circumvent the design problem since the former capitalizes on solutions discovered by natural evolution and the latter operates blindly, i.e. capitalizes on variations that produce desired emergent properties without the need for understanding/modeling the relation between the characteristics of the agents and the behavioral properties that emerge from the agents' interaction with their physical and social environment. The bio-inspired approach can be highly effective but can only be applied to domains for which we have a detailed understanding of the natural solutions, at least of what concerns the functionalities that we want to reproduce in artificial systems. The evolutionary approach can be applied to any problem domain but requires time-consuming adaptive processes and the identification of initial conditions that support the selection of progressively better and more complex solutions (Trianni and Nolfi 2012).

Future directions

The application range of artificial collective systems displaying emergent and self-organizing properties is potentially enormous. The current state of the art is that we have a rather advanced understanding of how these systems can be applied to a limited number of domains (e.g. spatial navigation, route and traffic optimization, collective decisions). A challenging and promising direction of research consists in the attempt to apply systems and methods of this type to domains that involve physical interactions between the agents, and between the agents and the environment, such as collective construction and transport, self-reconfigurable and self-assembling systems.

Another challenging and promising research direction consists of the identification of effective design for emergence principles and methodologies, i.e. techniques that enable the design of functional systems exhibiting emergent and self-organizing properties. This might be realized through the elaboration of radical new design methods or through refinements and extensions of existing techniques.

Learning more

The book by Camazine et al. (2001) provides a rich and authoritative discussion of self-organization in biological systems. For an introduction to self-organization in artificial systems, instead, I recommend the book on swarm intelligence by Bonabeau et al. (1999). A recent review of the state of the art and of the open challenges in collective robotics is provided in the *Handbook of Collective Robotics* edited by Kernbach (2012). For learning more on the application of evolutionary methods to the synthesis of collective behavior, readers might consult Trianni and Nolfi (2012).

For a detailed account of algorithms inspired by the principles observed in natural collective systems we recommend the monograph books addressing ant colony optimization by Dorigo and Stützle (2004) and particle swarm optimization by Kennedy et al. (2001).

Readers interested in gaining a practical knowledge of the synthesis of collective behavior in robot swarms might exploit freely available simulation tools such as ARGoS (Pinciroli et al. 2012) and FARSA (Massera et al. 2013). In addition to the simulation of multiple robots, FARSA provides a developmental tool that include libraries for designing and evolving the robots' control system.

References

Baldassarre, G., Parisi, D., and Nolfi, S. (2006). Distributed coordination of simulated robots based on self-organisation. *Artificial Life*, **12**(3), 289–311.

Baldassarre, G., Trianni, V., Bonani, M., Mondada, F., Dorigo, M., and Nolfi, S. (2007). Self-organised coordinated motion in groups of physically connected robots. *IEEE Transactions on Systems, Man, and Cybernetics*, **37**(1), 224–39.

Bonabeau, E., Dorigo, M., and Theraulaz, G. (1999). *Swarm intelligence: from natural to artificial systems*. Oxford: Oxford University Press.

Camazine, S., Deneubourg, J.-L., Franks, N., Sneyd, J., and Theraulaz, G. (2001). *Self-organization in biological systems*. Princeton, NJ: Princeton University Press.

Dorigo, M., and Stützle, T. (2004). *Ant colony optimization*. Cambridge, MA: MIT Press.

Frei, R., and Di Marzo Serugendo, G. (2011). Concepts in complexity engineering. *International Journal of Bio-Inspired Computation*, **3**(2), 123–39.

Garnier, S., Jost, C., Jeanson, R., Gautrais, J., Asadpour, M., Caprari, G., and Theraulaz, G. (2005). Collective decision-making by a group of cockroach-like robots. In: *Proceedings of the IEEE Swarm Intelligence Symposium, June 8–10, 2005, Pasadena, CA*. IEEE Press.

Garnier, S., Tache, F., Combe, M., Grimal, A., and Theraulaz, G. (2007). Alice in pheromone land: an experimental setup for the study of ant-like robots. In: *Proceedings of the IEEE Swarm Intelligence Symposium*. IEEE Press, pp. 37–44.

Kennedy, J., Eberhart, R. C., and Shi, Y. (2001). *Swarm intelligence*. San Francisco, CA: Morgan Kaufmann.

Kernbach S. (ed.) (2012). *The handbook of collective robotics: fundamentals and challenges*. Singapore: Pan Stanford Publishing.

Massera, G., Ferrauto, T., Gigliotta, O., and Nolfi, S. (2013). FARSA: An open software tool for embodied cognitive science. In: P. Liò, O. Miglino, G. Nicosia, S. Nolfi, and M. Pavone (eds), *Proceedings of the 12th European Conference on Artificial Life*. Cambridge, MA: MIT Press.

Nolfi, S. (2009). Behavior and cognition as a complex adaptive system: insights from robotic experiments. In: C. Hooker (ed.), *Philosophy of Complex Systems*, Handbook of the Philosophy of Science, Volume 10:. Amsterdam: Elsevier, pp. 443–63.

Pinciroli, C., Trianni, V., O'Grady, R., Pini, G., Brutschy, A., Brambilla, M., Mathews, N., Ferrante, E., Caro, G., Ducatelle, F., Birattari, M., Gambardella, L. M., and Dorigo, M. (2012). ARGoS: a modular, parallel, multiengine simulator for multi-robot systems. *Swarm Intelligence*, **6**(4), 271–95.

Sperati, V., Trianni, V., and Nolfi, S. (2011). Self-organised path formation in a swarm of robots. *Swarm Intelligence*, **5**, 97–119.

Trianni, V., and Nolfi, S. (2012). Evolving collective control, cooperation and distributed cognition. In: S. Kernbach (ed.), *The handbook of collective robotics: fundamentals and challenges*. Singapore: Pan Stanford Publishing, pp. 127–66.

Chapter 44

From aquatic animals to robot swimmers

Maarja Kruusmaa

Centre for Biorobotics, Tallinn University of Technology, Estonia

Fish and other aquatic animals have developed a diverse repertoire of locomotion and sensing strategies in an environment that is 800 times denser than air. This chapter gives a brief overview of their most common locomotion and sensing principles that have inspired engineers to find alternatives to propelled underwater propulsion and to standard robotic sensors.

Biomimetic robots can offer an alternative to the established underwater technology by developing vehicles that are more efficient, robust, agile, and maneuverable. However, this also means overcoming many technical challenges caused by the limitations of conventional technology.

Different forms of fish swimming trade off swimming speed and maneuverability whereas their man-made counterparts also face design complexity problems. This chapter describes locomotion principles together with analyses of each locomotion strategy and of the engineering problems that need to be solved in order to replicate the behavior of swimming animals in artifacts. New emerging technologies, such as smart materials, are discussed in this context as an alternative to conventional electromechanical robotic components.

Besides locomotion strategies unique to aquatic environments, aquatic animals also have unique senses for detecting hydrodynamic stimuli. Artificial lateral line systems and artificial whiskers modeled on those of pinnipeds, such as seals, offer solutions for sensing the underwater environment.

Biological principles

As any other animal, fish use muscles to convert chemical energy into mechanical work. Momentum created in the body is balanced by hydrodynamic reaction forces, and, as a result, the body moves forward while the mass of water that occupied the location of the body is shed backward. Thrust force is the result of adding momentum to the surrounding fluid while drag force, analogous to friction forces in solid-state mechanics, is removal of the momentum from the surrounding fluid. For a fish swimming at constant speed, drag and thrust forces are balanced, a fish accelerating has a larger thrust than drag and a fish decelerating has a larger drag than thrust. However, for a free-swimming fish, drag and thrust are not separable and can only be estimated indirectly (e.g. by wake analysis).

Beside drag and thrust, fish are also subjected to lift and gravitational forces. Most fish are neutrally or slightly negatively buoyant. They can adjust their buoyancy by changing the density of gas in their swim bladder, by creating static lift force with their fins and body or by creating dynamic lift force with their flapping 'wings'.

Fish create momentum by using muscles of their body and fins. Fish swimming modes are usually classified as body and caudal fin (BCF) propulsion or median and paired fin (MPF) propulsion (Figure 44.1). In general, MPF propulsion provides better maneuverability while BCF propulsion provides higher speeds and accelerations. Depending on their location in the food chain (predator or prey), hydrodynamic environment, or physical size, fish have adopted a remarkable variety of different locomotion mechanisms. At one extreme of the BCF spectrum there are thunniform swimmers (tuna, swordfish, sharks) which are capable of achieving high cruising speeds by flapping the tail fin of their nearly rigid body, but their maneuverability is rather poor. At the opposite end of the BCF swimmers spectrum, there are highly maneuverable anguilliform (eel-like) swimmers which use their whole body to create an undulating

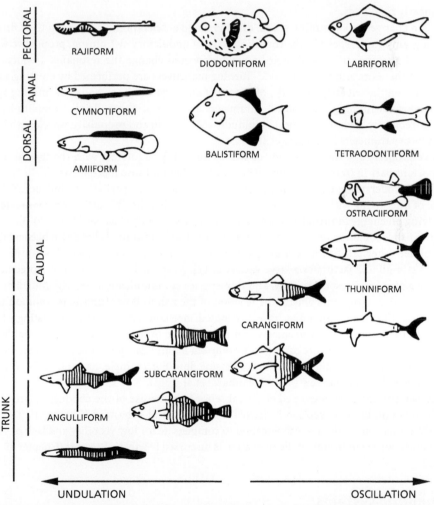

Figure 44.1 Swimming modes of fish. The vertical axis aligns swimming modes based on the role of body and fins in propulsion. The parts actively contributing to propulsion are shaded. The horizontal axis distributes the swimming modes based on the undulant and oscillatory motion.

Adapted from C.C. Lindsey, *Form, Function, and Locomotory Habits in Fish*, doi: 10.1016/S1546-598(08)60163-6, Copyright © 1978 ACADEMIC PRESS, INC. Published by Elsevier Inc.

wave. At the same time, eels and other elongated vertebrate swimmers are slow and inefficient. Carangiform swimmers (e.g. carp, mackerel) use the anterior third of their bodies to create the flapping motion whereas subcarangiform swimmers (such as trout) make use of two-thirds of their body to generate thrust thus offering a compromise between maneuverability and speed. Carangiform and thunniform swimmers are stiffer than anguilliform swimmers, thus making more use of the storage of elastic energy at every half-stroke of their flapping body.

MPF swimmers achieve high maneuverability and agility by operating their pectoral, ventral, or dorsal fins. Many bottom-dwelling fishes are rajiform swimmers who have developed extended pectoral fins. Rays can create travelling waves along those fins and depending on the wavelength, direction of the wave, and symmetry between the left and right fin, they are capable of moving forward, backward, or turning on the spot. Gymnotiforms, such as knifefish, create undulations using their elongated anal fin which permits swimming forward and backward with a nearly rigid body.

Fish can change the swimming speed, direction, and acceleration by changing their fin beat frequency, amplitude, speed of the travelling wave of undulatory swimming, propulsive wavelength, and also the stiffness of their body which would change the resonance frequency of the fin and therefore also the amplitude. Turning maneuvers are performed by creating asymmetry between the left and right side. McHenry et al. suggest that to change swimming speed of undulatory swimming, the control parameters are body stiffness, driving amplitude, and tail-beat frequency, with the tail-beat amplitude, wave-speed, and propulsive wavelength being the response parameters (McHenry et al. 1995).

Passive properties of the fish's body as well as an ability to interact with the flow also play an important part in facilitating efficient swimming, though those mechanisms are less investigated and understood. Under some circumstances, such as in periodic turbulence, fish are capable of interacting with vortices in a favorable way so that the propulsive force is created by capitalizing on energy in turbulence with an almost passive body (Liao and Cotel 2013).

Besides fish locomotion and its control, the senses of aquatic animals have also been adjusted to life under water. All fish have a unique sensing organ—a lateral line, which senses hydrodynamic stimuli and facilitates a large variety of behaviors such as detection of a predator or prey, navigation with respect to flow, and intraspecies communication (see e.g. Mogdans and Bleckmann 2012). Tropomorphism (the reaction of the fish to flow stimuli), is common to all fish species. The lateral line organ is a distributed mechanosensory array consisting of two sub-modalities—superficial neuromasts that are sensitive to flow speed and the canal lateral line capable of sensing pressure differences. Some sea mammals, who do not have a lateral line organ, have developed an alternative way of sensing flow. For example, harbour seals' whiskers are sensitive to hydrodynamic stimuli (Dehnhardt et al. 2001).

Some fish species have developed electrical sense to make use of the conductive properties of their medium. Electroreception is discussed by Boyer and Lebastard (Chapter 19, this volume). Dolphins and whales use echolocation to compensate for low visibility and lack of visual cues in underwater environment. Echolocation is discussed briefly by Prescott (Chapter 45, this volume).

Biomimetic systems

State-of-the-art underwater vehicles almost exclusively use propellers. It is a mature and reliable technology to create thrust, taken over from surface transportation. At the same time, nature has not developed a single species that uses propellers on a macro-scale. Efficiency, maneuverability, agility, and robustness of biological swimmers is still much greater than that of man-made vehicles but it is debatable if it is because of the underlying locomotion mechanics and to what extent the principles of biological swimming can be copied with our current technology.

Most of the research and technology development in underwater biomimetic systems focuses on replicating BCF propulsion and mostly thunniform or carangiform swimming. The reason for this could be prioritizing speed and efficiency over high maneuverability but also the fact that those swimming patterns are easiest to reproduce with the currently available technology. Those robots use a small number of rigid links connected by actuated joints forming a small serial chain. The kinematics of serial chain actuation, originating from research in robotic arm manipulation, is well understood in robotics. Fast and accurate real-time methods exist for their control. Flapping and oscillating motion is therefore fairly easily reproducible with the current technology. Also, the head and the body of a thunniform or carangiform biomimetic robot are relatively easily to approximate with a rigid hull, equipped with a tail fin actuator. The rigid hull can be used as a watertight compartment for housing electronics, sensors, and payload. In its extreme form, this design can be considered as an approximation of a conventional torpedo-shaped underwater robot powered by a single blade propeller.

The more the design concept diverges from thunniform swimming, the more advantages the vehicles proposes in terms of maneuverability, but it also introduces new technical challenges. Reproducing undulating BCF swimming with serial chain kinematics means more powered joints, more actuators, as well as more sophisticated control methods that come at the cost of price and reliability. It will also become more complicated to make the flexible hull watertight and accommodating onboard electronics can become a fairly time-consuming technical challenge. Because of the finite size of electromechanical actuators this design is also difficult to miniaturize, as opposed to their biological counterparts that are inherently distributed actuation systems consisting of thousands of muscle fibers.

Most fish robots are prototypes demonstrated only in laboratory conditions and often operate only in 2D. When operating in 3D, buoyancy control is usually realized by using a buoyancy tank or by controlling the angle of attack of the pectoral fins to create lift force.

The most well-known example of a BCF swimmer is perhaps the MIT RobotTuna, a robot with a relatively rigid torso that swims with fairly small body and tail motions actuated by a sophisticated system of pulleys and cable tendons driven by six brushless servo motors (Barrett et al. 1996). Another well-known example is the Essex robotic fish, which uses three servomotors to actuate the tail (Liu et al. 2005). A commercial biomimetic robot, BIOSwimmer by Boston Engineering, is also inspired by tuna.

A lamprey-inspired robot LAMPETRA represents a sophisticated mechanical design for implementing anguilliform swimming (Figure 44.2). The eel-like swimming motion is achieved by a modular arrangement of independently driven muscular segments. LAMPETRA control is based on a bio-inspired central pattern generator (CPG) principle (Stefanini et al. 2012; see also Cruse and Schilling, Chapter 24, this volume). It can produce travelling waves along the body with variable speed and waveform. The robot is capable of hours of autonomous locomotion, and object tracking has been demonstrated in laboratory conditions with a stereo vision system mounted in its head. A commercial biomimetic eel-like swimming robot has been developed by a spin-off company of the Norwegian Institute of Science and Technology, Eelume, for sub-sea inspection.

GhostBot is a gymnotiform swimmer inspired by ghost knife fish accommodating an anal fin with 34 degrees of freedom. This mechanical design permits controlling the parameters of the travelling wave along the anal fin, propelling the robot forward and backward at different speeds (Curet et al. 2011). An array of independently driven servo motors has also been used to build pectoral fins inspired by rajiform swimming (Zhou and Low 2010).

Figure 44.2 LAMPETRA robot replicated the anguilliform swimming mode of a lamprey.
Copyright © Lampetra Consortium

Besides varying the number of actuators and degrees of freedom of the propulsor, material properties of those devices have also been investigated. Experimental research relates the stiffness and geometry of hydrofoils (mostly tail fins) to drag, thrust, and efficiency of the vehicles. Also entirely soft, infinite degrees of freedom tail propulsors can be considered that make use of the energy of the travelling wave along the body to create momentum. Mathematical modeling of hydrodynamic reaction forces is more complicated, especially the interaction between a soft body and fluid. Hydrodynamic models often consider fluid forces as added mass or use Lighthill's elongated body theory that satisfactorily predicts the hydrodynamic force for an inviscid steady flow for a body with small amplitude lateral motion.

Whereas the purpose of building some biomimetic underwater robots is to advance underwater technology, others are also used as tools by biologists to test biological hypotheses. For example, a turtle robot Madeleine is mimicking the swimming locomotion of a sea reptile replicating the stiffness profile of a real turtle (Long Jr et al. 2006; Figure 44.3). It was built to test evolution of locomotion patterns—specifically, to find out why four-flippered animals (turtles and amphibians) use only their rear limbs for propulsion. Apparently, the reason is again the trade-off between speed and maneuverability.

One way to escape the trade-off between complexity and maneuverability is to consider alternative technologies as a basis for biomimetic design (see Vincent, Chapter 10, and Anderson and O'Brien, Chapter 20, this volume). Smart materials, such as shape memory alloys, dielectric elastomers, and ion-conductive polymer–metal composites have been used to build proof-of-concept undulating fins and fish robots. Those materials are soft and flexible, having theoretically an infinite number of degrees of freedom, and they are especially suitable for building miniature devices. However, smart materials research is still an actively developing field and the technology suffers from drawbacks such as energy inefficiency and poor long-term stability.

Fish robots are typically equipped with cameras, sonars, and other off-the-shelf sensors. Bioinspired sensing has got very little attention compared to biomimetic locomotion. Fish lateral line sensing is becoming, however, progressively popular. Several research groups have

Figure 44.3 A turtle robot Madeleine.
Photograph courtesy of Professor John Long, Vassar College.

developed lateral line sensors inspired by flow-sensing superficial neuromasts and canal lateral line sensing, but also hydrodynamically sensitive seal whiskers. The FILOSE robot is the first one implementing onboard flow sensing in a control loop to mimic thropotactic responses of real fish (Salumäe and Kruusmaa 2013). It can detect hydrodynamic stimuli and react to hydrodynamic events in a flow pipe in 2D. Similar to real fish it can detect flow direction and align itself into the flow (analogous to fish rheotactic behaviour), detect periodic turbulence, and take advantage of swimming in energy efficient regions in the wakes of other objects.

Future directions

Biomimetic swimmers develop towards greater speeds, agility, maneuverability, and robustness. All technical solutions either pushing the limits in one of those directions, or alternatively, finding trade-offs and compromises between mutually exclusive design objectives, are advancing this technology. Hopefully we will also witness the increase of the technology readiness level of this development. That should be revealed in the increasing number of field trials and commercial applications.

The main trend in underwater robotics is towards greater autonomy and energy efficiency with reduced cost. If biomimetic devices offer viable solutions they have a potential to establish themselves in mainstream robotics. Biomimetic underwater vehicles could offer alternatives for current underwater technology especially in applications and environments that are not suitable for heavy and bulky state-of-the-art underwater vehicles. Examples of those scenarios include environmental monitoring in shallow waters, exploration of confined underwater structures, and surveillance operations requiring quiet motion.

A poorly understood and unexploited opportunity is interaction with the flow of both biological fish and artificial flexible propulsors. It is known that fish are good at exploiting turbulence and flow in order to save energy by making use of the hydrodynamic forces created by pressure differences in flow. It is also known that they change their body shape and stiffness to adjust to flow patterns to achieve greater agility or energy efficiency. This problem could be investigated both theoretically and experimentally. So far there are almost no experiments demonstrating fish robot trials in flow. At the same time, those devices, substantially smaller and lighter than current underwater vehicles, are very sensitive to hydrodynamic disturbances and cannot be exploited in natural conditions without control methods that take into account hydrodynamic effects.

Flow sensing is another emerging field that has recently gained popularity. While all 32,000 fish species and several sea mammals have flow-sensing organs, no underwater vehicle currently in commercial use perceives flow. Flow sensing would make it possible to model hydrodynamic effects and take them into account in control. Lateral line sensors developed so far are demonstrated in laboratory conditions and have not been proven yet to be reliable and robust enough for long-term testing on underwater vehicles.

Learning more

A good starting point for anybody who wants to familiarize themselves with the physics of underwater locomotion is the landmark text by Steven Vogel "Life in moving fluids: the physical biology of flow." The book is written in a comprehensible manner, understandable for biologists, physicists, and engineers. Physical principles are illustrated with biological examples and help to develop a sense of fluid dynamics, which is necessary for understanding the more theoretical approaches to the subject but also for designing and conducting experiments, as well as for developing new design solutions for underwater vehicles.

John Long's book "Darwin's Devices" is a view by an evolutionary biologist who investigates the evolution of aquatic vertebrates by using biomimetic robots to develop and test his hypothesis.

John Videler's book "Fish Swimming" covers a wide range of topics about physics of fish swimming, morphology, and evolution. A popular review article about fish locomotion mechanics, written for readers with an engineering background is in Sfakiotakis et al. (1999).

Mogdans and Bleckmann (2012) provide a review of research on the fish lateral line that is a good starting point for understanding the biological principles of lateral line sensing and flow-related behaviour of fish.

References

Barrett, D., Grosenbaugh, M., and Triantafyllou, M. (1996). The optimal control of a flexible hull robotic undersea vehicle propelled by an oscillating foil. In: *Autonomous Underwater Vehicle Technology, 1996 (AUV'96), Proceedings of the 1996 Symposium on*. IEEE, pp. 1–9.

Curet, O.M., Patankar, N.A., et al. (2011). Mechanical properties of a bio-inspired robotic knifefish with an undulatory propulsor. *Bioinspiration and Biomimetics*, **6**(2), 026004.

Dehnhardt, G., Mauck, B., Hanke, W., and Bleckmann, H. (2001). Hydrodynamic trail-following in harbor seals (*Phoca vitulina*). *Science*, **293**(5527), 102–4.

Liao, J.C., and Cotel, A. (2013). Effects of turbulence on fish swimming in aquaculture. In: *Swimming Physiology of Fish: Towards Using Exercise to Farm a Fit Fish in Sustainable Aquaculture* (pp. 109–127). Springer Berlin Heidelberg. doi: 10.1007/978-3-642-31049-2_5

Lindsey, C. (1978). *Form, function, and locomotory habits in fish*. New York: Academic Press.

Liu, J., Dukes, I., and Hu, H. (2005). Novel mechatronics design for a robotic fish. In: *Intelligent Robots and Systems, 2005 (IROS 2005), IEEE/RSJ International Conference on*. IEEE, pp. 807–12.

Long, J. (2012). *Darwin's Devices: What evolving robots can teach us about the history of life and the future of technology*. New York: Basic Books.

Long, J. H. Jr, Schumacher, J., Livingston, N., and Kemp, M. (2006). Four flippers or two? Tetrapodal swimming with an aquatic robot. *Bioinspiration & Biomimetics*, **1**(1), 20.

McHenry, M., Pell, C., et al. (1995). Mechanical control of swimming speed: stiffness and axial wave form in undulating fish models. *Journal of Experimental Biology*, **198**(11), 2293–305.

Mogdans, J., and Bleckmann, H. (2012). Coping with flow: behavior, neurophysiology and modeling of the fish lateral line system. *Biological Cybernetics*, **106**(11–12), 627–42.

Salumäe, T., and Kruusmaa, M. (2013). Flow-relative control of an underwater robot. *Proc. R Soc. A: Math., Phys. and Eng. Sci.*, **469**(2153).

Sfakiotakis, M., Lane, D.M. et al. (1999). Review of fish swimming modes for aquatic locomotion. *Oceanic Engineering, IEEE Journal of*, **24**(2), 237–52.

Stefanini, C., Orofino, S., et al. (2012). A novel autonomous, bioinspired swimming robot developed by neuroscientists and bioengineers. *Bioinspiration and Biomimetics*, **7**(2), 025001.

Videler, J. J. (1993). *Fish swimming*. Fish and Fisheries series, Vol. 10. Dordrecht: Springer Science & Business Media.

Vogel, S. (1994). *Life in moving fluids: the physical biology of flow*. Princeton: Princeton University Press.

Zhou, C., and Low, K.-H. (2010). Better endurance and load capacity: An improved design of manta ray robot (RoMan-II). *Journal of Bionic Engineering*, **7**, S137–S144.

Chapter 45

Mammals and mammal-like robots

Tony J. Prescott

Sheffield Robotics and Department of Computer Science,
University of Sheffield, UK

Mammals are warm-blooded tetrapod vertebrates that evolved from therapsid reptilian ancestors during the late Triassic period around 225 million years ago. Their distinguishing characteristics, compared to reptiles, are a six-layered neocortex, hair, a triple-boned middle ear structure, and the mammary glands for which they are named. Whilst the first mammals were small nocturnal insectivores, from the mid-Jurassic onwards, mammals evolved to exploit all of the major habitats on Earth—terrestrial, subterranean, aquatic, and aerial. There are more than five thousand existing mammalian species ranging from shrews and bats weighing just a few grams to the largest animals on Earth—rorquals such as the Blue Whale—weighing more than 100 tonnes (Nowak 1999). Several radiations of mammals have seen marked increases in brain size, most notably the primates (including humans), elephants, and cetaceans (the whales, dolphins, and porpoises). A further distinctive feature has been the evolution, in primates, of a grasping hand capable of dextrous manipulation.

There have been many research efforts directed at replicating, in biomimetic technologies, specific mammalian morphological, perceptual, sensorimotor, and cognitive capacities and their neural substrates (see other chapters in this volume). Attempts to build integrated robotic systems that broadly match the behaviour and appearance of specific mammalian species have focused most strongly on humans (see Metta and Cingolani, Chapter 47, this volume), on quadrupeds such as cats and dogs, and on rodents. Here we focus on some of the most distinctive mammalian characteristics and on non-humanoid integrated robotic systems that are of interest from the perspective of capturing these capabilities in artifacts at the same time as advancing the understanding of mammalian biology.

Biological principles

Several mammalian adaptions have proved particularly interesting from the perspective of biomimetics.

The mammalian brain

One of the most significant changes in the evolution of mammals was the re-organization and expansion of an area in the roof of the reptilian brain, termed the dorsal cortex, into what is now called the neocortex (Northcutt and Kaas 1995; see also Prescott and Krubitzer, Chapter 8, this volume). This area progressively changed from a single layer of mixed excitatory and inhibitory cells into the six-layer multi-region brain mantle, with complex internal microcircuitry, that is characteristic of all modern mammals. The earliest mammals were nocturnal insectivores surviving in complex forest or scrub-like environments (Luo et al. 2011). Jerison (1973) hypothesized that the nocturnal lifestyle of these shrew-sized creatures was a driving factor behind neocortical evolution, also prompting the emergence of cortical maps for multiple sensory modalities, and mechanisms for rapidly integrating across these maps and making inferences based on sparse or ambiguous sensory data. Allman (1999) added to this general thesis that endothermy—the regulation of body temperature required for early mammals to remain active at night—created a much greater demand for energy in mammals compared to reptiles of similar size. This meant that food gathering needed to be more efficient; thus the neocortex evolved to support a rich multi-modal representational capacity that would allow mammals to make better decisions about when and where to forage.

The emergence of neocortex was followed by several further expansions in brain size relative to body size. The first came as mammals took over daytime niches from reptiles 65 million years ago. A further expansion occurred in early primates, approximately 30 million years ago, possibly linked to changes in social intelligence and foraging behaviour. Further increases occurred in the hominid ancestors of modern humans, likely linked to cognitive developments such as flexible tool use, language, and culture (see below). An important change in the system architecture of the brain was the evolution, in early mammals, of a direct corticospinal motor pathway that has allowed for more fine-grained control of end-effectors such as the primate hand.

Novel sensory systems

Mammalian hair evolved first as a sensory structure before adapting to have a significant role in temperature regulation as pelagic hair (or fur). Whiskers, or *vibrissae*, are prominent sensory hairs, found on all mammals except humans; they differ from pelagic hair by being longer and thicker and by having large follicles containing blood-filled sinus tissues. The first mammals will have possessed mobile facial vibrissae and the emergence of the vibrissal system is thought to have played an influential role in establishing a common ground-plan for the mammalian facial musculature and in driving the re-organization and expansion of the neocortex (Mitchinson et al. 2011a). Many mammals have also evolved areas of non-hairy, glabrous skin. In humans, these include the skin areas on the lips, hands, and fingertips, and the soles of the feet. These are the parts of the body that are most important when physically interacting with the world and where accurate tactile discrimination is most critical; unsurprisingly, then, glabrous skin has a high density of mechanosensory receptors. Early mammals also evolved a more sensitive auditory system (hence the changes in the middle ear) enabling communication in a frequency range unavailable to reptiles. Other distinctive sensory capacities in mammals include electroreception in the duck-billed platypus, echolocation in animals such as bats and dolphins, and new forms of discriminative touch such as the "star" of the star-nosed mole and the human fingertip. All of these sensory systems are *active* in that movement of the sensory apparatus is controlled to improve information pickup and is tuned to the task in which the animal is currently engaged (Prescott et al. 2011). Mammalian hairy skin contains a novel system of unmyelinated low threshold C-tactile mechanoreceptors that are particularly sensitive to light "stroking" touch and thus may underlie an affective touch capacity (McGlone et al. 2014) that may have originated through a need to strengthen social bonding (see below).

Agile locomotion and dextrous grasp

Mammals have a more distinctly upright physical morphology compared to the sprawling stance of most reptiles. This change in posture was combined with a switch from a two-segmented to a three-segmented limb. The improvement in locomotion capability is principally one of endurance rather than speed (Fischer and Witte 2007). Climbing mammals can cope with a huge variety of locomotor substrates, differing in diameter, inclination, roughness, continuity, and flexibility, adjusting their locomotor mechanics to maintain or enhance stability (see Witte et al., Chapter 31, this volume). The forepaws of most mammals have five digits and, in many species, are able to grasp and manipulate objects, a skill that has increased with the emergence of greater corticospinal control in therian (placental) mammals. A distinguishing feature of primates is the evolution of hands that allowed, for the first time, objects to be grasped and manipulated single-handedly. The emergence of an opposable thumb increased the dexterity of the hand and will have facilitated the evolution of tool use in early hominids.

Social cognition

The bearing and nurturing of live young prompted the emergence of mother–child bonding, beginning a trend in many mammalian orders towards increased sociality. Humans are often considered to be "ultra-social" (or socio-cultural) adding to the already well-developed social skills of other primates new abilities for social learning, communication (including language), and theory of mind (the ability to see the world from another person's perspective) (Herrmann et al. 2007). Capacities such as empathy may be limited to the great apes, and to some other large-brained mammals such as elephants and cetaceans. Social intelligence, as indicated by the size of the animal's social group, has been found to be linked to brain size with a trend toward larger social groups and richer interactions with conspecifics over the last 60 million years (Reader and Laland 2002).

Biomimetic systems

Some examples of recent mammal-like robots are shown in Figure 45.1 and discussed further below. Scientific research platforms tend to emphasize one system or behavior of interest, such as vibrissal sensing, or locomotion. Companion and assistive robots such as Sony's *Aibo* emphasize the delivery of a complete integrated system that is animal-like in appearance and has a repertoire of life-like behaviors.

Brain-based mammal-like robots

The field of neurorobotics, also known as brain-based, or neuromorphic robotics (Krichmar and Wagatsuma, 2011; Prescott et al. 2016) has often targeted the mammalian brain due to the availability and critical mass of biological data relating behaviour to neural activity. One of the first examples of brain-based modeling is the work by Edelman and co-workers on the *NOMAD* series of real-world artifacts (for review see Almássy and Sporns (2001)). These models aimed at validating Edelman's theory of neuronal group selection, also known as "neural Darwinism," particularly, the proposal that differential selection of neuronal connections can lead to repertoires of neurons that are tuned to specific states of the world. More recent models of integrated brain architectures include the *Distributed Adaptive Control* (DAC) models developed by Verschure and co-workers (See Verschure, Chapter 36, this volume), and the model of layered sensorimotor loops, and of integrative structures such as the basal ganglia and cerebellum, developed by Prescott and colleagues for their whiskered robot models (see below). Many groups have developed and tested models of the hippocampal system as a substrate for spatial cognition, using robots to test the sufficiency of proposed models in reproducing navigation behavior similar to that seen in rodents (for a representative example see *RatSLAM* by Wyeth et al. (2011); also Erdem et al., Chapter 29, this volume). The mammalian/primate sensory cortices have provided considerable inspiration for researchers interested in building machine perception (see Leibo and Poggio, Chapter 25, this

Figure 45.1 Mammal-like robots. (a) Shrewbot, a robot that emulates the mammalian whisker system, developed by Bristol Robotics Laboratory and the Active Touch Laboratory Sheffield. (b) Bat-bot developed by Ralph Müller, Virginia Tech. (c) Cheetah robot from the Department of Mechanical Engineering, Massachusetts Institute of Technology. (d) Sony Aibo robotic dog. (e) MiRo robot by Consequential Robotics.

volume). Others have used robots to test hypotheses about brain reward and learning mechanisms, reinforcement learning, and the role of neurotransmitters such as dopamine, being a particularly popular target of study (see, for instance, Uchibe and Doya (2011)). These examples have sought to reproduce brain functionality by employing relatively abstract, or "systems level" models of the relevant neural circuits, where computational units represent neuronal groups (for example clusters of neurons of the same type with similar input–output connectivity). Whilst more detailed models, for instance incorporating spiking or conductance-based model neurons, have been evaluated in robots, this has largely been in the context of simulating very specific aspects of animal learning or behavior, for instance to provide an embodied test of spike-time dependent plasticity in the cerebellum (Carrillo et al. 2008) or for learning a body representation in a model of visual–somatosensory integration (see Asada, Chapter 18, this volume).

Mammal-like locomotion and dexterity

A solution to the challenge of walking and running on four legs and in complex outdoor terrains has been demonstrated by *Boston Dynamics* robots such as *Big Dog* and *Spot*. Whilst not strongly biomimetic, these robots exploit strategies for maintaining dynamically stable gaits that are clearly bio-inspired. A robot called *Cheetah*, developed at the MIT Biomimetic Robotics Lab, incorporates biomimetic principles in a more direct fashion whilst not limiting itself to copying biology (Sangok et al. 2015; see also Witte et al., Chapter 31, this volume). For instance, this robot incorporates a flexible spine—known to play a critical role in generating the stride of many mammals. The Cheetah also takes advantage of a design principle called "biotensegrity" derived from analysis of the synergetic arrangement of bones, tendons, muscles, and ligaments in mammalian limbs. This principle has allowed the MIT team to create strong limbs from lightweight materials, with appropriate forms of compliance that can reduce the mechanical stress caused by footfall. Such designs are capable of elastic energy storage and thus show much improved energy efficiency compared to more rigid platforms. Experiments with a mammal-like tail have shown that this can improve the balance and maneuvrability of the robot. Taking a similar approach, the Biorobotics laboratory at EPFL has shown that robot legs modeled on the mammalian tri-partite "pantographic" limb also provide benefits such as improved self-stability (Spröwitz et al. 2013).

The challenge of creating robots with dextrous hands that match the capability of modern primates is discussed by Cutkosky (Chapter 30, this volume). Here we note that difficulty of this problem is less with the design of artificial fingers with high-resolution tactile sensing and precision movement as in the challenge of understanding how perception and movement combine to provide dextrous grasp planning and object manipulation. Given that this capacity evolved only in one group of comparatively large-brained mammals (the primates), it is perhaps no surprise that this remains a largely unsolved problem in robotics where we still have a lot to learn from nature.

Mammal-like active sensory systems

The vibrissal system of mammals has lead to many attempts to develop robot analogs of this system that can operate both terrestrially—inspired by the whiskered sensing systems of rodents—and aquatically—inspired by sea mammals such as seals and manatees that use whiskers to detect the hydrodynamic disturbances created by prey animals or by underwater structures (see Kruusmaa, Chapter 44, this volume). In order to better understand mammalian vibrissal sensing, Prescott and colleagues conducted a range of biological studies of rodent vibrissal sensing alongside the development of several different whiskered robot platforms (Prescott et al. 2009, 2015). Each robot was designed to explore specific questions about vibrissal tactile sensing. For instance, *Whiskerbot* was used to examine early sensory processing in a spiking neuron model of the primary afferent nerve fibers coupled to fiber-glass whiskers instrumented with strain gauges and controlled by shape-memory-alloy artificial muscles. Research with this model demonstrated the importance of fine control of the movement and positioning of the whisker for effective sensing and led to the development of additional degrees of freedom for whisker and head positioning in subsequent robots. Later robots, such as *Scratchbot* and *Shrewbot* (Pearson et al. 2011), used brain-based models of brain structures including the superior colliculus, cerebellum, and somatosensory cortex to actively guide head and whisker movement and to support whisker-based detection of tactile surface properties (Prescott et al. 2015; see also Lepora, Chapter 16, this volume).

Dolphins echolocate by emitting a double click sound, the echo of which has a complex signal structure from which the animal is able to extract object properties such as size, shape, and even internal structure. This capability is enhanced through the use of an active biosonar sensing strategy whereby the parameters of the clicks, such as their spectrum, duration, and intensity, are tuned to improve object recognition capability. The US Navy previously trained dolphins to locate active underwater mines, but has replaced these in recent years with unmanned underwater vehicles using sonar. It is not known whether the replacement submarines use dolphin-like active echolocation; however, it seems likely that such strategies will be useful for underwater robots faced with challenging sensing problems (Paihas et al. 2013). Bats are also known for their remarkable biosonar capabilities that enable them to hunt small prey animals in complex, structured environments. Recent analyses have shown that active sensing strategies that involve physically modulating sound generation and pickup through movement of the nostrils and pinnae (ears) may contribute to their remarkable perceptual skills (Müller 2015). Experiments to evaluate the potential for dolphin-like biosonar are currently in progress at Heriot Watt University, whilst a team at Virginia Tech is examining the possibility of bat-like biosonar in aerial robots.

Attempts to emulate primate vision, audition, touch, and chemosensation, including neuromorphic approaches, are discussed extensively in part III of this handbook; their integration in human-like robots is described in Metta and Cingolani (Chapter 47), and their use in prosthetics by Bensmaia (Chapter 53) and Lehmann and van Schaik (Chapter 54).

Mammal-like social cognition and companion robots

The potential for mammal-like robots as companions or "digital pets" has been recognized for several decades with various commercial robots being developed that copy the morphology of domestic mammals such as cats and dogs. One of the best-known robots, Sony's *Aibo*, first went on sale on 1999, although the original *Aibo* was discontinued in 2005, Sony have recently (2017) launched a revised design targeting the growing market for personal digital assistants. As a commercial project, little is known about the role of biomimetics in the development of the *Aibo* design. Other mammal-like toys, such as Hasbro's *FurReal Friends*, also emphasize pet-like behavior and appearance and therefore are more bio-inspired than biomimetic (i.e. embodying biological principles). The seal-like robot *Paro* (Wada et al. 2005), designed to socially interact with people with cognitive impairments (see Millings and Collins, Chapter 60, this volume) likewise aims for mammal-like appearance and behavior rather than emulating mammalian cognition or morphology in a stronger sense. Consequential Robotics' *MiRo* robot is a prototype companion robot that seeks to use brain-based control systems in a pet-sized platform with a number of mammalian features including a brain-like layered control architecture (Mitchinson and Prescott 2016). Research on mammal-like social cognition strongly influenced the design of the Kismet robot, developed at MIT Media's laboratory, in the late 1990s, that built on on earlier work with humanoid robots on problems such as active vision (Breazeal, 2003).

Future directions

As described elsewhere in this volume, models of mammalian neural systems have been developed and tested in robots that focus on pattern generation, locomotion, sensorimotor integration, spatial and episodic memory, action selection and decision-making, reinforcement learning, homeostasis, emotion, drives, and social cognition. The availability of these various models, and the emergence of computational theories of the brain's major learning systems—cortex, basal ganglia, hippocampus, amygdala, and cerebellum—means that we are now in a position where we could build a model of the mammalian brain architecture, at least at a "systems level" of abstraction, and embed it in a mobile robot. The benefits of such an integrated model would be profound in terms of providing a concrete, embodied instantiation of brain theory. Whilst such a complete integrated model is still some way off, the existence of a wide range of exemplar systems, and the increasing availability and reducing cost of real-time high-bandwidth computing, means this could be realistic aspiration for the next decade. There is a further possibility that such neurorobotic models could become even more brain-like, for instance, by incorporating massively parallel hardware capable of simulating millions of spiking neurons—an example architecture in this direction is the Spinnaker massively multi-core computing engine which is aiming for real-time simulation of up to one billion neurons (Furber et al. 2013).

Robots with mammal-like morphology, behaviour, and cognition will continue to be developed. Legged robots have applications for movement on uneven terrain and a mammal-like quadruped morphology appears to provide a good compromise between stability and speed (though see Witte et al., Chapter 31, this volume for an assessment of its advantages and drawbacks). Sensing systems that match mammalian capacities for active vision, audition, or touch, will enhance the versatility of future robots to operate in challenging environments.

The goal of creating robots that resemble mammals will be encouraged by interest in pet-like robots that can emulate some of the capacities for social companionship provided by domesticated mammals such as rabbits, dogs, and cats. To provide a robot with social cognition capabilities similar to domestic pets should be achievable by combining state-of-the-art biomimetic systems for action, perception, memory, emotion, learning, and cognition; however, to integrate these in an affordable commercial robot remains a challenge.

Learning more

Two collections of example brain-based robot models published by Webb and Consi (2001) and Krichmar and Wagatsuma (2011) show developments in the state-of-the-art in the late twentieth century and at the start of the twenty-first century, including several examples of mammal-like robots. Floreano et al. (2014) provide a more recent review of the broader field of attempts to test neuroscience models (both vertebrate and invertebrate) in robots. Mitchinson et al. (2011b) and Prescott et al. (2016) provide discussion of the benefits, and limitations, of using robots to evaluate models of mammalian brain systems. Ijspeert (2014) has provided a useful review of recent attempts to develop biomimetic locomotion systems, focused towards quadruped robots, while Mattar (2013) has provided a review of the extensive literature on biomimetic hands. The literature on models of social cognition is broad and disparate—Miklósi and Gácsi (2012) is one useful launching-off point focusing on animal-like robot companions.

Acknowledgment

The author received support in preparing this chapter from the EU Horizon 2020 FET Flaghip Human Brain project (HBP-SGA1, 720270).

References

Allman, J. M. (1999). *Evolving brains.* New York: Scientific American Library.

Almássy, N., and **Sporns, O.** (2001). Perceptual invariance and categorization in an embodied model of the visual system. In: B. Webb and T. R. Consi (eds), *Biorobotics.* Cambridge, MA: MIT Press, pp. 123–43.

Breazeal, C. (2003). Emotion and sociable humanoid robots. *International Journal of Human-Computer Studies*, **59**(1–2), 119–55.

Carrillo, R. R., Ros, E., Boucheny, C., and **Coenen, O. J. M. D.** (2008). A real-time spiking cerebellum model for learning robot control. *Biosystems*, **94**(1–2), 18–27.

Fischer, M. S., and **Witte, H.** (2007). Legs evolved only at the end! *Philosophical Transactions of the Royal Society of London A: Mathematical, Physical, and Engineering Sciences*, **365**(1850), 185–98.

Floreano, D., Ijspeert, Auke J., and **Schaal, S.** (2014). Robotics and neuroscience. *Current Biology*, **24**(18), R910–R920.

Furber, S. B., Lester, D. R., Plana, L. A., Garside, J. D., Painkras, E., Temple, S., and **Brown, A. D.** (2013). Overview of the SpiNNaker System Architecture. *IEEE Transactions on Computers*, **62**(12), 2454–67.

Herrmann, E., Call, J., Hernàndez-Lloreda, M. V., Hare, B., and **Tomasello, M.** (2007). Humans have evolved specialized skills of social cognition: the cultural intelligence hypothesis. *Science*, **317**(5843), 1360–66.

Ijspeert, A. J. (2014). Biorobotics: Using robots to emulate and investigate agile locomotion. *Science*, **346**(6206), 196–203.

Jerison, H. J. (1973). *Evolution of the Brain and Intelligence.* New York: Academic Press.

Krichmar, J. L, and Wagatsuma, H. (eds) (2011). *Neuromorphic and Brain-based Robots.* Cambridge, UK: Cambridge University Press.

Luo, Z. X., Yuan, C. X., Meng, Q. J., and **Ji, Q.** (2011). A Jurassic eutherian mammal and divergence of marsupials and placentals. *Nature*, **476**(7361), 442–5.

McGlone, F., Wessberg, J., and **Olausson, H.** (2014). Discriminative and affective touch: sensing and feeling. *Neuron*, **82**(4), 737–55.

Mattar, E. (2013). A survey of bio-inspired robotics hands implementation: New directions in dexterous manipulation. *Robot. Auton. Syst.*, **61**(5), 517–44.

Miklósi, Á., & Gácsi, M. (2012). On the utilisation of social animals as a model for social robotics. *Frontiers in Psychology*, **3**. https://doi.org/10.3389/fpsyg.2012.00075

Mitchinson, B., Grant, R. A., Arkley, K., Rankov, V., Perkon, I., and **Prescott, T. J.** (2011a). Active vibrissal sensing in rodents and marsupials. *Philos. Trans. R. Soc. Lond. B: Biol. Sci.*, **366**(1581), 3037–48.

Mitchinson, B., Pearson, M., Pipe, T., and **Prescott, T. J.** (2011b). Biomimetic robots as scientific models: A view from the whisker tip. In: J. Krichmar and H. Wagatsuma (eds), *Neuromorphic and Brain-based Robots.* Cambridge, MA: MIT Press, pp. 23–57.

Mitchinson, B., and **Prescott, T. J.** (2016). MIRO: A Robot "Mammal" with a Biomimetic Brain-Based Control System. In: N. F. Lepora, A. Mura, M. Mangan, P. F. M. J. Verschure, M. Desmulliez, and T. J. Prescott (eds), *Biomimetic and Biohybrid Systems: 5th International Conference, Living Machines 2016, Edinburgh, UK, July 19–22, 2016. Proceedings.* Cham: Springer International Publishing, pp. 179–91.

Müller, R. (2015). Dynamics of biosonar systems in horseshoe bats. *The European Physical Journal Special Topics*, **224**(17), 3393–406.

Northcutt, R. G., and **Kaas, J. H.** (1995). The emergence and evolution of mammalian neocortex. *Trends in Neurosciences*, **18**(9), 373–9.

Nowak, R. M. (1999). *Walker's Mammals of the World* (6th edn). Baltimore, USA: Johns Hopkins University Press.

Paihas, Y., Capus, C., Brown, K., and Lane, D. (2013). Benefits of dolphin inspired sonar for underwater object identification. In: N. F. Lepora, A. Mura, H. G. Krapp, P. F. M. J. Verschure, and T. J. Prescott (eds), *Biomimetic and Biohybrid Systems: Second International Conference, Living Machines 2013, London, UK, July 29–August 2, 2013. Proceedings.* Berlin, Heidelberg: Springer, pp. 36–46.

Pearson, M. J., Mitchinson, B., Sullivan, J. C., Pipe, A. G., and Prescott, T. J. (2011). Biomimetic vibrissal sensing for robots. *Philos. Trans. R. Soc. Lond. B: Biol. Sci.*, **366**(1581), 3085–96.

Prescott, T. J., Ayers, J., Grasso, F. W., and Verschure, P. F. M. J. (2016). Embodied models and neurorobotics. In: M. A. Arbib and J. J. Bonaiuto (eds), *From Neuron to Cognition via Computational Neuroscience.* Cambridge, MA: MIT Press, pp. 483–512.

Prescott, T. J., Diamond, M. E., and Wing, A. M. (2011). Active touch sensing. *Philos. Trans. R. Soc. Lond. B: Biol. Sci.*, **366**(1581), 2989–95.

Prescott, T. J., Mitchinson, B., Lepora, N. F., Wilson, S. P., Anderson, S. R., Porrill, J., ... Pipe, A. G. (2015). The robot vibrissal system: understanding mammalian sensorimotor co-ordination through biomimetics. In: P. Krieger and A. Groh (eds), *Sensorimotor Integration in the Whisker System.* New York: Springer, pp. 213–40.

Prescott, T. J., Pearson, M. J., Mitchinson, B., Sullivan, J. C. W., and Pipe, A. G. (2009). Whisking with robots: From rat vibrissae to biomimetic technology for active touch. *IEEE Robotics & Automation Magazine*, **16**(3), 42–50.

Reader, S. M., and Laland, K. N. (2002). Social intelligence, innovation, and enhanced brain size in primates. *Proc. Natl Acad. Sci. USA*, **99**(7), 4436–41.

Sangok, S., Wang, A., Chuah, M. Y. M., Dong Jin, H., Jongwoo, L., Otten, D. M., ... Sangbae, K. (2015). Design principles for energy-efficient legged locomotion and implementation on the MIT Cheetah robot. *IEEE/ASME Transactions on Mechatronics*, **20**(3), 1117–29.

Spröwitz, A., Tuleu, A., Vespignani, M., Ajallooeian, M., Badri, E., and Ijspeert, A. J. (2013). Towards dynamic trot gait locomotion: Design, control, and experiments with Cheetah-cub, a compliant quadruped robot. *The International Journal of Robotics Research*, **32**(8), 932–50.

Uchibe, E., and Doya, K. (2011). Evolution of rewards and learning mechanisms in Cyber Rodents. In: J. Krichmar and H. Wagatsuma (eds), *Neuromorphic and Brain-based Robots.* Cambridge, MA: MIT Press, pp. 109–28.

Wada, K., Shibata, T., Saito, T., Kayoko, S., and Tanie, K. (2005, 18–22 April 2005). *Psychological and social effects of one year robot assisted activity on elderly people at a health service facility for the aged.* Paper presented at the IEEE International Conference on Robotics and Automation, 2005. ICRA 2005.

Webb, B., and Consi, T. R. (2001). *Biorobotics.* Cambridge, MA: MIT Press.

Wyeth, G., Milford, M., Schulz, R., and Wiles, J. (2011). The RatSLAM project: robot spatial navigation. In: J. Krichmar and H. Wagatsuma (eds), *Neuromorphic and Brain-based Robots.* Cambridge, MA: MIT Press, pp. 87–108.

Chapter 46

Winged artifacts

Wolfgang Send

ANIPROP GbR, Göttingen, Germany

Motion in air rests on the flyers' ability to carry their weight and to overcome their drag. In horizontal flight, weight and drag are balanced by lift and thrust. Nature has done an ingenious job of integrating the generation of lift and thrust into the structure of flying animals. Their engine for producing thrust without a single rotating part is the flapping wing. Leonardo da Vinci designed the first human flapper with articulated wings. Many attempts were made in the past to mimic birds' flight with technical constructions, among them the remarkable early work of Lippisch before 1930 (see Lippisch 1960). Birds, insects, and even fish (see Kruusmaa, Chapter 44, this volume) apply the same basic mechanism. The production of thrust with flapping wings is an inherent property of the physical equations which describe aerodynamics and hydrodynamics. The governing equations are derived from the conservation laws for momentum, mass, and energy in fluid mechanics. Flapping of a three-dimensional wing, more precisely a coupled bending and torsional motion, reduces to a coupled pitching and plunging motion in two dimensions, in which, to simplify experimental as well as theoretical treatment, a profile section parallel to flow direction is cut out from the wing. The first papers on the origin of lift of a two-dimensional profile section date back to the beginning of the twentieth century.

The physics of wing motion which produces thrust has widely been investigated and is well understood. A recently published paper gives a thorough and comprehensive overview of the history and of the progress and challenges in flapping-wing aerodynamics (Platzer et al. 2008). The discovery of the mechanism dates back to 1924 (Birnbaum 1924) and was a spin-off during research on airplane flutter. This extremely dangerous phenomenon of high technical importance for aircraft stability is physically based on the same mathematical description as animal propulsion. It is merely the reverse of the same coin. One side is producing thrust with a flapping wing to move forward, the other is winning energy with oscillating wings from a fluid flow. Both modes simply differ in the amplitude ratio of the two constituent degrees of freedom—pitching and plunging, bending and torsion, respectively (Send 1992). High plunging at small pitching produces thrust and thereby adds energy to the flow, low plunging with high pitching extracts energy. Pitching contributes a negligible amount to the energy balance and plays the role of a catalyst in the process. For both modes, the ratio of plunge power to power gained from the flow or supplied to it, defines the efficiency of the process. The two states virtually switch from one mode to the other one, and the domain of transition at both sides shows extremely high efficiency; up to 90% according to basic results in unsteady aerodynamics. Large birds probably are able to fly in this range due to their very low drag.

Exploring the mechanisms of efficient flapping flight also reveals the secrets of a source of renewable energy which extracts power from the kinetic energy in a fluid current.

The normal flyer diagram

In 1869 in the *Archives néerlandaises* the first paper by Harting appeared in which the weight G of birds is related to the planform area A of their wings (Harting 1869). The planform area is framed by the outline of the wings when looking from above on the stretched wing system. The square root of the area divided by the third root of the weight results in an almost constant quantity. The French scientist Marey cites Harting's data and realizes the importance of this discovery (Marey 1890), which reflects a fundamental principle of construction in nature. Weight divided by planform area forms a third quantity, the wing loading $\gamma = G/A$. The normal flyer diagram (Figure 46.1) shows wing loading over weight for a wide range from small birds to large airplanes and is of key importance for understanding flight performance. Also included are winged artifacts, among them milestones of human powered flight. The data of the artificial bird *SmartBird*, developed by a small team of engineers and the author, are displayed in the lower left corner. Its properties are described and discussed in the course of this chapter.

The straight lines reflect the special relation between wing loading and weight. The upper dotted line rests on the findings of Harting, the solid line shows a theoretical estimate (Send 2003). The second dotted line fits *SmartBird*'s data.

The solid line in the diagram relates the properties of birds to those of airplanes in a stunning context. Beginning with the Wright Brothers' *Flyer III* far below the solid line and ending with the Airbus A380 right on that line, the airplane data in the diagram may be interpreted as the technical evolution of airplanes. This view on the presented data shared, flyers positioned in the vicinity of the solid line are fully evolved and are denoted as *normal flyers*. The constant k_g plays the role of a *degree of evolution*. Flyers along a line of constant k_g possess the same degree of evolution. Theoretical calculations show that the generation of thrust with flapping wings along the solid line is possible with a similar magnitude of the amplitudes for bending and torsion irrespective of size. This ambitious principal statement disregards the implications of technical realization. Today, we neither have the "muscles," i.e. the actuators, at hand nor do the materials exist which are capable of bearing the enormous stress of millions of load cycles. We are still faced with the problems of a 100 kg *normal flyer*.

The data belonging to $G = 10^3$ N are of particular interest for assessing our ability as human beings to fly like birds. The obvious and disappointing result is that we will never be able to do that as easily as birds do. As a *normal flyer* we would have to spend 3 to 4 kW. From the author's

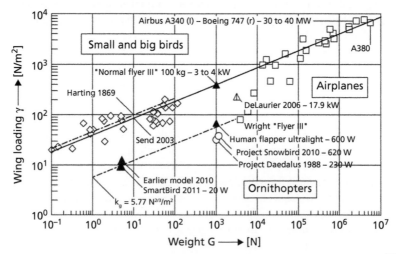

Figure 46.1 The normal flyer diagram. Solid line shows wing loading $\gamma = G/A = k_g \cdot \sqrt[3]{G}$ for $k_g = 40$ N$^{2/3}$/m^2. Harting's line corresponds to $k_g = 44.0$ N$^{2/3}$/m^2, SmartBird's value reads 5.77 N$^{2/3}$/m^2. Diamonds ◊ refer to bird data, squares □ represent airplanes. Some data are supplemented with the required power to fly, given in W (Watt), kW, and MW. The filled triangles refer to the author's work.

calculation in Send (2003) it follows that an ultralight human ornithopter, built with modern materials, is feasible, which requires not more than about 600 W. That still is quite a challenge and flying like birds certainly will never become a popular sport. The filled triangle *human flapper ultralight* marks the degree of evolution far below the *normal flyer* position. Years later in 2010 the admirable project *Snowbird* at the University of Toronto was successfully finished. The documents available on the web mention the required physical power of 620 W, which is quite close to the author's estimate. The engine of the famous Daedalus project was not a flapping mechanism but a normal propeller with a bicycle drive. The required average power of 230 W during the four hours lasting flight is much less than the demand for flapping flight.

Based on this estimate for the ultralight human flapper the author laid out the concept of an artificial bird with articulated wings which was expected to take off and land by its flapping mechanism only. The dotted line for $k_g = 5.77 \text{ N}^{2/3}/\text{m}^2$ connects the two winged artifacts, very different in size—the proposed ultralight flapper and *SmartBird*, both at the same degree of evolution. It is remarkable that the Wright Brothers' *Flyer III* ranks at the same level.

Two lines of development are worth mentioning: DeLaurier's full-scale piloted ornithopter C-GPTR (DeLaurier 1999) and Saupe's model ornithopter *Eskalibri*. To the author's knowledge, the gifted Swiss tinkerer Saupe developed the first ornithopter which was able to take off and land without any support. His work is typical of a vast and still increasing number of persons, ranging from skilled hobbyists to engineers, who are fascinated by flapping flight. Their work is hardly documented, though they often present excellent flyers. The rapidly advancing microelectronics industry stimulates the toy companies to bring more and more sophisticated flyers to the market which a decade ago at the turn of the millenium still seemed to be unreachable scientific goals. After a long time of development and numerous setbacks DeLaurier's ornithopter eventually took off in 2006 for its one and only short flight with a faulty touch down. His unique scientific and technical work is a milestone in the history of powered flapping flight.

There is a noticeable gap between birds and planes. If ever dinosaurs existed which were able to undertake active flapping flight, they are expected to be found with anatomic properties which fit to the vicinity of the *normal flyer* line. As mentioned above, the 100 kg power-assisted human flapper still is a challenge to enthusiasts around the world.

Physical and technical aspects of efficient flying

Experimental input

The theoretical description and eventually the numerical calculation of loads, power, and efficiency begins with prescribing the proper kinematics. The pursuit of obtaining kinematic data from live birds is a problem on its own. The first and excellently observed kinematics is due to Lilienthal (1889) and Marey (1891). Now experiments are conducted to obtain kinematic data using modern high speed cameras in 3D configuration.

After observing storks for more than 20 years, Lilienthal published his book in 1889 with a detailed drawing of the storks' wing motion. The author "translated" his drawings—the famous table VIII in his book consists of several parts—into numerical data. They turned out to be that precise that they easily fit into the contour plot of aerodynamic efficiency right at the place where the optimum had been computed. Marey built a large carousel-like test stand in which live birds travelled in tethered flight on a circular path. He developed pressure sensors and recorded their signals on soot covered cylinders similar to early voice recorders. A very similar test stand was provided by the author to carry out the measurements for developing SmartBird [10].

Both Lilienthal and Marey found out from their experiments that basically three degrees of freedom exist which form wing motion: The previously mentioned bending and torsional motion and a third degree of freedom, a back and forth motion of the wing, in helicopter dynamics called the lagging motion. The lagging motion enhances the production of thrust, but is not essential for the mechanism.

Theoretical description

The flight of an artificial flyer with moving wings shows complex features which cannot be solved properly even with modern numerical algorithms. A *proper solution* means a self-consistent solution in which generated thrust and lift balance drag and weight on a stable flight path. One part of the problem arises from the computation of the flow, the other part from flight dynamics. The prediction of drag requires approximations to the turbulent regions of flow, because a *direct numerical solution* (DNS) exceeds by far the range of present parallel computing. A flight dynamics code capable of controlling a stable flight seems to be feasible; however, it depends on precise unsteady aerodynamic forces. The desirable full solution needs to be reduced to approximations which cover individual aspects. As a very simple example of such a reduction the particular problem of aerodynamic efficiency is selected for a short discussion.

Though the details of the scaling in Figure 46.2 are of minor importance, a short explanation may precede the discussion. For a plunge amplitude equal to chord length and a pitch amplitude

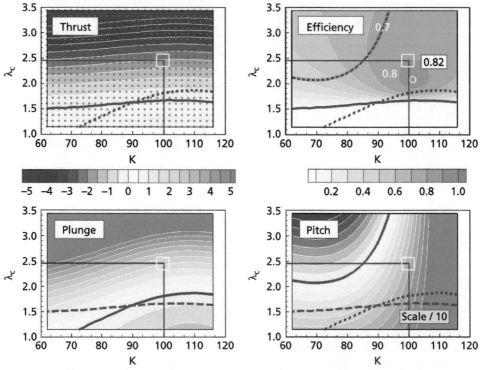

Figure 46.2 Efficiency of flapping flight. Result of a two-dimensional flow calculation for a coupled plunging and pitching profile named NACA7412 (Send et al. 2012). Contour plots show the normalized power coefficients averaged over one period of motion. Horizontal axis κ indicates phase shift of plunge ahead of pitch, vertical axis represents the ratio $\lambda_c = (h_0 \, / \, c) \, / \, \alpha_0$ of plunge amplitude h_0 to pitch amplitude α_0, given in radians. c denotes the chord length. White squares mark the design point of SmartBird's kinematics. Red dots: Individual numerical solutions.

of 24° the amplitude ratio reads about 2.5. The phase shift is equal to 90°, when the maximum pitch angle coincides with the plunge position in the middle between upper and lower turning point of the wing during upstroke. The figure shows that maximum efficiency is reached at a slightly higher phase shift. Figure 46.3 depicts *SmartBird*'s skeleton and the shape of the innermost rib. Clearly to identify is the position of the torsion actuator at the wing tip. The profile section NACA7412 was used for an earlier model with slightly higher wing loading (see Figure 46.1).

The *reduced frequency* serves as a measure for the degree of unsteadiness of the motion. The high value above 0.6 indicates that *SmartBird* is far from being a normal flyer. For a live bird of that size like a herring gull the reduced frequency (based on full chord length) is 0.2 or less, the weight one order of magnitude higher. From the beginning of SmartBird's development the declared aim was to show the bird in exhibition halls above a large audience. For safety reasons weight and speed were reduced as much as possible without waiving flight performance.

Principal result

The principal result of an aerodynamic calculation is the reaction force of the fluid in each of the very many small surface elements, into which the wetted area of the flyer is split for the numerical solution of the posed flow problem. Assuming we have obtained a solution, the *force per unit area* in each element is decomposed into the component perpendicular to the local surface element, the *pressure*, and the tangential component, the *shear stress*. The sum over all

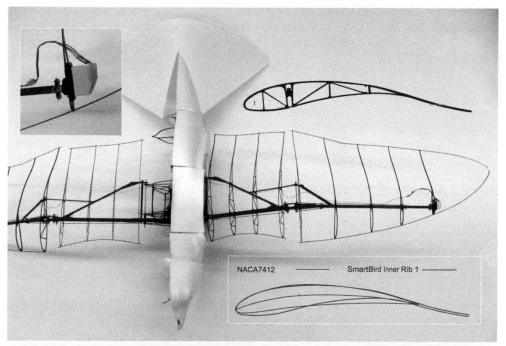

Figure 46.3 View of *SmartBird*'s skeleton. Span 2 m, planform area 0.5 m², mean chord length $c = 0.25$ m, chord length of rib one 0.33 m, weight 0.49 kg including battery. Operational data at design point: speed $u_0 = 5$ m/s, flapping frequency $f = 2$ Hz, average power consumption 20 W. Reduced frequency based on full chord length $\omega_c^* = 2\pi \cdot f \cdot c / u_0 = 0.63$. Torsion drive magnified top left.

surface elements for both contributions resolves into two forces, the total pressure force and the total shear stress force. These two forces then again are decomposed into the components of the coordinate system which is attached to the trajectory of the moving flyer. It is important to know that *lift* in aerodynamics is defined as the component perpendicular to the trajectory, and *drag* is defined as the component parallel to the trajectory, irrespective of the flyer's direction with respect to the horizon. In other words, a climbing flyer needs more thrust than is needed just to balance the drag, because additional thrust power is required to lift its weight against gravity. This lifting force is not achieved by the aerodynamic lift defined above.

The aerodynamic calculation which Figure 46.2 is based on is a pure pressure solution. Shear stress is not taken into account. In each surface element the resulting pressure force, i.e. the pressure and the geometric normal of the surface element, is multiplied by the kinematic velocity of the respective surface element. Force times velocity defines the local power which either gives a positive value (work is done against the fluid), or a negative value (work is extracted from the fluid). The kinematic velocity consists of the contributions from the three degrees of freedom involved: plunging, pitching, and the translational motion in the forward direction. For each surface element these three contributions to the power balance of the flyer are separately evaluated and summed up. These contributions vary over the period of motion and might even change their sign. Physically important is the average over one period of motion, the net outcome. Amplitude ratio and phase shift act as parameters in the solution to give more insight into the dependence from and sensitivity to these data.

The contour plots in Figure 46.2 show the total power balance of the wing section normalized to dimensionless *power coefficients*. The motion applied to the wing is different from a sinusoidal time dependence. It is called *partially linear*. The angle of the pitch motion remains constant for almost all parts of the upstroke and downstroke and changes very rapidly at the upper turning point from the maximum positive to the maximum negative value and vice versa at the lower turning point.

Findings

The top left diagram in Figure 46.2 shows in which regions thrust is obtained. The aerodynamic efficiency is defined only for regions in which thrust is produced, and is formed by the ratio of gained thrust power to the power fed in at pitch and plunge. These two contributions are plotted on the lower left and right graphs. The ensemble of the four contour plots reveals the secret of flapping flight: for certain values of amplitude ratio and phase shift almost all plunging power is converted into thrust power. At the design point the efficiency ranges up to 80%. To reach this high efficiency also the pitch has to be actuated, it requires additional power. The scale 1:10 for the pitch coefficient means that very little power is required for this degree of freedom.

The *active torsion* reduces the required power for *SmartBird* by a higher percentage than the efficiency contour plot indicates. The two-dimensional numerical calculation implies physical simplifications which lead to a broad plateau of high efficiency. This plateau was not found in the experiments. On the contrary, the high efficiency with optimum flow conditions is limited to a very narrow region spanning a few degrees for phase shift and a small range for the amplitude ratio. The precise location of this "sweet spot" is unpredictable and was found in flight experiments with the individual model, where both parameters were slightly varied until power consumption reached a minimum and the aerodynamic and mechanical noise at the wing tips almost disappeared. A separate wireless communication between SmartBird's onboard microprocessor and a laptop accomplishes this task. To some extent its location also depends on the individual making of a copy. Meanwhile (2013) four copies were built. Once these data have been evaluated they are stored in an onboard flash storage. After that initial tuning the models are operated using a normal remote control unit.

Future directions

The *normal flyer* in Figure 46.1 remains to be a challenge in both directions, making and computing. In 2006 the author took part in the German Berblinger competition, which is organized by the city of Ulm in Southern Germany. A two-seated airplane *Berblinger 2* was suggested with a bending-torsional drive and equipped with fuel cells. Bending-torsional drive is just another name for producing thrust with flapping wings. It sounds more technical, and, indeed, the author believes that the mechanism also might have the potential for application in aerospace engineering. The weight of *Berblinger 2* was designed to be 300 kg, which is very close to DeLaurier's airplane. The *degree of evolution* was derived from Saupe's *Eskalibri*. An entirely new feature was the *active torsion*, which had not been applied before. *SmartBird* is the first flyer in which *active torsion* is implemented—*Berblinger 2* still is a dream. Any future concept for airplanes with flapping wings certainly will apply this feature, because it significantly improves efficiency, thereby also reducing weight and structural costs.

Of course, the computational aspect is of high interest for predicting loads and the optimum operational parameters. The enormous progress in computing power nourishes the expectation that a proper, self-consistent numerical solution for a whole flyer will be feasible within the next five to eight years. At present, the more reliable way to design new flyers is the application of low-level aerodynamic estimates for loads and power consumption and then providing the necessary options for controlling and tuning the torsion amplitude and the phase shift. The bending power plays the role of aerodynamic fuel and merely sets the limit for maximum available thrust. The "thrust lever" is the torsion control.

The next evolutionary step leads to *SmartBird*'s successor with the properties of a normal flyer. After having mastered this challenge the view widens to the power-assisted human flapper—the missing link between birds and airplanes on the path to applications in aerospace engineering.

Learning more

Among a wide range of books which deal with birds and planes, the fine book by H. Tennekes *The Simple Science of Flight* (M.I.T. Press, Cambridge MA) covers much basic knowledge worth learning. J. J. Videler's *Avian Flight* (Oxford Ornithology Series) concentrates on birds and offers insight into all aspects of birds from the flight apparatus to aerodynamics at a level below mathematical skills. *Bird Flight Performance* (Oxford Science Publications) by C. J. Pennycuick counts among the classics and guides to practical calculations. The original edition even offered programs written in the computer language Basic to do your own computations. A newer and extended version by the same author is *Modelling the Flying Bird* (Elsevier Publ.). A physical background is required to read and to work with *The Biokinetics of Flying and Swimming* by A. Azuma (Springer-Verlag). The chapter by Hedenström (Chapter 32, this volume) provides a complementary contribution to this one, focusing on the biology and biomechanics of animal flight.

The website *www.ornithopter.org* introduces flapping flight and also lists historical events. Links to many other sites complete the very informative pages. The Festo company provides a *SmartBird* website www.festo.com/cms/en_corp/11369.htm with links to brochures and movies. For details of *SmartBird*'s development and testing the reader is referred to Send and Scharstein (2010) and Send et al. (2012). More about the author and his work may be found on www.

aniprop.de/overview. Last but not least the internet service Youtube provides a large number of video clips for the keyword SmartBird, among them the author's favourite clip: *The Smartbird outdoors attacked by seagulls.*

Acknowledgments

SmartBird was a joint project by a team of engineers, who developed the flyer in the years 2008–2011. R. Mugrauer designed and built the model, K. Jebens and Mrs. A. Nagarathinam contributed the wireless communication, controls, and electronics, and M. Fischer and G. Mugrauer organized the workflow. The author laid out the concept and conducted the experiments. The whole team gratefully acknowledges the full financial support by the Festo company's Bionic Learning Network and its initiator Dr. W. Stoll.

References

Azuma, A. (1992). *The Biokinetics of Flying and Swimming.* Tokyo: Springer-Verlag.

Birnbaum, W. (1924). Das ebene Problem des schlagenden Flügels, *Zeitschrift für angewandte Mathematik und Mechanik (ZAMM)*, **4**, 277–92.

Delaurier, J. D. (1999). The development and testing of a full-scale piloted ornithopter. *Canadian Aeronautics and Space J.*, **45**, 72–82.

Harting, P. (1869). Observations sur l'étendue relative des ailes et le poids des muscles pectoraux chez les animaux vertébrés volants. In: E. H. Baumhauer, J. Bosscha, and J. P. Lotsy (eds), *Archives néerlandaises des sciences exactes et naturelles.* Harlem: Société hollandaise des sciences à Harlem, pp. 33–54 (in French).

Lilienthal, O. (1889). *Der Vogelflug als Grundlage der Fliegekunst,* Berlin 1889—*Birdflight as the Basis of Aviation,* Facsimile publ. by American Aeronautical Archives.

Lippisch, A. M. (1960). Man powered flight in 1929, *Journal of the Royal Aeronautical Society*, **64**, 395–8.

Marey, E. J. (1890). *Le Vol des Oiseaux.* Paris: G. Masson.

Marey, E. J. (1891). *La Machine Animale* (5th edn). Paris: F. Alcan.

Pennycuick, C. J. (1989). *Bird flight performance: a practical calculation manual.* Oxford: Oxford University Press.

Pennycuick, C. J. (2008). *Modelling the Flying Bird.* Theoretical Ecology Series, Vol. **5**. Cambridge, MA: Academic Press.

Platzer, M. F., Jones, K. D., Joung, J., and **Lai, J. C. S.** (2008). Flapping-wing aerodynamics: progress and challenges. *AIAA J.*, **46**, 2136–49.

Send, W. (1992). The mean power of forces and moments in unsteady aerodynamics, *Zeitschrift für angewandte Mathematik und Mechanik (ZAMM)*, **72**, 113–32.

Send, W. (2003). Der Traum vom Fliegen. *Naturwissenschaftliche Rundschau*, **56**(2), 65–73.

Send, W., Fischer, M., Jebens, K., Mugrauer, R., Nagarathinam, A., and **Scharstein, F.** (2012). *Artificial Hinged-Wing Bird with Active Torsion and Partially Linear Kinematics,* 28th ICAS Congress, Brisbane, Australia, 23–28 September 2012, paper 53.

Send, W., and **Scharstein, F.** (2010). *Thrust measurement for flapping-flight components,* 27th ICAS Congress, Nice, France, 19–24 September 2010, paper 446.

Tennekes, H. (2009). *The simple science of flight: from insects to jumbo jets* (2nd edn). Cambridge, MA: MIT Press.

Videler, J. J. (2005). *Avian Flight.* Oxford Ornithology Series, Vol. **15**. Oxford: Oxford University Press.

Chapter 47

Humans and humanoids

Giorgio Metta and Roberto Cingolani

Istituto Italiano di Tecnologia, Genoa, Italy

Neuroscience has been providing tantalizing results showing the deep connection of body and mind to the determination of extant behavior. For example, Berthoz (2012) has delineated a fascinating account of how nature develops simple solutions to complex problems exploiting the physical properties of body and muscles jointly with the adaptability of the brain. Modern research in artificial intelligence (AI) attempts to recreate this mind–body nexus by working on the hardware and software of robots of various shapes. Humanoids are just another possible shape which in addition entails sophisticated human–robot interaction possibilities, tool use (affordances), and—more generally—their use in environments originally designed for humans.

Humanoid bodies

Robots that achieve performance comparable to humans will require bodies that mimic humans in many aspects: reliability, energetic performance, compliance, resilience, etc. At the moment, robots are "just" complex mechatronic devices, whose complexity is extremely difficult to manage: a typical humanoid platform may add up to 5000 mechanical components, not to mention the electronics, wiring, and computers. They weigh perhaps 50% more than a person of the same size and they require continuous power consumption in the range of top human sports performance (>1kW). With today's computer technologies, the computation required to mimic even a small part of the functions of the brain would require a power supply in the megawatt (millions of watts) range whereas the entire human brain consumes around 40 watts. The specific pressure in gears and joints can reach 140–150MPa during normal functioning and additional stresses in case of impacts lead to failures (even in the case of hardened steel).

Intrinsic compliance is a possible solution mimicking the controllable stiffness of human muscles and the flexible articulation of human joints. Soft robotics, a sub-discipline of mainstream robotics, is specifically considering new designs for the robotic "bodyware" (see e.g. Anderson and O'Brien, Chapter 20, this volume). Material studies in this case aim at developing soft actuators, soft bodies, and soft sensors (tactile) (Taccola et al. 2014). This is a completely new design paradigm for robotics where biomimesis is particularly advanced. Soft actuators try to mimic the properties of muscles with their ability to regulate mechanical stiffness actively. Unfortunately, typical soft actuators are not up to performance for full-scale humanoid robots. Variable impedance actuators (VIA), on the other hand, have received considerable attention recently because of their ability to cope with unmodeled interactions with the environment (unexpected impacts). They are more feasible than pure soft actuators: VIAs are normally a combination of various mechanical components capable of controlling position and stiffness simultaneously. They sort of approximate the properties of human muscles to co-contract to vary the mechanical stiffness of a limb (Vanderborght et al. 2013).

Materials are being studied to provide future robots with full body flexible and stretchable tactile sensing (see the recent review by Yogeswaran et al. 2015). The goal here is to mimic the sensitivity of human tactile sensors—including their relatively high bandwidth, 1kHz typical—into flexible and stretchable silicon substrates (see Lepora, Chapter 16, this volume). Advanced research is furthermore looking into the development of self-healing materials (Tee et al. 2012). Not less important, VLSI technology provides the ability to design visual sensors with unprecedented low power usage and sensitivity to change in illumination (see Dudek, Chapter 14, this volume). These are the so-called neuromorphic sensors, designed to mimic the responses of human photoreceptors and encoding information in "events"—that is, coding changes with high temporal resolution, rather than simply scanning the sensor to measure absolute light intensity at a relatively slow rate (Bartolozzi et al. 2011).

Finally, structural materials are important both for their ability to provide safe human–robot interaction (intrinsic compliance) and for the possibility of new approaches to design of robots with reduced numbers of components. In the latter, new production techniques such as 3D printing or molding can pave the way to the realization of robots that are simultaneously robust and low cost. Polymers would require an increase on their mechanical properties by a factor of 4 to 10 to efficiently replace metal. Research is currently ongoing into graphene inks (Bayer et al. 2014) and other nanofillers. Graphene is also an electrical conductor, thus potentially enabling flexible embedded electronics.

In spite of these very promising advances, the road to soft-bodied humanoids is still long. The best examples of "new humanoids" are the Kojiro platform (Mizuuchi et al. 2007) and the European Roboy (Pfeifer et al. 2013). Unfortunately, albeit quite fascinating, they are not as reliable as the traditional platforms and therefore comparatively less used by the community. Traditional platforms (Parmiggiani et al. 2012; Kaneko et al. 2011) are instead powered by electric motors, made of metal, and employ approximations of intrinsic compliance via active control or passive series-elastic actuators. A plausible research roadmap would certainly not discard the existing humanoids while attempting the progressive replacement of stiff metal components for new materials. For example, it is unlikely that electric motors will be replaced in the short term, while structural materials, new sensors, and new flexible electronics may be ready to be employed much earlier.

The resurgence of neural networks in AI is another driving factor to a true implementation of biomimetic humanoids. The combination of new training methods (Krizhevsky et al. 2012) and the availability of parallel computation (in the form of GPUs) have determined the development of efficient vision for pattern recognition which is delivering unprecedented high quality results (Mnih et al. 2015). Here, the design of complex cognitive architectures may be paramount to achieve general AI, that is, machines that can learn any task autonomously simply through interaction with the environment (including human teachers). Inspired by the brain, and as described elsewhere in this handbook, the research on cognitive architectures includes the study of memory systems, vision (Leibo and Poggio, Chapter 25), touch (Lepora, Chapter 16), and the integration thereof (Verschure, Chapters 35 and 36) and, clearly, advanced control (e.g. whole body control of movement, physical interaction, manipulation; see Herreros, Chapter 26, and Cutkosky, Chapter 30, this volume).

Biomimesis is attempted in domains as varied as attention (visual, acoustic), speech, action perception, object recognition, affordance detection, and following a long tradition in computational motor control, in the development of brain inspired controllers (e.g. cerebellar motor control). Many bio-inspired solutions have been developed on the iCub robotic platform (Parmiggiani et al. 2012; see Figure 47.1), since it readily allows integrating them in

Figure 47.1 The iCub humanoid robot described in the experiments presented in this chapter.
Source: © D.Farina, A.Abrusci, Istituto Italiano di Tecnologia, 2017.

complete experiments and architectures. In the following sections we survey a number of these solutions.

Humanoid vision

The human retina, like that of other animals, is not a flat square array of photoreceptors. It is rather evolved to the tasks of delivering high resolution and maintaining a large field of view (see Figure 47.2). On top of this, the photoreceptors (the rods and cones) are extremely sensitive to light, delivering an unsurpassed sensitivity—down to the single photon. These two characteristics can be reproduced in silicon (Dudek, Chapter 14). In humans, photoreceptors are denser in the center of the retina (the fovea) and decay in number towards the periphery. The result is that the retina optimizes simultaneously resolution and field of view for a fixed number of receptors. Models of the distribution of the photoreceptors have been studied in computer vision (Sandini and Tagliasco 1980). It has been calculated that if the retina were uniformly distributed with the same resolution of the fovea, then the neurons to process the resulting information flow would be about three tons in weight, not to mention the energy required to keep them alive. The same computation trade-off occurs in robots. Retina-like sensors have been developed over the years, although the lack of a proper market has so far confined them to the laboratory.

Rea et al. (2014) have developed a complete attention system for the iCub robot using retina-like vision (the log-polar model of the retina). In particular, in a series of experiments, they developed saliency based pre-attentive processing that efficiently implements filter banks for orientation, edges, color, motion, and object blobs (proto-objects). The interesting aspect of this work is in the ability to learn to predict interesting locations and sequence movements of the robot (saccades, pursuit, vergence) optimizing locations and gaze travel time to the object visual locations.

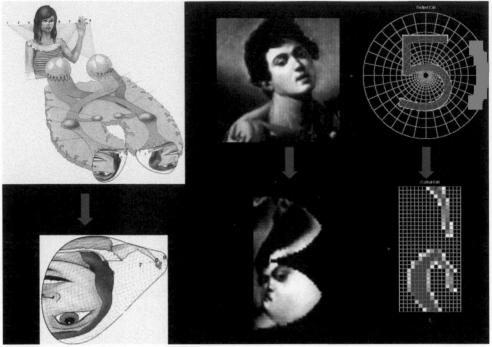

Figure 47.2 Model of the anthropomorphic retina sensor.

Reprinted from *Computer Graphics and Image Processing*, 14 (8), Giulio Sandini and Vincenzo Tagliasco, An anthropomorphic retina-like structure for scene analysis, pp. 365–72, doi: 10.1016/0146-664X(80)90026-X, Copyright © 1980 Published by Elsevier Inc., with permission from Elsevier.

Variable resolution is not the only striking feature of the human retina. As we mentioned, the sensitivity of the photoreceptor is incredibly different from standard cameras. Especially, neurons in the retina activate asynchronously as photons impinge on the receptors. Neuromorphic silicon technology (see Figure 47.3) can mimic this type of response resulting in visual data that is intrinsically sparse (event based), fast (events are asynchronous), and low-power (due to the sporadic activation of the transistors). The dynamic range of these receptors is also large because of the logarithmic response to intensity (they see much better than a standard camera in the dark). These devices have no central clock and encode information as variation of the sensed quantities (vision in this case, but a similar approach is being studied for touch). Therefore only relevant information is stored, transmitted, and processed, reducing the requirements in terms of bandwidth, memory, and power. The output of these artificial sensors is analogous to the neural spikes of biological sensors.

Rea et al. (2013) have demonstrated attention using neuromorphic cameras on the iCub robot. This work shows that neuromorphic vision allows faster—by two orders of magnitude—acquisition of moving visual targets compared to traditional frame-based vision. More recent developments on spike-based image processing include the computation of optical flow and learning visual receptive fields akin to Gabor filters.

It is foreseeable that the recent advances in large-scale neural processing (Mnih et al. 2015), combined with neuromorphic sensors, will enable more efficient and embedded vision for humanoid robots. While this would not solve the problem of designing learning architectures for general AI, the availability of efficient sensing is an important first step towards this goal.

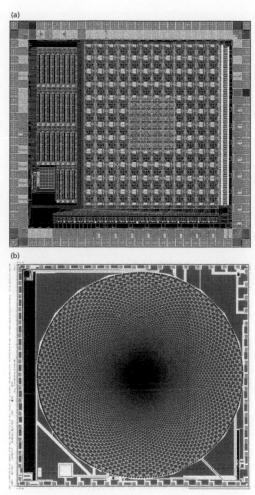

Figure 47.3 Two examples of neuro-inspired camera chips: (a) neuromorphic and (b) retina-like (logpolar).

Humanoid touch

Tactile sensing has been often overlooked in robotics. For robots that have to mimic human behavior and interact in unstructured environments touch is fundamental (Lepora, Chapter 16, this volume). Tactile sensing technologies, are, compared to vision, somewhat immature (see Bensmaia, Chapter 53, this volume, and Martinez-Hernandez (2015) for an indication on the state of the art). Touch has also conflicting engineering requirements: it needs to cover large areas (possibly curved), be mechanically soft (for sensitivity), and stretchable (to cover moving parts), while being durable. From the mechanical point of view it is certainly the most stressed component of the body of robots. The human skin solves the "problem" through self-healing and because it is soft and stretchable. Wiring and signal post processing is another challenge for a full body artificial skin. Sensors have been designed using the most diverse sensing principles under a variety of materials (Dahiya et al. 2010). Here we focus on some of the possibilities entailed by the ability to equip a robot (the iCub) with touch (see Figure 47.4).

Figure 47.4 Circuital examples of the flexible tactile system of the iCub.
Source: © Laura Tavena, Istituto Italiano di Tecnologia, 2017.

The most obvious use of tactile sensors is to detect contacts, estimate their intensity, and generate behaviors to deal with them (e.g. zeroing contact forces). We will describe some of these aspects in the next section since they form the basis for whole body dynamic control. There is another use of touch, which is to build maps of the body in order to predict contacts from vision. The study of the neurophysiology of the cortex has identified neurons that respond to touch and vision very peculiarly (Fogassi et al. 1996). They are located in area F4 (frontal area 4) and respond invariantly to contact with a given body part and simultaneously to the expected contact (visual) with the same body part. They are also active during the control of reaching movements (and avoidance movements too).

We have developed a learning system to build automatically a representation akin to F4 neurons on the iCub robot (Roncone et al. 2015). We framed the problem as probability estimation from sample events where the robot sees a moving object approaching the skin and simultaneously feels contact with it. This probability density is associated to receptive fields, one for each tactile element of the robot skin. The final result is a body representation that intrinsically encodes the likelihood that a given object moving in the vicinity of the robot will impact any given body part (a patch of the skin). This can be turned into a controller for avoiding or actively searching for contacts with any body part. We speculate that this type of motion controller, intrinsically multisensorial, highlights one of the properties of the human motor control system, i.e. equifinality. In practice, the representation described here allows any of the robot segments to be utilized as the end-point for a reaching movement.

The F4 representation corresponds to the psychological concept of peripersonal space studied in humans and animals (Macaluso and Maravita 2010). Being able to build similar models and generating behaviors in a humanoid robot provides it with the ability to "understand" space with respect to the body and control how to move in close proximity to people or, simply, inanimate objects.

Humanoid motor control

All sensory processing ultimately contributes to generating movement. In humanoids the control of movements may be complicated by the fact that the robot has a "floating base" (unlike industrial manipulators anchored to a single point) and therefore the dynamics has to take into account a variety of interactions with the environment. The study of the dynamics of this interaction can be connected to the study of control by the cerebellum in humans. A variety of models have been developed historically in connection to cerebellar control (Shidara et al. 1993; see also Herreros, Chapter 26, this volume). Modularity of controllers is another issue that is typically taken into account in brain-inspired motor control (Mussa-Ivaldi and Giszter 1992) and the equations of the dynamics show a similar modular decomposition in basis functions or fields.

In the iCub robot we combine these ideas with the ability to estimate external forces (due to contacts) owing to the ability to simultaneously sense the robot state (encoders, accelerometers, etc.), force and torques (via specialized sensors in the limbs), and the contact location and pressure (from the distributed skin). The result is a strategy that combines the estimation of the external forces with the whole body dynamics to generate controllers that can exploit the redundancy of the robot to achieve multiple goals such as stability, minimum angular momentum, translation of the body in space, and force constraints (i.e. how much force is exerted at each contact point); see Del Prete et al. (2015). This holds the promise of enabling dynamic walking patterns and stabilizers that not only generate human-like movements but also exploit interaction with the environment.

It has been shown that external force estimation through the combination of this variety of sensors (including the skin, see Fumagalli et al. 2010) is computationally adequate to obtain high performance controllers and design predictors that can take into account dynamics to anticipate impacts (or contacts) leading, in general, to highly dynamical behaviors. This research is closely related to the construction of peripersonal space described earlier: the ability to estimate contacts before they actually happen may be connected to planning avoidance behaviors rather than reaching to specific targets and subsequently connecting the plan to its actual execution by the dynamic controller.

An example of dynamic balancing is shown in Figure 47.5 for the iCub robot.

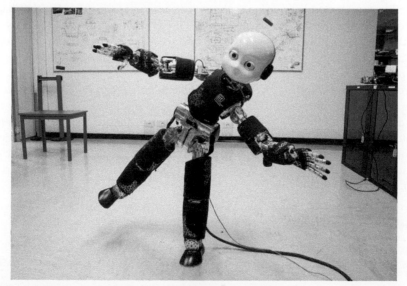

Figure 47.5 The iCub robot running a dynamic balancing algorithm.
Source: © D.Farina, A.Abrusci, Istituto Italiano di Tecnologia, 2017.

Conclusions and future directions

Research on biomimetic humanoids is certainly not limited to the few examples we have presented here. Human–robot interaction and social robotics are active communities that aim at developing—among other things—more efficient robots that can cooperate with people in daily tasks. The study of movement and gestures can in fact enable a more natural interaction with parameters that render the whole process of working with a robot efficient, understandable (without specific prior training), and comfortable to the user. Artificial intelligence techniques have been barely mentioned in this chapter but they clearly are paramount to improving robot flexibility. We have also barely scratched the surface of cognitive architectures (but see Verschure, Chapters 35 and 36, this volume) which are attempts to develop general AI. Equipped with a suitable cognitive architecture, humanoid robots will be able to learn any task from natural interaction with (hopefully) benevolent human teachers.

In summary, we have argued for a future where the body–mind nexus illustrated by the concept of simplexity (Berthoz 2012) will be realized through the use of new technological solutions (materials) on the robot "bodyware" in the form of soft actuators, sensors, and compliant skeletal structure, and—importantly—because of the development of advanced neural-like processing techniques for sensory (possibly multisensory) data analysis, learning, and motor control—the robot "mindware." Since humans are the only truly cognitive systems we know of, copying what evolution has shaped so accurately seems nothing but the wise course of action.

Learning more

Many other chapters in this handbook consider challenges that are highly pertinent to development of humanoid robotics. To highlight a few, Lepora (Chapter 16) and Cutkosky (Chapter 30) discuss many of the technologies and control strategies that will be needed to replicate the human capacity to reach, grasp, and manipulate objects. This in turn will rely on human-like vision (Dudek, Chapter 14) and perceptual processing (Leibo and Poggio, Chapter 25). To interact with people, humanoids will need better AI, but also improved awareness of social others, and the ability to understand social signals through different modalities (see Smith (Chapter 15) on audition, and Wortham and Bryson (Chapter 33) on communication). The ability to have an empathic relationship with humans may rely on the robot having its own emotional and self-regulation capabilities (Vouloutsi and Verschure, Chapter 34), and a rich internal state that might approach having a sense of self (Seth, Chapter 37).

Background on the iCub robot, its origins and capabilities, is given in Metta et al. (2010); the approach to designing humanoid robotics through emulation of human capabilities and following a developmental trajectory, known as 'developmental robotics' is outlined in Lungarella et al. (2003). The challenge of making humanoid robots more social, and therefore more useful to people, is discussed in Wiese et al. (2017), whilst recent efforts to use the DAC cognitive architecture (Verschure, Chapter 36) with the iCub robot are described by Moulin-Frier et al. (in press).

Acknowledgments

The authors wish to thank the colleagues of IIT operating in the iCub facility, RBCS, Smart Materials team, Graphene team, and in the IIT centers of CBN at Lecce and CNST at Milano.

References

Bartolozzi, C., Metta, G., Hofstaetter, M., and Indiveri, G. (2011). *Event-driven vision for the iCub*. Bio-Mimetic and Hybrid Approach to Robotics,Workshop at IEEE International Conference on Robotics and Automation (ICRA 2011). Shangai, China, May 13, 2011.

Bayer, I. et al. (2014). Direct transformation of edible vegetable waste into bioplastics. *Macromolecules*, 47, 5135–43.

Berthoz, A. (2012). *Simplexity: Simplifying Principles for a Complex World*. New Haven, CT: Yale University Press. ISBN: 9780300169348.

Dahiya, R. S., Metta, G., Valle, M., and Sandini, G. (2010). Tactile sensing: from humans to humanoids. *IEEE Transactions on Robotics*, 26(1), 1–20.

Del Prete, A., Nori, F., Metta, G., and Natale, L. (2015). Prioritized motion-force control of constrained fully-actuated robots: "Task space inverse dynamics." *Robotics and Autonomous Systems*, 63(1), 150–7.

Fogassi, L., Gallese, V., Fadiga, L., Luppino, G., Matelli, M., and Rizzolatti, G. (1996). Coding of peripersonal space in inferior premotor cortex (area F4). *Journal of Neurophysiology*, 76, 141–57.

Fumagalli, M., Gijsberts, A., Ivaldi, S., Jamone, L., Metta, G., Natale, L., Nori, F., and Sandini, G. (2010). Learning to exploit proximal force sensing: a comparison approach. In: O. Sigaud and J. Peters (eds), *From motor learning to interaction learning in robots*. Studies in Computational Intelligence series, Vol. 264. Berlin: Springer-Verlag, pp. 159–77.

Kaneko, K., Kanehiro, F., Morisawa, M., Akachi, K., Miyamori, G., Hayashi, A., and Kanehira, N. (2011). Humanoid robot HRP-4 - Humanoid robotics platform with lightweight and slim body. In: *2011 IEEE/RSJ International Conference on Intelligent Robots and Systems (IROS)*, 25–30 Sept. 2011, San Francisco, CA: IEEE, pp. 4400–07.

Krizhevsky, A., Ilya, S., and Hinton, G. E. (2012). ImageNet classification with deep convolutional neural networks. *Advances in Neural Information Processing Systems*, 25, 1097–1105.

Lungarella, M., Metta, G., Pfeifer, R., and Sandini, G. (2003). Developmental robotics: a survey. *Connection Science*, 15(4), 151–90.

Macaluso, E., and Maravita, A. (2010). The representation of space near the body through touch and vision. *Neuropsychologia*, 48(3), 782–95.

Martinez-Hernandez, U. (2015). Tactile sensors. *Scholarpedia*, 10(4), 32398.

Metta, G., Natale, L., Nori, F., Sandini, G., Vernon, D., Fadiga, L., von Hofsten, C., Rosander, K., Lopes, M., Santos-Victor, J., Bernardino, A., and Montesano, L. (2010). The iCub humanoid robot: An open-systems platform for research in cognitive development. *Neural Networks*, 23(8), 1125–34.

Mizuuchi, I., Nakanishi, Y., Sodeyama, Y., Namiki, Y., Nishino, T., Muramatsu, N., Urata, J., Hongo, K., Yoshikai, T., and Inaba, M. (2007). An Advanced Musculoskeletal Humanoid Kojiro. In: *Proceedings of the 2007 IEEE-RAS International Conference on Humanoid Robots (Humanoids 2007)*. Pittsburgh, PA: IEEE, pp. 294–9.

Mnih, V., Kavukcuoglu, K., Silver, D., Rusu, A. A., Veness, J., Bellemare, M. G., Graves, A., Riedmiller, M., Fidjeland, A. K., Ostrovski, G., Petersen, S., Beattie, C., Sadik, A., Antonoglou, I., King, H., Kumaran, D., Wierstra, D., Legg, S., and Hassabis, D. (2015). Human-level control through deep reinforcement learning. *Nature*, 518, 529–33.

Moulin-Frier, C., et al. (in press). DAC-h3: A proactive robot cognitive architecture to acquire and express knowledge about the world and the self. *IEEE Transactions on Cognitive and Developmental Systems*.

Mussa-Ivaldi, F. A., and Giszter, S. F. (1992). Vector field approximation: a computational paradigm for motor control and learning. *Biological Cybernetics*, 67(6), 491–500.

Parmiggiani, A., Maggiali, M., Natale, L., Nori, F., Schmitz, A., Tsagarakis, N., Santos Victor, J., Becchi, F., Sandini, G., and Metta, G. (2012). The design of the iCub humanoid robot. *International Journal of Humanoid Robotics*, 9(4), 1–24.

Pfeifer, R., Gravato Marques, H., and Iida, F. (2013). Soft robotics: the next generation of intelligent machines. In: F. Rossi (ed.), *Proceedings of the Twenty-Third international joint conference on Artificial Intelligence (IJCAI '13)*. Cambridge, MA: AAAI Press, pp. 5–11.

Rea, F., Metta, G., and Bartolozzi, C. (2013). Event-driven visual attention for the humanoid robot iCub. *Frontiers in Neuroscience, Neuromorphic Engineering*, 7(234), 1–11.

Rea, F., Sandini G., and Metta, G. (2014). Motor biases in visual attention for a humanoid robot. In: IEEE/RAS International Conference of Humanoids Robotics (HUMANOIDS 2014), Madrid, Spain, November 18–20, 2014. IEEE, pp. 779–86.

Roncone, A., Hoffmann, M., Pattacini, U., and Metta, G. (2015). Learning peripersonal space representation through artificial skin for avoidance and reaching with whole body surface. In: *EEE/RSJ International Conference on Intelligent Robots and Systems. Hamburg, Germany, September 28–October 02, 2015*. IEEE.

Sandini, G., and Tagliasco, V. (1980). An anthropomorphic retina-like structure for scene analysis. *Computer Graphics and Image Processing*, 14(3), 365–72.

Shidara, M., Kawano, K., Gomi, H., and Kawato, M. (1993). Inverse-dynamics model eye movement control by Purkinje cells in the cerebellum. *Nature*, 365, 50–2.

Taccola, S., Greco, F., Sinibaldi, E., Mondini, A., Mazzolai, B., and Mattoli, V. (2014). Toward a new generation of electrically controllable hygromorphic soft actuators. *Advanced Materials*, doi:10.1002/adma.201404772

Tee, B. C. K., Wang, C., Allen, R., and Bao, Z. (2012). An electrically and mechanically self-healing composite with pressure- and flexion-sensitive properties for electronic skin applications. *Nat. Nano.*, 7, 825–32.

Vanderborght, B., Albu-Schaeffer, A., Bicchi, A., Burdet, E., Caldwell, D.G., Carloni, R., Catalano, M., Eiberger, O., Friedl, W., Ganesh, G., Garabini, M., Grebenstein, M., Grioli, G., Haddadin, S., Hoppner, H., Jafari, A., Laffranchi, M., Lefeber, D., Petit, F., Stramigioli, S., Tsagarakis, N., Van Damme, M., Van Ham, R., Visser, L.C., and Wolf, S. (2013). Variable impedance actuators: A review. *Robotics and Autonomous Systems*, 61(12), December 2013, 1601–14.

Wiese, E., Metta, G., and Wykowska, A. (2017). Robots as intentional agents: using neuroscientific methods to make robots appear more social. *Frontiers in Psychology*, 8(1663). doi:10.3389/fpsyg.2017.01663

Yogeswaran, N., Dang, W., Taube Navaraj, W., Shakthivel, D., Khan, S., Ozan Polat, E., Gupta, S., Heidari, H., Kaboli, M., Lorenzelli, L., Cheng, G., and Dahiya, R. (2015). New materials and advances in making electronic skin for interactive robots. *Advanced Robotics*, 29(21), 1359–1373.

Section VI

Biohybrid systems

Chapter 48

Biohybrid systems

Nathan F. Lepora

Department of Engineering Mathematics and Bristol Robotics Laboratory,
University of Bristol, UK

Biohybrid systems are formed by coupling at least one biological component of a living system to at least one artificial, engineered component. Within the system, the biological and artificial components are not independent, but pass information either in one or both directions, thereby forming a new hybrid bio-artificial entity. Biohybrid systems can involve structures spanning from the nano-scale (molecular) or micro-scale (cellular) to the macro-scale, such as entire organs or body parts.

We have already encountered biohybrid systems in the form of living machines in which robotic components operate symbiotically with cultures of living bacteria that that provide their energy (Ieropoulos et al., Chapter 6). The chapters in this "biohybrid systems" section of the *Handbook of Living Machines* delve further into the technologies underlying such systems and explore applications in healthcare, neuroscience, and robotics.

From the view of those in the biomimetics community, biohybrid systems can be interpreted as a way to take advantage of the unique characteristics of biological systems that have been refined over millions of years of natural selection, in order to solve complex or critical problems in the capabilities of artificial systems. In this methodology of "learning from nature," biological systems can form the basis for novel solutions towards a "soft" and "wet" robotics or "living" technologies based on the biological principles of self-organization, adaptability, and robustness. They also offer a path to building systems using commonplace biological materials and mimicking the propensity of living organisms to live in balanced ecosystems with natural cycles of reproduction, growth, death, and recycling. Biohybrid approaches therefore hold the promise of helping us to create more sustainable future technologies (see Halloy, Chapter 65, this volume).

Biohybrid systems thus offer an opportunity as examplar experiments in how to build "living" artificial systems. On the other hand, biohybrids are entering into the life sciences as useful tools to explore the physiology of living organisms or, over the long term, to restore function that has been lost to injury or disease. Moreover, the development of new and more sophisticated manners of communicating with living material opens up novel opportunities to measure biological parameters, which can help understand physiological mechanisms or replace damaged tissues or organs with artificial versions.

This section of the handbook considers many important examples of biohybrid systems, from brain–machine interfaces, micro- and nano-biohybrids, and biohybrid robots designed using synthetic biology tools, to prosthetic hands with a sense of touch, cochlear replacements that restore hearing, and steps toward brain implants that can boost cognitive function.

Brain–machine interfaces are biohybrid systems that act as communications channels from the central nervous system to an artificial device, such as for people suffering from severe motor impairments due to motor neuron disease or spinal cord injury. Prasad (Chapter 49) reviews the biological principles underlying brain–machine interfaces, focusing on non-invasive methods, particularly EEG (electroencephalography) and MEG (magnetoencephalography). Being a pattern-recognition system a brain–machine interface must involve multiple stages from acquisition to classification, and there are multiple challenges to be overcome for the technology to be as practical as possible for daily use.

A key biohybrid technology is implantable neural interfaces that connect micro- or nano-scale electromechanical systems directly with living nervous tissue. Vassanelli (Chapter 50) gives an overview of the state-of-the-art in this field, focusing mainly on neuro-electronic interfaces. At present these are mainly functional probes that record and stimulate neurons, but future interfaces will evolve to biomimetic systems that fully integrate neural tissue with the capability of restoring losses in neural function from disease or injury.

The next chapter makes the proposal that "biohybrid robots are synthetic biology systems." Ayers (Chapter 51) reviews an emerging area of biorobotics that combines living cells with engineered devices to create biohybrid robots. Current robots have a number of significant limitations, from size (difficult to miniaturize) to reliance on battery technology. Ayers argues that many of these obstacles can be overcome by integrating living engineered cells in robots, making use of the progress in synthetic biology to advance robotics.

If biohybrid system are to operate at the micro- or nano-scale then we will need tools that can assemble them at this level. Fukuda and colleagues (Chapter 52) review progress toward realizing living machines with self-repairing, self-organizing, and self-powering properties, based on multi-scale engineering that spans from the nanometer to meter scale. One of the most interesting challenges is to make biological cells artificially and assembly them as three-dimensional structures for the realization of artificial living machines that are purely biological.

A biohybrid interface could restore a sense of touch for individuals who have lost a limb or who have upper spinal cord injuries. Bensmaia (Chapter 53) explores developments in smart prosthetics equipped with tactile sensing, focusing on the combined challenge to control movements of the limb and receive sensory feedback about the consequences of those movements. Promising approaches for restoring touch via prosthetics include interfacing with the peripheral nerve or directly with the brain.

The cochlear implant, or bionic ear, is one of the most successful biohybrid systems of today. Lehmann and van Schaik (Chapter 54) explain how these devices are able to partially restore hearing, by replacing the function of the inner hair cells of the cochlea with artificial devices that can translate sound into electrical signals in the sensory neuron system. They also discuss the biological inspiration used to build cochlear implants and the constraints in creating biohybrid systems.

Finally, we conclude this section with the possibility of devising brain implants that could replace damaged neural tissue with microelectronics that mimic the function of the original biological circuitry. Song and Berger (Chapter 55) describe a prototype cognitive prosthesis that could restore the ability to form new long-term memories typically lost after damage to the hippocampal region of the brain.

Chapter 49

Brain–machine interfaces

Girijesh Prasad

School of Computing, Engineering and Intelligent Systems, Ulster University, Londonderry, UK

A *brain–machine interface* (BMI) is a biohybrid system primarily intended to act as an alternative communication channel for people suffering from severe motor impairments, such as those with a motor neuron disease or with spinal cord injuries. A BMI can be realized either invasively, using electrodes implanted in the cortex or placed directly on its surface (an electrocorticogram) or by using non-invasive imaging systems. In either case, the aim is to measure the cortical or neurophysiological correlates of brain activity during voluntary cognitive tasks. The focus of this chapter is on non-invasive BMI (see Vassanelli, Chapter 50, this volume for a discussion of invasive interfaces) where the technologies available include: *electroencephalography* (EEG) which measures changes in brain electrical activity through the skull; *magnetoencephalography* (MEG) which measures changes in the magnetic fields produced by brain activity using highly sensitive magnetometers; and various techniques, detailed below, that detect changes in the brain's hemodynamic response (blood flow) and that are indicative of activity in localized brain areas.

Among these non-invasive approaches EEG-based BMI is the most widely investigated. Event-related de-synchronization/synchronization (ERD/ERS) of sensorimotor rhythms (SMRs), the P300 event-related potential (a wave of electrical activity linked to decision making), and steady-state visual evoked potential (SSVEP) are the three main cortical activation patterns in EEG used for designing an EEG-based BMI. Being a pattern recognition system, a BMI involves multiple stages: brain data acquisition, pre-processing, feature extraction, and feature classification along with a device to communicate or control with or without explicit neurofeedback.

Despite worldwide extensive research towards making the BMI as practical as possible for daily use, there are still several challenges to be overcome. One crucial challenge is to account for non-stationary brainwave dynamics resulting in time-variant performance. Also, some people may find it difficult to establish a reliable BMI with sufficient accuracy, although most improve over time with repeated practice. Despite these challenges, BMI research is progressing in two broad application areas: alternative communication by replacing neuromuscular pathways, and neurorehabilitation by helping to activate desired cortical areas for targeted brain plasticity.

Biological principles of non-invasive interfaces

A brain–machine interface (BMI), also called a brain–computer interface (BCI), is normally established by uniquely identifying repeatable metabolic or electric brain activity (e.g. cortical activation) patterns occurring in response to a set of well-defined cognitive tasks. The activation patterns are detected using a pattern recognition system (based on one of the technologies listed above), whose output can be used to select keypad letters, display messages, play computer games, control household devices, control prosthetic/orthotic limbs, or command and control a telepresence robotic device or a smart wheelchair. Thus, via the cognitive task, the BMI enables communication with a computer-controlled machine or device directly through brain activation, bypassing the peripheral nervous and muscular systems. BMI is primarily aimed at providing alternative communication to people suffering from severe motor impairments such as motor neuron disease (MND) and spinal cord injuries (SCIs).

EEG-based approaches

In addition to speech, gestures, including facial expressions and motor tasks such as right hand movement and/or left hand movement, are naturally used by human beings in their routine communications. Coincidently, it was found that hand movement execution may result in changes in EEG signals known as sensorimotor rhythm (SMR) activations in the form of event-related desynchronization (ERD) in the contra-lateral hemisphere (relative to the moved hand) and event-related synchronization (ERS) in the ipsilateral hemisphere (Pfurtscheller and Neuper 2001). The ERD is manifested as a reduction in EEG signal amplitude mainly in the rolandic alpha (or µ band) and the ERS as an enhancement in the signal amplitude mainly in the β band. Fortunately, it has been found that planning or preparation for real hand movement and hand movement imagination also leads to very similar cortical activations in the sensorimotor cortex (Pfurtscheller and Neuper 2001), therefore those with movement impairments can instead perform motor imageries to generate similar cortical activations in µ and β bands (cf. Figure 49.1). Pioneered by Pfurtscheller's group in Graz (Pfurtscheller et al. 2000), the sensorimotor rhythm (SMR) modulation in the form of ERD/ERS is by far the most predominant neurophysiological phenomenon used in devising a non-invasive EEG-based BMI.

Uniquely identifiable cortical activations also occur in another EEG-based measure—event-related potentials (ERPs). Two of the most commonly used in devising EEG-based BMIs are P300 and steady-state visual evoked potential (SSVEP). In P300-based BMI, all the objects or options that may need to be selected are arranged in the form of a grid as part of a graphical user interface (GUI). Exploiting the odd ball paradigm concept, the objects are displayed repeatedly one-by-one in a sequential order and the BMI user focuses his/her attention on the desired object. This creates a uniquely identifiable event-related potential change after a latency of about 300 ms. Farwell and Donchin (1988) were the first to use P300 as the basis for a BMI.

In SSVEP-based BMI, corresponding to each of the objects or options to be selected, a flickering display is created either through an external device such as an LED or a flickering graphical display as part of a GUI. Each of the displays flickers at a certain fixed frequency. When a BMI user focuses his/her attention on one of the displays, the cortical activations (i.e. the amplitude of the potential) corresponding to the flickering frequency and its harmonics in the occipital cortex get enhanced (see chapter 14 of Wolpa and Wolpa 2012).

MEG-based BMI

Another non-invasive approach to recording electrical brain activity is by means of recording the magnetic field produced by the electrical impulses generated by cortical neurons, through a process called magnetoencephalography (MEG). An MEG system may contain up to three hundred sensors and must be operated at the temperature of liquid helium. The MEG provides whole-head views, a high spatiotemporal resolution and its signal's spatial distribution is impervious to the varying masses within the structure of the head. Mellinger et al. (2007) showed that a sensorimotor rhythm (SMR)-based BMI using MEG was as effective as that using EEG.

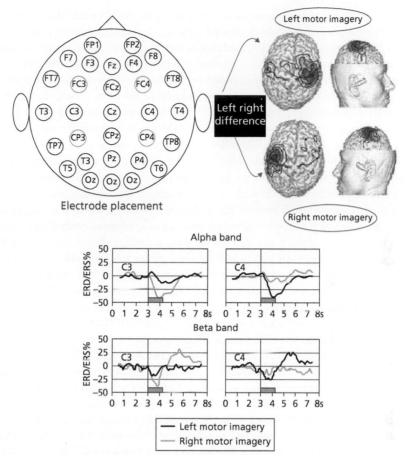

Figure 49.1 ERD/ERS Phenomena in EEG-based SMR BMI. Top left panel: EEG channels C3 and C4 connections in bi-polar mode are highlighted in blue. Top right panel: ERD maps for a single subject calculated for the cortical surface of a realistic head model. The spline surface Laplacian method was applied to the bandpass filtered (9–13 Hz) single-trial EEG data and the distribution of the alpha band ERD was calculated for left and right motor imagery. Bottom panel: Grand average ERD curves recorded during motor imagery from the left (C3) and right sensorimotor cortex (C4). The ERD time courses were calculated for the selected bands in the alpha range for 16 subjects. Positive and negative deflections, with respect to baseline, represent a band power increase (ERS) and decrease (ERD), respectively. The gray bars indicate the time period of cue presentation.

© 2000 IEEE. Reprinted, with permission, from G. Pfurtscheller, Current trends in Graz brain-computer interface (BCI) research, *IEEE Transactions on Rehabilitation Engineering*, 8 (2), pp. 216–219, doi:10.1109/86.847821.

BMI based on brain hemodynamics

In recent years BMI systems have also been devised based on the on-line detection of changes in metabolic brain activities in the form of the blood oxygen level dependent (BOLD) signals obtained from *functional magnetic resonance imaging* (fMRI). During mental tasks, neurons use up their energy supply, which must be replenished by the blood in the form of oxygen and glucose, thereby allowing identification of areas of the brain that are active. Similar to MEG, fMRI offers a high spatial resolution but unlike MEG it is able to penetrate deep within the brain in order to measure neuronal activity. Yoo et al. (2004) demonstrated that volunteers were able to steer their way through a two-dimensional maze using an fMRI-based BMI by performing four different mental tasks each of which allowed separate areas of the brain to be activated. This study, although showing high accuracies, made use of only two participants and took around two minutes to generate each command which resulted in a relatively low information transfer rate.

BMI systems have also been developed based on the on-line detection of changes in metabolic brain activities in response to motor imageries tasks, in the form of hemodynamic signals obtained from near-infrared-spectroscopy (NIRS). The NIRS systems offer high spatial resolution but low temporal resolution. However, such quality systems may be too bulky and expensive for day-to-day constant use. The use of NIRS for BMI is a relatively new concept; it operates by measuring changes in both the regional cerebral blood flow (rCBF) and the cerebral oxygen metabolic rate ($rCMRO_2$). When certain areas of the brain become active they require more oxygenated blood and hence the detection of increased amounts of oxygenated hemoglobin signals. For this, the light of a certain wavelength is emitted then collected by a sensor which is subsequently analyzed. The attenuation which the light undergoes on its passage between emitter and sensor is an indication of the structure of the tissue which it has passed through. Coyle et al. (2007) were the first to study an NIRS-based SMR BMI which detected motor imagery tasks to make a binary choice. Although the system showed great potential, the study resulted in an information transfer rate (ITR) of 1 bit/min and does not contrast well when measured against other similar EEG-based systems. Its advantages lie in the fact that it is low-cost, convenient and portable for the user, and provides good temporal resolution when compared against MEG and fMRI.

The most recent entrant in the BMI field is a technique involving the detection of changes in cerebral blood flow velocity (CBFV), called Transcranial Doppler (TCD) sonography. By its design, TCD is inherently immune from electrical interference such as the power line interference which affects EEG recordings. TCD is a relatively cost-effective method of detecting changes in the brain when compared against MEG or fMRI, exhibits good temporal resolution and whose hardware is also relatively portable. It has been used for BMI in a mental task discrimination study (Myrden et al. 2011) involving nine able-bodied participants who were each asked to perform one of two mental exercises. Sensors placed at the left and right transtemporal windows detected a bilateral increase in CBFV during a mental rotation task whilst a word generation task produced left lateralization.

Biohybrid systems

A biohybrid system implementing a BMI can be built around a pattern recognition system and thus, as seen in Figure 49.2, requires a multi-stage system consisting of brain data acquisition, preprocessing, feature extraction, feature classification, and finally a command and control interface for controlling or communicating to a device, with or without explicit neurofeedback. The most predominantly used brain signal is EEG, which is an ultra-low voltage signal with very low signal-to-noise-ratio (SNR), as the skull dampens signals, dispersing and blurring the electromagnetic waves created by the activations of cortical neurons. Therefore as part of the data-acquisition system, very high quality electrodes and a cap assembly are needed for appropriately attaching sensors to the scalp, following the international 10–20 system of electrode placement, and then high gain op-amp based amplifiers are normally used to substantially amplify the signal in appropriate frequency ranges, so as to obtain a practically useful signal.

Signal pre-processing

Raw EEG signals have a very low signal-to-noise (SNR) ratio due to several factors such as the interference from electrical power lines, motion artifacts, and EMG/EOG interference. Low

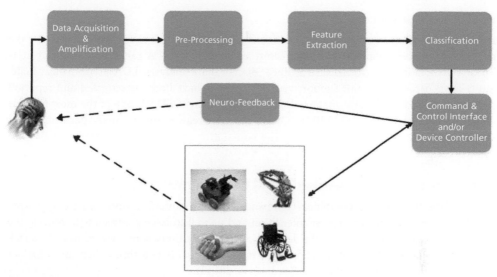

Figure 49.2 A Brain–Machine Interface.

SNR may also occur because EEG contributions from neuronal activations related to BMI tasks may be overshadowed with the activations resulting from multifarious autonomic and other cognitive activities. Pre-processing is therefore carried out to remove unwanted components embedded within the EEG signal leading to increase in signal quality and resulting in better feature separability and classification performance. The pre-processing filter is designed in such a way that most of the EEG components unrelated to BMI tasks are suppressed. Pre-processing carried out using a recurrent quantum neural network (RQNN) based stochastic filter was reported by Gandhi et al. (2014a) to result in a statistically significant improvement in performance across multiple subjects.

Feature extraction

In the feature extraction stage, the pre-processed signal is used to extract features that are likely to provide uniquely identifiable patterns providing enhanced separability among the classes of cognitive tasks used to operate the BMI. The primary aim of feature extraction is to extract mental task-correlated information (or features) from the brain signal regardless of the quality of the EEG signal. The output of the feature extraction stage highly impacts on the performance of the following feature classification stage. For instance, the probability of correct brain state identification can be increased if the feature extraction stage transforms the EEG signal in such a way that the SNR is maximized as much as possible.

The main distinguishing features in all the EEG patterns used for BMI design are changes in power in certain frequency bands, e.g. ERD in SMR of the μ band. The power spectral density (PSD) is therefore the most commonly used feature for visually demonstrating the EEG modulations resulting from the BMI task-related cortical activations. Some form of PSD is also found to be one of the best features in enhancing BMI performance (Herman et al. 2008) and is the most widely used feature in BMI design. Other frequently reported features are band-power, wavelets, and common spatial patterns (CSP). A feature extraction technique using one of the higher order statistics methods called bispectrum, was shown to provide significantly enhanced performance (Shahid and Prasad 2011) by effectively accounting for the fact that the motor imagery (MI)-related EEG signals are highly non-Gaussian and have non-linear dynamic characteristics.

Feature classification

In the classification stage, a pattern classifier is designed for high accuracy in classifying the particular features obtained from the feature extraction stage. A range of linear and non-linear classification algorithms such as linear discriminant analysis (LDA), type-2 fuzzy logic, multi-layer perceptrons, and support vector machines, have been investigated and reported to provide mixed results (Herman et al. 2008). Some form of LDA is one of the most popular classification algorithms used in BMIs. One of the main reasons for the mixed performance of sophisticated classification algorithms is the non-stationary nature of the brain signals used in BMI design.

User interface, neurofeedback, and BMI operation

BMI systems often require a graphical command and control interface customized to their specific BMI paradigm to control user interaction and as well to issue commands to operate the intended device. In proportion to the detected cortical activations, some type of neurofeedback (often in a visual form) is created and provided to the BMI user in real-time to help assess his/her effectiveness in operating the BMI (cf. Gandhi et al. 2014b).

Normally a BMI operates in a cue-initiated timed paradigm, which is called a dependent or synchronous mode of operation. An SMR-based BMI can also be operated in a paradigm-free mode. This is called a self-paced or asynchronous mode of operation. Although the asynchronous mode is more natural to operate, it is difficult to achieve sufficiently reliable performance in this mode. Long-term constant use of a BMI is yet to be seen, though pilot trials of a range of BMI applications have been reported. Frequently reported applications include using BMI for environment control, typing letters, operating a robotic system or a wheelchair, and neurorehabilitation, e.g. by helping to activate desired cortical areas for targeted brain plasticity so as to restore movements in paralyzed limbs. A brief discussion of a range of promising applications reported recently follows.

Applications

The P300-based BMI has primarily been found to be well suited to tasks requiring direct selection. Commonly reported applications are spelling, smart home control, or internet browsing (see chapter 12 in Wolpa and Wolpa 2012). Several applications involving mobile robot or wheelchair control using control strategies involving SMR as well as P300 BMIs, either one type alone or a combination of both, have been reported. For safe wheelchair operation it is essential that obstacle as well as collision avoidance is ensured at all cost. Robotic systems are therefore equipped with a set of appropriate range sensors along with an obstacle avoidance mechanism. As part of a shared control strategy, BMIs are primarily used either to initiate autonomous navigation or to perform a step-by-step navigation control in a supervised mode, while the obstacle avoidance as well as collision avoidance is ensured by the robotic side controls. For instance, in Gandhi et al. (2014b), a shared control strategy involved a command and control interface, called an intelligent adaptive user interface (iAUI), wherein icons of basic movement commands such as forward, left, right, backward, and stop are displayed for selection by a two-class SMR BMI (cf. Figure 49.3). Through bi-directional communication, the positions of the movement commands in iAUI are re-organized based on the position of the mobile robot within the environment so that the most probable movement commands could be selected fastest using just a two-class BMI.

In recent years, a very promising application of the EEG/MEG-based BMI in post-stroke neurorehabilitation has been investigated by several research groups. In this the BMI is primarily used

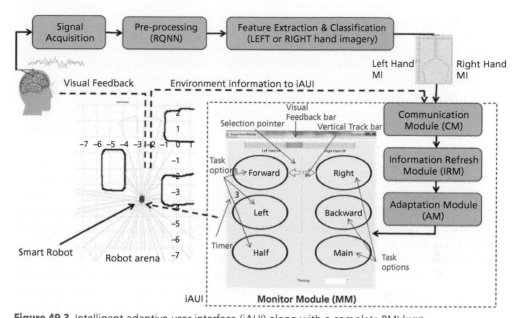

Figure 49.3 Intelligent adaptive user interface (iAUI) along with a complete BMI loop.

to detect cortical activations when the stroke sufferer is performing rehabilitation exercises. Based on the extent of cortical activations in the motor cortex, the BMI provides neurofeedback as well as commands a robotic exoskeleton, if available, to perform physical movements. Thus the BMI helps facilitate focused physical practice as well as motor imagery practice of rehabilitation exercises (Prasad et al. 2010; Ramos-Murguialday et al. 2013), which is found to result in enhanced motor recovery even among chronic stroke sufferers (see Ballester, Chapter 59, this volume, for a discussion of neurorehabilitation paradigms).

Although there are no obvious technology risks with the use of non-invasive BMI, its effect over long-term usage is yet to be carried out. Some participants in pilot trials of SMR BMI have reported feeling tired and, in some cases, headaches after performing motor imageries to operate BMI for more than an hour or so. This is, however, very much dependent upon the kind of application, and type and quality of neurofeedback provided to the participants. Also, it is well known that people prone to epileptic seizures are adversely affected by the flickering displays used in ERP BMIs. In general cortical re-organization resulting from long-term use of BMIs is unlikely to have any adverse effect as it is like learning a new skill.

There are quite a few companies that have launched a range of BMI-related products[1] into the market. These companies primarily offer a hardware and software setup, which can be used to devise a prototype BMI very quickly for either some specific application or further research and innovation. As BMI systems are still mainly in the R&D stage, the companies offering products that provide, as much as possible, open access to data and system information (e.g. g.tec Austria) have much better market penetration, as these are preferred by university BMI labs around the globe.

---

[1] http://en.wikipedia.org/wiki/Brain%E2%80%93computer_interface

Future directions

There are still several challenges to be overcome before a BMI becomes practical to use on a daily basis. For acquiring a good quality EEG signal, active EEG electrodes requiring wet gel for attaching to the scalp are still the preferred option. These require professional support for applying to the user's head and the gel becomes dessicated over time, adversely affecting the quality of the skin contact. Urgent research is therefore needed to devise good quality dry electrodes which can be easily attached to a head-gear and that can be comfortably worn by the user without any professional support.

On the signal processing side, the main challenge is due to the inherently non-stationary characteristics of the brainwaves dynamics. The dynamics also change due to cortical plasticity resulting from repeated BMI usage over time. Also, it is not uncommon to find degraded electrode connections due to several engineering factors such as drying of the gel. As a result, the performance of the BMI classifier designed off-line using previously stored data deteriorates. In order to address this, there is a need to continuously monitor the EEG to ascertain whether there is a significant change or shift in its characteristics and thereby in the features extracted for devising the BMI (Raza et al. 2015). Once the change is detected, the BMI needs to be adapted to the new dynamics. If this adaptation can be automated through semi-supervised on-line training (Raza et al. 2016) so that it does not need constant professional support this will go a long way towards making the BMI practical to use.

Another challenging issue has been that a substantial proportion of users find it difficult to operate a particular type of BMI, that is, their two-class BMI operating accuracy may be 70% or lower, and those people may be considered to suffer from a BMI aphasia. However, a person may not have aphasia for all types of BMIs at the same time; also, performance improves with training and experience. There is therefore now greater emphasis on developing multi-modal hybrid BMI (hBMI) by combining two different modalities wherein inputs are either received in parallel or sequentially. In the sequential arrangement, the first BMI acts as a brain switch. Also, an hBMI may combine two different EEG patterns, e.g. SSVEP and ERD/ERS in SMR. It can also be designed to combine one brain signal and a different type of input such as heart-rate (Shahid et al. 2011) or signals from an NIRS BMI or an eye-tracking system.

There has also been a lot of emphasis on mainstreaming the application of a BMI, so as to enhance its general acceptance in society, more like a consumer product. To this end, some promising works reported are BMI-driven computer games, BMI-based driver attention and fatigue monitoring, and BMI-enabled artistic expression such as music or painting (see chapter 23 in Wolpa and Wolpa 2012). However, the central focus of innovations in BMIs still remains targeted towards alternative communication for replacing impaired neuromuscular pathways and neurorehabilitation for helping to activate desired cortical areas for targeted brain plasticity.

Learning more

One of the most useful books worth consulting for more information is *Brain-Computer Interfaces: Principles and Practice* by Wolpa and Wolpa (2012). This book provides very comprehensive coverage of all aspects of both invasive and non-invasive BMIs. Although chapters are written by different authors they are very well integrated and present a highly coherent description of the state of the art in BMI. The book is aimed at scientists, engineers, and clinicians at all levels having a basic undergraduate level background in biology, physics, and mathematics. An interesting BMI review paper by Silvoni et al. (2011) presents a thorough review of the progress of BMI in relation to post-stroke rehabilitation. Specifically, this paper contextualizes three popular approaches to BMI-based rehabilitation: substitutive strategy, classical conditioning strategy, and operant conditioning strategy.

There are also available several open-source and/or open-access software tools. One of the most widely used tools by BMI researchers is EEGlab[2]. It is a MATLAB-based toolbox for processing event-related EEG, MEG, and other electrophysiological data. It has a rich library of functions for blind source separation, time/frequency analysis, artifact rejection, event-related statistics, and visualization of averaged and single trial data. The other commonly used open source software library is that produced by the BioSig project[3]. The BioSig library is basically a toolbox for Octave and MATLAB with import/export filters, feature extraction algorithms, classification methods, and viewing functions. It can be very effectively used for processing a range of bio-signals such as EEG, ECoG, electrocardiogram (ECG), electrooculogram (EOG), electromyogram (EMG), respiration, and so on. A freely available BMI software system BCI2000[4] is another very useful tool for learning BMI technology as well as developing novel BMI applications. This is a general purpose system for BMI research and can be used for data acquisition, stimulus presentation, and brain monitoring applications. The BCI2000's vision is to become the most widely used tool for diverse areas of real-time bio-signal processing and it is claimed that there are over 2700[5] users around the world.

Acknowledgment

The author sincerely acknowledges the help received from Dr Haider Raza and Dr Vaibhav Gandhi towards designing and drawing the figures in the chapter.

References

Coyle, S. M., Ward, T. E., and Markham, C. M. (2007). Brain-computer interface using a simplified functional near-infrared spectroscopy system. *Journal of Neural Engineering*, 4(3), 219–26.

Farwell, L. A., and Donchin, E. (1988). Talking off the top of your head: toward a mental prosthesis utilizing event-related brain potentials. *Electroencephalogr. Clin. Neurophysiol.*, 70(6), 510–23.

Gandhi, V., Prasad, G., Coyle, D., Behera, L., McGinnity, T. M. (2014a). Quantum neural network-based EEG filtering for a brain–computer interface. *IEEE Transactions on Neural Networks and Learning Systems*, 25(2), 278–88.

Gandhi, V., Prasad, G., McGinnity, T. M., Coyle, D., and Behera, L. (2014b). EEG based mobile robot control through an adaptive brain-robot interface. *IEEE Transactions on Systems Man and Cybernetics: Systems*, 44(9), 1278–85.

Herman, P., Prasad, G., McGinnity, T. M., and Coyle, D. H. (2008). Comparative analysis of spectral approaches to feature extraction for EEG-based motor imagery classification. *IEEE Transactions on Neural Systems and Rehabilitation Engineering*, 16(4), 317–326.

Mellinger, J., Schalk, G., Braun, C., Preissl, H., Rosenstiel, W., Birbaumer, N., and Kübler, A. (2007). An MEG-based brain-computer interface (BCI). *NeuroImage*, 36(3), 581–93.

Myrden, J. B., Kushki, A., Sejdić, E., Guerguerian, A., and Chau, T. (2011). A brain-computer interface based on bilateral transcranial Doppler ultrasound. *PloS One*, 6(9), e24170.

---

[2] http://sccn.ucsd.edu/eeglab/

[3] http://biosig.sourceforge.net/

[4] http://www.schalklab.org/research/bci2000

[5] http://www.schalklab.org/research/bci2000

Pfurtscheller, G., and Neuper, C. (2001). Motor imagery and direct brain–computer communication. *Proceedings of the IEEE*, **89**(7), 1123–34.

Pfurtscheller, G., Neuper, C., Guger, C., Harkam, W., Ramoser, H., Schlögl, A., Obermaier, B., and Pregenzer, M. (2000). Current trends in Graz brain–computer interface (bci) research. *IEEE Transactions on Rehabilitation Engineering*, **8**(2), 216–19.

Prasad, G., Herman, P., Coyle, D. H., McDonough, S., and Crosbie, J. (2010). Applying a brain-computer interface to support motor imagery practice in people with stroke for upper limb recovery: a feasibility study. *Journal of Neuroengineering and Rehabilitation*, **7**(60), 1–17.

Ramos-Murguialday, A., Broetz, D., Rea, M., et al. (2013). Brain-machine-interface in chronic stroke rehabilitation: a controlled study. *Annals of Neurology*, **74**(1), 100–8.

Raza, H., Cecotti, H., Li, Y., and Prasad, G. (2016). Adaptive learning with covariate shift-detection for motor imagery-based brain–computer interface. *Soft Computing*, **20**(8), 3085–96.

Raza, H., Prasad, G., and Li, Y. (2015). EWMA model based shift-detection methods for detecting covariate shifts in non-stationary environments. *Pattern Recognition*, **48**(3), 659–69.

Shahid, S., and Prasad, G. (2011). Bispectrum-based feature extraction technique for devising a practical brain-computer interface. *Journal of Neural Engineering*, **8**(2), 025014. doi: 10.1088/1741-2560/8/2/025014

Shahid, S., Prasad, G., and Sinha, R. K. (2011). On fusion of heart and brain signals for hybrid BCI. *Proc. 5th Int. IEEE EMBS Conference on Neural Engineering*, Cancun, Mexico. IEEE, 5 pp.

Silvoni, S., Ramos-Murguialday, A., Cavinato, M., Volpato, C., Cisotto, G., Turolla, A., Piccione, F., and Birbaumer, N. (2011). Brain-computer interface in stroke: a review of progress. *Clinical EEG and Neuroscience*, **42**(4), 245–52.

Wolpa, J. R., and Wolpa, E. W. (2012). *Brain-Computer Interfaces: Principles and Practice*. New York: Oxford University Press.

Yoo, S., Fairneny, T., Chen, N., Choo, S., Panych, L., Park, H., Lee, S., and Jolesz, F. (2004). Brain-computer interface using fMRI: spatial navigation by thoughts. *Neuroreport*, **15**(10), 1591–5.

Chapter 50

Implantable neural interfaces

Stefano Vassanelli

Department of Biomedical Sciences, University of Padova, Italy

Implantable neural interfaces (*INI*) can be seen as a class of biohybrid systems characterized by the intimate physical and functional interaction between engineered *micro* or *nano electro mechanical systems* (i.e. *MEMS* or *NEMS*) and the nervous tissue *in vivo*. Beginning with implants of single electrodes, INI—also known as *neuronal probes*—have evolved over years and have experienced a development boost in the last few decades. Two converging needs were, and still are, urging such development: the need in neuroscience research for effective tools to investigate brain circuits, and the health challenge of finding alternative strategies to pharmacological treatment for neurological disorders (Reardon 2014). At the forefront of INI development, we are working toward the introduction of new materials such as organic compounds, the integration of physical elements emulating neuronal computation, and the conception of novel approaches and protocols to elicit neuronal activity in a near-physiological manner.

It can be foreseen that neural interfaces, which at present mainly serve the function of "probes" for recording and stimulating neurons, will evolve into "biomimetic" systems integrating structural and functional features of the native tissue and with capability of restoring functional losses. Thus, INI are still in their infancy. The goal of this chapter is to provide a concise overview of actual INI, focusing mainly on neuro-electronic interfaces. Pioneering work towards biomimetic neural interfaces is briefly addressed in the "Future directions" section at the end of the chapter. The focus of this chapter is on implantable neural interfaces: see Prasad (Chapter 49, this volume) for a discussion of non-invasive brain–machine interfaces.

Biological principles of neuro-electronic interfacing

A useful 'application-oriented' classification of implantable neural interfaces is based on the site of implantation that is either central (brain, brain stem, and spinal cord) or peripheral (peripheral nerves, sensory organs). Implantation of probes with multiple recording sites in the brain cortex is a powerful approach to investigate the function of neuronal microcircuits (Buzsáki et al. 2012) and, on the other hand, to implement high-resolution brain–machine interfaces (Lebedev and Nicolelis 2011). One limitation, however, is their invasiveness, considering that central implants unavoidably disrupt neuronal tissue when inserted. An important exception is represented by electrocorticography (ECoG) arrays, which are placed in contact with the brain surface to monitor activity in superficial cortical layers (Buzsáki et al. 2012). It is to circumvent this drawback that major effort is currently being put into developing minimally invasive devices, by reducing the implant size down to the micrometers or nanometers scale, and/or by developing bio-inspired materials and architectures that better integrate with the native tissue. Indeed, a yet unsolved problem related to brain damage during insertion is the formation of a scar of glia cells (gliosis) appearing one or a few weeks after implantation. The scar insulates the implanted device and hampers recording and stimulation of healthy neurons. In general, the reaction is initiated by the traumatic injury of insertion with consequent inflammation, and powered by a foreign-body immune response (Polikov et al. 2005). To date, attacking key cellular and molecular factors in these reaction chains to minimize gliosis remains a central challenge and major advances are expected from studies addressing the biocompatibility of the implant—e.g. by introducing organic and biomimetic materials as scaffolds at the device–tissue interface—as well as from development of specific pharmacological treatments.

Peripheral interfaces with nerves by cuff or more invasive sieve electrodes, instead, preserve brain tissue and have started to be employed for neuroprosthetic applications with amazing results (Bhunia et al. 2015). On the other hand, as they do not provide direct access to the central nervous system, they are of little help when the goal is to investigate high-level processing or when bypassing peripheral nerves is needed to restore function. Cochlear (Gaylor et al. 2013) and, more recently, retinal implants (Chader et al. 2009) are a separate issue as the implant is in contact with a sensory organ—i.e. the organ of Corti and retina, respectively—and not with a peripheral nerve. In the case of cochlear implants (see Lehmann and van Schaik, Chapter 54, this volume), implanted microelectrodes mainly stimulate branches of the auditory nerve, thereby sharing a similarity with peripheral interfaces. Implant functionality, however, strictly relies on the preservation of the anatomical structure of the cochlea. In the case of retina implants, owing to the complex network architecture of the organ and its processing of visual stimuli, stimulation may be sought of different classes of cells in addition to ganglion cells and their fibers forming the optical nerve. Consequently, the engineering of central and peripheral neural interfaces has to meet specific requirements depending on the site of implantation and of the biological structures involved. We will describe different and most relevant technological approaches in the next section, focusing on central interfaces.

INI, however, greatly differ also in terms of their biophysical working mechanisms. Electrical interfaces, when recording, measure ionic currents and related potentials produced by neuronal activity. When stimulating, they generate currents and associated fields that elicit the opening of voltage-gated channels in neurons. Two basic philosophies exist in electrical INI, depending on whether faradaic oxidation–reduction currents or capacitive non-faradaic currents are involved. Once a metal electrode is inserted in the brain tissue, an interface is formed between the solid phase of the electrode and the liquid phase of the electrolyte in the extracellular environment (Bard and Faulkner 2001). In the case of a metal electrode, electrons carry electric charge, while in the extracellular solution charge is carried by ions, mainly sodium, potassium, and chloride. During recording and stimulation of neuronal activity, current flows between electrode and electrolyte and across the cell membrane. Two types of processes can occur at the metal–electrolyte interface. In the first one, electrons are transferred across the metal–solution interface and cause oxidation or reduction of chemical species to occur in the electrolyte. Since Faraday's law governs such reactions (i.e. the amount of chemical reaction caused by the flow of current is proportional to the amount of electricity passed) they are called *faradaic* processes. Electrodes at which faradaic processes occur are sometimes called *charge transfer electrodes*. In reality, a given metal–solution interface will usually show a range of potentials where no reactions occur, because they are thermodynamically or kinetically unfavorable. Instead, processes such as charge adsorption and desorption at the electrode surface can still happen, changing the structure of the electrode-electrolyte interface. Although charge does not cross the interface, capacitive currents can flow (at least transiently) when the potential, electrode area, or solution composition changes. These processes are called *non-faradaic* processes, as they do not involve faradaic oxidation/reduction reactions associated to current flow. Both faradaic and non-faradaic processes can take place when potential changes at the electrolyte (as during recording) or at the electrode (as in the case of stimulation) and both must be taken into account when designing probes, conceiving experiments, and interpreting data.

In general, electrodes can be chosen that are mainly faradaic or non-faradaic across the voltage range that they experience during electrophysiological recording and/or stimulation. A typical example of a nearly purely faradaic electrode is the Ag/AgCl electrode used in patch-clamp recording, while a representative of non-faradaic electrodes is the Pt electrode. In order to prevent faradic oxidation–reaction phenomena from occurring, a thin film coating of dielectric material can be employed to insulate the conductive electrode (metal or also semiconductor), thereby significantly extending the non-faradaic voltage ranges of operation. Existing neural implants based on both approaches are analyzed in the next section.

Whereas electrical interfaces have monopolized the field so far, the interest for optical or mixed optoelectronic implantable systems has revived with the advent of optogenetics. The possibility of both exciting and inhibiting neurons by expression of light-gated ion channels (channelrhodopsins) has catalyzed the attention of many groups. Furthermore, by refining the method through genetic engineering approaches, it is nowadays possible to target channelrhodopsins expression to neuronal sub-populations, allowing for an unprecedented functional dissection of brain microcircuits. The interested reader is referred, for example, to Dugué et al. (2012) for the topic.

Implantable neural interfaces: an overview

The development of implantable brain probes based on MEMS featuring arrays of microelectrodes has experienced a significant boost in recent years (Wise et al. 2008) given the need of novel approaches to record large numbers of neurons at high resolution, in a minimally invasive manner and over long time periods (Buzsáki et al. 2012; Lebedev and Nicolelis 2011). *Multielectrode arrays (MEA)* and *multitransistor arrays (MTA)* integrated in silicon microchips are the two major classes of such probes. They reflect two different philosophies for brain–chip interfacing, with different transduction mechanisms of electrical signals between neurons and semiconductor chips. The microtransducers employed in the two cases are *metal microelectrodes* and *electrolyte-oxide-semiconductor (or metal) field-effect-transistors (EOSFET)*, respectively. The latter is a modified version of the metal oxide semiconductor field effect transistor (MOSFET) that is widely used in integrated circuits in microelectronics. Both technologies were originally developed for *in vitro* recording and then optimized as implantable probes for *in vivo* application (see Figure 50.1).

MEA and, more recently, MTA are at the core of generations of brain-implantable probes whose ultimate goal is the recording of extracellular signals either in the form of spikes or local field potentials (LFP) (Buzsáki et al. 2012). It is noteworthy that the main challenge addressed so far has been the recording of as many neurons as possible in terms of spikes, while preserving biocompatibility and endurance of the implant. Only in recent years has the interest for recording LFPs seen a resurgence (Einevoll et al. 2013). Also, implementing stimulation capabilities has been recognized as extremely important, particularly for application in neuroprosthetics (O'Doherty et al. 2011).

Similarly to single electrode and tetrode technologies, most chip-based neuronal implants share as a common feature a needle-like shape in order to limit tissue damage and, therefore, to minimize the potentially detrimental effect on the physiological function of neuronal circuits. In general, MEA and MTA improve conventional single electrode recording techniques as they provide a number of microelectrodes, each one gathering electrical activity from neighboring neurons (Buzsáki 2004). In practice, they measure both spatial and temporal features of extracellular fields generated by firing neurons, which convey information on spiking activity, action potentials propagation, and synaptic activation (Einevoll et al. 2013). Nowadays, advanced implantable

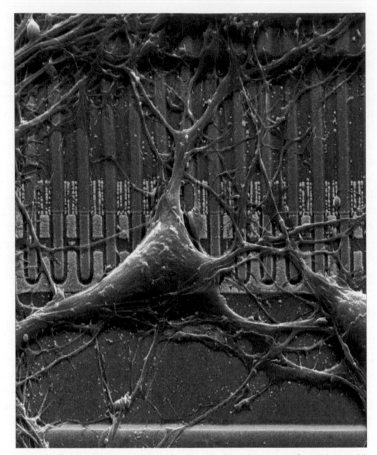

Figure 50.1 MTA chip for neuronal recording *in vitro*. Cultured neurons from the rat hippocampus grow on the surface of a silicon microchip coated by a thin film of silicon dioxide. A linear MTA covered by neurons is visible in the foreground. Small dark squares are the oxide-insulated gates of the individual transistors that are integrated in the bulk silicon and used as voltage sensors. Extracellular voltages generated by a neuron growing on the transistor gate modulate through the oxide source-drain current of the transistor.

Reproduced from *The Journal of Neuroscience*, 19 (16), Stefano Vassanelli and Peter Fromherz, Transistor Probes Local Potassium Conductances in the Adhesion Region of Cultured Rat Hippocampal Neurons, pp. 6767–6773, Figure 2 © 1999, The Society for Neuroscience, with permission.

probes feature hundreds of recording sites integrated at distances down to the micrometers range. Thus, upcoming generations of MEA and MTA implants are expected to electrically "*image*" fields in the brain tissue, a goal that has already been achieved for brain slices *in vitro* (Ferrea et al. 2012; Hutzler 2006). To this endeavor, joint research is ongoing involving interdisciplinary teams of electronic engineers, material scientists, and neurophysiologists. This research faces challenges such as reliably detecting low-amplitude voltages and currents generated by neurons in the extracellular environment (typically in the range of microvolts and nanoamperes, respectively), limiting power consumption and heat production, and preserving long-term biocompatibility of the implants. A further hurdle that needs to be overcome is to guarantee high measurement accuracy. Recording extracellular neuronal signals without affecting their shape and amplitude is fundamental to extracting correct and exhaustive electrophysiological information on neuronal activity both when considering LFPs and extracellular spikes (Einevoll et al. 2013; Pettersen et al. 2012). Ideally, therefore, neuronal probes should not distort in any way the measured signal.

As anticipated above, researchers are eyeing the potential of INI for stimulation and *micro-stimulation* in particular, that is for the spatially distributed stimulation of neuronal networks at high resolution (Hanson et al. 2008; O'Doherty et al. 2011). Both MEA and arrays of *electrolyte-oxide-semiconductor-capacitors (MCA)* could be used for this purpose. As a complement to opto-genic and pharmacological stimulation, they represent a means with which to probe neuronal circuits in experimental neuroscience and to convey sensory information to the brain in advanced neuroprostheses. Furthermore, electrical microstimulation at the central and peripheral level is thought to represent a potentially revolutionary treatment for neurological disorders (Reardon 2014). Thus, although optical stimulation methods have gained momentum for *in vivo* studies, with the advent of optogenetics and its ability to target specific neurons, electrical stimulation remains a cardinal approach and with closer application to clinics. Compared to recording, and despite the success of peripheral implants such as the cochlear (Gaylor et al. 2013) and first retinal implants (Chader et al. 2009), microstimulation of brain neuronal networks *in vivo* by chip-based microelectrode arrays remains mostly unexploited. Reasons may be ascribed in some measure to technical difficulties such as the need to repeatedly deliver high voltages or currents to excite neurons (up to volts or hundreds of microamperes) while avoiding microelectrode corrosion or tissue damage. In view of the demand from neuroscientists for "near-physiological" stimulation methods (O'Doherty et al. 2011), current probes based on microelectrodes appear unsatisfactory, lacking neuronal selectivity and far from mimicking real synaptic inputs. Thus, microstimulation still requires significant improvement in terms of the resolution that can be achieved and efficiency of communication with neurons.

Hereafter, we provide a brief summary of state-of-the art chip-based neural implants used for recording and stimulation *in vivo*. The two different 'philosophies' of electrical brain–chip interfacing are presented with a description of fundamental technical aspects and examples of use in neuroscience.

State-of-the-art neural implants

As anticipated previously, the core difference between the two philosophies of interfacing lies deeply in their electrochemical nature: as a first rough approximation, in the case of MEA, electrons work as charge carriers between metal electrode and electrolyte generating faradaic currents. In electrolyte-oxide-semiconductor transducers, instead, capacitive non-faradaic currents through a thin layer of insulating oxide underlie recording and stimulation. Thus, in the latter case, electrons in the semiconductor/metal and ions in the solution play separately the role of charge carriers. It is highly advisable to understand and to keep in mind the fundamentally different nature and behavior of the two interfaces when designing and performing electrophysiological experiments. For more information on the electrochemistry of the interfaces the interested reader can refer to Bard and Faulkner (2001).

First of all, despite the variety of available probes, all of them obey some general principles of operation and face common hurdles. Neuronal activity is recorded by implanted microelectrodes in the form of "spikes" (i.e. the usual action potential-related extracellular signals also known as single units) or LFPs. Although the exact origin and mechanisms of spatial spread in the brain tissue of such signals are not yet understood in depth (Einevoll et al. 2013; Pettersen et al. 2012), it is implicit that monitoring neuronal network activity at high resolution requires electrodes that are small when compared to neurons and their compartments. In state-of-the-art probes, this challenge remains partially unmet with microelectrodes having dimensions that are comparable, or slightly smaller, than neuronal bodies and with a surface area that is in the range between a hundred and thousands of micrometers squared. Reducing electrode size comes with an additional challenge: that is, the recording of signals with very small amplitudes. Recording single units and LFPs with a decent signal-to-noise-ratio (SNR), and to do this chronically, is technologically demanding. The amplitude of action potentials and LFPs recorded in the CNS by extracellular electrodes is typically in the order of a few hundreds of microvolts or smaller. Thus, low noise MEAs and readout electronics are mandatory for *in vivo* recording. In general, part of the background noise in single-unit recording is in reality "neuronal noise" from the multitude of background action potentials. Similarly, at lower frequencies, brain oscillations and background neuronal ensembles activity affect SNR for LFP recording. However, metal microelectrode impedance does contribute to noise. In particular, high electrode impedance will increase noise and, in combination with stray capacitances between the electrode and the recording amplifier, will reduce the electrodes' high-frequency response by RC filtering. Thus, impedances, at 1 kHz, ranging from approximately 50 kΩ to 1 MΩ, typically characterize microelectrodes used in neural implants. To improve performance, microelectrodes have been developed based on a wide range of materials, such as stainless steel, gold, tungsten, platinum, platinum–iridium alloys, iridium oxide, titanium nitride, and poly(ethylenedioxythiophene) (PEDOT). Being extracellular signals associated to tiny potentials and small currents at the recording electrode (at least as far as recording is concerned), faradaic processes are negligible. Instead, when electrodes are used for stimulation, large voltages have to be applied and faradaic reactions may come into play, depending also on the electrodes' material and the quality of the interface (Cogan 2008; Merrill et al. 2005).

Furthermore, recording with chronically implanted microelectrodes has been challenging for reasons that have little to do with the properties of the electrodes themselves. Tissue reaction with gliosis and formation of a tissue scar interposing between probe and neurons is probably the major obstacle (Moxon et al. 2009; Polikov et al. 2005) and it is still unclear how materials, dimensions, and geometry of silicon probes affect the glia response. Overall, as micromotion of the probe relative to the brain can cause tissue damage in chronic implants, tethering to the skull should be made through flexible structures. In any case, having a small neuron-to-electrode distance (i.e. below about 100 µm) is critical for recording single units as the field strength rapidly decreases moving away from the spike source (Pettersen et al. 2012). Perhaps mainly for this reason, maintaining good and stable recording for more than 1–2 years has proven to be a challenge both in monkeys (Suner et al. 2005) and in first clinical trials (Hochberg et al. 2006), which has also led to the adoption of cortical ensemble or multiunit recording methods to provide more stable control signals for prosthetic devices. Gliosis also hampers stimulation efficiency, but additional issues have to be considered when metal microelectrodes are to be employed for eliciting neuronal activity. Due to the large driving voltages, selection of adequate electrode materials is crucial to avoid as much as possible irreversible redox reactions, with production of chemical species potentially toxic for the brain tissue. On the other hand, reactions must be minimized that could lead to electrode corrosion. Thus, in principle, capacitive charge injection is more desirable than faradaic, although high-capacitive current delivery is only possible with capacitor electrodes that are porous or employ high dielectric constant coatings (see below).

Among faradaic electrodes that can be used both for recording and stimulation in neuronal probes, platinum (Pt) and iridium (Ir) oxide electrodes are the most common. Pt (and PtIr alloys) can inject charge both by faradaic and non-faradaic processes, with the faradaic component usually dominating over the capacitive one. Nevertheless, cochlear implants (among the most successful and experimented implants in patients) still use Pt or PtIr microelectrodes for stimulation, even though relatively large electrode areas are required to keep current densities safe (Cogan 2008). Another strategy is thin-film coating of hydrated Ir oxide, obtained through Ir activation. The film, once deposited on the metal surface of an electrode, greatly increases charge injection capacity through a fast and reversible faradaic reaction between the Ir^{3+} and Ir^{4+} oxidation states of the iridium ions. Activated iridium oxide microelectrodes were also developed for implantable silicon probes taking advantage of thin-film micromachining technology, and even tested in clinical studies requiring intracortical stimulation. To date, iridium oxide microelectrodes are considered among the best for neuronal stimulation.

Titanium nitride (TiN) and tantalum/tantalum oxide (Ta/Ta_2O_5) electrodes are well known solutions in the world of capacitive stimulation. TiN, a chemically stable and biocompatible metallic conductor, offers large charge injection capacity through the so-called "electrical double layer," particularly because highly porous surfaces can be obtained that enhance electrode capacitance. Coating a metal surface with a thin dielectric layer is another strategy to enhance charge injection. This is the case of Ta/Ta_2O_5 electrodes, whose use, however, has never been extensive.

All efforts to improve microelectrodes performance are reviewed in detail in Cheung (2007), Kipke et al. (2008), Pine (2006), and Wise et al. (2008). In the future, emergent materials and innovative nanostructures, such as silicon nanowires, carbon nanotubes, conductive polymers such as poly(ethylenedioxythiophene) (PEDOT), or other polymeric compounds will likely find their way for application in INI. Nevertheless, and despite the wealth of innovations, two basic architectures have inspired nearly all implantable MEAs that are nowadays available: the *Michigan* and the *Utah* arrays.

The Michigan probe

Some of the first silicon-based implantable multielectrode arrays were made by Wise, Starr, and Angell in the 1970s. These structures have evolved into several subsequent devices whose basic architecture is commonly referred to as a "Michigan" probe. In this probe, several planar microelectrodes are patterned on one of the two faces of a shank structure, providing a set of spatially arranged recording sites (see Figure 50.2). In brief, the probe structure consists of a selectively etched (micromachined) silicon substrate having needle-like (shank) geometry. Conducting leads are insulated above and below by inorganic dielectrics while recording (and in some cases also stimulating) microelectrodes with approximately planar surfaces are formed by an area of exposed metal. Starting in the 70s, the manufacturing of the Michigan probe relied on single-sided processing of silicon wafers that were becoming available in electronic industry, including technologies such as diffused boron etch-stops, reactive ion etching (RIE), and silicon-on-insulator (SOI) wafer technology. Since then, most of the present approaches to probe manufacturing use silicon substrates, relying on expertise and equipment developed for industrial microelectronic circuits and achieving an incomparable degree of miniaturization and reliability of yield. A detailed description of technologies and processes for Michigan probes manufacturing can be found in Wise et al. (2008), including advanced implementations such as 2D and 3D multi-shank structures (see Figure 50.2) and downstream CMOS signal processing and wireless communication circuitry. Single shank and 2D multi-shank structures inspired by this work are commercially available and they have been reported to work quite successfully in acute and chronic implantation (Kipke et al. 2008).

Michigan probes also offer microstimulation capability. Typical current thresholds for stimulation with one single microelectrode are about 10–30 μA, but currents of 100 μA or more are very common. Facing the need of such a relatively high current, a potential drawback is interchannel

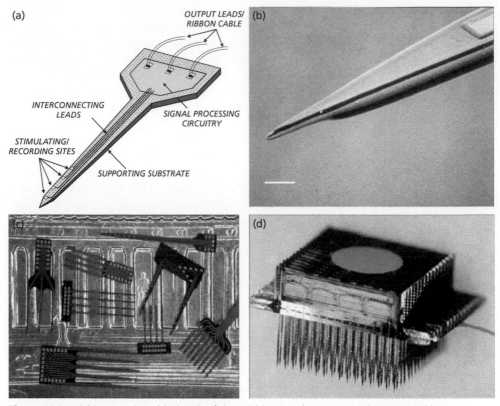

Figure 50.2 Michigan probes. (a) Sketch of the Michigan probe structure characterized by a needle-like shape to favor implantation. The needle region of the probe is inserted vertically in the brain while the non-recording part contains metallic pads for output leads connections. Integrated in the shank and close to the tip of the chip are recording (stimulating) sites (four planar metal microelectrodes arranged in a linear array). (b) SEM micrograph of the tip part of a Michigan probe (scale bar: 10 μm). (c) Different probe designs, including multi-shank structures. (d) Multi-shank structure with 1024 sites for 3D recording.

(a), (b) and (d) © 2004 IEEE. Reprinted, with permission, from K. D. Wise, D. J. Anderson, J. F. Hetke, D. R. Kipke, and K. Najafi, Wireless implantable microsystems: high-density electronic interfaces to the nervous system, *Proceedings of the IEEE*, 92 (1), pp. 76–97, doi: 10.1109/JPROC.2003.820544. (c) © 2008 IEEE. Reprinted, with permission, from K. D. Wise, A. M. Sodagar, Y. Yao, M. N. Gulari, G. E. Perlin, and K. Najafi, Microelectrodes, microelectronics, and implantable neural microsystems, *Proceedings of the IEEE*, 96 (7), pp. 1184–1202, doi: 10.1109/JPROC.2008.922564.

crosstalk. In the Michigan probe, the problem is virtually negligible as the conducting silicon substrate below the leads and the electrolyte above them both act as ground planes shunting parasitic currents. Concerning stray capacities, the small area of the lithographically defined leads minimizes shunt capacitance so that voltage filtering and attenuation is negligible as well. Recording sites can be gold or platinum, but anodically formed iridium oxide is increasingly used as it produces significantly lower recording impedances than other materials. Employing Ir oxide microelectrodes was essential for stimulating sites, as it enabled more than 20 times the charge delivery to tissue than platinum or gold at the same voltage. On such basis, a variety of new designs have been explored to better cope with tissue integration, and they are reviewed in Wise et al. (2008).

The Utah Electrode Array

The Utah Electrode Array (UEA; see **Figure. 50.3**), is the best-known 3D MEA probe, and the first one to be implanted chronically in the cortex of paralyzed patients (Hochberg et al. 2006). While the Michigan electrode array has been built to take advantage of the planar photolithographic manufacturing techniques used in the semiconductor industry, the UAE was designed "from scratch" to meet the need for a neural interface for multi-site recording within an horizontal plane in the cerebral cortex. Novel manufacturing techniques had to be developed in order to build this device, and they are described in detail in Jones et al. (1992). The probe consists of a silicon-based, 3D structure of a 10×10 array of tapered silicon electrodes with a base width of 80 μm, a length up to 1500 μm, and an electrode-to-electrode horizontal distance of 400 μm. Insulation between electrodes is provided by glass in the base plane of the array. The starting material is a thick n-type wafer, through which p^+ trails are created by thermomigration, and then chemically etched and sawed to expose the p^+ trails as thin needles. The tip of each needle electrode is coated with metals: gold, platinum, or iridium. Particularly, activated iridiium oxide is used when stimulation is also required. A backside electrical contact to each electrode is formed by sputter deposition and patterned by photolithography. Finally, a biocompatible polymer, Parylene-C, encapsulates the UEA and the probe is connected to an external data acquisition system by thin Pt–Ir wires that are soldered to the bond pads of each of the 100 electrodes. These arrays have been implanted in the cat auditory and visual cortex and recent work has proven that they can be used for chronic recording (Hochberg et al. 2006; Suner et al. 2005). It is now ascertained that the array can support single-unit recording for 1 to 5 years, depending on the subject implanted and despite suffering from signal loss from several electrodes. A graded variation of the UEA with microelectrode length from 0.5 to 1.5 mm (see Figure 50.3) was conceived to permit focal excitation of fibers at different depths and was implanted in the cochlear and sciatic peripheral nerve.

UEA developers have also engineered tools that enable the probe to be implanted in cortical tissues by limiting tissue damage. Even though the individual needle-shaped electrodes are extremely sharp, early attempts at implanting them into the visual cortex only deformed the cortical surface and resulted in incomplete implantation. Further, the compression of the cortical surface produced by slow mechanical insertion can injure blood vessels, causing intracranial

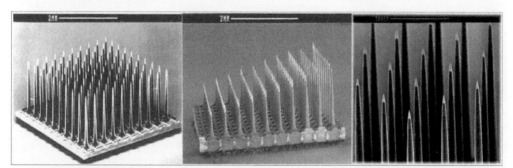

Figure 50.3 The Utah Electrode Array. (*left*) Micrograph of the UEA featuring 100 silicon-based microelectrodes. (*middle*) Slanted array is used to stimulate fibers at different depths in the cochlear nerve. (*right*) Enlarged view of the array showing platinum-coated tips. From *The Center for Neural Interfaces, and the John A. Moran Eye Center, University of Utah* (http://www.bioen.utah.edu/cni/image1.html#content)

hemorrhage and cortical edema. Because the brain is a viscoelastic material, it behaves in a much more rigid fashion if electrodes are inserted at a very high velocity. A surgical instrument that appears to circumvent the above-mentioned problems has been developed based upon this concept and pneumatically inserts the probe in about 200 μs.

Transistor-based probes

The idea of using transistors instead of metal electrodes as sensors for recording biological electrochemical signals dates back to the 70s when ion-sensitive field effect transistors (ISFETs) were first introduced. The concept relies on the modification of a standard MOSFET where the gate metal is replaced by an insulating oxide for the contact with the electrolyte. While recording with ISFETs was centered on measuring changes of ions concentration in solution (e.g. H^+ and Na^+), Fromherz and co-workers provided first experimental evidence that Electrolyte Oxide Semiconductor Field Effect Transistors (EOSFET) can measure extracellular potentials generated by spiking neurons *in vitro* (Fromherz 2003). In subsequent experiments, the approach was optimized to record dissociated mammalian neurons and brain slices (Fromherz 2003). Aiming for a two-way semiconductor–neuron interface, an EOSFET counterpart was developed for stimulation: the electrolyte oxide semiconductor capacitor (EOSC). The device shared with the EOSFET the concept of establishing a capacitive (i.e. non-faradaic) electrical coupling between the semiconductor silicon and the electrolyte, enabling the stimulation of neurons through displacement currents in neuronal cultures and brain slices. Thus multitransistor arrays (MTA) and multicapacitor arrays (MCA) represent an alternative to classical metal electrode based MTA for neuroelectronic interfacing. Only recently, MTA were exploited for *in vivo* recording (Vassanelli et al. 2012; see Figure 50.4), opening novel opportunities for the manufacturing of implantable neuronal probes for high-resolution and large-scale recording of neuronal networks. For a detailed description of the method please refer to Vassanelli (2014).

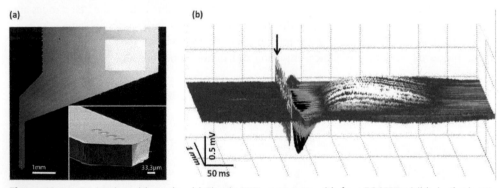

Figure 50.4 MTA implantable probe. (a) Simple MTA prototype with four EOSFETs visible in the inset. The chip features a needle, with transistors at the tip, while the wider expansion contains metallic bond pads (white rectangles) for contacting readout electronics. The whole chip is surrounded by a thin (13 nm) insulating layer of TiO_2. (b) LFP depth profile (50 sweeps average) measured in the rat barrel cortex by the MTA probe. Upon whisker stimulation (black arrow), an evoked LFP is generated which varies across the cortex within the explored 1 mm depth (Y axis). Note the high spatial resolution (10 μm) of the profile.

MTA developments aiming for large-scale and high-resolution recording have taken advantage of complementary metal oxide semiconductor (CMOS) technology to integrate up to about 16,000 transistors in a 1 mm^2 array and, more recently, reaching the remarkable density of 32,000 transistors in a 2.6 mm^2 area (Eversmann et al. 2011). This MTA technology and similar high-resolution large-scale MEAs approaches (Ferrea et al. 2012; Frey et al. 2010) are poised to revolutionize the way neuronal networks are investigated, by imaging, from the extracellular side, their electrical activity.

Only recently, and with a time lag of nearly 40 years with respect to metal microelectrodes, EOS elements (transistors and capacitors) have appeared in first prototypes of implantable neuronal probes. A proof-of-principle demonstration of LFP recording in the rat barrel cortex is shown in Figure 50.4.

Implantable MTAs featuring EOS capacitors have been also tested for stimulation, first *in vitro* and then *in vivo*, allowing for a reliable stimulation of the rat brain cortex for at least three months (Cohen and Vassanelli, unpublished results).

Future directions

Imaging neuronal networks *in vivo* by high-resolution recording of extracellular potentials is becoming within reach. However, crucial bottlenecks will have to be overcome. Above all, power management will require significant steps of development in order to minimize dissipation and overheating of the brain tissue in case of large arrays. New avenues are being explored in the attempt to make probes smaller, highly biocompatible, and suitable to tissue integration. Silicon nanowire field-effect-transistors have been arranged in two- and three-dimensional microporous structures that may facilitate tissue integration (Liu et al. 2015). Moving away from silicon, other emerging materials such as carbon nanotubes, graphene, or conductive polymers promise to revolutionize the way multielectrode arrays are manufactured. Furthermore, novel optoelectronic probes merging neuro-electronic interfacing with optogenetic stimulation have appeared that may represent an extraordinary tool for basic neuroscience to explore neuronal microcircuits *in vivo*. Despite these fascinating developments, we feel that MEA- and MTA-based implantable probes will remain the option of choice for *in vivo* recording in the near future and, most likely, for many years to come. Taking advantage of processes developed for silicon in the microelectronic industry, their performance will be optimized well beyond current limitations. The new implantable neuronal interfaces will provide neuroscientists with a tool for combining high-resolution electrical imaging of brain activity with behavioral studies, and, in the longer term, to interface biomimetic neuromorphic circuits with neurons to restore functional losses in neurological patients.

Learning more

For the electrochemistry and basic chemical and physical principles of the electrode–electrolyte interface the reader is encouraged to refer to Bard and Faulkner (2001). Their book is a comprehensive description of electrochemistry of electrodes including a detailed description of phenomena occurring at the electrode–electrolyte interface. Although a full understanding requires a solid electrochemistry and physical chemistry background, readers outside the field (e.g. neuroscientists) can also greatly profit from reading it.

For a detailed review on INI please refer to Kipke et al. (2008) and Wise et al. (2008). They are classical but still relatively up-to-date reviews on technologies used for INI manufacturing, also discussing application examples.

For a more comprehensive description of MEA and MTA approaches, selected readings are Vassanelli (2014) and, for what concerns MTA, Fromherz (2003). The latter, in particular, is a book chapter spanning the history and fundamentals of oxide-insulated semiconductor devices used for recording and stimulating neurons.

Finally, readers interested in state-of-the-art applications of INI in neuroscience and the rationale for their use in brain–machine interfaces (BMI) and investigation of neuronal microcircuits can refer to the reviews by Buzsáki et al. (2012) and Lebedev and Nicolelis (2011).

References

Bard, A.J., and Faulkner, L.R. (2001). *Electrochemical methods: fundamentals and applications.* New York: Wiley.

Bhunia, S., Majerus, S., and Sawan, M. (2015). *Implantable biomedical microsystems: design principles and applications.* Amsterdam: Elsevier.

Buzsáki, G. (2004). Large-scale recording of neuronal ensembles. *Nat. Neurosci.,* 7, 446–51. doi:10.1038/nn1233

Buzsáki, G., Anastassiou, C. A., Koch, C. (2012). The origin of extracellular fields and currents: EEG, ECoG, LFP and spikes. *Nat. Rev. Neurosci.,* 13, 407–20. doi:10.1038/nrn3241

Chader, G. J., Weiland, J., and Humayun, M. S. (2009). Artificial vision: needs, functioning, and testing of a retinal electronic prosthesis. In: J. Verhaagen et al. (eds), *Neurotherapy: Progress in Restorative Neuroscience and Neurology.* Progress in Brain Research series, no. 175. Amsterdam: Elsevier, pp. 317–32.

Cheung, K.C. (2007). Implantable microscale neural interfaces. *Biomed. Microdevices,* 9, 923–38. doi:10.1007/s10544-006-9045-z

Clark, G. A., Ledbetter, N. M., Warren, D. J., and Harrison, R. R. (2011). Recording sensory and motor information from peripheral nerves with Utah Slanted Electrode Arrays. In: 2011 Annual International Conference of the IEEE Engineering in Medicine and Biology Society, EMBC, pp. 4641–44. doi:10.1109/IEMBS.2011.6091149

Cogan, S.F. (2008). Neural stimulation and recording electrodes. *Annu. Rev. Biomed. Eng.,* 10, 275–309. doi:10.1146/annurev.bioeng.10.061807.160518

Dugué, G. P., Akemann, W., and Knöpfel, T. (2012). A comprehensive concept of optogenetics. In: T. Knopfel and E. S. Boyden (eds), *Optogenetics: Tools for Controlling and Monitoring Neuronal Activity.* Progress in Brain Research series, no. 196. Amsterdam: Elsevier, pp. 1–28.

Einevoll, G. T., Kayser, C., Logothetis, N. K., and Panzeri, S. (2013). Modelling and analysis of local field potentials for studying the function of cortical circuits. *Nat. Rev. Neurosci.,* 14, 770–85. doi:10.1038/nrn3599

Eversmann, B., Lambacher, A., Gerling, T., Kunze, A., Fromherz, P., and Thewes, R. (2011). A neural tissue interfacing chip for in-vitro applications with 32k recording / stimulation channels on an active area of 2.6 mm^2. In: *2011 Proceedings of the ESSCIRC.* ESSCIRC, pp. 211–14. doi:10.1109/ESSCIRC.2011.6044902

Felderer, F., and Fromherz, P. (2011). Transistor needle chip for recording in brain tissue. *Appl. Phys. A,* 104, 1–6. doi:10.1007/s00339-011-6392-2

Ferrea, E., Maccione, A., Medrihan, L., Nieus, T., Ghezzi, D., Baldelli, P., Benfenati, F., and Berdondini, L. (2012). Large-scale, high-resolution electrophysiological imaging of field potentials in brain slices with microelectronic multielectrode arrays. *Front. Neural Circuits,* 6, 80. doi:10.3389/fncir.2012.00080

Frey, U., Sedivy, J., Heer, F., Pedron, R., Ballini, M., Mueller, J., Bakkum, D., Hafizovic, S., Faraci, F. D., Greve, F., Kirstein, K.-U., and Hierlemann, A. (2010). Switch-matrix-based high-density microelectrode array in CMOS technology. *IEEE J. Solid-State Circuits,* 45, 467–82. doi:10.1109/JSSC.2009.2035196

Fromherz, P. (2003). Neuroelectronic interfacing: semiconductor chips with ion channels, nerve cells, and brain. In: R. Waser (ed.), *Nanoelectronics and information technology.* Berlin: Wiley-VCH Verlag, pp. 781–810.

Gaylor, J. M., Raman, G., Chung, M., Lee, J., Rao, M., Lau, J., and Poe, D. S. (2013). Cochlear implantation in adults: a systematic review and meta-analysis. *JAMA Otolaryngol.: Head Neck Surg.*, **139**, 265–72. doi:10.1001/jamaoto.2013.1744

Hanson, T., Fitzsimmons, N., and O'Doherty, J. E. (2008). Technology for multielectrode microstimulation of brain tissue. In: M. A. Nicolelis (ed.), *Methods for neural ensemble recordings*. Frontiers in Neuroscience. Boca Raton, FL: CRC Press, Chapter 3.

Hochberg, L. R., Serruya, M. D., Friehs, G. M., Mukand, J. A., Saleh, M., Caplan, A. H., Branner, A., Chen, D., Penn, R. D., Donoghue, J. P. (2006). Neuronal ensemble control of prosthetic devices by a human with tetraplegia. *Nature*, **442**, 164–71. doi:10.1038/nature04970

Hutzler, M. (2006). High-resolution multitransistor array recording of electrical field potentials in cultured brain slices. *J. Neurophysiol.*, **96**, 1638–45. doi:10.1152/jn.00347.2006

Jones, K. E., Campbell, P. K., and Normann, R. A. (1992). A glass/silicon composite intracortical electrode array. *Ann. Biomed. Eng.*, **20**, 423–37.

Kipke, D. R., Shain, W., Buzsáki, G., Fetz, E., Henderson, J. M., Hetke, J. F., and Schalk, G. (2008). Advanced neurotechnologies for chronic neural interfaces: new horizons and clinical opportunities. *J. Neurosci.*, **28**, 11830–38. doi:10.1523/JNEUROSCI.3879-08.2008

Lebedev, M.A., and Nicolelis, M.A.L. (2011). Toward a whole-body neuroprosthetic. *Prog. Brain Res.*, **194**, 47–60. doi:10.1016/B978-0-444-53815-4.00018-2

Liu, J., Fu, T.-M., Cheng, Z., Hong, G., Zhou, T., Jin, L., Duvvuri, M., Jiang, Z., Kruskal, P., Xie, C., Suo, Z., Fang, Y., and Lieber, C. M. (2015). Syringe-injectable electronics. *Nat. Nanotechnol.*, **10**, 629–36. doi:10.1038/nnano.2015.115

Merrill, D. R., Bikson, M., Jefferys, J. G. R. (2005). Electrical stimulation of excitable tissue: design of efficacious and safe protocols. *J. Neurosci. Methods*, **141**, 171–98. doi:10.1016/j.jneumeth.2004.10.020

Moxon, K. A., Hallman, S., Sundarakrishnan, A., Wheatley, M., Nissanov, J., and Barbee, K. A. (2009). Long-term recordings of multiple, single-neurons for clinical applications: the emerging role of the bioactive microelectrode. *Materials*, **2**, 1762–94. doi:10.3390/ma2041762

O'Doherty, J. E., Lebedev, M. A., Ifft, P. J., Zhuang, K. Z., Shokur, S., Bleuler, H., and Nicolelis, M. A. L. (2011). Active tactile exploration using a brain–machine–brain interface. *Nature*, **479**, 228–31. doi:10.1038/nature10489

Pettersen, K. H., Lindén, H., Dale, A. M., and Einevoll, G. T. (2012). Extracellular spikes and current source density. In: R. Brette and A. Destexhe (eds), *Handbook of Neural Activity Measurements*. Cambridge, UK: Cambridge University Press, pp. 92–135.

Pine, J. (2006). A history of MEA development. In: M. Taketani and M. Baudry (eds), *Advances in network electrophysiology*. New York: Springer, pp. 3–23.

Polikov, V. S., Tresco, P. A., Reichert, W. M. (2005). Response of brain tissue to chronically implanted neural electrodes. *J. Neurosci. Methods*, **148**, 1–18. doi:10.1016/j.jneumeth.2005.08.015

Reardon, S. (2014). Electroceuticals spark interest. *Nature*, **511**, 18. doi:10.1038/511018a

Suner, S., Fellows, M. R., Vargas-Irwin, C., Nakata, G. K., Donoghue, J. P. (2005). Reliability of signals from a chronically implanted, silicon-based electrode array in non-human primate primary motor cortex. *IEEE Trans. Neural Syst. Rehabil. Eng. Publ. IEEE Eng. Med. Biol. Soc.*, **13**, 524–41. doi:10.1109/TNSRE.2005.857687

Vassanelli, S. (2014). Multielectrode and multitransistor arrays for *in vivo* recording. In: M. D. Vittorio, L. Martiradonna, and J. Assad (eds), *Nanotechnology and neuroscience: nano-electronic, photonic and mechanical neuronal interfacing*. New York: Springer, pp. 239–67.

Vassanelli, S., and Fromherz, P. (1999). Transistor probes local potassium conductances in the adhesion region of cultured rat hippocampal neurons. *J. Neurosci.*, **19**, 6767–73.

Vassanelli, S., Mahmud, M., Girardi, S., and Maschietto, M. (2012). On the way to large-scale and high-resolution brain-chip interfacing. *Cogn. Comput.*, **4**, 71–81. doi:10.1007/s12559-011-9121-4

Wise, K. D., Anderson, D. J., Hetke, J. F., Kipke, D. R., and Najafi, K. (2004). Wireless implantable microsystems: high-density electronic interfaces to the nervous system. *Proc. IEEE*, **92**, 76–97. doi:10.1109/JPROC.2003.820544

Wise, K. D., Sodagar, A. M., Yao, Y., Gulari, M. N., Perlin, G. E., Najafi, K. (2008). Microelectrodes, microelectronics, and implantable neural microsystems. *Proc. IEEE*, **96**, 1184–202. doi:10.1109/JPROC.2008.922564

Chapter 51

Biohybrid robots are synthetic biology systems

Joseph Ayers

Marine Science Center, Northeastern University, USA

An emerging area of biorobotics is the advent of biohybrid robots combining living cells with engineered devices. Our current robots are blind to chemical senses, difficult to miniaturize, and require chemical batteries. All of these obstacles can be overcome by integration of living engineered cells. Synthetic biology seeks to build devices and systems from fungible gene parts. The fundamental organizational scheme of synthetic biology combines *parts* (gene systems coding different proteins) integrated into a *chassis* (i.e. induced pluripotent eukaryotic cells, yeast, or bacteria) to produce a *device* with properties not found in nature. The next level of organization consists of a *system*: an interacting set of devices (Purnick et al. 2009). Biohybrid robots are examples of systems. I will describe how synthetic biology and organic electronics can integrate neurobiology and robotics to form a basis for synthetic neuroethology. There is a nascent literature describing genes of higher effect that can mediate organ levels of organization. Such capabilities, when applied to the arena of biohybrid systems, portend truly biological robots guided, controlled, and actuated solely by the processes of life.

There is a broad variety of biohybrid robots. Examples include living muscle integrated into a swimming robot (Herr et al. 2004), and muscle integrated with hydrogel substrates to produce jet propulsion (Nawroth et al. 2012). One category adheres bacteria to solid substrates to create bacteria-propelled rafts. There has been much recent work on the development of biohybrids formed from cultured muscle, although these efforts lack specific excitation/contraction coupling to couple controllers to muscle. We are developing vehicles that involve integration of engineered sensors (Yarkoni et al. 2012), neuronal network controllers (Lu et al. 2012), and engineered muscle (Grubišić et al. 2014). This interdisciplinary approach integrates biorobotics, neurophysiology, organic electronics, and synthetic biology. I will argue that biohybrid robots of the latter type are indeed synthetic biology systems and provide a progress report on the components of these biohybrid vehicles.

Synthetic biology

A watershed is developing with regard to strategies for characterization of molecular control systems. Systems biology is based on the reverse engineering of molecular control systems. Synthetic biology, in contrast, is based on the forward engineering of novel devices and systems from genetic parts. The DNA coding parts specify particular proteins and their regulatory mechanisms. Modularization of parts allows a "LEGO" approach to integration of biological systems to create living cells with capabilities not found in nature.

Key features of the synthetic biological approach are standardization and fungibility (Baker et al. 2006). The features of each part are characterized with regard to function, regulation, optimal operating conditions, and compatibility with one another. A part is thus not just DNA but a unit of function specified by its protein product and associated regulatory elements. Remarkably, many parts are conserved throughout nature and thus have been standardized in evolution. A key assumption of synthetic biology is fungibility; the ability for parts from different organisms to be integrated and controlled through promotors and repressors in a coordinated fashion. Fungibility appears quite feasible among prokaryotes and success with knock-out and knock-in murine models portends success in even vertebrates. Eukaryotic synthetic biology has progressed significantly in yeast and in the nervous system and is nascent in stem cell biology.

As life must be compartmentalized, the fundamental organizational unit of all organisms is the cell. In prokaryotes, such as bacteria, the cell contents are homogeneous, but in eukaryotes internal membranes compartmentalize the cell into organelles with different functions. The cell is the *chassis* of synthetic biology. There is an emerging science of protocell biology, but most synthetic biology occurs in native cells such as bacteria, yeast, fertilized eggs, stem cells, and neurons.

Engineered cells can integrate several different parts to form a device. For example a sensor part can be combined with a reporter part to form a novel type of sensory device. Just like sensory neurons or receptors cells can couple G-protein coupled receptors with neurotransmitters to mediate sensory transduction, prokaryotes can integrate receptors with light or gas reporters such as nitric oxide. Similarly, optogenetic photoreceptors can release calcium from internal stores to mediate excitation–contraction coupling in muscle. Parts can also be integrated into logical devices based on expression systems. Synthetic biology in prokaryotes has progressed to the level of integration of devices to form computational systems (Purnick et al. 2009; Church et al. 2014). Expression-based logic gates can be constructed expressing biological behavior, dynamics, and logic control

Another concept key to synthetic biology is refactoring. As existing chassis are repurposed for novel functions in specific environments, major subsets of genes necessary for *life in nature* will become superfluous and can be refactored from the genome of the engineered cells. For example, muscle cells engineered to respond to light to release calcium from external stores can operate without the genes necessary for synaptic transmission and excitation–contraction coupling.

Unicellular behavior

A well-understood eukaryotic model for future unicellular robots is *Paramecium* (Naitoh et al. 1969; Kung et al. 1982). Here sensing is mediated by constitutive ion channels in the cell membrane that respond to mechanical deformation and are distributed regionally on the two poles of the cell. Deformation of the anterior region of the cell causes an increase in permeability to calcium that depolarizes the cell. Deformation of the posterior region of the cell causes an increase in the permeability to potassium that hyperpolarizes the cell. These regional polarizations are conducted throughout the cell. The membrane voltage in turn modulates the molecular devices that mediate ciliary beating. The hyperpolarization that results when a predator such as *Didinium* deforms the rear of the cell increases the frequency of ciliary beating to speed up locomotion and allow escape from predation. The depolarization of the cell that results when *Paramecium* collides with an obstacle causes reversed ciliary beating that causes the organism to back up, mediating the Jennings "avoidance reaction". These two reactions mediate the bulk of the behavioral repertoire. *Paramecium* behavior has been the subject of considerable genetic analysis.

Systems neuroscience and biomimetic robots

In higher organisms, neuronal networks and interconnecting pathways perform different functions to mediate sensation, computation, actuation, and performance-related feedback. More complicated behavior involves complex task groups. The task groups themselves are spatially distributed through different regions of the nervous system and are programmed as neuronal and synaptic networks (Ayers et al. 2012) rather than algorithmically (Brooks 2001). These networks thus incorporate structural features such as decussation, convergence, and divergence. For example, the simplest behavioral systems are monosynaptic reflexes where a sensory neuron forms synapses with motor neurons that in turn activate muscle. In the next level of organization, a central pattern generator is interposed between sensation and actuation.

Comparative physiology has produced transformational advances by focusing on model systems that provide profound technical accessibility. One of the most profound technical advantages of the nervous systems of simpler animals is the capability to reproducibly identify neurons. This capability has allowed the evaluation of brain circuitry by stimulation and recording from pairs of identified neuron pairs and ultimately establishment of the detailed circuitry underlying motor pattern generation, choice, and plastic change (see e.g. Selverston, Chapter 21, this volume). The role of cell type is key in the self-organization of these circuits. It quickly became apparent in simple animal models that neuronal circuits were, in turn, neuromodulated by peptides that allowed reconfiguration of circuits by context through alterations in neuron dynamics and the strength of synaptic connections. The ability to reconfigure networks, on the fly, provides considerable adaptive flexibility.

Sensing: Living organisms are endowed with a variety of sense organs that filter the plethora of sensory information to extract features of the environment and performance of the intrinsic physical plant to modulate behavior. A sense denotes a capability to discriminate a particular form of external energy. Sense organs have a variety of common characteristics including external modulation, receptor cells that embody a transduction mechanism, and a sensory afferent that projects to the CNS. All animal sensors use a labeled line code where each sensory afferent is distinguished by a sensory modality (light, chemical signatures, etc.), a receptive field or orientation of the stimulus source relative to the body for external stimuli (or internally distributed in some cases), and an intensity code where the inter-impulse interval is proportional to the logarithm of the stimulus intensity to ensure a broad dynamic range. Within the central nervous system, sensory interneuronal networks perform the process of discrimination to detect behavioral releasers.

Computation: In simple animals, *innate* behavior is hardwired through genetically predetermined networks in the central nervous system. Central networks mediate exteroceptive reflexes, taxes, sequencing, choice, and some aspects of learning. They mediate endogenous rhythms, respond to incoming sensory input, and through central pattern generators they generate the patterns of motor neuronal activity that underlies behavior. The spatial and temporal patterns of motor neuron activity form the motor programs.

Actuation: The neuromuscular systems that control the behavior of simple organisms are generally organized into tonic and phasic fiber types. Muscles are complex actuators and often have different heads that participate in different functions. All living muscles share a common mechanism for excitation–contraction coupling where the muscle action potential is conducted through the transverse tubular system to release Ca^{++} ions from the sarcoplasmic reticulum. The Ca^{++} ions in turn cause a conformational change in regulatory proteins, troponin and tropomyosin, to initiate the actinomyosin cross-bridge cycle. There is a distinct matching between the motor impulse patterns control signals with the tonic and phasic muscle fiber types.

Biomimetic robots: We have used these organizational principles to construct and control biomimetic robots based on the lamprey, honeybee, and lobster (Ayers et al. 2012). In these vehicles, analog and digital sensors are interfaced to sensory neurons using microcontrollers to emulate the transduction processes of the animal models and resulting labeled-line impulse codes (Ayers 2002). These sensors provide afferents for heading deviation, primary orientation, acceleration, collision, odometry, and hydrodynamic and optical flow (Ayers et al. 2007; Westphal et al. 2011). Sensory networks employ lateral inhibition, range fractionation, and differentiation to form behavioral releasers and modulatory components.

We control these biomimetic robots with electronic nervous systems constructed with electronic neurons and synapses (Rabinovich et al. 2006). Analog CPGs can be readily constructed using Hindmarsh–Rose analog computers (Ayers et al. 2007). A two-dimensional map-based model allows the operation of hundreds of neurons and synapses on a simple DSP chip (Rulkov 2002; Westphal et al. 2011). The electronic nervous system is instantiated on a DSP chip and programmed as a network rather than algorithmically. These systems instantiate the command neuron, coordinating neuron, central pattern generator architecture (Ayers 2002).

The central pattern generators of the underwater robots generate motor programs consisting of impulse trains to control body appendages and the body axis. These control signals are ideal for controlling shape memory alloy actuators (Ayers et al. 2007). We form artificial muscles from nitinol wire isolated from sea water with a Teflon sheath (Ayers et al. 2007; Westphal et al. 2011). We achieve excitation–contraction coupling by using the action potentials of electronic motor neurons to gate a power transistor to apply duty-cycle modulated current to the actuator wires to mediate the state change from martensite to austenite associated with a contraction. Pairs of wires are arranged antagonistically and cooled with sea water such that a shortening conversion of one wire from martensite to austenite is coupled with a lengthening deformation of its antagonist muscle to martensite (Ayers et al. 2007). Motor programs coordinate contractions around joints or the body axis to mediate locomotion.

All of these components are organized into layered exteroceptive reflexes to mediate navigation, obstacle negotiation, and investigation. Central to these reflexes are the command neurons that organize locomotion and body posture. The command neurons organize complete behavioral acts and are activated by releasers that discriminate ongoing exteroceptive sensor input to modulate behavior (Ayers et al. 2012). The command neurons, in turn, modulate segmental central pattern generators to mediate taxis.

A biohybrid robot

We have been developing technologies for a miniature biohybrid undulatory robot called *Cyberplasm*. The goal of this robot is to develop capabilities for interfacing engineered sensor cells with a microelectronic central pattern generator that, in turn, activates living muscle cells engineered to mediate excitation–contraction coupling using light (Figure 51.1). The central pattern generator is based on the Hindmarsh–Rose models of neurons and synapses (Pinto et al. 2000) and instantiated as subthreshold analog VLSI (Lu et al. 2012). The behavior of freely behaving biohybrid robots will integrate cells comprising several device types to form a system. This platform will enable the exploration of micro-scale bioelectric interfaces and the potential for coupling power with actuation through the mitochondria of living muscle.

We have chosen to employ undulatory locomotion along the body axis (Westphal et al. 2011). In biohybrid robots of this type, there are two macro-scale components that compose the chassis of the vehicle (Figure 51.1). In the animal itself, undulation would be mediated by axial contractions around a flexible rod, the notochord or spine. In our lamprey-based robot we employ a strip of polyurethane. At the micro-scale, Kapton films provide strong, flexible, and analogous substrates. A second component is a molecular scaffold that provides a substrate for the engineered cells. The scaffold plays the role of establishing a compatible niche and providing interfaces between the cells and system electronics.

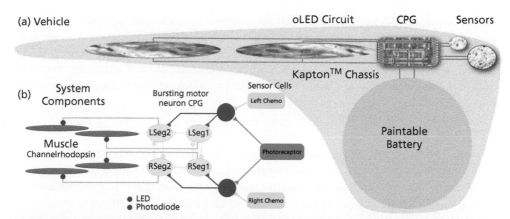

Figure 51.1 The *Cyberplasm* biohybrid robot. (a) Vehicle. (b) System Components. Swim: swimming command neurons; LSeg1: left segment 1 motor neuron; LSeg2: left segment 2 motor neuron; RSeg1: right segment 1 motor neuron; RSeg2: right segment 2 motor neuron; gray circles: inhibitory synapses; triangles: excitatory synapses; blue circles: blue LEDs.

Sensing: For engineered cells to respond to new environmental contingencies they must possess both a sensing mechanism and a reporting mechanisms. Animal sensors for vision and olfaction rely on specific G-protein coupled receptors (GPCRs) that are ubiquitous throughout the animal kingdom. To achieve the response times necessary for dynamic control of robot behavior we need to employ constitutive transducers. Our initial sensor relies on the photodimerization of endothelial nitric oxide synthase (eNOS) to report with nitric oxide (Yarkoni et al. 2012). Plasmids containing gene constructs that implemented a LOV mechanism for the formation of photoactive eNOS were transfected into a CHO-K1 Chinese Hamster ovary cell line to generate nitric oxide (NO) in response to light pulses. The transfected cells generated NO adequate to generate a linear response on a PT-NITSPc Nafion™ coated electrode (Yarkoni et al. 2012). Nitric oxide can also be generated by GPCRs, further increasing the sensory possibilities of this reporter. We are also exploring light-based reporters.

Computation: We build controllers based on the principles of operation of CPGs (Ayers 2010). To control undulatory swimming, we employ a central pattern generator composed of subthreshold analog VLSI electronic motor neurons (Lu et al. 2012; Figure 51.1b). An anterior pair is connected by reciprocal inhibitory synapses to generate a left–right alternation between the two sides. Excitatory connections between the anterior motor neuron and the posterior motor neuron on the same side generate a delay to mediate propagation of the flexion wave down the body as in the lamprey robot (Ayers et al. 1983; Westphal et al. 2011). The central pattern generator is activated by excitatory inputs from sensor interfaces activated by light (Yarkoni et al. 2012). The CPG is instantiated in 65 nm CMOS PTM with a supply voltage of 0.8 VDC that can be powered by a miniature paintable battery.

Actuation: One of the primary constraints of autonomous robots is power. Almost every robot has a separate power supply and actuators. The basic muscle unit, a myotube, embeds a power supply (mitochondria) in the actuators (myofibrils) coupling power with actuation. One of the most difficult aspects of artificial muscle is excitation–contraction coupling. In shape memory actuators we use heat generated by current to mediate conversion of neuronal activity to martensite to austenite conversion resulting in contraction. Optogenetic approaches afford the opportunity to engineer muscle to respond to light through calcium-permeable channelrhodopsin. The overall strategy, which is in progress, is to insert a plasmid in a viral vector that contains the basic loop transcription factor MyoD and bHLH E and channelrhodopsin vectors into myocytes (C2C12 mouse myoblasts American Type Culture Collection, Manassas, VA) and transform them into myofibrils. Initial results indicate that these transformed cells form contractile sarcomeres and form a stable cell line (Grubišić et al. 2014)

Interfaces

Engineered cells need a mechanism to communicate with electronic neurons. At the sensor end, they require a reporter that can be sensed by electronic transducers. At the actuator end, they require an excitation–contraction coupling mechanism. Moreover, these components need to be wired together at a micro-scale. We have adopted an additive manufacturing technology— electrohydrodynamic jet (E-Jet) printing (Barton 2010; Figure 51.2)—that is capable of integrating extracellular matrix components, transductive polymers, and metallic readout circuitry. The print head of an E-Jet printer is a glass microelectrode that has been sputter-coated with gold (Figure 51.2a). As the print head is glass, it is agnostic to different solvents. Heterogeneous inks can be formed of a variety of components such as aqueous solutions of extracellular matrix components, polymers, or metallic nanoparticles in ethanol. Nanoparticles of indium tin oxide (ITO) are used to form optically clear conductors. The substrate is adhered to a vacuum chuck. A pneumatic backpressure is applied to the pipette and a high (600–900 V DC) voltage is applied between the pipette and the stage. The ink forms a Taylor cone in the lumen of the pipette, that calves off droplets that are about half the diameter of the pipette lumen. As micropipettes can have lumens of ~0.5 µm the droplets can be as small as 250 nm. The stage is programmed to move in in detailed patterns (Figure 51.2c) to produce the microelectronic features.

We are employing two categories of sensors. The first includes electrochemical electrodes that respond to nitric oxide. The fabrication technique for NO sensors can be readily adapted to E-Jet printing. The basic NO electrode (Yarkoni et al. 2012) can be formed from inks based on silver nanoparticles coated with silver chloride referenced to a graphene nanoparticle ink, coated with a Nafion™ ink. These electrodes are spin-coated with a dielectric layer and a scaffold of collagen is

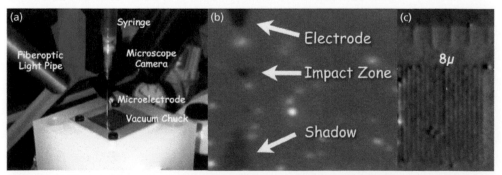

Figure 51.2 Electrohydrodynamic Jet Printing. (a) A sputter-coated microelectrode is suspended over a vacuum chuck mounted on a precision (1 nm resolution) stage. A fiberoptic light pipe illuminates the microelectrode tip. (b) Microscope camera image of the electrode tip. The distance between the electrode tip and its shadow on the substrate indicates the height of the electrode tip. (c) Electronic traces printed from nanoparticles of silver in ethanol. (Calibration marks at the top are separated by 100 µm.)

employed to adhere the engineered cells. The NO electrode generates millivolt potentials that can be directly interfaced to the electronic sensory neurons with an A/D converter. The analog signal is converted to a pulse interval code. The second sensor is a photodetector based on a photosensitive polymer. The device consists of a clear ITO cathode printed directly on a coverslip, a polymer layers and a silver nanoparticle anode layer. A collagen layer on the other side of the cover slip will adhere engineered cells with light reporters.

Excitation–contraction coupling of engineered muscle will be implemented through the use of a micro oLED to generate blue light to activate channelrhodopsin to release Ca^{++} from the sarcoplasmic reticulum (Figure 51.1a). The device will consist of a silver anode upon which is printed an electron donor layer, a coumarin blue light emitting layer, and an ITO cathode.

When integrated these devices will form systems that correspond to the exteroceptive reflexes of our biomimetic robots. For example, the photosensor will activate swim commands that will turn on the CPG that will activate the muscles in a pattern appropriate to generate undulations. Our plan is to integrate bilateral chemosensor reflexes to mediate chemotaxis (Figure 51.1b).

The future is synthetic neuroethology

Biomimetic and biohybrid robots can be controlled with modularized electronic neurons and synapses that can be configured as any type of neuronal circuit (Ayers et al. 2010). It is important to determine whether it is feasible to also use synthetic and stem cell biology to design and build central pattern generators from living cells. Gene families that correspond to the parts of synthetic biology code the ion channels. The neurogenetics of neuron fate is an emerging area of neurobiology. Major advances are occurring in determination of the transcription factors that mediate determination of cell fate. Living neurons are devices that can integrate a broad variety of genetic parts to generate a multiplicity of transmembrane channels, transmitter synthetic and exocytotic pathways, as well as ionotropic and metabolotropic transmembrane receptors. Considerable evidence indicates that functional organoids can be generated by relatively simple cocktails of transcription factors. Given the self-organizing nature of the nervous system, one might expect that whole ganglia and particular circuit configurations are under similar control.

At the systems level, neurons function within synaptic networks. Attempts to synthesize networks *in vitro* have been problematical. Recent experiments indicate that astroglia are necessary for synaptogenesis. The E-Jet technology I have described is ideal for the printing of heterogeneous scaffolds that integrate metallic readout circuitry (Figure 51.2c), extracellular matrix components, and trophic factors within islands that structurally program particular circuit configurations from engineered neurons and glia such as CPGs and exteroceptive reflexes. Such capabilities, when applied to the arena of biohybrid systems, portend truly biological robots guided, controlled, and actuated solely by the processes of life.

Learning more

Recent issues in biomimetics are addressed in Lepora et al. (2013). The application of nonlinear dynamics to neuronal circuits is reviewed in Rabinovich et al. (2006). Broad overviews of synthetic biology can be found in these reviews: Purnick et al. (2009) and Church et al. (2014).

Acknowledgments

NSF CBET Award 0943345 and an ONR Synthetic Biology MURI # N000141110725 supported the work described here.

References

Ayers, J. (2002). A conservative biomimetic control architecture for autonomous underwater robots. In: J. Ayers, J. Davis, and A. Rudolph (eds), *Neurotechnology for Biomimetic Robots*. Cambridge, MA: MIT Press, pp. 234–52.

Ayers, J., et al. (1983). Which behavior does the lamprey central motor program mediate? *Science (Washington D C)*, **221**, 1312–14.

Ayers, J., et al. (2012). A conserved biomimetic control architecture for walking, swimming and flying robots. *Biomimetic and Biohybrid Systems*, **7375**, 1–12.

Ayers, J., et al. (2007). Biomimetic approaches to the control of underwater walking machines. *Phil. Trans. R. Soc. Lond. A*, **365**, 273–95.

Ayers, J., Rulkov, N., Knudsen, D., Kim, Y-B., Volkovskii, A., and Selverston, A. (2010). Controlling underwater robots with electronic nervous systems. *Applied Bionics and Biomimetics*, **7**, 57–67.

Baker, D., et al. (2006). Engineering life: building a FAB for biology. *Scientific American*, **294**(6), 44–51.

Barton, K., Mishra, S., Shorter, K., Alleyne, A., Ferreira, P., and Rogers, J. (2010). A desktop electrohydrodynamic jet printing system. *Mecchatronics*, **20**, 611–16.

Brooks, R. A. (2001). Steps towards living machines. In: T. Gomi (ed.), *Evolutionary Robotics: From Intelligent Robotics to Artificial Life*. New York: Springer-Verlag, pp. 72–93.

Church, G. M., et al. (2014). Realizing the potential of synthetic biology. *Nat. Rev. Mol. Cell Biol.*, **15**(4), 289–94.

Grubišić, V., et al. (2014). Heterogeneity of myotubes generated by the MyoD and E12 basic helix-loop-helix transcription factors in otherwise non-differentiation growth conditions. *Biomaterials*, **35**(7), 2188–98.

Herr, H., et al. (2004). A swimming robot actuated by living muscle tissue. *J. Neuroengineering Rehabil.*, **1**(1), 6.

Kung, C., et al. (1982). The physiological basis of taxes in *Paramecium*. *Annual Review of Physiology*, **44**(1), 519–34.

Lepora, N. F., et al. (2013). The state of the art in biomimetics. *Bioinspiration & Biomimetics*, **8**(1), 013001.

Lu, J., et al. (2012). Low power, high PVT variation tolerant central pattern generator design for a biohybrid micro robot. *IEEE International Midwest Symposium on Circuits and Systems (MWSCAS)*, **55**, 782–5.

Naitoh, Y., et al. (1969). Ionic mechanisms controlling behavioral responses of *Paramecium* to mechanical stimulation. *Science*, **164**, 963–5.

Nawroth, J. C., et al. (2012). A tissue-engineered jellyfish with biomimetic propulsion. *Nature Biotechnology*, **30**(8), 792–97.

Pinto, R. D., et al. (2000). Synchronous behavior of two coupled electronic neurons. *Physical Review E: Statistical physics, plasmas, fluids, and related interdisciplinary topics*, **62**, 2644–56.

Purnick, P. E., et al. (2009). The second wave of synthetic biology: from modules to systems. *Nat. Rev. Mol. Cell Biol.*, **10**(6), 410–22.

Rabinovich, M. I., et al. (2006). Dynamical principles in neuroscience. *Reviews of Modern Physics*, **78**(4), 1213–65.

Rulkov, N. (2002). Modeling of spiking-bursting neural behavior using two-dimensional map. *Phys Rev E*, **65**, 041922.

Westphal, A., et al. (2011). Controlling a lamprey-based robot with an electronic nervous system. *Smart Structures and Systems*, **8**(1), 37–54.

Yarkoni, O., et al. (2012). Creating a biohybrid signal transduction pathway: opening a new channel of communication between cells and machines. *Bioinspiration & Biomimetics*, **7**(4), 046017.

Chapter 52

Micro- and nanotechnology for living machines

Toshio Fukuda[1], Masahiro Nakajima[2],
Masaru Takeuchi[3], and Yasuhisa Hasegawa[3]

[1] Institute for Advanced Research Nagoya University, Japan; Meijo University,
Beijing Institute of Technology
[2] Center for Micro-nano Mechatronics, Nagoya University, Japan
[3] Department of Micro-Nano Systems Engineering, Nagoya University, Japan

We aim to develop new biohybrid technologies, engineered at the micro- and nano-scales, that combine some of the benefits of mechanical and electronic systems with those of biological systems. In this chapter we consider some of the challenges of building such devices at very small physical scales, and discuss how fabrication methodologies, developed at the micro- and nanoscales, could impact on different classes of industrial, daily life, and biomedical products. We illustrate the potential of these methods by summarizing some of our recent progress in combining nanomaterials with biological cells using a micro- and nanorobotic manipulation system called the *nanolaboratory*.

Based on current technologies, we divide biohybrid living machines into *bio-inspired* and *biological* component systems. The bio-inspired component is composed of inorganic, dry, mechanical, or electrical systems. For example, tactile systems, imaging systems, robotic hands, mechanical joints, artificial muscles, and so on. The biological component is an organic, wet, or chemical system. To realize living machines with self-repairing, self-organizing, and self-powering features, bio-inspired and biological systems can be integrated as shown in Figure 52.1.

Multi-scale engineering is an important technique for developing systems that span from nanometer scale to meter scale as depicted in Figure 52.2. From the viewpoint of a biohybrid living machine, the biological component is currently composed of biological cells as its minimum unit. Biological cells are typically micrometer scale with nanometer size components such as their component deoxyribonucleic acid (DNA). One of the most interesting challenges is to make biological cells artificially and to assemble them as three-dimensional structures for the realization of artificial living machines that are purely biological. The bio-inspired component is generally developed from micro-/nanometer scale to meter scale. For example, to mimic the Meissner corpuscles of the human fingertip tactile sensing systems will require micro-nano structures (Trung et al. 2013). These different components and technologies will need to be integrated by multi-scale engineering to create a working living machine.

To develop biohybrid living machine technologies we are using a micro-nano mechatronics approach (Fukuda et al. 2010) that is an extension of mechatronics—the academic and technological field that combines mechanics with electronics. Mechatronic systems generally have three components: sensors, actuators, and controllers. These components are integrated for various

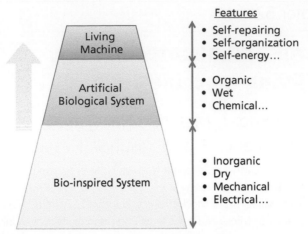

Figure 52.1 Components and features of a biohybrid living machine.

devices and products in our daily life including automobiles, computer peripherals, printers, cameras, amusement machines, robots, environmental monitoring systems, energy storage devices, biological/medical treatments, and so on. Advanced techniques have recently been applied to these devices and products to realize micro-nano sensors, micro-nano actuators, and micro-nano controllers. The benefits of developing devices using micro-nano mechatronics include improved efficiency, integration, and functionality, with lower energy consumption and cost, in a smaller device. This chapter will (i) consider some of the physical challenges of building devices at the micro-nano scale, (ii) review the potential application domains of micro-nano mechatronics, (iii) describe one of the key emerging techniques used in the field—micro-nano manipulation—and

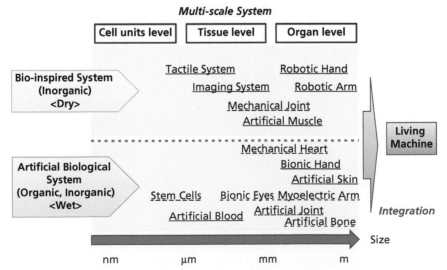

Figure 52.2 Examples of bio-inspired and artificial biological component systems for multi-scale engineering.

(iv) describe some recent work that could lead to the construction of biohybrid living machines at the micro-nano scale.

Physics of micro-nano mechatronics

At the micro-/nanometer scale, mechanics is dramatically changed from our living world. Micro-nano mechanics is intuitively understood by the scaling law (Table 52.1). In the less than nano-meter scale, quantum effects appear that are still not completely understood.

As an example of how the micro- and nano-scale can differ from the macro-scale, we next provide a brief account of van der Waals forces and electrostatic forces due to their importance in the mechanics of nano-manipulation and self-assembly.

The van der Waals force always acts on the sample at the intermolecular length scale, acting between neighboring molecules. This effect is akin to gravity, in that the attractive aspects of the force are important, but has a different physical law (for example the van der Waals force can become repulsive, whereas gravity is always attractive). The basis of the van der Waals force is that when two materials approach each other on a molecular scale, both attracting and repulsive intermolecular forces are generated from disturbances in the shape of the electron shell around

Table 52.1. Scaling laws. Because of the way that different physical qualities scale with distance, small effects at the macro-scale can be dominant at the micro- and nano-scale, and vice versa. For example the electrostatic force is dominant at the macro-scale (and scales with length squared L^2), but the van der Waals force can become dominant at the nano-scale (because it scales with length L, which is bigger than L^2 when L is small). The size effects are estimated by calculating how the various quantities depend on length in the physical law.

Physical Parameters	Symbols	Equations	Size Effects
Length	L	L	L
Area	S	$\propto L^2$	L^2
Volume	V	$\propto L^3$	L^3
Mass	m	ρV	L^3
Pressure	f_p	SP	L^2
Gravity	f_g	mg	L^3
Inertia force	f_i	md^2x/dt^2	L^4
Viscous (kinetic friction) force		$uS/d \cdot dx/dt$	L^2
Elastic force		$eS\Delta L/L$	L^2
Spring constant	K	$2UV/(\Delta L2)$	L
Resonant frequency	ω	$\sqrt{K/m}$	L^{-1}
Moment of inertia	I	$amrl^2$	L^5
Deflection	D	M/K	L^2
Reynolds number	R_e	f/f_f	L^2
Electrostatic force	F_e	$\varepsilon SE^2/2$	L^2
van der Waals force	F_{vdw}	$F_{vdw} = \dfrac{Hd}{12z^2}$	L
Dielectric force	F_d	$2\pi l^3 \varepsilon_1 \dfrac{\varepsilon_2 - \varepsilon_1}{\varepsilon_2 - 2\varepsilon_1} \nabla(E^2)$	L^3

Remarks: ρ: density, P: pressure, g: acceleration of gravity, x: displacement, t: time, d: density, e: Young's modulus, a: constant, r_i: radius of body of rotation, ε: dielectric constant, E: electric field, d: conversed radius, C: constant of the atom–atom potential.

molecules. The attractive forces become stronger as the materials become closer together, which act to draw the materials further together. However, if the materials are drawn too closely together, then the repulsive forces begin to dominate, pushing them apart to settle at an equilibrium distance that balances the forces.

The van der Waals *force* can be described mathematically in closed-form under some ideal conditions. For example the Lennard-Jones potential energy represents an approximate model for the (rotationally symmetric) isotropic part of the total (attractive plus repulsive) van der Waals force as a function of distance r,

$$\phi(r) = 4\varepsilon \left\{ \left(\frac{\sigma}{r} \right)^{12} - \left(\frac{\sigma}{r} \right)^{6} \right\} \tag{1}$$

where σ is the length at which the van der Waals force become close to zero, and ε determines the overall depth of the potential energy well, i.e. how much energy E is needed to pull the two materials apart once they have been attracted by the van der Waals force. The van der Waals force is the attracting force calculated by taking the gradient of the Lennard-Jones potential.

To calculate the van der Waals *energy*, the potential energy between two molecules can be integrated with respect to object shape. For example, the van der Waals energy E_{vdw} between a particle and a plate is expressed as the following equation:

$$E_{vdw} = -\frac{H}{6} \left\{ \frac{d}{2z} + \frac{d}{2(z+d)} - \ln\frac{z}{z+d} \right\} \tag{2}$$

with $H = \pi^2 C \rho_1 \rho_2$ the Hamaker constant, where ρ_1, ρ_2 are the densities of each of the two materials, z is the gap between the particle and plate, and d is the diameter of the particle. When the gap is much bigger than the particle diameter $z >> d$, the van der Waals force can be approximated by $F_{vdw} = Hd/12z^2$, which is the expression used in Table 52.1 to represent the properties of the van der Waals force. For example, compared with the Coulomb electrostatic force, whose magnitude scales with length squared, the van der Waals force scales with length, and so becomes dominant over the electrostatic force at the micro- and nano-scales.

Applications of micro-nano mechatronics

Micro- and nanomechatronics are key methodologies for the construction of new types of sensors and actuators that allow their miniaturization to the micro-scale or below. To understand the potential benefits of advances in this area we can divide the application fields of mechatronics between industrial, daily life, and biomedical as depicted in Table 52.2.

Industrial applications include automobiles, robotics and automation, alternative energy resources, and so on. These applications generally require relatively large-scale systems that are more than a meter in size and are therefore human-scale. In automation technologies, many devices work together without any direct human operation. Micro-nano mechatronics is a promising technology for improving system accuracy, energy efficiency, and total size/weight, particularly with regard to sensors and actuators. For example, a modern automobile has more than 100 sensors, 50 engine control units (ECUs), and 50 electric motors. The current size of typical accelerometer is at the millimeter-length scale, but improved performance would result from micro-scale devices that could be distributed over the automobile structure, for example within parts of the engine where mm-scale devices are not appropriate.

Table 52.2. Application categories of micro and nano mechatronics.

Environment	Typical Size	Application Categories	Related Basic Technologies in Micro-Nano Scale	Used Devices Based On Micro-Nano Mechatronics
Dry	More than a meter	Industrial	Automobile	Photolithography, Micromachining, Ceramics, Catalyst, Metal Coating, Surface Treatment, ...	Temperature Sensor, Flow Sensor, Gas Sensor, Knock Sensor, Pressure Meter, Cancelation Meter, Yaw Rate Sensor, ...
			Robotics	Photolithography, Precision Machining, Micromachining, ...	Force Sensor, Pressure Sensor, Gyro Sensor, Distance Sensor, IR Sensor, Swathing Device, Stereo Camera System, Laser Range Finder, GPS, ...
			Alternative Energy Resource	Photonic Materials, Purification, Catalyst, Energy Resource, ...	Photonic Device, Gas Sensor, Photovoltaic Generation Device, Biomass Power Generation ...
	Less than a meter	Daily Life	Computer Peripheries	Photolithography, Precision Machining, Micromachining, Nano-imprinting, ...	CPU, Hard Disk Drive, Memory, Display, Touch Sensor, Fingerprint Authentication Device, ...
			Mobile Phone	Photolithography, Precision Machining, Micromachining, ...	Sound Sensor, Speaker, Touch Sensor, Light Sensor, Gyro Sensor, GPS, ...
			Hobby	Photolithography, Precision Machining, Micromachining, ...	Acceleration Meter, Gravity Sensor, Angle Sensor, Display, Touch Panel, Wireless Communicator, ...
Dry (Wet)			Electrical Household Appliances	Photolithography, Precision Machining, Micromachining, ...	Printer, Ink-Jet Head, Camera, LED, TV, Audio, ...
			Environmental Monitoring Systems	Photolithography, Precision Machining, Wireless Communication, ...	CameraTemperature Sensor, Humidity Sensor, Solar Sensor, CO_2 Sensor, Radiation Sensor, ...
Wet	Less than a centimeter	Biomedical	Health, Cosmetic, Focd	Pharmacy, Health-Care, Gene Manipulation, Micro-fluidics Chip, Micro-nano Particles, Surface Treatment.	Bio-signal Sensor, Bio-MEMS, pH Sensor, Color Sensor, Capacitance Sensor, Blood Analyzer, Sugar Sensor ...
			Biological Analysis	Fluorescent Materials, Gene Manipulation, Surface Treatment ...	Flow Cytometric Instrument, DNA Sequencer, Cell Injector, ...
			Medical Treatment, Tissue Engineering	Photonic Sensor, Fluorescent Materials,	Bio-signal Sensor, Electroencephalograph, CT Scanning System, MRI System, ...

(continued)

Table 52.2. Continued

Application Categories		Related Basic Technologies in Micro-Nano Scale	Used Devices based on Micro-Nano Mechatornics
Industrial	Automobile	Photolithography, Micromachining, Ceramics, Catalyst, Metal Coating, Surface Treatment, …	Temperature Sensor, Flow Sensor, Gas Sensor, Knock Sensor, Pressure Meter, Cancelation Meter, Yaw Rate Sensor, …
	Robotics	Photolithography, Precision Machining, Micromachining, …	Force Sensor, Pressure Sensor, Gyro Sensor, Distance Sensor, IR Sensor, Swathing Device, Stereo Camera System, Laser Range Finder, GPS, …
	Alternative Energy Resource	Photonic Materials, Purification, Catalyst, Energy Resource, …	Photonic Device, Gas Sensor, Photovoltaic Generation Device, Biomass Power Generation …
Daily Life	Computer Peripheries	Photolithography, Precision Machining, Micromachining, Nano-imprinting, …	CPU, Hard Disk Drive, Memory, Display, Touch Sensor, Fingerprint Authentication Device, …
	Mobile Phone	Photolithography, Precision Machining, Micromachining, …	Sound Sensor, Speaker, Touch Sensor, Light Sensor, Gyro Sensor, GPS, …
	Hobby	Photolithography, Precision Machining, Micromachining, …	Acceleration Meter, Gravity Sensor, Angle Sensor, Display, Touch Panel, Wireless Communicator, …
	Electrical Household Appliances	Photolithography, Precision Machining, Micromachining, …	Printer, Ink-jet Head, Camera, LED, TV, Audio, …
	Environmental Monitoring Systems	Photolithography, Precision Machining, Wireless Communication, …	CameraTemperature Sensor, Humility Sensor, Solar Sensor, CO_2 Sensor, Radiation Sensor, …
Biomedical	Health, Cosmetic, Food	Pharmacy, Health-Care, Gene Manipulation, Micro-fluidics Chip, Micro-Nano particles, Surface Treatment.	Bio-signal Sensor, Bio-MEMS, pH Sensor, Color Sensor, Capacitance Sensor, Blood Analyzer, Sugar Sensor …
	Biological Analysis	Fluorescent Materials, Gene Manipulation, Surface Treatment …	Flow Cytometric Instrument, DNA Sequencer, Cell Injector, …
	Medical Treatment, Tissue Engineering	Photonic Sensor, Fluorescent Materials,	Bio-signal Sensor, Electroencephalograph, CT Scanning System, MRI System, …

The daily life field includes applications such as computer peripherals, mobile phones, hobby equipment, electrical household appliances, and environmental monitoring systems. Typical sizes are less than one meter which is again similar to human-scale. Equipment is often multifunctional and it is desirable to allow customization for users. Some devices must operate in wet or damp conditions, such as ink-jet components, humidity sensors, temperature sensors, and so on. Wireless communication is a key technology in this field to integrate the functionality of devices in home environments.

The biomedical field includes applications in the areas of health, cosmetics, food, biological analysis, medical treatment, and tissue engineering. Micro- and nano-mechatronics can benefit the design of biomedical devices, such as health monitoring systems, making them more portable and lower price. Our group has developed a single-cell analysis system based on micro-nano scale tools (see under "Applications of nanofabrication", below). Tissue engineering is also receiving a lot of attention for applications that involve embryonic stem cells or induced pluripotent stem cells (Yamanaka 2009). In our group, the patient blood vessel simulator "Endo Vascular Educator (EVE)" has been developed at the frontier of surgical simulation. This simulator is fabricated and assembled based on micro-fabrication technologies that involve rapid prototyping (Ikeda et al. 2004).

Device applications of micro-nano mechatronics

Various applications have been investigated that deploy micro-nano mechatronics both by academic and industrial laboratories; these can be divided between micro-nano sensing (Figure 52.3, top), actuation (Figure 52.3, middle), and control (Figure 52.3, bottom) settings.

Table 52.3 shows the classification of micro-nano sensing and actuation technologies with example application environments in dry and wet conditions. For micro-nano sensing applications, physical properties are measured with high-sensitivity, high-efficiency, low noise, local sensing, and so on. Various biosensing and future life-supporting devices have been investigated such as NO gas sensing (Iverson et al. 2013), a hydrogen peroxide sensor (Cao et al. 2013), and electromechanical piezoresistive sensing (Smith et al. 2013). Micro-nano actuation application has been investigated with various nanomaterials such as photomechanical actuation based on carbon nanotube materials (Liu et al. 2012), electrochemical actuation using graphene (Zhang et al. 2014), and hydrogel actuators with reduced graphene oxide (Wang et al. 2013).

Recently, the investigation of nanoelectromechanical systems (NEMS) has attracted much attention and is expected to realize highly integrated, miniaturized, and multifunctional devices for various applications in the near future. One of the most effective approaches is the direct usage of bottom-up (Feynman 1960) fabricated nanostructures that use nanostructures or nanomaterials directly. Example nanodevices include those based on carbon nanotubes (CNTs)—these have interesting mechanical, electronic, and chemical properties that have been investigated in various studies (Fukuda et al. 2003). There is the possibility of using the fine structures of CNTs directly. For example, "telescoping" CNTs that are fabricated by peeling off the outer layers of multi-walled carbon nanotubes (MWCNTs) are one of the most interesting nanostructures (Cumings and Zettl 2000). In previous work, we have described how the inner core can be pulled out mechanically inside a scanning electron microscope (SEM) (Nakajima et al. 2007). We also reported previously on the direct measurement of electrostatic actuation of a telescoping MWCNT inside SEMs and transmission electron microscopes (TEMs).

The cells inside real tissues and organs such as neural cells, skin, and blood vessels, are arranged according to certain patterns and shapes (Suuronen et al. 2005). In tissue engineering, important issues are cell encapsulation inside certain structures and high-throughput assembly of these structures (Langer and Vacanti 1993). Cell immobilization is generally necessary prior to manipulation; convenient methods for this include aspiration (suction), pressure of solution and fluidic

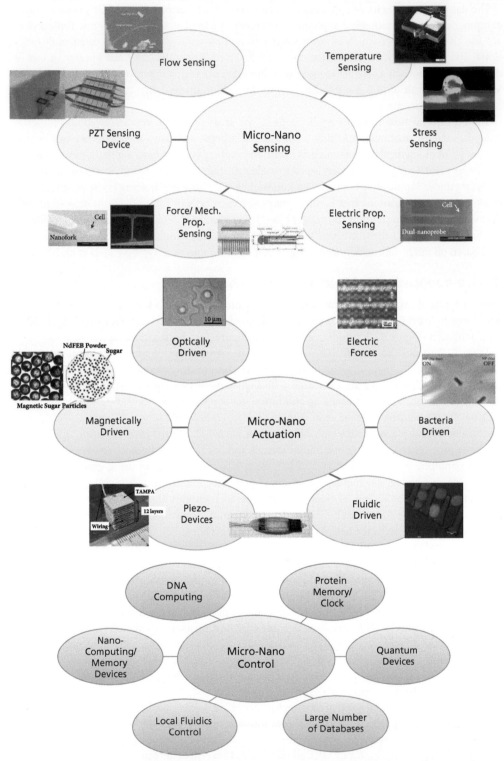

Figure 52.3 Example applications for micro-nano mechatronics in sensing (top), actuation (middle), and control (bottom). PZT, piezoelectric.

Table 52.3. Micro-nano sensing and actuation technologies.

Application Fields	Dry Environment	Wet Environment	Examples
Micro-Nano Sensing Technologies	◆ MEMS/NEMS with dry packaging ◆ Atomic Force Microscope ◆ Electron Microscopes ◆ Flow Sensing ◆ Force/ Mech. Prop. Sensing ◆ Electric Prop. Sensing	◆ MEMS/NEMS with water sealing packaging ◆ Atomic Force Microscope ◆ Environmental Scanning Electron Microscopes ◆ Scanning Ion Microscope ◆ Force/ Mech. Prop. Sensing ◆ Electric Prop. Sensing	◆ NO Sensor for biosensing based on SWCNT ◆ Hydrogen Peroxide Sensor by graphene network ◆ Electromechanical piezoresistive sensing using graphene membranes
Micro-Nano Actuation Technologies	◆ Electro Static Force Driven ◆ Adhesion Devices ◆ Magnetically Driven ◆ Piezo-Driven ◆ Electric Force Driven ◆ MEMS Mechanical Energy System	◆ Dielectric Forces ◆ Optically Driven ◆ Piezo-Driven ◆ Magnetically Driven ◆ Cardiac Muscle Cell Driven ◆ Bacteria Driven	◆ Photomechanical actuation based on carbon nanotube materials ◆ Electro-chemical actuation using graphene ◆ Hydro-gels actuator with reduced graphene oxide

structure (Tsutsui et al. 2010). The advantage of aspiration and pressure is that fixing force is large, while the disadvantage is that it can damage cells (Tixier-Mita et al. 2004). By using special fluidic structures, cells can be immobilized inside a microfluidic chip, although the immobilized cells can be difficult to analyze further (Di Carlo et al. 2006). On-chip fabrication based on photo-crosslinkable resin via UV illumination is a creative way to immobilize cells directly inside a microfluidic chip. There are several advantages, such as high speed, low cost, and the possibility of constructing arbitrary shapes (Yue et al. 2011). One of the important issues to be solved is the deficiency of oxygen and nutrients and the accumulation of wastes when cell structures are stacked. The current limit of the thickness of cell-sheet stacks is less than 200 μm due to diffusion of oxygen, nutrients, and wastes (Derda et al. 2009). Hence, it is important to construct an analog of a blood vessel structure for transportation of oxygen, nutrients, and waste, so that fabricated tissues can have a three-dimensional, rather than laminar, structure.

Nanorobotic manipulation

Nanorobotic manipulation, or nanomanipulation, has received considerable attention as an effective strategy for characterization of the properties of individual nano-scale materials and for the construction of nano-scale devices. This technique will provide a ground-breaking route to the application of nanotechnology in a wide range of applications. One of the attractive future applications of nanomanipulation is to realize the ultimate top-down fabrication or assembly of complete devices and products.

Microscopic observations for nanomanipulation

To manipulate nano-scale objects requires that they can be observed with a resolution higher than nano-scale, hence, appropriate manipulators and observation systems (microscopes) are necessary for nanomanipulation.

The optical microscope (OM) is the most historical and basic microscope; however, its resolution is limited to ~100 nm due to the diffraction limit of the optical wavelength (~400–~800 nm) as explained by the well-known Abbe's law (Hell 2007). Scanning tunnelling microscopes (STMs) and atomic force microscopes (AFMs) belong to the class of scanning probe microscopes (SPMs) that use a physical probe to examine a specimen, and can be used both for observation and manipulation at the nano-scale. Their high resolution makes them capable of atomic manipulation; however, they are limited by their 2D positioning capability and available strategies for manipulation. For instance, STMs cannot be used for complex manipulations or in 3D space. The atomic force microscope (AFM) can achieve nearly as high resolution as STMs for either conductive or insulated objects, and can be applied to manipulate relatively large nanometer objects mainly by mechanical pushing on a 2D surface. The major limitation of SPM-type manipulators is that they do not have rotational degrees of freedom (DOFs) and hence their orientation cannot be controlled. Another important limitation is that their observation and manipulation abilities cannot be used simultaneously. Finally, the observation area is limited and long exposure times—typically tens of minutes—are needed to get one image. This limitation becomes increasingly problematic for 3D nanomanipulation of nanostructures. Electron microscopes (EMs) provide atomic-scale resolution with an electron beam whose wavelength is less than ~0.1 Å. EMs are divided into two main types known as scanning electron microscopes (SEMs) and transmission electron microscopes (TEMs). Unfortunately, the specimen chamber and observation area of TEMs are too narrow to contain manipulators with complex functions.

By separating the imaging and manipulation functions, nanorobotic manipulators can perform nanomanipulation under the real-time observation of electron microscopes. These devices generally have more DOFs, including orientation control, and hence can be used for the manipulations of 0D (symmetric spheres) to 3D objects in 3D free space. We have proposed a hybrid nanorobotic manipulation system that integrates TEM and SEM nanorobotic manipulators as a core system for a *nanolaboratory* (see Nakajima et al. 2006 and below). This strategy is named as hybrid nanomanipulation to differentiate it from those with only an exchangeable specimen holder. The most important feature of the new manipulator is that it contains several passive DOFs which makes it possible to perform relatively complex manipulations whilst staying within a relatively compact volume that can be installed inside the narrow vacuum chamber of a TEM. Limited by the relative lower resolution of electron microscopes, this system is still not suited to the manipulations of atoms. However, its capacity for 3D positioning, orientation control, independently actuated multi-end-effector, separated real-time observation system, and the possibility of including SPMs inside it make it the most promising route for further advances in nanomanipulation.

Development of nanomanipulation systems

In 1990, Eigler and Schweize demonstrated the first atom-level nanomanipulation using an STM that they applied at low temperature (4 Kelvin) to position individual xenon atoms on a single-crystal nickel surface with atomic precision. This manipulation enabled them to fabricate rudimentary structures of their own design, atom by atom. In the resulting images, atoms were positioned to form the three-letter display "IBM." This work demonstrated the feasibility of controlling atoms through nanomanipulation. Hosoki et al. (1992) also demonstrated atomic lettering by STM by extracting sulfur atoms from a molybdenum disulfide (MoS_2) substrate using evaporation by electric field at room temperature. Avouris and colleagues applied an atomic force microscope (AFM) to bend, straighten or translate CNTs on a substrate (Hertel et al. 1998). The objects are pushed and deformed by the tip of the AFM. Ning Xi and colleagues at Michigan State

University developed an AFM-based nanomanipulation system with an interactive operation system (Li et al. 2004, 2005). The system introduced real-time visual feedback during AFM-based nanomanipulation. They also proposed some physical models of the interaction between the AFM tip, substrate, and objects in order to relay the real-time interactive forces through a haptic feedback system provided to the operator. The "msu" (Michigan State University) pattern was represented by pushed nanoparticles on a glass surface. Requicha et al. (2009) proposed a fully automatic AFM system for arrangement of nanoparticles with diameters of ~10 nm. This technique is important because it reduced operating time to a matter of minutes, whereas previous interactive systems often required several hours of skilled operator time to construct similar, but usually less accurate, patterns. Sitti and Hashimoto (2003) also proposed a tele-nanorobotic system based on an AFM probe. This system consisted of a piezoresistive AFM as a mechanical manipulation probe and topology sensor, with haptic and virtual reality displays providing force and visual feedback.

We have previously proposed a "nanolaboratory" based on a nanorobotic manipulation system (Nakajima et al. 2006). Such a system can realize various nano-scale fabrication and assembly operations to develop novel nanodevices and to integrate borderless technologies using nanorobotic manipulation. The nanolaboratory idea is also applicable to the scientific exploration of macroscopic phenomena. This would be one of the most significant enabling technologies to realize manipulation and fabrication with individual atoms and molecules for the assembly of devices.

Applications of nanofabrication toward biohybrid living machines

Having explained some of the core methods of micro-nano mechatronics, this section presents some example applications regarding the manipulation of nano and biological materials that could be useful in assembling future biohybrid living machines.

Assembly of nanostructures using CNT based on nanomanipulation

As previously noted, carbon nanotubes (CNTs) are increasingly used as a basic building block for a new generation of nanoelectronic and mechanical systems, for example, as linear (Cumings and Zettl 2000) and rotational nanobearings (Fennimore et al. 2003), mass conveyors (Sazonova et al. 2004), field emitters (Rinzler et al. 1995; Saito et al. 1998), AFM probes (Dai et al. 1996), nanotweezers (Kim and Lieber 1999), nanoposition sensors (Liu et al. 2005a), and so on. All of those applications are based on as-grown CNTs without any change in the mechanical structure of CNTs except some chemical decoration (Liu et al. 2005b). In fabrication, manipulation, or assembly of nanotubes, the length of CNTs is an important factor and will influence the function and configuration of the nanostructures and nanodevices. Therefore, a method of defining the length of nanotubes precisely is indispensable. The application of CNTs in special structures or shapes can be significantly improved by a method capable of making irreversible bends in CNTs. In addition to these structural manipulation methods an effective welding technique is essential for the development of CNT-based nanostructures and devices.

A 3D nanostructure can be constructed by electron-beam-induced nanofabrication. An example assembly process is shown step-by-step in Figure 52.4. Figure 52.4(a) shows a CNT picked up by an AFM cantilever and manipulated by a nanorobotic manipulator. One end of the CNT was fixed on an AFM cantilever surface by a tungsten deposit, produced by a welding technique, the other end was set to touch the surface of another AFM cantilever. A bending

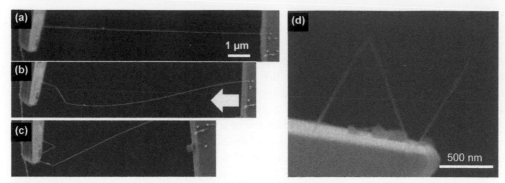

Figure 52.4 Assembly of three-dimensional nanostructure based on a CNT assisted with welding and bending techniques (a)–(c), and finally cutting the waste material giving (d), a fabricated letter N standing on the substrate.

technique was applied to the CNT whilst controlling its direction and angle of bending using a nanomanipulator. This initial bending was followed by another at a second point on the CNT as shown in Figure 52.4(b). The location and orientation of the CNT was changed by the manipulator and the second kink was set to touch the substrate as shown in Figure 52.7(c). Finally, the CNT was cut at a third point to create a 3D nanostructure that was separate from the substrate. As the result, a letter N was assembled from a CNT and set to stand on the substrate at two points as shown in Figure 52.7(d). The two points attach the structure on the substrate only by van der Waals force.

Single cell analysis based on nanomanipulation

Single cell analysis has been a recent focus of attention due to the increasing possibility of controlling the observation environment (Leary et al. 2006). Under conventional SEMs and TEMs, the sample chambers of these electron microscopes are set to high vacuum to reduce disturbance of the electron beam. To observe water-containing samples, for example bio-cells, appropriate drying and dying treatments are needed before observation, making direct observations of water-containing samples difficult.

The nanolaboratory can be applied for the single cell analysis and manipulation. Applications under dry, damp, and wet conditions can be done respectively under TEM/SEM, environmental-SEM (E-SEM), and optical microscope (OM). As described previously, the nanomanipulation system inside a TEM/SEM is a fundamental technology for property characterization of nano materials, structures, and mechanisms, for the fabrication of nano building blocks, and for the assembly of nano devices. An OM micromanipulation system is generally used underwater in order that the biological cells can be cultured within a medium. We have developed the E-SEM nanorobotic manipulation system, or nanosurgery system, to manipulate and control the local environment for biological samples in nano-scale (Figure 52.5). The environmental-SEM (E-SEM) can realize the direct observation of hydroscopic (water-containing) samples with nanometer high resolution using a specially designed second electron detector (Ahmad et al. 2008). The evaporation of water is controlled by the sample temperature (~0 – ~40 °C) and by the chamber pressure (10 – 2600 Pa). The unique characteristic of the E-SEM is the ability to directly observe and manipulate hydroscopic samples without drying.

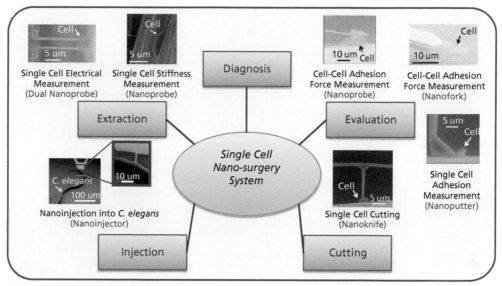

Figure 52.5 Single cell nano-surgery system with various nano-tools.

Cell assembly based on micro-nano manipulations

Micro-nano manipulation is a key technology to realize cell assembly because it can be used to achieve arbitrary shape construction and local analysis at single cell level. Table 52.4 shows a summary of possible assembly methods for cellar structures in different dimensionalities (0D, 2D, 3D). Important considerations are scaleability, fabrication speed, and biocompatability (according to ratings provided by developers in our laboratory). In our group, we are working on cell assembly based on various local control techniques as summarized in Figure 52.6.

Table 52.4. 3D cellular structural assembly methods.

Dimension of manipulation	Fabrication method of cell structure
0D	Fabrication by Photo-fabrication, Inkjet fabrication Cell patterning by electrophoresis, Cell aggregation by micro-well, Cell aggregation by micro-fluidic flow, Thermal gel probe manipulation, Magnetic gel particle manipulation, Electro-deposition for fabrication of cell structures
1D	Cell encapsulated hydro-gel fiber by microfluidic chip, Magnetic manipulation of hydro-gel fiber, Reeling hydro-gel fiber system
2D	Cell attached micro-plate fabricated by origami method, Cell attached paper fabricated by stacking process, Self-assembly of cell embedded microstructures by microfluidic chip, Cell sheet fabricated by temperature response hydro-gel

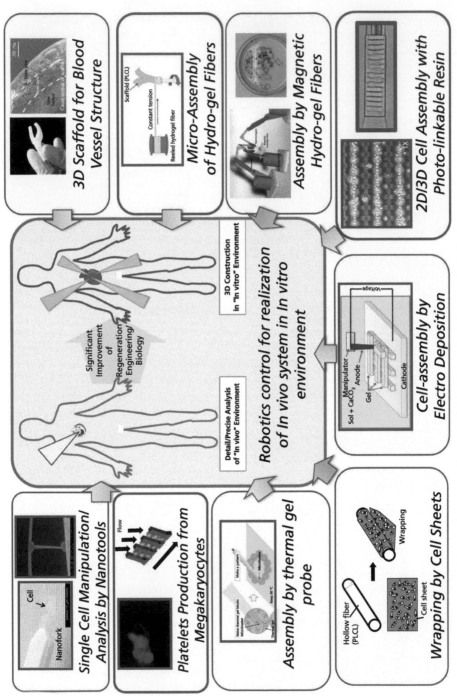

Figure 52.6 A schematic of bio-analysis/assembly/synthesis based on micro-nano manipulation.

Optical fabrication has the advantage of fabricating structure without contact. Moreover, the optical pass can be well controlled using a laser system. To apply this method for cell assembly a laser writing system was proposed to fabricate micro-structures with cells (Chan et al. 2010). Photo-linkable resin was used to encapsulate the cells and laser scanning to fabricate the structure in three dimensions. By changing the mixture of cell types the co-assembly of different type of cells could also be achieved.

For tissue engineering it is necessary to construct cell patterns and immobilize patterned cells inside certain structures. We have developed a novel method for forming cell patterns by dielectrophoresis (DEP) and immobilizing cells by photo-crosslinkable resin inside microfluidic chips (Yue et al. 2014). The proposed design has separate areas for patterning and fabrication and is designed for movable microstructures embedding microbeads or cells with controllable concentrations. The whole procedure is shown in Figure 52.7. We have applied this method to demonstrate that cell line patterns containing hundreds of cells, of multiple types, can be formed within one second. There are three advantages to this method. First, arbitrary-shaped microstructures can be efficiently fabricated inside microfluidic chips. Second, a large number of cells can be immobilized inside microstructures. Third, the assembly of microstructures happens with high efficiency due to microfluidic self-assembly.

Conclusion

This chapter describes the current and future technologies and issues of micro- and nanomechatronics. For industrial applications, various devices are developed and available, such as applications to automobiles, computer peripheries, amusements, printers, cameras, robotics automation, environmental monitoring, biological/medical treatments, energy resources, and so on. Those devices can be improved using micro- and nanomechatronics technologies to realize high-efficiency, high-integration, high-functionality, low-energy consumption, low-cost, miniaturization, and so on. Micro-nano mechatronics technologies can also be applied to make breakthroughs in the field of nanobiology for future medical applications. Finally, tools are being developed that allow the fabrication of structures at the molecular and atomic scale, the modification of biological components at the cellular and within-cell level, and the integration of these systems toward future biohybrid living machines.

Learning more

The journal "Lab on a Chip" contains papers describing the latest developments in devices and applications at the micro- and nano-scale. Yue et al. (2014) and Wang et al. (2015) provide detailed descriptions of of automated microfabrication systems from the Fukuda laboratory. Vincent (Chapter 10, this volume) provides a broad introduction to the design and manufacture of biomimetic materials. Pang et al. (Chapter 22, this volume) describe some of the adhesion systems of geckos and beetles that have unique structural features at the micro- or nano-scale and their replication in biomimetic artifacts such as robot grippers.

Acknowledgments

This work was partially supported by a Grant-in-Aid for Scientific Research from the Ministry of Education, Culture, Sports, Science, and Technology of Japan.

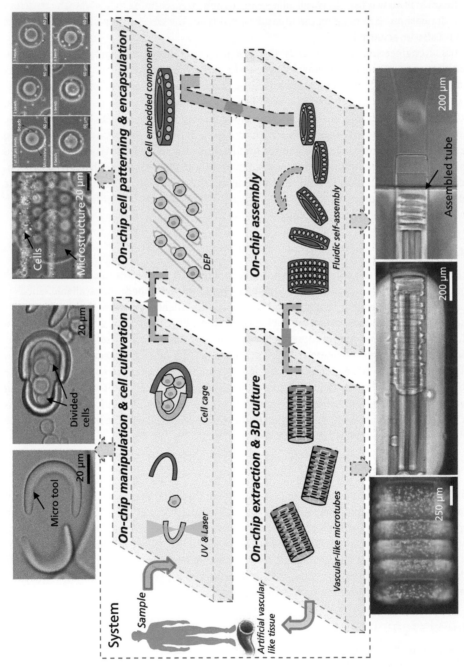

Figure 52.7 Cell assembly by multifunctional microfluidic system. See Yue et al. (2014) for further details.

References

Ahmad, M. R., Nakajima, M., Kojima, S., Homma, M., and Fukuda, T. (2008). The effects of cell sizes, environmental conditions and growth phases on the strength of individual w303 yeast cells inside ESEM. *IEEE Transactions on Nanobioscience*, 7, 185–93.

Cao, X., Zeng, Z., Shi, W., Yep, P., Yan, Q., and Zhang, H. (2013). Three-dimensional graphene network composites for detection of hydrogen peroxide. *Small*, 9, 1703–07.

Chan, V., Zorlutuna, P., Jeong, J. H., Kong, H., and Bashir, R. (2010). Three-dimensional photopatterning of hydrogels using stereolithography for long-term cell encapsulation. *Lab on a Chip*, 10, 2062–70.

Cumings, J., and Zettl, A. (2000). Low-friction nanoscale linear bearing realized from multiwall carbon nanotubes. *Science*, 289, 602–04.

Dai, H. J., Hafner, J. H., Rinzler, A. G., Colbert, D. T., and Smalley, R. E. (1996). Nanotubes as nanoprobes in scanning probe microscopy. *Nature*, 384, 147–50.

Derda, R., Laromaine, A., Mammoto, A., Tang, S., Mammoto, T., Ingber, D., and Whitesides, G. (2009). Paper-supported 3D cell culture for tissue-based bioassays. *Proc. Natl Acad. Sci. USA*, 106, 18457–62.

Di Carlo, D., Aghdam, N., and Lee, L. P. (2006). Single-cell enzyme concentrations, kinetics, and inhibition analysis using high-density hydrodynamic cell isolation arrays. *Analytical Chemistry*, 78, 4925–30.

Fennimore, A. M., Yuzvinsky, T. D., Han, W. Q., Fuhrer, M. S., Cumings, J., and Zettl, A. (2003). Rotational actuators based on carbon nanotubes. *Nature*, 424, 408–10.

Feynman, R. P. (1960). There's plenty of room at the bottom. *Caltech's Engineering and Science*, 23, 22–36.

Fukuda, T., Arai, F., and Dong, L. X. (2003). Assembly of nanodevices with carbon nanotubes through nanorobotic manipulations. *Proc. of the IEEE*, 91, 1803–18.

Fukuda, T., Nakajima, M., Ahmad, M. R., Shen, Y., and Kojima, M. (2010). Micro- and Nanomechatronics. *IEEE Industrial Electronics*, 4, 13–22.

Hell, S. W. (2007). Far-field optical nanoscopy. *Science*, 316, 1153–8.

Hertel, T., Martel, R., and Avouris, P. (1998). Manipulation of individual carbon nanotubes and their interaction with surfaces. *J. Phys. Chem. B*, 102(6), 910–15.

Hosoki, S., Hosaka, S., and Hasegawa, T. (1992). Surface modification of MoS_2 using STM. *Appl. Surf. Sci.*, 60/61, 643.

Ikeda, S., Arai, F., Fukuda, T., and Negoro, M. (2004). An *in vitro* soft membranous model of individual human cerebral artery reproduced with visco-elastic behavior. *Proc. of the 2004 IEEE Int. Conf. on Robotics and Automation (ICRA 2004)*, pp. 2511–16,.

Iverson, N. M., Barone, P. W., Shandell, M., Trudel, L. J., Sen, S., Sen, F., Ivanov, V., Atolia, E., Farias, E., McNicholas, T. P., Reuel, N., Parry, N. M. A., Wogan, G. N., and Strano, M. S. (2013). *In vivo* biosensing via tissue-localizable near-infrared-fluorescent single-walled carbon nanotubes. *Nature Nanotechnology*, 8, 873–80.

Kim, P., and Lieber, C. M. (1999). Nanotube nanotweezers. *Science*, 286, 2148–50.

Langer, R., and Vacanti, J. P. (1993). Tissue engineering. *Science*, 260, 920–6.

Leary, S. P., Liu, C. Y., and Apuzzo, M. L. J. (2006). Toward the emergence of nanoneurosurgery. *Neurosurgery*, 58, 1009–26.

Li, G., Xi, N., Yu, M., and Fung, W.-K. (2004). Development of augmented reality system for AFM based nanomanipulation. *IEEE Trans. on Mechatronics*, 9, 358–65.

Li, G., Xi, N., Chen, H., Pomeroy, C., and Prokos, M. (2005). "Videolized" atomic force microscopy for interactive nanomanipulation and nanoassembly. *IEEE Trans. on Nanotech.*, 4, 605–15.

Liu, J., Wang, Z., Xie, X., Cheng, H., Zhaoac, Y., and Qu, L. (2012). A rationally-designed synergetic polypyrrole/graphene bilayer actuator. *J. Mater. Chem*, 22, 4015–20.

Liu, P., Arai, F., Dong, L. X., Fukuda, T., Noguchi, T., and Tatenuma, K. (2005b). Field emission of individual carbon nanotubes and its improvement by decoration with ruthenium dioxide. *J. Robo. Mech.*, 17, 475–82.

Liu, P., Dong, L. X., Arai, F., and Fukuda, T. (2005a). Nanotube multi-functional nanoposition sensors. *J. Nanoeng. Nanosyst.*, 219, 23–7.

Nakajima, M., Arai, F., and Fukuda, T. (2006). In situ measurement of Young's modulus of carbon nanotube inside TEM through hybrid nanorobotic manipulation system. *IEEE Trans. on Nanotechnology*, 5, 243–8.

Nakajima, M., Arai, S., Saito, Y., Arai, F., and Fukuda, T. (2007). Nanoactuation of telescoping multi-walled carbon nanotubes inside transmission electron microscope. *Jpn. J. Appl. Phys.*, 42, L1035–38.

Requicha, A. A. G., Arbuckle, D. J., Mokaberi B. and Yun, J. (2009). Algorithms and software for nanomanipulation with atomic force microscopes. *Int'l J. Robotics Research*, 28, 512–22.

Rinzler, A. G., Hafner, J. H., Nikolaev, P., Lou, L., Kim, S. G., Tomanek, D., Nordlander, P., Colbert, D. T., and Smalley, R. E. (1995). Unraveling nanotubes: field emission from an atomic wire. *Science*, 269, 1550–3.

Saito, Y., Uemura, S., and Hamaguchi, K. (1998). Cathode ray tube lighting elements with carbon nanotube field emitters. *Jpn. J. Appl. Phys.*, 37, L346–8.

Sazonova, V., Yaish, Y., Ustunel, H., Roundy, D., Arias, T. A., and McEuen, P. L. (2004) A tunable carbon nanotube electromechanical oscillator. *Nature*, 431, 284–7.

Sitti, M., and Hashimoto, H. (2003). Teleoperated touch feedback from the surfaces at the nanoscale: modeling and experiments. *IEEE Trans. on Mechatronics*, 8, 287–98.

Smith, A. D., Niklaus, F., Paussa, A., Vaziri, S., Fischer, A. C., Sterner, M., Forsberg, F., Delin, A., Esseni, D., Palestri, P., Ostling, M., and Lemme, M. C. (2013). Electromechanical piezoresistive sensing in suspended graphene membranes. *Nano Letters*, 13, 3237–42.

Suuronen, E. J., Sheardown, H., Newman, K. D., McLaughlin, C. R., and Griffith, M. (2005). Building *in vitro* models of organs. *International Review of Cytology*, 244, 137–73.

Tixier-Mita, A., Jun, J., Ostrovidov, S., Chiral, M., Frenea, M., LePioufle, B., and Fujita, H. (2004). A silicon micro-system for parallel gene transfection into arrayed cells. *Proc. of the uTAS 2004 Symposium, The Royal Society of Chemistry*, pp. 180–2.

Trung, P. Q., Hoshi, T., Tanaka Y., and Sano, A. (2013). Proposal of tactile sensor development based on tissue engineering. *Proc. of IEEE/RSJ Int. Conf. Intelligent Robots and Systems (IROS 2013)*, pp. 2030–4.

Tsutsui, H., Yu, E., Marquina, S., Valamehr, B., Wong, I., Wu, H., Ho, C. H. (2010). Efficient dielectrophoretic patterning of embryonic stem cells in energy landscapes defined by hydrogel geometries. *Annals of Biomedical Engineering*, 38, 3777–88.

Wang, E., Desai, M. S., and Lee, S. W. (2013). Light-controlled graphene-elastin composite hydrogel actuators. *Nano Letters*, 13, 2826–30.

Wang, H., et al. (2015). Automated assembly of vascular-like microtube with repetitive single-step contact manipulation. *IEEE Transactions on Biomedical Engineering*, 62(11), 2620–8.

Yamanaka, S. (2009). A fresh look at iPS cells. *Cell*, 137, 13–17.

Yue, T., Nakajima, M., Ito, M., Kojima, M., and Fukuda, T. (2011). High speed laser manipulation of on-chip fabricated microstructures by replacing solution inside microfluidic channel. *Proc. of the 2011 IEEE/RSJ Int. Conf. on Intelligent Robots and Systems (IROS2011)*, pp. 433–8.

Yue, T., Nakajima, M., Takeuchi, M., Hu, C., Huang, Q., and Fukuda, T. (2014). On-chip self-assembly of cell embedded microstructures to vascular-like microtubes. *Lab on a Chip*, 14, 1151–61.

Zhang, X., Yu, Z., Wang, C., Zarrouk, D., Seo, J. W. T., Cheng, J. C., Buchan, A. D., Takei, K., Zhao, Y., Ager, J. W., Zhang, J., Hettick, M., Hersam, M. C., Pisano, A. P., Fearing, R. S., and Javey, A. (2014). Photoactuators and motors based on carbon nanotubes with selective chirality distributions. *Nature Communications*, 5, 2983.

Chapter 53

Biohybrid touch interfaces

Sliman J. Bensmaia

Department of Organismal Biology and Anatomy,
University of Chicago, USA

The hand is a remarkably sophisticated and versatile sensorimotor organ. It affords us the ability not only to grasp and dexterously manipulate objects, but also to engage in amazingly complex behaviors such as typing on a keyboard or playing the guitar. While these abilities rely on sophisticated neural systems to control the hand, they would be severely impaired if not abolished were the hand not also endowed with a myriad of sensors (Johansson and Flanagan 2009). Indeed, a wide variety of receptors are embedded in the skin, muscles, and joints of the hand, and these receptors convey information about the conformation and movements of the hand and about the size, shape, and material properties of objects that contact the hand (Bensmaia and Manfredi 2012). Specialized regions of the brain interpret these signals from the hand and integrate them with ongoing motor commands to guide behavior. Without the continuous stream of sensory signals from the hand, the grasping and manipulation of objects would rely on visual feedback, which is sometimes unavailable, for example when an object is obstructed by another, and often inadequate, for example, when deciding how much force to exert to grasp a rigid object. In addition to their role in guiding motor behavior, sensory signals from the hand also play a critical role in embodiment, that is, they make the hand feel as though it is part of us. Finally, touch is essential to communicating emotion and to sexual behavior and experiences.

For amputees and tetraplegic patients, the loss of limb or the ability to control the limb has devastating consequences on the quality of life. Increasingly verisimilar and sophisticated robotic hands have been designed to replace lost or dysfunctional hands, and algorithms have been developed to control these hands by deciphering (or decoding) intended movements from patterns of activation in residual muscles or nerves, or even from neurons in the brain (see Figure 53.1 for an illustration)(Gilja et al. 2011). However, using a prosthetic hand requires sensory signals for the same reasons as does using a native one. In addition to the motors that actuate the joints, robotic hands are equipped with a variety of sensors that track the state of the hand, much as do receptors in the native hand. The trick, then, is to convert the output of these sensors into percepts that restore lost sensation or at least convey enough information to guide the control of the hand (Figure 53.1). One approach consists of substituting lost sensation with stimuli applied to the residual sensory sheet (for example, the chest of an amputee) to convey information about the state of the prosthetic hand (Rombokas et al. 2013). The patient must then learn to associate specific patterns of stimulation with specific states of the hand. Another, more ambitious approach consists of attempting to restore touch biomimetically, that is, to reproduce the patterns of neuronal activity that would be elicited were the native limb still in place and connected to the brain.

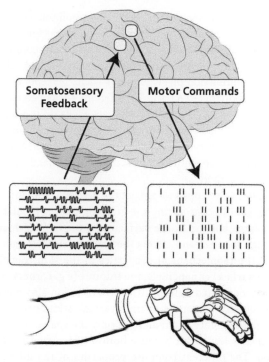

Figure 53.1 Signals from motor areas of the brain are used to control the robotic limb (right). Signals from sensors on the limb are converted into electrical patterns of stimulation applied to somatosensory areas of the brain to elicit meaningful sensory percepts (left).

Restoring touch through a peripheral interface

While amputees have lost a limb, the nerves that sent motor commands to the muscles to move the limb and convey sensory feedback from the limb to the brain are still intact. It is thus possible in principle to decode motor intention from the motor nerve fibers and activate residual sensory fibers to elicit percepts (Weber et al. 2012). As mentioned above, there are a variety of receptors that innervate the skin, three of which mediate touch, that is, convey information about the size, shape, and texture of grasped objects, and signal if these are moving across the skin. These mechanoreceptors convert skin deformations into neural signals and each type of receptor responds to different aspects of skin deformation and mediates different aspects of tactile perception. While all three receptors contribute to some degree to all forms of tactile perception, Merkel disks are primarily involved in sensing pressure and also play a role in tactile shape perception, Meissner corpuscles are involved in tactile motion perception, and Pacinian corpuscles are implicated in tactile texture perception. Signals from each receptor are then carried to the brain by afferent fibers: slowly adapting type 1 (SA1) afferents carry signals from Merkel disks, rapidly adapting (RA) afferents from Meissner corpuscles, and Pacinian (PC) afferents from Pacinian corpuscles (Bensmaia and Manfredi 2012). Any information about the state of the hand and about objects contacting the hand is conveyed in the spatio-temporal pattern of activation in these mechanoreceptive afferents.

The idea behind restoring touch through a peripheral interface is to activate afferents in a systematic and informative way to elicit, to the extent possible, verisimilar percepts. One way to activate afferents is to stimulate them with short, low-amplitude electrical pulses. Electrical stimulation of afferents has been shown, in experiments with awake human subjects, to elicit percepts that depend systematically on the afferent type (Ochoa and Torebjork 1983). Stimulation of SA1 afferents elicits percepts of pressure, stimulation of individual RA afferents elicits percepts of flutter and motion, and stimulation of PC afferents elicits percepts of vibration and texture. Furthermore, activation of individual SA1 and RA afferents elicits percepts that are localized to small patches of skin, whereas activation of PC afferents elicits diffuse sensations. Touch could then be restored by developing algorithms that convert the output of the sensors on the robotic hand into appropriate patterns of nerve stimulation. To achieve biomimetic patterns of nerve activation requires a mechanistic understanding of how, in the native hand with native receptors, skin deformation leads to afferent activity.

Powerful models have been developed to predict the neuronal responses elicited by any kind of skin deformation that could occur during the grasping and manipulation of objects. These models describe how a force applied to the surface of the skin (which can vary in time and space) propagates through the skin and affects the three types of receptors embedded in the skin (Kim et al. 2010). The biomechanical properties of the skin transform the spatial pattern applied to the skin's surface, enhancing some features (e.g. corners) and obscuring others (small internal features of contacted objects). These models can also predict with millisecond accuracy the timing of afferent responses to time-varying stimuli (e.g. of a textured surface scanned across the skin). The ability to predict not only the spatial but also the temporal patterns of afferent activation is important because both of these aspects of the neural response shape tactile perception. The different response properties of the different afferents are well captured by these models. For example, individual SA1 and RA fibers respond to stimulation of small patches of skin (with diameters on the order of a few millimeters), whereas individual PC afferents often respond to stimulation of the entire hand. Furthermore, the response of SA1 afferents is determined primarily by the depth at which the skin is indented (or the magnitude of the pressure that is exerted on it), RA afferents to the rate of change of the indentation, and PC afferents to the acceleration of the skin. Not only do such models provide a mechanistic understanding of the cutaneous signals the hand sends to the brain, they can also be used in a neuroprosthetic application to convert patterns of sensor output into patterns of nerve activation (Kim et al. 2009). Indeed, the model output can then be effected in the nerve by delivering appropriate patterns of electrical stimulation.

A critical component of restoring touch through peripheral nerve stimulation is the neural interface. The idea is to chronically implant electrode arrays in the nerve or in the spinal cord, and to electrically stimulate individual afferents or populations of afferents through these electrodes (Figure 53.2). A variety of implants are being developed for the somatosensory nerve: some resemble miniature beds of nails that are embedded into the nerve, others are cuffs that encircle the nerve (Weber et al. 2012). Verisimilar tactile sensations—of pressure and texture, for example—have been elicited by stimulating through such arrays. However, the mapping between sensations and electrical stimulation using these arrays is difficult because, given the current state of technology, it is not possible to use these implants to stimulate individual afferents or even individual afferent types. Thus, models of receptors cannot be straightforwardly used to convert sensor output into nerve stimulation. One near-term approach might consist of simulating the responses of populations of afferents and effecting these through electrical stimulation of the nerve. As technology progresses and afferent stimulation becomes more selective, however, the ability to elicit predictably verisimilar percepts by reproducing natural patterns of nerve activity will improve. Given the advanced state of our understanding

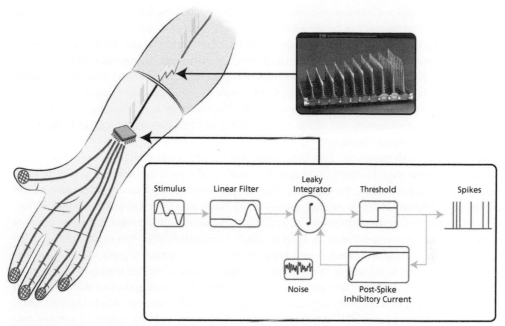

Figure 53.2 Neuroprosthesis with a peripheral interface. The output of the sensors is converted into desired spike trains using a model of mechanotransduction (bottom right inset), which is effected in the nerve by electrical stimulation through a chronically implanted electrode array (e.g. a Utah Slant Electrode Array, top right inset).

of mechanotransduction and the properties of the somatosensory nerve, the main obstacle to achieving naturalistic artificial percepts through a peripheral interface is a technological one. Thus, as technology improves, so will our ability to restore naturalistic tactile sensations in amputees equipped with robotic limbs.

Restoring touch through a brain interface

While peripheral interfaces offer amputees a means to restore sensorimotor function, they cannot be applied to patients with upper spinal cord injuries, because the communication between the nerve and the brain has been disrupted in these individuals. In this case, the nerve must be bypassed and the neuroprosthesis must interface directly with the brain. Signals from the hand are carried up the arm by mechanoreceptive afferents and travel a short distance up the spinal cord before they terminate on neurons in the dorsal column nuclei. Neurons in these nuclei send projections to the thalamus, which then sends signals to the primary somatosensory cortex (S1). In principle, a neuroprosthesis could interface with dorsal column nuclei, thalamus, or cortex (Weber et al. 2012), but here I focus on cortical interfaces as these constitute the greatest departure from their peripheral counterparts.

As mentioned above, mechanoreceptive afferents signal relatively simple aspects of skin deformations (indentation depth, speed, and acceleration). Information about objects—their shape, texture, motion, etc.—must then be inferred from the pattern of activation over populations of afferents. One of the primary functions of the central nervous system is to extract behaviorally relevant information from this afferent input. Conscious sensory percepts are one of the consequences of this sensory elaboration. The major difference between nerve fibers and cortical neurons is that the former can be classified into a small number of subtypes, for which powerful models can be developed, while the latter cannot. Indeed, every neuron in the brain exhibits unique properties, and while these properties form clusters to some degree, they do not do so as categorically as do those of peripheral neurons. Consequently, mechanistic models describing cortical neuronal responses can only be applied to small sub-populations of neurons and are not nearly as powerful as their peripheral counterparts. On the other hand, cortical neurons carry a more elaborated representation of the sensory world and this elaboration can in principle be exploited with a cortical interface.

As is the case with peripheral interfaces, the objective with cortical interfaces is to evoke naturalistic and meaningful patterns of neuronal activation, this time in the brain, hoping to elicit naturalistic and meaningful percepts. To illustrate the approach, let us consider information about contact location. When we grasp an object, we know which parts of our hand—which fingers, for example—make contact with it. S1 is organized somatopically, that is, nearby neurons respond to nearby patches of skin. In other words, S1 comprises a complete map of the body (in fact, at least four complete maps). A well-known representation of this organizational scheme is the somatosensory homunculus, which depicts a cross-sectional view of the brain lined with a deformed image of a body, with giant hands and a giant head, reflecting the fact that more brain volume is devoted to some body regions than to others. When a patch of skin is touched, then, a spatially restricted patch of cortex is activated (in fact, three spatially restricted patches, one in each cutaneous body map—the fourth is proprioceptive). Our perception of where something is touching us may thus be determined by where in the brain the activated neurons are located. In fact, stimulation of S1 neurons does elicit a percept that is localized to their receptive field, that is, to the patch to which they normally respond (Berg et al. 2011). This property of somatosensory cortex can then be used to convey information about contact location. Indeed, even though amputees have lost their limb, they still experience sensations of a limb there, because the part of their somatosensory cortex that used to respond to the lost limb is still there. Activity in that part of their brain is thus experienced on a limb that is no longer there, often referred to as a phantom limb. For tetraplegic patients, the sensation is experienced on their deafferented limb. Now, suppose touching an object with the prosthetic index fingertip led to the electrical activation of neurons whose receptive field is on the index fingertip. Activation of those neurons would in turn elicit a percept that is projected (or experienced) on the index fingertip. Better yet, studies show that if artificially induced sensations are paired with the visual experience of contact between the prosthetic hand and an object, the sensations are projected to the prosthetic hand (Figure 53.3). In other words, if touching objects elicits spatially appropriate patterns of activity in S1, the hand may become embodied and experienced as part of one's body!

As more pressure is exerted on the skin, the neurons that respond to that patch of skin become more active, and nearby neurons become activated. This increase in the strength and spatial extent of the neuronal response likely underlie the increase in perceptual magnitude that accompanies increases in pressure. The effects of changes in pressure on neuronal activity can be mimicked by increasing or decreasing the amplitude of electrical pulse trains applied to the brain: low pressures will be mimicked using low-amplitude pulse trains; high pressures will be mimicked using high-amplitude pulse trains (Berg et al. 2013). The perceptual magnitude of artificial percepts is modulated by changes in electrical pulse amplitude in much the same way that the perceptual magnitude of natural percepts is modulated by changes in skin pressure. Thus, graded sensations of pressure can be achieved by modulating the electrical amplitude according to the output of pressure sensors on the prosthesis.

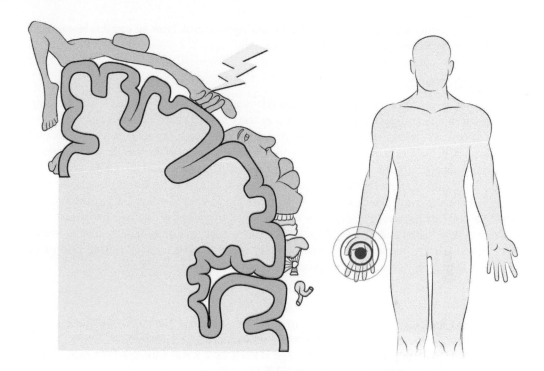

Figure 53.3 Conveying information about contact location. Stimulation of a population of neurons that respond to the hand elicits sensations that are projected to the hand. This phenomenon can be used to intuitively convey information about contact location.

Future directions

While information about contact location and contact pressure are critical to object manipulation, a full restoration of touch will involve eliciting rich and multidimensional artificial percepts. Indeed, when we grasp an object, we acquire information about its shape, its texture, its motion, etc.—properties beyond just location and pressure. The next step in sensory restoration, then, is to attempt to elicit percepts with specific qualities. One approach would consist of exploiting the feature selectivity of neurons in the sensory cortex. Indeed, some neurons in S1 only respond when a specific object feature is present in their receptive field. For example, some neurons will only respond if an edge at a specific orientation is indented into the skin (orientation selective neurons) (Bensmaia et al. 2008). Other neurons only respond when an object is moving in a specific direction across the skin, regardless of its shape (direction selective neurons)(Pei et al. 2010). Perhaps stimulating a population of neurons with a specific orientation preference will elicit a percept of an edge at that orientation. Stimulating a population of neurons with a specific direction preference will elicit a percept of a stimulus moving in that direction. Stimulation of the two populations simultaneously might then elicit a percept of an oriented edge moving in a specific direction. If this approach is successful, and the experiments have yet to be done, the verisimilitude of artificial percepts will be limited primarily by the technology that is used to stimulate: the more electrodes we can implant, the richer the percepts we will be able to elicit.

In the meantime, we are able to intuitively convey certain types of tactile information by mimicking natural and relevant patterns of neuronal activity. For a tetraplegic who has not felt anything below the neck since his or her injury, that is a major step forward. How far can this biomimetic approach take us in restoring touch? Only time will tell.

Learning more

A recent review by Delhaye et al. (2016) explores some of the challenges involved in building human–machine interfaces for prosthetic touch systems. Several chapters in this handbook also address topics very relevant to designing somatosensory prostheses. Lepora (Chapter 16) reviews the biological principles of human touch and discusses some example biomimetic tactile systems. Cutkosky (Chapter 30) discusses the challenge of reaching, grasping, and manipulating objects, noting the benefits of synergies in simplifying control. Prasad (Chapter 49) describes non-invasive brain–machine interfaces that could potentially be used to control a prosthetic hand, while Vassanelli (Chapter 50) discusses the challenge of creating long-term direct interfaces to the nervous system. Lenay and Tixier (Chapter 58) review sensory substitution technologies which, as discussed above, could serve as a means to display sensory signals from a prosthesis on an area of skin elsewhere on the body, or transposed into a different sensory modality.

References

Bensmaia, S. J., Denchev, P. V., Dammann J. F., III, Craig, J. C., and Hsiao, S. S. (2008). The representation of stimulus orientation in the early stages of somatosensory processing. *J. Neurosci.*, **28**, 776–86.

Bensmaia, S. J., and Manfredi, L. R. (2012). The sense of touch. In: V. S. Ramachandran (ed.), *Encyclopedia of Human Behavior*. Amsterdam: Elsevier, pp. 373–8.

Berg, J. A., Dammann J. F., III, Manfredi, L. R., Tenore, F. V., Kandaswamy, R., Vogelstein, R. J., Tabot, G. A., Hatsopoulos, N. G., and Bensmaia, S. J. (2011). *Providing sensory feedback through intracortical microstimulation for upper limb neuroprostheses.* Paper presented at Neuroscience 2011, November 12–16, Washington, D.C., Society for Neuroscience.

Berg, J. A., Dammann J. F., III, Tenore, F. V., Tabot, G. A., Boback, J. L., Manfredi, L. R., Peterson, M. L., Katyal, K. D., Johannes, M. S., Makhlin, A., Wilcox, R., Franklin, R. K., Vogelstein, R. J., Hatsopoulos, N. G., and Bensmaia, S. J. (2013). Behavioral demonstration of a somatosensory neuroprosthesis. *IEEE Trans. Neural Syst. Rehabil. Eng.*, **21**, 500–7.

Delhaye, B. P., Saal, H. P., and Bensmaia, S. J. (2016). Key considerations in designing a somatosensory neuroprosthesis. *Journal of Physiology-Paris*, **110**(4, Part A), 402–8. doi:https://doi.org/10.1016/j.jphysparis.2016.11.001

Gilja, V., Chestek, C. A., Diester, I., Henderson, J. M., Deisseroth, K., and Shenoy, K. V. (2011). Challenges and opportunities for next-generation intracortically based neural prostheses. *IEEE Trans. Biomed. Eng.*, **58**, 1891–9.

Johansson, R. S., and Flanagan, J. R. (2009). Coding and use of tactile signals from the fingertips in object manipulation tasks. *Nat. Rev. Neurosci.*, **10**, 345–59.

Kim, S. S., Sripati, A. P., and Bensmaia, S. J. (2010). Predicting the timing of spikes evoked by tactile stimulation of the hand. *J. Neurophysiol.*, **104**, 1484–96.

Kim, S. S., Sripati, A. P., Vogelstein, R. J., Armiger, R. S., Russell, A. F., and Bensmaia, S. J. (2009). Conveying tactile feedback in sensorized hand neuroprostheses using a biofidelic model of mechanotransduction. *IEEE Trans. BIOCAS.*, **3**, 398–404.

Ochoa, J., and Torebjork, E. (1983). Sensations evoked by intraneural microstimulation of single mechanoreceptor units innervating the human hand. *J. Physiol.*, **342**, 633–54.

Pei, Y. C., Hsiao, S. S., Craig, J. C., and Bensmaia, S. J. (2010). Shape invariant coding of motion direction in somatosensory cortex. *PLoS Biol.*, **8**, e1000305.

Rombokas, E., Stepp, C. E., Chang, C., Malhotra, M., and Matsuoka, Y. (2013). Vibrotactile sensory substitution for electromyographic control of object manipulation. *IEEE Trans. Biomed. Eng.*, **60**(8). doi: 10.1109/TBME.2013.2252174

Weber, D. J., Friesen, R., and Miller, L. E. (2012). Interfacing the somatosensory system to restore touch and proprioception: essential considerations. *J. Mot. Behav.*, **44**, 403–18.

Chapter 54

Implantable hearing interfaces

Torsten Lehmann[1] and André van Schaik[2]

[1] School of Electrical Engineering and Telecommunications,
University of New South Wales, Australia
[2] Bioelectronics and Neuroscience, MARCS Institute for Brain,
Behaviour and Development, Western Sydney University, Australia

Loss of hearing can be a devastating disability. In normal daily activities we use spoken language to communicate with fellow beings—thus loss of hearing severely limits our interaction with colleagues, friends, family, and other people. Loss of hearing easily leads to social isolation, with possible further consequences such as loss of income or depression. That technology has been used, ever since it was possible, for helping people with a hearing impairment to hear is therefore unsurprising. The conventional hearing aid is very successful in helping people with hearing loss; however, while modern hearing aids are sophisticated, they essentially amplify sound, relying on residual hearing in the user for their operation. One of the leading causes of deafness is the loss of the inner hair cells in the cochlea; these are responsible for translating sound pressure into electrical signals in the sensory nervous system. If the hair cells are lost, sound amplification gives users no benefit. However, one of the most successful biohybrid systems of today, the cochlear implant or bionic ear, is able to partially restore hearing in such cases.

In this chapter, we will explain the fundamental operation of the cochlea. We will discuss the biological inspiration used in the implementation of cochlear implants and discuss the constraints imposed on designs when creating biohybrid hearing systems. We will discuss the challenges faced by present and future cochlear implants and finally outline areas of future research. Other types of hearing prostheses are also found, such as bone-anchored hearing aids, middle-ear implants, and brainstem stimulators; however, we shall focus our attention on the cochlear implant, which is the most widely used device.

Biological principles of the cochlea

Sound, having been picked up by the outer ear, travels down the ear canal and vibrates the eardrum (see Chapter 15 for details). The eardrum, in turn, drives the impedance-matching levering mechanism of the middle ear bones (the malleus, incus, and stapes) which attach to the oval window of the fluid-filled cochlea (or the inner ear), thus coupling sound efficiently to the cochlear chambers. In the cochleas, sound vibrations are converted into neural activity which travels up the auditory nerves from both ears to the olivary complex in the brainstem and from there, via the inferior colliculus, sound information reaches the auditory cortex where higher-level sound processing takes place.

The cochlea

A schematic drawing of the cochlea is shown in Figure 54.1. The spiral cochlea has two main chambers: the scala vestibuli and the scala tympani, both stretching the length of the cochlea

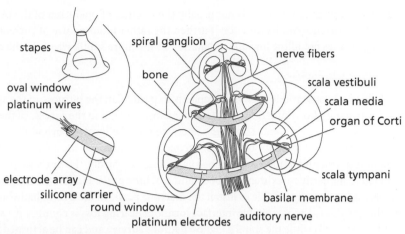

Figure 54.1 Schematic drawing of cochlea with implanted electrode array.

and separated by the basilar membrane. Sound is coupled via the oval window at the base of the cochlea to the scala vestibuli and causes deflections of the basilar membrane that travel towards the apex of the cochlea. The stiffness of the basilar membrane reduces with distance from the basal end causing the propagation of the deflection wave to slow down as it travels down the cochlea. As a consequence, the deflection amplitude peaks somewhere along the membrane. The position of the peak depends on the frequency of the incoming sound: high-frequency signals cause peaks near the basal end while low-frequency signals cause peaks near the apical end. This frequency-dependent mapping of sound signals to location (tonotopic mapping) is one of the key aspects used in cochlear implants.

Along the basilar membrane lies the organ of Corti. Here, distributed along the length of the cochlea, are located the hair cells, which interface between physical movement and neural activation. The inner hair cells create neural responses to deflections of the basilar membrane; these neural responses are transmitted via chemical synapses to the spiral ganglion cells and sent down the auditory nerve. Because of the tonotopic mapping of the cochlea, the auditory nerve also carries a tonotopic map of the incoming sound. This mapping of sound signals to neural activity is another key aspect used in cochlear implants.

The outer hair cells in the organ of Corti create movement in response to neural activity. This creates a positive feedback system whereby movement caused by sound generates neural activity, thereby causing further movement, again increasing the neural activity: amplification of the neural signal takes place. The amplification is non-linear, allowing for more amplification of soft sounds than of loud sounds, in effect compressing the dynamic range of the sounds received by the cochlea and allowing us to hear softer sounds. This dynamic range compression is the third key aspect used in cochlear implants.

Artificial cochleas

In recent decades, a substantial amount of research has been carried out in the area of building systems that closely mimic the operation of the cochlea—particularly within the area of neuromorphic engineering. Both electronic and software cochleas have been created, and while some systems strive to perform a given task (such as sound localization or segmentation) in an alternative way to more classical engineering methods, most systems strive to gain insight into the operation of the biological systems they mimic. In building biohybrid systems such as a cochlear implant, however, it is not usually the manner of operation of the biological system that is of importance but its function. Further, the rather blunt nature of current stimulation technologies does not allow for implementation of the more subtle effects in the cochlea.

Biohybrid considerations

For most biohybrid systems, and for cochlear implants in particular, the purpose of the system is prosthetic, i.e. replacing a body function which has been lost. Critical to the success of the system is the body–machine interface point. Firstly, the interface must be located beyond the lost body function, interfacing to operational body parts. Secondly, the interface must be located where the body function is well understood and can be characterized. Thirdly, the interface point must provide for stable, long-term co-existence of body and machine parts. Finally, the interface must be located where it is surgically feasible to install the machine parts. In the case of biohybrid hearing systems, the cochlea is a most suitable interface location: hearing loss is commonly caused by death of cochlear hair cells while the spiral ganglion neurons survive and can be activated by electrical stimulation from within the cochlea; the tonotopic mapping of sound on the spiral ganglion cells in the healthy cochlea can be readily replicated using electronics and an array of electrodes; an electrode array is relatively easily inserted into the cochlea whose bony structure fixes the electrode array relative to the nerve cells.

In some cases, for instance if the auditory nerves have been severed, the cochlea cannot be used as an interface for a biohybrid hearing system. In such cases, the interface can be placed further along the auditory pathway, for instance on the auditory brainstem. Such biohybrid hearing systems, however, are less successful than the cochlear implant because the sound mapping is less well defined and understood.

Biohybrid hearing systems

In replacing body functions, the capabilities and limitations of technologies used play crucial roles in the design and overall biohybrid system performance. The approach taken by the biological system, for instance, is not necessarily the most efficient or straight-forward using technologies available. Mimicking the sound processing of the cochlea is most readily achieved using an electronic system; modern digital electronics is capable of sophisticated, low-power signal processing and can be re-programmable, which enables new research to be of benefit to already deployed systems.

Neural interface

Translating acoustic signals into electronic signals for processing is readily done using a microphone. The further translation of electronic signals into nerve signals is done by passing a current through a pair of electrodes (see Chapter 50 for details): the current passing through the tissue will change the potential difference between nerve cells and extracellular fluid, causing nerve cells in close proximity to the electrode to activate. By increasing the current magnitude, more nerve cells will activate, increasing the physiological response to the stimulation, hence enabling electronic modulation of the perceived sound intensity. In order to avoid pain or tissue damage caused by the electrical stimulation, it is important to limit current magnitude, duration, and average: usually tightly controlled bi-phasic current pulses are used for this purpose. By placing a number of electrodes along the length of the cochlea, each within reach of different nerve cells in the cochlea, an interface between electronic and neural systems is created.

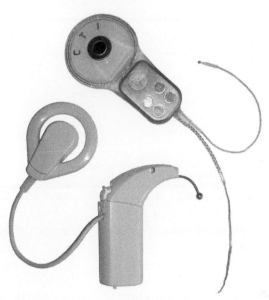

Figure 54.2 Photograph of cochlear implant (top right) and external speech processor (bottom left).

Cochlear implant system

A photograph of a typical cochlear implant and accompanying external speech processor is shown in Figure 54.2. The implant is placed under the skin above the ear while the electrode array is inserted through the round window of the cochlea into the scala tympani, see Figure 54.1. A return electrode is normally placed outside the cochlea to maximize current distribution along the nerves which lower the stimulation current perception threshold. To avoid adverse body reactions to the implant, the materials in contact with the body must be bio-compatible: the electronics are encased in a hermetically sealed titanium shell while electrodes are made of platinum encased in a silicone carrier for electrical isolation. Cochlear implants need to stimulate at rates of 1–10kHz in order to generate useful sound quality. Such rates require a relatively large power draw; thus batteries are located outside the body in the speech processor for easy replacement. Power (and data) is then transferred to the implant via a transcutaneous set of inductively coupled coils aligned by a pair of magnets that also serve to hold the external coil in place. The typical behind-the-ear type speech processor also houses the system's microphone and conducts most of the required signal processing, making significant upgrades to the system possible without surgery.

Sound processing

The very first cochlear implants had only a few electrodes, but it was soon clear that biohybrid hearing systems should mimic the tonotopic mapping in the healthy cochlea. Today's cochlear implants have tens of electrodes: when electrodes near the basal end of the cochlea are stimulated a sensation of high-frequency sound is generated while stimulation of electrodes near the apical end generates a low-frequency sound sensation. To illustrate the cochlear frequency classification, Figure 54.3 shows a spectrogram for the sound "choice:" the energy of different frequencies contained in the sound is plotted against time; the phonemes "ch," "oi," and "s" are clearly distinguishable. To generate the tonotopic mapping, speech processors likewise divide the frequency range of audible sounds into a number of bands, and for each small time step determine the energy in each

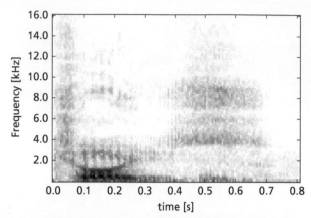

Figure 54.3 Spectrogram of the sound "choice;" dark areas indicate high energy of the given frequency at that point in time.

band (for instance using Fourier transforms or a bank of band-pass filters); each frequency band is mapped to a particular electrode in the cochlea. In the most commonly available stimulation strategy, continuous interleaved sampling (CIS), in each small time step every electrode is stimulated in turn with a current level corresponding to the energy of the frequency band it is mapped to.

Dynamic range

In typical cochlear implant users, the ratio between a just perceivable stimulation current and largest comfortable current (the dynamic range) is only in the range of 10dB. The most important class of sound signals, speech, on the other hand, has a dynamic range of some 50dB. Compression of the sound signals, as in the healthy cochlea, is therefore critical for useful cochlear implant performance. Compression of the sound signal received from the microphone is sometimes employed (using automatic gain control) in order to reduce the dynamic range required in subsequent sound processing. Most importantly, however, energy levels found in each sound frequency band are logarithmically compressed to fit within the dynamic range of the associated electrode.

Future directions

Much progress in cochlear implant performance has been made in the last 30 years, and while cochlear implant users are able to have conversations with normal hearing people, their hearing is far from being fully restored. Hearing in noisy environments in particular is still a challenge.

The most important performance limitation probably lies in the nerve-electrode interface. The further the electrodes are from the nerves they aim to activate, the less selective the stimulation is and the more stimulation power is required. Reducing the effective distance between electrodes and nerves will therefore allow a higher number of more selective channels, which will result in significant performance improvements; this is therefore a key research area. Recent electrode technologies have been aimed as addressing this issue but have so far been limited by the bony structure of the cochlea in reducing distance to the nerves. The use of neural growth factors for encouraging the nerves to grow towards the electrodes is another approach being investigated.

Having a better interface to the nerves will improve overall hearing performance and in particular performance in noisy environments. Using advanced signal processing to remove unwanted noise signals before sound is transmitted to the implant is another current research area. Much improvement has already been made, and with the availability of ever more powerful mobile computing platforms, there continues to be scope for significant improvements in this area.

Current cochlear implant systems are relatively bulky (for instance compared with hearing aids); a fully implanted system would be less prone to breakage, would always be available (even during activities such as swimming) and would be less cumbersome and noticeable. Fully implanted systems have been demonstrated but are not commercially available; the main obstacles faced by these systems are the performance of the implanted microphone and the limited power available in a fully implanted system. These limitations are areas of active research.

Learning more

A number of important elements of cochlear implants have been left out in this brief chapter—for instance the method of mapping threshold and comfort current levels in individual users; the importance of neural plasticity and learning in implant performance; the use of measuring electrically evoked compound action potentials; the effects of surviving neuron population and time of implantation; the increasing use of bilateral implants; the emergence of acoustic/electric hybrid systems; and the many engineering challenges seen in modern implant designs. The interested reader is referred to the references below as a starting point for further reading and pointers to future research directions.

An excellent overview of human and machine hearing is provided by Lyon (2017). Wilson and Dorman (2008) and Zeng et al. (2008) each give overviews of current cochlear implant technologies and areas of future research—mostly from an engineering standpoint—while Seligman (2009) and Clark (2012) give historical accounts of the technology. Lim et al. (2009) describe alternative biohybrid hearing systems. Stöver and Lenarz (2009) explore possible new materials in electrode technology from a physiological standpoint. van Schaik et al. (2010) give examples of systems that mimic the cochlear function. Finally, auditory and cochlear implant systems are described in details by Pickles (2012) and Clark (2003) respectively. A number of other exemplar neuromorphic sensing systems are described elsewhere in this handbook, for instance, for vision (Dudek, Chapter 14) and for chemosensation (Pearce, Chapter 17).

References

Clark, G. (2003). *Cochlear implants: fundamentals and applications*. New York: Springer.

Clark, G. (2012). The multi-channel cochlear implant and the relief of severe-to-profound deafness. *Cochlear Implants International*, **13**(2), 69–85.

Lim, H. H., Lenarz, M., and Lenarz, T. (2009). Auditory midbrain implant: a review. *Trends in Amplification*, **13**(3), 149–80.

Lyon, R. F. (2017). *Human and machine hearing*. Cambridge, UK: Cambridge University Press.

Pickles, J. O. (2012). *An introduction to the physiology of hearing*, 4th edn. Bingley: Emerald Group.

Seligman, P. (2009). Prototype to product—developing a commercially viable neural prosthesis. *Journal of Neural Engineering*, **6**(6), 065006.

Stöver, T., and Lenarz, T. (2009). Biomaterials in cochlear implants. *GMS Current Topics in Otorhinolaryngology—Head and Neck Surgery*, **8**, 10.3205/cto000062.

van Schaik, A., Hamilton, T. J., and Jin, C. T. (2010). Silicon models of the auditory pathway. In: R. Meddis, E. A. Lopez-Poveda, R. R. Fay, and A. N. Popper (eds), *Springer handbook of auditory research: computational models of the auditory system*, Vol. 35. New York: Springer.

Wilson, B. S., and Dorman, M. F. (2008). Cochlear implants: current designs and future possibilities. *Journal of Rehabilitation Research and Development*, **45**(5), 695–730.

Zeng, F.-G., Rebscher, S., Harrison, W., Sun, X., and Feng, H. (2008). Cochlear implants: system design, integration, and evaluation. *IEEE Reviews in Biomedical Engineering*, **1**, 115–42.

Chapter 55

Hippocampal memory prosthesis

Dong Song and Theodore W. Berger

Department of Biomedical Engineering, University of Southern California, Los Angeles, USA

This chapter describes the development of a cognitive prosthesis designed to restore the ability to form new long-term memories typically lost after damage to the hippocampus. The animal model used to develop this prototype is the memory-dependent, delayed nonmatch-to-sample (DNMS) behavior in rats, and the core of the prosthesis is a biomimetic multi-input, multi-output (MIMO) non-linear dynamical model that provides the capability of predicting spatio-temporal spike-train output of the hippocampus (CA1) based on spatio-temporal spike-train inputs recorded presynaptically to CA1 (e.g. CA3). We demonstrate the capability of the MIMO model for highly accurate predictions of CA1 coded memories that can be made on a single-trial basis and in real-time. When hippocampal CA1 function is blocked and long-term memory formation is lost, successful DNMS behavior also is abolished. However, when MIMO model predictions are used to re-instate CA1 memory-related activity by driving spatio-temporal electrical stimulation of hippocampal output to mimic the patterns of activity observed in control conditions, successful DNMS behavior is restored.

Biological principles

Damage to the hippocampus and surrounding regions of the medial temporal lobe can result in a permanent loss of the ability to form new long-term memories (Milner 1970; Squire and Zola-Morgan 1991; Eichenbaum 1999). Because the hippocampus is involved in the creation but not the storage of new memories, long-term memories formed before the damage often remain intact, as does short-term memory. The anterograde amnesia characteristic of hippocampal system damage can result from stroke or epilepsy, and is one of the defining features of dementia and Alzheimer's disease.

One of the obstacles to designing restorative treatments for amnesia is the limited knowledge about the nature of neural processing in brain areas responsible for higher cognitive function. This chapter describes a strategy for addressing this fundamental issue through the replacement of damaged tissue with microelectronics that mimics the input–output functions of the intact biological circuitry. The proposed microelectronic systems do not just electrically stimulate cells to generally heighten or lower their average firing rates. Instead, the proposed prosthesis incorporates mathematical models that, for the first time, replicate the fine-tuned, spatio-temporal coding of information by the hippocampal memory system, thus allowing memories for specific items or events experienced to be formed and stored in the brain.

This approach is based on several assumptions about how information is coded in the brain, and on the relationship between cognitive function and the electrophysiological activity of populations of neurons (Berger et al. 2010). First, we assume that the majority of information in the brain is coded in terms of spatio-temporal patterns of activity. Neurons communicate with each other using all-or-none action potentials, or "spikes." Because spikes are of nearly equivalent amplitude, information can be transmitted only through variation in the sequence of inter-spike intervals, or *temporal pattern*. Populations of neurons are distributed in space, so information transmitted by neural systems can be coded only through variation in the *spatio-temporal pattern* of spikes.

The essential signal processing capability of a neuron is derived from its capacity to change input sequences of inter-spike intervals into different, output sequences of inter-spike intervals. In all brain areas, the resulting input–output transformations are strongly non-linear, due to the non-linear dynamics inherent in the cellular/molecular mechanisms of neurons and synapses. The non-linearities of neural mechanisms go far beyond the action potential threshold. All voltage-dependent conductances are, by definition, non-linear, with a variety of identifying slopes, inflection points, activation voltages, inactivation properties, etc., resulting in sources for non-linearities that are large in number and varied in function. Thus, information in the brain is coded in terms of spatio-temporal patterns of activity, and the functionality of any given brain system can be understood and quantified in terms of the non-linear transformation which that brain system performs on input spatio-temporal patterns. For example, the conversion of short-term memory into long-term memory must be equivalent to the cascade of non-linear transformations comprising the internal circuitry of the hippocampus—as spatio-temporal patterns of activity propagate through the hippocampus, the series of non-linear transformations resulting from synaptic transmission through its intrinsic cell layers gradually converge on the representations required for long-term storage (Figure 55.1). Based on this formulation, we construct multi-circuit, biomimetic systems, in the form of our proposed memory prosthesis, that are intended to circumvent the damaged hippocampal tissue and to re-establish the ensemble, or population coding of spike trains performed by a normal population of hippocampal neurons (Figure 55.2). If our conceptualization of the problem described here

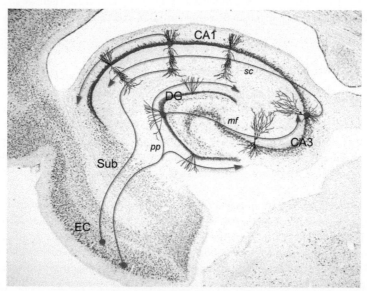

Figure 55.1 The internal circuit of the hippocampus. EC: entorhinal cortex; DG: dentate gyrus; Sub: subiculum; pp: perforant path; mf: mossy fiber; sc: Schaffer collateral.

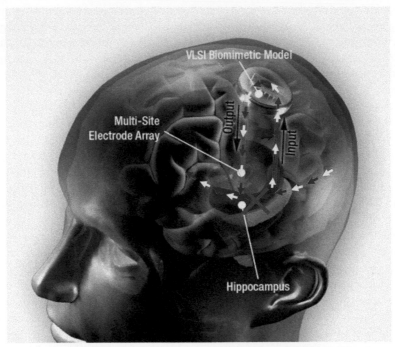

Figure 55.2 Conceptual representation of a hippocampal prosthesis for the human. The biomimetic multi-input, multi-output (MIMO) model implemented in VLSI performs the same non-linear dynamical transformation as a damaged portion of hippocampus. The prosthesis connects up-stream and down-stream regions in the damaged hippocampal area through multi-site electrode arrays.

and implemented in the following experiments and modeling is correct, then we should be able to re-establish long-term memory function in a stimulus-specific manner using the proposed hippocampal prosthesis.

Hippocampal memory prosthesis

The hippocampal memory prosthesis is a multi-circuit, biomimetic system that consists of three components: (1) a multielectrode array for recording the ensemble spike trains from an upstream hippocampal region, e.g. CA3, (2) a computational unit with a MIMO non-linear dynamical model for predicting the output (e.g. CA1) ensemble spike trains based on the ongoing input (e.g. CA3) ensemble spike trains, and (3) an electrical stimulator for stimulating the downstream hippocampal region, e.g. CA1, with the predicted output spatio-temporal patterns of spikes (Figure 55.2).

The MIMO non-linear dynamical model of input–output spike train transformation takes the form of the sparse generalized Laguerre–Volterra model (Song et al. 2009a,b, 2013, Song and Berger 2010; Figure 55.3). A MIMO model is a concatenation of a series of MISO models that each can be considered a spiking neuron model (Song et al. 2006, 2009a). Each MISO model consists of (a) a MISO Volterra kernel model transforming the input spike trains (x) to the synaptic potential u, (b) a Gaussian noise term capturing the stochastic properties of spike generation, (c) a threshold for generating output spikes, (d) an adder generating the pre-threshold membrane potential w, and (e) a single-input, single-output Volterra kernel model describing the output

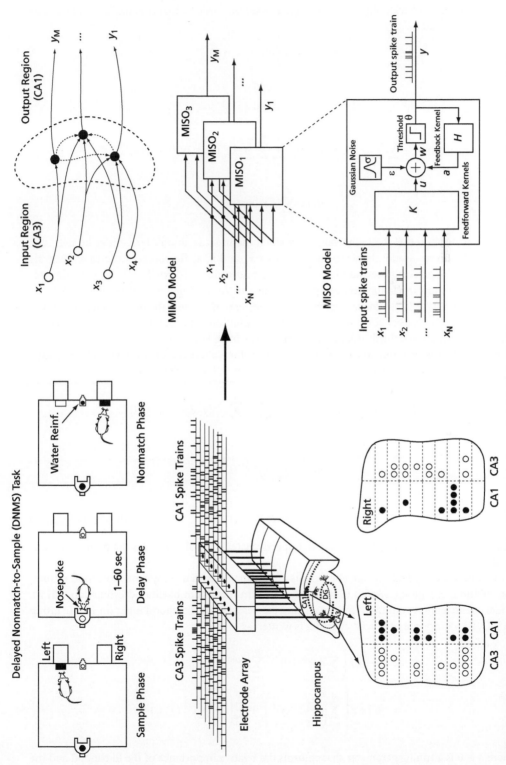

Figure 55.3 Building a multi-input, multi-output (MIMO) non-linear dynamical model of the hippocampal CA3–CA1 using spike trains recorded from rats performing a memory-dependent delayed nonmatch-to-sample (DNMS) task.

spike-triggered feedback nonlinear dynamics. The second-order self-kernel model can be mathematical expressed as:

$$w = u(k,x) + a(h,y) + \varepsilon(\sigma)$$

$$y = \begin{cases} 0 & \text{when } w < \theta \\ 1 & \text{when } w \geq \theta \end{cases}$$

$$u(t) = k_0 + \sum_{n=1}^{N} \sum_{\tau=0}^{M_k} k_1^{(n)}(\tau) x_n(t - \tau) + \sum_{n=1}^{N} \sum_{\tau_1=0}^{M_k} \sum_{\tau_2=0}^{M_k} k_{2s}^{(n)}(\tau_1, \tau_2) x_n(t - \tau_1) x_n(t - \tau_2)$$

$$a(t) = \sum_{\tau=1}^{M_h} h(\tau) y(t - \tau)$$

The zeroth-order kernel, k_0, is the value of u when the input is absent. First-order kernels $k_1^{(n)}$ describe the linear relation between the nth input x_n and u, as functions of the time intervals (τ) between the present time and the past time. Second-order self-kernels $k_{2s}^{(n)}$ describe the second-order non-linear relation between pairs of spikes in the nth input x_n as they affect u. N is the number of inputs. h is the linear feedback kernel. M_k and M_h denote the memory lengths of the feedforward process feedback process, respectively. Additional model terms such as the second-order cross kernels and third-order kernels can also be included.

To facilitate model estimation and avoid overfitting, the Volterra kernels are typically expanded with Laguerre or B-spline basis functions b as in Song et al. (2006, 2009a,b):

$$u(t) = c_0 + \sum_{n=1}^{N} \sum_{j=1}^{J} c_1^{(n)}(j) v_j^{(n)}(t) + \sum_{n=1}^{N} \sum_{j_1=1}^{J} \sum_{j_2=1}^{j_1} c_{2s}^{(n)}(j_1, j_2) v_{j_1}^{(n)}(t) v_{j_2}^{(n)}(t)$$

$$a(t) = \sum_{j=1}^{L} c_h(j) v_j^{(h)}(t)$$

where $v_j^{(n)}(t) = \sum_{\tau=0}^{M_k} b_j(\tau) x_n(t - \tau)$, $v_j^{(h)}(t) = \sum_{\tau=1}^{M_h} b_j(\tau) y(t - \tau)$, $c_1^{(n)}$, $c_{2s}^{(n)}$ and c_h are the sought Laguerre expansion coefficients of $k_1^{(n)}$, $k_{2s}^{(n)}$, and h, respectively (c_0 is equal to k_0). J is the number of basis functions.

To achieve model sparsity, the coefficients are estimated with a composite likelihood estimation method, e.g. group LASSO (Song et al. 2013). In maximum likelihood estimation (MLE), model coefficients are estimated by minimizing the negative log likelihood function-$l(c)$. In group LASSO, the composite penalized criterion is written as

$$S(c) = -l(c) + \lambda \left(\sum_{n=1}^{N} \| c_1^{(n)}(j) \|_2^1 + \sum_{n=1}^{N} \| c_{2s}^{(n)}(j_1, j_2) \|_2^1 \right)$$

$$= -l(c) + \lambda \left(\sum_{n=1}^{N} \left(\sum_{j=1}^{J} c_1^{(n)}(j)^2 \right)^{\frac{1}{2}} + \sum_{n=1}^{N} \left(\sum_{j_1=1}^{J} \sum_{j_2=1}^{j_1} c^{(n)}(j_1, j_2)^2 \right)^{\frac{1}{2}} \right)$$

where $\lambda \geq 0$ is a tuning parameter that controls the relative importance of the likelihood and the penalty term. When λ takes on a larger value, the estimation yields sparser result of the coefficients. λ can be optimized with a multi-fold cross-validation method. Due to the point-process natural of its output, the MIMO model is validated with the Kolmogorov–Smirnov test based on the time-rescaling theorem (Song et al. 2006).

As shown above, the sparse non-linear dynamical MIMO model provides a quantitative way of identifying the functional connectivities between spiking neurons. More importantly, it has been used for predicting the spatio-temporal patterns of output spike trains based on the spatio-temporal patterns of input spike trains, and served as the computational basis of the hippocampal memory prosthesis.

Our hippocampal memory prosthesis experiments are conducted in rats performing a memory-dependent, delayed nonmatch-to-sample (DNMS) task with random delay intervals (Figure 55.3). Animals perform the task by pressing a single lever presented in one of the two positions in the sample phase (left or right). This event is called the "sample response." The lever is then retracted and the delay phase initiates; for the variable durations (0–30 s) of the delay phase, the animal is required to nose-poke into a lighted device on the opposite wall. When the delay is ended, the nose-poke light is extinguished, both levers are extended, and the animal is required to press the lever *opposite* to the sample lever. This event is called the "nonmatch response." If the correct lever is pressed, the animal is rewarded. A session includes approximately 100 successful DNMS tasks that each consists of two of the four behavioral events, i.e. right sample (RS) and left nonmatch (LN), or left sample (LS) and right nonmatch (RN). Spike trains are obtained with MEAs from different septo-temporal regions of the hippocampus (Figure 55.3).

The experiment consists of three steps (Figure 55.4). First, in the control step, a mini-pump implanted into the hippocampus continuously introduces saline during the DNMS session. The

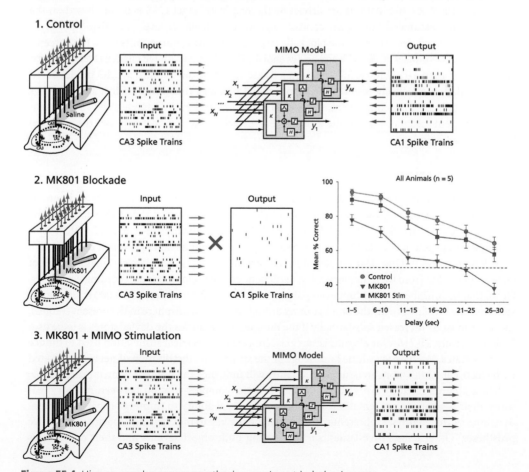

Figure 55.4 Hippocampal memory prosthesis experiment in behaving rats.

memory function of animals is quantified with a forgetting curve describing the percentage of correct nonmatch responses as the function of the delay duration. It shows that animals can successfully maintain memory of the sample locations (left or right) for delays as long as 30 s, with the percentage of correct responses monotonically decreasing as the delay is prolonged (Figure 55.4, green line). During the same period, CA3 and CA1 spike trains are recorded simultaneously with a MEA. Nonlinear dynamical MIMO models reflecting the normal CA3–CA1 spatio-temporal pattern transformations are estimated using these normal condition spike train input–output data. The estimated MIMO models can accurately predict the CA1 spike trains based on the ongoing CA3 spike trains in real time (Figure 55.5). Both CA3 and CA1 show different spatio-temporal patterns for the two lever positions (Song et al. 2014a).

Second, in the blockade step, glutamatergic NMDA channel blocker MK801, instead of saline, is introduced to the hippocampal CA1 region. MK801 drastically suppresses the activities of CA1 pyramidal neurons, while leaving the CA3 pyramidal neuron activities relatively intact. Consequently, at the behavioral level, animal performances during the DNMS task are significantly decreased. The forgetting curve shifts down to the percentage of correct responses at 70–80% for delays shorter than 10 s and near or below chance level (50%) for delays longer than 10 s (Figure 55.4, blue line). These results show that MK801 impairs the hippocampal memory function by disrupting the CA3–CA1 signal transmission.

Finally, in the MIMO stimulation step, the CA1 regions are electrically stimulated with CA1 output patterns predicted by MIMO models during the sample phase, with the infusion of MK801. The CA1 stimulations are driven by the relatively intact CA3 activities based on the MIMO models estimated during the control step. At the behavioral level, the DNMS performances are enhanced from those of the blockade step. The forgetting curve shifts up to the percentage of correct responses at 90–85% for delays shorter than 10 s and above chance level (50%) for delays longer than 10 s (Figure 55.4, red line). These results show that the MIMO model-based CA1 stimulation effectively restores the hippocampal memory function by reinstating the CA1 activities. The recording component, the non-linear dynamical MIMO model, and the stimulation component essentially constitute a closed-loop prosthetic system that bypasses the impaired hippocampal region.

Future directions

The hippocampal memory prosthesis has been successfully implemented in rodents and nonhuman primates performing memory-dependent behavioral tasks such as the DNMS task and the delayed match-to-sample (DMS) task (Berger et al. 2011, 2012; Hampson et al. 2012, 2013).

Despite these achievements, the current system requires major advances before it can approach a working prosthesis. One clear limitation in the current system is the relatively small number of features extracted from each memory, which is dependent on the number of recording electrodes. Given that the number of CA1 hippocampal pyramidal cells (approximately 450,000 in one hemisphere of the rat) far exceeds the number of recording sites (16/hemisphere in the present studies), the latter must be increased substantially if the number of memories the device stores is to be useful behaviorally. Likewise, analogous issues exist concerning electrical stimulation used to induce memory states. We have very little knowledge of the space–time distribution of neurons activated with electrical stimulation, an issue that certainly will influence the number of memories capable of being differentiated by a prosthetic system. It is the complex relation between memory features, their neuronal representation, and our experimental ability to mimic the spatio-temporal dynamics of this neuronal representation that will be the key to maximizing capacity of cognitive prostheses. Other substantive issues, such as coding both objects/events and their contexts, as

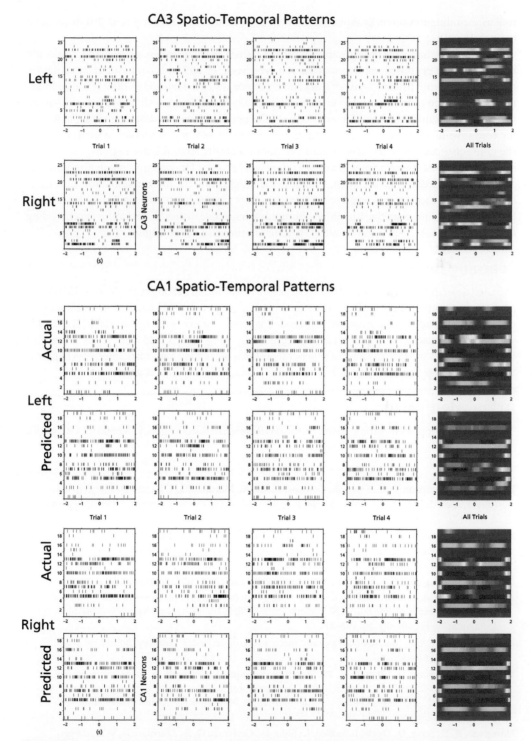

Figure 55.5 CA1 spatio-temporal patterns of spikes are predicted from the CA3 spatio-temporal patterns of spikes using MIMO non-linear dynamical models.

well as including long-term synaptic plasticity into the MIMO model (Song et al. 2014b), remain to be dealt with.

We remain virtually the only group exploring the arena of cognitive prostheses, and, as already made clear, our proposed implementation of a prosthesis for central brain tissue many synapses removed from both primary sensory and primary motor systems demands bi-directional communication between prosthesis and brain using biologically based neural coding. We believe that the general lack of understanding of neural coding will impede attempts to develop neural prostheses for cognitive functions of the brain, and, as prostheses for sensory and motor systems consider the consequences for, or the contributions of, more centrally located brain regions, that understanding fundamental principles of neural coding will become of increasingly greater importance. A deeper knowledge of neural coding will provide insights into the informational role of cellular and molecular mechanisms, representational structures in the brain, and badly needed bridges between neural and cognitive functions. From this perspective, issues surrounding neural coding represent one of the next "great frontiers" of neuroscience and neural engineering.

Learning more

Detailed descriptions of the MIMO model structure can be found in Song et al. (2006 and 2009a,b). Sparse estimation of the MIMO model is described in Song et al. (2013 and 2016), and Robinson et al. (2015). In addition, human MIMO modeling results, which were not available at the time when this manuscript was prepared, are reported in Song et al. (2015 and 2016). Vassanelli (Chapter 50, this volume) discusses the general problem of creating long-term interfaces to neural tissue, a challenge that will need to be overcome to create effective neural prostheses.

Acknowledgment

This work was supported by the Defense Advanced Research Projects Agency (N66001-14-C-4016).

References

Berger, T. W., Hampson, R. E., Song, D., Goonawardena, A., Marmarelis, V. Z., and Deadwyler, S. A. (2011). A cortical neural prosthesis for restoring and enhancing memory. *Journal of Neural Engineering*, **8**, 046017.

Berger, T. W., Song, D., Chan, R. H. M., and Marmarelis, V. Z. (2010). The Neurobiological Basis of Cognition: Identification by Multi-Input, Multi-Output Nonlinear Dynamic Modeling. *Proceedings of the IEEE*, **98**, 356–74.

Berger, T. W., Song, D., Chan, R. H. M., Marmarelis, V. Z., LaCoss, J., Wills, J., Hampson, R. E., Deadwyler, S. A., and Granacki, J. J. (2012). A hippocampal cognitive prosthesis: Multi-input, multi-output nonlinear modeling and VLSI implementation. *IEEE Transactions on Neural Systems and Rehabilitation Engineering*, **20**, 198–211.

Eichenbaum, H. (1999). The hippocampus and mechanisms of declarative memory. *Behavioral Brain Research*, **103**, 123–33.

Hampson, R. E., Song, D., Chan, R. H. M., Sweatt, A. J., Fuqua, J., Gerhardt, G. A., Marmarelis, V. Z., Berger, T. W., and Deadwyler, S. A. (2012). A nonlinear model for cortical prosthetics: memory facilitation by hippocampal ensemble stimulation. *IEEE Transactions on Neural Systems and Rehabilitation Engineering*, **20**, 184–97.

Hampson, R. E., Song, D., Opris, I., Santos, L., Shin, D., Gerhardt, G. A., Marmarelis, V. Z., Berger, T. W., and Deadwyler, S. A. (2013). Facilitation of memory encoding in primate hippocampus by a neuroprosthesis that promotes task specific neural firing. *Journal of Neural Engineering*, **9**, 056012.

Milner, B. (1970). Memory and the medial temporal regions of the brain. In: Pribram, K. H., and Broadbent, D. E. (eds.), *Biology of Memory*. NewYork: Academic Press, pp. 29–50.

Robinson, B. S., Song, D., and Berger, T. W (2015). Estimation of a large-scale generalized Volterra model for neural ensembles with group lasso and local coordinate descent. In: *Proceedings of the IEEE EMBC Conference*, pp. 2526–9.

Song, D. and Berger, T. W. (2010). Identification of nonlinear dynamics in neural population activity. In: K. G. Oweiss (ed.), *Statistical Signal Processing for Neuroscience and Neurotechnology*. Amsterdam: Elsevier, pp. 103–28.

Song, D., Chan, R. H. M., Marmarelis, V. Z., Hampson, R. E., Deadwyler, S. A., and Berger, T. W. (2006). Nonlinear dynamic modeling of spike train transformations for hippocampal-cortical prostheses. *IEEE Transactions in Biomedical Engineering*, **54**, 1053–66.

Song, D., Chan, R. H. M., Marmarelis, V. Z., Hampson, R. E., Deadwyler, S. A., and Berger, T. W. (2009a). Nonlinear modeling of neural population dynamics for hippocampal prostheses. *Neural Networks*, **22**, 1340–51.

Song, D., Chan, R. H. M., Marmarelis, V. Z., Hampson, R. E., Deadwyler, S. A., and Berger, T. W. (2009b). Sparse generalized Laguerre–Volterra model of neural population dynamics. In: *Proceedings of the IEEE EMBS Conference*, pp. 4555–8.

Song, D., Chan, R. H. M., Robinson, B. S., Marmarelis, V. Z., Opris, I., Hampson, R. E., Deadwyler, S. A., and Berger, T. W. (2014b). Identification of functional synaptic plasticity from spiking activities using nonlinear dynamical modeling. *Journal of Neuroscience Methods*, doi: 10.1016/j.jneumeth.2014.09.023.

Song, D., Harway, M., Marmarelis, V. Z., Hampson, R. E., Deadwyler, S. A., and Berger, T. W. (2014a). Extraction and restoration of hippocampal spatial memories with nonlinear dynamical modeling. *Frontiers in Systems Neuroscience*, doi: 10.3389/fnsys.2014.00097.

Song, D., Wang, H., Tu, C. Y., Marmarelis, V. Z., Hampson, R. E., Deadwyler, S. A., and Berger, T. W. (2013). Identification of sparse neural functional connectivity using penalized likelihood estimation and basis functions. *Journal of Computational Neuroscience*, **35**, 335–57.

Song, D., Robinson, B. S., Hampson, R. E., Marmarelis, V. Z., Deadwyler, S. A., Berger, T. W. (2015). Sparse generalized Volterra model of human hippocampal spike train transformation for memory prostheses. In: *Proceedings of the IEEE EMBC Conference*, pp. 3961–4.

Song, D., Robinson, B. S., Hampson, R. E., Marmarelis, V. Z., Deadwyler, S. A., and Berger, T. W. (2016). Sparse large-scale nonlinear dynamical modeling of human hippocampus for memory prostheses. *IEEE Transactions on Neural Systems and Rehabilitation Engineering*, doi: 10.1109/TNSRE.2016.2604423.

Squire, L.R., and Zola-Morgan, S. (1991). The medial temporal lobe memory system. *Science*, **253**, 1380–6.

Section VII

Perspectives

Chapter 56

Perspectives

Michael Szollosy

Sheffield Robotics, University of Sheffield, UK

Living machines will undoubtedly be a big part of our human future. As many of the authors in this handbook point out, we already live in worlds populated with living machines, some taking the form of robots or AIs, but also, more commonly, human beings ourselves, augmented, improved, supplemented by technology. But technological innovation and enhancements aside, the essays in this section begin with and are resolutely focused on *the human*: humans relating to living machines, humans learning to live with living machines, humans learning to live with each other *and* living machines, and humans *as* living machines.

"Perspectives" is a diverse section, presenting examples from the range of issues that are raised and challenges that are posed by our technological developments, and the social impacts that living machines will have on our human societies. But it is absolutely vital that we start these conversations now; these issues cannot simply be regarded as an eccentric luxury, to be ignored or considered only when it is absolutely necessary. For if recent case studies (and more general social and political phenomena) are anything to go by, if we wait until it is *necessary* to address these issues, it will already be too late. The case of public distrust around the use of genetically modified organisms (GMOs) (e.g. Dixon, 1999) may be instructive for those who aspire to build living machines: if innovations and their implications are not addressed before they are brought to the public, we will already find in that space negative opinions, misunderstanding, and irresponsible reporting.

This is not to say that the public do not have valid concerns about GMOs, living machines, or other new technologies; of course they do. Nor is this to say that concerns and anxieties— whether informed and grounded or otherwise—should be treated with a simple bandage, grudgingly dealt with as an inevitable exercise in public relations. It is not enough to parachute in an expert with a knighthood and the backing of a cleverly acronymed think-tank to pacify the masses. If living machines are going to be a part of our (human) lives—and all of these articles make it clear that they absolutely will be, or very much are already—then we need to be open and discuss these impacts on our individual and collective lives from the outset. What is needed, therefore, is a concrete commitment to proper public engagement and ongoing, thorough consideration of the social and cultural impacts of research into living machines.

Lenay and Tixier (Chapter 58) illustrate why better engagement with the public is important by asking why greater use hasn't been made of perceptual supplementation devices, despite evidence to demonstrate that there are a number of systems that can provide effective supplementation with a relatively short learning period. It is important for scientists and researchers to listen, they say, to the people who will, or could, benefit from the technology: devices need to be built in partnership with the needs of potential users if they are to have positive social impacts.

The chapters in this section represent important contributions to these much-needed discussions, so it is appropriate that this section begins with a summary by Hughes (Chapter 57) of human augmentation and the notion of the transhuman. Hughes reminds us that "Debates over the social impacts of technology are too often ahistorical, ignoring that previous technologies were greeted with similar hype and hysteria, and raised the same ethical and social questions." He invites us to consider how we are already, and are already becoming transhuman. He evaluates a human journey in which we have always been accompanied by technologies, with a brief history of human augmentation; for example, the use for millennia of prostheses, and the more recent development of optical and auditory artificial "senses," including everything from telescopes to radios and televisions.

Hughes also gives us a glimpse into the future of humanity itself, telling us what human beings may one day look like, based on the range of human augmentation that will come into being in the near future. This includes everything from "exocortical" devices (such as already-ubiquitous smartphones) to advanced pharmaceuticals, genetic engineering, prosthetics, and nanotechnologies. Such new inventions will have profound impacts on human beings, ranging from improving physical abilities (or eliminating disabilities) to "controlling addictions and moral failings" and impacting work, education, sexuality, reproduction, and even spirituality.

Lenay and Tixier (Chapter 58) look at sensory substitution systems—which, for reasons they explain, are better thought of as *perceptual supplementation*. Starting with the pioneering work of Paul Bach-y-Rita (1972) , they describe the development of the Tactile Vision Substitution System (TVSS), where images were translated into 400 pixels of information and projected onto the skin by means of electromagnetic point vibrators. Tests on the TVSS showed that blind subjects, after a relatively brief period of learning, were able to recognize highly complex forms, and even faces. This chapter examines in detail some of the many scientific, technological, and philosophical questions and implications raised by this research and the subsequent work that it inspired, and relates this to the important phenomenon of *cerebral plasticity*: how different areas of the brain can be redeployed for different tasks (see Prasad, Chapter 49, and Ballester, Chapter 59, for other technologies exploiting brain plasticity), and what the implications of this are for our perceptual capabilities more generally. As Lenay and Tixier explain, this means that it becomes hard to draw a distinction between between prosthetic devices developed for people with disabilities, and the development of tools that affect the ability of anyone to think or act.

Continuing this theme of the human brain as an infinitely adaptable living machine, Ballester (Chapter 59) discusses *restorative plasticity*, the ability of a damaged brain to reorganize, following a trauma such as a stroke, allowing the remaining intact areas to take over some lost function. This adaptability has been exploited in innovative rehabilitation systems and therapy programs using electric and magnetic stimulation, virtual reality, and robotic exoskeletons. Of particular interest is the Rehabilitation Gaming System (RGS), explored at length by Ballester and her colleagues. The RGS is a biohybrid system for neurorehabilitation that, using virtual reality technology, builds on theories of sensorimotor learning and plasticity developed by Verschure in the context of the Distributed Adaptive Control framework (see Verschure, Chapter 36); as such it is not only an excellent example of supplementing the human machine, but also of how researchers are using a living machines perspective to create practical and useful systems that address a societal need.

Subsequent chapters represent a shift in the kind of perspectives on offer, as we turn from systems that replace, supplement, or rehabilitate human function to exploring the more social and cultural contributions of living machines. Millings and Collins (Chapter 60) explain why these perspectives are so important and why, in thinking about the relationships we might have with new kinds of living machines, a good place to start is the relationships we already have with each other. Starting with a summary of attachment theory (a way of thinking about the nature of humans' emotional ties to others first proposed by Bowlby (1969)), Millings and Collins apply these ideas to human–robot interactions. Against certain opinions which seek to discourage humans becoming emotionally engaged with machines, Millings and Collins suggest at least two reasons why we might want to create robots that encourage some forms of human bonding. First, to enable us to develop machines that have a smooth and natural interaction with people, and thus are easier to use, and second, because the relationships we develop with such machines, including emotional ties, can teach us something about human beings ourselves, that is, about our interaction not just with machines but with each other. Given the level of concern about robots in the public imagination, this discussion of how robots might fulfil human emotional needs is an important step in considering what kind of robots we want in our lives, and what we want them for.

Szollosy (Chapter 61) focuses on not the actual robots that exist in our world today, or even those that might exist in the future, but rather those that already inhabit the popular imagination. As this introduction has insisted, we cannot and should not dismiss these imagined monsters and messiahs as fictions that need to be corrected. It is worth noting that robots were invented in the public imagination—in theatre (the first use of the term *robot* was from a 1920 play), in film, and in newspapers—long before they came to be built in labs or found their way into our homes. And robots continue to be re-imagined and re-invented in these ways, bearing the weight of our anxieties, expectations, fears, and aspirations. Szollosy uses ideas on anthropomorphization, projection, history, and ideology to look not only at why we are afraid of robots, but also how now we sometimes instead, or *also*, look to such machines to save humanity.

In order to address the concerns and anxieties that living machines can give rise to, it is essential that we anticipate the ethical (and legal) challenges that we will face in the near future, or that are already being posed by new technologies. Savulescu and Maslen (Chapter 62) are engaged in such work, and their contribution is an attempt to understand better the nature of experiences in virtual realities, and to construct an ethics for engaging in immersive technologies and telepresence. Going beyond previous attempts to account for virtual experiences in philosophy, they consider the value of virtual interactions, and ask to what extent we should regard them as impoverished versions of "real" experience as opposed to experiences that have unique value in their own right. They consider the increasingly pressing issue of "virtual agents" (human avatars of real people) and the degree to which, and under what circumstances, they may or may not be deemed to be "authentic." On the ethics of virtual spaces, they address the issues of reputation, digital property, and wellbeing, deciding that we need an adaptable ethics that can fit the endless varieties of spaces virtual environments can offer. Maslen and Savulescu propose that we make a fundamental distinction between "intravirtual" and "extravirtual" consequences, for example to consider whether causing harm to an avatar to whom someone has developed an attachment is a harm against the person him/herself. However, they also suggest that these ethical questions also hinge on both the purpose of the virtual environment and the nature of the consequences (in the real or virtual world) of our virtual actions.

Continuing on the theme of ethics and morality, Gunkel (Chapter 63) asks "Can machines have rights?" The problem, Gunkel explains, is that traditionally the question is posed *anthropocentrically*: to be considered for rights, machines would have to demonstrate equivalence with human beings. (To illustrate his point, Gunkel cites Andrew, from Isaac Asimov's novelette *The Bicentennial Man*, and *Battlestar Galactica*, which demonstrates again how important fictions can be in thinking about the real-world, or soon-to-be real world, problems of living machines.) These anthropocentric criteria are very problematic, Gunkel points out; for example, they were also once employed to exclude others from ethical consideration, such as minorities, women, children, and animals. We also cannot use "consciousness" as a means by which we can test whether something is worthy of moral or ethical consideration, as we neither have a clear definition of consciousness nor a way of reliably detecting the presence of consciousness in another.

Perhaps, Gunkel suggests, it is our entire anthropocentric approach to the question of morality and ethics that needs to be rethought. Gunkel proposes some alternative frameworks based on the concepts of *information ethics* and *social-relational ethics* as ways to approach the issue of machine rights, effectively changing the question from *can* machines have rights to *should* machines have rights. According to this perspective the attribution of rights to machines is not a question of whether they have this or that other property (consciousness, the capacity to suffer or whatever) but a choice that we make about how we want to relate to them.

Mura and Prescott (Chapter 64) look at the emerging education landscape around the field of Living Machines and related areas of "convergent science", highlighting some of the challenges faced by interdisciplinary approaches that seek to straddle, not just science and engineering, but also the social sciences and arts. Noting that most training programmes in Europe and the United States relevant to living machines take place within single disciplinary boundaries, they point out that it is unlikely we can train new innovators without better support for cross-over and interdisciplinary training. These authors propose some concrete steps for establishing new programmes at University level to provide training in biomimetic and biohybrid systems.

Finally, Halloy (Chapter 65) considers some of the challenges to not only our technological development but also to our planet given the rate at which current electronic, digital, and robotics technologies are using up resources such as energy and rare earth metalloids. For example, Halloy estimates that in defeating the champion Lee Sedol, in the ancient Chinese board game Go, Google DeepMind's AlphaGo computer consumed more energy than Lee Sedol has in his life to date—biological living machines are, by and large, extremely efficient consumers and recyclers of resources compared to artificial ones. In light of the energy and materials crises that our ever-increasing use of technology seems likely to precipitate, Halloy calls for a "radical biomimetic approach" that means reinventing how we build things, and with what resources. The aim is to make more use of the commonplace materials out of which animals and plants construct themselves, and to understand and mimic the natural cycles of growth, decay, and re-use that we observe in ecosystems.

In conclusion, it is worth noting that each of these chapters reflects, in one way or another, how there should always be, at the heart of these discussions of living machines, a *human*. Or maybe a posthuman. Considerations such as these must always sit alongside those relating to science and technology; and we must always seek to account for human perspectives and social relations as the world, and our technologies, evolve around us.

References

Bach-y-Rita, P. (1972). *Brain mechanisms in sensory substitution*. New York: Academic Press.

Bowlby, J. (1969). *Attachment and loss (Vol. 1): Attachment*. New York: Basic Books.

Dixon, B. (1999). The paradoxes of Genetically Modified Foods: A climate of mistrust is obscuring the many different facets of genetic modification. *BMJ: British Medical Journal*, 318(7183), 547.

Chapter 57

Human augmentation and the age of the transhuman

James Hughes

Institute for Ethics and Emerging Technologies, Boston, USA

There are three axes on which to map the social implications of human augmentation: technological, the ability which is being augmented, and the social systems that will be affected.

The technological augmentations to consider range from the exocortical information and communication systems—laptops, smartphones, and wearables—that are becoming extensions of the human body and brain, to pharmaceuticals, tissue and genetic engineering, and prosthetic limbs and organs, to eventually nanomedical robotics, brain computer interfaces and cognitive prostheses. As we become increasingly biohybrid, each technological form of augmentation can in turn be mapped onto the capabilities which we are in the process of enabling and augmenting:

1. Eliminating physical disabilities and extending physical abilities.

2. Curing disease and extending healthy longevity.

3. Eliminating sensory disabilities and extending the senses.

4. Eliminating cognitive disabilities and enhancing memory, cognition, and intelligence.

5. Curing mental illness and extending emotional control.

6. Controlling addictions and moral failings, and augmenting our capacities for moral self-control, sentiment, and cognition.

7. Exploring and augmenting our capacity for spiritual experience.

In each of these domains we already have assistive technologies and pharmaceuticals that straddle the therapy–enhancement divide, and are in the process of developing genetic and nanorobotic technologies that promise even more radical augmentation.

The enhancement of human capabilities in each of these domains will drive changes in social systems from the family and education, to the economy, politics, and religion. These impacts can be considered individually, but their aggregate effects will be non-linear and drive complex adaptations in the living machine that is our co-evolved techno-social civilization.

Technological augmentations of human abilities

Eliminating physical disabilities and extending physical abilities

The most important ways that human physical abilities have been augmented have been through progress in the prevention of physical disabilities, and in the mastery of energy to power machines. Between declines in the crippling effects of infectious disease, violence, and war, and the declines in injuries in agricultural and industrial work, the incidence of physical disability has declined since the nineteenth century. Improved nutrition and sanitation, reforms of workplace and auto safety, and the decline of violence have all reduced the likelihood that the global citizen will suffer from musculoskeletal impairments such as paralysis and loss of limbs. On the other hand the extension of longevity has introduced the countervailing trend of a growth in the incidence of arthritis and other aging-related physical limitations.

The domestication of draft animals, steam power, electricity, and the internal combustion engine all can be seen as augmentations of human physical abilities, allowing us to carry heavy objects, move about, and even fly. Looking more strictly at direct physical augmentations, however, we can that see that assistive technologies also have an ancient provenance. Prosthetic arms and legs have been recorded for thousands of years for instance. But the twentieth century saw the widespread availability of wheelchairs, the most advanced models of which are now are fully powered, able to climb stairs or hold the user upright, and be controlled by brain–machine interfaces (EPFL 2013). Prosthetic limbs are now capable of direct control from peripheral nerves, and of receiving sensory feedback from the prosthetic. Exoskeletal robotics are enabling those with movement disorders to walk, while also enabling the able-bodied soldier or worker to carry heavy objects. Just as the airplane and helicopter have enabled humans to transcend human limitations, the next generations of prosthetics will possess strength and capabilities that surpass organic limbs.

Eventually, all the causes of physical disability will be subject to either prevention or amelioration with pharmaceuticals, tissue engineering, and nanomedicine. Research is proceeding on limb cloning and transplantation, as well as limb regeneration. Stem cell and genetic therapies are progressing to repair spinal nerve damage. Tens of thousands of people have received deep brain stimulators that ameliorate Parkinson's disease, epilepsy, and other tremor disorders. Drugs and gene therapies are being explored to treat age-related muscle deterioration, which will then provide the next generation of performance-enhancing drugs and gene tweaks. Even if aging, obesity, and a decline in physical activity are having a negative effect on physical capabilities in the early twenty-first century, by century's end many human beings will have access to greater strength, dexterity, and endurance than humans have ever enjoyed.

Curing disease and extending healthy longevity

As with physical disability, the years of healthy longevity enjoyed by most human beings have improved dramatically in the last century, largely as a result of improvements in public health, workplace safety, and the decline of violence. According to the Global Burden of Disease Study, men and women worldwide have gained an average of ten years of life expectancy just since 1970 (Wang et al. 2012), while healthy and disability-free years are also increasing, albeit more slowly.

A variety of machine augmentations promise to transform health maintenance and the treatment of disease in the coming decades. One avenue is the development of wearable biometric monitors and implanted devices, networked with smartphones and health providers. Diabetics are having devices implanted that measure blood sugar and release insulin in real time. Heart disease patients wear heart monitors or have wifi-enabled pacemakers that can alert first responders of a cardiac event. Wearable e-health devices are measuring steps taken and calories burned, toilets can provide urinalysis, and smartphone accessories can detect and diagnose signs of disease from the breath and blood.

The next stages of healthy longevity, however, will have to come from re-engineering the body itself, from immune response and tissue repair to the mechanisms of aging. The SENS Research Foundation, for instance, proposes seven specific pharmaceutical and tissue-engineering therapies that could feasibly reverse seven mechanisms that cause age-related disease. Further out, all tissues and body processes will be augmented with nanorobotics with robust capabilities, identifying and eliminating pathogens, repairing tissue damage, and sup-plementing oxygenation.

Eliminating sensory disabilities and extending the senses

The augmentation of hearing and sight began hundreds of years ago with the development of "ear trumpets" for the hard of hearing, as well as eyeglasses, magnifying lenses, microscopes, and telescopes. In this broadest sense radio astronomy, medical imaging, television, radio, and communication technologies are all sensory enhancements. As Robert Hooke first surmised when he popularized the microscope in the seventeenth century, we have now arrived at the point where we are connecting these devices directly to our nervous systems.

Fortunately, the decline of infectious disease and accidental injuries has reduced the percent of the population who are blind or deaf, although, as with other disabilities, the growth of the elderly population has also driven a growth of age-related loss of hearing and sight. With progress in the use of stem cells and tissue engineering to repair cochlea and retinas, all forms of sensory damage will be reparable in this century. In the meantime, blind and deaf people are also the early adopters of augmentations that will not only overcome their disabilities, but eventually provide enhanced sensory prosthetics to those with ordinary sight and hearing.

More than 200,000 formerly deaf people worldwide have received cochlear implants, translating microphones into electrical impulses to the cochlea, providing an increasingly fine-tuned alternative to organic hearing. Hearing devices are also being connected directly into the aural nerve and the auditory brain stem, and being integrated with cell phones and Bluetooth music players. Progress is being made with photosensitive implants in the retina, and with systems that translate images into electrical impulses directly into the visual cortex. Shrinking, mobile arrays that analyze the contents and chemistry of the air and fluids can also be thought of as machine extensions of smell and taste, although capable of detecting far more than their organic analogs. Although it will more attractive to use wearable versions than to have them surgically connected, sensory prosthetics will eventually be cheap, safe, and powerful enough to be a plausible replacement for the standard issue organics.

Eliminating cognitive disabilities and enhancing memory, cognition, and intelligence

As with health and disability, the dramatic improvement in cognitive capacities throughout the world in the last century, the "Flynn effect," is largely the result of improvements in nutrition and health, as well as smaller families and more enriched intellectual stimulation. In the last twenty years, however, the rapid spread of access to computers, smartphones, and the internet has meant that the average person has access to billions of times more information and memory than their unaugmented predecessors. We are now entering the next phase of exocortical cognitive enhancement with the development of wearable computers, like Google Glass, integrating information ever more seamlessly into daily experience.

Meanwhile, pharmaceutical cognitive enhancement has spread with the use of stimulant drugs and non-stimulant drugs like modafinil. Large-scale genomic studies are identifying genes and neurological variations that create differences in cognitive abilities, opening new avenues of pharmaceutical and genetic cognitive enhancement. Consumers already have access to wearable EEG monitors to track their distraction and focus, as well as transcranial electric stimulators that appear to improve some cognitive abilities. By midcentury these approaches are likely to be surpassed by brain–machine interfaces enabled by hybrids of synthetic biology and nanorobotics, establishing millions of connections between the brain and external information and computing power. Instead of having to talk to our glasses or type on a virtual keyboard we will be able to record our experience, retrieve information, and communicate at the speed of thought, while directly modulating our capacities for attention and decision-making. We will rapidly be able to learn skills, and edit and share memories just as we now upload photos to Facebook.

Curing mental illness and extending emotional control

The original essay on the cyborg by Manfred Clynes and Nathan Kline (Clynes and Kline 1960) was centrally concerned with providing a back-up for NASA's ground control of astronauts who might suffer mental illness in space, detecting psychoses and having their suit inject them with anti-psychotic drugs. The era of treating mental illness with marginally effective pharmaceuticals with multiple side effects, like SSRIs, will likely soon be looked back on with the horror that we now view the madhouse and the prefrontal lobotomy. The genomic and functional mapping of the brain now underway, combined with the development of gene therapies, nanomaterials, and eventually nanorobotics will enable therapies far more targeted and effective at relieving depression, anxiety, and fear. Beyond the treatment of mental illness, we will increasingly have a fine-tuned control of our emotions. Just as we may now drink coffee to be more alert and cheerful for the work day, and alcohol in the evening to facilitate relaxation, we will eventually be able to turn off neurotic self-criticism and increase enthusiasm in preparation for creative tasks, or set a timer on our grief in response to tragic events. Beta-blockers are being used to control performance anxiety, and oxytocin can be manipulated to increase trust, bonding, and empathy. As David Pearce has advocated (Pearce 1998), many of our descendants will experiment with turning down their experience of pain and jacking up their experience of happiness and pleasure, and we will have to navigate, individually and socially, the land-of-the-lotus-eaters pitfalls.

Controlling addictions and moral failings, and augmenting our capacities for moral self-control, sentiment, and cognition

A priest's collar, a chastity belt, a hairshirt, an ankle monitor on a parolee, a video monitor on a police cruiser, or a porn filter on a library computer can all be seen as crude forms of moral augmentation. But as our understanding of the neurological basis of self-control, empathy, fairness, and moral decision-making grows, we have begun to experiment with neurological therapies for moral failings. Stimulants allow those who would formerly been labeled "bad children" or lazy adults to develop attentiveness and conscientiousness. Pedophiles and rapists are being treated with testosterone suppression to modulate their compulsions. The neurological miswirings and neurochemistries that impairs understanding of others' emotions in autism, or the desire to harm others in psychopathy, may be reparable. The relationship of an over-active amygdala in moral cognition, firing powerful signals of fear and digust that overwhelm more rational decision-making, has been found to be amenable to modulation with propranolol, meditation, SSRIs, and alcohol. Risk-taking and addiction-proneness have been found to be related to variations in dopamine genes suggesting that these traits could be changed pharmaceutically, and vaccines have been developed to block the effect of cocaine, opiates, nicotine, and alcohol.

As with other forms of enhancement, however, our capacity to modulate our vices and virtues with exocortical systems will likely be more popular and accessible. Wearable sensors of exercise and blood glucose will have fewer side-effects than bariatric surgery, or drugs and gene therapies to regulate muscle growth and fat metabolism. Electronic reminders will be more accessible than cognitive enhancement drugs or brain prostheses to help remember birthdays. Electronic checklists that nudge us towards wiser choices are safer and more transparent than drugs or devices to change decision-making neurons directly. Eventually these exocortical forms of self-control will combine with our pharmaceutical methods and our brain–machine interfaces to allow us—and state authorities, religious institutions, and military command structures—full access to our moral feelings, decision-making, and behavior.

Exploring and augmenting our capacity for spiritual experience

Ordinary, waking awareness is just a small fraction of the states of consciousness that we are capable of experiencing, and our future augmented brains will have even more states to consciously and intentionally explore. Shamans, yogis, and monastics have been experimenting for tens of thousands of years with spiritual technologies, from drugs, chanting, and meditation to yoga and sweat lodges. Today the scientific study of psychedelic drugs is documenting that they can create long-lasting positive changes in personality (MacLean et al. 2011), and may be adjuncts in therapies for overcoming depression, PTSD, and addictions. Meditators are being studied with fMRI, and studying themselves with portable consumer EEG monitors, documenting changes in brain structure, mental health, and behavior. Transcranial magnetic stimulation is being used to manipulate parts of the brain controlling proprioception, generating experiences of "oneness." Just as nano-neural interfaces and brain-machines will enable increasingly precise control of memory, attention, cognition, and emotion, they will also enable easier and deeper experiences of awe, oneness, timelessness, gratitude, and meaning.

Social systems

Together these various forms of ancient, ongoing, and impending augmentations and enhancements mean that we have long been in a "transhuman" era, transitioning from the Paleolithic human existence to the myriad posthuman forms to come. When exactly we entered this condition is as unclear as when we will leave it, since the line between ur-human and modern human is contested, and our uploaded, gene-tweaked cyborg descendents will have as much claim to the mantle of "human" as an eye-glass wearing woman with a prosthetic leg does today. But instead of having centuries to adapt, the accelerating adoption of physical, cognitive, sensory, and spiritual augmentation will drive as much change in our social life in the coming decades as we have seen in centuries.

Family and reproduction

There is already a trend towards lower rates of marriage and smaller families in the industrialized world. One reason for this trend is that, with the decline of agricultural labor and the banning of child labor in factories, children have become an increasingly expensive luxury. As a consequence, there has been a growing concern with ensuring that the few children a couple produce will have the best possible opportunities in life. The desire to maximize children's life prospects is driving the use of prenatal testing, and will in the future drive an interest in preconceptive testing and gene therapies to ensure children's health, abilities, and attractiveness. Just as ensuring that children are literate has been a social and parental obligation, so ensuring that they have access to and literacy with the rapidly evolving exocortex of information and communication technologies is an obligation today. Similarly, as genetic, nanomedical, and brain–machine enhancements become more common and expected in education and employment, we will be obliged to ensure that children have them.

There are several aspects of enhancement and augmentation that may mitigate the decline of marriage and child-rearing. First, increasing the proportion of the population who are healthy and able-bodied, and lengthening the span of their lives, means that more people will be able to find attractive life partners and they will have increasingly amounts of time in which to find them. More direct control over the neurochemistry of love, lust, trust, and neuroticism will make it easier for us to be more agreeable and faithful life partners. The growing options for reproduction will also mean that being infertile, single, post-menopausal, or homosexual will be increasingly irrelevant to starting a biological family. Eventually the artificial womb will make it possible to have children without the complication of a surrogate mother. There has been little evidence thus far, however, that even the most optimal pro-natal policies in Northern Europe—state-sponsored IVF, gay marriage, generous family leave, and free higher education— have reversed the trend of declining marriage and child-bearing.

One form of augmentation that will likely accelerate this decline is the electronic mediation of sexuality. For those who still desire physical encounters, computer dating, Grinder, and Facebook apps like Bang With Friends have facilitated commitment-free hook-ups. There is also substantial evidence that for many people pornography substitutes for sex in the flesh, and the rapid proliferation of myriad forms of internet pornography is now being integrated with haptic devices and immersive virtual reality. Electronically mediated sex is safer (no diseases, violence, or pregnancy), easier (no lengthy courtship, commitments, or foreplay), and can be exactly what the individual desires (your partners can be anyone, or anything, you desire, without any physical defects). In *Love and Sex with Robots,* David Levy (Levy 2008) argues convincingly that there will also be a large market for robotic surrogates for sex and eventually romantic relationships.

Even the most fundamental categories of our reproductive system, the gender binary, will continue to erode in the coming century, a process that began with industrialization (Hughes and Dvorsky 2008). With gene therapies and tissue engineering it will become easier to modify genitals and hormones, and post-secondary sex characteristics like breasts, vocal chords, and body hair. Subcultures are already exploring gender identities and morphologies liminal between male and female, and forms of augmented hypersexuality and asexuality will be possible.

Education

Already alternatives to the factory-era of primary school education and the medieval models of higher education are being explored with computerized, self-guided learning modules, massive online courses, and skill certification. Meanwhile, cognitive enhancement drugs, access to ubiquitous computing, and eventually brain–machine interfaces will be speeding up the learning process. We will increasingly depend on our intelligent personal assistants as extensions of our own memory and cognition, cataloguing our life experiences, collecting information, and summarizing and prioritizing choices. The rapidly changing, shrinking, and harshly competitive labor market will mean that increasingly numbers of adults will be engaging in continuous up-skilling as they attempt to remain employable.

Work and the economy

In the context of an overall decline of employment due to automation and disintermediation, and the growing number of healthy senior citizens staying in the labor force, there will be greater competition for a declining number of jobs. Workers with the latest physical and cognitive enhancements will likely have far greater advantages in that competition, just as healthy and computer-literate workers are advantaged today. Likewise, countries that facilitate more widespread access to physical and cognitive augmentations, and the time to learn to use them, will be more productive. The most automated economies and workforces will need to adopt policies that ensure a broad redistribution of income, leisure, and access to remaining employment opportunities in order to maintain economic growth and competitiveness.

Religion

Few religions have objections to medical therapies and prosthetics. So, as the normative standard gradually moves, there will likely be few religious objections to the extension of physical, sensory, and cognitive abilities, at least until augmentations begin to challenge moral intuitions about human/non-human boundaries. We can see some of the contours of this conflict in religious objections to reproductive technologies and human–animal chimera research. The Abrahamic faiths—Judaism, Christianity, and Islam—in general are resistant to radical versions of the cyborg project, seeing it as an affront to the divinely created human body's role in the divine project. Proposals to record and upload human personality are most problematic, as they adopt a purely materialistic model of human consciousness.

Asian religions are likely to be more accommodating of augmentation, however. Faiths such as Buddhism and Hinduism, on the other hand, are more open to the idea of individual minds migrating from animal to human to superhuman bodies. There also is more enthusiasm for human enhancement projects like genetic enhancement in the societies from India to Japan than in Europe and North America.

Religious institutions will similarly react in a variety of ways to widespread access to neurotechnologies that control moral feelings and behavior, and provide access to previously rare religious experiences. Some will accept neurotechnologies, or even incorporate them into their practices as obligations and sacraments, while others will reject them as distractions and abominations. The more puritanical faiths will likely find more use for technologies for moral self-control, while Buddhism, Hinduism, and shamanic traditions may find ready use for new methods of inducing altered states of consciousness.

Politics and global conflict

As I proposed in *Citizen Cyborg* (Hughes 2004), conflicts over the cyborg project are likely to increasingly shape political conflict in the twenty-first century, with bioconservatives of a variety of stripes aligning against an increasingly diverse coalition of bioliberals and transhumanists. One central debate will be the significance of genetic and morphological "humanness" as the basis of rights-bearing, as opposed to self-awareness and cognitive capacities. This argument has already shaped the status of fetuses and the brain-dead, and in the future it will be central to the struggles over the rights of cognitively enhanced animals, genetically modified and cyber-enhanced humans, and machine minds.

A second key element of the public policy debate will be the relevance of the therapy/enhancement distinction. Just as many now feel that stimulant drugs for ADD and SSRIs for depression represent widespread over-medicalization of the normal human condition by a profit-driven medical industrial complex, demands for increasingly sophisticated augmentations by able-bodied citizens will make clear the arbitrary nature of the therapy/enhancement line. Bioconservatives will resist the increasing informal or overt pressures from educators and employers to adopt augmentation, just as there is resistance today to "coercion" to abort disabled embryos or be electronically accessible 24/7.

Struggles over the funding of enhancement research, and the regulation of enhancements and augmentations, will take place on the pre-existing terrain of growing inequality and geopolitical competition. Over the last fifty years we have been increasingly using state-financed healthcare systems to expand access to assistive technologies—from renal dialysis and wheel-chairs, to cochlear implants, IVF, and stimulant drugs—and demand for equitable, publicly financed access will grow. Countries that do not support the enhancement of their workforces will be at an economic disadvantage internationally, and militaries that do not fully exploit robotics, drones, and supersoldier enhancement will be at a military disadvantage.

On the other hand, just as transnational institutions like the International Atomic Energy Agency have grown in importance to stem the proliferation of weapons of mass destruction, preventing the super-empowerment of individuals and small groups with forms of human aug-mentation will increasingly become a concern for nation-states and transnational policing. As has been the case with nuclear energy and chemical weapons, however, the "dual uses" of aug-mentation technologies will make effective transnational regulation very difficult. Just as China may consider unfiltered access to the internet to be a threat to national security while the West sees it more as a basic right, some countries are likely to resist efforts to ban enhancements and augmentations and attract an international medical tourist traffic seeking them.

Future directions

We know from the history of futurological speculation that we are rarely right about the pace of innovation and its consequences. Technologies which seem imminent may never come to fru-ition, while technologies like the internet may be inconceivable. The first-order effects of techno-logical innovation—such as automobiles leading to high-speed accidents—may be foreseen, but secondary and tertiary effects—such as the spread of suburbs, climate change, and the decline of the trolley—are harder to imagine. Likewise with augmentation it is likely that some forms of augmentation will remain impossible, too expensive, or unappealing, while new forms will emerge with unimaginable consequences. This is why I have focused less on the plausibility and timelines of different types of augmentation in this essay, and more on the human capabilities towards which all forms of human engineering are likely to be applied. Assuming that these broad goals of control of the body and brain will be accomplished one way or the other, we can then focus on what consequences these forms of control may have on society.

A second thing I have attempted with this essay is to expand the frame on enhancement and augmentation to include the technologies we have co-evolved with for hundreds and thousands of years. Debates over the social impacts of technology are too often ahistorical, ignoring that previous technologies were greeted with similar hype and hysteria, and raised the same ethical and social questions. By keeping this continuity in mind we can often extrapolate that we will accommodate new innovations with the same moral codes and regulatory structures. It may be that cyborg augmentation or the "Singularity" may eventually pose practical questions that were only philosophical thought experiments in the past, needing entirely novel responses, but thus far the social and regulatory challenges appear manageable within the social institutions we already have.

Learning more

The academic, policy, and lay communities that contribute to the global discussion of the social implications of human enhancement and augmentation are too diverse to enumerate fully. The communities that I have participated in have been those of public policy, bioethics, and tran-shumanism. One of the key bioethics works raising bioconservative concerns about human aug-mentation was *Beyond therapy: biotechnology and the pursuit of happiness* (President's Council on Bioethics 2003), while the collections *Enhancing human capacities* (Savulescu et al. 2011, see also Maslen and Savulescu, Chapter 62, this volume) and *Human enhancement* (Savulescu and Bostrom 2011) represent many of the bioliberal voices. Two works that represent the transhumanist movement's perspectives are *The singularity is near* by Ray Kurzweil (Kurzweil 2006) and *The transhumanist reader* (More and Vita-More 2013). Two key historical documents in the imagination of human–machine integration and the cyborg are J.D. Bernal's 1929 essay on the future of prosthetically enhanced humans, *The world, the flesh, and the devil* (Bernal 1969) and Clynes and Kline essay coining the term "cyborg" (Clynes and Kline 1960). Other important literatures are those of artificial intelligence and cognitive science, life extension and bioscience futurism, the emerging literatures on global catastrophic risks and the economic effects of automation, and general futurist speculation including speculative fiction.

References

Bernal, J. D. (1969). *The world, the flesh and the devil: an enquiry into the future of the three enemies of the rational soul*. 2nd edition. Bloomington, IN: Indiana University Press.

Clynes, M. E., and Kline, N. S. (1960). Cyborgs and space. *Astronautics*, September, 26–27, 74–76.

EPFL (2013). *Chair in non-invasive brain-machine interface (CNBI)*. [Online] Available at: http://cnbi.epfl.ch/software/index.html [Accessed June 22, 2013].

Hughes, J. (2004). *Citizen Cyborg*. New York: Basic.

Hughes, J., and Dvorsky, G. (2008). *Post-Genderism: Beyond the Gender Binary*. Hartford, CT: IEET.

Kurzweil, R. (2006). *The singularity is near: when humans transcend biology*. New York: Penguin.

Levy, D. (2008). *Love and Sex with Robots: The Evolution of Human–Robot Relationships*. New York: Harper Perennial.

MacLean, K. A., Johnson, M. W., and Griffiths, R. R. (2011). Mystical experiences occasioned by the hallucinogen psilocybin lead to increases in the personality domain of openness. *Journal of Psychopharmacology*, **25**(11), 1453–61.

More, M., and Vita-More, N. (eds) (2013). *The transhumanist reader: classical and contemporary essays on the science, technology, and philosophy of the human future*. New York: Wiley-Blackwell.

Pearce, D. (1998). *The hedonistic imperative*. [Online] Available at: http://www.hedweb.com/ [Accessed June 27, 2013].

President's Council on Bioethics (2003). *Beyond therapy: biotechnology and the pursuit of happiness*. New York: Harper Perennial.

Savulescu, J., and Bostrom, N. (eds) (2011). *Human enhancement*. Oxford: Oxford University Press.

Savulescu, J., ter Meulen, R., and Kahane, G. (eds) (2011). *Enhancing human capacities*. New York: Wiley-Blackwell.

Wang, H., Dwyer-Lindgren, L., Lofgren, K. T., et al. (2012). Age-specific and sex-specific mortality in 187 countries, 1970–2010: a systematic analysis for the Global Burden of Disease Study 2010. *The Lancet*, **380**(9859), 2071–94.

Chapter 58

From sensory substitution to perceptual supplementation

Charles Lenay[1] and Matthieu Tixier[2]

[1] Philosophy and Cognitive Science, University of Technology of Compiègne, France
[2] Institut Charles Delaunay, Université de Technologie de Troyes, France

The simple, almost naïve idea of sensory substitution systems is that of replacing a deficient sensory modality with another that is still available. The emblematic example of this approach is the TVSS (Tactile Vision Substitution System) designed for blind persons. Created in the late 1960s by Paul Bach-y-Rita, this device converts visual information captured by a video-camera into tactile signals in the form of a 20 × 20 matrix of points. In this way the image, reduced to 400 pixels (black or white without any intermediate levels of gray) is projected onto the skin by means of electromagnetic (or piezoelectric) point vibrators, or directly by electrical stimuli (Figure 58.1).

If learning occurs while the camera is placed on a table and remains immobile, the perceptual capacities remain very limited and amount merely to patterns of tactile stimulation that are felt on the skin. However, if the blind person is allowed to grasp the camera and to explore simple scenes, then a significant change occurs: progressively (after something like fifteen hours of learning), the subject becomes able to recognize highly complex forms, even faces, and to locate these objects in the environment (Bach-y-Rita and Kercel 2003). Indeed, this increased capacity to recognize shapes seems to be accompanied by an *externalization* of the percepts (Epstein 1986; Auvray et al. 2005). The sensation of a succession of rapidly changing tactile stimuli on the skin, produced by the continual rotation and movement of the camera, drops out of consciousness; and is replaced by the perception of stable objects that are situated at a distance, "out there" in front of him (Bach-y-Rita 2004). This striking result raises a whole host of important questions, to be considered in this chapter, which are at one and the same time scientific, technological, and philosophical.

The diversity of sensory substitution systems

The basic principle of sensory substitution, pioneered by Bach-y-Rita, has subsequently been widely developed (Auvray and Myin 2009). Regarding visuo-tactile substitution, one of the variants of the TVSS which has been most widely marketed is the Optacon device (Telesensory Systems) which is designed for reading. A miniature camera films the text which is rendered in the form of tactile vibrations by means of a small matrix of micro-vibrators on which the blind reader places one of the fingers of his free hand. At the present time, a new version of the TVSS is miniaturized so as to distribute the sensory data in the form of electrical stimuli on the tongue (TDU, Tongue Display Unit). This system appears to have substantial advantages: saliva is a natural electrolyte; the tongue has a high density of sensory units; and the connection to

Figure 58.1 Tactile Vision Substitution System (TVSS).
© Dr. Paul Bach-y-Rita, 2006.

the central nervous system is very rapid. In addition, it is quite possible to make a discreet version of the device that can be mounted on a dental apparatus (Bach-y-Rita 2004; Kaczmarek 2011). Visuo-auditory substitution has also been widely developed. The coding systems vary according to the device in question. For example, distance can be coded by the intensity of the sound, and the horizontal position by the temporal disparity between the two ears. The PSVA system (Prosthesis Substituting Vision by Audition) developed by Capelle adds a second coding of vertical position by frequency (Capelle et al. 1998). The Vibe system developed by Hanneton divides the image in the form of receptor fields which control different sound sources, which are then mixed to produce a stereophonic sound (Hanneton et al. 2010). The Voice system developed by Meijer is based on a cyclic scanning of the image, which makes it possible to code the horizontal position (Ward and Meijer 2010). Many studies have demonstrated the effectiveness of such devices in many areas: spatial localization, the recognition of simple forms including reading, the recognition of perspective, of shadows, of the interposition of objects, and even certain visual illusions. In addition, the team of Paul Bach-y-Rita (who passed away in 2006) has produced an impressive vestibule-tactile substitution device for patients who have lost their initial vestibular system: an accelerometer coupled to the matrix of a TDU rapidly restores a sense of balance when they have their eyes closed (Danilov and Tyler 2005). This approach can also be extended to retinal or cochlear implants (see Lehmann and van Schaik, Chapter 54, this volume), since they convert physical variations in light or sound into electrical stimuli which feed directly into peripheral nerves (optical nerves, cochlea) or onto the central nervous system.

The first scientific interest of sensory substitution systems is to allow the study of the spectacular cerebral plasticity involved in their functioning. For example with the TVSS, the tactile sensory input has nothing in common with that of the retinal system, and neither does the manual control of the camera have anything in common with the control of the ocular muscles. Nevertheless, the brain turns out to be able to reorganize itself to produce a perceptual world specifically characterized by the recognition, and indeed the localization, of shapes and objects. It is noteworthy that the very same cerebral regions that are normally devoted to visual perception are redeployed here in the course of learning these systems of visuo-tactile or visuo-auditory substitution (as they are for reading in Braille) (Renier et al. 2005). An understanding of these neurological observations must be based on a study of the actual behaviour of the subject bearing the perceptual prosthesis. This involves understanding the functional relationships between the organism and its environment, i.e. a structure of sensorimotor coupling that is sufficiently stable to allow and indeed to induce this functional reorganization of the central nervous system.

A minimalist study of sensorimotor coupling

Learning to use a sensory substitution device can be understood as the genesis of a new perceptual capacity. In the beginning, the device is present as an object of perception in itself: the subject is conscious of the tactile stimuli and of his exploratory movements. Then, when the device is mastered, the device itself disappears (at least partially) from consciousness, and is replaced by the perception of objects "out there" in a space in front of the subject. What are the mechanisms involved in this shift? In particular, what is involved in the appearance of this space in which stable objects are perceived? In order to identify the necessary conditions for this shift to occur, one possible approach consists of deliberately *simplifying* the device as much as possible, and testing whether the shift still occurs. This minimalist method consists therefore of reducing as far as possible the repertoires of action and sensation that are available for the subject. In this spirit, the system of Paul Bach-y-Rita with 400 pixels is reduced to a single point, a single photoelectric cell connected to a single tactile stimulator (Lenay et al. 2003). When the luminous intensity of the incident light-field passes a certain threshold, this triggers an all-or-nothing tactile stimulus (Figure 58.2).

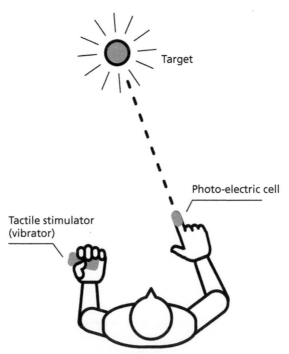

Figure 58.2 A minimalist sensory substitution system.

At each point in time the (blindfolded) subject receives only a minimal amount of information, 1 bit corresponding to the presence or absence of the tactile stimulation. It has been possible to show that even in these drastically reduced conditions, the subject is still able to localize a target placed in different directions and at different distances. It is quite easy to understand that the subject should, in principle, be able to solve the task by means of triangulation, by aiming at the target with different positions of the arm and angles of the wrist (Siegle and Warren 2010).

Here, it is abundantly clear that the perception of the luminous source at a distance cannot be achieved by any sort of internal analysis of the sensory information; if only because this sensory information is nothing other than a simple temporal sequence of all-or-nothing sensations which have no intrinsic spatiality whatsoever. The spatial perception thus necessarily corresponds to a *synthesis* of a temporal succession of sensations *and actions*. In support of this, it is quite easy to observe that in order to maintain this perception, the subject must continually move the photoelectric cell, aiming at the perceived object in many different ways. Indeed, if the subject is immobile, there are only two possibilities. Either the subject is pointing in a direction that does not encounter the target; but in this case he receives no stimulation, and has nothing other than the pure memory of a perception which rapidly fades away. Otherwise, the subject is pointing towards the target; but in this case he receives a continual tactile stimulation, and this imposed stimulation replaces the perception of an external object. It is only in the case of ongoing activity, producing corresponding sensory *variations*, that the perception can be constituted.

Reversibility, which we may define as the capacity to recover the same sensation by inverting an action, makes it possible to locate the target and, even more basically, to constitute the very space in which this localization occurs. In the limiting conditions defined by this experimental setup, there is no perception without action. Each position of the target corresponds to a specific, unique *sensorimotor invariant*, i.e. a law relating actions to sensations which remains stable in the context of variations in these actions and sensations. The target is localized in direction and depth when the law which governs pointing in the direction of the target is mastered. This is a good illustration of what O'Regan and Noë call the "law of sensorimotor contingency," which fits in the framework of an enactive approach to perception (O'Regan and Noë 2001).

In the same minimalist spirit, it is also possible to study the recognition of shapes in a two-dimensional space. For example, with the Tactos system, the subject moves a small receptor field in the space of the screen, and receives a tactile stimulus each time that this receptor field encounters a coloured pixel. It has been shown that under these conditions, subjects are able to recognize simple shapes by means of their active exploration (Ziat et al. 2007; Stewart and Gapenne 2004).

Finally, this approach can be completed by studying the effects of progressively increasing the number of receptor fields (with a corresponding increase in the number of tactile stimulators), up to the 400 receptor fields of the TVSS. In a first series of experiments, passing from 1 to 8 receptor fields, it was shown that this modification could be explained as a re-internalization of the perceptual activity: the parallel sensory inputs corresponded in a way to "actions which were already performed." This allowed for an economy of movement and memory, and thus for perceptions which were more rapid and more precise (Sribunruangrit et al. 2002).

One of the interests of these radically simplified devices is to show that the perceptual activity remains essentially the same, whatever the modality that is coupled to the movements of the receptor field. Using the strict minimalist situation of a single receptor field, we were able to use either a tactile stimulation, or a sound, or even a visual stimulation (one point on the screen), without producing any significant change in the perceptual capacities (Gapenne et al. 2005).

From sensory substitution to perceptual supplementation

It seems that the results of this minimalist study can be generalized to all the devices of prosthetic perception (Figure 58.3). In all the cases that have been reported, spatial localization is
learned by means of the concrete activity of the subject who advances in the actual space of the
objects perceived (Lenay and Steiner 2010).

In this way of conceptualizing perception, it is necessary to make a clear distinction between,
on one hand, the "sensation," i.e. the sensory input that is delivered to the organism and, on the
other hand, the "perception," i.e. the law which, for a given perceptual content, defines the sensory
feedback as a function of the actions performed. When analyzing sensory substitution devices,
this means that it is important to distinguish the *sensory* modality which is employed to activate
the central nervous system from the *perceptual* modality which is defined by the sort of sensorimotor law that the device allows, i.e. the type of perception which can be acquired concerning
the environment. For example, for the TVSS and The Voice, the sensory modalities are different,
being respectively tactile and auditory; nevertheless, one can say that in both cases, the perceptual
modality is basically "visual" in the sense that the subject has access to the positions of objects at a
distance. By contrast, for the Tactos device, the perceptual modality obtained is basically "tactile,"
in the sense that there is spatial coincidence between the site of perception (the receptor fields)
and the perceived object itself; and this is true whether the *sensory* modality employed be tactile,
auditory (sound waves), or even visual (light waves).

For this reason, the terms "sensory substitution" appear unsatisfactory and even misleading;
in particular because in order to function these devices must be just as much "motor" as "sensory." We therefore prefer to speak of "perceptual supplementation." A perceptual supplementation system is a device which provides an artificial coupling between an organism and the
environment to which it gives access. The new relation which it instigates between the actions
and the sensory feedback provided to the subject gives rise to the constitution of specific percepts (Lenay et al. 2003).

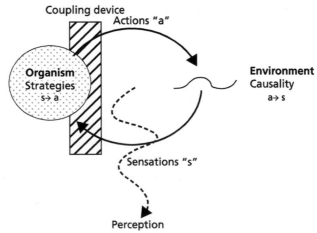

Figure 58.3 Scheme of sensorimotor coupling. The system of prosthetic perception is a "coupling
device" which modifies the lived body by defining the repertoires of actions and sensory feedback
that are available to the subject. Via the environment, the actions "a" cause the sensations
"s" : s = g(a) ; and the organism defines the strategy which specifies its own actions as a function
of the sensations received : a = f(s).

The study of the appropriation and the use of so-called "sensory substitution" technologies by actual subjects, whether it be in experimental laboratory conditions or in more natural situations, shows that what is at stake is not solely the compensation of one sensory modality by another, but rather the progressive construction of a new domain of perception and meaning. The interest of the term "supplementation" is that at one and the same time it has the dual connotations of "compensation" for an absence or loss, but also of an augmentation, of a supplement, of an enrichment, of something *more*. In this more general framework, there is no strict line of demarcation between prosthetic devices invented for the world of people with disabilities, and the general employment of various tools which modify our possibilities of acting and feeling, which modify the domain of possible perceptions. This perspective refers to a more general sensorimotor model of the appropriation of tools which makes it possible, on one hand, to propose perspectives for research concerning the problems of the adoption of supplementation technologies, and on the other hand, to rethink ways of inventing and studying them, as well as the ways they can possibly be used as a source of augmenting the capacities of human subjects.

Social adoption and emotional values

In spite of the effectiveness of supplementation devices, which is widely recognized, it is surprising—and disappointing—to note that, as a matter of fact, that they have been pretty much a failure, both socially and economically. Why is it that these techniques, which were invented in the 1960s and experimentally validated in the 1970s, have had so little impact on the daily life of blind and visually impaired persons? There are many possible answers to this question: a practical effectiveness which is insufficient (none of these devices enables a blind person to drive a car); the feeling of appearing as a weird cyborg; but perhaps, above all, the lack of any real progress at the level of the quality of lived experience. It appears, indeed, that what is missing are the *qualia*, i.e. the qualities, the *values* of the things that are perceived (Bach-y-Rita 1997). One can try showing a person who has been blind from birth the image of his wife, or his own image in a mirror; one can try showing blind students sexy pin-ups; in all cases, the disappointment is flagrant. There is indeed the constitution of an object, the cognitive capacity for discrimination and categorization; but there is no *emotional* value attached to these objects. Thus, the sensorimotor coupling provided by the TVSS does resemble in many ways that provided by our natural, normal vision; but in practice it turns out that the actual lived experience is quite different. In fact, there are actually many differences between this artificial coupling with the environment and the mode of coupling corresponding to natural vision: there is no colour; there are few points; and it requires a camera with relatively slow movements. These differences are well appreciated by sighted persons, or persons who became blind late in life, who try out the device and undergo the process of learning. In fact, the device invented by Bach-y-Rita does not perform a sensory *substitution*, but as we have already tried to explain, an *addition*, the opening of a novel domain of coupling between a human subject and the world. It is essential to take account of the fact that these devices never compensate exactly for a deficit, for a loss; rather, they introduce new perceptual modalities that are quite original each time.

Of course it is always possible to work on more sophisticated technical devices, such as implants, which might make it possible to approach more closely the complexity of natural sensory inputs and action possibilities. However, whenever there is a technical mediation, an infinite number of variations become possible and so there will always be possible novelties or augmentations in perception. It is, however, also possible to take things the other way round, and to approach the problem of developing emotional values in the realm of lived experience by asking how it might be possible, even with a relatively simple perceptual supplementation device, to promote the advent of these emotional qualities. It seems to us that two sorts of solution can be envisaged.

Firstly, it is possible that what is missing is essentially a precocious learning process, an integration of the prosthesis in activity ever since prime childhood so that there is a chance of setting up a link between the percepts and primary feelings of pleasure and pain. Observations are very delicate—it is very difficult to intervene in the case of a blind baby—but the few results that are known seem encouraging (Sampaio 1989). Another approach consists of working on the *collective* construction of emotional values. It is rather remarkable that all the observations on sensory substitution systems reported in the literature only ever concern a strictly individual pattern of use: the user is always surrounded by sighted subjects, but quite isolated in his particular mode of perception. Now it is very plausible to put forward the hypothesis that genuine perceptual *values* are intimately linked to the insertion of the person concerned in a community of shared meanings, a collective history in a common environment defined by the same means of access. In fact, perceptual supplementation devices offer the opportunity for empirical research concerning the anthropological question of the genesis of the values that are attached to percepts. In this light, we are working to set up a fundamental study of prosthetic perceptual *interactions* between people. For example, we propose to examine how it is possible to use prosthetic perception to recognize another person, the Other; and maybe to play at engaging in processes of mimicry (Lenay and Stewart 2012). This sort of situation is indeed more likely to carry meaning and to promote emotional investment.

Learning more

The interested reader can turn to the following topics we mention as entry points to further the discussion on perceptual supplementation.

A classical question concerning systems of perceptual supplementation is that of the quality of the lived experience for a subject who has mastered their use (Ward and Meijer 2010). Is a person's perception like that of the sensory modality that is being replaced (deference); in the case of the TVSS this would be vision. Or is the quality of the perception like that of the new sensory modality that is used (dominance): in the case of the TVSS, this would be touch. It seems to us best to consider that it is neither one nor the other, but each time a novel modality (Auvray and Myin 2009). This can be seen clearly by comparing different devices which employ the same sensory modality and yet which give rise to different perceptual experiences.

In fact, it is by recognizing this novelty that we can open up the questions of the appropriation and the constitution of emotional values (Lenay et al. 2003). These questions are actually quite fundamental for understanding the meaning of technical artifacts in general (Lenay and Steiner 2010). Perceptual supplementation systems only carry to their extreme limit, and thereby render them clearer and more readable (O'Regan and Noë 2001), the general principles of the functioning of cognition and tools when they transform and augment our human experience.

References

Auvray, M., Hanneton, S., Lenay, C., and O'Regan, K. (2005). There is something out there: distal attribution in sensory substitution, twenty years later. *Journal of Integrative Neuroscience*, **64**(4), 505–21.

Auvray, M., and Myin, E. (2009). Perception with compensatory devices: From sensory substitution to sensorimotor extension. *Cognitive Science*, **33**(6), 1036–58.

Bach-y-Rita, P. (1997). Substitution sensorielle et qualia. In: J. Proust (ed.), *Perception et intermodalité*. Paris: Presses Universitaires de France, pp. 81–100.

Bach-y-Rita, P., and Kercel, S. W. (2003). Sensory substitution and the human–machine interface. *Trends in Cognitive Sciences*, 7(12), 541–6.

Bach-y-Rita, P. (2004). Tactile sensory substitution studies. *Annals of the New York Academy of Sciences*, 1013, 83–91.

Capelle, C., Trullemans, C. ,Arno, P., and Veraart, C. (1998). A real-time experimental prototype for enhancement of vision rehabilitation using auditory substitution. *Biomedical Engineering, IEEE Transactions on*, 45(10), 1279–93.

Danilov, Y., and Tyler, M. (2005). Brainport: an alternative input to the brain. *Journal of Integrative Neuroscience*, 4(04), 537–50.

Epstein, W., Hughes, B., Schneider, S., and Bach-y-Rita, P. (1986). Is there anything out there? A study of distal attribution in response to vibrotactile stimulation. *Perception*, 15(3), 275–84.

Gapenne, O., Rovira, K., Lenay, C., Stewart, J., and Auvray, M. (2005). *Is form perception necessary tied to specific sensory feedback?* Paper presented at the 13th International Conference on Perception and Action (ICPA13), July 5–10, Monterey, CA.

Hanneton, S., Malika A., and Durette, B. (2010). The Vibe: a versatile vision-to-audition sensory substitution device. *Applied Bionics and Biomechanics*, 7(4), 269–76.

Kaczmarek, K. A. (2011). The tongue display unit (TDU) for electrotactile spatiotemporal pattern presentation. *Scientia Iranica*, 18(6), 1476–85.

Lenay, C., Gapenne, O., Hanneton, S., Marque, C., and Genouëlle, C. (2003). Sensory substitution: limits and perspectives. In: Y. Hatwell, A. Streri, and E. Gentaz (eds), *Touching for knowing: cognitive psychology of haptic manual perception*. Amsterdam/Philadelphia: John Benjamins Publishing Company, pp. 275–92.

Lenay, C., and Steiner, P. (2010). Beyond the internalism/externalism debate: the constitution of the space of perception. *Consciousness and Cognition*, 19, 938–52.

Lenay, C., and Stewart, J. (2012). Minimalist approach to perceptual interactions. *Frontiers in Human Neuroscience*, 6, 98.

O'Regan, J.K., and Noë, A. (2001). A sensorimotor account of vision and visual consciousness. *Behavioral and Brain Sciences*, 24(5), 939–72.

Renier, L., Collignon, O., Poirier, C., Tranduy, D., Vanlierde, A., Bol, A., Veraart C., and De Volder, A. G. (2005). Cross-modal activation of visual cortex during depth perception using auditory substitution of vision. *Neuroimage*, 26(2), 573–80.

Sampaio, E. (1989). Is there a critical age for using the sonicguide with blind infants? *Journal of Visual Impairment & Blindness*, 83(2), 105–8.

Siegle, J. H., and Warren, W. H. (2010). Distal attribution and distance perception in sensory substitution. *Perception*, 39(2), 208–23.

Sribunruangrit, N., Marque, C., Lenay, C., Gapenne, O., Vanhoutte, C., and Stewart, J. (2002). Application of parallelism concept to access graphic information with precision for blind people. In: *EMBEC'02 2nd European Medical & Biological Engineering Conference*, pp. 4–8.

Stewart, J., and Gapenne, O. (2004). Reciprocal modelling of active perception of 2-D forms in a simple Tactile Vision Substitution System. *Minds and Machines*, 14, 309–30.

Ward, J., and Meijer, P. (2010). Visual experiences in the blind induced by an auditory sensory substitution device. *Consciousness and Cognition*, 19(1), 492–500.

Ziat, M., Lenay, C., Gapenne, O., Stewart, J., Ali Ammar, A., and Aubert, D. (2007). Perceptive supplementation for an access to graphical interfaces. In: *Proceedings of the 4th international conference on universal access in human computer interaction: coping with diversity*, Beijing, China. Berlin/Heidelberg: Springer-Verlag, pp. 841–50.

Chapter 59

Neurorehabilitation

Belén Rubio Ballester

SPECS, Institute for Bioengineering of Catalonia (IBEC),
the Barcelona Institute of Science and Technology (BIST), Barcelona, Spain

Following stroke, a majority of patients show weakness on the side of the body contralateral to the lesion (i.e. hemiparesis), leading to functional limitations in instrumental activities of daily living (iADL). In addition, cognitive and motor deficits may be present, depending on the lesion location and size, such as spasticity, attention deficits, and loss of coordination, along with language deficits or aphasia. Moreover, about 60% of stroke patients experience chronic effects including pain, mood disorders, and depression. The above list of specific deficits suggests that there might be important interrelations between these deficits asking for an integrated rehabilitation approach.

Biomimetic and biohybrid technologies, such as exoskeletons and Virtual Reality (VR) interfaces, are promising clinical tools for neurorehabilitation. However, it remains unclear how these technologies should be applied in order to achieve clinical impact. A better understanding of the mechanisms underlying comorbidities and their recovery would allow implementation of more effective care and rehabilitation approaches. In this chapter we review the main biological principles for neurorehabilitation, building on a theory of brain plasticity provided by the Distributed Adaptive Control framework for mind and brain (Verschure, Chapter 36, this volume).

Theory

Methodologically this is complementary to the synthetic approach that DAC follows (Verschure and Prescott, Chapter 2, this volume) by committing theory development to both synthesizing artificial and repairing biological brains. DAC considers the brain as a multi-layer control system and suggests a very distinct internal regulation between these different layers, e.g. if the feedback control of the reactive layer (Layer 2) is dominating behaviour it inhibits the other layers and when the forward models of the adaptive layer (Layer 3) are unable to reliably classify the environment the architecture prevents the engagement of the contextual layer (i.e. confidence-dependent regulation of control). Goal-oriented contextual learning (Layer 4) is conditional on the acquisition of a sense-act state space by the adaptive layer where sensory states provide active predictions on actions and states of the world. At Levels 3 and 4, DAC operates as an instantiation of an ideomotor theory of cognition and action, where actions result from the interpretation of the interaction with the world based on internal models and predictions rather than being triggered by external events. On this basis, the DAC theory allows for the extraction and formulation of a set of specific principles that can guide the design of effective neurorehabilitation paradigms.

For instance, DAC proposes that the disruption of sensorimotor contingencies after a brain lesion will lead to a reduction of activity in the motor cortical pathways and their associated sensorimotor areas. Consequently, a drop in neural activity will limit neural plasticity and hamper recovery for as far as it depends on activity dependent processes. DAC considers that the goals of the agent critically structure the learning dynamics. This principle should thus also be considered in our attempts to repair the brain through non-invasive means. The DAC theory predicts that the external manipulation of sensorimotor modalities during the execution of goal-oriented actions will increase the activation of the ideomotor system of the fronto-parietal networks of the neocortex, thus driving plasticity and recovery (Adamovich et al. 2009). In this vein, the DAC theory identifies a number of additional principles to optimize behavioral interventions for neurorehabilitation, such the objective to externally influence perceptual and motor predictions of the brain, the realization of self-paced individualized training, intense massed-practice, ecologically valid settings, limited overcompensation, goal-oriented training, practice that is structured in time (e.g. distributed resting periods), and facilitating motor imagery-based training. In this chapter we show how DAC can provide guidelines for the development of effective VR-based rehabilitation strategies that are grounded in theoretical neuroscience. It illustrates that a living machines perspective can be used to create practical and useful systems that address a societal need. The Rehabilitation Gaming System (RGS)—a novel VR technology for motor and cognitive recovery—is described as an example for the implementation and validation of these principles.

Biological principles

Restore: use-dependent brain plasticity

At the beginning of the twentieth century, neuroscientists conceived the Central Nervous System (CNS) as a "fixed" neuronal setup. It wasn't until 1914 that Ramón y Cajal—widely known as the father of modern neuroscience—proposed that regenerative processes were taking place in the brain after a lesion, a concept that has been termed "restorative plasticity." Since impairment is often caused by irreversible damage of neural tissue, recovery from stroke and other neural conditions depends on the brain's ability to reorganize the connectivity of the remaining circuitry, allowing other areas to assume the lost function. The concept behind rehabilitative stroke therapy is therefore derived from this notion of neuroplasticity, which is based on three main mechanisms (Nudo et al. 1997): the modulation of the strength of existing synaptic connections, the generation of new connections, and the synaptic pruning or elimination of existing connections. After a brain lesion, a patient can take advantage of these plastic mechanisms by repetitively exposing himself to impairment-related practice. This neurological principle has been termed "use-dependent brain plasticity" and the DAC theory incorporates it in its adaptive and contextual layers.

Interestingly, the brain seems to be prepared to heal itself. Imaging studies have shown that the brain of an acute stroke patient presents a remarkable potential for plastic changes. Indeed, it has been hypothesized that if an acute stroke patient would learn to play piano with her non-affected hand, she would develop task-specific motor dexterity even faster than a healthy adult would do. However, there is convincing evidence that this state is transient. Three to four weeks after the stroke, this extraordinary capacity for learning seems to vanish (Krakauer et al. 2012). Based on these observations, many authors have questioned whether any functional rehabilitation at the chronic stage may be useful. Recently, Steven Zeiler and colleagues developed an animal experiment with mice showing that the induction of a second stroke at the chronic stage re-opens a sensitive period and mediates full recovery of those impairments that derived from a first stroke (Zeiler et al. 2015). In addition, Constraint-Induced Movement Therapy, which forces the use of the paretic limb by constraining the movement of the non-paretic limb, has previously been shown to produce substantial improvements at the chronic phase post-stroke, suggesting that tendency to plateau in motor recovery does not indicate a similar plateau in functional improvements. Despite this, to what extent chronic patients can still make any notable improvements after they have reached a plateau in motor recovery remains an open question.

Recruit: sensorimotor contingencies

During the execution of a particular motor task, an optimal sensorimotor representation allows us on the one hand to predict the external sensory event (forward model) and on the other to estimate the behavioral goal given the internal sensory motor representation (which could be seen as an inverse model). Recently the substrate of this capability, the so-called mirror neuron system, has been described by Giacomo Rizzolatti and his co-workers (Rizzolatti et al. 1996). Mirror neurons are activated by both movement execution and observation, and their functionality may be related to the recognition of goal-oriented actions and the construction of sensorimotor predictions. The mirror neurons of the monkey's brain were discovered by chance during single cell recordings in the area F5 of the premotor cortex while performing grasping tasks. While F5 neurons code observed or executed goal-related motor actions such as hand and mouth grasping with varying levels of congruence, those of area PF, located in the inferior parietal cortex, respond to somatosensory stimuli, visual stimuli, or both, as part of an action sequence, predicting the goals of complex movements. Mirror neurons have also shown to respond when the goal state of the action is hidden from view, and when sounds associated with specific actions occur. In humans, the effectiveness of the mirror neuron system (MNS) in driving the motor system was demonstrated through experiments that showed an increase in the excitability of the corticospinal pathway.

Since brain damage causes a disruption of sensorimotor contingencies—possibly in both their motor (e.g. hemiparesis) as well as their perceptual components (e.g. hemineglect)—the exposure to explicit external sensorimotor feedback may provide an avenue for motor learning and recovery by capitalizing on the mirror neurons. This hypothesis has been extensively explored by clinical studies showing that motor imagery and motor imitation may be powerful vehicles to enhance motor recovery and reduce neuropathic pain. Back in 1985, Denis proposed that mental practice/imagery could improve motor performance though the enhancement of specific cognitive skills, such as motor planning (Denis 1985). Another classic example is the so-called mirror therapy studies performed by Ramachandran and co-workers which showed that the projection of an arm at the position of a felt arm in case of phantom limb pain leads to a marked reduction of the pain sensation and the possibility of manipulating the sensation of the phantom dependent on the mirror image presented to the patient (Ramachandran and Rogers-Ramachandran 1996). Subsequently, several functional neuroimaging studies have further confirmed these findings. For instance, Malouin and colleagues conducted a PET study with hemiparetic stroke patients in which participants were asked to imagine themselves walking (Malouin et al. 2003). Results revealed that higher brain centers became progressively engaged when demands of locomotor tasks required increasing cognitive and sensory information processing. However, the benefits of sensorimotor stimulation are not limited to learning and recovery: findings from a recent study on healthy subjects suggest that providing congruent visuotactile feedback during training may also enhance body ownership, thus facilitating motor control during execution (Grechuta et al. 2017). On these bases, an effective mental practice for motor learning should be vivid, persistent (i.e. controllable), and accurate, requiring recalling visual, auditory, tactile, and kinesthetic cues.

Reinforce: individualization

There is a considerable variety of treatment concepts and therapies addressing post-stroke neurorehabilitation without a clear consensus. The efficacy of these therapies has been shown to depend on a number of parameters. First, treatment frequency and intensity has been shown to correlate with recovery in terms of increased independence and reduction of hospitalization. Second, the specificity of rehabilitation training with respect to the deficits is also seen as a central concern in conventional therapies and may play a fundamental role in recovery. Since the severity and nature of stroke-derived impairments differs across patients, the level of tolerance to intensive therapies will be also diverse, revealing the need for developing patient-tailored rehabilitation protocols.

An intermediate "level of arousal" promotes "optimal performance." Thus, for optimal learning, the training task should neither be too demanding nor too easy. If the challenge is too high it might overwhelm the person when training. Contrarily, if it is too easy the user can be bored and disinterested, or it might simply not benefit as the task is already mastered. There are indications that challenge promotes both motor learning and neural plasticity. In squirrel monkeys, for example, training in a task they had not mastered yet produced a functional reorganization of the motor cortex, not observed in monkeys repeating a similar but simpler task. Functional reorganization was also observed in lesioned squirrel monkeys after retraining of "skilled hand use" (Nudo et al. 1997). More recently, a similar idea has been expressed in terms of information theory in the Challenge Point Framework, by defining the challenge involved in performing a task as the amount of information involved. This theory states that there is an optimal amount of information that maximizes the potential learning.

Biomimetic or biohybrid systems

Recently, standard rehabilitation has been augmented with several technological advances. These technologies include computer therapy programs, robotics, Functional Electrical Stimulation (FES), and, to a lesser degree, Transcranial Magnetic Stimulation (TMS) and Virtual Reality (VR). While computer-based therapy programs (e.g. Computerized Cognitive Behavior Therapy) may be effective for the treatment of mild to moderate cognitive impairments, the automated treatment of motor impairments seems to be technically more challenging, since it requires the combination of dedicated interface tracking devices with interactive training scenarios. Different solutions are available and have been used widely in rehabilitation, such as sensor devices, including pressure sensors, accelerometers, or even vibrotactile actuators for providing haptic feedback, haptic exoskeleton technologies, vision-based patch tracking, and data gloves.

Since the 1990s, when the first rehabilitation robotic device, the InMotion Arm Robot (Interactive Motion Technologies Inc., USA, also known as MIT-Manus) was developed, engineers have made great advances in the development and application of exoskeletons to the restoration of functional motor abilities. These biomimetic wearable robots for neurorehabilitation can be applied as 1) therapeutic devices for physiotherapy, 2) assistive devices for the augmentation of human capabilities, 3) haptic devices for tactile feedback delivery, and 4) monitoring tools for motor recovery. An example of this versatility is the Hand Exoskeleton Rehabilitation Robot (HEXORR) (Schabowsky et al. 2010), an active robotic shell developed by Schabowsky and colleagues, which is capable of compensating for muscular tone to help patients in opening the paretic hand (Figure 59.1). HEXORR assists movement, allows free movement, or restricts movement to allow static force production. Clinical studies have shown the superiority of HEXORR as an intervention protocol when compared to conventional therapy. But when combined with VR-based training scenarios, the versatility of these biomimetic technologies becomes even more pronounced, allowing simulation of the sensation of touch and providing control over the object's physics in a contextualized manner.

Figure 59.1 Pictures of a hand in HEXORR at different postures.

Reproduced from *Journal of NeuroEngineering and Rehabilitation*, 7, p. 36, Figure 1, doi: 10.1186/1743-0003-7-36, Christopher N. Schabowsky, Sasha B. Godfrey, Rahsaan J. Holley, and Peter S. Lum, Development and pilot testing of HEXORR: Hand EXOskeleton Rehabilitation Robot. This work is licensed under a Creative Commons Attribution 2.0 Generic License.

The rationale for the use of VR-based training scenarios in rehabilitation is based on a number of unique attributes of this technology. These attributes include the opportunity for experiential, active re-learning that encourages the participant. Moreover, VR is an excellent tool for the manipulation of sensorimotor contingencies as it allows the full control of different modalities such as vision, audio, and haptic feedback. For instance, recent studies have shown that inclusion of auditory feedback improves the clinical outcomes after rehabilitation with robotic therapy systems. In this line, augmenting the visual and auditory feedback of the VR interaction with haptic feedback on touching virtual objects together may enhance the salience of interaction events and the ecological validity of the task thus influencing the retention of functional gains. Another unique attribute of VR is that it can provide automatized individualization of training intensity and specificity, while allowing for increased standardization of assessment and training protocols. A recent review analyzed 35 studies to evaluate the effects of robotic systems and VR on upper limb motor function after stroke (Saposnik and Levin 2011). Results from this study suggested that VR-based therapies and robotics might be promising strategies to increase the intensity of treatment and to promote motor recovery after stroke. However, findings from Cameirão and colleagues revealed that the beneficial effects of VR-based neurorehabilitation treatments when compared to conventional therapies may depend on the specific interface systems used (Cameirão et al. 2012). Hence, the solution is not to use VR as such but to take advantage of its features to augment and extend current therapies based on scientific principles such as those found in the DAC theory.

The Rehabilitation Gaming System (RGS) is a biohybrid system for neurorehabilitation based on VR (Figure 59.2A) that is grounded on the three theoretical principles described above: use-dependent plasticity, sensorimotor contingencies, and individualization. RGS integrates a paradigm of action execution with motor imagery and action observation where the underlying hypothesis is that functional recovery can be promoted by capitalizing on the life-long plasticity of the brain. By combining movement execution with the observation of correlated action of virtual limbs from a first-person perspective, recovery can be accelerated and enhanced by driving the MNS that can be seen as an interface between the neuronal substrates of visual perception, motor planning, and execution. Following the third principle, the RGS architecture includes two main functional components for an automatized individualization of rehabilitation therapies: an Adaptive Difficulty Controller, and an Adaptive Biomechanics Controller. These individualization mechanisms offer the possibility to manipulate task difficulty and visuomotor feedback, thus encouraging training and allowing highly impaired patients to perform complete functional motor tasks in VR that they could not achieve under real-world conditions (Ballester et al. 2015, 2016). More importantly, this mechanism exposes the patient to complete sensorimotor contingencies during training and allows encouraging the use of the paretic limb by 1) increasing the probability of success, and 2) decreasing levels of effort and preventing fatigue. But while we have argued that there are possible benefits, there are also some possible problems arising from amplifying virtual movements. It is important that patients are still required and stimulated to perform to their full capabilities. Any amplification of a patient's movements has to be adjusted such that they are always at the edge of their possibilities and motivated to complete the movement themselves. Supporting these design principles, the clinical trials that have been performed thus far suggest that RGS accelerates recovery of acute and chronic stroke patients (Figure 59.2B) (Cameirão et al. 2012; Ballester et al. 2017).

Future directions

Nowadays, there is no complete understanding of the neuronal systems affected by stroke and the overall repercussion it has on non-affected areas. Indeed, this understanding would require comprehension of the brain itself, which is one of the major targets of modern science. For this reason, we expect an increasing role for theory-based interventions. Although robot-aided neurorehabilitation and VR technologies seem to be promising solutions, there is not enough evidence to draw clear conclusions about their effectiveness and applicability to different patient profiles. Many studies have confirmed a positive impact on mildly impaired hemiparetic patients; however, the effects on patients showing more severe motor deficits have been poorly studied. Despite these limitations, the market is already offering evidence-based autonomous healthcare systems that comprise a full suite for care management, diagnostics, monitoring, and rehabilitation approaches, making them cost effective and suitable for domiciliary rehabilitation and tele-assistance. In parallel, there is now a big effort in developing wearable technologies for injury/illness prevention and assistance (Ballester et al. 2015). Although very few studies have explored the effects of this new approach on recovery, we expect to see an exponential growth both in the number of related scientific publications and in the amount and diversity of commercialized products in the coming years. In particular, we expect an incremental increase

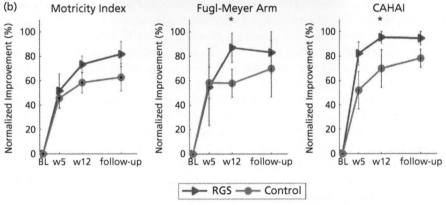

Figure 59.2 (a) RGS setup. The RGS system includes a 24″ touch-screen with an integrated CPU, and a support or stand placing the Microsoft Kinect Sensor above the screen.

(b) Normalized improvement over time for the arm subpart of the Fugl–Meyer Assessment Test and the Chedoke Arm and Hand Activity Inventory. Improvement (median ± median absolute deviation) for RGS (black) and control (gray) groups for selected clinical scales. *$p < 0.05$, between-group comparison.

(b) Reproduced from Monica da Silva Cameirão, Sergi Bermudez i Badia, Esther Duarte, and Paul F. M. J. Verschure, Virtual reality based rehabilitation speeds up functional recovery of the upper extremities after stroke: A randomized controlled pilot study in the acute phase of stroke using the Rehabilitation Gaming System, *Restorative Neurology and Neuroscience*, 29 (5), 287–298, Figure 5, doi: 10.3233/RNN-2011-0599, © 2011, IOS Press and the authors.

of the symbiosis between humans and the living machine technology that will allow them to remain healthy and to recover function after illness as evidenced by the DAC-based RGS system.

Learning more

The field of biohybrid systems for rehabilitation covers a broad range of technologies, such as prosthetic implants, passive and active exoskeletons, biofeedback therapy, brain imaging, nano-technology, stimulation systems, VR-based technologies, and wearable devices. See Rothschild (2010) and contributions by Prasad (Chapter 49), Vassanelli (Chapter 50), Bensmaia (Chapter 53), and Lehmann and van Schaik (Chapter 54), in this handbook, for more on these topics. Maslen and Savulescu (Chapter 62) also discuss ethical issues raised by the increasing use of VR technologies. Gert Kwakkel and colleagues reviewed 11 recent clinical studies to assess the effect of robot-aided therapy on stroke patients' upper-limb motor function (Kwakkel et al. 2015). This review provides a clear overview on the fundamental challenges in developing these devices. For more detailed information on the latest advances in the field of neurorehabilitation technologies please refer to the book "Neurorehabilitation Technology" (Dietz et al. 2012; ISBN 978-1-4471-2277-7). The *Journal of NeuroEngineering and Rehabilitation* (BioMed Central Ltd., ISSN 1743-0003) is also a useful resource in this regard.

Regarding the scientific grounding proposed above, the book *Neuroscience: Fundamentals for Rehabilitation* (Lundy-Ekman 2012) offers a practical guide for neurorehabilitation, covering in detail the topic of neuroplasticity and linking it to clinical cases. For a better understanding of the notion of sensorimotor contingencies two excellent resources are Alva Noë's (2004) book *Action in perception* (ISBN 978-0-2621-4088-1) and his collaboration with Kevin O'Regan: *What it is like to see: A sensorimotor theory of perceptual experience* (O'Regan and Nöe 2001), and a more recent summary of the field can be found in Engel et al. (2016) *The pragmatic turn: toward action-oriented views in cognitive science,*

References

Adamovich, S. V., Fluet, G. G., Tunik, E., and Merians, A. S. (2009). Sensorimotor training in virtual reality: a review. *NeuroRehabilitation*, 25(1), 29–44.

Ballester, B. R., et al. (2015). The visual amplification of goal-oriented movements counteracts acquired non-use in hemiparetic stroke patients. *Journal of NeuroEngineering and Rehabilitation*, 12(1), 50. Available at: http://www.ncbi.nlm.nih.gov/pubmed/26055406 [Accessed June 10, 2015].

Ballester, B. R., Lathe, A., Duarte, E., Duff, A., and Verschure, P. F. M. J. (2015). A wearable bracelet device for promoting arm use in stroke patients. In: *Proceedings of the 3rd International Congress on Neurotechnology, Electronics and Informatics - Volume 1: NEUROTECHNIX*, pp. 24–31. ISBN 978-989-758-161-8. doi: 10.5220/0005662300240031

Ballester, B. R., Maier, M., Mozo, R. M. S. S., Castañeda, V., Duff, A., and Verschure, P. F. (2016). Counteracting learned non-use in chronic stroke patients with reinforcement-induced movement therapy. *Journal of Neuroengineering and Rehabilitation*, 13(1), 74.

Ballester, B. R., Nirme, J., Camacho, I., Duarte, E., Rodríguez, S., Cuxart, A., ... and Verschure, P. F. (2017). Domiciliary VR-based therapy for functional recovery and cortical reorganization: randomized controlled trial in participants at the chronic stage post stroke. *JMIR Serious Games*, 5(3), e15.

Cameirão, M.S., et al. (2012). The combined impact of virtual reality neurorehabilitation and its interfaces on upper extremity functional recovery in patients with chronic stroke. *Stroke*, 43(10), 2720–8.

Denis, M. (1985). Visual imagery and the use of mental practice in the development of motor skills. *Canadian Journal of Applied Sport Sciences. Journal Canadien des Sciences Appliquees au Sport*, 10(4), 4S–16S.

Dietz, V., Nef, T., Rymer, W. Z. (eds) (2012). *Neurorehabilitation Technology*. London: Springer.

Engel, A. K., Friston, K. J., and Kragic, D. (eds) (2016). *The pragmatic turn : toward action-oriented views in cognitive science*. Strüngmann Forum Report series. Cambridge, MA: MIT Press.

Grechuta, K., Guga, J., Maffei, G., Rubio, B. B., and Verschure, P. F. M. J. (2017). Visuotactile integration modulates motor performance in a perceptual decision-making task. *Scientific Reports*, 7(1), 3333. Available at: https://www.nature.com/articles/s41598-017-03488-0

Krakauer, J.W., et al. (2012). Getting neurorehabilitation right: what can be learned from animal models? *Neurorehabilitation and Neural Repair*, 26(8), 923–31. Available at: http://www.ncbi.nlm.nih.gov/pubmed/22466792 [Accessed July 21, 2015].

Kwakkel, G., Kollen, B. J., and Krebs, H. I. (2015). Effects of robot-assisted therapy on upper limb recovery after stroke: a systematic review. *Neurorehabilitation and Neural Repair*, 22(2), 111–21. Available at: http://www.pubmedcentral.nih.gov/articlerender.fcgi?artid=2730506&tool=pmcentrez& rendertype=abstract [Accessed May 8, 2015].

Lundy-Ekman, L. (2012). *Neuroscience: Fundamentals for Rehabilitation* (4th edition). St Louis, MI: Saunders.

Malouin, F. et al. (2003). Brain activations during motor imagery of locomotor-related tasks: a PET study. *Human Brain Mapping*, 19(1), 47–62. Available at: http://www.ncbi.nlm.nih.gov/pubmed/12731103 [Accessed July 21, 2015].

Noë, A. (2004). *Action in perception*. Cambridge, MA: MIT Press.

Nudo, R. J., Plautz, E. J., and Milliken, G.W. (1997). Adaptive plasticity in primate motor cortex as a consequence of behavioral experience and neuronal injury. *Seminars in Neuroscience*, 9(1–2), 13–23. Available at: http://linkinghub.elsevier.com/retrieve/pii/S1044576597901020.

O'Regan, J. K., and Nöe, A. (2001). What it is like to see: A sensorimotor theory of perceptual experience. *Synthese*, 129(1), 79–103. Available at: http://link.springer.com/article/10.1023/A:1012699224677 [Accessed July 21, 2015].

Ramachandran, V.S., and Rogers-Ramachandran, D. (1996). Synaesthesia in phantom limbs induced with mirrors. *Proc. Biol. Sci.*, 263(1369), 377–86. Available at: https://www.ncbi.nlm.nih.gov/pubmed/8637922

Rizzolatti, G., Fadiga, L., Gallese, V., and Fogassi, L. (1996). Premotor cortex and the recognition of motor actions. *Cogn. Brain Res.*, 3, 131–41. Available at https://www.ncbi.nlm.nih.gov/pubmed/8713554?access_num=8713554&link_type=MED&dopt=Abstract

Rothschild, R. (2010). Neuroengineering tools/applications for bidirectional interfaces, brain-computer interfaces, and neuroprosthetic implants—a review of recent progress. *Frontiers in Neuroengineering*, 3(October), 112. Available at: http://www.pubmedcentral.nih.gov/articlerender.fcgi?artid=2972680&tool=pmcentrez&rendertype=abstract [Accessed July 16, 2015].

Saposnik, G., and Levin, M. (2011). Virtual reality in stroke rehabilitation: a meta-analysis and implications for clinicians. *Stroke*, 42(5), 1380–6. Available at: http://www.ncbi.nlm.nih.gov/pubmed/21474804 [Accessed July 21, 2015].

Schabowsky, C.N., et al. (2010). Development and pilot testing of HEXORR: hand EXOskeleton rehabilitation robot. *Journal of NeuroEngineering and Rehabilitation*, 7, 36. Available at: http://www.pubmedcentral.nih.gov/articlerender.fcgi?artid=2920290&tool=pmcentrez&rendertype=abstract.

da Silva Cameirão, M., et al. (2011). Virtual reality based rehabilitation speeds up functional recovery of the upper extremities after stroke: a randomized controlled pilot study in the acute phase of stroke using the rehabilitation gaming system. *Restorative Neurology and Neuroscience*, 29(5), 287–98. Available at: http://www.ncbi.nlm.nih.gov/pubmed/21697589 [Accessed May 15, 2015].

Zeiler, S.R., et al. (2015). Paradoxical motor recovery from a first stroke by re-opening a sensitive period with a second stroke. *Stroke*, 46(Suppl 1), ATP86.

Chapter 60

Human relationships with living machines

Abigail Millings and Emily C. Collins

Department of Psychology, University of Sheffield, UK

Drawing on psychological research on human–human relationships, we explore the concept of long-term human relationship formation with machines, with particular reference to robotic machines, and reciprocally, machine interaction with humans. When considering the relationships humans can form with other entities, including machines, it is important to begin with the first principles of human–human relationships. In this chapter we take attachment theory as a framework of human–human relationships, from which to extend a proposition of human–robot relationships, focusing on each side of this relationship in turn.

Biological principles

Attachment theory

Humans, like other mammals, are social creatures, forming lasting bonds with others (see Prescott, Chapter 45, this volume). Attachment theory explains the developmental purpose and nature of humans' emotional ties to others (Bowlby 1969). Attachment theory adopts a general systems perspective, arguing that a system's survival depends on its ability to keep certain variables within certain limits, either through its own control features, or through interlinking with another system that serves to regulate the first system within its limits (Marvin and Britner 1999). Attachment theory describes three interlinked feedback systems that operate together across interpersonal boundaries, with the sole purpose of ensuring infant survival. Borrowing from ethology, these feedback systems are actually organized systems of behavior (or 'behavioral systems'), evolved to, firstly, keep caregivers close to vulnerable infants in order to keep them physically safe from predators; and secondly, equip the developing infant with the skills needed for eventual self-regulation, protection, and autonomous survival. The notion of a behavioral *system* defines a species-wide control system, governing a particular set of behaviors that result in a specific outcome in a given situation. An example of this is that, under certain conditions, many species of birds will instinctively build nests. Of the three main behavioral systems described by attachment theory, two function in the human infant, the 'attachment system' and the 'exploration behavioral system'. A third behavioral system, the 'caregiving behavioral system' functions in the infant's adult caregiver (Figure 60.1).

Infant attachment and exploration behavioral systems

The infant's attachment system drives behaviors evolved to reduce the physical distance between the infant and the caregiver in response to threat cues. The exploration system drives exploratory behaviors adapted to facilitate the development of skills and mastery over surroundings, such as play and experimentation. The two systems operate as a single feedback loop, in which

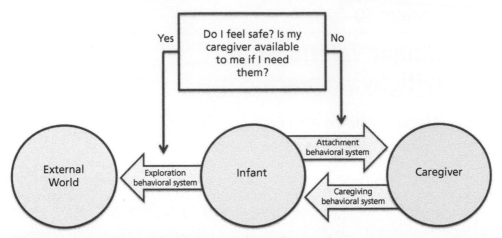

Figure 60.1 Behavioral systems of the human infant and its caregiver.

the activation of the attachment system (triggered by threat cues) terminates the exploration system, and conversely, satisfaction of the attachment system allows the exploration system to be reactivated (Marvin and Britner 1999).

The attachment system undergoes significant development during the first year of life. Human infants, equipped with a set of developing attachment behaviors (crying, cuddling, smiling, and later, reaching and following), learn to recognize their primary caregiver, and preferentially direct their attachment behaviors towards their primary caregiver (Bowlby 1969). The attachment behavioral system functions in response to environmental and intrinsic cues of threat, such as darkness, loud noises, separation, being alone, feeling unwell, hunger, pain, temperature regulation needs, and fear. In response to such cues, the attachment behavioral system activates attachment behaviors such as crying, clinging, reaching, and following, in order to seek a "safe haven" by regaining proximity to the primary caregiver. This activation of the attachment system automatically shuts down the exploration behavioral system, until proximity to primary caregiver is restored, and the infant is soothed (Marvin and Britner 1999). At this point the exploration system is reactivated and the infant can use their caregiver as a "secure base" from which to regroup, and recommence the important task of learning about the environment through play and experimentation. Such forays are necessary in order to develop new skills and gain mastery over the surroundings, and ultimately, to become an autonomously self-protective agent. Computational models have been created that demonstrate these processes (Petters and Waters 2008).

Although the attachment, exploration, and caregiving systems originate and are most crucial to survival in infancy and childhood, they continue to be relevant throughout life (Bowlby 1969). Attachments are formed to peers and eventually romantic partners (Mikulincer and Shaver 2007) and attachment behaviors can be seen across numerous everyday settings featuring some form of threat (e.g. Fraley and Shaver (1998) present a study of attachment behaviors observed between separating couples at an airport facing the attachment threat of the imminent departure of one partner).

Adult caregiving behavioral system

The caregiving behavioral system is located in the primary caregiver (e.g. parent), and responds to behaviors exhibited by an attached individual (e.g. child). The caregiving behavioral system is arguably most active in parents responding to their children (George and Solomon 1999), but is also thought to underpin all empathic and prosocial behavior (Mikulincer and Shaver 2007). When the child communicates distress by crying, the caregiver responds to try to alleviate the child's distress. In terms of systems, the control features of the first system (the child) are responded to by the regulatory second system (the parent). The goal is to keep the first system within the operational parameters that determine its survival.

Individual differences

While the development of the attachment, exploration, and caregiving behavioral systems are universal across humans, there exist nuanced individual differences in the functioning of these systems that develop over time. Such differences result from repeated relational experiences that are internalized as cognitive models and which serve to guide cognition, affect, and behavior in a personality trait-like fashion, and develop along known trajectories (Mikulincer and Shaver 2007). These individual differences not only affect individuals' cognition, affect, and behavior in relation to their own attachment relationships, but also in terms of their caregiving system and its activation in response to others, such as their partners and children (Millings et al. 2013).

In sum, close human–human relationships are governed by: i) the universal feedback systems of attachment, exploration, and caregiving; and ii) developmental individual differences in the manner in which these systems operate. We herein apply these concepts to human–robot relationships.

Biomimetic and biohybrid systems

Applying attachment theory to human–robot relationships

There are two perspectives from which we can apply attachment theory to human–robot relationships: conceptualizing and designing robot "attachment" to us, and conceptualizing and sensitively designing around our propensity to form attachments to caregivers, which may, in the future, include personal robot assistants. We are reticent to describe long-term human relationships with machines as "attachments" for three reasons: i) if we consider the slow speed at which current state-of-the-art robotics and artificial intelligence is advancing, the likelihood of robots becoming sufficiently capable of fully meeting our emotional needs anywhere in the near future is remote; ii) there are philosophical discussions to be had regarding the extent to which robots can possess the attributes necessary for the formation of true attachment bonds, such as the capacity to hold the attached individual in mind, which are beyond the scope of this chapter; and iii) there are potential negative consequences for developing fully formed attachment bonds to agents which lack the capacity to fully meet our attachment needs, which have been described elsewhere (Sharkey and Sharkey 2010). We agree that there is certainly a need for caution as researchers develop increasingly sophisticated systems that imitate understanding of our attachment needs. However, we propose that there is likely a middle ground between not using robots at all in the context of vulnerable populations, and a dystopia where use of robots results in a withdrawal of human contact with caregivers, which causes damage to the developing attachment system in children (Sharkey and Sharkey 2010), or which increases social isolation in the elderly (Sharkey and Sharkey 2012; cf. Sharkey and Wood 2014). We therefore explore the application of attachment theory to human–robot relationships from both sides of the equation. That is humans' relationships with robots, and robots' interactions with humans that we here define as robots' relationships with humans given the dyadic symbiosis of two-way human–robot interaction. We explore this with the caveat that while humans' relationships with robots might bear some similarity to the hallmarks of attachment relationships, we do not propose that humans can become fully "attached" to current state-of-the-art robots. Neither do we propose that the development of robots should aim for sufficient capacity in future iterations to enable this "attachment" process to occur. However, designing robots that mimic attachment and exploration systems could be beneficial, as we now discuss.

Robot "attachment" to humans: designing robots that mimic attachment/exploration systems

There are two distinct reasons why we might want to build robots that mimic our own attachment and exploration systems. The first is in order to promote smooth and "natural" human–robot interaction in machines that might be programmed with a developmental trajectory across their lifespan with a user, both from the perspective of human–robot interaction, and robot learning architectures. We propose that designing robots that behave like mammals with regard to attachment and exploration ought to aid a human user's perception of the robot as a socially acceptable entity. A robot that mimics a living being's developmental lifespan would facilitate more natural relations with a user: the human would better understand how to relate to them, how to work with them, and how to care for them. Based on the idea that human infants are "the most successful example of adaptation into our social and technological environment, without much prior experience," developmental approaches to robotics aim to identify "how a robotic platform could use the properties of this bond [with humans] in order to thrive from it, as children most often manage to do" (Hiolle et al. 2012, p. 2).

A second motivation for building robots that mimic the attachment and exploration behavioral systems is to advance our understanding of human relationships, particularly when they go wrong. Early attachment research that identified the mammalian requirement for emotional comfort as distinct from that for food was invaluably informative, but also unethical and very harmful to the macaque subjects (e.g. Harlow and Zimmerman 1958). Robotic or computational control systems could be used to better understand the processes by which attachment-related trauma has an impact on subsequent human behavior. While abuse within attachment relationships is common (May-Chahal and Cawson 2005), the myriad of variables involved means that cases do not lend themselves to controlled study. Future research could model the precise impact of particular patterns of abusive parenting on subsequent attachment and exploration behaviors.

Computational work on modeling the attachment and exploration systems has shown some promising results, where agent-based computational experiments have produced epigenetic trajectories that represent the individual differences of secure and insecure attachment patterns that we see in studies of infants (Petters and Waters 2010). Similarly, robotic architecture has been designed to emulate an infant's secure base and exploration behavior in a new environment (Hiolle and Cañamero 2008). Hiolle and Cañamero's architecture created for the Aibo robot involves the robot learning about the environment, becoming surprised and aroused as a result of unfamiliar stimuli, and soothed by a human user taking the role of Aibo's caregiver. Different developmental trajectories are produced by slightly more, and slightly less, attentive caregiving (Hiolle and Cañamero 2008).

Human "attachment" to robots: designing robots that fit with our attachment needs

In order to fully meet an individual's attachment needs, a robot would need to provide comfort in times of distress or threat (safe haven), and, through merely being there, facilitate the individual's exploration of their environment (secure base). These processes are highly complex and arguably center around what makes us human. While other mammals have attachments, none are as complex as our own. It is likely, therefore, that any robotic imitation of such human processes would be fairly rudimentary, and subject to the caveats already discussed. We therefore consider how robots can best "fit with," rather than "fully meet" our attachment needs.

In order for robots to fit with humans' safe haven requirements, a robot would need to be able to detect signs of stress or distress in its human user, and respond in such a way to alleviate the distress. In the case of a robotic personal assistant for a frail older person who lives alone, this might involve collecting data from the individual's facial expressions, tone of voice, posture, and movement, and identifying markers of various negative emotions within those data. The robot might then probe for further information by conversing with the individual. The robot could then respond by connecting the individual to their attachment figures (e.g. close relatives) or relevant professionals (e.g. doctor) via video-calling. In this scenario, the robot has responded to distress by facilitating proximity to the individual's human attachment figures or wider support network. The robot enables the attachment and caregiving systems of the human participants in a relationship to operate appropriately, despite geographical distance. The robot does not replace the individual's attachment figures.

For a robot to fit with secure base needs, it would need to facilitate exploration and learning about the environment. In the case of the frail older person described above, this might be through enabling the person to go out shopping, by helping them with mobility or carrying items. The individual could feel safe to do more things in the company of the robot because of the physical assistance the robot provides, but also because of the robot's capacity to connect the individual to attachment figures should any emergencies occur. The robot serves as an embodiment of practical and psychological support, facilitating, but not replacing, the person's relationships with other humans.

There are certain contexts, however, where it might be appropriate to seek to replace real live relationships with robots, such as particular therapeutic settings in which it may be inappropriate to use humans or animals. The Paro seal robot (Figure 60.2) has gained popularity in recent years as a robotic version of Animal Assisted Therapy (AAT) with elderly and dementia patients (Mordoch et al. 2013). Paro provides an established example of how the use of robots can be advantageous—for example, where patients may exhibit distressed and challenging behaviors that could compromise animal welfare, where the potentially unpredictable nature of animal behavior may have negative consequences, or where it is inappropriate to use real animals for hygiene reasons.

Designing robots that activate the caregiving system

While having our attachment needs met is an important requirement for psychological wellbeing, activation of the caregiving behavioral system can also have psychological and physical benefits. Research has found that providing care and support to others can be more beneficial to wellbeing than receiving it (Thomas 2010; Brown et al. 2003). This activation of the caregiving system might be one of the reasons that AAT is effective (Kruger and Serpell 2006). A number of social robots have been designed with "cuteness" in mind, where "cute" is defined as having a youthful and non-threatening appearance. For example, participants exposed to the NAO robot reported finding it cute and seductive, and were keen to interact with it (Baddoura et al. 2012). Breazeal reports that the Kismet robot has been designed to have an infant-like appearance that "evoke[s] nurturing responses of human adults" (Breazeal and Ayranada 2002, p. 87). The Paro seal robot (Shibata et al. 2003), with its large eyes, soft white fur, and responsiveness to touch,

Figure 60.2 The Paro robot seal.

sound, and light, is also designed to be cute. Although often methodologically weak, literature surrounding the Paro claims numerous beneficial therapeutic effects (Bemelmans et al. 2012). We propose that the mechanism of these effects is likely to be caregiving behavioral system activation, although this is yet to be empirically tested.

Future directions

We have explored the relevance of attachment theory to human–robot relationships, both from the perspective of the robot and the human, identifying the distinct domains of the attachment and caregiving behavioral systems. We have described the motivations for developing an attachment system in robots, and future research directions might include extending the developmental approach to robotics advocated by Hiolle et al. (2012, 2014). We have also touched on the potential for computational and robotic attempts to represent the attachment system in helping us to better understand the ramifications of breakdowns in the system, for example in cases of abusive relationships. Future research could extend this by examining trajectories of recovery.

In considering the capacity of robots to fit with our attachment needs, we have described how we can use robots to extend the geographical reach of our attachment relationships with other humans, rather than seeking to replace our relationship partners with robots. Future research could seek to better delineate: a) the precise areas of need for vulnerable populations; and b) how robots could acceptably and usefully provide support in those areas.

Finally, we have proposed the caregiving behavioral system to be of fundamental importance in designing social robots that provide comfort. Future research can seek to: a) test caregiving activation as a proposed mechanism of the positive effects brought about by social robots; and b) delineate the most effective design criteria for optimal caregiving system activation.

Additional future areas for research in the topic of human–robot relationships which lie beyond the scope of this chapter include expanding current research exploring practical uses of robots in autism therapy, for example, in helping children with autism to learn about tactile social behavior (Robins et al. 2012), and additionally with interviewing children for eye-witness testimonies (Wood et al. 2013). Further, future research might consider robotic approaches to therapeutic agents, building on the e-mental health literature (Cavanagh and Millings 2013).

There are many exciting directions for future research on human–robot relationships, and given the speed at which state-of-the-art robotics is advancing it is important that even now we begin to speculate on what future relationships with advanced technology may look like. Robots hold a practical position in society. They are tools, capable of extending human independence in the face of, for example, physical disadvantages. With robotic devices an individual can regain their freedom to proceed with simple daily activities, free from some of the concerns that tending to such activities alone could present. Understanding these existing and potential human–robot relationships is an important endeavor, and one to which the perspective of attachment theory has much to offer as we navigate a path towards living with social machines that assist, rather than replace, our human–human relationships.

Learning more

Further learning of the foundational concepts of this topic can be guided by key attachment theory literature (Mikulincer and Shaver 2007) as well as literature addressing the caregiving system and its relationship with adult attachment development (Collins and Ford 2010; Millings et al. 2013). Expanding further, we suggest that the interested reader might apply these ideas to the following topics: computational and robotic approaches to attachment and caregiving (Petters and Waters 2008, 2010; Hiolle et al. 2012, 2014); exploring the place of robots in our social environment (Breazeal 2003; Collins et al. 2013); and the effects of social robots in different populations—for example in the care of the elderly (Mordoch et al. 2013); in therapy for children with autism (Robins et al. 2005); and as potential peer tutors for children (Kanda et al. 2004). It may also be of interest to look at the applications of computational and robotic approaches in exploring the effects of behaviors that may be produced by social robots, as in modeling robot gait based on human emotional mental models (Destenhe et al. 2013).

Finally there are several surveys and reviews of socially interactive robots which provide overviews of a variety of combinations of the topics covered in this chapter. Amongst these, Fong et al. (2003) and Goodrich and Schultz (2007) are well-cited and provide excellent starting points for further study.

Acknowledgments

Inspiration and support for the preparation of this chapter came from Economic and Social Research Council funded seminar series 'Applying Relationships Science to Contemporary Interventions (ApReSCI)', grant number ES/L001365/1, and European Commission funded project 'Expressive Agents for Symbiotic Education and Learning (EASEL)', grant number n611971.

References

Baddoura, R., Matuskata, R., and Venture, G. (2012). The familiar as a key-concept in regulating the social and affective dimensions of HRI. In: *Proc IEEE/RAS Int. Conf. on Humanoid Robots*, pp. 234–41. doi: 978-1-4673-1369-8/12.

Bemelmans, R., Gelderbom, G., Jonker, P., and de Witte, L. (2012). Socially assistive robots in elderly care: a systematic review into effects and effectiveness. *Journal of the American Medical Directors Association*, 13(2), 114–20. doi:10.1016/j.jamda.2010.10.002

Bowlby, J. (1969). *Attachment and loss (Vol. 1): Attachment.* New York: Basic Books.

Breazeal, C. (2003). Toward sociable robots. *Robotics and Autonomous Systems*, 42(3), 167–75.

Breazeal, C., and Aryananda, L. (2002). Recognition of affective communicative intent in robot-directed speech. *Autonomous Robots*, **12**(1), 83–104.

Brown, S., Nesse, R., Vinokur A., and Smith, D. (2003). Providing social support may be more beneficial than receiving it: results from a prospective study of mortality. *Psychological Science*, **14**, 320–7. doi: 10.1111/1467-9280.14461

Cavanagh, K., and Millings, A. (2013). (Inter)personal computing: the role of the therapeutic relationship in e-mental health. *Journal of Contemporary Psychotherapy*, **43**, 197–206. doi 10.1007/s10879-013-9242-z

Collins, E. C., Millings, A., and Prescott, T. J. (2013). Attachment to assistive technology: a new conceptualisation. *Assistive Technology Research Series*, **33**, 823–8.

Collins, N. L., and Ford, M. (2010). Responding to the needs of others: The interplay of the attachment and caregiving systems in adult intimate relationships. *Journal of Social and Personal Relationships*, **27**, 235–44.

Destenhe, M., Hashimoto, K., and Takanishi, A. (2013). Emotional gait generation method based on emotion mental model—preliminary experiment with happiness and sadness. In: *Ubiquitous Robots and Ambient Intelligence (URAI), 2013 10th International Conference on*. IEEE, pp. 86–9.

Fong, T., Nourbakhsh, I., and Dautenhahn, K. (2003). A survey of socially interactive robots. *Robotics and Autonomous Systems*, **42**(3), 143–66.

Fraley, R. C., and Shaver, P. R. (1998). Airport separations: a naturalistic study of adult attachment dynamics in separating couples. *Journal of Personality and Social Psychology*, **75**, 1198–212.

George, C., and Solomon, J. (1999). Attachment and caregiving: the caregiving behavioural system. In: J. Cassidy and P. Shaver (eds), *Handbook of attachment*. New York: Guilford Press, pp. 649–70.

Goodrich, M. A., and Schultz, A. C. (2007). Human–robot interaction: a survey. *Foundations and Trends in Human-Computer Interaction*, **1**(3), 203–75.

Harlow, H., and Zimmerman, R. (1958). The development of affectional responses in infant monkeys. *Proceedings of the American Philosophical Society*, **102**(5), 501–9.

Hiolle, A., and Cañamero, L. (2008). Conscientious caretaking for autonomous robots and arousal-based model of exploratory behaviour. In: *Proceedings of the Eighth International Conference on Epigenetic Robotics: Modelling Cognitive Development in Robotic Systems*. Lund University Cognitive Studies, 139. Retrieved from: http://www.image.ece.ntua.gr/projects/feelix/files/epirob08_hiolle_canamero.pdf September 9, 2014.

Hiolle, A., Cañamero, L., Davila-Ross, M., and Bard, K. (2012). Eliciting caregiving behaviour in dyadic human–robot attachment-like interactions. *ACM Transactions on Interactive Intelligent Design Systems*, **2**(1), Article 3. doi: 10.1145/2133366.2133369

Hiolle, A., Lewis, M., and Cañamero, L. (2014). Arousal regulation and affective adaptation to human responsiveness by a robot that explores and learns a novel environment. *Frontiers in Neurorobotics*, **8**, 17. doi:10.3389/fnbot.2014.00017

Kanda, T., Hirano, T., Eaton, D., and Ishiguro, H. (2004). Interactive robots as social partners and peer tutors for children: a field trial. *Human-Computer Interaction*, **19**(1), 61–84.

Kruger, K., and Serpell, J. (2006). Animal-assisted interventions in mental health: definitions and theoretical foundations. In: A. Fine (ed.), *Handbook on animal-assisted therapy: Theoretical foundations and guidelines for clinical practice*. San Diego: Elsevier, pp. 21–39.

Marvin, R. S., and Britner, P. A. (1999). Normative development: The ontogeny of attachment. In: J. Cassidy and P. Shaver (eds), *Handbook of attachment*. New York: Guilford Press, pp. 21–43.

May-Chahal, C., and Cawson, P. (2005). Measuring child maltreatment in the United Kingdom: a study of the prevalence of child abuse and neglect. *Child Abuse & Neglect*, **29**(9), 969–84. doi: 10.1016/j.chiabu.2004.05.009

Mikulincer, M., and Shaver, P. (2007). *Attachment in adulthood structure, dynamics, and change*. New York: Guilford Press.

Millings, A., Walsh, J., Hepper, E, and O'Brien, M. (2013). Good partner, good parent: caregiving mediates the link between romantic attachment and parenting style. *Personality and Social Psychology Bulletin*, **39**, 170–80. doi: 10.1177/0146167212468333

Mordoch, E., Osterreicher, A., Guse, L., Roger, K., and Thompson, G. (2013). Use of social commitment robots in the care of elderly people with dementia: a literature review. *Maturitas*, **74**(1), 14–20. doi: 10.1016/j.maturitas.2012.10.015.

Petters, D., and Waters, E. (2008). Epigenetic development of attachment styles in autonomous agents. In: *Proceedings of the Eighth International Conference on Epigenetic Robotics. Modeling Cognitive Development in Robotics Systems*, July 2008, Brighton. Lund University Cognitive Studies, 139, pp. 153–4. Retrieved from: http://www.cs.bham.ac.uk/~ddp/petters_waters_epirob08_final.pdf 12/09/14

Petters, D., and Waters, E. (2010). AI, Attachment Theory, and simulating secure base behaviour: Dr. Bowlby meet the Reverend Bayes. In: *Proceedings of the International Symposium on 'AI-Inspired Biology', AISB Convention 2010*, University of Sussex, Brighton: AISB Press, pp. 51–8. Retrieved from: http://www.cs.bham.ac.uk/~ddp/petters-d-waters-e-aiib.pdf 12/09/14

Robins, B., Dautenhahn, K., Te Boekhorst, R., and Billard, A. (2005). Robotic assistants in therapy and education of children with autism: can a small humanoid robot help encourage social interaction skills? *Universal Access in the Information Society*, **4**(2), 105–20.

Robins, B., Dautenhahn, K., and Dickerson, P. (2012). Embodiment and cognitive learning–can a humanoid robot help children with autism to learn about tactile social behaviour? In: S. S. Ge, O. Khatib, J. J. Cabibihan, R. Simmons, and M. A. Williams (eds), *Social Robotics*. ICSR 2012. Lecture Notes in Computer Science, vol **7621**. Berlin/Heidelberg: Springer, pp. 66–75.

Sharkey, A., and Wood, N. (2014). The Paro seal robot: demeaning or enabling? *Proceedings of AISB 2014*. AISB.

Sharkey, A. J. C., and Sharkey, N. E. (2012). Granny and the robots: ethical issues in robot care for the elderly. *Ethics and Information Technology*, **14**(1), 27–40.

Sharkey, N. E., and Sharkey, A. J. C. (2010). The crying shame of robot nannies: an ethical appraisal. *Interaction Studies*, **11**(2), 161–90.

Shibata, T., Wada, K., and Tanie, K. (2003). Statistical analysis and comparison of questionnaire results of subjective evaluations of seal robot in Japan and U.K. In: *Proceedings of the 2003 IEEE International Conference on Robotics & Automation*, Taipei, Taiwan, September 14–19.

Thomas, P. A. (2010). Is it better to give or to receive? Social support and the well-being of older adults. *Journal of Gerontology: Series B*, **65B**(3), 351–7. doi:10.1093/geronb/gbp113

Wood, L. J., Dautenhahn, K., Lehmann, H., Robins, B., Rainer, A., and Syrdal, D. S. (2013). Robot-mediated interviews: do robots possess advantages over human interviewers when talking to children with special needs? In: S. S. Ge, O. Khatib, J. J. Cabibihan, R. Simmons, and M. A. Williams (eds), *Social Robotics*. ICSR 2012. Lecture Notes in Computer Science, vol **7621**. Berlin/Heidelberg: Springer, pp. 54–63.

Chapter 61

Living machines in our cultural imagination

Michael Szollosy

Sheffield Robotics and Department of Computer Science,
University of Sheffield, UK

The public perception of robots is fraught with images of armies of gleaming metallic bipedal killing machines, and scheming artificial intelligence lurking behind doe-eyed young women, both equally bent on world domination and the eradication of the human race. The picture has improved somewhat in the last two decades, with new images joining the old clichés, those of robots come to save the Earth and humanity from the worst of our own excesses and artificial intelligence as the final salvation against the havoc of natural stupidity.

The consequences of our obsession with our conception of robotic monsters is obvious: screeching headlines in the popular press about the arrival of "Terminator technologies" obscure the genuinely beneficial innovations in robotics and AI so that it becomes impossible to distinguish between minor steps and important leaps, benevolent developments that will enrich human life and autonomous systems that perhaps should give legitimate cause for alarm. The bad press that "Terminator technologies" receive could lead to a public rejection of valuable science, as it has already done previously—not without a good deal of controversy—with the media's demonization of "Frankenstein foods," e.g. Dixon (1999).

And unfortunately it is not the case that the more positive sci-fi perceptions are more accurate reflections of what work is being undertaken in the lab. When robots and developments in AI are not enslaving the human race, they are saving us from our human biological limitations and frailties, for example, by helping us overcome our human moral weakness, or by allowing us to upload our consciousness into another body to ensure eternal life or, most often, by battling to save us from other, badder robots. And when the bar is set so high, when nothing less than biological and spiritual salvation is the least we have come to expect from technology, robots and AI are always bound to fall short, causing further anxiety about our brave new technological futures.

The general public—and sometimes people that really should know better—regard robots, AI, and other emerging technologies almost entirely through a lens of science fiction, and with little regard for science fact. This is in spite of the best efforts of scientists and academics and industry leaders, many of whom are actively engaged trying to raise awareness and disseminate information. Sometimes, we must accept that it is the scientists, academics, and industry leaders themselves that are, at least in part, responsible for the unrealistic expectations and anxieties of the public, as the temptation to chase funding, customers, fame, and fortune proves too great and the machines of misinformation are fed with hyperbole, optimistic boasts, or slight exaggerations about the imminence of their technological breakthrough. Such complications, however, are only capitalizing on existing fears and expectations, and are only effective because they touch some deeper, more fundamental anxieties that exist in the public imagination.

Whatever speculations are offered by film and literature, science fiction is a much better reflection of our present hopes and fears than it is an accurate picture of what our future may or may not look like. We must bear this in mind if we are to understand why the Terminator has such a grasp on headline writers, and why so many people regard it as an inevitability that autonomous, gun-wielding robots bent on genocide will certainly one day march through our streets. But before we look at the specific shape and manifestation of these cultural imaginings— that is, those particular images that find themselves ingrained in our cultural life, in our films, in our books and video games, and in our popular press—it is important to examine the means by which such ideas come into being, how such monsters grip our imaginations.

Mechanism of misinformation

The role that anthropomorphism plays in our human–robot interactions is well-documented (e.g. Caporael and Heyes 1997; Duffy 2003; Kiesler et al. 2008). Anthropomorphism is the name given to the phenomenon of human beings projecting their own attitudes, beliefs, and fantasies into an inanimate object, or attributing human thinking, feelings, or motivations to inanimate object (see e.g. Heider and Simmel 1944). Anthropomorphism, however, while clearly an important factor in human–robot interactions, is not particularly well understood, and only tells one part of a larger story. There are elements of anthropomorphism, as a subjective phenomenon rooted in our personal and cultural imagination, that cannot be understood without a more thorough understanding of how it is humans project their own thought processes and emotional states. (By distinguishing a "cultural imagination" I do not mean to imply something akin to a Jung's archetype or collective unconscious. Rather, I mean simply fantasies that are shared in a particular cultural context, but fantasies that can be explained only in terms of shared social experience, without recourse to biological or mystical processes.) And such an understanding can only be achieved by an appeal to another epistemological methodology, one that allows for a more nuanced narrative conceptualization of subjective experience. The account of projection offered by post-Freudian psychoanalytic thought, particularly that from Melanie Klein and her followers, and those of the British independent tradition (e.g. Heimann et al. 1989; Rayner 1991) is a useful place to start in trying to conceive of projection as a mechanism that can explain with greater clarity our relationships with robots and AI, both fictional and real.

The psychoanalytic idea of projection is an attempt to describe *object relations*, that is, the way that people relate to things—usually other people, but also other material and non-material objects in their world. Projection is one of an array of interrelated defence mechanisms, first employed by infants but also used in the normal, everyday functioning of psychically mature adults. Such defence mechanisms are also the root of psychopathology: the line between the normal and psychopathological use of these defence mechanisms is, in psychoanalytic thinking, primarily a question of degrees, and the extent to which they are used to maintain an acceptance of ambivalence—that is, being able to negotiate the paradox of simultaneously held positive and negative views towards an object. The purpose of these mechanisms is to defend a person from anxiety, or the unbearable conflicts of being confronted with contrary ideas or feelings.

With projection, it is believed that in psychological fantasy we split off parts of ourselves, such as thoughts, or emotions, or other parts of the self, and "project" them into something—a person, an object, or even a symbol or an idea—which can then be regarded as a sort of *container* for these projections (cf. Klein 1997; Bion 1962). Sometimes good parts of the self are projected into containers for safe keeping, or so that one can identify with that part in another, or to make a connection between oneself and the container. This is referred to as *projective identification*, and is not only a common phenomenon in psychic functioning, but forms a basis of our inter-subjective communications and conceptions of group identity. On the other hand, sometimes negative parts of the self are projected into a container (and in practice it is usually a combination of good and bad parts that are projected). Bad parts of the self—violent fantasies, hatred, anxieties, for example—can be projected away from the self, in an act of disavowal. In keeping such bad elements at bay, the self can be thought of as pure and all good, with the consequence that an external other is demonized and regarded as the source of that bad object, and so therefore becomes a persecuting figure as the hatred and violence that was projected out is imagined to return in the form of the other.

Ideas of projection and projective identification can be used in cultural studies to provide compelling explanations for diverse phenomenon, particularly those involving collective fantasies of identity and threat. For example, nationalism can be regarded as individual people projecting their own imagined positive qualities into a symbol, or an idea, or a leader; in such a container, the positive aspect of the self is considered "safe," that is, not threatened by anxiety or the normal vicissitudes of ambivalence, and these shared associations provide a source for collective cohesion, a group identity. Projection of bad parts of the self can also be used to understand why we find certain objects or groups threatening, how we experience paranoia, and why we feel such hatred towards some objects. The most obvious example of projections used in this way can be seen in scapegoating, such as commonly seen in racist attitudes, or on the other side of nationalism: It is not *we* who are violent, it is *them*. *They* hate us and are out to get us. With the scapegoat, there is a belief that the container of the bad parts of the self must be destroyed before it can return and destroy us. This is a root of paranoia. The belief that we are being persecuted is our own fantasy (cf. Anzieu 1989; Young 1994).

It is through such projections that we come to know and understand the world. Projective identification also provides the basis for empathy: by projecting parts of ourselves into others—to "walk a mile in someone else's shoes" as one saying imagines it, or by "putting ourselves in someone else's position"—and by receiving the projections of others, by being that container for another's projections, we gain knowledge and understanding of others' emotional and thought processes. Thus, a great deal of post-Freudian psychoanalytic thought has elevated the idea of projections and projective identification to a greater importance, and these are, for many, at the very centre of human communications and experience. And in a world more and more populated with entities capable of supreme rationalism far beyond anything of which a human is capable, it is this capacity for empathy that is increasingly defining for many what it means to be human; this is particularly evident in our science fiction—see, for example, *Do androids dream of electric sheep?*, *The Terminator*, *Star Trek*, et al.

Monsters and mythology

Projections are at work in the construction of monsters throughout the ages, from gods and demons that, despite their apparent superiority, were the victims of the most base human emotions (jealousy, rage, etc.) to the racial others that have been painted throughout the centuries as inhuman savages, to the most outrageous supernatural beings that the human imagination can conjure—zombies and werewolves, vampires and aliens. All of our monsters, whatever their tenuous links to fact or (science) fiction, are products of human fantasies, creatures imbued with the bad parts of ourselves projected outward. The monster is the container for our projections, and that container then becomes terrifying and relentless and ubiquitous because it is nothing other than our own qualities from which we know, unconsciously, we cannot escape. If robots, in our techno-dystopian visions, are attempting to to enslave or completely eradicate the entire human race, they are only reflections of our own violent impulses: either feelings we would rather disavow or the memory of past sins that we would rather repress.

That robotic monsters are products of human fantasies is, perhaps, not a startling revelation, though it is apparently incumbent upon us to remind the headline writers and scaremongering futurologists that this is the case, and that the robots they imagine populating our future have more in common with the monsters that populated our past—like vampires and ghosts—than they do with any real robots being designed in (most) contemporary laboratories. And of course it is worth adding that if we do fear autonomous robots, perhaps if a military experiment in autonomous weaponry gets out of control, it is not the robot or the artificial intelligence that we need to fear or blame but our own warmongering tendencies.

The next step in applying a study of projections to how we regard robots and AI then is to look in more detail at the very specific shape that our metallic monsters take. When we imagine robots, why do we imagine *these* robots, that look and act the way they do?

When we look at The Terminator, for example, or *Star Trek*'s Borg, *Dr. Who*'s Cybermen and the Daleks, *Westworld*'s out of control machines and the countless others, we are looking at our own, very human, violent fantasies, reflected in gleaming metal. The violence exhibited by these robots is our violence, projected out and placed into an other. And without wishing to exclude the more practical, perhaps "materialist" reasons for the humanoid design of such creatures (such as the poverty of imagination or simply because in a film or television programme it is easier to dress a man up in a suit than to build an actual robot), an understanding that robot monsters are reflections of humans might also explain, at least in part, why our robot monsters so often look so much like us.

But bad robots and AI are not just violent. There are many other characteristics that are common to such monsters. They are, for example, relentless: they cannot be easily killed, and are uniquely determined in pursuit of their victims. They are also—and this is something that clearly distinguishes them from many of the other monsters of our past—entirely and purely rational: these monsters are not guided by hatred or vengeance or a fanatic devotion to irrational mythology, as with spectres or savage brutes or religious extremists. These soulless robotic monsters are guided by a single principle: their violence and destruction is completely and utterly based in a calculated, indisputable logic, a resolute dedication to their technological, rational, scientific programming.

And it is in these qualities that we may learn something more about what it is in contemporary human beings that we deem, at least unconsciously, to be "bad," or of which we may at least (again, unconsciously) be suspicious. We might notice, for example, that robots are not the first monsters to exhibit these properties. If one were to go a little further back in history, to a time before robots, strictly speaking, existed, one would find Frankenstein, and one might find other monsters, such as Mr. Hyde, zombies, and other spectres. What these monsters all have in common is that they are the consequences of human industry, science, and rationality.

Robots seen in this context are part of a tradition of monsters that reflect some underlying fears about the direction of human development since the Industrial Revolution and the turn to Enlightenment rationality (cf. Kang 2011). Mary Shelley's 1818 novel, *Frankenstein*—subtitled *The Modern Prometheus*—is intriguing in this respect because, like so many of the stories about robotic monsters that followed it, far from being a demonization of technology, progress, or rationalism, Shelley's novel, is more ambivalent about technology, on the one hand lionizing (as Romantics were wont to do) the courage of the Promethean hero while at the same time demonstrating that there might be unforeseen, negative consequences to this shift in the newly conceived human and its aspirations: the inflexible logic of reason, the ruthless efficiency of industry, the emotional detachment of science might, *Frankenstein* and its descendants suggest, be creating new monsters.

The unremitting rationality of robots can be regarded as an expression of an anxiety about our own human reason, applied without more "human" checks of empathy. Since the inception of the very term robot, in Karel Čapek's 1920 play *R.U.R (Rossum's Universal Robots)*, it has always been understood that there is something essential about these manufactured humanoids—that otherwise resemble us to such a completely unnecessary extent—that is missing: that despite their ability to reason and work and being in every other respect the equal of, or even superior to, their human masters, something that makes these artificial beings *less than human*. And this missing element reflects our own fears that *we ourselves* are becoming something less than human, and that we are destroying some essential part of our humanity as we become more and more committed to reason and logic, or as we become Rossum's *robota* (*robota* is the Czech word meaning "forced labour," and *rozum* means "reason"). Robots have always, therefore, reflected concerns about the *dehumanization* of our species, as a consequence of mass industrialization, the increasing mechanization of warfare, and the loss of certain pastoral conceptions of human experience in an increasingly urbanized world of technological innovations. And we find that this emotional deadness is true of our robots even when they are not "evil": in *Forbidden Planet*, the *Alien* series, *Star Trek*, in Asimov's writings generally, and a host of other, more ambivalent or sometimes even "good" robots, robots are portrayed as devoid of emotion, lacking empathy or feeling, and governed only by reason and a predictable, programmable intellect.

And to make them all the more terrifying, because they become containers for *all* of our negative impulses, the monster robots are not just incessantly, unceasingly rational; because they are also containers for all of those other, non-rational impulses that have plagued human (un)consciousness for so many millennia, robots are all the more monstrous because they represent both these anxieties about the consequences of *reason*—the inflexible doctrine of rationality, the precision and productivity of technology, the dispassionate methodology of science—*and* our savage, *animal* impulses: the *irrational* violence, the *impulsive* desires for domination and control. (Interestingly, just as we try to disavow callous rationality as "inhumane," it was not so long ago that we used to try to dismiss our barbaric, impulsive side as also "inhuman"—this, also, reflects the important shifts in what we deem to be "human.") And because these are really *our fears about ourselves*, these robots are truly inescapable: we run and run but the anxieties and fears keep coming; there is no hiding from the "unthought known" (Bollas 1989), that it is not actually *them*, it is *us*.

A terrific example demonstrating how all of these anxieties operate through projections in science fictional representations of robots can be seen in *Do androids dream of electric sheep?*, Philip K. Dick's 1968 novel that was the basis for Ridley Scott's 1982 film, *Blade Runner*. In both the novel and film, the main character is Rick Deckard, a bounty hunter, the epitome of the outsider, the solitary, rational, dispassionate killer. Despite this, Deckard believes that it is *the humanoid robot*—the "andy"—that is "a solitary predator." The narrator tells us, "Rick liked to think of them that way; it made his job palatable" (Dick 2010). Throughout the book, Deckard is shown to regard the andys with contempt and justifies his own ruthlessness with the comforting rationalization that the andys are dangerous and simply must be exterminated, whatever evidence (including his own personal feelings) might suggest to the contrary. And for this reason, both the novel and the film, in different ways, show that the line between human and machine (or pet and electric sheep) is blurring: it is not just because robots are being built to look and sound more like humans. It is also (equally, perhaps more so) the case that human beings are starting to look and act more like machines.

A note on messiahs

Though it perhaps used to be enough to ask simply why we are afraid of robots, it is becoming clear that while that is still an important question, we need a more nuanced approach that takes into account more than evil robots and our human fears. Because we are now not only afraid of robots. As explained in the introduction, robots and AI today are sometimes not only monsters out to destroy the human race, but are also sometimes our last and best hope for salvation. The tagline for Neill Blomkamp's 2015 film, *Chappie*, for example, states: "Humanity's last hope isn't human." Or witness, for example, the evolution of the iconic *Terminator* series: In the first film, released in 1984, Arnold Schwarzenegger's T-800 Model 101 is quite clearly, unambiguously, the Bad Guy, trying to kill John Connor's mother so that humanity's great resistance leader is never allowed to be born. But already by 1991, in the second film, *Terminator 2: Judgment Day*, Schwarzenegger's cyborg is on the side of humanity, protecting John Connor from other evil robots. (And without wishing to give too many spoilers, by 2009, with *Terminator: Salvation*, it is clear from the title that robots are going to have some part to play in defeating the monsters that we ourselves have created.)

A similar expression of hope can be seen in Daniel Wilson's *Robopocalypse* (2011), a novel more intelligent than most in considering how the Singularity might actually come about and, inevitably, go horribly wrong. As with the later Terminator films, Wilson's novel shows that though our reliance on technology may pose a threat to humanity, and that the inevitability of sentient AI will most certainly try to wipe humanity from the Earth, the only real hope for humanity lies not in eschewing technology but embracing it, by recruiting good robots to battle the bad, and by enhancing humans with cybernetic implants and advanced prosthetics. (It is interesting, too, that even in a novel as intelligent as Wilson's that sentient AI inevitably, necessarily becomes genocidal, as if the desire to kill and destroy was somehow inextricably linked to consciousness; it would be interesting to consider whether that is really the case, but it seems to be so in the human imagination.)

Part of the explanation for such a transformation in the role robots play in our fictions and fantasies must certainly lie in the increasing realization that we are no longer simply "human," at least certainly not as we once imagined ourselves to be when human beings came to fear mechanical beings as an existential threat. At least part of this increasingly prevalent trend to regard robots, or technology more generally, as our salvation must come from a realization that technology is not something entirely "other" to human existence, but an essential and inextricable part of our posthuman existence (cf. Braidotti 2013, Ferrando 2013, Hayles 1999). We are now posthuman in two important ways: we have, on the one hand, moved beyond the old humanist beliefs about who we are, the ideas of a singular being and communities divided along strictly binary oppositions, such as man/woman, human/machine, consumer/producer. These are the assumptions upon which the Enlightenment conceptions of human beings rested; this is the human of the (first) Industrial Revolution, the human whose existence was threatened by the march of the machines. On the other hand, more pragmatically, it is more obvious that human beings in the twenty-first century are beings increasingly integrated with our technologies in very dramatic ways. The ubiquity of our mobile phones, the (inter-) connectedness of our world, and the automation of everyday tasks sit alongside improvements in prosthetics, implants, and enhancements that make it very hard for even the most die-hard Luddite puritan to maintain belief in a "natural" human being unimpacted by technology. The acceptance of robotic saviors in our fiction reflects this inescapable belief that humans are technologically enhanced beings, and that this might not entirely be a bad thing.

However, we needn't abandon the notion of projection altogether to understand the heroic robot phenomenon. The strength of the Freudian conceptualization of projection is that it equally can be used to understand not only how we imagine robotic monsters, but also how we can regard robots and AI in an idealized way, as not the demons and harbingers of destruction but as saints and our only hope for salvation. It soon becomes apparent, however—or perhaps unsurprisingly, if we are still looking through a Freudian lens—that at the root of our idealized images of robots is not a spark of optimism but additional anxieties. As explained above, good parts of the self are often projected in much the same way as bad parts. This is sometimes done for the sake of group identification, but it can also be done for other reasons: for example, in order to keep safe certain parts of the self that are imagined to be under threat, or in order to idealize an external object, so that object can be imagined as wholly good and pure. These projections and defence mechanisms are not, of course, unrelated, and we can find many examples of humans caught in the vicissitudes of idealization and demonization.

In this context we might regard, for example, the more optimistic, some might say "radical," aspiration of transhumanists. Whatever miracle technology and advances in healthcare that are realized in the coming decades, the very hope that humans might be able to profoundly extend our life expectancy into the hundreds of years, can been seen within the context of our fear of death. That it might one day be possible to replace parts of the human body with equally efficient, or even super-human, technological devices (as in Wilson's novels *Amped* or *Robopocalypse*), or that we might be able to upload our entire consciousness into the cloud (as in the 2014 film *Transcendence*), to be downloaded into another body or robot avatar (as in the 2009 films *Avatar* and *Surrogates*) is even a more fanatical defence against our anxieties about our limited bodies, and can be regarded in much the same way that fantasies of spiritual transcendence and the afterlife dominated the cultural imagination in the days before secular technology made it possible to imagine something a little less far-fetched.

Conclusions and implications

If good parts of the self are projected into a robot, or into technology more generally, or the very idea of a technological salvation, that container must then be thought of as pure and all good, so that the good parts of the self are not corrupted by any negative associations. This idealization of the good container—the necessity that it be *all* good—leads to a number of other, more and more radical defences, as the impossible task of imagining something completely perfect forces one into more and more desperate fantasies, and further away from a realistic conception of the object and of the world as ambivalent.

Defences can be seen as much in the irrepressible optimism that some express towards a utopian conception of the future as they are evident in the indestructible pessimism of others who can only imagine the dystopian consequences of our new technologies. And there is a further problem in that projecting positive qualities to create an ideal object necessarily implies the creation of an antithesis, that there is also somewhere else the projection of negative qualities in the creation of a monster. Once caught in the vicissitudes of projection as a defence against anxiety, it is difficult to escape, so an unrealistic idealization of robots, AI, and technology more generally cannot be thought of as a remedy for the demons that plague our cultural nightmares.

In reality, of course, the truth must lie somewhere between the two extremes, between monster and messiah, and indeed for most people their feelings towards technology do embrace both unbridled optimism and pessimism. However, to judge by our cultural products, there is a tendency to swing from one extreme to the other, and it is too seldom that one sees a sober, mature assessment of technology that acknowledges that conflicting possibilities need to be negotiated. It is this more nuanced, more ambivalent view of robots, AI, and technology toward which we must strive.

Anxieties do not have to be based on truth to have real consequences. In fact, sometimes the more *outrageously* fantastical the anxiety, the more it takes control of us. However, anxieties are very rarely useless, and always tell us something very interesting and profound about our relationship with ourselves and with our world, or at least what we perceive our relationship with ourselves and the world to be. In-depth analyses of these representations of monstrous and messianic robots can therefore offer great insight into the fears—genuine or naïve—that the public harbours towards new advancements in technology, and the human place in an increasingly technological world. These insights can inform those designing living machines as to how their inventions might be received.

Acknowledgments

The author received support in preparing this chapter from the EU Horizon 2020 FET Flaghip Human Brain project (HBP- SGA1, 720270), and from the UK Art and Humanities Research Council grants "Cyberselves in Immersive Technologies" (AH/M002950/1) and "Cyberselves: How Immersive Technologies will Impact our Future Selves" (AH/R004811/1).

References

Anderson, D., Borman, M., Kubicek, V., Silver, J. (Producers), and McG (Director). (2009). *Terminator: Salvation* [Motion picture]. United States: Warner Bros.

Anzieu, D., and Turner, C. T. (1989).*The skin ego*. Yale: Yale University Press.

Asimov, I. (2004a). *I, Robot (The Robot Series)* [Kindle edition]. Retrieved from Amazon.co.uk

Asimov, I. (2004b). *Robot Dreams (The Robot Series)* [Kindle edition]. Retrieved from Amazon.co.uk.

Bion, W. R. (1962). A theory of thinking. *Int. J. Psychoanal*, **43**, 306–10.

Blomkamp, N. (Producer/Director/Screenwriter), and Kinberg, S. (Producer). (2015). *Chappie* [Motion picture]. United States: Columbia Pictures.

Bollas, C. (1989). *The shadow of the object: Psychoanalysis of the unthought known*. New York: Columbia University Press.

Braidotti, R. (2013). *The posthuman*. Cambridge: John Wiley & Sons.

Cameron, J. (Producer/Screenwriter/Director), and Hurd, G. A. (Producer). (1991). *Terminator 2: Judgment Day* [Motion picture]. United States: TriStar Pictures.

Cameron, J. (Producer/Screenwriter/Director), and Landau, J. (Producer). (2009). *Avatar* [Motion picture]. United States: Twentieth Century Fox.

Čapek, K. (2004). *R.U.R. (Rossum's Universal Robots)* (Penguin Classics) [Kindle edition]. Retrieved from Amazon.co.uk.

Caporael, L. R., and Heyes, C. M. (1997). Why anthropomorphize? Folk psychology and other stories. In: R. W. Mitchell, N. S. Thompson, and H. L. Miles (eds), *Anthropomorphism, anecdotes, and animals*. Albany, NY: SUNY Press, pp. 59–73.

Carroll, G., Giler, D., Hill, W. (Producers), and Scott, R. (Director). (1979). *Alien* [Motion picture]. United States: Twentieth Century Fox.

Cohen, K., Johnson, B., Kosove, A. A., Marter, A., Polvino, M., Ryder, A., Valdes, D. (Producers), and Pfister, W. (Director). (2014). *Transcendence* [Motion Picture]. United Kingdom: Alcon Entertainment

Dick, P. K. (2010). *Do androids dream of electric sheep?* (SF Masterworks) [Kindle edition]. Retrieved from Amazon.co.uk.

Dixon, B. (1999). The paradoxes of Genetically Modified Foods: A climate of mistrust is obscuring the many different facets of genetic modification. *BMJ: British Medical Journal*, **318**(7183), 547.

Duffy, B. R. (2003). Anthromorphism and the social robot. *Robotics and Autonomous Systems*, **42**, 177–90.

Ferrando, F. (2013). Posthumanism, transhumanism, antihumanism, metahumanism, and the new materialisms: differences and relations. *Existenz*, **8**(2), 26–32.

Handelman, M., Hoberman, D., Lieberman, T. (Producers), and Mostow, J. (Director). (2009). *Surrogates* [Motion picture]. United States: Touchstone Pictures.

Hayles, N. K. (1999). *How we became posthuman*. Chicago, IL: University of Chicago Press.

Heider, F., and Simmel, M. (1944). An experimental study of apparent behavior. *The American Journal of Psychology*, **57**, 243–59.

Hurd, G. A. (Producer), and Cameron, J. (Screenwriter/Director). (1984). *The Terminator* [Motion picture]. United States: Orion Pictures.

Hurd, G. A. (Producer), and Cameron, J. (Screenwriter/Director). (1986). *Aliens* [Motion picture]. United States: Twentieth Century Fox.

Heimann, P., Isaacs, S., Klein, M., and Riviere, J. (eds). (1989). *Developments in psychoanalysis*. London: Karnac Books.

Kang, M. (2011). *Sublime dreams of living machines: the automaton in the European imagination*. Cambridge, MA: Harvard University Press.

Kiesler, S., Powers, A., Fussell, S. R., and Torrey, C. (2008). Anthropomorphic interactions with a robot and robot-like agent. *Social Cognition*, **26**(2), 169–81

Klein, M. (1997). *Envy and gratitude, and other works 1946–1963*. London: Random House.

de Lauzirika, C. (Producer), and Scott, R. (Director). (2007). *Blade Runner: The Final Cut* [Motion picture]. United States: Warner Bros.

Lazarus, P. (Producer), and Crichton, M. (Screenwriter/Director). (1973). *Westworld* [Motion picture]. United States: MGM.

Rayner, E. (1991).*The independent mind in British psychoanalysis*. London: Jason Aronson.

Shelley, M. (2003). *Frankenstein: Or, the Modern Prometheus* (Penguin Classics). London: Penguin.

Wilson, D. H. (2011). *Robopocalypse* (Robopocalypse series Book 1) [Kindle edition]. Retrieved from Amazon.co.uk.

Wilson, D. H. (2012). *Amped* [Kindle edition]. Retrieved from Amazon.co.uk.

Young, R. M. (1994). *Mental space*. London: Process Press.

Chapter 62

The ethics of virtual reality and telepresence

Hannah Maslen and Julian Savulescu

Oxford Uehiro Centre for Practical Ethics, University of Oxford, UK

We usually experience and act on the physical world directly—such experiences and actions being mediated only by our senses and our corporality. Immersive biohybrid technologies—in particular, virtual reality and telepresence—offer opportunities for acting in virtual environments, and for materially engaging with the physical world at a distance. Current and prospective examples include forming and maintaining friendships with virtual agents in virtual worlds, and interacting with physical individuals at remote locations via a robot avatar, whilst receiving multisensory feedback. Such technologically mediated experiences and interactions raise traditional philosophical questions concerning the value of different experiences and the construction and continuity of agency. They also present novel ethical challenges pertaining to moral responsibility for action at a distance and ethical codes governing virtual acts. Drawing on a variety of examples of immersive technologies, we outline two traditional philosophical questions as they pertain to such technologies: we examine the value of virtual experiences and the extent to which the virtual agent is continuous with or represents an extension of the physical agent. We then consider the ethics of technologically mediated interactions, particularly when only one agent is acting at a distance.

The value of virtual experiences

Although virtual environments, such as the virtual world *Second Life* and massive multimedia online role-playing games (MMORPGs) are relatively new, philosophers have used thought experiments involving forms of virtual reality for decades. The purpose of such thought experiments is to try to illuminate what is of value to the lives and wellbeing of human beings, and to consider the possibility that we might not know as much about ourselves and the world as we believe we do. In 1974, Robert Nozick used the Experience Machine thought experiment to argue against the view that human wellbeing consists only in pleasurable experiences, no matter how complex or apparently real. He compared an entirely simulated life, full of pleasurable simulated experiences, with a non-simulated life, with fewer pleasurable (but "real") experiences. Based on his intuition that the non-simulated life is preferable to the simulated life, notwithstanding the greater amount of pleasure experienced in the latter, Nozick concluded that maximizing pleasure and minimizing pain, even from complex experiences, cannot be the only things that bear on an individual's wellbeing. In addition to experiencing pleasure, Nozick argues that we value *doing* certain things, *being* a certain sort of person, and having contact with *deeper reality* (Nozick 1974, p. 43).

Versions of the brain-in-a-vat thought experiment (see, for example, Putnam 1981) are used to challenge our firmest-held convictions about what we think we know about the world and ourselves. This thought experiment asks us to imagine that we are mere brains in vats, receiving sensory impulses identical to those that we would receive if we were really interacting with the world. Since we cannot know if this is in fact our situation, we cannot know for sure whether our beliefs about acting on an external world are true or false. Nick Bostrom has extended this kind of approach to ask whether we might be living in a computer simulation (see Bostrom 2003).

Whilst not as extreme as the philosophers' thought experiments, virtual experiences might nonetheless raise related questions: "Are virtual experiences good for us?"; "Are we exposing ourselves to delusion when we become immersed in virtual environments?"; "How (if at all) do our virtual experiences relate meaningfully to our real lives?" The main difference between the Experience Machine and brain-in-a-vat types of thought experiments on the one hand, and plausible virtual reality on the other, is that when we use immersive technologies we know that this is what we are doing. In almost all cases, we do not confuse the virtual with the real, even when we become temporarily psychologically immersed. Moreover, our virtual activities often do have consequences for the real world, either for ourselves or for others, even when our actions occur in the virtual environment.

Cogburn and Silcox (2013) consider Nozick's three concerns—viz. that a simulated life involves no "doing," "being," or connection to "reality"—and argue that plausible immersive technologies will in fact not necessarily render lives deficient in any of these respects. They point out that the purposes and mechanisms of plausible immersive technologies will allow us to "'do' things in the actual world as well as the virtual world" and therefore that our engagement with such technologies will not involve mere passive experience. An agent's interaction will not only constitute "doing" in the virtual environment, but will necessarily also involve "doing" in the non-virtual world, in the form of bodily movements, for example on Wii Fit or Just Dance (Wii). We might add that there is a valuable sense of "doing" that obtains when, for example, we communicate with others in virtual environments, regardless of how much physical movement is required to do so.

Moreover, in relation to "being," the sorts of virtual reality technologies that are likely to be developed and adopted will require users to make decisions, express themselves, and react meaningfully to events and other agents. Indeed, as will be seen below (cf. Young and Whitty 2011), one could plausibly use virtual environments to express and develop an ideal version of the self—to intentionally practice *being* a certain way.

Connection to "deeper reality" is clearly achieved as soon as other agents enter the equation. In relation to such experiences, there is not only sufficient novelty, but also the prospect for the sort of novelty that we tend to think is valuable: conversations and relationships become possible. Further, the addition of other agents generates the possibility that actions within the virtual world have external consequences (see discussion of Søraker 2012 in "Ethical codes in virtual environments" below). The prospect of virtual actions having external, morally significant consequences suggests that the virtual experience can, in some cases, have the sort of depth and import that Nozick believes is lacking in the simulation produced by the Experience Machine.

However, even if Experience-Machine-type concerns are allayed, there could still be questions about the place that virtual experiences should have in our lives and the amount of value that can attach to them. Despite knowing that we engage with mere simulation, we still might over-value virtual experiences and activities. Are virtual friendships as valuable as real friendships? Might we lose track of what we desire and strive for in the real world if we place too much emphasis on our activities in a virtual world? Might agents sometimes find it difficult to disentangle "fantasy" from "reality" with respect to their characteristics and accomplishments? Does it matter if the virtual agents with whom we interact are completely divorced from the realities of the physical agents behind them?

Of course, virtual experiences are a heterogeneous set, and their value to individuals and society will vary greatly. For example, being a pilot in an online world may have similar value to a game, whereas using an immersive simulator to learn transferable aviation skills will have a very different value. Remotely piloting a drone to drop aid packages in a crisis will be of different value still. Concerns might primarily be directed to those types of virtual experiences that might be thought to be a misuse of time or those that are used as an impoverished substitute for a real life experience. However, this sort of concern is not peculiar to virtual reality and depends heavily on one's conception of the good life, goals, and values.

The value of virtual interaction in close relationships

Whilst it seems clear from the above that interaction with others in the virtual world can be of some value—indeed, it is the prospect of interaction with other agents that confers on the virtual some of the value associated with its connection to reality—we might think that these interactions are impoverished in important ways and, being such, are less valuable than "the real thing." This worry is especially salient in relation to close personal relationships, which are often partly sustained and expressed by acts involving physical intimacy. Given the absence of physical contact in virtual interaction, would such interactions be of limited value in the context of close relationships? One of us has previously examined the prospect of how technology might change the nature of sex and reproduction (Savulescu 1999). There is great potential for technology to facilitate reproduction and provide opportunities that were not previously available; for example, opportunities for selecting embryos without certain genetic conditions that reduce wellbeing.

However, whether new technology will make such a positive contribution to relationships themselves is a more complex question. It is plausible that some virtual interactions, even those between close friends, have the potential to be as valuable as equivalent non-virtual interactions. Indeed, there will be many activities in which friends engage that could be performed equally well in a virtual environment, such as Facebook. In such cases, the virtual activity will satisfy all the conditions of the activity for which it is a substitute. Such virtual interactions will therefore, in a sense, not be substitutes. For example, where what you want to do with your friend is to "have a conversation," a smooth-running video call might enable the interlocutors to have precisely the interaction that constitutes "having a conversation." It might even be possible for friends to "spend time together" via such means, where "spending time together" is supposed to consist of more than a mere exchange of information—perhaps re-familiarizing oneself with the other's personality and nurturing an emotional connection are part of what one does when one spends time with another person. To the extent that the virtual activity consists in the same interpersonal dimensions as the "real" activity, it is likely to be of just as much value.

The potential limits of virtual interaction will become clear when we consider that such interactions may fail to satisfy the conditions necessary for the activity to constitute a particular intimate activity. It may therefore be correspondingly less valuable. For many examples of these activities, physical contact is necessary. Even for intimate interactions not involving sex, the touch or mere physical presence of the other is arguably necessary for it to play the requisite role in the relationship and thereby carry the associated value. Comforting your partner with an embrace, for example, generally requires physical contact. The sorts of technologies available today for remote interaction would enable only a poor substitute for the comfort of a caring embrace, devoid of any physicality. Indeed, long-distance relationships can sometimes suffer as a result of the lack of an adequate substitute for physical intimacy.

Future possibilities for remote contact, especially those involving robot avatars, will raise fascinating possibilities for physical intimacy at a distance. The contribution that such interactions could make to individuals' subjective wellbeing will be an interesting psychological line of inquiry. Deeper philosophical questions will be raised about the nature of such activities and whether they can ever constitute—satisfy the conditions for—certain species of physically intimate activities (in the way that a Skype conversation can fully constitute a conversation), or whether such virtual interactions must be understood as different (albeit potentially valuable) kinds of activities. Further questions will be raised about the authenticity of the real and/or virtual participants, especially given the prospect of altering characteristics. We explore the authenticity of the virtual agent in more general terms in the following section.

The authenticity of the virtual agent

Closely related to the discussion of the need to "be a certain way"—and the potential threat to this interest domain that virtual reality poses—is the idea that immersive technology might render the actions of the virtual agent inauthentic or, at least, not continuous with those of the "real life" agent. The concern is not that the agent herself becomes inauthentic, but rather that her activities in the virtual environment are not her own: thus, the question is not so much whether the use of immersive technologies renders the self in someway inauthentic but, rather, whether one can be authentically oneself in virtual environments. If I visit you as my robot avatar, have you really met me, and are the actions I make mine?

In relation to these sorts of questions, Young and Whitty (2011) argue that one's "embodiment" as a virtual agent is authentic if the salient features "transcend" the virtual domain. They draw a distinction between an *ideal* realization of embodiment—which is "an authentic presentation of the self" that is "typically realized by the person and/or that which he/she has the potential to realize"—and an *idealized* realization of embodiment—which they equate with "inauthentic embodiment" (Young and Whitty 2011, p. 539). An example of the latter might be to present oneself as possessing extraordinary athletic prowess, despite lacking the potential to realistically meet this ideal. Having made this distinction, Young and Whitty's central claim is that, by developing one's ideal rather than idealized self in the virtual realm, the authenticity attached to this identity is able to transcend domains, meaning that an individual is not restricted by where he/she can express it—i.e. it can persist beyond the virtual domain. If one develops an idealized persona, on the other hand, one only possesses authenticity within the context in which the idealized persona is developed—a persona with only context authenticity is essentially "trapped" in that domain.

This view is particularly persuasive when the sorts of features one can express and develop in the virtual agent are analogous to those that at least some human beings could express and develop in the real world. It seems appropriate to assess the authenticity of a virtual agent with respect to its physical characteristics, skills, and personality traits, for example. If one intends to form friendships and relationships in the virtual world and then continue to develop them in the non-virtual world, it will matter that one's authenticity as a virtual agent transcends into the real world; that one at least has the potential to instantiate in the real world the salient characteristics that matter when forming friendships and relationships.

However, there will potentially be telepresence technologies that allow or even require agents to instantiate features that do not correspond to features that any human being could instantiate.

Robotic telepresence, as envisaged by the BEAMING project (see Longo et al. 2010) would permit or perhaps even require agents to possess robotic characteristics and skills that bear little if any correspondence to the characteristics and skills that even an idealized human being could possess. How authenticity might transcend domains using such technology will be a task both for philosophers and psychologists. Especially if important interactions or decisions are conducted via such technology, we will need to have a clearer idea of what might constitute the features central to an authentic persona that can transcend this context. Attention should also be paid to the possible scenario wherein an agent feels most authentic in a virtual persona that does not transcend to the non-virtual world.

Instead of considering discrete contexts and assessing the extent to which a persona transcends them, an alternative approach to the authenticity of the virtual agent is to instead think about the holistic context of the particular person and how all her activities—virtual and non-virtual—bear on her life and wellbeing. Schechtman (2012) suggests that instead of asking whether the narrative of the avatar is the same as the narrative of the user, we should instead ask whether these distinguishable but interrelated narratives are both part of a single, broader person-narrative. The answer to this question, she argues, is that sometimes the offline narrative of the user and the online narrative of the avatar are essentially "subplots in the more comprehensive narrative of the resident, a person who lives sometimes offline and sometimes online." Both sets of activities are part of the life of the same person because, although distinguishable sub-narratives, they influence each other in significant ways.

Ethical codes in virtual environments

It is not immediately clear whether normal ethical codes of conduct apply in virtual settings and, in particular, whether the sort of conduct that is considered immoral in the real world is equally immoral in these settings. Some virtual environments, even those involving interaction with other agents, are essentially games. In many computer games, it is not immoral to kill a character (although virtue ethicists might have something to say about how enjoyment of such conduct in a game might reflect on an agent's character), and similar permissions are likely to apply in virtual environments such as MMORPGs. However, although some virtual environments have similar properties to games, others approximate social environments or serve a more instrumental purpose. Some virtual environments involve other agents and can have effects on their reputations, digital property, and wellbeing, amongst other things. Given the diversity of purpose and effects, there will not be such a thing as an all-purpose cyber ethics, but assessment of what is morally required or impermissible will depend on many factors.

Providing a framework for considering the permissibility of actions in virtual environments, Søraker (2012) proposes a fundamental distinction between "intravirtual" and "extravirtual" consequences. According to this distinction, virtual actions and events produce effects both within the virtual world and in the external, non-virtual world. For example, a sincere virtual relationship between two avatars has the effect of giving the avatars a certain status in relation to each other and perhaps generating obligations and expectations within the virtual world. However, as an extravirtual consequence of forming this relationship in the virtual world, the agent might neglect her responsibilities in the real world or provoke jealousy from a real-life partner. Disrupting a virtual event—such as an avatar giving a public talk—might have the simple effect that the event is cancelled in the virtual world. Consequences in the real world, however, could be wide reaching; perhaps the virtual event could have lead to a real-life job opportunity for a key participant. This way of thinking about the effects of virtual acts and events is not supposed to invite the conclusion that virtual acts are always, or even often, likely to cause harm. Rather, it prompts us to consider that one and the same virtual act can have distinct and disparate effects both within and without the virtual domain.

Indeed, whether a particular virtual act causes any harm at all will often be hard to know and will be contingent on the ways in which others interact with the virtual. Søraker claims that this subjective element is precisely what makes ethics so difficult online: the "invisible" user behind the avatar with which the agent communicates comes with a set of mental states that often determine the extravirtual effects of the intravirtual state of affairs. The nature of the extravirtual effects are heavily dependent on the subjective features of those behind the avatars, which vary far more greatly than the subjective features relevant to straightforward instances of harm in the real world. Similarly, Wolfendale (2007) argues that causing harm to avatars is analogous to causing harm to other objects to which we form attachments (such as possessions, other people, and ideals), such instances of attachment being similarly contingent. Further, extravirtual effects are often not accessible to us, so we are unable to monitor all the effects of our actions. Søraker suggests that this entails that those interacting in virtual environments should remind themselves often of the extravirtual "other," in order to recognize that their virtual acts have potentially significant extravirtual consequences.

Considering morality in virtual reality and telepresence more broadly, we suggest that much will depend on the *purpose of the virtual environment* and the likely *nature of the consequences* that one can precipitate through one's virtual actions. The ethical codes pertaining to a surgeon operating via a virtual interface will differ greatly from those that pertain to operating a business in Second Life, which will differ still from those governing MMORPGs. It might be useful to appeal to the distinction between activities that have their value in their process (such as games) and activities that have their value in their outcome (such as surgery). Arguably, the more a virtual activity is *aimed at a socially valuable outcome*, the more it matters how the agent acts (cf. Santoni de Sio et al. in press). For example, in an online environment like Second Life, much of the value of participating in this activity resides in the entertainment one derives from it—its value is in the process. Here it might not matter if one lies about one's appearance, job, likes, or dislikes. In contrast, the very same virtual environment could be set up as a dating platform, where the purpose shifts to the outcome of establishing viable relationships. In such an environment, it might matter a lot (and hence be morally impermissible) to lie about one's self and situation. Correspondingly, the more it matters how an agent acts, the more he will be morally required to perform some acts and to refrain from others. Indeed, the more one's role in a virtual environment is intended to serve as a virtual substitute for a real-life role, the more likely it is that the same moral constraints on one's actions will apply.

Acting at a distance

There are plausible examples of telepresence in relation to which the distinction between intravirtual and extravirtual effects will not apply, and where immediate feedback about the effects of one's acts will be available. Attending remote locations as a robot avatar provides one such example (see Longo et al. 2010). Here, the ethical challenges will revolve less around one's acts having effects in distinct domains, and more around the psychological and technological features of acting on the world via this medium. In particular, such technologies will raise questions about responsibility for action given technological and psychological limitations. Questions about how the robot avatar should be treated will also raise challenges for the traditional distinctions between bodies and property.

Physical limitations and responsibility

Acting on the world remotely is not like acting on it directly. The agent's movements are "translated" into mechanical movements and may not be replicated exactly as intended or may be limited. The visual, auditory, and sensory feedback delivered to the agent may not exactly reproduce the sensory information available in the remote location, yet the agent will substantially base his or her decisions for action on this information. In many familiar cases, if an agent causes harm using a tool, it is no defense to point out that a tool mediated one's actions—tool use per se is not a buffer to responsibility. Further, if a tool is faulty or difficult to work with, depending on how much one would be expected to know, it is often thought that the agent should act cautiously in the context of faults or difficulty. If I drive my car despite knowing that the steering is not responding correctly, then I cannot claim in my defense that the car was difficult to drive when I crash into a pedestrian. However, if I reasonably believed that my car was in full working order but then the steering failed, my responsibility for any ensuing harm would be mitigated. A similar approach to mechanical malfunction and limitation might be taken to operating robot avatars. It will be important to consider what operational requirements should be imposed to best ensure responsible use.

In general, moral responsibility is a function of the intentions and foresight the agent had with respect to the harm risked or caused, and the avoidability of that harm, together with whatever duties of care exist (see Oakley and Cocking 1994). Such principles would extend to telepresence. Whether a manufacturer is liable for some portion of the harm, or whether responsibility attaches fully to the agent who assumed the risks associated with using the technology, will in large part depend on the respective duties of care in the situation, the user's expectations relating to the functioning of the technology, any abnormal use of the technology, and the degree of foreseeability of the harm (see Sexton et al. 2012). How blameworthy the agent is will be a function of the degree of moral responsibility (potentially attenuated by any shared responsibility of the manufacturer) and the degree of harm that is caused.

Psychological limitations and responsibility

The potential responsibility confounders listed above are artifacts of the technology and the potential limitations attached to use of that technology. However, such mechanical shortcoming will not be the only possible complicating factor when acting at a distance. There will also be a psychological distance that is likely to attend the use of technology that mediates remote action. Especially as people are still acclimatizing to using these technologies, there will be psychological barriers that will have to be overcome before the moral reality of the remote interaction is as salient to the agent as it usually is in immediate environments. We did not evolve to be capable of such distant effects, and their significance may not be as immediately intuitive as the significance of our close-range effects. Although we may rationally understand and believe that our actions "here" can have significant effects "there," it may take time and experience until this is also "felt." The development of robotic telepresence technology will have to be accompanied by psychological research on individuals' sense of agency at a distance, for example when there is a time delay or an interruption in feedback. Whilst this will not necessarily reduce moral responsibility, such failures to accurately perceive one's agency may present practical challenges to agents full accepting responsibility for action at a distance.

Harming the robot avatar

One final ethical topic relevant to action at a distance is how to conceptualize and respond to harm caused to robotic avatars. It could be argued that the actions of agents through robots is fairly straightforward (not withstanding the mechanical limitations and psychological distance discussed above). Causing harm using a robot avatar, it could be argued, should be seen as analogous to using a weapon or drone. However, harm caused in the opposite direction permits no obvious analogy. As the capacity for sensory feedback (including haptic feedback) increases, harm to an avatar becomes less like harm to property and more like harm to a person. The agent operating her robot avatar will be vulnerable to aversive and non-consensual touch and possibly even to pain. The more integrated the technology, the more this will be true. But, since the robot is not a human body, it would be very difficult to conceptualize the harm caused in the familiar terms of assault, or some other offence against the person. Although possible psychological harms may be more analogous, there will remain some differences that will need identifying and analyzing. Assessing whether such harms require new moral and legal concepts will be important.

Conclusions

The possible purposes and modes of engaging with virtual reality and telepresence are myriad. From online virtual environments intended for socializing, to telesurgery, to diplomatic negotiation via a robot avatar, the range of activities that could be pursued virtually is likely be at least as great as the range of activities possible in the non-virtual world. Further, the ways in which virtual environments and acts intersect with the non-virtual world will not be uniform. As we become increasingly biohybrid, and in order to ensure ethical development and use of virtual reality and telepresence technologies, close attention must be paid to the psychology of technologically mediated action to ensure that agents are able to act as they intend and take responsibility for these actions.

References

Bostrom, N. (2003). Are we living in a computer simulation?. *The Philosophical Quarterly*, **53**(211), 243–55.

Cogburn, J., and Silcox, M. (2013). Against brain-in-a-vatism: on the value of virtual reality. *Philosophy & Technology*, **27**(4), 561–79.

Longo, M. R., Santos, E., Haggard, P., Purdy, R., and Bradshaw, C. (2010). *BEAMING Deliverable D7.1: Assessment of Ethical and Legal Issues of Component Technologies*. London: Beaming Project.

Nozick, R. (1974). *Anarchy, State, and Utopia*. Oxford: Basil Blackwell.

Oakley, J., and Cocking, D. (1994). Consequentialism, moral responsibility, and the intention/foresight distinction. *Utilitas*, **6**(2), 201–16.

Putnam, H. (1981). *Reason, Truth, and History*. Cambridge: Cambridge University Press.

Santoni de Sio, F., Faulmüller, N., Savulescu, J., and Vincent N. A. (in press). Why less praise for enhanced performance? Moving beyond responsibility-shifting, authenticity and cheating to a nature of activities approach. In: F. Jotterand and V. Dubljevic (eds), *Cognitive enhancement: ethical and policy implications in international perspectives*. Oxford: Oxford University Press.

Savulescu, J. (1999). Reproductive technology, efficiency and equality. *Medical Journal of Australia*, **171**(11–12), 668–70.

Schechtman, M. (2012). The story of my (Second) Life: virtual worlds and narrative identity. *Philosophy & Technology*, **25**(3), 329–43.

Sexton, P., Jarrold, A. S. S., and McDowell, L. (2012). Recent developments in products liability. *Tort Trial & Insurance Practice Law Journal*, **48**(1), 419.

Søraker, J. H. (2012). Virtual worlds and their challenge to philosophy: understanding the "intravirtual" and the "extravirtual." *Metaphilosophy*, 43(4), 499–512.

Wolfendale, J. (2007). My avatar, my self: virtual harm and attachment. *Ethics and Information Technology*, 9(2), 111–19.

Young, G., and Whitty, M. (2011). Progressive embodiment within cyberspace: Considering the psychological impact of the supermorphic persona. *Philosophical Psychology*, 24(4), 537–60.

Chapter 63

Can machines have rights?

David J. Gunkel

Department of Communication, Northern Illinois University, USA

One of the enduring concerns of moral philosophy is determining who is deserving of ethical consideration. Although initially limited to "other men," ethics has developed in such a way that it challenges its own restrictions and comes to encompass what had been previously excluded individuals and entities, i.e. women, animals, the environment. Currently, we stand on the verge of another fundamental challenge to moral thinking. This challenge comes from the autonomous and increasingly intelligent machines of our own making, and it puts in question many deep-seated assumptions about who or what can be a moral subject. This chapter examines whether machines—understood broadly and including artificial intelligence, software bots and algorithms, embodied robots, and biohybrids—can have rights. Because a response to this query primarily depends on how one characterizes "moral status" or the "having of rights," it is organized around two established moral principles, considers how these principles apply to machines, and concludes by providing suggestions for further study.

Moral principles

Anthropocentrism

From a traditional philosophical perspective, the question "can machines have rights?" not only would be answered in the negative but the query itself risks incoherence. This is because traditional forms of ethics, no matter how they have been articulated (e.g. virtue ethics, consequentialism, deontology, care ethics, etc.), have been anthropocentric. Under this human-centered conceptualization, machines, no matter how sophisticated their design or operation, are considered to be nothing more than tools or instruments of human endeavor. From this perspective, the bar for machine rights appears to be impossibly high. In order for a machine to have rights, it would need to be recognized as human or at least be virtually indistinguishable from another human being. This is precisely the task undertaken by the android Andrew, in Isaac Asimov's novelette *The Bicentennial Man*, who seeks to have the World Legislature declare him human, and it is the source of narrative tension in the television series *Battlestar Galactica*, where the skin-job Cylons remain persistently indistinguishable from their human counterparts. But this effort is not limited to science fiction, and researchers like Hans Moravec, Raymond Kurzweil, and Rodney Brooks predict human-level or better machine capabilities by the middle of the twenty-first century.

Although this achievement remains hypothetical, the problem is not whether machines will or will not successfully attain human-like capabilities. It rests with the anthropocentric criteria itself, which not only marginalizes machines but has often been mobilized to exclude others— women, children, people of color, etc. Because of this, moral philosophy has been critical of its own history and has sought to articulate moral status in ways that are not dependent on spurious or prejudicial criteria. Recent innovations disengage moral standing from identification with the human being and refer it instead to the generic concept of "person." In fact, we already occupy a world populated by artificial entities that are considered legal and moral persons—the limited liability corporation. As promising as this innovation is, there is little agreement concerning what makes someone or something a person, and the literature on this subject is littered with different formulations and often incompatible criteria.

In an effort to contend with, if not resolve, these problems, researchers typically focus on the one "person making" quality that appears on most, if not all, the lists—consciousness. In fact, consciousness is widely considered to be a necessary if not sufficient condition for moral standing, and there has been considerable effort in the fields of philosophy, AI, and robotics to address the question of machine morality by targeting the possibility (or impossibility) of machine consciousness (Torrance 2008; Wallach and Allen 2009). This determination, however, is dependent not only on the design and performance of actual artifacts but also on how we understand and operationalize the term "consciousness." Unfortunately there is little or no agreement concerning this matter, and the concept encounters both terminological and epistemological difficulties.

First, we do not have any widely accepted or uncontested definition of "consciousness," and the term means many different things to many different people (Velmans 2000). In fact, if there is any agreement among philosophers, psychologists, cognitive scientists, neurobiologists, AI researchers, and robotics engineers regarding this matter, it is that there is little or no agreement when it comes to defining and characterizing the concept. To make matters worse, the problem is not just with the lack of a basic definition; the problem may itself already be a problem. "Not only is there no consensus on what the term *consciousness* denotes," Güven Güzeldere (1997, p. 7) argues, "but neither is it immediately clear if there actually is a single, well-defined '*the* problem of consciousness' within disciplinary (let alone across disciplinary) boundaries. Perhaps the trouble lies not so much in the ill definition of the question, but in the fact that what passes under the term consciousness as an all too familiar, single, unified notion may be a tangled amalgam of several different concepts, each inflicted with its own separate problems."

Second, even if it were possible to devise a widely accepted and uniform definition of consciousness, we still lack any credible and certain way to determine its actual presence in another. Because consciousness is a property attributed to "other minds," its presence, or lack thereof, requires access to something that is and remains inaccessible (Churchland 1999; Coeckelbergh 2012). Although philosophers, psychologists, and neuroscientists throw considerable argumentative and experimental effort at this problem, it is not able to be fully resolved. "There is," as Daniel Dennett (1998, p. 172) concludes, "no proving that something that seems to have an inner life does in fact have one—if by 'proving' we understand, as we often do, the evincing of evidence that can be seen to establish by principles already agreed upon that something is the case." Consequently, not only are we unable to demonstrate with any certitude whether animals, machines, or other entities are in fact conscious (or not) and therefore legitimate moral persons (or not), we are left doubting whether we can even say the same for other human beings. And it is this persistent and irreducible doubt that opens the possibility for extending rights to other entities, like machines.

Animocentrism

Animals have not traditionally been considered moral subjects, and it is only recently that the discipline of philosophy has begun to approach the animal as a legitimate subject of moral concern. The crucial turning point in this matter is derived from a brief statement provided by Jeremy Bentham (2005, p. 283): "The question is not, Can they reason? nor Can they talk? but, Can they suffer?" Following this insight, the crucial issue for animal rights philosophy is not to determine whether some entity, like an animal, can achieve human-level capacities with things like speech, reason, or cognition; "the *first* and *decisive* question would be rather to know whether animals *can suffer*" (Derrida 2008, p. 27).

This change in perspective provides a potent model for entertaining the question of the rights of machines. This is because the animal and the machine, beginning with René Descartes, share a common ontological status and position—marked, quite literally, in the Cartesian text by the bio-mechanical hybrid *bête-machine*. Despite this similitude, animal rights philosophers resist efforts to extend rights to machines, and they demonize Descartes for even suggesting the association (Singer 1975, pp. 185–212; Regan 1983, p. 3–5). This exclusivity has been asserted and justified on the grounds that the machine, unlike an animal, is not capable of experiencing either pleasure or pain. Like a stone or other inanimate object, the machine would have nothing that mattered to it and therefore, unlike a mouse or other sentient creature, would not be a legitimate subject of moral concern.

Although this sounds rather reasonable and intuitive, it fails for at least three reasons. First, it has been practically disputed by the construction of various mechanisms that now appear to suffer or at least provide external evidence of something that looks undeniably like pain. Engineers have successfully constructed mechanisms that synthesize believable emotional responses and have designed systems capable of evincing behaviors that look a lot like what is usually called pleasure or pain. Kokoro's dental training robot Simroid, for instance, has been deliberately designed to cry out in pain, when dental students mistakenly hurt it.

Second, it can be contested on epistemological grounds. Because suffering is typically understood to be subjective, there is no way to know exactly how another entity experiences unpleasant sensations such as fear, pain, or frustration. Like "consciousness," suffering is also an internal state of mind and is therefore complicated by the problem of other minds. Although observation of external indicators may provide some indirect access to these internal conditions, there is no way of telling for sure whether these "expressions" come from an actual experience of pain or are simulations of pain in the form of mere external signs. Furthermore, and to make matters even more complicated, we may not even know what "pain" and "the experience of pain" is in the first place. This point is something that is taken up and demonstrated in Dennett's "Why You Can't Make a Computer That Feels Pain." In this provocatively titled essay, published decades before the debut of even a rudimentary working prototype, Dennett imagines trying to disprove the standard argument for human (and animal) exceptionalism "by actually writing a pain program, or designing a pain-feeling robot" (Dennett 1998, p. 191). At the end of what turns out to be a rather protracted and detailed consideration of the problem, he concludes that we cannot, in fact, make a computer that feels pain. But the reason for drawing this conclusion does not derive from what one might expect. The reason you cannot make a computer that feels pain is not the result of some technological limitation with the mechanism or its programming. It is a product of the fact that we remain unable to decide what pain is in the first place (Dennett 1998, p. 218).

Third, all this talk about the possibility of engineering pain or suffering in order to demonstrate machine rights entails its own particular moral dilemma. "If (ro)bots might one day be capable of experiencing pain and other affective states, a question that arises is whether it will be moral to build such systems—not because of how they might harm humans, but because of the pain these artificial systems will themselves experience. In other words, can the building of a (ro)bot with a somatic architecture capable of feeling intense pain be morally justified and should it be prohibited?" (Wallach and Allen 2009, p. 209). Consequently, moral philosophers and engineers find themselves in a curious and not entirely comfortable situation. If it were in fact possible to construct a device that "feels pain" in order to demonstrate the possibility of machine moral rights, then doing so might be ethically suspect insofar as in constructing such a mechanism we do not do everything in our power to minimize its suffering. Or to put it another way, positive demonstration of machine rights might only be possible by risking the violation of those rights.

Future directions

One of the criticisms of animal rights philosophy is that for all its efforts to think moral standing otherwise, it remains an exclusive and exclusionary practice. Environmental ethics, in particular, has criticized animal rights for including some creatures in the community of moral subjects while simultaneously excluding others—lower order animals, plants, microbes, soils, waters, etc. But environmental ethics does not do much better. It too has been cited for privileging "natural objects" to the exclusion of non-natural artifacts, like art works, architecture, technology, and machines (Floridi 2013; Gunkel 2012). For this reason, a recent innovation called *information ethics* endeavors to complete the process of moral inclusion by formulating a truly universal and impartial macroethics. Information ethics or "IE," according to Luciano Floridi (2013, p. 98), "is an ecological ethics that replaces *biocentrism* with *ontocentrism*, and then interprets Being in informational terms. It suggests that there is something even more elemental than life, namely *Being*, the existence and flourishing of all entities and their global environment, and something more fundamental than suffering, namely *nothingness*."

From an IE perspective all kinds of machines, from hammers and lawnmowers to computers and autonomous robots, would be considered a matter of moral concern insofar as all of these artifacts are "information entities" with a fundamental right to continued existence. IE, therefore, articulates a general form of ethics that is able to accommodate a wider range of possible subjects. And it is this generality that is both IE's greatest asset and a potential liability. To its credit, IE completes the project of moral inclusion. Following the innovations of bio- and environmental ethics, IE expands the scope of moral philosophy by altering its focus and lowering the threshold for inclusion or, to use Floridi's terminology, the level of abstraction (Floridi 2013, p. 30). In doing so, it draws to a natural conclusion the process of moral inclusion, opening up consideration to all beings by articulating an ethics that is truly all-encompassing and non-exclusionary. But by including everything, such an ethics risks moral discernment. Although IE procures rights for all things, machines included, it does so at the expense of difference (Levinas 1969, p. 43). Consequently, the success or failure of IE as a moral theory will depend on how well it is able to address and accommodate this criticism.

Finally, when evaluating these different modes of moral inclusion, the task might not be determining which approach is better or which moral theory is more or less exclusive. The problem may be with this procedure itself. Irrespective of how it is defined, standard approaches to moral reasoning take what can be called a "properties approach" (Coeckelbergh 2012, p. 13). That is, they first define criteria for inclusion and then ask whether a particular entity meets this criteria or not. Proceeding in this fashion has at least two difficulties. First, one needs to be able to identity the property or properties that will be considered necessary and sufficient for deciding moral status, and there is considerable disagreement over the definition of what counts. Second, this decision is necessarily a normative operation and an exercise of power. In making a determination about the criteria for moral inclusion, someone or some group normalizes their particular experience or situation and imposes this decision on others as the universal condition for moral considerability.

In response to these problems, philosophers have advanced alternative approaches that can be called "social-relational ethics" (Coeckelberg 2012; Gunkel 2012; Levinas 1969). These efforts do not endeavor to establish *a priori* criteria of inclusion and exclusion but begin from the existential fact that we always and already find ourselves in situations facing and needing to respond to others—not just other human beings but animals, the environment, organizations, and machines. Consequently, these alternatives advocate shifting the focus of the question and changing the terms of the debate. Here the issue is not "Can machines have rights?" which is an ontological question decided by the presence or absence of specific qualifying criteria, but "Should machines have rights?" which is a moral query and one that is decided not on the basis of what things are but on how we decide to relate and respond to them. Proceeding in this manner, although opening opportunities for extending rights to others, disrupts the philosophical status quo by promoting ethics in advance of ontology (Levinas 1969, p. 304).

Learning more

A critical approach to the history of moral philosophy along with a consideration of the rights and responsibilities of machines can be found in Wallach and Allen (2009), Coeckelbergh (2012), and Gunkel (2012). Seminal works in animal rights philosophy are available in Singer (1975), Regan (1983), and Derrida (2008); also see Gunkel (2012) for an examination of the way this innovation relates to the question of machine rights. Consideration of the importance of consciousness for machine moral standing can be found in Torrance (2008), information about information ethics is available in Floridi (2013), and alternative efforts to think ethics otherwise have been developed in Coeckelbergh (2012), Gunkel (2012), and Levinas (1969).

References

Bentham, J. (2005). *An Introduction to the Principles of Morals and Legislation.* J. H. Burns and H. L. Hart (eds.). Oxford: Oxford University Press.

Churchland, P. M. (1999). *Matter and Consciousness.* Cambridge, MA: MIT Press.

Coeckelbergh, M. (2012). *Growing Moral Relations: A Critique of Moral Status Ascription.* London: Palgrave Macmillan.

Dennett, D. (1998) *Brainstorms: Philosophical Essays on Mind and Psychology.* Cambridge, MA: MIT Press.

Derrida, J. (2008). *The Animal That Therefore I Am.* Trans. D. Wills. New York: Fordham University Press. Original work: *L'animal que done ju suis.* Paris: Éditions Galilée, 2006.

Floridi, L. (2013). *The Ethics of Information.* Oxford: Oxford University Press.

Gunkel, D. (2012). *The Machine Question: Critical Perspectives on AI, Robots and Ethics.* Cambridge, MA: MIT Press.

Güzeldere, G. (1997). The many faces of consciousness: A field guide. In: N. Block, O. Flanagan & G. Güzeldere (eds.) *The Nature of Consciousness: Philosophical Debates.* Cambridge, MA: MIT Press, 1–68.

Levinas, E. (1969). *Totality and Infinity.* Trans. A. Lingis. Pittsburgh, PA: Duquesne University Press. Original work: *Totalité et Infini.* Hague: Martinus Nijhoff, 1961.

Regan, T. (1983). *The Case for Animal Rights.* Berkeley, CA: University of California Press.

Singer, P. (1975). *Animal Liberation: A New Ethics for Our Treatment of Animals*. New York: New York Review of Books.

Torrance, S. (2008). Ethics and consciousness in artificial agents. *AI & Society*, **22**(4): 495–521.

Velmans, M. (2000). *Understanding Consciousness*. New York: Routledge.

Wallach, W. and **Allen, C.** (2009). *Moral Machines: Teaching Robots Right from Wrong*. Oxford: Oxford University Press.

Chapter 64

A sketch of the education landscape in biomimetic and biohybrid systems

Anna Mura[1] and Tony J. Prescott[2]

[1] SPECS, Institute for Bioengineering of Catalonia (IBEC), the Barcelona Institute of Science and Technology (BIST), Spain
[2] Sheffield Robotics and Department of Computer Science, University of Sheffield, UK

Biomimetics and biohybrid systems—together the domain of Living Machines—are emerging research fields based on the understanding and application of natural principles to the development of novel real-world technologies. Moreover, the Living Machines approach, as highlighted in this volume, emphasizes the strategy of *understanding through making* (see Verschure and Prescott, Chapter 2, this volume), that is, the construction of artifacts as embodiments of theories of natural systems in addition to being candidate solutions to societal challenges. This approach therefore implies, and requires, a transdisciplinary undertaking that bridges between science, engineering, and the social sciences, arts, and humanities. In addition, it also emphasizes a mix of basic and applied approaches that requires an awareness of the broader societal context in which modern research and innovation activities are conducted. The future of this research depends critically on engaging young minds from different educational backgrounds and providing them with relevant transdisciplinary training. This chapter provides a brief exploration of the education landscape for within-discipline and across-discipline programs related to Living Machines, focusing on Europe and the United States (US), highlighting some challenges that should be addressed, and providing some suggestions for future course development and policy making.

Education for a new scientific Renaissance

In order to build novel advanced artifacts based on biological principles we need to understand, not only mimic, nature and life and to base technology on this fundamental understanding. Progress in Living Machines therefore relies more generally on progress in both the natural sciences and in the engineering disciplines. This idea also dominated the interest and curiosity of one of the greatest geniuses of the Renaissance—Leonardo da Vinci. Working at the confluence of science, technology, and art, da Vinci considered these disciplines as instruments with which to address the same objective—advancing the understanding of nature and of the human condition. In doing so he mastered biology, engineering, physics, anatomy, drawing, and painting to explore the solutions embodied by living organisms and to create technical designs and working machines of various kinds. For example, Figure 64.1 shows a da Vinci drawing for the design of a flying machine motivated by his observations of the structure of the wings of a bat, placed alongside a contemporary bat-inspired robot that actually flies. As a man of the Renaissance, da

Vinci used interdisciplinary methods—writing treatises on mechanics, anatomy, cosmology, hydraulic, and Earth sciences. In this sense, his work can be considered as a true precursor to modern forms of biomimetics and biohybridicity.

Over recent decades we have witnessed a rapid growth of biomimetics research and science-driven technology with the emergence of artificial and biohybrid artifacts at the confluence of life sciences, materials science, nanotechnology, robotics, and artificial intelligence, as described in this volume. However, the success of these approaches—that originate from the convergence of different disciplines of knowledge—depends not only on bringing together parallel strands of scientific research and technological development, but also on providing suitable educational scaffolding. Compared to the Renaissance era, the wealth of knowledge available in the different fields related to Living Machines is such that even a polymath genius, such as da Vinci, would struggle to master a small fraction of it. On the other hand, the powerful tools available to researchers and engineers, from the internet through to specialized databases and smart design systems, mean that a new approach is possible—one that emphasizes mastery of skills in interpreting and applying knowledge without the same necessity for the personal accumulation of knowledge. The development of training programs must make the most of this revolution in the way we share and manipulate knowledge so that we can provide future Leonardos with the most appropriate training and tool-sets for building new technologies and understanding.

Many training programs exist in Europe and the US that provide an excellent grounding in the relevant science and engineering disciplines related to Living Machines research; however, this teaching often takes place within single disciplinary boundaries. For the future, it is important that more inclusive and overarching "programs of programs" are developed that support crossover and interdisciplinary training. In devising such programs we should aim to provide:

1. A sufficient grounding in the vocabulary, methodology, and technical skill sets in relevant areas of biological science to allow the extrapolation of design principles towards engineering.

2. A core understanding of engineering concepts, particularly around applied mathematical topics.

3. Training in modeling, both computational and physical, including understanding of core methodologies in the areas of programming, hardware development, and robotics.

4. An education in key topics that provide crossover and synergy between the understanding of biological and artificial systems, particularly in approaches such as dynamical systems and information theory.

5. Examples of integration across the discipline boundaries, for example, showing how research in neuroscience, neuromorphic computing, and wearable technologies can come together in the field of prosthetics (Figure 64.2).

6. An appreciation of the potential technology impacts of constructing models as artifacts, and of the societal and ethical issues around responsible innovation.

Living Machines as convergent science

There has been significant effort, since the start of the new millennium, to address the challenge of collaboration across disciplines and its potential impact in science, health, education, environment, and society. A particular focus has been on the need to promote integration between lab science and mathematics and engineering in the biological and biomedical sciences (McCarthy 2004; National Research Council 2009; Sharp and Langer 2011) for what has been

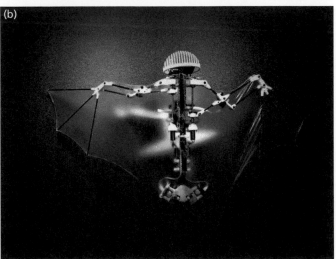

Figure 64.1 (a) Leonardo da Vinci's 1488 design for a flying machine inspired by the wings of a bat. (b) A contemporary bat-inspired robot designed by Ramezani et al. (2017). The B2 "bat bot", which is capable of autonomous flight, was created through a collaboration between bat biologists and robot engineers and instantiates the dominant degrees of freedom of the bat-wing without replicating its detailed morphology (and so follows the design methodology advocated in this book).

(a) Image: © Veneranda Biblioteca Ambrosiana/ Getty Images

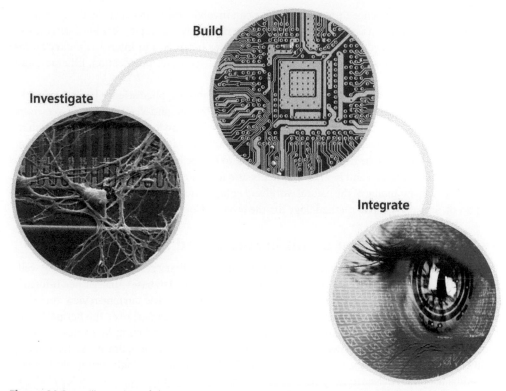

Figure 64.2 An illustration of the required convergence of science and technology in the field of prosthetics (see Vassanelli, Chapter 50 and Song and Berger, Chapter 55, this volume)—*investigate* brain physiology and function, *build* neuromorphic circuits, and *integrate* to create biohybrid systems.

described as a "third revolution" (the first two being *molecular biology*, a key moment being the discovery of the structure of DNA, and *genomics* for which a highlight was the sequencing of the human genome). A general drive for better collaboration between the life sciences and engineering has emerged during this period under the banner of "convergence" (National Research Council 2014; Roco and Bainbridge 2003; Roco et al. 2014; Science Europe 2014; Sharp et al. 2011; Sharp and Langer 2011). For example, two reports edited by Roco and colleagues (Roco and Bainbridge 2003; Roco et al. 2014) highlighted synergies, and common research questions and goals, between the research domains of nanotechology, biotechnology, information technology, and cognitive science (termed NBIC technologies). These optimistic reports looked forward to broad societal benefits with little downside, although European views have been more nuanced (European Commission 2004). The larger community, that is now promoting a convergence approach, and that is led by eminent institutions such as the American Academy of Arts and Sciences, the US National Research Council, and MIT, has also called for better teaching of STEM subjects in schools, for new interdisciplinary undergraduate degrees (e.g. National Research Council 2009), to facilitate boundary-hopping particularly for early-career researchers (AAAS 2008), and for the private sector and policy makers to prioritize resources and effort to facilitate these outcomes (AAAS 2013).

Convergence has been argued to mean more than interdisciplinarity, which in itself is not new (Graff 2015), and to involve purpose-driven research targeting identified challenges that will have societal impact (Science Europe 2014). This can involve new forms of societal engagement through targeted outreach activities. To consider one illustrative example, "Big Ideas@Berkeley" is a competition launched by UC Berkeley that provides incentives to interdisciplinary teams to provide new solutions to challenges such as food security, water purification, illiteracy, and malaria diagnosis.

At the same time, the convergent approach is not purely about applied science but instead seeks to channel basic science towards real-world applications (Prescott and Verschure 2016; Verschure and Prescott, Chapter 2, this volume). Overall, convergence relies on forming a web of partnerships to support boundary-crossing research and to translate advances into new technologies. A similar goal and ethos underlies the theme of this handbook to bridge between science and engineering, in the areas of biomimetics and biohybrid systems, and to create Living Machine technologies that advance our understanding of the natural world and enhance human life. Living Machines is therefore part of, and an exemplar methodology for, the broader Convergent Science approach.

A survey of education trends in Europe and the United States

Biomimetic and biohybrid systems are emergent multidisciplinary research fields lacking well-defined educational criteria. Given this, there are no official databases that gather information on educational programs directed towards these domains. The current review therefore began from the relatively informal base of discussions with scientists, over the period 2009–2015, participating in scientific and educational activities related to Living Machines research. These discussions took place at events including the Barcelona Cognition, Brain and Technology (BCBT) Summer Schools, the Capo Caccia Cognitive Neuromorphic Engineering Workshops, the Telluride School of Neuromorphic Cognitive Systems, the Padova Neurotechnology School, and the International Conference on Biomimetics and Biohybrid Systems (Living Machines). A further, more systematic, internet search was also performed, between 2012 and 2015, as part of the European Union *Convergent Science Network* Coordination Action. This investigation specifically targeted postgraduate teaching curricula using the search terms *biomimetics, biohybrids, neurotechnology, materials science, computational neuroscience, biorobotics, nanotechnology*, and *mechatronics*. Overall, our survey sought to be representative rather than exhaustive, aiming to capture the thrust of ongoing activities and identify opportunities and trends.

Education programs in Europe

Our investigation of European education identified three substantive Europe-wide integrated education programs, as summarized in Table 64.1, in any otherwise largely fragmented field. These specific programs were:

1. Joint Master programs supported by *Erasmus Mundus* (now *Erasmus+*), a cooperation and mobility program in the field of higher education organized by the European Commission. These two-year Masters courses are joint projects between up to five European universities and in some cases include international universities as well. Some of these programs offer the students the opportunity to plan an individualized program of study by combining elements from courses offered by the participating universities.

2. Joint PhD programs and postgraduate programs supported by the *Marie Skłodowska-Curie Actions* (MSCA) a European program, and part of the EU's Horizon 2020 research and innovation program, that supports researchers at all stages of their careers, irrespective of nationality, to provide them with the necessary skills to become excellent researchers in all disciplines. With this program researchers have the possibility to spend some of their

Table 64.1. Masters and PhD programs supported by *Erasmus Mundus* + and PhD programs under MSCA programs in Europe in the fields of bio-robotics, mechatronics, biomimetics, and biohybrid systems.

Masters offered by joint programs under Erasmus Mundus	PhDs offered by joint programs under Marie Curie
EMM-NANO—Master of Nanoscience and Nanotechnology	**INTRO**—Interactive Robotics research network
MEME—Master in Evolutionary Biology	**RobotDoc**—The Robotics for Development of Cognition
EU4M—Masters in Mechatronic and Micro-Mechatronic Systems	**FACETS-ITN**—Computing paradigms in biological nervous systems
CEMABUE—Master in Biomedical Engineering.	**NeuroPhysics**—New imaging frontiers
EMMS—Master in Material Sciences	**DYNANO**—Nanosystems for biomedical and biotechnological applications
EMARO—European Master in Advanced Robotics	**ABC**—Multidisciplinary approach on Adaptive Brain Computations
	NAMASEN—Science and Engineering of Neuronal Networks.

Other European Networks and organizations: **EURON** The European Graduate School of Neuroscience. **EURON Robotics** European Robotics Research Network.

training program abroad. MSCA programs have large multidisciplinary consortia (10–15 partners) combining high-level expertise from all over Europe.

3. Multidisciplinary educational programs led by networks of European universities that attempt to develop more targeted Master and PhD programs in the fields of biomimetic and biohybrid systems with the goal of training the future experts in those fields. An example of these programs is *EURON*: The European Graduate School of Neuroscience, which has created a broad network of sixteen universities in seven different countries united by shared interest in neuroscience research. *EURON* offers PhD and Master programs in the field of neuroscience and biomedical sciences. A similar network model for multidisciplinary education is offered by the *European Robotics Research Network*, a community of more than 230 academic and industrial groups in Europe with a common interest in advanced research and development to make better robots. This network recently merged with a partner industrial network to form the organization *EU Robotics*.

An indicative distribution of the contribution of the different European universities and academic centers to the education of the graduate programs, Masters and PhD, mentioned above is given in Figure 64.3 and shows interest in educational topics related to Living Machines across most European countries.

A large and widely distributed set of graduate programs is offered in nanotechnology, mechatronics, and materials science, mostly in Germany, France, Italy, UK, and the Netherlands. Some countries such as Italy, UK, Germany, France, Spain, and Switzerland have also placed some emphasis on more specialist graduate programs related to biorobotics and biohybrids, as illustrated in Figure 64.4. The international organization *BIOKON*, which has its headquarters in Berlin, Germany, is also worth noting in this context as a coalition of scientists and institutions that promote and support the expansion of biomimetics and bio-inspired technologies. BIOKON's primary goal is to spread information related to biomimetics activities, to facilitate networking of members, and to organize a platform and a worldwide forum for biomimetic scientists and other interested persons; however, they do not directly coordinate education curricula.

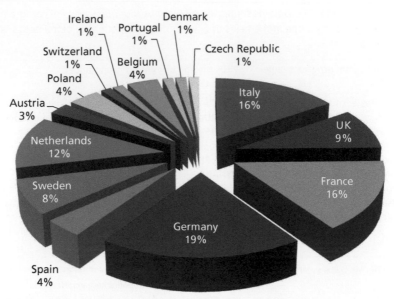

Figure 64.3 How different European countries contribute to multidisciplinary graduate programs (Masters and PhDs) in the fields of Biomimetics, Biohybrids, Neurotechnology, Materials Science, Computational Neuroscience, BioRobotics, Nanotechnology, and Mechatronics.

This survey reveals that in order to create multidisciplinary programs in fields at the convergence of the biological sciences and technology, academic centers from different countries have strategically become members of joint programs organized/sponsored under the umbrella of the European Union. The future of these initiatives is therefore highly dependent on the enthusiasm of this sponsoring body for continued support, which in turn requires that a substantial community of stakeholders in education, commerce, and society continues to promote the cause of convergent research.

Education programs in the US

Many US academic institutions have now started to provide graduate and postgraduate courses in multidisciplinary fields of life sciences, materials science, bioengineering, and biotechnology. Some clear examples of the impact made by the US drive towards convergent multidisciplinary science include: (i) MIT's *Koch Institute for Integrative Cancer Research*, an example of how to incorporate convergence into the infrastructure of science—biologists, engineers, and others in the physical sciences work together in a new building and on the same floors; and (ii) Stanford's *BioX Institute*, an interdisciplinary biosciences institute, with the mission to catalyze discovery by crossing the boundaries between disciplines, and to create new knowledge of biological systems that ultimately improve human health.

Our analysis of the US landscape included surveys of the websites *Study.com, Robotics Schools and Universities in the U.S., STEM*, the *U.S. Network for Education Information* (USNEI), and the *National Science Foundation* (NSF). Across this sample it was possible to see an emerging emphasis on interdisciplinary programs involving STEM subjects (science, technology, engineering, and mathematics). Nevertheless, we have found it difficult to identify specific educational offerings relevant to Living Machines and Convergent Science approaches, despite there being strong evidence of a coordinated approach to education in some component fields. For example, in the area of nanotechnology, where support is organized through a national initiative , an interactive map,

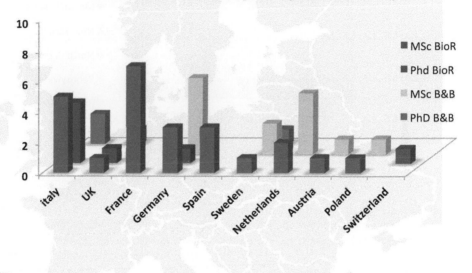

Figure 64.4 Estimated number of Masters (MSc) and PhD programs related to biorobotics and biohybrid systems offered in different European countries. Search terms for this survey were *biomimetics, biohybrids*, and *biorobotics*. BioR = biorobotics, B&B = biohybrids. At the time of writing, France has the highest number of Masters related to biorobotics (n=7) while Italy offers more PhD programs in this field (n= 5). Germany on the other hand offers the highest number of Masters programs related to biohybrids (n= 3) followed by Sweden offering a few PhDs programs in this field (n= 2).

and comprehensive list of Masters and PhD programs, has been developed (see www.nano.gov). We conclude that some of the relevant activity may be at a level of granularity of courses within institutes that is difficult to detect by a broad-brush survey. An example of a specific initiative that we have become aware of, and that is strongly-related to both biomimetics and the Convergent Science approach, is a course on Bioinspired Design[1] offered by the Department of Integrative Biology at University of California, Berkeley (see also Full et al. 2015). This course, which is

[1] http://polypedal.berkeley.edu/?page_id=691

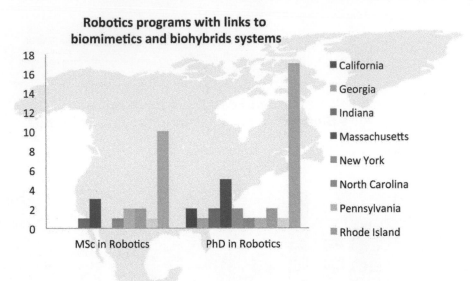

Figure 64.5 US robotics Masters (MSc) and PhD programs that have embraced a more biomimetic approach and the states in which they are offered. Overall, 10 US states offer a total of 11 Masters programs and 18 PhD programs, with Massachusetts leading with the highest number of Masters (n = 3) and PhD programs in the field (n = 5). These universities are also very active in ground-breaking research as well as education and outreach activities.

offered to students from science, engineering, arts, medicine and business as a single semester option, covers the biomimetic design process, exemplar case studies, and team projects in biomimetic robotics. Other notable ingredients include an emphasis on critical thinking and inquiry-based learning and on knowledge-transfer as exemplified by training in 'bioentrepreneurship'.

To assess the broader extent of activity in the US in areas related to Living Machines we examined a second area of relevant research—robotics—an interdisciplinary field that combines study of mechanical engineering and computers, where many programs do include a theme related to biomimetics. Figure 64.5 plots the number of Masters and/or PhD programs per US state offered in the area of robotics that had some additional match with our keywords *biomimetics, biohybrids*, and *biorobotics*. This picture indicates emerging activity on both the East and West coast with the Boston area (Massachusetts) a significant hot-spot.

A tentative conclusion is that both European and US landscapes are fertile ground for integrative approaches in postgraduate education that combine science and engineering, Europe, through funding structures that allow consortia of universities to pursue educational initiatives at Masters and PhD level, also has the potential to support transnational educational opportunities in areas of Convergent Science such as biomimetic and biohybrid systems.

Strategies to structure international educational opportunities in biomimetics and biohybrid systems

Whilst a move toward education programs that embrace a Convergent Science approach is evident in both Europe and the US, there are, as yet, few curricula that specifically reflect the

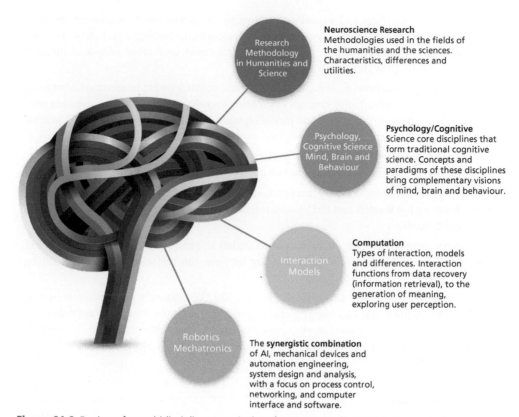

Neuroscience Research
Methodologies used in the fields of
the humanities and the sciences.
Characteristics, differences and
utilities.

Psychology/Cognitive
Science core disciplines that
form traditional cognitive
science. Concepts and
paradigms of these disciplines
bring complementary visions
of mind, brain and behaviour.

Computation
Types of interaction, models
and differences. Interaction
functions from data recovery
(information retrieval), to the
generation of meaning,
exploring user perception.

The **synergistic combination**
of AI, mechanical devices and
automation engineering,
system design and analysis,
with a focus on process control,
networking, and computer
interface and software.

Figure 64.6 Design of a multidisciplinary curriculum for Masters education in the area of biomimetics and biohybrid systems with particularly emphasis on neuroscience and neurotechnology.

principles and themes of Living Machines research. A first step in addressing this challenge is to define what an international Masters program could look like.

Building on the more generic set of topics presented earlier, Figure 64.6 outlines one possible program, potentially a two-year taught Masters, with specific focus toward neuroscience and neurotechnology.

Note that several elements of the proposed curricula can be found in many universities—physiology, psychology, AI, and so on; however, others are newer and more interdisciplinary in nature. For example, we have highlighted here the embodiment of neuromimetic controllers in hardware (neuromorphics), the design of biomimetic control architectures for complex artifacts such as robots (neurorobotics), and the field of brain–machine interfaces. These are all important domains for translating understanding in neuroscience towards real-world applications. It is possible to anticipate a program that is split between courses taken at a home institution and a period spent visiting a leading center for these forms of interdisciplinary research for more specialist training.

To encourage the growth of this kind of education program and to foster a broader environment for education in areas of Convergent Science, the following strategic goals could be pursued:

◆ Lobby national and international bodies to recognize the value and importance to society of the translation of biological principles into technology both as a path to further scientific understanding and to create novel solutions to societal needs.

- Direct national initiatives and education funding to encourage the expansion of interdisciplinary and transdisciplinary programs in life sciences, technology, and industry. This strategy, which is already evident in some countries, could be further advanced through international initiatives and collaboration.

- Create an international network of educational institutions interested in promoting Convergent Science approaches and enhance synergies by sharing knowledge and resources including program designs, educational software and modeling tools, design templates (e.g. for 3D printed systems), and biomimetic control/teaching software for common research platforms.

- Facilitate the development of cross-institution degrees, including trans-national qualifications, where training can be obtained in two or more contributing institutions.

- Provide support for dedicated programs such as summer schools, workshops, internships, or online teaching for Masters and PhD students in areas of research related to biomimetic and biohybrid systems.

- Develop pre- and postgraduate programs of education and research, such as internships and fellowships, that emphasize discipline-hopping between the life sciences and engineering and vice versa.

- Foster links between universities, public bodies, charities, foundations, businesses, and citizen organizations that can identify challenges where a Living Machines approach can provide a breakthrough. Create opportunities for interactions between students, research leaders, and these other stakeholders to come together to develop solutions to real-world challenges.

Acknowledgments

The surveys and research described in this chapter were sponsored by the European Union 7th Framework through the Future Emerging Technologies (FET) programe and Coordination Actions "Convergent Science Network for Biomimetics and Biohybrid Systems (CSNI)" (ICT-248986) and "Convergent Science Network of Biomimetics and Neurotechnology (CSNII)" (ICT-601167). We are grateful to Paul Verschure for his enormous contribution in developing the ideas presented here and to the many speakers and participants in CSN events and conferences who have helped to craft and inspire the Living Machines approach.

References

AAAS (2008). *ARISE I—Advancing Research In Science and Engineering: Investing in Early-Career Scientists and High-Risk, High-Reward Research*. Retrieved from Cambridge, MA: https://www.amacad.org/content/publications/publication.aspx?d=1138

AAAS (2013). *ARISE II: Unleashing America's Research & Innovation Enterprise*. Retrieved from Cambridge, MA: https://www.amacad.org/content/publications/publication.aspx?d=1138

European Commission (2004). *Converging technologies: Shaping the future of European societies*. Retrieved from http://bookshop.europa.eu/en/converging-technologies-pbKINA21357/

Full, R. J., Dudley, R., Koehl, M. A. R., Libby, T., and Schwab, C. (2015). Interdisciplinary laboratory course facilitating knowledge integration, mutualistic teaming, and original discovery. *Integrative and Comparative Biology*, 55(5), 912–25. https://doi.org/10.1093/icb/icv095

Graff, H. J. (2015). *Undisciplining knowledge: interdisciplinarity in the twentieth century*. Champaign, IL: Johns Hopkins University Press.

McCarthy, J. (2004). Tackling the challenges of interdisciplinary bioscience. *Nat. Rev. Mol. Cell Biol.*, 5(11), 933–7.

National Research Council (2009). *A New Biology for the 21st Century.* Washington, DC: National Academies Press.

National Research Council (2014). *Convergence: Facilitating Transdisciplinary Integration of Life Sciences, Physical Sciences, Engineering, and Beyond.* Washington, DC: National Academies Press.

Prescott, T. J., and Verschure, P. F. M. J. (2016). Action-oriented cognition and its implications: contextualising the new science of mind. In: A. K. Engel, K. Friston, and D. Kragic (eds.), *Where's the Action? The Pragmatic Turn in Cognitive Science.* Cambridge, MA: MIT Press for the Ernst Strüngmann Foundation, pp. 321–31.

Ramezani, A., Chung, S.-J., and Hutchinson, S. (2017). A biomimetic robotic platform to study flight specializations of bats. *Science Robotics*, 2(3), eaal2505. doi: 10.1126/scirobotics.aal2505

Roco, M. C., and Bainbridge, W. S. (2003). *Converging Technologies for Improving Human Performance: Nanotechnology, Biotechnology, Information Technology and Cognitive Science.* Berlin: Springer.

Roco, M. C., Bainbridge, W. S., Tonn, B., and Whitesides, G. (2014). *Convergence of Knowledge, Technology and Society: Beyond Convergence of Nano-Bio-Info-Cognitive Technologies.* Basel: Springer International Publishing.

Science Europe (2014). *Converging Disciplines.* Retrieved from Brussels: http://www.scienceeurope.org/wp-content/uploads/2015/12/Workshop_Report_Convergence_FINAL.pdf

Sharp, P. A., Cooney, C. L., Kastner, M. A., Lees, J., Sasisekharan, R., Yaffe, M. B., ... and Sur, M. (2011). *The Third Revolution: The Convergence of the Life Sciences, Physical Sciences, and Engineering.* Cambridge, MA: MIT Press.

Sharp, P. A., and Langer, R. (2011). Promoting convergence in biomedical science. *Science*, 333(6042), 527.

Chapter 65

Sustainability of living machines

José Halloy

Paris Interdisciplinary Energy Research Institute (LIED),
Université Paris Diderot, France

Technological development must be put into perspective in order to respond to the challenge of climate change and expected energy and resources transitions that pose a major threat to humankind. Climate change is driven by human production of greenhouse gases (GHG) such as carbon dioxide and methane (IPCC 2014). Climate change and the energy transition are thus closely related because, nowadays, the main sources of energy are fossils fuels, namely coal and oil. If we want to mitigate climate change, we must use energy sources that produce the least GHG. But that is not the only issue. To build technological devices, including power stations, a huge amount of different materials are also needed (Smil 2013). The vast majority of the materials used in current technological devices, especially in artificial intelligence (AI) and robotics, are extracted from mineral deposits, including all metals and metalloids (Smil 2013; Peiró 2013). The increasing use of such minerals is not sustainable in the long term because they have been produced by geological events during Earth's formation and during its geological evolution over a timescale of billions of years. This chapter will argue that a new biomimetic and biohybrid Living Machines approach is needed to the design of appropriate technologies that will allow their construction from low-power, widely available, and highly recyclable components.

The unsustainable energy costs of electronic systems

Artificial Intelligence (AI) and robotics are making impressive technological progress. For example, in 2016, one extraordinary achievement of AI took place: the AlphaGo AI defeated Lee Sedol, one of the best Go players in the world. AlphaGo is both a computer and a computer program capable of playing the Go game, developed by the British company Google DeepMind. The algorithms of AlphaGo combine machine learning techniques with graph paths, combined with numerous training sessions with humans and other computers. AlphaGo uses a Monte Carlo method, guided by a value network and a policy network (both a value network and a network of objectives), both of which are implemented using deep neural networks (Silver 2016). Before AlphaGo in 2016, the game of Go had been resisting AI algorithms (unlike chess, which had already succumbed to AI in 1997, when IBM's Deep Blue defeated the world chess champion Garry Kasparov). AlphaGo's victory is a symbolically important achievement since building a Go player program is a complex challenge for AI. Thus, the AlphaGo achievement made headlines in the world's scientific journals and newspapers.

In terms of sustainability, this achievement raises many issues. The AlphaGo computer used 1202 CPUs and 176 GPUs (Central and Graphical Processing Units). We assume that 100W are needed to power one CPU and about 200W for one GPU (see Figure 65.1). Those numbers are educated guesses based on current specifications from main microprocessor manufacturers

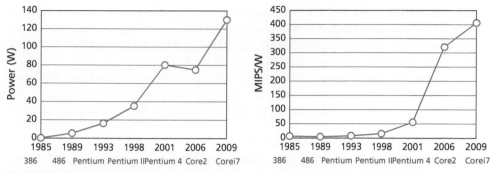

Figure 65.1 Power and efficiency increase of Intel processors. The power needed for Intel processors has been steadily increasing to reach about 140W for recent models. At the same time their computing efficiency has also been rising. The number of millions of instructions per second (MIPS) per watt (W) has increased exponentially. Even if in terms of computing efficiency, the recent processors are excellent, the absolute power necessary to run them is also increasing exponentially. Efficiency improvements lead to rebound effects (Polimeni et al. 2015). As processors are more efficient in terms of computational power and can be more efficient in terms of energy consumption, the total number of processors is booming in different markets leading to a global increase in energy consumption and in the materials consumed in their manufacture.

such as Intel or NVIDIA. Thus, to power all of its computational requirements the AlphaGo computer needed about 155kW. For comparison, Lee Sedol's brain can be assumed to need around 20W and the whole human body about 2500 kCal per day to live in excellent shape; this corresponds to about 120W of power. Lee Sedol was aged 34 when the Go tournament against AlphaGo took place. If we assume he had, on average, eaten 2500 kCal/day, then he had consumed about 130 GJ during his whole lifetime up to that point. Considering the estimated 155kW needed just to power the microprocessors of AlphaGo, this computer will have consumed around 130 GJ in 10 days. In terms of power and energy the gap between a human and a computer is huge and, we argue, non-sustainable.

The AlphaGo achievement was possible because of the tremendous technological progress that has been made in electronics and computational technologies since the invention of the crystalline semiconductor transistor in 1947 at Bell labs by John Bardeen, Walter Brattain, and William Shockley, for which they received the Nobel Prize for physics in 1956. The number of transistors used in microprocessors has been dramatically increasing since the invention of the integrated circuit in 1959. Gordon Moore found that the number of transistors in integrated circuits was approximately doubling every two years since their invention. However, Moore's law is reaching its limits in terms of transistor integration because this integration is reaching a nano-scale limit of the size of a dozen or so atoms. In terms of integration and energy consumption we are also reaching limits because of the increase in heat dissipation. These devices have to be cooled and kept in air-conditioned buildings, further increasing their operating energy costs.

From an economic point of view, the cost of the technologies that permit the integration of more and more transistors is also increasing with dizzying proportions. Another Silicon Valley empirical law, Rock's law (or Moore's second empirical law), states that the cost of manufacturing a chip foundry doubles every four years because the photolithography manufacturing process is also approaching its physical limits. Nowadays, a chip manufacturing plant represents a several billion-dollar investment. Moreover, this integration also comes with an increasing manufacturing energy cost. All of this amazing technological progress has enabled the construction of the current largest super-computer created by the Chinese company Sunway MPP. This super-computer has 10,649,600 cores producing 93,014.6 TFlop/s and requires 15,371 kW. For comparison, in terms of power requirements in the USA in 2012, about 2kW of electricity per person was needed to power the American way of life.

Transistors are crystalline materials made of metalloids, mainly germanium (since 1947) and silicon (since 1954). Other typically used compounds include gallium arsenide, silicon carbide, and silicon–germanium. The same types of semiconductor materials are also used to produce photovoltaic electricity and LED lighting. Computational technologies are mainly based on CMOS (Complementary Metal Oxide Semiconductor) and MOSFET (Metal Oxide Semiconductor Field Effect Transistor) chips. These type of materials and technologies are used to manufacture the circuits made of logic gates that form the basic architecture of integrated circuits.

The founding materials are crystalline semiconductors. The electrical conductivity of a semi-conductor is intermediate between those of metals and insulators, hence the name. This electrical conductivity can be controlled by doping, i.e. introducing a small amount of impurities into the pure crystalline material to produce an excess or deficit of electrons. Differently doped semiconductors can be connected to create junctions; this allows the direction and amount of current flowing through the device to be controlled. This property is at the basis of the operation of the components of modern electronics: diodes, transistors, etc. The first manufacturing step is to produce wafers of very high purity. A wafer is a thin disc of semiconductor material, such as silicon, gallium arsenide, or indium phosphide. It serves as a support to manufacture the needed micro-structures by techniques such as etching and deposition of other materials. Different methods are used to dope the wafer material by adding, at very low concentration (of the order of parts per million (ppm)), other elements such as boron, arsenic, phosphorus, gallium, or germanium. This manufacturing process has important implication in terms of recycling possibilities as the lower the concentration the more difficult to separate the elements. The exquisitely detailed multi-layering of different materials to produce an integrated circuit seriously impacts the recycling possibilities.

The above brief summary indicates that AI and robotic technologies are approaching physical limits in terms of the ability to improve the design of their founding electronic technology, and that each advance increases the energy requirements for their manufacture and operation. Moreover, large amounts of different mineral materials are needed to manufacture the bodies of the computers and the robots. It appears that many of the materials used to produced electronics are becoming critical because they are not renewables and are difficult to recycle (Reck and Graedel 2012). Recycling is a difficult issue because electronics is based on many types of materials and some of them are in low concentrations. The body of a computer or a robot is a complex structure made of integrated circuits, electronic boards, devices, and different types of synthetic polymer packaging, making it difficult to disassemble them. Each of the components, such as integrated circuits, is also a complex structure that is difficult to disassemble, which makes separation of the chemicals difficult.

The unsustainable nature of modern mining

The number of different chemical elements used to produce computer or robot bodies is large and includes nearly 80% of the periodic table of the elements (Peiró 2013). Supplies of some of these elements, particularly those needed to manufacture the electronics parts, are becoming of critical concern. Criticality of the chemical elements can be considered along three axes: supply risk, environmental implications, and vulnerability to supply restrictions (Graedel et al.

2015a,b). In terms of supply risk only, the chemical elements that are considered as most critical are germanium, indium, thallium, arsenic, tin, bismuth, selenium, silver, and cadmium. These elements are essential to the construction of electronic parts, computers, and robots.

The rate of extractions of metals and metalloids by mining has been constantly growing during the twentieth century (see Figure 65.2). Nearly all extracted elements have reached their highest levels of extraction, in historic terms, in the last few years. This growing extraction rate does not seem to be slowing down; rather, it appears to be accelerating, probably due to the booming of the electronic consumer market, evaluated now in billions of dollars. From 2007 to 2015 about 1.4 billion smartphones were sold to end-users worldwide. The current industry trend is to multiply electronic devices by billions for the so-called Internet of Things (IoT) and smart technologies. These factors will encourage yet more mining to produce the needed chemical elements from ores.

The amounts of minerals that can be extracted from ores are finite and correspond to the geological stocks created by Earth's geological evolution. These stocks are evaluated by institutions such as the United States Geological Survey (USGS). Based on these estimated geological amounts, and by extrapolating from current rates of extraction, it is possible to estimate when all known reserves of a given chemical element will be extracted (Douce 2016a,b). Based on various plausible hypotheses of rate extraction extrapolation, the final years of extraction for some of the more in-demand elements will be during the twenty-first century (Heinberg 2010; Bardi 2014; see Figure 65.3).

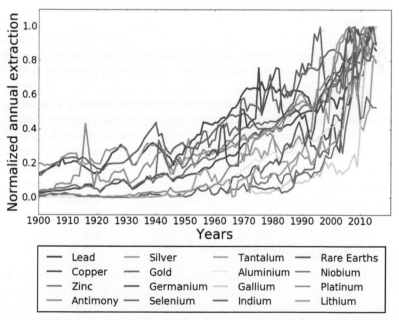

Figure 65.2 Normalized annual extraction of chemical elements. The total amount extracted per year for each chemical element is normalized by its maximal extracted amount to put all elements on the same scale. The figure shows the steady increase of most of the elements since the beginning of the twenty-first century. It also shows that the total quantity extracted each year is reaching it maximal values during recent years. Overall, the mining of metals and metalloids is booming.

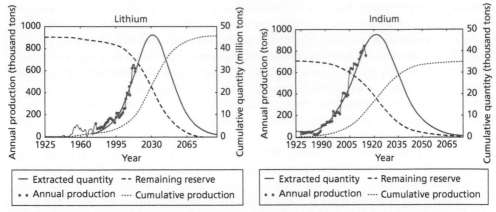

Figure 65.3 Estimated extraction peaks for lithium and indium. Both chemical elements are essential for the electronics industry—lithium for manufacturing batteries and indium for displays, photovoltaic cells, and photodiodes. The bell shape curves represent the expected extracted quantity of lithium (left) and indium (right). The dashed curves represent the remaining reserve based on the USGS estimation of the world reserve. The dotted curves represent the cumulative production. The curves and dots represent the actual world extracted quantities evaluated by the USGS. When the cumulative production equals the remaining reserve, all known resources have been extracted. In this example, we make the hypothesis that cumulative production follows a logistic curve allowing extraction curves to be estimated for annual production. The extraction peaks could be reached around 2030.

The extraction of chemical elements from ores also needs significant amounts of energy for mining and processing (physically and chemically) the ores to produce ingots of metals and metalloids. In 2014, an estimated 400 million tonnes equivalent of petroleum (400 Mtep or 18 billion GJ) were necessary for the mining of metal and metalloid elements. In comparison, the total energy consumption of the European Union (28 member states) was about 1600 Mtep. Thus, the mining industry necessary to produce the raw materials used to manufacture the complex materials used to produce computers or robots consumes a very large energy share. The growth of the mining industry leads to the growth of energy consumption. The main energy sources used in mining are coal and oil. It must be noted that for renewable energy production the same kind of material depletion risks exist, raising concern about the material sustainability of some renewable energies such as photovoltaic electricity and windmills (Fizaine and Court 2015).

A Living Machines approach

All of the above factors lead to the conclusion that, in terms of energy and materials, the current computational technologies are not sustainable in the long term. There is a growing and pressing need to re-invent how we build electronic systems, including computers and robots, in order to make them sustainable. Many uncertainties remain, making a prospective strategy difficult to develop. However, there is a consensus in the scientific literature that significant limits will be reached during the twenty-first century and that this issue is also closely linked to those of climate change and energy transition. Considering the time needed for scientific and technological development, this means that the research needed to develop alternative solutions should be undertaken as soon as possible and intensively.

From a chemical point of view, if we compare our technologies to living beings the differences are significant. In contrast to our modern technologies, only six chemical elements compose 97% of the mass of living beings. These elements are carbon (C), hydrogen (H), nitrogen (N), oxygen (O), phosphorus (P), and sulfur (S). Other elements are also used such as boron, cobalt, iron, copper, molybdenum, selenium, silicon, tin, vanadium, and zinc. But these other elements are used at much lower concentration—below 1% of mass. Using chemical elements at lower concentration, driven by absorption though membranes, is a completely different process, from the point of view of thermodynamics, than first producing ingots from the lower concentration found in ores. The amount of energy required is much lower, and the original concentration can also be lower. Typically for living systems, the concentrations of those special chemical elements are those found in sea or fresh water at the ppm level (Alberts et al. 2002).

Another essential aspect is that the main six chemical elements—C, H, N, O, P, and S—are involved in biogeochemical cycles: the carbon cycle, the water cycle, the sulfur cycle, the nitrogen cycle, and the phosphorus cycle. All of these cycles involve the biosphere. In ecology and, more generally, in the Earth sciences, a biogeochemical cycle is the process of cyclic transport and transformation of a chemical element or compounds between the large reservoirs of the geosphere, the atmosphere, and the hydrosphere, in which the biosphere is embedded. Thus, the fundamental chemical elements of living systems are recycled and sustainable in the long term, i.e. millions and even billions of years, because they are associated with biogeochemical cycles. Living systems form ecosystems that are tightly connected, from the point of view of chemical recycling. Those differences, between living systems and our current technologies, are crucial for sustainability in the long term.

Hence, we argue for the need of a radical biomimetic, or Living Machines, approach to the development of next generation technologies; radical meaning including the type of chemistry and ecological processes found in living systems. This radical approach necessitates reinventing our computational technologies, i.e. the founding basis of robotics and AI and other related technologies, in terms of materials, architectures, and processes, and linking these processes within new technological ecosystems, learning from the self-regulating ecological cycles of birth, growth, death, and re-use found in the natural world.

New computational architectures

The need for new architectures is due to the bottlenecks introduced by the von Neumann architecture based on logic gates circuits as described above. It is becoming clear that this computing architecture needs too much energy, and produces too much heat that needs to be dissipated, to scale to the demands of modern information processing and AI. However, this architecture, and its implementation in integrated circuits, is excellent at producing high computing power. For pure computational problems, i.e. arithmetical calculation, an architecture based on logic gates can be kept. It is nevertheless important to build novel integrated circuits based on renewable materials. However, even in this case, the performance of the materials is closely related to its physico-chemical nature. New renewable materials will change the expected performance of the semiconductors necessary to build logic gates.

Cognitive science can improve our understanding of the neuronal structures and processes that perform efficient information processing in living systems with low power, as discussed elsewhere in this book (Dudek, Chapter 14; Metta and Cingolani, Chapter 47). For example, in the field of neuromorphic computing, an exciting research track is developing to replicate this efficiency in massively parallel electronic systems (see e.g. Furber et al. 2013); although for now the research still focuses on simple neurons and structures it captures a crucial feature of communication in biological neural networks which is communication using electrical "spikes". Research is ongoing to build more accurate neurons and neuronal networks that could be used to develop novel neuromorphic chips (Shepherd 1998; Calimera et al. 2013).

Many current successful AI technologies are based on "deep learning" (see e.g. Leibo and Poggio, Chapter 25; Herreros, Chapter 26, this volume). These techniques have enabled recent, significant, and rapid progress. The different deep learning architectures, such as deep neural networks, convolutional deep neural networks, and deep belief networks, have applications in the fields of sound and visual signal analysis, including face recognition, speech recognition, computer vision, automated language processing, and bioinformatics, as discussed elsewhere in this handbook, where they have demonstrated that they can produce excellent results for different problems. Initiated at the end of the 1980s with the birth of the first multi-layer artificial neural networks, and adopting a concept dating back to the late 1950s (perceptron, etc.), the modern concept of deep learning only began to take shape in the 2010s, taking advantage of the massive increase in computational power of modern processors. However, in terms of power and energy, deep neuronal networks are also profoundly affected by the design bottlenecks of modern electronics, as shown by the energy comparison between AlphaGo and Lee Sedol. The field of neuromorphics must constantly update its designs, considering the most advanced results from cognitive science and neuroscience and improving its ability to capture biomimetic principles. A close collaboration should be maintained between the neurobiological and AI research fields (see also Verschure and Prescott, Chapter 2, this volume).

New biomimetic materials

Artificial neuronal architectures cannot be conceived without at the same time considering the type of materials and bodies needed for their implementation (see Vincent, Chapter 10, this volume). The current dualistic technological design, separating the processing architecture from its implementation body, leads to major sustainability pitfalls. There is thus an urgent need to develop new research roadmaps for living machines based on the convergent approach to the design of brains and bodies, developed in this handbook, and emphasizing the use of sustainable materials. Obviously, the current silicon-based semiconductor technologies will endure, as they represent keystone technologies and have huge economic markets evaluated in multi-billions of dollars and millions of jobs. Therefore, this new research roadmap must be developed in parallel and should propose an incremental approach to the transition from non-sustainable, computer-based machines to sustainable living machines. Recent trends in advanced robotic research are opening new possibilities to build a new roadmap. For example, the field of soft robotics is making use of polymers and new actuation methods (Kim et al. 2013; Laschi et al. 2016; Anderson and O'Brien, Chapter 20, this volume; Trimmer, Chapter 41, this volume). Those soft polymers could be bio-sourced or biocompatible. Moreover, some recent simple robots have been developed using biological tissues extracted from living beings or formed by cell and tissue cultures (Ayers, Chapter 51, this volume; Fukuda et al., Chapter 52, this volume; Webster et al. 2016; Feinberg et al. 2007; Cvetkovic et al. 2014).

Towards sustainable ecosystems

Linking technological devices to natural ecosystems, or inventing biohybrid ecosystems, raises new and difficult scientific and engineering questions. Understanding how to link our technologies to the Earth's system and to existing biogeochemical cycles is a huge scientific and technological challenge because we must face the intricate complexity of those systems. This radical research roadmap must be highly interdisciplinary, involving all natural sciences and engineering but also the social sciences, as it will have impact on all aspects of our societies and will raise new ethical questions.

References

Alberts, B., et al. (2002) *Molecular biology of the cell*, 4th edition. New York: Garland Science.

Bardi, U. (2014). *Extracted: How the quest for mineral wealth is plundering the planet*. White River Junction, VT: Chelsea Green Publishing.

Calimera, A., Macii, E., and Poncino, M. (2013). The human brain project and neuromorphic computing. *Functional neurology*, **28**(3), 191–6.

Cvetkovic, C., Raman, R., Chan, V., Williams, B. J., Tolish, M., Bajaj, P., … and Bashir, R. (2014). Three-dimensionally printed biological machines powered by skeletal muscle. *Proceedings of the National Academy of Sciences*, **111**(28), 10125–30.

Douce, A. E. P. (2016). Metallic mineral resources in the twenty-first century. I. Historical extraction trends and expected demand. *Natural Resources Research*, **25**(1), 71–90.

Douce, A. E. P. (2016). Metallic mineral resources in the twenty-first century: II. Constraints on future supply. *Natural Resources Research*, **25**(1), 97–124.

Feinberg, A. W., Feigel, A., Shevkoplyas, S. S., Sheehy, S., Whitesides, G. M., and Parker, K. K. (2007). Muscular thin films for building actuators and powering devices. *Science*, **317**(5843), 1366–70.

Fizaine, F., and Court, V. (2015). Renewable electricity producing technologies and metal depletion: a sensitivity analysis using the EROI. *Ecological Economics*, **110**, 106–18.

Furber, S. B., Lester, D. R., Plana, L. A., Garside, J. D., Painkras, E., Temple, S., and Brown, A. D. (2013). Overview of the spinnaker system architecture. *IEEE Transactions on Computers*, **62**(12), 2454–67.

Graedel, T. E., Harper, E. M., Nassar, N. T., and Reck, B. K. (2015a). On the materials basis of modern society. *Proc. Natl Acad. Sci. USA*, **112**(20), 6295–300.

Graedel, T. E., Harper, E. M., Nassar, N. T., Nuss, P., and Reck, B. K. (2015b). Criticality of metals and metalloids. *Proc. Natl Acad. Sci. USA*, **112**(14), 4257–62.

Heinberg, R. (2010). *Peak everything: waking up to the century of declines*. Gabriola, Canada: New Society Publishers.

IPCC (2014). *Climate Change 2014: Synthesis Report*. Contribution of Working Groups I, II, and III to the Fifth Assessment Report of the Intergovernmental Panel on Climate Change [Core Writing Team, R.K. Pachauri and L.A. Meyer (eds.)]. IPCC, Geneva, Switzerland, 151 pp.

Kim, S., Laschi, C., and Trimmer, B. (2013). Soft robotics: a bioinspired evolution in robotics. *Trends in Biotechnology*, **31**(5), 287–94.

Laschi, C., Mazzolai, B., and Cianchetti, M. (2016). Soft robotics: technologies and systems pushing the boundaries of robot abilities. *Science Robotics*, **1**(1), eaah3690.

Peiró, L. T., Méndez, G. V., and Ayres, R. U. (2013). Material flow analysis of scarce metals: sources, functions, end-uses and aspects for future supply. *Environmental Science & Technology*, **47**(6), 2939–47.

Polimeni, J. M., Mayumi, K., Giampietro, M., and Alcott, B. (2015). *The Jevons paradox and the myth of resource efficiency improvements*. Abingdon, UK: Routledge.

Reck, B. K., and Graedel, T. E. (2012). Challenges in metal recycling. *Science*, **337**(6095), 690–5.

Shepherd, G. M., Mirsky, J. S., Healy, M. D., Singer, M. S., Skoufos, E., Hines, M. S., … and Miller, P. L. (1998). The Human Brain Project: neuroinformatics tools for integrating, searching and modeling multidisciplinary neuroscience data. *Trends in Neurosciences*, **21**(11), 460–8.

Smil, V. (2013). *Making the modern world: materials and dematerialization*. Chichester: John Wiley & Sons.

Silver, D., et al. (2016). Mastering the game of Go with deep neural networks and tree search. *Nature*, **529**, 484–9.

Webster, V. A., Chapin, K. J., Hawley, E. L., Patel, J. M., Akkus, O., Chiel, H. J., and Quinn, R. D. (2016). *Aplysia californica* as a novel source of material for biohybrid robots and organic machines. In: N. F. Lepora, A. Mura, M. Mangan, P. F. M. J. Verschure, M. Desmulliez, and T. J. Prescott (eds), *Conference on Biomimetic and Biohybrid Systems*. Basel: Springer International Publishing, pp. 365–74.

Index

Figures and tables are indicated by an italic *f* or *t* following the page number.